Geophysical Monograph Series

Geophysical Monograph Series

240 **Chemostratigraphy Across Major Chronological Eras** *Alcides N. Sial, Claudio Gaucher, Muthuvairavasamy Ramkumar, and Valderez Pinto Ferreira (Eds.)*
241 **Mathematical Geoenergy: Discovery, Depletion, and Renewal** *Paul Pukite, Dennis Coyne, and Daniel Challou (Eds.)*
242 **Ore Deposits: Origin, Exploration, and Exploitation** *Sophie Decree and Laurence Robb (Eds.)*
243 **Kuroshio Current: Physical, Biogeochemical and Ecosystem Dynamics** *Takeyoshi Nagai, Hiroaki Saito, Koji Suzuki, and Motomitsu Takahashi (Eds.)*
244 **Geomagnetically Induced Currents from the Sun to the Power Grid** *Jennifer L. Gannon, Andrei Swidinsky, and Zhonghua Xu (Eds.)*
245 **Shale: Subsurface Science and Engineering** *Thomas Dewers, Jason Heath, and Marcelo Sánchez (Eds.)*
246 **Submarine Landslides: Subaqueous Mass Transport Deposits From Outcrops to Seismic Profiles** *Kei Ogata, Andrea Festa, and Gian Andrea Pini (Eds.)*
247 **Iceland: Tectonics, Volcanics, and Glacial Features** *Tamie J. Jovanelly*
248 **Dayside Magnetosphere Interactions** *Qiugang Zong, Philippe Escoubet, David Sibeck, Guan Le, and Hui Zhang (Eds.)*
249 **Carbon in Earth's Interior** *Craig E. Manning, Jung-Fu Lin, and Wendy L. Mao (Eds.)*
250 **Nitrogen Overload: Environmental Degradation, Ramifications, and Economic Costs** *Brian G. Katz*
251 **Biogeochemical Cycles: Ecological Drivers and Environmental Impact** *Katerina Dontsova, Zsuzsanna Balogh-Brunstad, and Gaël Le Roux (Eds.)*
252 **Seismoelectric Exploration: Theory, Experiments, and Applications** *Niels Grobbe, André Revil, Zhenya Zhu, and Evert Slob (Eds.)*
253 **El Niño Southern Oscillation in a Changing Climate** *Michael J. McPhaden, Agus Santoso, and Wenju Cai (Eds.)*
254 **Dynamic Magma Evolution** *Francesco Vetere (Ed.)*
255 **Large Igneous Provinces: A Driver of Global Environmental and Biotic Changes** *Richard. E. Ernst, Alexander J. Dickson, and Andrey Bekker (Eds.)*
256 **Coastal Ecosystems in Transition: A Comparative Analysis of the Northern Adriatic and Chesapeake Bay** *Thomas C. Malone, Alenka Malej, and Jadran Faganeli (Eds.)*
257 **Hydrogeology, Chemical Weathering, and Soil Formation** *Allen Hunt, Markus Egli, and Boris Faybishenko (Eds.)*
258 **Solar Physics and Solar Wind** *Nour E. Raouafi and Angelos Vourlidas (Eds.)*
259 **Magnetospheres in the Solar System** *Romain Maggiolo, Nicolas André, Hiroshi Hasegawa, and Daniel T. Welling (Eds.)*
260 **Ionosphere Dynamics and Applications** *Chaosong Huang and Gang Lu (Eds.)*
261 **Upper Atmosphere Dynamics and Energetics** *Wenbin Wang and Yongliang Zhang (Eds.)*
262 **Space Weather Effects and Applications** *Anthea J. Coster, Philip J. Erickson, and Louis J. Lanzerotti (Eds.)*
263 **Mantle Convection and Surface Expressions** *Hauke Marquardt, Maxim Ballmer, Sanne Cottaar, and Jasper Konter (Eds.)*
264 **Crustal Magmatic System Evolution: Anatomy, Architecture, and Physico-Chemical Processes** *Matteo Masotta, Christoph Beier, and Silvio Mollo (Eds.)*
265 **Global Drought and Flood: Observation, Modeling, and Prediction** *Huan Wu, Dennis P. Lettenmaier, Qiuhong Tang, and Philip J. Ward (Eds.)*
266 **Magma Redox Geochemistry** *Roberto Moretti and Daniel R. Neuville (Eds.)*
267 **Wetland Carbon and Environmental Management** *Ken W. Krauss, Zhiliang Zhu, and Camille L. Stagg (Eds.)*
268 **Distributed Acoustic Sensing in Geophysics: Methods and Applications** *Yingping Li, Martin Karrenbach, and Jonathan B. Ajo-Franklin (Eds.)*
269 **Congo Basin Hydrology, Climate, and Biogeochemistry: A Foundation for the Future (English version)** *Raphael M. Tshimanga, Guy D. Moukandi N'kaya, and Douglas Alsdorf (Eds.)*
269 **Hydrologie, climat et biogéochimie du bassin du Congo: une base pour l'avenir (version française)** *Raphael M. Tshimanga, Guy D. Moukandi N'kaya, et Douglas Alsdorf (Éditeurs)*
270 **Muography: Exploring Earth's Subsurface with Elementary Particles** *László Oláh, Hiroyuki K. M. Tanaka, and Dezsö Varga (Eds.)*
271 **Remote Sensing of Water-Related Hazards** *Ke Zhang, Yang Hong, and Amir AghaKouchak (Eds.)*
272 **Geophysical Monitoring for Geologic Carbon Storage** *Lianjie Huang (Ed.)*
273 **Isotopic Constraints on Earth System Processes** *Kenneth W. W. Sims, Kate Maher, and Daniel P. Schrag (Eds.)*
274 **Earth Observation Applications and Global Policy Frameworks** *Argyro Kavvada, Douglas Cripe, and Lawrence Friedl (Eds.)*
275 **Threats to Springs in a Changing World: Science and Policies for Protection** *Matthew J. Currell and Brian G. Katz (Eds.)*
276 **Core-Mantle Co-Evolution: An Interdisciplinary Approach** *Takashi Nakagawa, Madhusoodhan Satish-Kumar, Taku Tsuchiya, and George Helffrich (Eds.)*
277 **Compressional Tectonics: Plate Convergence to Mountain Building (Tectonic Processes: A Global View, Volume 1)** *Elizabeth J. Catlos and İbrahim Çemen (Eds.)*
278 **Extensional Tectonics: Rifting and Continental Extension (Tectonic Processes: A Global View, Volume 2)** *İbrahim Çemen and Elizabeth J. Catlos (Eds.)*
279 **Strike-Slip Tectonics: Oceanic Transform Faults to Continental Plate Boundaries (Tectonic Processes: A Global View, Volume 3)** *İbrahim Çemen and Elizabeth J. Catlos (Eds.)*
280 **Landscape Fire, Smoke, and Health: Linking Biomass Burning Emissions to Human Well-Being** *Tatiana V. Loboda, Nancy H. F. French, and Robin C. Puett (Eds.)*
281 **Clouds and Their Climatic Impacts: Radiation, Circulation, and Precipitation** *Sylvia Sullivan and Corinna Hoose (Eds.)*
282 **Fast Processes in Large-Scale Atmospheric Models: Progress, Challenges, and Opportunities** *Yangang Liu and Pavlos Kollias (Eds.)*
283 **Helicities in Geophysics, Astrophysics, and Beyond** *Kirill Kuzanyan, Nobumitsu Yokoi, Manolis K. Georgoulis, and Rodion Stepanov (Eds.)*
284 **Noisy Oceans: Monitoring Seismic and Acoustic Signals in the Marine Environment** *Gaye Bayrakci and Frauke Klingelhoefer (Eds.)*
285 **Alfvén Waves Across Heliophysics: Progress, Challenges, and Opportunities** *Andreas Keiling (Ed.)*
286 **Salt in the Earth Sciences: Evaporite Rocks and Salt Deposition** *Webster Mohriak*
287 **Salt in the Earth Sciences: Basin Analysis and Applications** *Webster Mohriak*
288 **Microanalysis of Atmospheric Particles: Techniques and Applications** *Joseph M. Conny and Peter R. Buseck (Eds.)*
289 **Distributed Acoustic Sensing in Borehole Geophysics** *Yingping Li, Robert Mellors, and Ge Zhan (Eds.)*
290 **Tectonics and Seismic Structure of Alaska and Northwestern Canada: EarthScope and Beyond** *Natalia A. Ruppert, Margarete A. Jadamec, and Jeffrey T. Freymueller (Eds.)*

Geophysical Monograph 278

Extensional Tectonics

Rifting and Continental Extension

Tectonic Processes: A Global View, Volume 2

Editors

İbrahim Çemen
Elizabeth J. Catlos

This Work is a co-publication of the American Geophysical Union and John Wiley and Sons, Inc.

AGU
ADVANCING EARTH AND SPACE SCIENCES

WILEY

This edition first published 2025

Published under the aegis of the AGU Publications Committee

Matthew Giampoala, Vice President, Publications
Steven A. Hauck, II, Chair, Publications Committee
For details about the American Geophysical Union visit us at www.agu.org.

Registered Office
John Wiley & Sons, Inc., 111 River Street, Hoboken, NJ 07030, USA

Editorial Office
111 River Street, Hoboken, NJ 07030, USA

For details of our global editorial offices, customer services, and more information about Wiley products visit us at www.wiley.com.

Wiley also publishes its books in a variety of electronic formats and by print-on-demand. Some content that appears in standard print versions of this book may not be available in other formats.

Library of Congress Cataloging-in-Publication Data
Names: Çemen, Ibrahim, 1951- editor. | Catlos, Elizabeth J., 1971- editor.
Title: Extensional tectonics : rifting and continental extension / editors Ibrahim Çemen, Elizabeth J. Catlos.
Description: Hoboken, NJ : Wiley, 2025. | Series: Geophysical monograph series | Includes index.
Identifiers: LCCN 2024039058 (print) | LCCN 2024039059 (ebook) | ISBN 9781119773740 (hardback) | ISBN 9781119773771 (adobe pdf) | ISBN 9781119773764 (epub)
Subjects: LCSH: Geology, Structural. | Rifts (Geology)
Classification: LCC QE601. E98 2025 (print) | LCC QE601 (ebook) | DDC 551.8–dc23/eng/20240929
LC record available at https://lccn.loc.gov/2024039058
LC ebook record available at https://lccn.loc.gov/2024039059

Cover design: Wiley
Cover image: © Gerald Corsi/Getty Images

Set in 10/12pt Times New Roman by Straive, Pondicherry, India

CONTENTS

LIST OF CONTRIBUTORS

Ian D. Bastow
Department of Earth Science and Engineering
Imperial College London
London, United Kingdom

Elizabeth A. Bell
Department of Earth and Space Sciences
University of California
Los Angeles, California, USA

Erin Beutel
Department of Geology and Environmental Geosciences
College of Charleston
Charleston, South Carolina, USA

Cathy J. Busby
Department of Earth and Planetary Sciences
University of California at Davis
Davis, California, USA

Elizabeth J. Catlos
Jackson School of Geosciences
Department of Earth and Planetary Sciences
University of Texas at Austin
Austin, Texas, USA

İbrahim Çemen
Department of Geological Sciences
University of Alabama
Tuscaloosa, Alabama, USA

James A. Conder
School of Earth Systems and Sustainability
Southern Illinois University
Carbondale, Illinois, USA

Zuze Dulanya
Geography and Earth Sciences Department
University of Malawi
Zomba, Malawi

Thomas M. Etzel
Jackson School of Geosciences
Department of Earth and Planetary Sciences
University of Texas at Austin
Austin, Texas, USA

and

ExxonMobil
Parkway Spring, Texas, USA

Irina Filina
Department of Earth and Atmospheric Sciences
University of Nebraska–Lincoln
Lincoln, Nebraska, USA

Karen Fontijn
Department of Geosciences, Environment and Society
Université Libre de Bruxelles
Brussels, Belgium

James Granath
Granath and Associates Consulting Geology
Highland Ranch, Colorado, USA

Valeria Jaramillo
Department of Earth and Space Sciences
University of California, Los Angeles
Los Angeles, California, USA

Derek Keir
Department of Ocean and Earth Science
University of Southampton
Southampton, United Kingdom

and

Department of Earth Sciences
University of Florence
Florence, Italy

Folarin Kolawole
Department of Earth and Environmental Sciences
Columbia University
New York, USA

Rezene Mahatsente
School of Geosciences
University of Louisiana at Lafayette
Lafayette, Louisiana, USA

Nicholas Mariita
Geothermal Energy Training and Research Institute
Dedan Kimathi University of Technology
Nyeri, Kenya

Osman Merey
Department of Geological Sciences
University of Alabama
Tuscaloosa, Alabama, USA

James D. Muirhead
School of Environment
University of Auckland
Auckland, New Zealand

Keith Putirka
Department of Earth and Environmental Sciences
California State University
Fresno, California, USA

Tyrone O. Rooney
Department of Earth and Environmental Sciences
Michigan State University
East Lansing, Michigan, USA

Axel K. Schmitt
Institute of Earth Sciences
Heidelberg University
Heidelberg, Germany

and

John de Laeter Centre
Curtin University
Bentley, Australia

Daniel Stockli
Jackson School of Geosciences
Department of Earth and Planetary Sciences
University of Texas at Austin
Austin, Texas, USA

Berry H. Tew, Jr.
Department of Geological Sciences
University of Alabama
Tuscaloosa, Alabama, USA

and

Geological Survey of Alabama
Tuscaloosa, Alabama, USA

Samuel R. Walker
Department of Geological Sciences
University of Alabama
Tuscaloosa, Alabama, USA

Bernhard Weise
School of Geography, Geology and the Environment
University of Leicester
Leicester, United Kingdom

Matthew W. Wielicki
Department of Geological Sciences
University of Alabama
Tuscaloosa, Alabama, USA

Liang Xue
Department of Earth and Environmental Sciences
Syracuse University
Syracuse, New York, USA

An Yin
Department of Earth and Space Sciences
University of California
Los Angeles, California, USA

PREFACE

TECTONIC PROCESSES: A GLOBAL VIEW

Tectonic processes control the shape and structure of the Earth, and these processes affect the Earth's climate, geomorphology, magmatism, geochemistry, sedimentary environments, and economic resources. The evolution of these features through geologic times can be explained within the framework of "plate tectonics," which is the overwhelmingly accepted unified theory of Earth sciences. This makes plate tectonics the central core discipline in geoscience research.

The Earth's lithosphere is divided into major plates and microplates that interact with each other along divergent (extensional), convergent (compressional), and transform (strike-slip) plate boundaries. Within the past few decades, Earth sciences have made tremendous advances in our understanding of plate tectonics processes along these three types of plate boundaries.

The overarching objective of the three-volume series *Tectonic Processes: A Global View* is to present an up-to-date compendium and valuable reference for students of Earth sciences at all levels, from advanced undergraduate and graduate to doctoral and postdoctoral researchers, as well as for educators, policymakers, and research professionals in academia and industry. The collection contains three volumes:

- Volume 1: *Compressional Tectonics: Plate Convergence to Mountain Building*
- Volume 2: *Extensional Tectonics: Rifting and Continental Extension*
- Volume 3: *Strike-Slip Tectonics: Oceanic Transform Faults to Continental Plate Boundaries*

EXTENSIONAL TECTONICS

Extensional Tectonics: Rifting and Continental Extension reviews present-day knowledge of extensional tectonic processes and explores examples from around the world. It starts with an introductory chapter outlining extensional tectonics processes from the initiation of continental rifting to oceanic basin formation. The remainder of the book is divided into three parts.

Part I: Large-Scale Extensional Tectonics

The four chapters in Part I review large-scale extensional terranes. Chapter 2 by Rooney et al. presents an overview of the East African Rift System. Chapter 3 by Filina and Beutel suggests a new tectonic reconstruction of the Gulf of Mexico based on geological and geophysical constraints. In Chapter 4, Busby and Putirka introduce a new concept for active obliques rifting for the extensional process from Walker Lane to the Gulf of California. And in Chapter 5, Catlos et al. investigate extensional tectonics in Western Anatolia, Turkey, as part of the Aegean Extension.

Part II: Extensional Tectonic Processes

The three chapters in Part II explore processes affecting rifting that may apply broadly to other extensional terranes. Chapter 6 by Conder explains the relationship between isostasy and seafloor spreading as a self-sustaining process. Chapter 7 by Weise and Mahatsente explores the rigidity-seismicity relation in the East African Rift System. To conclude, Chapter 8 by Walker et al. examines the evolution of the Mississippi Embayment and associated alkaline magmatism based on an edge-driven convection-based model.

Part III: Case Studies of Continental Extensional Tectonics

In Part III, four regional case studies are presented. Extensional processes in these regions can be applied to other extended terrains. Chapter 9 by Kolawole et al. explores the evolution of the Middle Shire and Nsanje rift interaction zone located along three contiguous nonvolcanic propagating rift segments. In Chapter 10, Jaramillo et al. tackle the root problem of the mid-tertiary Cordilleran detachment faults. In Chapter 11, Merey et al. suggest a new model for the Büyük Menderes Graben in Western Turkey within the framework of the Cenozoic back-arc extension setting. Finally, in Chapter 12, Catlos and Çemen emphasize the importance of the transfer zones of extended terranes with examples from the Aegean microplate and Western Anatolia.

ACKNOWLEDGMENTS

We appreciate the authors' contributions to this volume and acknowledge the time, effort, and diverse perspectives of a large number of insightful reviewers. This volume could not be realized without the persistent efforts of the authors and reviewers. Thank you all.

İbrahim Çemen
Department of Geological Sciences
The University of Alabama, USA

Elizabeth J. Catlos
Jackson School of Earth Sciences
The University of Texas at Austin, USA

1

Processes and Perspectives in Extensional Tectonics

Elizabeth J. Catlos[1] and İbrahim Çemen[2]

ABSTRACT

The study of extensional tectonics harkens to the early days of plate tectonics, as the search for mechanisms driving large continental blocks to drift apart led to advances in paleomagnetism and geochronology. Divergent plate boundaries form extensive and continuous volcanic systems covering large portions of the Earth and are fundamental to understanding how plates form. The study of extensional dynamics transects traditional geoscience disciplinary boundaries and is critical in the search for the origin and evolution of life. Extensional and normal faults are not restricted to divergent plate boundaries but are also located in regions of plate convergence and within plates. Failed rifts pose significant seismic hazards. This chapter aims to identify the importance of the study of extension and dispel common misconceptions regarding the process. Stress is a critical factor in explaining why extension occurs in the lithosphere and how magma focuses in regions of extensional tectonics.

1.1. INTRODUCTION

Divergent plate boundaries are places where pieces of Earth's lithosphere (crust + upper mantle) drift away from each other (Figures 1.1 and 1.2). They are spreading boundaries, where, in some cases, new igneous crust fills space between them or sediments deposit within new accommodation space. Divergent plate boundaries can separate ocean, continental, or transitional lithosphere and have been reviewed elsewhere (Wyllie, 1988; Sibuet & Tucholke, 2013; Acocella, 2014; Philippon & Corti, 2016; Peron-Pinvidic et al., 2019; Acocella, 2021; Zwaan & Schreurs, 2022; Olive, 2023). Divergent plate boundaries separate plates that move in opposite directions due to an extensional stress regime. Plates can extend, stretch out, enlarge in breadth, continue in length, or expand.

Despite an apparent simple observation (two plates move apart), developments in our understanding of plate tectonics have revealed that some common generalizations about these regions are incorrect. At divergent plate boundaries, lithospheric extension is often perceived to be narrow and localized between specific plates. However, the process can also occur within plates due to diverse activity at plate boundaries and mantle dynamics. In addition to its well-known roles at divergent boundaries, extension occurs in both transform and convergent plate boundaries and plate interiors. Misconceptions arise regarding extension because it is usually introduced during introductory geoscience courses without a complete understanding of the fundamentals of stress, strain, and rock physiochemical properties. These misconceptions include discriminating between tension versus extension and the role of magma as a passive outcome versus an active driver. Because introductory geoscience students are exposed to hazards early on at convergent and strike-slip

[1] *Jackson School of Geosciences, Department of Earth and Planetary Sciences, University of Texas at Austin, Austin, Texas, USA*

[2] *Department of Geological Sciences, University of Alabama, Tuscaloosa, Alabama, USA*

Extensional Tectonics: Rifting and Continental Extension, Geophysical Monograph 278, First Edition.
Edited by İbrahim Çemen and Elizabeth J. Catlos.

DOI: 10.1002/9781119773757.ch01

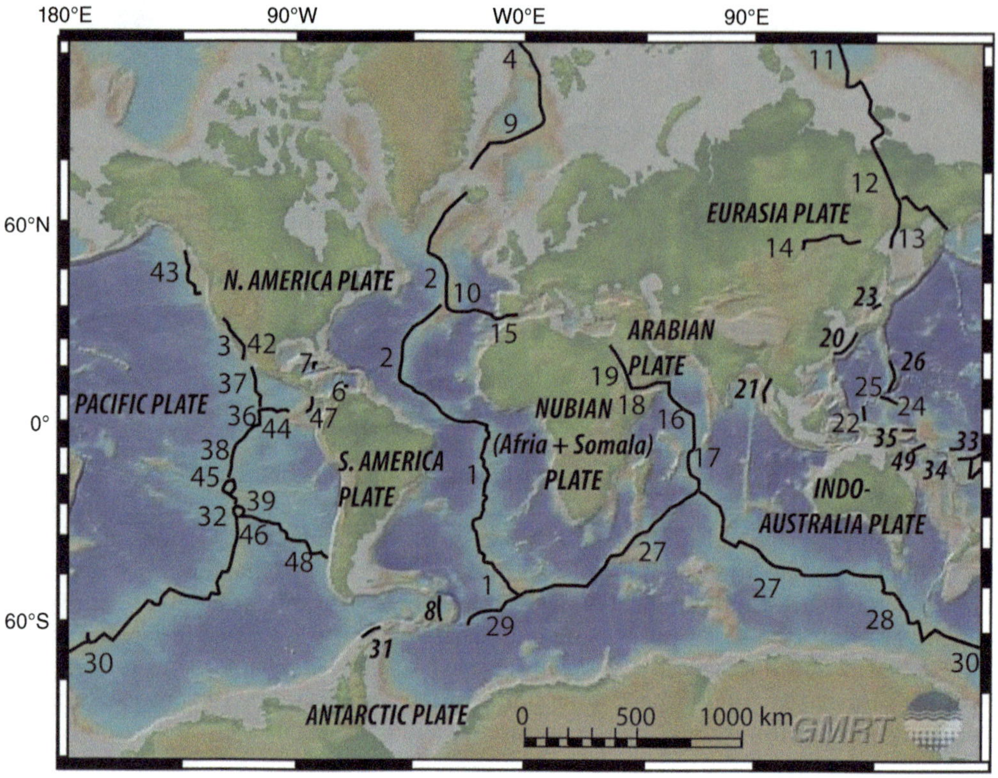

Figure 1.1 Global multiresolution topography map. *Source:* Adapted from Ryan et al. (2009). Divergent plate boundaries and extensional back-arc regions are numbered. Numbers in italics are related to subduction systems. *Source:* Adapted from Coffin et al. (1998). See Table 1.1 and Table 1.2 for information.

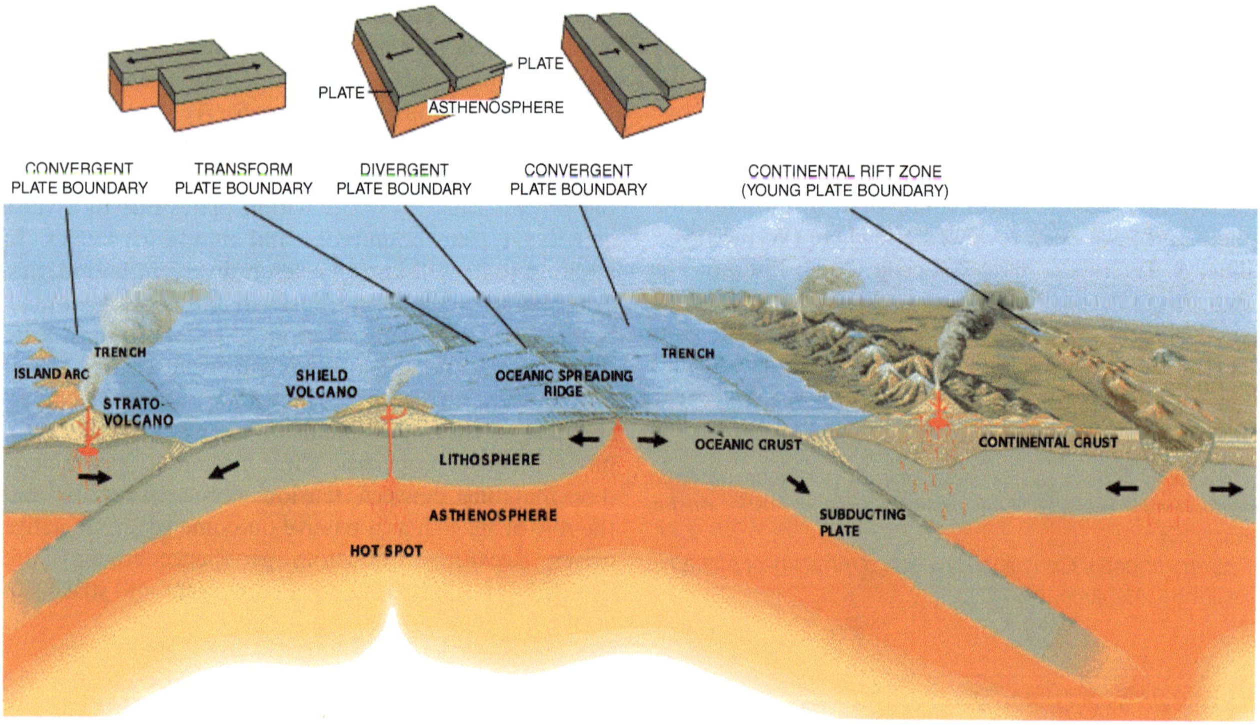

Figure 1.2 The cartoon cross-section illustrates the main plate boundary types. *Source:* Adapted from Kious and Tilling (2016) / U.S. Department of the Interior (USGS) / Public Domain.

Table 1.1 Extensional plate boundaries.

No.	Plate boundary name	Plate separation
1	South Atlantic Ridge (part of Mid Atlantic Ridge)	S. America–Africa
2	North and Central Atlantic Ridge Axis (also Aegir Ridge) (part of the Mid Atlantic Ridge)	N. America–Eurasian
3	Gulf of California	N. America–Pacific
4	Eurasian Basin Spreading Ridge	N. America–Eurasia
6	Falcon Basin Fault	Maricaibo–Perija
7	Cayman Trough (Trench, Spreading Center, Rise)	N. America–Caribbean Plate
9	Reykjanes-Nansen Ridge	N. America–Eurasia
10	Azores Spreading Center and Triple Junction	N. America–Eurasian–Africa
11	Ridge north of Taimyr, Arctic Ocean	N. America–Eurasia
12	Ridge in northeast Asia	N. America–Eurasia
13	Ridge near Okhotsk Sea	Okhotsk–N. America
14	Baikal Rift Zone	Amur–Eurasia
15	Iberia-Africa Plate Boundary	Iberia–Africa
16	Carlsberg Ridge	Atrica–Indo–Australia
17	Central Indian Ridge	Africa–Indo–Australia
18	Gulf of Aden Spreading Ridge	Africa (Somali)–Arabian
19	Red Sea Ridge Axis	Africa (Somali)–Arabian
22	Ayu Trough	Philippine Sea–Caroline
24	Sorol Trough NE of Eauripik Rise	Pacific–Caroline
25	Caroline Ridge	East–West Caroline
27	Southeast Indian Ridge	Australia–Antarctic
28	Australia-Antarctic Spreading Ridge	Australia–Antarctic
29	East American-Antarctic Ridge	S. America–Antarctic
30	Pacific Antarctic Spreading Ridge	Pacific–Antarctic
32	Juan Fernandez-Antarctic Spreading Center (part of Chile Rise)	Juan Fernandez–Antarctic
36	Pacific-Galapagos Spreading Center	Pacific–Galapagos
37	Pacific-Cocos Spreading Center	Pacific–Cocos
38	Pacific-Nazca Spreading Center	Pacific–Nazca
39	Juan Fernandez-Pacific S Spreading Center	Juan Fernandez–Pacific
42	Pacific-Rivera Ridge	Pacific–Rivera
43	Explorer Ridge	Juan de Fuca–Gorda
44	Galapagos-Nazca Spreading Center	Galapagos–Nazca
45	Easter-Nazca Spreading Center	Easter–Nazca
46	Juan Fernandez-Nazca Spreading Center (part of Chile Rise)	Juan Fernandez–Nazca
47	Cocos-Nazca Spreading Center (Cocos Ridge)	Cocos–Nazca
48	Nazca-Antarctic Spreading Center (part of Chile Rise)	Nazca–Antarctic

Adapted from Coffin et al. (1998).

plate boundaries (e.g., the Himalaya, Andes, and San Andreas Fault dominate the coverage in introductory geoscience textbooks), they may not recognize the inherent nature of seismic risk associated with divergent plate boundaries. For example, failed rift systems pose some of the highest risks in intraplate tectonic settings.

Advances in our understanding of lithospheric extension have also led to significant debate about the processes involved in developing regions of extended lithosphere and new and exciting ideas about how the Earth works. This chapter aims to clarify some of the basics of extensional tectonics, emphasize the importance of the study regions of crustal extension, dispel some common misconceptions, and set the stage for the chapters that focus on extensional tectonics in this volume.

1.2. THE IMPORTANCE OF EXTENSIONAL TECTONICS

Extensional tectonics plays a vital role in understanding an array of Earth processes that range from fundamental tectonics to the evolution of life. Continental drift requires extension and the rifting apart of large land masses (Wegener, 1912; Davies, 2022). These were the first plate boundaries used as evidence for plate tectonic theory, which also led to the application and further development of the fields of paleomagnetism and radiometric dating (Hess, 1962; Vine & Matthews, 1963; Wilson, 1965; LePichon, 1968; McKenzie, 1977; Palin & Santosh, 2021; Olive, 2023). Continental breakup and basin formation are fundamental and vital processes in Earth tectonics (e.g., Pagli et al., 2015; Wang et al., 2019; Buck, 2004; Palin et al., 2020; Bonifacio et al., 2023).

The lack of acceptance of plate tectonics in the geoscience community, despite ample evidence of continental connections and shapes, paleontology, and similar rock stratigraphy and types, was in part due to the unknown mechanisms driving the process of drifting and extension (McKenzie, 1977; Frisch et al., 2011; Nyarko & Rudge, 2022; Bonifacio et al., 2023). Modern-day plate reconstructions rely on extensional dynamics to shape and develop supercontinents. Continental extension is often restricted to occurring along narrow zones in the geological past (Dunbar & Sawyer, 1987; Seton et al., 2012). Although infrequent in the modern day, active continental extension is an important process in regulating intraplate stresses (Dalmayrac and Molnar, 1981; Rey et al., 2001; others). This volume includes a chapter on understanding reconstructions of the Gulf of Mexico (Filina and Beutel, "Geological and Geophysical Constraints Guide New Tectonic Reconstruction of the Gulf of Mexico"). The "push" from mid-ocean ridges is invoked as a force that separates landmasses (e.g., Vink, 1982; Dunbar & Sawyer, 1987; Seton et al., 2012, 2023). Ridge push is a sliding force driven by horizontal density contrasts that develop from cooling and thickening oceanic lithosphere and is thought to contribute about 10% to plate driving forces (Hales, 1969; Solomon et al., 1980; Hager & O'Connell, 1981; Spence, 1987; Turcotte & Schubert, 2014; Mulyukova & Bercovici, 2019; Lu et al., 2021).

Divergent plate boundaries form extensive and continuous volcanic systems covering large portions of the Earth. Mid-ocean ridges are Earth's most extensive continuous volcanic systems (Figures 1.1 and 1.2) (Perfit & Chadwick, 1998; MacDonald et al., 1991). The ridge system extends 67,338 km but is only 5–30 km wide (Macdonald, 2001; Bird, 2003). Mid-ocean ridges are also considered the longest mountain ranges on Earth, separating plates at a rate of 10 mm/yr (Gakkel Ridge) to 170 mm/yr (East Pacific Rise) (e.g., MacDonald et al., 1991; Püthe & Gerya, 2014). Approximately 60% of Earth's surface comprises oceanic crust created at mid-ocean ridges (Cogley, 1984; Searle, 2013), and these regions are of primary importance in modeling how new oceanic plates develop (Haymon et al., 1991; Purdy et al., 1992). The average crust production at mid-ocean ridges is estimated at ~18 km^3/year. Assuming the global ridge system was constant in length at 67,338 km, the average crust production rate is ~233–326 km^3.

Oceanic crust is basaltic in composition and is further classified as mid-ocean ridge basalt (MORB), oceanic island basalt (OIB), intraoceanic arc basalt (IOAB), island arc basalt (IAB) and continental arc basalt (CAB), depending on chemistry (see review in Xia & Li, 2019). Compositional information gleaned from basalts in MORB, OIB, IOAB, and CAB settings inform the types of tectonic processes that operate within the framework of extensional terranes. Due to their origin, MORB is Earth's most voluminous igneous phenomenon (Brooks et al., 1991; Gannoun et al., 2016). The composition of MORB is a cornerstone of discrimination diagrams used to constrain the tectonic setting of numerous regions (e.g., Floyd, 1995; Pearce, 1996; Grimes et al., 2015). The composition of minerals within basaltic rocks has also proved helpful in determining their conditions of crystallization and formation (e.g., Bryndzia et al., 1989; Ernst & Liu, 1998). Basaltic melts are generated beneath mid-ocean ridges via decompression melting of upwelling mantle and focus beneath a ridge axis (Perfit & Chadwick, 1998; Soule, 2015). As such, basalts are used to examine processes such as mantle composition and partial melting, crust–mantle interactions, and convection (Boutilier & Keen, 1999; Gannoun et al., 2016). Differences in the morphology of fast (>80 mm/yr total spreading rate) and slow-spreading (<55 mm/yr) mid-ocean ridges may be due to the periodicity of magmatic heat input, which controls lithospheric rheology (e.g., Small, 1998; Dick et al., 2003; Buck et al., 2005; Püthe & Gerya, 2014). Other factors that can control mid-ocean ridge morphology include the thermal regime, offsets in thermal perturbations, crustal thickness, and the influence of rheology on fault systems (see discussion in Püthe & Gerya, 2014). Submarine lava flows that characterize mid ocean ridges also influence morphology (e.g., Kennish & Lutz, 1998).

Divergent plate boundaries that separate continental lithosphere create rift zones (Figure 1.2) (e.g., Milanovsky, 1972; Kinabo et al., 2008, Zou et al., 2013). The active African Rift Valley is 3,500 km long but only 50–150 km wide (Figure 1.3) (Davies, 2008; Scoon, 2018). This volume includes two chapters dedicated to the East African Rift Valley (Weise and Mahatsente, "Rigidity–Seismicity Relation in the East African Rift System"; and Rooney et al., "The East African Rift System"). Due to the difference in the degree of contamination by the continental lithosphere, melt chemistry in regions of continental rifting differ from mid-ocean ridges (e.g., Bailey, 1977; Wyllie, 1988), and the processes involved in generating melts in these regions are under active investigation (Neugebauer, 1983; Brune, 2016; Chiasera et al., 2021; Brune et al., 2023). Besides their origin, magma systems beneath divergent plate boundaries are studied for how

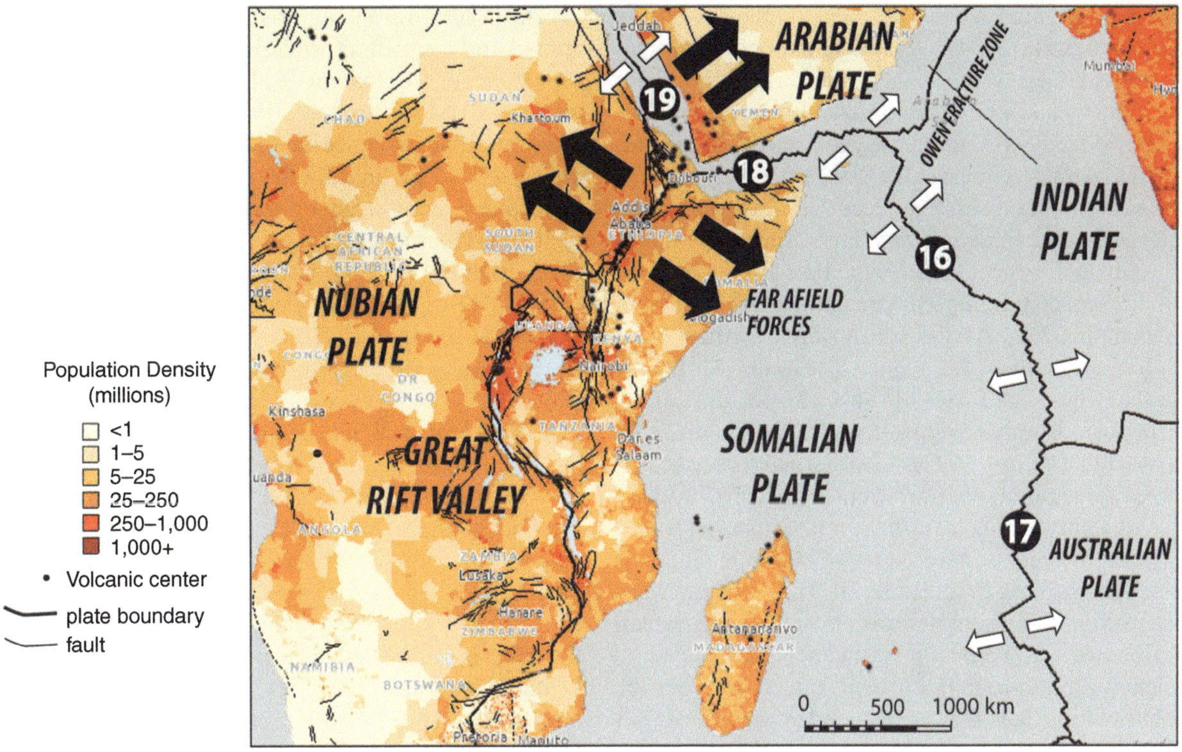

Figure 1.3 Map of the Great African Rift Valley. Numbers on the plate boundaries are listed in Table 1.1. Arrows indicate the approximate motion of rifting. The map was developed using ArcGIS. The layers are the Global Faults Layer from ArcAtlas (Esri) (USGS), Gridded Population of the World v4 2020 (Center for International Earth Science Information Network, Columbia University, World Wide Fund for Nature), Tectonic Plates and Boundaries (Peter Bird, Hugo Ahlenius/Nordpil), and Volcanoes (Smithsonian in conjunction with the USGS, value added by TERC's earth exploration team and simplified by Roger Palmer).

they facilitate the rifting process through dike emplacement and thermal weakening (e.g., Acocella, 2014, Daniels et al., 2014; Urbani et al., 2015; Sigmundsson et al., 2015; Acocella & Trippanera, 2016) and triggering low-angle detachment faulting (Ebinger & Casey, 2001; Bialas et al., 2010). Fundamentally, how melts are produced in the mantle, migrate, pond in the lithosphere, and are injected into the upper crust relates to constraints on the processes involved in creating volcanoes (e.g., Pagli et al., 2015).

Lithospheric extension also creates topographic depressions and basins. A supradetachment basin forms above a low-angle (<30°) normal fault system that experienced large (>5 km) displacement (e.g., Friedmann & Burbank, 1995; Muñoz-Barrera et al., 2022) and is rarely preserved as exhumation, uplift, and erosion proceeds (Jia et al., 2021). However, these are found worldwide (e.g., Basin and Range Province, Davis & Friedmann, 2005; North American Cordillera, Whitney et al., 2013; western Anatolia and the Aegean region, van Hinsbergen, 2010, Oner & Dilek, 2011; Norwegian Sea, Muñoz-Barrera et al., 2022; South China Sea, Jia et al., 2021; Sun et al., 2022). These basins also occur throughout geological time, documented as early as the Archean (Zametzer et al., 2023). Because they are closely related to the tectonic processes that developed them, they provide critical constraints on the kinematic evolution of the extensional system (e.g., Milia & Torrente, 2015).

Volcanism at continental rifts poses hazards when intersecting large population centers, affecting the quality of air, soil, and ground and surface hydrology, which are reflected in water and food quality (Davies, 2008). Today, East Africa's rift hosts the largest urban populations and critical infrastructure, including transportation and energy and water supplies (Figure 1.3) (Hearn, 2022). Hearn (2022) identifies several of its geological hazards, including earthquakes and volcanism, landslides, ground fissuring and cavity collapse, flooding, and sedimentation problems. Significant quantities of CO_2 are emitted at continental rifts, which may have influenced deep carbon fluxes and contributed to climate change (Brune et al., 2017). Lakes and rivers that develop in continental rift zones are critical for understanding climate change (e.g., Ayenew & Legesse, 2007; Olaka et al., 2010; Scholz et al., 2011). The geomorphology and active tectonics of the African rift may have controlled human evolution (Abdelsalam et al., 2004). Volatile emissions in submarine and subaerial gas releases at extensional settings have played a role in killing humans and livestock and are often overlooked as significant geohazards (Edmonds et al., 2015; Annunziatellis et al., 2003).

Fluid-driven processes that occur at divergent plate boundaries have diverse impacts. Hydrothermal systems that develop at divergent plate boundaries also have implications for understanding climate change via degassing (Poulsen et al., 2003). At mid-ocean ridges, hydrothermal circulation and alteration regulate seawater chemistry and global chemical fluxes (Früh-Green et al., 2022). These regions are also important in understanding the origin of life (Holm, 1992). Geothermal processes at mid-ocean ridges today support abundant marine life in the absence of light, so much so that an investigation found new species in every square meter of seafloor sediment (Grassle & Maciolek, 1992; Rona, 2003). Ocean gateways develop during continental breakup, influencing ocean circulation and life (Peron-Pinvidic et al., 2019).

Ancient and modern mid-ocean ridges are sites of valuable ore resources that appear under conditions that facilitate their development and preservation (e.g., Rona, 1984, 2003). Metalliferous muds with large amounts of Cu, Zn, Pb, Fe, Au, and Ag were first noted in the Red Sea rift (Figure 1.1 and Figure 1.3) (Laurila et al., 2015). The high concentrations are related to new oceanic crust developing in smaller, isolated deeps with high-saline brines that favor preservation (Scholten et al., 2017). Massive sulfide deposits are in mounds along the East Pacific Rise near the Gulf of California, the Galapagos Rift, the Guaymas Basin, and the Juan de Fuca Ridge (Bischoff et al., 1983). The International Sea Bed Authority (https://www.isa.org.jm/) estimates the extensional systems at ridge, arc, and backarc spreading centers host $\sim 3 \times 107$ tons of Cu and Zn. Economic deposits can be hosted within the hydrothermal volcanic system (volcanogenic massive sulfide [VMS] deposits and seafloor massive sulfide [SMS]) or within associated sedimentary rocks (Herzig & Hannington, 1995; Humphris et al., 1995; Knight et al., 2018). Gold occurs in sulfides from the Southern Explorer Ridge and Axial Seamount (Northeast Pacific), the Trans-Atlantic Geotraverse (TAG) hydrothermal field and Snakepit vent field (Mid-Atlantic Ridge) (Hannington et al., 1991). Gold enrichment in these regions is associated with late-stage, low-temperature (<300 °C) vents (Hannington et al., 1991; Herzig & Hannington, 1995). Slower-spreading ridges appear more enriched in gold (Knight et al., 2018). Demand for economic deposits found at mid-ocean ridges continues to increase (Murton et al., 2019), and the International Sea Bed Authority oversees the regulation of seafloor to minimize the impact of mining on the benthic communities associated with these deposits.

Geothermal resources exist beneath regions of widespread and incipient crustal extension, even in amagmatic regions (e.g., Coolbaugh et al., 2002; Banks, 2012; Jolie et al., 2021; English et al., 2023). The Basin and Range physiographic province in the western United States is characterized by broadly distributed normal faults, thin lithosphere, long-lived episodic magmatism, and high heat flow (Figure 1.4a,b) (e.g., Eaton, 1982). Figure 1.4b shows that the region hosts a significant fraction of geothermal resources and electricity generation capacity (Sifford & Bloomquist, 2000; Wisian & Blackwell, 2004). Extensional geothermal systems also comprise a significant portion of the geothermal installed capacity in the region (Wisian & Blackwell, 2004). Normal faults associated with extended lithosphere are critical portions of these geothermal systems and facilitate heat transfer from deep to shallow levels (e.g., Faulds et al., 2010). Two chapters in this volume focus on the extensional dynamics of the U.S. Basin and Range (Jaramillo et al., "Root Problem of Mid-Tertiary Cordilleran Detachment Faults: Deciphering the Evolution of the Whipple Mountains Detachment Shear Zone in Southeastern California"; and Busby and Putirka, "Introduction to an Active Oblique Rift: Walker Lane to Gulf of California").

Hot springs and geothermal activity also occur during incipient continental extension, even in areas associated with flat-slab subduction. Back arc extension associated with flat-slab subduction zones has been shown to lead to significant continental hydration, resulting in a reduction of lithospheric strength, mantle delamination, the development of ignimbrites, and increased lithospheric buoyancy (e.g., Scott et al., 2020; Hiett et al., 2021). This hydrothermal activity can lead to the development of significant ore deposits (e.g., Landis and Rye, 1974; Bissig et al., 2001; Newell et al., 2015).

Ore deposits are a signature of post-collisional extensional environments. For example, Pb–Zn–Ag, Cu–Mo, and Au–Ag deposits are found in association with the development of metamorphic core complexes in the Rhodope Massif of Bulgaria (e.g., Marchev et al., 2005; Rohrmeier et al., 2013). Groves et al. (1998) document several regions with significant Au deposition that were sometimes developed during compression but facilitated and exhumed during lithospheric extension. The combination of high metamorphic temperatures, fluid infiltration, magmatism, and tectonic exhumation identifies regions of extension ideal for developing and providing access to ore minerals.

Rifted margin research has classically been related to the search for hydrocarbons (oil, gas, and coal) (e.g., Peron-Pinvidic et al., 2019; Alves et al., 2020; Li et al., 2023). Rifts with post-rift sag basins and rift basins with marine fill host the most known reserves of recoverable oil and gas (Lambiase & Morely, 1999). Rift basins in the Newark region (northeastern New Jersey,

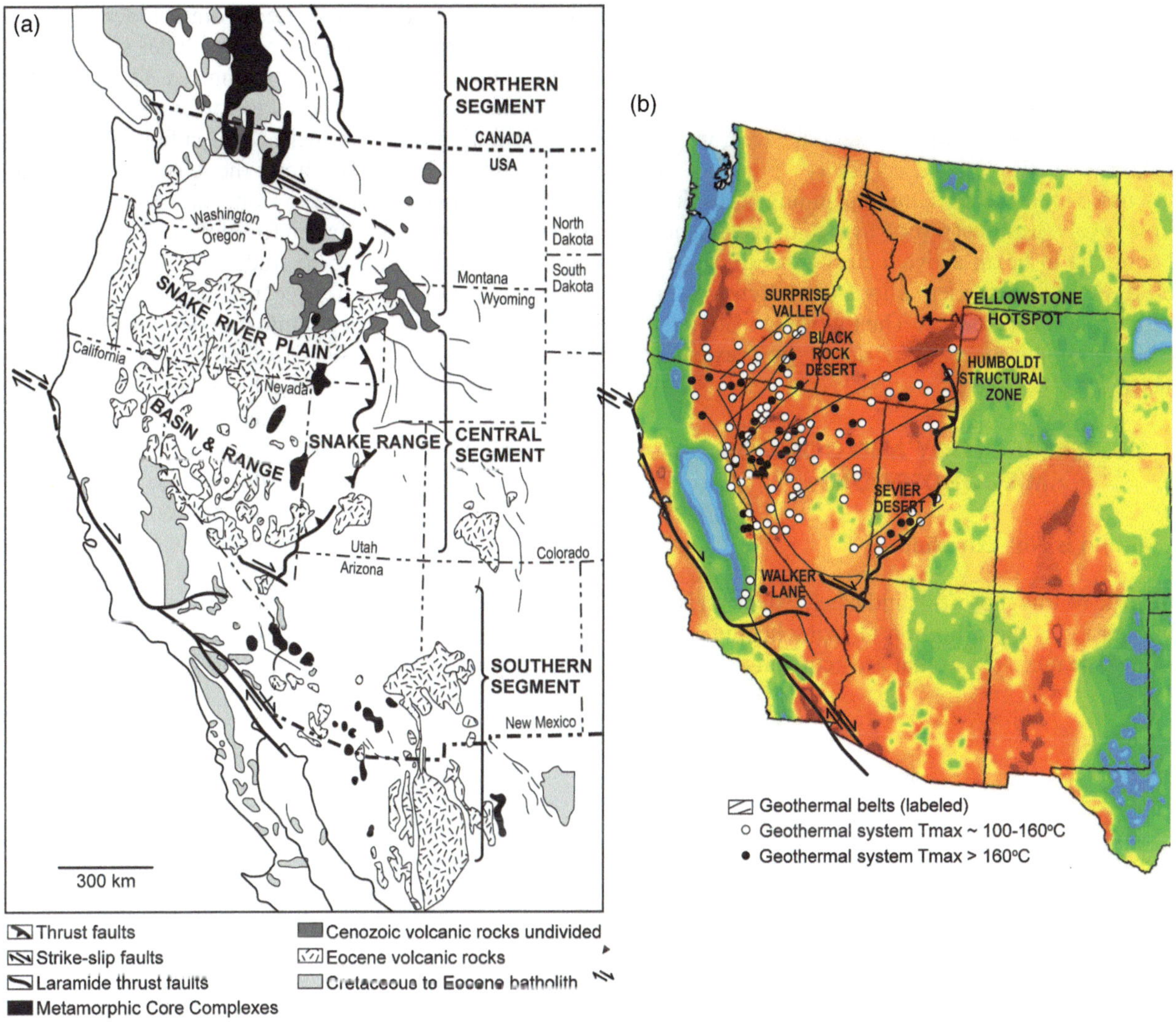

Figure 1.4 (a) Map of the western portion of the United States, showing the location of several metamorphic core complexes. *Source:* Adapted from Gébelin et al. (2011) / with permission of John Wiley & Sons. (b) Geothermal heat flow map of the same region. *Source:* Blackwell et al. (2011). We include geothermal belts and systems. *Source:* Adapted from Faulds et al. (2010).

United States) developed during the Mesozoic and host bituminous coal that played a role in the early economic history of the United States (Robbins et al., 1988). Today, they are evaluated for their potential for CO_2 sequestration (Zakharova et al., 2020).

Determining the origin and driving forces of rifting and how it evolves tectonically directly correlates to understanding its hydrocarbon potential (e.g., Rop et al., 2022). Rifts can encompass a variety of geometries as sediments within a rift system are hosted on a range of basement compositions, and architectures develop over geological time (e.g., Alves et al., 2020; Brune et al., 2023). Critical archives of environmental and climate changes are recorded within rift-related sediments (e.g., McNeill et al., 2019; Hou et al., 2020). Deeper structural levels of highly magmatic active rifts are largely inaccessible, but ancient rifts exhumed in continental interiors can be used to understand parts of these systems and their dynamics (e.g., Oslo rift, Neumann et al., 1992). Rifts can also fail, providing additional snapshots of the extensional process (Elling et al., 2022). In this volume, we include a chapter that discuss failed rifts (Walker et al., "Edge-Driven Convection-Based Models for the Evolution of the Mississippi Embayment and Associated Alkaline Magmatism"). Eventually, some rift systems will interact with other plate boundaries, including convergent ones. The internal architecture of some collisional orogens is controlled by rift inheritance (e.g., Jourdon et al., 2019; Peron-Pinvidic et al., 2019). Intraoceanic settings are one of the central regions for subduction development as convergent systems initiate at or near mid-ocean ridges (e.g., Barth et al., 2008; Wu et al., 2023). Extinct mid-ocean ridges may subduct, creating seismic hazards (e.g., Zhu et al., 2023) and the development of flat-slab (low-angle) subduction zones (Gutscher et al., 2000; Horton et al., 2022). Subduction of these features may also drive extension within the continental crust (e.g., George et al., 2022).

As discussed in this section, understanding lithospheric extension involves almost every aspect of geosciences. It will continue to be an active focus of investigation as we seek to develop new ideas about plate tectonics and search for economic resources. The following section lists a few common misconceptions about how extension operates in the lithosphere.

1.3. TERMINOLOGY AND COMMON MISCONCEPTIONS

1.3.1. Tension Is Not Extension

As the title of this subsection indicates, extension is not tension. Tension is the condition of being held in a state between two or more forces that are acting in opposition to each other. Extension is used from a kinematic perspective to represent tectonic processes under stretching in Earth's crust and lithosphere, giving rise to a range of phenomena, such as extensional tectonics and extensional basin formation (e.g., Wernicke & Burchfiel, 1982).

For example, Figure 1.3 shows the African Rift, where the Nubian plate separates from the Somalia plate, breaking up the larger African (or Nubia–Somalia) plate. In introductory courses, some students imagine a scenario where the plates separate due to imaginary forces far afield from the rift valley. Unfortunately, this common misconception may persist throughout a student's geoscience education as this plate boundary is often used as the type locality for introducing continental lithosphere extension. Students need clarification, as the arrows representing motion along the East African rift appear at odds with the sense of motion indicated by mid-ocean ridges. However, the arrows in Figure 1.3 represent tension, not stress. Rocks respond to stress (σ), an internal force resulting from an applied load that acts along a cross-section of a mechanical or structural component, by changing volume or form (Figure 1.5, inset).

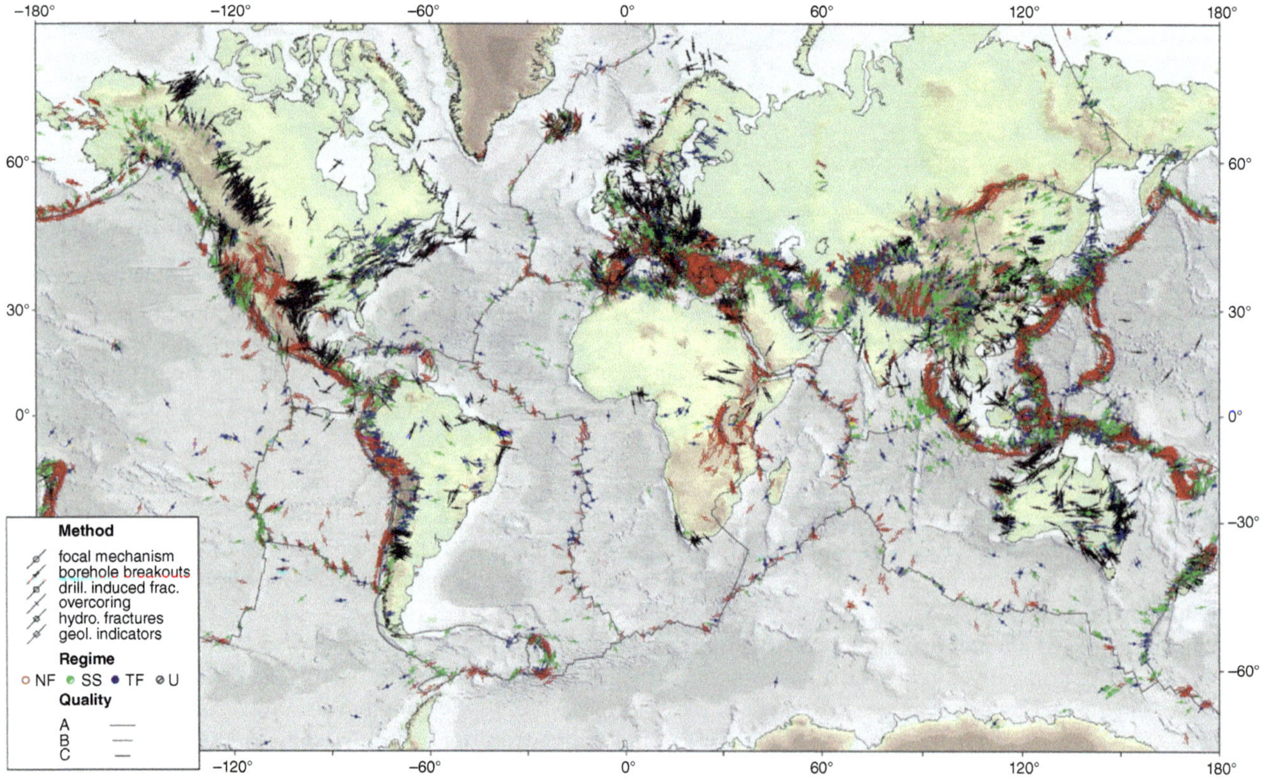

Figure 1.5 World Stress Map (2016). The stress map shows the maximum horizontal stress (SHmax) orientation in the Earth's crust. Symbols denote the SHmax orientation from various stress indicators. Colors denote the stress regime: normal faulting (NF), strike-slip (SS), thrust faulting (TF), and unknown (U). The inset shows the method used to determine the stress orientation. Rocks are displaced by normal faulting, strike-slip, and thrust faulting. The principal stress axes are identified. *Source:* Adapted from Heidbach et al. (2018).

Stress is defined as force/area and has units of pressure (N/m^2 or lb/in^2 or Pa, pascals). Stress can be normal (perpendicular to the surface) or shear (parallel). Anderson (1905, 1951) linked the orientation of the causative stress tensor relative to the Earth's surface relation to fault types in the upper, shallower levels of the crust (Figure 1.5) (Simpson, 1997; Healy et al., 2012; Sorkhabi, 2012). The magnitude of stress is defined as maximum $\sigma 1$ > intermediate $\sigma 2$ > minimum $\sigma 3$. Figure 1.5 shows the World Stress Map (Heidbach et al., 2018), indicating that if the largest principal stress is vertical, then high-angle extensional normal faults develop. Brittle rocks can fail in tension even when all three principal stresses are compressive (Barton & Shen, 2018). Extension occurs when the maximum stress direction acts normally or vertically on the rock mass. All three principal stresses in the Earth are nearly always compressional or inward-directed. A tensional state of stress is rare.

1.3.2. Magma Does Not Force Open Divergent Plate Boundaries

Divergent plate boundaries are classified based on the extent of their igneous activity and spreading rate. Those that do not contain evidence of igneous activity are nonmagmatic, nonvolcanic, amagmatic, or nontectonic (Acocella, 2021; Gordon et al., 2013; Bickert et al., 2023). Relationships between mid-ocean ridge segmentation and structure, spreading rate, and igneous activity have been extensively explored (e.g., Schouten et al., 1985; Sinton et al., 1991; Purdy et al., 1992). Typically, slow-spreading ridges (half-spreading rates <30 mm/yr) have more significant variability in magmatism, lithology, and vent chemistry as detachment faults expose lower-crustal and serpentinized mantle rocks (Früh-Green et al., 2022). Mid-ocean ridges with slower spreading rates are also characterized by deep axial valleys bounded by rugged flanking topography, whereas fast-spreading ridges have more continuous axial ridges with smoother flanking topography (e.g., Small, 1998; Phipps Morgan & Chen, 1993). Ultraslow spreading ridges (<12 mm/yr) have linked amagmatic and magmatic ridge segments (Dick et al., 2003).

Schematic diagrams like those in Figure 1.2 are commonly used in introductory geoscience courses to demonstrate that the flow of the mantle upwellings forces plates to move apart. They also suggest a directionality of forces acting on plate boundaries, allowing some confusion regarding the difference between plate motion vectors and force tensors. However, weak gravity anomalies associated with ridges suggest that any upwelling is shallow (e.g., Acocella, 2021). Plate divergence also contributes to decompression melting as the mantle rises and volcanic activity is focused. Anderson (2002) indicates that the state of stress in the lithosphere is more important than the temperature in localizing volcanism. Two chapters in this volume (Conder, "Oceanic Isostasy: Self-Sustaining Seafloor Spreading and Rift Localization"; and Kolawole et al., "Rapid Versus Delayed Linkage and Coalescence of Propagating Rift Tips") discuss the process of how rifting develops and the role of magma as it enters divergent systems.

1.3.3. Extension and Normal Faulting Are Part of Collisional Orogens and Plate Boundaries

Divergent plate boundaries are often linked with high-angle normal faults, whereas convergent plate boundaries show reverse or thrust faults. However, the World Stress Map demonstrates that divergent plate boundaries are not the only areas associated with normal faulting (Figure 1.5). Subduction zone systems (convergent plate boundaries) can display extensional dynamics and develop normal faults. Worldwide, several back-arc basins are associated with extensional dynamics and the development of normal faults, enough to be designated plate boundaries by Coffin et al. (1998) (Figure 1.1, Table 1.2). The development of extensional systems in convergent settings has been explored using physical and numerical modeling (e.g., Andrews & Sleep, 1974; Chemenda et al., 1995; Toussaint et al., 2004; Granado & Ruh, 2019).

Figure 1.6a shows a cartoon demonstrating the expected relationship between subduction and the development of extensional regimes in the back-arc region. Within the subduction system, normal faults can also develop within the accretionary prism of subduction zones and near or within the subducting plate (Figures 1.6b and c). The World Stress Map (Figure 1.5) shows a large red zone in the Aegean region, indicative of back-arc extension related to the subduction of the

Table 1.2 Back-arc extensional systems related to subduction.

No.	Name	Related Subduction system
8	South Sandwich Arc Spreading Ridge	South Sandwich
20	Okinawa Trough Spreading Center	Ryukyu
21	Andaman Sea Spreading Ridge	Sumatra
23	Yamato Basin Spreading Center	Japan
26	Mariana Spreading Center	Mariana
31	Bransfield Basin Spreading Ridge	Phoenix–Antarctic
33	Lau Basin Fracture Zone	Northern Tonga
34	Fiji Basin	Solomon–Vitiaz
35	Manus Spreading Ridge	Solomon Sea Plate
49	Woodlark Basin Spreading Ridge	Solomon Sea Plate

Adapted from Coffin et al. (1998).

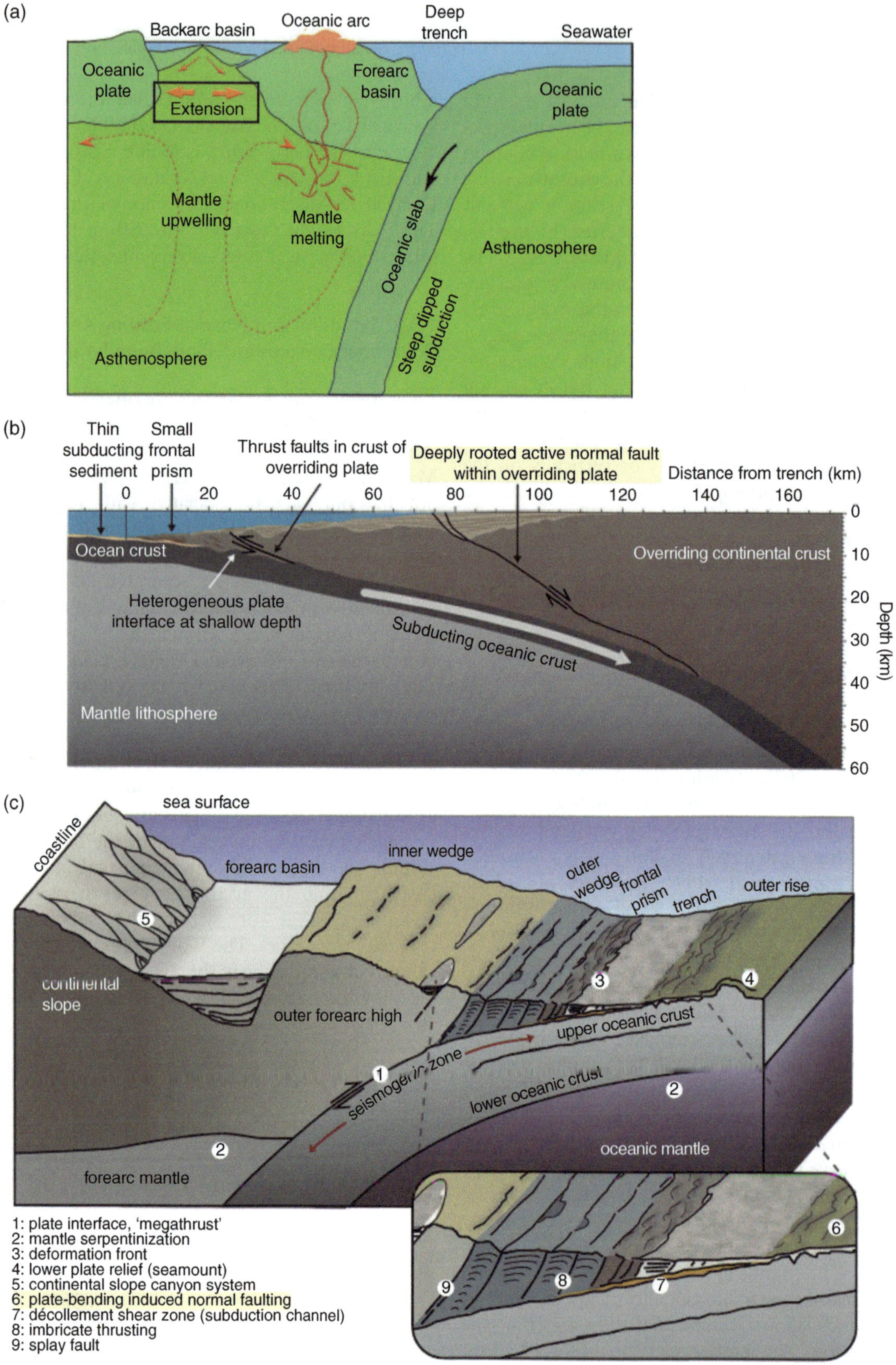

Figure 1.6 Examples of extension and normal fault tectonics in subduction zone systems. (a) Cartoon showing the components of a subduction zone with the back-arc basin highlighted. *Source:* Adapted from Zheng et al. (2015) / with permission of Springer Nature. (b) Cartoon of a structure of the subduction zone off the Alaska Peninsula. A large normal fault cuts through the overriding North American plate at a roughly 45° angle and connects to the megathrust fault along which the North American and Pacific plates move against each other. *Source:* K. Cantner, AGI, based on data adapted from Bécel et al. (2017). (c) Cartoon of a marine forearc of a convergent margin. In this scenario, plate-bending-induced normal faults are located between the outer rise and trench regions. *Source:* Adapted from Kopp (2013) / with permission of Elsevier.

African (Nubian) slab beneath the Aegean and Anatolian microplates. This volume discusses and reviews this area with three chapters (Catlos et al., "Extensional Tectonics in Western Anatolia, Turkey: Eastward Continuation of the Aegean Extension"; Catlos & Çemen, "Transfer Zones in Extended Terranes: Examples from Western Anatolia"; and Merey et al. "Büyük Menderes Graben in Western Turkey: An Example of a Rift Basin Developed in Back-Arc Extension"). Western Anatolia and the Aegean region are known for low-angle detachment systems associated with metamorphic core complexes (e.g., Catlos & Çemen, 2005; Çemen et al., 2006).

Extension can also play a significant role in the development and ultimate collapse of mountain belts (see review by Rey et al., 2001). For example, orogen-parallel extension has long been noted to play a significant role in the exhumation of the Himalayan metamorphic core (Coleman, 1996; Jessup et al., 2008; Xu et al., 2013), and extensional faulting in the range may have operated contemporaneously with thrusting (e.g., Burchfiel et al., 1992; Hodges et al., 1998). Orogen-parallel extension also accompanies the development of oroclinal bends (i.e., the portion of the Southern Carpathians that links the Carpathian and Balkan Mountains around the western edge of the Moesian platform in Romania [Schmid et al., 1998] and the Ibero-Armorican Arc [Authemayou et al., 2019]). Orogen-perpendicular extension is also recorded within the Himalaya (Coleman, 1996; Chen et al., 2022) and Alps (Champagnac et al., 2004) at specific locations and time frames. Extension also occurs perpendicular to thickened crust in the Basin and Range Province (Wernicke, 1992; Sonder & Jones, 1999). Structures associated with these extensional orientations are often explored to understand the development of ore deposits (e.g., Sánchez et al., 2014; Occhipinti et al., 2016).

Oblique extension might assist the development of rifting in nonmagmatic settings and subsequent continental break-up (Brune et al., 2012). Trench-parallel extension also occurs in subduction zone settings in the sub-Andes (e.g., Dalmayrac & Molnar, 1981; Suárez et al., 1983) and in the southern Ryukyus arc south of Taiwan (Lallemand et al., 1999). As orogens form barriers to atmospheric circulation, the extensional dynamics of large collisional and contractional mountain belts are inextricably linked to climate.

1.3.4. Regions of Extension Pose Seismic Risk

Continental normal fault-related earthquakes are thought never to reach the maximum moment magnitudes recorded by continental strike-slip and reverse faults (>Mw 8.5; Doglioni et al., 2015; Xu et al., 2018; Neely & Stein, 2021). Earthquakes along the oceanic spreading ridges show smaller magnitudes and shallower depths than convergent plate boundaries (e.g., Simkin, 1993). Moment release rates along mid-ocean ridges correlate with relative velocity between plates, and the stress drop is lower than compared to convergent boundaries (Frohlich & Wetzel, 2007). The deepest earthquakes associated with mid-ocean ridges occur at those with slow spreading rates (32 km, Southwest Indian Ridge; Grevemeyer et al., 2019). Earthquakes >Mw4.5 are absent in mid-ocean ridges with fast and intermediate spreading rates (Grevemeyer et al., 2019). The Arctic mid-ocean ridge system is the most active source of earthquakes in the northern polar region, with an Mw 6.3 earthquake occurring on the Gakkel Ridge (Schlindwein et al., 2015).

Understanding the stress field is crucial to interpreting seismic activity at mid-ocean ridge settings (e.g., Sen et al., 2023). Stresses arise at these plate boundaries from extensional kinematic boundary conditions and thermal effects as young oceanic crust cools (e.g., Choi et al., 2008; Lin & Parmentier, 1989). Topography is also a source of ridge-parallel tensile stress (e.g., Neves et al., 2004). Flexural stresses develop as spreading directions change (Pockalny et al., 1996). Rundquist and Sobolev (2002) find that relationships between the seismic moment release, fault length, and spreading rate differ between the transform and rift portions of mid-ocean ridges and indicate that lithospheric thermal structure primarily controls seismicity.

Earthquakes within stable continental regions (intraplate earthquakes) concentrate along preexisting zones of weakness, including failed rifts (Sykes, 1978; Mooney et al., 2012; Gangopadhyay & Talwani, 2003; Talwani, 2017). Although continental normal fault earthquakes are thought to have smaller maximum magnitudes than continental earthquakes associated with other fault geometries (Neely & Stein, 2021), some structures are a source of significant concern. For example, the Wasatch Fault Zone (WFZ) at the eastern boundary of the Basin and Range Province is the longest continuous, active normal fault (343 km) in the United States (Machette et al., 1991). This structure has the potential for more significant magnitude earthquakes, depending on models for its segmentation (Valentini et al., 2019) or historical seismicity (McCalpin & Nishenko, 1996). If unsegmented, larger ruptures with earthquakes of 6.9–7.9 magnitudes are possible but would be infrequent (Valentini et al., 2019). The Teton normal fault is one of the most active faults within the western interior of the United States, extending 55 km along the eastern front of the Teton Range in Wyoming as seen by prominent postglacial fault scarps (e.g., Byrd et al., 1994; DuRoss et al., 2020). This structure has the potential for large-magnitude earthquakes, as evidenced by historical seismicity (e.g., White et al., 2009; DuRoss et al., 2021). A magnitude 7.5 earthquake is anticipated for the Teton fault using a predicted return period of 800–1,800 yr (Doser & Smith, 1983). Paleoseismic events along the Teton normal fault and its associated structures may be related to the melting of glacial ice (Hampel et al., 2007).

Stresses at the zones of weakness associated with the failed rifts are similar to those at plate boundaries (hundreds of MPa, Talwani, 2017). Some of the more notable earthquakes with large magnitudes that occurred at failed rifts include the New Madrid earthquake sequence of 1811–1812 (Mw 7–8) on the Cambrian Reelfoot rift (Csontos & Arsdale, 2008; Pollitz & Mooney, 2014). The region is considered the most seismically active in the eastern United States (Csontos & Arsdale, 2008). The Kachchh Rift Basin on the western margin of peninsular India formed due to rifting of Eastern Gondwanaland during the Late-Triassic to Early Jurassic (Shaikh et al., 2020; Kothyari et al., 2022). This rift basin is considered one of the most seismically active intraplate terrains in the world, recorded by the 1819 Allah Bund earthquake (Mw 6.0), 1956 Anjar earthquake (Mw 7.8), and 2001 Bhuj earthquake (Mw 7.7) (Shaikh et al., 2020; Kothyari et al., 2022). The 1303 Hongdong earthquake occurred along the Shanxi rift system, one of the most active intraplate tectonic zones in the North China Block, with at least eight earthquakes with Mw >7 (Li et al., 2015). Failed rifts pose a significant earthquake hazard to unprepared populations and are a concern in planning infrastructure (e.g., Johnston & Kanter, 1990).

Extensional faults are observed in contractional settings and can develop and activate after large thrust events (e.g., Jara-Muñoz et al., 2022). For example, trench-parallel normal faults off the coast of central Chile were discovered after the 27 February 2010 Mw 8.8 Maule earthquake (Lange et al., 2012; Costa et al., 2020). However, some events with an extensional moment tensor can occur with no clear association with a preceding thrust event associated with the convergent margin. For example, the 1946 Ancash earthquake (Mw 6.8) occurred along the Quiches normal-slip fault (Bellier et al., 1991). The stress regime here is extensional, not compressional, and highlights the seismic risk inherent to syn-convergent extensional stress regimes (Dalmayrac & Molnar, 1981; Schwartz, 1988; Deverchère et al., 1989; Bellier et al., 1991; Wise & Noble, 2003; Siame et al., 2006).

Trench-parallel aftershocks with normal sense of motion can occur after large subduction zone earthquakes because the outer rise is temporarily in tension due to the under-thrusting motion of the oceanic plate (e.g., Christensen & Ruff, 1988; Lange et al., 2012). Surface normal faults developed after the Ancash (Peru) earthquake of 10 November 1946 (Mw 6.8) (Hodgson & Bremner, 1953; Doser, 1987; Bellier et al., 1991). This earthquake is considered the largest, historically recorded, intracontinental normal-fault earthquake that was documented to occur along the Andean margin (Doser, 1987).

1.4. OBJECTIVES AND ORGANIZATION OF THE BOOK

The objective of this volume was to compile a series of chapters that would be valuable references for audiences interested in extensional tectonics. The target audience is advanced undergraduate and graduate students, postdocs, educators, research professionals, and policymakers in academia and industry. The topics covered by the chapters in this volume synthesize current knowledge about global extension and divergent plate boundaries. Several reviewers helped critically evaluate the research presented in the chapters and appraised the current literature. The topics covering extensional tectonics are described in the preface of this volume and highlighted in the volume's table of contents. The chapters are also mentioned in this chapter of the introduction.

ACKNOWLEDGMENTS

We appreciate drafting assistance from Jeffrey S. Horowitz. No real or perceived financial conflicts of interest exist for any author. We appreciate constructive comments from Dr. Tyler Grambling and an anonymous reviewer.

AVAILABILITY STATEMENT

No new data was generated, analyzed, or presented in the manuscript.

REFERENCES

Abdelsalam, M. G., Atekwana, E. A., Keller, G. R., & Klemperer, S. L. (2004). The life cycle of continental rifting as a focus for U.S.-African scientific collaboration. *Eos, Transactions American Geophysical Union*, *85*(47), 500–500. doi: 10.1029/2004EO470004

Acocella, V. (2014). Structural control on magmatism along divergent and convergent plate boundaries: Overview, model, problems. *Earth-Science Reviews*, *136*, 226–288. doi: 10.1016/j.earscirev.2014.05.006

Acocella, V. (2021). Volcanoes at Divergent Plate Boundaries. In V. Acocella (Ed.), *Volcano-Tectonic Processes* (pp. 407–447). Cham: Springer International Publishing. doi: 10.1007/978-3-030-65968-4_11

Acocella, V., & Trippanera, D. (2016). How diking affects the tectonomagmatic evolution of slow spreading plate boundaries: Overview and model. *Geosphere*, *12*(3), 867–883. doi: 10.1130/GES01271.1

Alves, T., Fetter, M., Busby, C., Gontijo, R., Cunha, T. A., & Mattos, N. H. (2020). A tectono-stratigraphic review of continental breakup on intraplate continental margins and its impact on resultant hydrocarbon systems. *Marine and Petroleum Geology*, *117*, 104341. doi: 10.1016/j.marpetgeo.2020.104341

Anderson, D. L. (2002). Plate tectonics as a far-from-equilibrium self-organized system. *Geodynamics Series*, *30*, 411–425.

Anderson, E. M. (1905). The dynamics of faulting. *Transactions. Edinburgh Geological Society*, *8*, 387–402.

Anderson, E. M. (1951). *The dynamics of faulting and dyke formation with application to Britain*. Edinburgh: Oliver & Boyd.

Andrews, D. J., & Sleep, N. H. (1974). Numerical modelling of tectonic flow behind island arcs. *Geophysical Journal International*, *38*(2), 237–251. doi: 10.1111/j.1365-246X.1974.tb04118.x

Annunziatellis, A., Ciotoli, G., Lombardi, S., & Nolasco, F. (2003). Short- and long-term gas hazard: the release of toxic gases in the Alban Hills volcanic area (central Italy). *Journal of Geochemical Exploration*, *77*(2–3), 93–108. doi: 10.1016/S0375-6742(02)00272-8

Authemayou, C., Le Gall, B., Caroff, M., & Bussien Grosjean, D. (2019). Wrench-related dome formation and subsequent orogenic syntax bending in a hot orogen (Variscan Ibero-Armorican Arc, the Ouessant Island, France). *Tectonics*, *38*(10), 3563–3585. doi: 10.1029/2018TC005189

Ayenew, T., & Legesse, D. (2007). The changing face of the Ethiopian rift lakes and their environs: call of the time. *Lakes & Reservoirs: Science, Policy and Management for Sustainable Use*, *12*(3), 149–165. doi: 10.1111/j.1440-1770.2007.00332.x

Bailey, D. K. (1977). Lithosphere control of continental rift magmatism. *Journal of the Geological Society*, *133*(1), 103–106. doi: 10.1144/gsjgs.133.1.0103

Banks, D. (2012). *An introduction to thermogeology: Ground source heating and cooling* (1st ed.). Chichester, West Sussex: Wiley-Blackwell. doi: 10.1002/9781118447512

Barth, M. G., Mason, P. R. D., Davies, G. R., & Drury, M. R. (2008). The Othris Ophiolite, Greece: a snapshot of subduction initiation at a mid-ocean ridge. *Lithos*, *100*(1–4), 234–254. doi: 10.1016/j.lithos.2007.06.018

Barton, N., & Shen, B. (2018). Extension strain and rock strength limits for deep tunnels, cliffs, mountain walls and the highest mountains. *Rock Mechanics and Rock Engineering*, *51*(12), 3945–3962. doi: 10.1007/s00603-018-1558-2

Bécel, A., Shillington, D. J., Delescluse, M., Nedimović, M. R., Abers, G. A., Saffer, D. M., et al. (2017). Tsunamigenic stsructures in a creeping section of the Alaska subduction zone. *Nature Geoscience*, *10*(8), 609–613. doi: 10.1038/ngeo2990

Bellier, O., Dumont, J. F., Sébrier, S., & Mercier, J. L. (1991). Geological constraints on the kinematics and fault-plane solution of the quiches fault zone reactivated during the 10 November 1946 Ancash earthquake, northern Peru. *Bulletin of the Seismological Society of America*, *81*(2), 468–490.

Bialas, R. W., Buck, W. R., & Qin, R. (2010). How much magma is required to rift a continent? *Earth and Planetary Science Letters*, *292*(1–2), 68–78. doi: 10.1016/j.epsl.2010.01.021

Bickert, M., Cannat, M., & Brunelli, D. (2023). Hydrous fluids down to the semi-brittle root zone of detachment faults in nearly amagmatic ultra-slow spreading ridges. *Lithos*, *442–443*, 107084. doi: 10.1016/j.lithos.2023.107084

Bird, P. (2003). An updated digital model of plate boundaries. *Geochemistry, Geophysics, Geosystems*, *4*(3), 2001GC000252. doi: 10.1029/2001GC000252

Bischoff, J. L., Rosenbauer, R. J., Aruscavage, P. J., Baedecker, P. A., & Crock, J. G. (1983). Sea-floor massive sulfide deposits from 21 degrees N East Pacific Rise, Juan de Fuca Ridge, and Galapagos Rift; bulk chemical composition and economic implications. *Economic Geology*, *78*(8), 1711–1720. doi: 10.2113/gsecongeo.78.8.1711

Bissig, T., Lee, J. K. W., Clark, A. H., & Heather, K. B. (2001). The Cenozoic history of volcanism and hydrothermal alteration in the central Andean flat-slab region: new ^{40}Ar-^{39}Ar constraints from the El Indio–Pascua Au (-Ag, Cu) Belt, 29°20′–30°30′ S. *International Geology Review*, *43*(4), 312–340. doi: 10.1080/00206810109465016

Blackwell, D., Richards, M., Frone, Z., Batir, J., Ruzo, A., Dingwall, R., & Williams, M. (2011). Temperature-at-depth maps for the conterminous U. S. and geothermal resource estimates. *GRC Transactions*, *35*, 1545–1550.

Bonifacio, J. F., Ganade, C. E., Santos, A. C. D., & Trindade, R. I. F. D. (2023). Review and critical assessment on plate reconstruction models for the South Atlantic. *Earth-Science Reviews*, *238*, 104333. doi: 10.1016/j.earscirev.2023.104333

Boutilier, R. R., & Keen, C. E. (1999). Small-scale convection and divergent plate boundaries. *Journal of Geophysical Research: Solid Earth*, *104*(B4), 7389–7403. doi: 10.1029/1998JB900076

Brooks, C. K., Larsen, L. M., & Nielsen, T. F. D. (1991). Importance of iron-rich tholeiitic magmas at divergent plate margins: A reappraisal. *Geology*, *19*(3), 269. doi: 10.1130/0091-7613(1991)019<0269:IOIRTM>2.3.CO;2

Brune, S. (2016). Rifts and rifted margins: a review of geodynamic processes and natural hazards. In J. C. Duarte & W. P. Schellart (Eds.), *Geophysical monograph series* (1st ed., pp. 11–37). Wiley. doi: 10.1002/9781119054146.ch2

Brune, S., Kolawole, F., Olive, J.-A., Stamps, D. S., Buck, W. R., Buiter, S. J. H., et al. (2023). Geodynamics of continental rift initiation and evolution. *Nature Reviews Earth and Environment*, *4*(4), 235–253. doi: 10.1038/s43017-023-00391-3

Brune, S., Popov, A. A., & Sobolev, S. V. (2012). Modeling suggests that oblique extension facilitates rifting and continental break-up. *Journal of Geophysical Research: Solid Earth*, *117*(B8), 2011JB008860. doi: 10.1029/2011JB008860

Brune, S., Williams, S. E., & Müller, R. D. (2017). Potential links between continental rifting, CO_2 degassing and climate change through time. *Nature Geoscience*, *10*(12), 941–946. doi: 10.1038/s41561-017-0003-6

Bryndzia, L. T., Wood, B. J., & Dick, H. J. B. (1989). The oxidation state of the Earth's sub-oceanic mantle from oxygen thermobarometry of abyssal spinel peridotites. *Nature*, *341*(6242), 526–527. doi: 10.1038/341526a0

Buck, W. R. (2004). Consequences of asthenospheric variability on continental rifting. In G. D. Karner, B. Taylor, N. W. Driscoll, & D. L. Kohlstedt (Eds.), *Rheology and deformation of the lithosphere at continental margins* (pp. 1–30). Columbia University Press. doi: 10.7312/karn12738-002

Buck, W. R., Lavier, L. L., & Poliakov, A. N. B. (2005). Modes of faulting at mid-ocean ridges. *Nature*, *434*(7034), 719–723. doi: 10.1038/nature03358

Burchfiel, B. C., Zhiliang, C., Hodges, K. V., Yuping, L., Royden, L. H., Changrong, D., & Jiene, X. (1992). The South Tibetan Detachment system, Himalayan orogen: extension contemporaneous with and parallel to shortening in a collisional mountain belt. In *Geological Society of America Special Papers* (Vol. *269*, pp. 1–41). Geological Society of America. doi: 10.1130/SPE269-p1

Byrd, J. O. D., Smith, R. B., & Geissman, J. W. (1994). The Teton fault, Wyoming: topographic signature, neotectonics, and mechanisms of deformation. *Journal of Geophysical Research: Solid Earth*, *99*(B10), 20095–20122. doi: 10.1029/94JB00281

Catlos, E. J., & Çemen, I. (2005). Monazite ages and the evolution of the Menderes Massif, western Turkey. *International Journal of Earth Sciences*, *94*(2), 204–217. doi: 10.1007/s00531-005-0470-7

Çemen, I., Catlos, E. J., Göğüs, O., & Özerdem, C. (2006). Postcollisional extensional tectonics and exhumation of the Menderes massif in the Western Anatolia extended terrane, Turkey. In Y. Dilek & S. Pavlides (Eds.), *Postcollisional tectonics and magmatism in the Mediterranean region and Asia*. Geological Society of America. doi: 10.1130/2006.2409(18)

Champagnac, J., Sue, C., Delacou, B., & Burkhard, M. (2004). Brittle deformation in the inner NW Alps: from early orogen-parallel extrusion to late orogen-perpendicular collapse. *Terra Nova*, *16*(4), 232–242. doi: 10.1111/j.1365-3121.2004.00555.x

Chemenda, A. I., Mattauer, M., Malavieille, J., & Bokun, A. N. (1995). A mechanism for syn-collisional rock exhumation and associated normal faulting: results from physical modelling. *Earth and Planetary Science Letters*, *132*(1–4), 225–232. doi: 10.1016/0012-821X(95)00042-B

Chen, S., Zhang, B., Zhang, J., Wang, Y., Li, X., Zhang, L., et al. (2022). Tectonic transformation from orogen-perpendicular to orogen-parallel extension in the North Himalayan Gneiss Domes: evidence from a structural, kinematic, and geochronological investigation of the Ramba gneiss dome. *Journal of Structural Geology*, *155*, 104527. doi: 10.1016/j.jsg.2022.104527

Chiasera, B., Rooney, T. O., Bastow, I. D., Yirgu, G., Grosfils, E. B., Ayalew, D., et al. (2021). Magmatic rifting in the Main Ethiopian Rift began in thick continental lithosphere; the case of the Galema Range. *Lithos*, *406–407*, 106494. doi: 10.1016/j.lithos.2021.106494

Choi, E., Lavier, L., & Gurnis, M. (2008). Thermomechanics of mid-ocean ridge segmentation. *Physics of the Earth and Planetary Interiors*, *171*(1–4), 374–386. doi: 10.1016/j.pepi.2008.08.010

Christensen, D. H., & Ruff, L. J. (1988). Seismic coupling and outer rise earthquakes. *Journal of Geophysical Research: Solid Earth*, *93*(B11), 13421–13444. doi: 10.1029/JB093iB11p13421

Coffin, M. F., Gahagan, L. M., & Lawver, L. A. (1998). *Present-day plate boundary digital data compilation* (Progress Report No. 174) (p. 5). Austin, Texas: The University of Texas Institute of Geophysics. http://hdl.handle.net/2152/65603

Cogley, J. G. (1984). Continental margins and the extent and number of the continents. *Reviews of Geophysics*, *22*(2), 101–122. doi: 10.1029/RG022i002p00101

Coleman, M. E. (1996). Orogen-parallel and orogen-perpendicular extension in the central Nepalese Himalayas. *Geological Society of America Bulletin*, *108*(12), 1594–1607. doi: 10.1130/0016-7606(1996)108<1594:OPAOPE>2.3.CO;2

Coolbaugh, M. F., Taranik, J. V., Raines, G. L., Shevenell, L. A., Sawatzky, D. L., Bedell, R., & Minor, T. B. (2002). *A geothermal GIS for Nevada: Defining regional controls and favorable exploration terrains for extensional geothermal systems*. Paper presented at the Geothermal Resources Council: Annual Meeting, Reno, NV. https://pubs.usgs.gov/publication/70023917

Costa, C., Alvarado, A., Audemard, F., Audin, L., Benavente, C., Bezerra, F. H., et al. (2020). Hazardous faults of South America; compilation and overview. *Journal of South American Earth Sciences*, *104*, 102837. doi: 10.1016/j.jsames.2020.102837

Csontos, R., & Van Arsdale, R. (2008). New Madrid seismic zone fault geometry. *Geosphere*, *4*(5), 802. doi: 10.1130/GES00141.1

Dalmayrac, B., & Molnar, P. (1981). Parallel thrust and normal faulting in Peru and constraints on the state of stress. *Earth and Planetary Science Letters*, *55*(3), 473–481. doi: 10.1016/0012-821X(81)90174-6

Daniels, K. A., Bastow, I. D., Keir, D., Sparks, R. S. J., & Menand, T. (2014). Thermal models of dyke intrusion during development of continent–ocean transition. *Earth and Planetary Science Letters*, *385*, 145–153. doi: 10.1016/j.epsl.2013.09.018

Davies, G. F. (2022). *Stories from the deep Earth: how scientists figured out what drives tectonic plates and mountain building*. Cham: Springer International Publishing. doi: 10.1007/978-3-030-91359-5

Davies, T. C. (2008). Environmental health impacts of East African Rift volcanism. *Environmental Geochemistry and Health*, *30*(4), 325–338. doi: 10.1007/s10653-008-9168-7

Davis, G., & Friedmann, S. (2005). Large-scale gravity sliding in the Miocene Shadow Valley supradetachment basin, Eastern Mojave Desert, California. *Earth-Science Reviews*, *73*(1–4), 149–176. doi: 10.1016/j.earscirev.2005.04.008

Deverchère, J., Dorbath, C., & Dorbath, L. (1989). Extension related to a high topography: results from a microearthquake survey in the Andes of Peru and tectonic implications. *Geophysical Journal International*, *98*(2), 281–292. doi: 10.1111/j.1365-246X.1989.tb03352.x

Dick, H. J. B., Lin, J., & Schouten, H. (2003). An ultraslow-spreading class of ocean ridge. *Nature*, *426*(6965), 405–412. doi: 10.1038/nature02128

Doglioni, C., Carminati, E., Petricca, P., & Riguzzi, F. (2015). Normal fault earthquakes or graviquakes. *Scientific Reports*, *5*(1), 12110. doi: 10.1038/srep12110

Doser, D. I. (1987). The Ancash, Peru, earthquake of 1946 November 10: evidence for low-angle normal faulting in the high Andes of northern Peru. *Geophysical Journal International*, *91*(1), 57–71. doi: 10.1111/j.1365-246X.1987.tb05213.x

Doser, D. I., & Smith, R. B. (1983). Seismicity of the Teton-southern Yellowstone region, Wyoming. *Bulletin of the*

Seismological Society of America, *73*(5), 1369–1394. doi: 10.1785/BSSA0730051369

Dunbar, J. A., & Sawyer, D. S. (1987). Implications of continental crust extension for plate reconstruction: an example from the Gulf of Mexico. *Tectonics*, *6*(6), 739–755. doi: 10.1029/TC006i006p00739

DuRoss, C. B., Gold, R. D., Briggs, R. W., Delano, J. E., Ostenaa, D. A., Zellman, M. S., et al. (2020). Holocene earthquake history and slip rate of the southern Teton fault, Wyoming, USA. *GSA Bulletin*, *132*(7–8), 1566–1586. doi: 10.1130/B35363.1

DuRoss, C. B., Zellman, M. S., Thackray, G. D., Briggs, R. W., Gold, R. D., & Mahan, S. A. (2021). Holocene paleoseismology of the Steamboat Mountain Site: evidence for full-length rupture of the Teton Fault, Wyoming. *Bulletin of the Seismological Society of America*, *111*(1), 439–465. doi: 10.1785/0120200212

Eaton, G. P. (1982). The basin and range province: origin and tectonic significance. *Annual Review of Earth and Planetary Sciences*, *10*(1), 409–440. doi: 10.1146/annurev.ea.10.050182.002205

Ebinger, C. J., & Casey, M. (2001). Continental breakup in magmatic provinces: an Ethiopian example. *Geology*, *29*(6), 527. doi: 10.1130/0091-7613(2001)029<0527:CBIMPA>2.0.CO;2

Edmonds, M., Grattan, J., & Michnowicz, S. (2015). Volcanic gases: silent killers. In C. J. Fearnley, et al. (Eds.), *Observing the volcano world* (pp. 65–83). Cham: Springer International Publishing. doi: 10.1007/11157_2015_14

Elling, R., Stein, S., Stein, C., & Gefeke, K. (2022). Three major failed rifts in central North America: similarities and differences. *GSA Today*, *32*(6), 4–11. https://doi.org/GSATG518A.1

English, J. M., English, K. L., Dunphy, R. B., Blake, S., Walsh, J., Raine, R., et al. (2023). An overview of deep geothermal energy and its potential on the Island of Ireland. *First Break*, *41*(2), 33–43. doi: 10.3997/1365-2397.fb2023009

Ernst, W. G., & Liu, J. (1998). Experimental phase-equilibrium study of Al- and Ti-contents of calcic amphibole in MORB; a semiquantitative thermobarometer. *American Mineralogist*, *83*(9–10), 952–969. doi: 10.2138/am-1998-9-1004

Faulds, J., Coolbaugh, M., Bouchot, V., Moek, I., & Oguz, K. (2010). *Characterizing structural controls of geothermal reservoirs in the Great Basin, USA, and Western Turkey: developing successful exploration strategies in extended terranes*. Paper presented at the World Geothermal Congress, Bali, Indonesia. https://brgm.hal.science/hal-00495884

Floyd, P. A. (1995). Igneous activity. In R. D. Dallmeyer, W. Franke, & K. Weber (Eds.), *Pre-Permian geology of central and eastern Europe* (pp. 59–81). Berlin, Heidelberg: Springer Berlin Heidelberg. doi: 10.1007/978-3-642-77518-5_6

Friedmann, S. J., & Burbank, D. W. (1995). Rift basins and supradetachment basins: intracontinental extensional end-members. *Basin Research*, *7*(2), 109–127. doi: 10.1111/j.1365-2117.1995.tb00099.x

Frisch, W., Meschede, M., & Blakey, R. (2011). Subduction zones, island arcs and active continental margins. In W. Frisch, M. Meschede, & R. C. Blakey (Eds.), *Plate tectonics* (pp. 91–122). Berlin, Heidelberg: Springer Berlin Heidelberg. doi: 10.1007/978-3-540-76504-2_7

Frohlich, C., & Wetzel, L. R. (2007). Comparison of seismic moment release rates along different types of plate boundaries. *Geophysical Journal International*, *171*(2), 909–920. doi: 10.1111/j.1365-246X.2007.03550.x

Früh-Green, G. L., Kelley, D. S., Lilley, M. D., Cannat, M., Chavagnac, V., & Baross, J. A. (2022). Diversity of magmatism, hydrothermal processes and microbial interactions at mid-ocean ridges. *Nature Reviews Earth and Environment*, *3*(12), 852–871. doi: 10.1038/s43017-022-00364-y

Gangopadhyay, A., & Talwani, P. (2003). Symptomatic features of intraplate earthquakes. *Seismological Research Letters*, *74*(6), 863–883. doi: 10.1785/gssrl.74.6.863

Gannoun, A., Burton, K. W., Day, J. M. D., Harvey, J., Schiano, P., & Parkinson, I. (2016). Highly siderophile element and os isotope systematics of volcanic rocks at divergent and convergent plate boundaries and in intraplate settings. *Reviews in Mineralogy and Geochemistry*, *81*(1), 651–724. doi: 10.2138/rmg.2016.81.11

Gébelin, A., Mulch, A., Teyssier, C., Heizler, M., Vennemann, T., & Seaton, N. C. A. (2011). Oligo-Miocene extensional tectonics and fluid flow across the Northern Snake Range detachment system, Nevada. *Tectonics*, *30*(5), 2010TC002797. doi: 10.1029/2010TC002797

George, S. W. M., Perez, N. D., Struble, W., Curry, M. E., & Horton, B. K. (2022). Aseismic ridge subduction focused late Cenozoic exhumation above the Peruvian flat slab. *Earth and Planetary Science Letters*, *600*, 117754. doi: 10.1016/j.epsl.2022.117754

Gordon, A. C., Mohriak, W. U., & Barbosa, V. C. F. (2013). Crustal architecture of the Almada Basin, NE Brazil: an example of a non-volcanic rift segment of the South Atlantic passive margin. *Geological Society, London, Special Publications*, *369*(1), 215–234. doi: 10.1144/SP369.1

Granado, P., & Ruh, J. B. (2019). Numerical modelling of inversion tectonics in fold-and-thrust belts. *Tectonophysics*, *763*, 14–29. doi: 10.1016/j.tecto.2019.04.033

Grassle, J. F., & Maciolek, N. J. (1992). Deep-sea species richness: Regional and local diversity estimates from quantitative bottom samples. *The American Naturalist*, *139*(2), 313–341. doi: 10.1086/285329

Grevemeyer, I., Hayman, N. W., Lange, D., Peirce, C., Papenberg, C., Van Avendonk, H. J. A., et al. (2019). Constraining the maximum depth of brittle deformation at slow- and ultraslow-spreading ridges using microseismicity. *Geology*, *47*(11), 1069–1073. doi: 10.1130/G46577.1

Grimes, C. B., Wooden, J. L., Cheadle, M. J., & John, B. E. (2015). "Fingerprinting" tectono-magmatic provenance using trace elements in igneous zircon. *Contributions to Mineralogy and Petrology*, *170*(5–6), 46. doi: 10.1007/s00410-015-1199-3

Groves, D. I., Goldfarb, R. J., Gebre-Mariam, M., Hagemann, S. G., & Robert, F. (1998). Orogenic gold deposits: a proposed classification in the context of their crustal distribution and relationship to other gold deposit types. *Ore Geology Reviews*, *13*(1–5), 7–27. doi: 10.1016/S0169-1368(97)00012-7

Gutscher, M., Spakman, W., Bijwaard, H., & Engdahl, E. R. (2000). Geodynamics of flat subduction: seismicity and tomographic constraints from the Andean margin. *Tectonics*, *19*(5), 814–833. doi: 10.1029/1999TC001152

Hager, B. H., & O'Connell, R. J. (1981). A simple global model of plate dynamics and mantle convection. *Journal of Geophysical Research: Solid Earth*, *86*(B6), 4843–4867. doi: 10.1029/JB086iB06p04843

Hales, A. L. (1969). Gravitational sliding and continental drift. *Earth and Planetary Science Letters*, *6*(1), 31–34. doi: 10.1016/0012-821X(69)90156-3

Hampel, A., Hetzel, R., & Densmore, A. L. (2007). Postglacial slip-rate increase on the Teton normal fault, northern Basin and Range Province, caused by melting of the Yellowstone ice cap and deglaciation of the Teton Range? *Geology*, *35*(12), 1107. doi: 10.1130/G24093A.1

Hannington, M., Herzig, P., Scott, S., Thompson, G., & Rona, P. (1991). Comparative mineralogy and geochemistry of gold-bearing sulfide deposits on the mid-ocean ridges. *Marine Geology*, *101*(1–4), 217–248. doi: 10.1016/0025-3227(91)90073-D

Haymon, R. M., Fornari, D. J., Edwards, M. H., Carbotte, S., Wright, D., & Macdonald, K. C. (1991). Hydrothermal vent distribution along the East Pacific Rise crest (9°09′–54′N) and its relationship to magmatic and tectonic processes on fast-spreading mid-ocean ridges. *Earth and Planetary Science Letters*, *104*(2–4), 513–534. doi: 10.1016/0012-821X(91)90226-8

Healy, D., Sibson, R. H., Shipton, Z., & Butler, R. (2012). Stress, faulting, fracturing and seismicity: the legacy of Ernest Masson Anderson. *Geological Society, London, Special Publications*, *367*(1), 1–6. doi: 10.1144/SP367.1

Hearn, G. J. (2022). Some of the geological challenges and opportunities associated with the dynamics of the Cenozoic East African Rift System. *Quarterly Journal of Engineering Geology and Hydrogeology*, *55*(3) qjegh2021-060. doi: 10.1144/qjegh2021-060

Heidbach, O., Rajabi, M., Cui, X., Fuchs, K., Müller, B., Reinecker, J., et al. (2018). The World Stress Map database release 2016: crustal stress pattern across scales. *Tectonophysics*, *744*, 484–498. doi: 10.1016/j.tecto.2018.07.007

Herzig, P. M., & Hannington, M. D. (1995). Polymetallic massive sulfides at the modern seafloor a review. *Ore Geology Reviews*, *10*(2), 95–115. doi: 10.1016/0169-1368(95)00009-7

Hess, H. H. (1962). History of ocean basins. In A. E. J. Engel, H. L. James, & B. F. Leonard (Eds.), *Petrologic studies* (pp. 599–620). USA: Geological Society of America. doi: 10.1130/Petrologic.1962.599

Hiett, C. D., Newell, D. L., & Jessup, M. J. (2021). The evidence for fluid transfer and continental hydration above a flat slab. *Earth and Planetary Science Letters*, *556*, 116722. doi: 10.1016/j.epsl.2020.116722

Hodges, K., Bowring, S., Davidek, K., Hawkins, D., & Krol, M. (1998). Evidence for rapid displacement on Himalayan normal faults and the importance of tectonic denudation in the evolution of mountain ranges. *Geology*, *26*(6), 483. doi: 10.1130/0091-7613(1998)026<0483:EFRDOH>2.3.CO;2

Hodgson, J. H., & Bremner, P. C. (1953). Direction of faulting in the Ancash, Peru, earthquake of November 10, 1946, from teleseismic evidence. *Bulletin of the Seismological Society of America*, *43*(2), 121–125. doi: 10.1785/BSSA0430020121

Holm, N. G. (1992). Why are hydrothermal systems proposed as plausible environments for the origin of life? In N. G. Holm (Ed.), *Marine hydrothermal systems and the origin of life* (pp. 5–14). Dordrecht: Springer Netherlands. doi: 10.1007/978-94-011-2741-7_1

Horton, B. K., Capaldi, T. N., & Perez, N. D. (2022). The role of flat slab subduction, ridge subduction, and tectonic inheritance in Andean deformation. *Geology*, *50*(9), 1007–1012. doi: 10.1130/G50094.1

Hou, Y., Wang, H., Fan, T., Zhang, H., Yang, R., Li, Y., & Long, S. (2020). Rift-related sedimentary evolution and its response to tectonics and climate changes: A case study of the Guaizihu sag, Yingen-Ejinaqi Basin, China. *Journal of Asian Earth Sciences*, *195*, 104370. doi: 10.1016/j.jseaes.2020.104370

Humphris, S. E., Herzig, P. M., Miller, D. J., Alt, J. C., Becker, K., Brown, D., et al. (1995). The internal structure of an active sea-floor massive sulphide deposit. *Nature*, *377*(6551), 713–716. doi: 10.1038/377713a0

Jara-Muñoz, J., Melnick, D., Li, S., Socquet, A., Cortés-Aranda, J., Brill, D., & Strecker, M. R. (2022). The cryptic seismic potential of the Pichilemu blind fault in Chile revealed by off-fault geomorphology. *Nature Communications*, *13*(1), 3371. doi: 10.1038/s41467-022-30754-1

Jessup, M. J., Newell, D. L., Cottle, J. M., Berger, A. L., & Spotila, J. A. (2008). Orogen-parallel extension and exhumation enhanced by denudation in the trans-Himalayan Arun River gorge, Ama Drime Massif, Tibet-Nepal. *Geology*, *36*(7), 587. doi: 10.1130/G24722A.1

Jia, J., Wu, Y., Miao, C., Fu, C., Xie, W., Qin, J., & Wang, X. (2021). Tectonic controls on the sedimentation and thermal history of supra-detachment basins: a case study of the Early Cretaceous Fuxin Basin, NE China. *Tectonics*, *40*(5), e2020TC006535. doi: 10.1029/2020TC006535

Johnston, A. C., & Kanter, L. R. (1990). Earthquakes in stable continental crust. *Scientific American*, *262*(3), 68–75.

Jolie, E., Scott, S., Faulds, J., Chambefort, I., Axelsson, G., Gutiérrez-Negrín, L. C., et al. (2021). Geological controls on geothermal resources for power generation. *Nature Reviews Earth and Environment*, *2*(5), 324–339. doi: 10.1038/s43017-021-00154-y

Jourdon, A., Le Pourhiet, L., Mouthereau, F., & Masini, E. (2019). Role of rift maturity on the architecture and shortening distribution in mountain belts. *Earth and Planetary Science Letters*, *512*, 89–99. doi: 10.1016/j.epsl.2019.01.057

Kennish, M. J., & Lutz, R. A. (1998). Morphology and distribution of lava flows on mid-ocean ridges: a review. *Earth-Science Reviews*, *43*(3–4), 63–90. doi: 10.1016/S0012-8252(98)00006-3

Kinabo, B. D., Hogan, J. P., Atekwana, E. A., Abdelsalam, M. G., & Modisi, M. P. (2008). Fault growth and propagation during incipient continental rifting: insights from a combined aeromagnetic and shuttle radar topography mission digital elevation model investigation of the Okavango Rift Zone, northwest Botswana. *Tectonics*, *27*(3), 2007TC002154. doi: 10.1029/2007TC002154

Kious, W. J., & Tilling, R. I. (2016). *This dynamic earth: the story of plate tectonics*. Online Edition. U.S. Geological Survey. https://pubs.usgs.gov/gip/dynamic/dynamic.html

Knight, D., R Roberts, S., & P Webber, A. (2018). The influence of spreading rate, basement composition, fluid chemistry and chimney morphology on the formation of gold-rich SMS deposits at slow and ultraslow mid-ocean ridges. *Mineralium Deposita*, *53*(1), 143–152. doi: 10.1007/s00126-017-0762-4

Kopp, H. (2013). Invited review paper: the control of subduction zone structural complexity and geometry on margin segmentation and seismicity. *Tectonophysics*, *589*, 1–16. doi: 10.1016/j.tecto.2012.12.037

Kothyari, G. C., Kandregula, R. S., Dumka, R., Chauhan, G., & Taloor, A. K. (2022). Quaternary tectonic history of seismically active intraplate Kachchh Rift Basin, western India: a review. *Geodesy and Geodynamics*, *13*(3), 192–204. doi: 10.1016/j.geog.2021.09.011

Lallemand, S., Liu, C., Dominguez, S., Schnürle, P., & Malavieille, J. (1999). Trench-parallel stretching and folding of forearc basins and lateral migration of the accretionary wedge in the southern Ryukyus: a case of strain partition caused by oblique convergence. *Tectonics*, *18*(2), 231–247. doi: 10.1029/1998TC900011

Lambiase, J. J., & Morley, C. K. (1999). Hydrocarbons in rift basins: the role of stratigraphy. *Philosophical Transactions of the Royal Society of London, Series A: Mathematical, Physical and Engineering Sciences*, *357*(1753), 877–900. doi: 10.1098/rsta.1999.0356

Landis, G. P., & Rye, R. O. (1974). Geologic, fluid inclusion, and stable isotope studies of the Pasto Buena tungsten-base metal ore deposit, Northern Peru. *Economic Geology*, *69*(7), 1025–1059. doi: 10.2113/gsecongeo.69.7.1025

Lange, D., Tilmann, F., Barrientos, S. E., Contreras-Reyes, E., Methe, P., Moreno, M., et al. (2012). Aftershock seismicity of the 27 February 2010 Mw 8.8 Maule earthquake rupture zone. *Earth and Planetary Science Letters*, *317–318*, 413–425. doi: 10.1016/j.epsl.2011.11.034

Laurila, T. E., Hannington, M. D., Leybourne, M., Petersen, S., Devey, C. W., & Garbe-Schönberg, D. (2015). New insights into the mineralogy of the Atlantis II Deep metalliferous sediments, R ed S ea. *Geochemistry, Geophysics, Geosystems*, *16*(12), 4449–4478. doi: 10.1002/2015GC006010

Le Pichon, X. (1968). Sea-floor spreading and continental drift. *Journal of Geophysical Research*, *73*(12), 3661–3697. doi: 10.1029/JB073i012p03661

Li, B., Atakan, K., Sørensen, M. B., & Havskov, J. (2015). Stress pattern of the Shanxi rift system, North China, inferred from the inversion of new focal mechanisms. *Geophysical Journal International*, *201*(2), 505–527. doi: 10.1093/gji/ggv025

Li, Y., Wang, S., Ye, Z., Lin, S., Wang, Y., & Zuo, S. (2023). The sedimentary filling differences of subsags and their impacts on oil shale and coal accumulation in a rift lacustrine sag: A case study from the Lower Cretaceous in the Zhangqiang sag, southern Songliao Basin, Northeast China. *Marine and Petroleum Geology*, *155*, 106354. doi: 10.1016/j.marpetgeo.2023.106354

Lin, J., & Parmentier, E. M. (1989). Mechanisms of lithospheric extension at mid-ocean ridges. *Geophysical Journal International*, *96*(1), 1–22. doi: 10.1111/j.1365-246X.1989.tb05246.x

Lu, G., Zhao, L., Chen, L., Wan, B., & Wu, F. (2021). Reviewing subduction initiation and the origin of plate tectonics: what do we learn from present-day Earth? *Earth and Planetary Physics*, *5*(2), 123–140. doi: 10.26464/epp2021014

Macdonald, K. C. (2001). Mid-ocean ridge tectonics, volcanism and geomorphology. In *Encyclopedia of ocean sciences* (pp. 1798–1813). Elsevier. doi: 10.1006/rwos.2001.0094

MacDonald, K. C., Scheirer, D. S., & Carbotte, S. M. (1991). Mid-ocean ridges: Discontinuities, segments and giant cracks. *Science*, *253*(5023), 986–994. doi: 10.1126/science.253.5023.986

Machette, M. N., Personius, S. F., Nelson, A. R., Schwartz, D. P., & Lund, W. R. (1991). The Wasatch fault zone, Utah—segmentation and history of Holocene earthquakes. *Journal of Structural Geology*, *13*(2), 137–149. doi: 10.1016/0191-8141(91)90062-N

Marchev, P., Kaiser-Rohrmeier, M., Heinrich, C., Ovtcharova, M., Von Quadt, A., & Raicheva, R. (2005). 2: Hydrothermal ore deposits related to post-orogenic extensional magmatism and core complex formation: the Rhodope Massif of Bulgaria and Greece. *Ore Geology Reviews*, *27*(1–4), 53–89. doi: 10.1016/j.oregeorev.2005.07.027

McCalpin, J. P., & Nishenko, S. P. (1996). Holocene paleoseismicity, temporal clustering, and probabilities of future large (M > 7) earthquakes on the Wasatch fault zone, Utah. *Journal of Geophysical Research: Solid Earth*, *101*(B3), 6233–6253. doi: 10.1029/95JB02851

McKenzie, D. (1977). Surface deformation, gravity anomalies and convection. *Geophysical Journal International*, *48*(2), 211–238. doi: 10.1111/j.1365-246X.1977.tb01297.x

McNeill, L. C., Shillington, D. J., Carter, G. D. O., Everest, J. D., Gawthorpe, R. L., Miller, C., et al. (2019). High-resolution record reveals climate-driven environmental and sedimentary changes in an active rift. *Scientific Reports*, *9*(1), 3116. doi: 10.1038/s41598-019-40022-w

Milanovsky, E. E. (1972). Continental Rift zones: their arrangement and development. *Tectonophysics*, *15*(1–2), 65–70. doi: 10.1016/0040-1951(72)90052-2

Milia, A., & Torrente, M. M. (2015). Rift and supradetachment basins during extension: insight from the Tyrrhenian rift. *Journal of the Geological Society*, *172*(1), 5–8. doi: 10.1144/jgs2014-046

Mooney, W. D., Ritsema, J., & Hwang, Y. K. (2012). Crustal seismicity and the earthquake catalog maximum moment magnitude (Mcmax) in stable continental regions (SCRs): correlation with the seismic velocity of the lithosphere. *Earth and Planetary Science Letters*, *357–358*, 78–83. doi: 10.1016/j.epsl.2012.08.032

Mulyukova, E., & Bercovici, D. (2019). The generation of plate tectonics from grains to global scales: a brief review. *Tectonics*, *38*(12), 4058–4076. doi: 10.1029/2018TC005447

Muñoz-Barrera, J. M., Rotevatn, A., Gawthorpe, R. L., Henstra, G., & Kristensen, T. B. (2022). Supradetachment basins in necking domains of rifted margins: insights from the Norwegian Sea. *Basin Research*, *34*(3), 991–1019. doi: 10.1111/bre.12648

Murton, B. J., Lehrmann, B., Dutrieux, A. M., Martins, S., De La Iglesia, A. G., Stobbs, I. J., et al. (2019). Geological fate of seafloor massive sulphides at the TAG hydrothermal field (Mid-Atlantic Ridge). *Ore Geology Reviews*, *107*, 903–925. doi: 10.1016/j.oregeorev.2019.03.005

Neely, J. S., & Stein, S. (2021). Why do continental normal fault earthquakes have smaller maximum magnitudes? *Tectonophysics*, *809*, 228854. doi: 10.1016/j.tecto.2021.228854

Neugebauer, H. J. (1983). Mechanical aspects of continental rifting. In *Developments in geotectonics* (Vol. *19*, pp. 91–108). Elsevier. doi: 10.1016/B978-0-444-42198-2.50013-4

Neumann, E.-R., Olsen, K. H., Baldridge, W. S., & Sundvoll, B. (1992). The Oslo Rift: a review. *Tectonophysics*, *208*(1–3), 1–18. doi: 10.1016/0040-1951(92)90333-2

Neves, M. C., Bott, M. H. P., & Searle, R. C. (2004). Patterns of stress at midocean ridges and their offsets due to seafloor subsidence. *Tectonophysics*, *386*(3–4), 223–242. doi: 10.1016/j.tecto.2004.06.010

Newell, D. L., Jessup, M. J., Hilton, D. R., Shaw, C. A., & Hughes, C. A. (2015). Mantle-derived helium in hot springs of the Cordillera Blanca, Peru: implications for mantle-to-crust fluid transfer in a flat-slab subduction setting. *Chemical Geology*, *417*, 200–209. doi: 10.1016/j.chemgeo.2015.10.003

Nyarko, S. C., & Rudge, D. W. (2022). Using the history of plate tectonics to teach the nature of science. *International Journal of Science Education*, *44*(12), 1958–1977. doi: 10.1080/09500693.2022.2105977

Occhipinti, S. A., Metelka, V., Lindsay, M. D., Hollis, J. A., Aitken, A. R. A., Tyler, I. M., et al. (2016). Multicommodity mineral systems analysis highlighting mineral prospectivity in the Halls Creek Orogen. *Ore Geology Reviews*, *72*, 86–113. doi: 10.1016/j.oregeorev.2015.07.003

Olaka, L. A., Odada, E. O., Trauth, M. H., & Olago, D. O. (2010). The sensitivity of East African rift lakes to climate fluctuations. *Journal of Paleolimnology*, *44*(2), 629–644. doi: 10.1007/s10933-010-9442-4

Olive, J.-A. (2023). Mid-ocean ridges: geodynamics written in the seafloor. In *Dynamics of plate tectonics and mantle convection* (pp. 483–510). Elsevier. doi: 10.1016/B978-0-323-85733-8.00018-4

Oner, Z., & Dilek, Y. (2011). Supradetachment basin evolution during continental extension: the Aegean province of western Anatolia, Turkey. *Geological Society of America Bulletin*, *123*(11–12), 2115–2141. doi: 10.1130/B30468.1

Pagli, C., Mazzarini, F., Keir, D., Rivalta, E., & Rooney, T. O. (2015). Introduction: anatomy of rifting: tectonics and magmatism in continental rifts, oceanic spreading centers, and transforms. *Geosphere*, *11*(5), 1256–1261. doi: 10.1130/GES01082.1

Palin, R. M., & Santosh, M. (2021). Plate tectonics: what, where, why, and when? *Gondwana Research*, *100*, 3–24. doi: 10.1016/j.gr.2020.11.001

Palin, R. M., Santosh, M., Cao, W., Li, S.-S., Hernández-Uribe, D., & Parsons, A. (2020). Secular change and the onset of plate tectonics on Earth. *Earth-Science Reviews*, *207*, 103172. doi: 10.1016/j.earscirev.2020.103172

Pearce, J. A. (1996). A user's guide to basalt discrimination diagrams. In D. A. Wyman (Ed.), *Trace element geochemistry of volcanic rocks: applications for massive sulphide exploration* (Vol. *12*, pp. 79–113). Geological Association of Canada, Short Course Notes.

Perfit, M. R., & Chadwick, W. W. (1998). Magmatism at mid-ocean ridges: constraints from volcanological and geochemical investigations. In W. Roger Buck, P. T. Delaney, J. A. Karson, & Y. Lagabrielle (Eds.), *Geophysical Monograph Series* (pp. 59–115). Washington, DC: American Geophysical Union. doi: 10.1029/GM106p0059

Peron-Pinvidic, G., Manatschal, G., & "IMAGinING RIFTING" Workshop Participants. (2019). Rifted margins: state of the art and future challenges. *Frontiers in Earth Science*, *7*, 218. doi: 10.3389/feart.2019.00218

Philippon, M., & Corti, G. (2016). Obliquity along plate boundaries. *Tectonophysics*, *693*, 171–182. doi: 10.1016/j.tecto.2016.05.033

Phipps Morgan, J., & Chen, Y. J. (1993). Dependence of ridge-axis morphology on magma supply and spreading rate. *Nature*, *364*(6439), 706–708. doi: 10.1038/364706a0

Pockalny, R. A., Gente, P., & Buck, R. (1996). Oceanic transverse ridges: a flexural response to fracture-zone–normal extension. *Geology*, *24*(1), 71. doi: 10.1130/0091-7613(1996)024<0071:OTRAFR>2.3.CO;2

Pollitz, F. F., & Mooney, W. D. (2014). Seismic structure of the Central US crust and shallow upper mantle: uniqueness of the Reelfoot Rift. *Earth and Planetary Science Letters*, *402*, 157–166. doi: 10.1016/j.epsl.2013.05.042

Poulsen, C. J., Gendaszek, A. S., & Jacob, R. L. (2003). Did the rifting of the Atlantic Ocean cause the Cretaceous thermal maximum? *Geology*, *31*(2), 115. doi: 10.1130/0091-7613(2003)031<0115:DTROTA>2.0.CO;2

Purdy, G. M., Kong, L. S. L., Christeson, G. L., & Solomon, S. C. (1992). Relationship between spreading rate and the seismic structure of mid-ocean ridges. *Nature*, *355*(6363), 815–817. doi: 10.1038/355815a0

Püthe, C., & Gerya, T. (2014). Dependence of mid-ocean ridge morphology on spreading rate in numerical 3-D models. *Gondwana Research*, *25*(1), 270–283. doi: 10.1016/j.gr.2013.04.005

Rey, P., Vanderhaeghe, O., & Teyssier, C. (2001). Gravitational collapse of the continental crust: definition, regimes and modes. *Tectonophysics*, *342*(3–4), 435–449. doi: 10.1016/S0040-1951(01)00174-3

Robbins, E. I., Wilkes, G. P., & Textoris, D. A. (1988). Coal deposits of the Newark rift system. In *Developments in Geotectonics* (Vol. *22*, pp. 649–682). Elsevier. doi: 10.1016/B978-0-444-42903-2.50032-4

Rohrmeier, M. K., Von Quadt, A., Driesner, T., Heinrich, C. A., Handler, R., Ovtcharova, M., et al. (2013). Post-orogenic extension and hydrothermal ore formation: high-precision geochronology of the central Rhodopian metamorphic core complex (Bulgaria-Greece). *Economic Geology*, *108*(4), 691–718. doi: 10.2113/econgeo.108.4.691

Rona, P. A. (1984). Hydrothermal mineralization at seafloor spreading centers. *Earth-Science Reviews*, *20*(1), 1–104. doi: 10.1016/0012-8252(84)90080-1

Rona, P. A. (2003). Resources of the Sea Floor. *Science*, *299*(5607), 673–674. doi: 10.1126/science.1080679

Rop, B. K., Anyona, S., & Anyona, M. (2022). *The East African rift system: does it have oil and gas potential?* Paper presented at the 2016 Sustainable Research & Innovation (SRI) Conference. https://sri.jkuat.ac.ke/jkuatsri/index.php/sri/article/view/299

Rundquist, D. V., & Sobolev, P. O. (2002). Seismicity of mid-oceanic ridges and its geodynamic implications: a review. *Earth-Science Reviews*, *58*(1–2), 143–161. doi: 10.1016/S0012-8252(01)00086-1

Ryan, W. B. F., Carbotte, S. M., Coplan, J. O., O'Hara, S., Melkonian, A., Arko, R., et al. (2009). Global multi-resolution

topography synthesis. *Geochemistry, Geophysics, Geosystems, 10*(3), 2008GC002332. doi: 10.1029/2008GC002332
Sánchez, M. G., Allan, M. M., Hart, C. J. R., & Mortensen, J. K. (2014). Extracting ore-deposit-controlling structures from aeromagnetic, gravimetric, topographic, and regional geologic data in western Yukon and eastern Alaska. *Interpretation, 2*(4), SJ75–SJ102. doi: 10.1190/INT-2014-0104.1
Schlindwein, V., Demuth, A., Korger, E., Läderach, C., & Schmid, F. (2015). Seismicity of the Arctic mid-ocean Ridge system. *Polar Science, 9*(1), 146–157. doi: 10.1016/j.polar.2014.10.001
Schmid, S. M., Berza, T., Diaconescu, V., Froitzheim, N., & Fügenschuh, B. (1998). Orogen-parallel extension in the Southern Carpathians. *Tectonophysics, 297*(1–4), 209–228. doi: 10.1016/S0040-1951(98)00169-3
Scholten, J. C., Staffers, P., Garbe-Schdnberg, D., & Moammar, M. (2017). Hydrothermal mineralization in the Red Sea. In D. S. Cronan (Ed.), *Handbook of marine mineral deposits* (1st ed., pp. 369–395). Routledge. doi: 10.1201/9780203752760-14
Scholz, C. A., Cohen, A. S., Johnson, T. C., King, J., Talbot, M. R., & Brown, E. T. (2011). Scientific drilling in the Great Rift Valley: The 2005 Lake Malawi Scientific Drilling Project — an overview of the past 145,000 years of climate variability in Southern Hemisphere East Africa. *Palaeogeography, Palaeoclimatology, Palaeoecology, 303*(1–4), 3–19. doi: 10.1016/j.palaeo.2010.10.030
Schouten, H., Klitgord, K. D., & Whitehead, J. A. (1985). Segmentation of mid-ocean ridges. *Nature, 317*(6034), 225–229. doi: 10.1038/317225a0
Schwartz, D. P. (1988). Paleoseismicity and neotectonics of the Cordillera Blanca Fault Zone, northern Peruvian Andes. *Journal of Geophysical Research: Solid Earth, 93*(B5), 4712–4730. doi: 10.1029/JB093iB05p04712
Scoon, R. N. (2018). The East African Rift system. In I. R. N. Scoon (Ed.), *Geology of national parks of central/southern Kenya and northern Tanzania* (pp. 19–28). Cham: Springer International Publishing. doi: 10.1007/978-3-319-73785-0_3
Scott, B. E., Newell, D. L., Jessup, M. J., Grambling, T. A., & Shaw, C. A. (2020). Structural controls on crustal fluid circulation and hot spring geochemistry above a flat-slab subduction zone, Peru. *Geochemistry, Geophysics, Geosystems, 21*(7), e2020GC008919. doi: 10.1029/2020GC008919
Searle, M. (2013). Crustal melting, ductile flow, and deformation in mountain belts: cause and effect relationships. *Lithosphere, 5*(6), 547–554. doi: 10.1130/RF.L006.1
Sen, J., Sarkar, S., & Mandal, N. (2023). Complexity in the 3D stress fields of mid-ocean ridge tectonics: a manifestation of sub-crustal mush dynamics. Preprint. doi:10.31223/X5GD5S
Seton, M., Müller, R. D., Zahirovic, S., Gaina, C., Torsvik, T., Shephard, G., et al. (2012). Global continental and ocean basin reconstructions since 200Ma. *Earth-Science Reviews, 113*(3–4), 212–270. doi: 10.1016/j.earscirev.2012.03.002
Seton, M., Williams, S. E., Domeier, M., Collins, A. S., & Sigloch, K. (2023). Deconstructing plate tectonic reconstructions. *Nature Reviews Earth and Environment, 4*(3), 185–204. doi: 10.1038/s43017-022-00384-8
Shaikh, M. A., Maurya, D. M., Mukherjee, S., Vanik, N. P., Padmalal, A., & Chamyal, L. S. (2020). Tectonic evolution of the intra-uplift Vigodi-Gugriana-Khirasra-Netra Fault System in the seismically active Kachchh rift basin, India: implications for the western continental margin of the Indian plate. *Journal of Structural Geology, 140*, 104124. doi: 10.1016/j.jsg.2020.104124
Siame, L. L., Sébrier, M., Bellier, O., & Bourles, D. (2006). Can cosmic ray exposure dating reveal the normal faulting activity of the Cordillera Blanca Fault, Peru? *Revista de la Asociación Geológica Argentina, 61*(4), 536–544.
Sibuet, J.-C., & Tucholke, B. E. (2013). The geodynamic province of transitional lithosphere adjacent to magma-poor continental margins. *Geological Society, London, Special Publications, 369*, 429–452. doi: 10.1144/SP369.15
Sifford, A., & Bloomquist, R. G. (2000). Geothermal electric power production in the United States: a survey update for 1995–1999. In *Proceedings world geothermal congress* (pp. 389–396). Japan: Kyushu - Tohoku. www.geothermal-energy.org/pdf/IGAstandard/WGC/2000/R0243.PDF
Sigmundsson, F., Hooper, A., Hreinsdóttir, S., Vogfjörd, K. S., Ófeigsson, B. G., Heimisson, E. R., et al. (2015). Segmented lateral dyke growth in a rifting event at Bárðarbunga volcanic system, Iceland. *Nature, 517*(7533), 191–195. doi: 10.1038/nature14111
Simkin, T. (1993). Terrestrial volcanism in space and time. *Annual Review of Earth and Planetary Sciences, 21*(1), 427–452. doi: 10.1146/annurev.ea.21.050193.002235
Simpson, R. W. (1997). Quantifying Anderson's fault types. *Journal of Geophysical Research: Solid Earth, 102*(B8), 17909–17919. doi: 10.1029/97JB01274
Sinton, J. M., Smaglik, S. M., Mahoney, J. J., & Macdonald, K. C. (1991). Magmatic processes at superfast spreading mid-ocean ridges: glass compositional variations along the East Pacific Rise 13°–23°S. *Journal of Geophysical Research: Solid Earth, 96*(B4), 6133–6155. doi: 10.1029/90JB02454
Small, C. (1998). Global systematics of mid-ocean ridge morphology. In W. Roger Buck, P. T. Delaney, J. A. Karson, & Y. Lagabrielle (Eds.), *Geophysical monograph series* (pp. 1–25). Washington, DC: American Geophysical Union. doi: 10.1029/GM106p0001
Solomon, S. C., Richardson, R. M., & Bergman, E. A. (1980). Tectonic stress: models and magnitudes. *Journal of Geophysical Research: Solid Earth, 85*(B11), 6086–6092. doi: 10.1029/JB085iB11p06086
Sonder, L. J., & Jones, C. H. (1999). Western United States extension: how the west was widened. *Annual Review of Earth and Planetary Sciences, 27*(1), 417–462. doi: 10.1146/annurev.earth.27.1.417
Sorkhabi, R. (2012). Know your faults! *GeoExPro, 9*(6), 72–76.
Soule, S. A. (2015). Mid-ocean ridge volcanism. In *The encyclopedia of volcanoes* (pp. 395–403). Elsevier. doi: 10.1016/B978-0-12-385938-9.00021-3
Spence, W. (1987). Slab pull and the seismotectonics of subducting lithosphere. *Reviews of Geophysics, 25*(1), 55–69. doi: 10.1029/RG025i001p00055
Suárez, G., Molnar, P., & Burchfiel, B. C. (1983). Seismicity, fault plane solutions, depth of faulting, and active tectonics of the Andes of Peru, Ecuador, and southern Colombia. *Journal of Geophysical Research: Solid Earth, 88*(B12), 10403–10428. doi: 10.1029/JB088iB12p10403

Sun, S., Qiu, L., Yan, D.-P., Zhou, Z., Zhang, J., Wang, X., et al. (2022). Formation and evolution of supradetachment basins during continental extension: insights from the Fuxin Basin in NE China. *Frontiers in Earth Science*, *10*, 845812. doi: 10.3389/feart.2022.845812

Sykes, L. R. (1978). Intraplate seismicity, reactivation of preexisting zones of weakness, alkaline magmatism, and other tectonism postdating continental fragmentation. *Reviews of Geophysics*, *16*(4), 621–688. doi: 10.1029/RG016i004p00621

Talwani, P. (2017). On the nature of intraplate earthquakes. *Journal of Seismology*, *21*(1), 47–68. doi: 10.1007/s10950-016-9582-8

Toussaint, G., Burov, E., & Avouac, J. -P. (2004). Tectonic evolution of a continental collision zone: a thermomechanical numerical model. *Tectonics*, *23*(6), 2003TC001604. doi: 10.1029/2003TC001604

Turcotte, D., & Schubert, G. (2014). *Geodynamics* (3rd ed.). Cambridge University Press. doi: 10.1017/CBO9780511843877

Urbani, S., Trippanera, D., Porreca, M., Kissel, C., & Acocella, V. (2015). Anatomy of an extinct magmatic system along a divergent plate boundary: Alftafjordur, Iceland. *Geophysical Research Letters*, *42*(15), 6306–6313. doi: 10.1002/2015GL065087

Valentini, A., DuRoss, C. B., Field, E. H., Gold, R. D., Briggs, R. W., Visini, F., & Pace, B. (2019). Relaxing segmentation on the Wasatch fault zone: impact on seismic hazard. *Bulletin of the Seismological Society of America*, *110*(1), 83–109. doi: 10.1785/0120190088

Van Hinsbergen, D. J. J. (2010). A key extensional metamorphic complex reviewed and restored: the Menderes Massif of western Turkey. *Earth-Science Reviews*, *102*(1–2), 60–76. doi: 10.1016/j.earscirev.2010.05.005

Vine, F. J., & Matthews, D. H. (1963). Magnetic anomalies over oceanic ridges. *Nature*, *199*(4897), 947–949. doi: 10.1038/199947a0

Vink, G. E. (1982). Continental rifting and the implications for plate tectonic reconstructions. *Journal of Geophysical Research: Solid Earth*, *87*(B13), 10677–10688. doi: 10.1029/JB087iB13p10677

Wang, P., Huang, C.-Y., Lin, J., Jian, Z., Sun, Z., & Zhao, M. (2019). The South China Sea is not a mini-Atlantic: plate-edge rifting vs intra-plate rifting. *National Science Review*, *6*(5), 902–913. doi: 10.1093/nsr/nwz135

Wegener, A. (1912). Die entstehung der kontinente. *Geologische Rundschau*, *3*(4), 276–292.

Wernicke, B. (1992). Cenozoic extensional tectonics of the U.S. Cordillera. In B. C. Burchfiel, P. W. Lipman, & M. L. Zoback (Eds.), *The Cordilleran Orogen* (pp. 553–517). Boulder, Colorado: Geological Society of America. doi: 10.1130/DNAG-GNA-G3.553

Wernicke, B., & Burchfiel, B. C. (1982). Modes of extensional tectonics. *Journal of Structural Geology*, *4*(2), 105–115. doi: 10.1016/0191-8141(82)90021-9

White, B. J. P., Smith, R. B., Husen, S., Farrell, J. M., & Wong, I. (2009). Seismicity and earthquake hazard analysis of the Teton–Yellowstone region, Wyoming. *Journal of Volcanology and Geothermal Research*, *188*(1–3), 277–296. doi: 10.1016/j.jvolgeores.2009.08.015

Whitney, D. L., Teyssier, C., Rey, P., & Buck, W. R. (2013). Continental and oceanic core complexes. *Geological Society of America Bulletin*, *125*(3–4), 273–298. doi: 10.1130/B30754.1

Wilson, J. T. (1965). A new class of faults and their bearing on continental drift. *Nature*, *207*(4995), 343–347. doi: 10.1038/207343a0

Wise, J. M., & Noble, D. C. (2003). Geomorphic evolution of the Cordillera Blanca, Northern Peru. *Boletín de La Sociedad Geológica Del Perú*, *96*, 65–85.

Wisian, K. W., & Blackwell, D. D. (2004). Numerical modeling of basin and range geothermal systems. *Geothermics*, *33*(6), 713–741. doi: 10.1016/j.geothermics.2004.01.002

Wu, Y., Liao, J., Qing, J., & Shen, Y. (2023). Subduction development along extinct mid-ocean ridges versus weakened passive continental margins. *Gondwana Research*, *117*, 243–260. doi: 10.1016/j.gr.2023.02.001

Wyllie, P. J. (1988). Magma genesis, plate tectonics, and chemical differentiation of the Earth. *Reviews of Geophysics*, *26*(3), 370–404. doi: 10.1029/RG026i003p00370

Xia, L., & Li, X. (2019). Basalt geochemistry as a diagnostic indicator of tectonic setting. *Gondwana Research*, *65*, 43–67. doi: 10.1016/j.gr.2018.08.006

Xu, Y., He, H., Deng, Q., Allen, M. B., Sun, H., & Bi, L. (2018). The CE 1303 Hongdong earthquake and the Huoshan Piedmont Fault, Shanxi Graben: implications for magnitude limits of normal fault earthquakes. *Journal of Geophysical Research: Solid Earth*, *123*(4), 3098–3121. doi: 10.1002/2017JB014928

Xu, Z., Wang, Q., Pêcher, A., Liang, F., Qi, X., Cai, Z., et al. (2013). Orogen-parallel ductile extension and extrusion of the Greater Himalaya in the late Oligocene and Miocene. *Tectonics*, *32*(2), 191–215. doi: 10.1002/tect.20021

Zakharova, N. V., Goldberg, D. S., Olsen, P. E., Collins, D., & Kent, D. V. (2020). Reservoir and sealing properties of the Newark rift basin formations: implications for carbon sequestration. *The Leading Edge*, *39*(1), 38–46. doi: 10.1190/tle39010038.1

Zametzer, A., Reimann, S., Heubeck, C. E., Thomsen, T. B., & Serre, S. H. (2023). Tectonic and sedimentary evolution of a potential supra-detachment fault basin, Archaean Moodies Group (~3.22 Ga), central Barberton Greenstone Belt. *South African Journal of Geology*, *126*(3), 235–260. doi: 10.25131/sajg.126.0016

Zheng, Y., Chen, Y., Dai, L., & Zhao, Z. (2015). Developing plate tectonics theory from oceanic subduction zones to collisional orogens. *Science China Earth Sciences*, *58*(7), 1045–1069. doi: 10.1007/s11430-015-5097-3

Zhu, G., Yang, H., Yang, T., & Zhang, G. (2023). Along-strike variation of seismicity near the extinct mid-ocean ridge subducted beneath the manila trench. *Seismological Research Letters*, *94*(2A), 792–804. doi: 10.1785/0220220304

Zou, C., Zhang, G., Zhu, R., Yuan, X., Zhao, X., Hou, L., et al. (2013). Formation and distribution of volcanic rock. In *Volcanic reservoirs in petroleum exploration* (pp. 11–29). Elsevier. doi: 10.1016/B978-0-12-397163-0.00002-6

Zwann, F., & Schreurs, G. (2022). *Continental rifted margins 1: definition and methodology* (1st ed.). Wiley. doi: 10.1002/9781119986928

Part I
Large-Scale Extensional Tectonics

2
The East African Rift System

Tyrone O. Rooney[1], Ian D. Bastow[2], Karen Fontijn[3], Nicholas Mariita[4], James D. Muirhead[5], and Derek Keir[6,7]

ABSTRACT

The East African Rift System (EARS) is the iconic continental rift and is now recognized as a profoundly interconnected system, extending from the core-mantle boundary to Earth's surface. Markedly broad, low seismic wavespeeds extend from the core-mantle boundary into the upper mantle and reveal thermochemically anomalous material of the African Superplume. This dominates the upper mantle and impacts melt generation and rifting processes. Associated flood basalt volcanism commenced during the Eocene and was accompanied by rifting. Rift maturity varies from embryonic fault-controlled rifting in the south with classic graben and half graben geomorphology. Further north the locus of extension has migrated to narrower zones of magma intrusion that may eventually mark a new mid-ocean ridge. The EARS shows us that magmatic extension commences well before the development of an ocean basin, with important implications for stretching factors and subsidence at rifted margins. With increasing urbanization, research of rifting in East Africa has implications for volcanic and seismic hazards and climate change through CO_2 release. Geothermal energy too is a growing field with important socio-economic implications for a fast-growing population. This contribution provides a succinct review of the EARS in the context of the multi-scale interconnections of the rift system.

2.1. INTRODUCTION

The East African Rift System (EARS) is perhaps the most iconic example of a continental rift (Figure 2.1). The relatively narrow rift valleys, flanked by border faults with significant throw and associated relief, prompted early geoscientists to provide the first definition of a rift valley as a "linear valley with parallel almost vertical sides, which has fallen due to a series of vertical faults" (Gregory, 1896). New insights into the topography of East Africa that were developed throughout the 1800s resulted in the understanding of an interconnected rift system (Sueß, 1891). Further insights in our understanding of rift development followed in the 20th century, particularly during the advent of plate tectonic theory, with the recognition of a coincidence between continental flood basalts and continental rifts, which resulted in the hypothesis that thermochemical anomalies rising from the deep mantle may interact with and weaken the continental lithosphere, resulting in rift development (Morgan, 1971). East Africa was not among the examples initially posed in this model; however, the region was so central to our understanding of continental rifting that Morgan (1971) lamented that a "search should be made for such

[1] *Department of Earth and Environmental Sciences, Michigan State University, East Lansing, Michigan, USA*

[2] *Department of Earth Science and Engineering, Imperial College London, London, United Kingdom*

[3] *Department of Geosciences, Environment and Society, Université Libre de Bruxelles, Brussels, Belgium*

[4] *Geothermal Energy Training and Research Institute, Dedan Kimathi University of Technology, Nyeri, Kenya*

[5] *School of Environment, University of Auckland, Auckland, New Zealand*

[6] *Department of Ocean and Earth Science, University of Southampton, Southampton, United Kingdom*

[7] *Department of Earth Sciences, University of Florence, Florence, Italy*

Extensional Tectonics: Rifting and Continental Extension, Geophysical Monograph 278, First Edition.
Edited by İbrahim Çemen and Elizabeth J. Catlos.

DOI: 10.1002/9781119773757.ch02

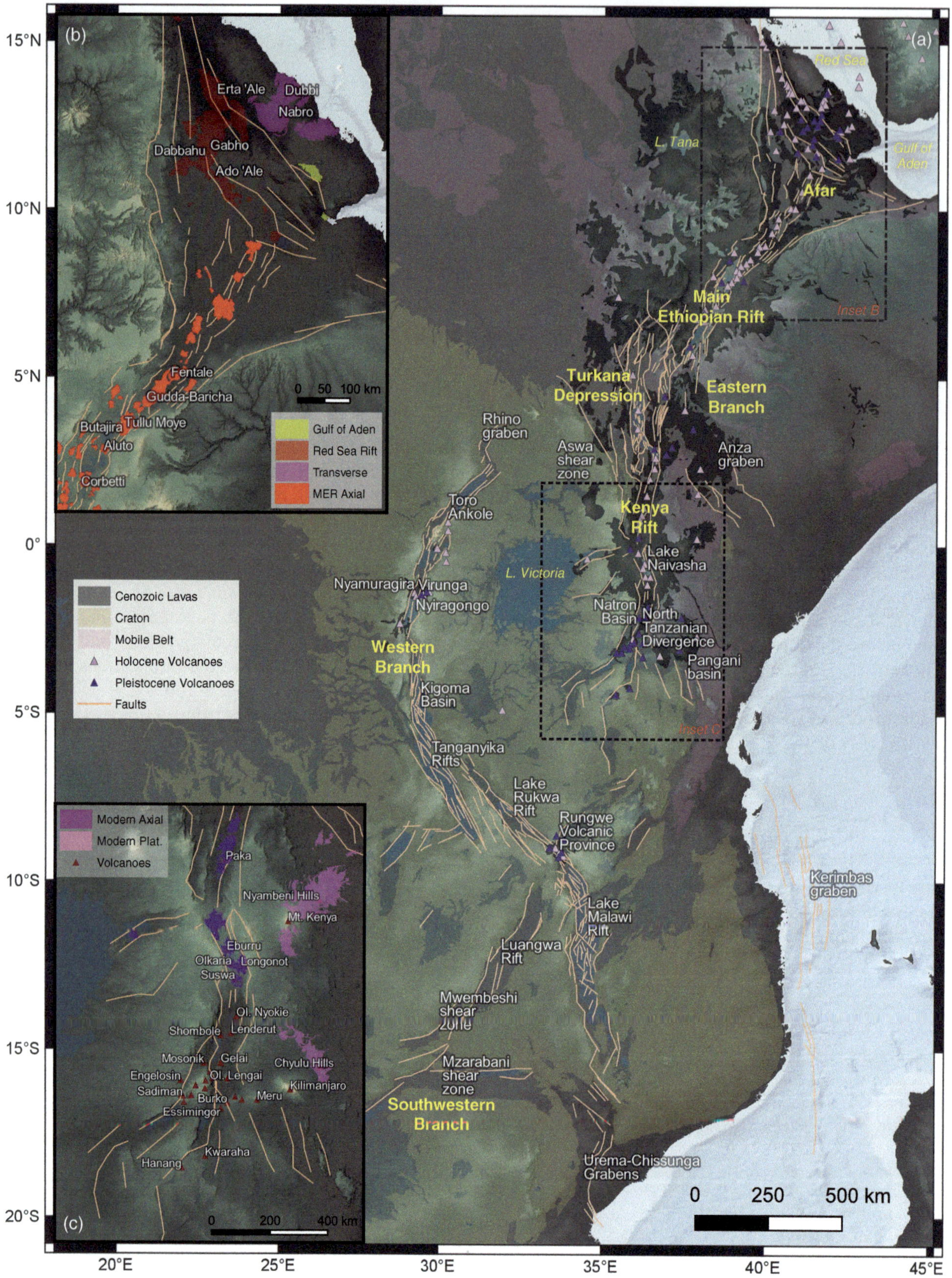

Figure 2.1 Generalized map of the East African Rift System. Major faults are derived from MacGregor (2015) / with permission of Elsevier. The extent of Cenozoic magmatism in East Africa, in addition to the Craton/Mobile belt division, is derived from the USGS Surficial Geology of Africa (geo7_2ag). Volcano locations are derived from the Smithsonian Volcano Database. (a) Map showing the broad structure of the East African Rift. (b) Inset of the Northern Main Ethiopian Rift and Afar Depression. The shaded regions represent Quaternary volcanic units within this region using the divisions described in Rooney (2020d) / with permission of Elsevier. (c) Inset of the Kenya Rift and Northern Tanzania Divergence. The shaded regions represent Quaternary volcanic units within this region using the divisions described in Rooney (2020c) / with permission of Elsevier. Note that "Modern Plat." are units erupted on the plateaus surrounding the Kenya Rift. Major volcanoes and basins are labeled.

continental [magmatic] activity in East Africa." Subsequent studies indeed revealed the presence of continental flood basalts in East Africa (e.g., Davidson & Rex, 1980; Mohr & Zanettin, 1988); however, the age, distribution, and composition of these flood basalts are far more complex than the initial simple mantle plume models predict (e.g., Davidson & Rex, 1980; Ebinger & Sleep, 1998; Furman et al., 2016; George et al., 1998; Hofmann et al., 1997; Montagner et al., 2007).

The advent of the digital era in the 1980s–1990s made it possible, for the first time, for large volumes of geophysical data to be collected along the EARS. Wide-angle seismic studies and satellite data began to reveal the structural complexities of the crust, including the unexpected observation that the EARS' pronounced surface morphology is in many places not mirrored at depth by markedly thinned crust. Contrary to expectations of traditional rifting models, significant extension was instead found to be accommodated by magma intrusion long before the birth of a new ocean basin (e.g., Keranen et al., 2004; Mackenzie et al., 2005). Deeper still, mantle seismic tomography began to allow geophysicists to comment on debates concerning the thermal and chemical evolution of the EARS: specifically, whether or not it was underlain by a mantle plume. There, too, a surprising result emerged: rather than imaging narrow plume tails below the uplifted Ethiopian and East African plateaus, a broad (>500 km wide), inclined, low-wavespeed anomaly was illuminated, emanating from the core-mantle boundary below southern Africa, impinging on the base of the African plate somewhere below Ethiopia and Kenya (e.g., Ritsema et al., 1999).

In recent years, the rapidly growing human population of East Africa has prompted geoscientific focus in East Africa to engage with sustainable energy and hazard research. Satellite radar interferometry (InSAR) and other remote sensing tools, allied with more intensive geophysical, geological, and geochemical investigations of active volcanic systems, have generated insights into volcanic cyclicity and flux of CO_2 into the atmosphere (e.g., Hunt et al., 2017). In parallel with these new insights on volcanism, East Africa has become a leader in geothermal exploration with mature systems proving the utility of this sustainable energy source (e.g., Omenda, 1998). Advances from the initial observations of the digital era have resulted in improved deep mantle tomographic imaging (e.g., Boyce et al., 2021; Chang et al., 2020), more advanced geodynamic models (e.g., Garnero et al., 2016; Koptev et al., 2018; Moucha & Forte, 2011), and region-wide studies of magmatism (e.g., Rooney, 2017). Through this research, it has become apparent that the East African upper mantle is largely contaminated by thermochemically anomalous material rising from one of two antipodal deep mantle anomalies: the African Large Low Shear Velocity Province (LLSVP). As research progresses during this modern era, early insights into an interconnected rift system have relevance once more; the initial view of a *laterally* interconnected rift system has evolved to demonstrate that the EARS is also *vertically* interlinked and extends from the very deepest mantle anomalies through to the surface manifestations of rifting evident in geothermal fields.

This chapter examines advances in our understanding of the origin and evolution of the EARS from a multidisciplinary perspective. After describing the crustal architecture of the EARS, we examine how thermochemical mantle anomalies have influenced the development of rifting and magmatism. Finally, we explore volcanism, hazards, and geothermal power associated with the East African Rift. Given the volume of the existing scholarship on the EARS, we aim to provide a succinct synthesis of the key results.

2.2. STRUCTURE OF THE EARS

2.2.1. Regional-Scale Structural Patterns in the EARS

The EARS comprises a series of segmented half- and full-graben rift basins that have gradually developed since ~45 Ma (e.g., Boone et al., 2019). We focus primarily on the onshore surface manifestation of Cenozoic rift activity on the eastern African continent, which occurs as far north as the Afar Triple Junction in Ethiopia and southward into Okavango graben in Botswana (e.g., Macgregor, 2015; Scholz et al., 1976). However, geodetic, reflection seismic, and seismicity data reveal that Cenozoic rifting associated with the EARS also extends into offshore Mozambique/Tanzania (e.g., Stamps et al., 2008; Franke et al., 2015), and GPS observations indicate a ~600 km-wide zone of diffuse deformation and extension across Madagascar (Stamps et al., 2010). These many branches of modern-day rifting associated with the EARS dissect the Nubian and Somalian plates to form the Victoria, Rovuma, Lwandle, and San microplates (Jestin et al., 1994; Stamps et al., 2008; Saria et al., 2013; Daly et al., 2020).

The EARS is divided into several rift branches containing multiple basin segments (Figure 2.1); the most prominent branches, the Eastern and Western Branches, are separated by ~700 km where the EARS bifurcates around the Tanzania Craton (e.g., Daly et al., 1989; Ebinger, 1989). The Eastern Branch can be traced from the southern end of the Afar Triple Junction in a SSW direction through the Main Ethiopian and Kenya Rifts and from the Natron-Manyara segments of the North Tanzanian Divergence, with the possibility of extending southward toward the Rungwe Volcanic Province (e.g., Mulibo & Nyblade, 2016) (Figure 2.1). Additionally, the Eastern Branch splits at the North Tanzanian Divergence, creating a SSE-trending rift arm from the Pangani basin in north Tanzania southward to offshore rift segments, such as the Kerimbas graben in offshore Mozambique (e.g., Franke et al., 2015; Vormann & Jokat, 2021). The northernmost expression of the Western Branch is observed immediately northwest of the Tanzania Craton in the Rhino graben, Uganda (Figure 2.1). From here, the Western Branch continues southward around the edge of the Tanzania Craton, forming a crude SSE trend as the branch merges into the Lake Malawi Rift and then continues farther south into Urema-Chissunga grabens as it approaches offshore Mozambique (Steinbruch, 2010). In southern sectors of the EARS, another potential branch of Neogene rifting (the Southwestern Branch) occurs south of the Rungwe Volcanic Province and runs from the Luangwa Rift to the Okavango graben (Macgregor, 2015; Pastier et al., 2017; Scholz et al., 1976). Of note, evidence for significant Neogene rift faulting (e.g., displaced syn-rift sediments) is lacking in some of these basins, such as the Luangwa Rift (e.g., Banks et al., 1995); however, their inclusion here is based on the present distribution of seismicity (Craig et al., 2011; Delvaux & Barth, 2010; Mulibo & Nyblade, 2016; Scholz et al., 1976) and evidence of Quaternary-age faulting in the southernmost sector in the Okavango Rift (e.g., Modisi et al., 2000). Considering all the various rift branches outlined, the cumulative length of rift basins associated with onshore Neogene rifting in East Africa exceeds 5,000 km.

To a first order, the distribution of rifting since the Late Paleogene is influenced by basement terranes and the many suture zones dissecting these, as well as Mesozoic-age rifting that predates the onset of the EARS (Banks et al., 1995; Chorowicz et al., 1987; Daly et al., 1989; Kolawole et al., 2018; Scholz et al., 1976; Smith & Mosley, 1993). Overall, rifting has nucleated preferentially in Proterozoic lithosphere, not Archean terranes, as exemplified by the bifurcation of the EARS around the Tanzania Craton (Figure 2.1). Additionally, continental-scale shear zones (e.g., Aswa shear zone) facilitate abrupt deviations in the general trend of some rift sectors (Smith & Mosley, 1993) (Figure 2.1). Similarly, earthquake locations and the general orientation of rift basins in the Southwestern Branch follow the NE-SW tends of major Proterozoic shear zones (e.g., Mwembeshi and Mzarabani shear zones; Banks et al., 1995; Modisi et al., 2000) while also avoiding cratonic lithosphere (e.g., Calais et al., 2006; Kolawole et al., 2017). Mesozoic rifts are also important in controlling the general pattern of younger rifting events. The Lake Rukwa and Tanganyika Rifts nucleated along a zone of Karoo-age rifting (Klerkx et al., 1998; Wright et al., 2020) and in some cases appear to have reactivated Mesozoic faults (Kilembe & Rosendahl, 1992). Finally, the initial (~45 Ma) locus of EARS rifting in the Turkana Depression localized within a site of Cretaceous to Early Paleogene rifting in the Anza graben (e.g., Ebinger et al., 2000; Morley et al., 1999).

2.2.2. Rift Basin Structure and Segmentation

Rift sectors throughout the EARS are typically expressed as a series of half-graben basins, with rare full grabens, which are kinematically linked by transform or accommodation zones (Mohr, 1974; Morley et al., 1999) (Figures 2.2 and 2.3). These basins are bounded by large border faults, which can be up to ~120 km long (Ebinger et al., 1999) and typically accrue 1–10 km of normal dip-slip displacement over the lifetime of the rift (Morley, 1988; Morley et al., 1999; Wright et al., 2020) (Figure 2.3). Along-strike displacement profiles gradually decrease away from the fault center toward the fault tip (Ebinger et al., 1987, 1999; Shillington et al., 2020). At the distal ends of border faults, strain is typically transferred to an overlapping border fault segment, forming an adjacent basin (Ebinger et al., 1999), often resulting in soft-linked basin segments (e.g., Rowland & Sibson, 2001) (Figure 2.3).

Border fault traces have been previously depicted with arcuate shapes and facing directions that alternate along strike between basins (e.g., Bosworth et al., 1986; Rosendahl et al., 1986). However, structural mapping utilizing suites of field, satellite, seismicity, and reflection seismic data (e.g., Agostini et al., 2011; Ebinger, 1989; Mohr, 1974; Morley, 1988) suggests these were idealized depictions in accordance with preconceived models of rift architecture. Earlier studies also favored dominantly listric fault hypotheses, with border faults manifesting at depth as lower crustal detachments (e.g., Bosworth et al., 1986; Morley, 1989). However, seismic reflection profiles reveal dominantly steep faults (~45–70° dips) and structural restorations across these profiles in the Lake Tanganyika Rift balance with steep planar faults (e.g., Muirhead et al., 2019; Wright et al., 2020). Subsurface projections of border faults to regions of deep seismicity (20–27 km) in early-stage basins in the Eastern Branch (e.g., Natron-Manyara basins) also support minimum fault dips of ~50°, consistent with their corresponding earthquake fault plane solutions (Weinstein et al., 2017). Seismicity in the Main Ethiopian Rift (MER) similarly reveals steep fault dips (Keir et al., 2006). A notable exception is the broadly rifted Turkana Depression, where reflection seismic observations reveal both high- and low-angle border fault segments in Late Oligocene to Miocene-age basins (Morley et al., 1999).

The opposing sides of half-graben basins in the EARS are defined by a flexural margin (Figure 2.2). This margin represents the crest of a monoclinal fold resulting from the characteristic deflection profile of the hanging wall in response to vertical displacements along the fault plane (Bott, 1996). The width of this zone of flexure, and hence

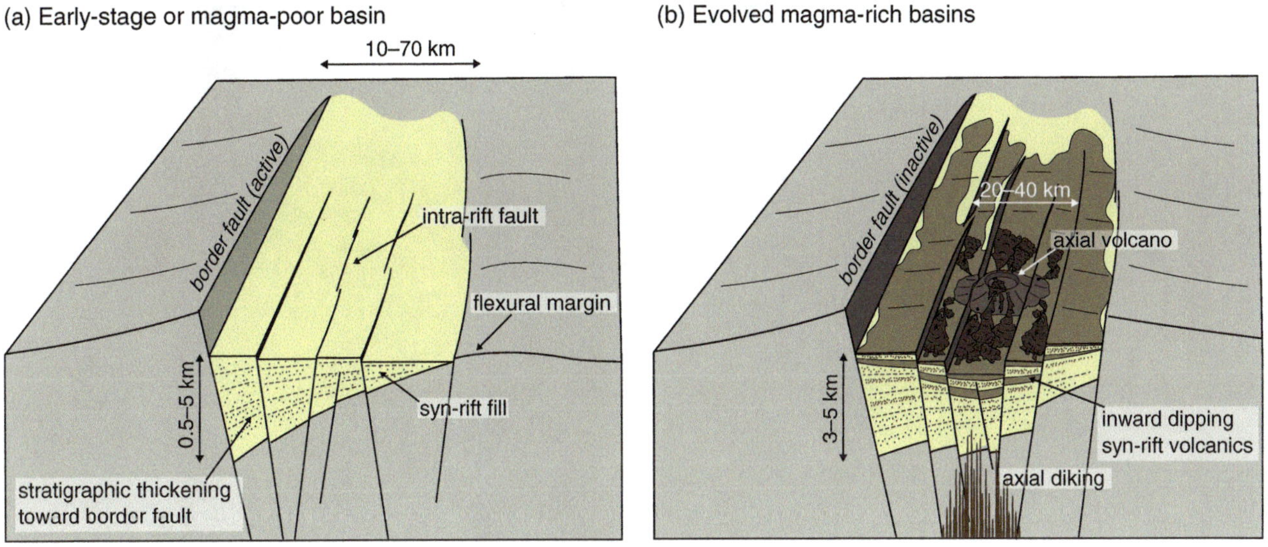

Figure 2.2 Simplified conceptual illustration of the shallow crustal structure (less than 10 km depth) of evolving rift basins in the East African Rift System. The model is adapted from Ebinger et al. (2013). (a) The structure of early-stage basins (<5 Ma) in the Eastern Branch and the generally magma-poor basins of the Western Branch, where extension is largely accommodated by large vertical displacements on border faults. These border fault displacements result in half-graben basin morphologies with typical basin widths of ~50 km. (b) As rift basins evolve (>10 Ma basins) in the Eastern Branch, magmatism becomes a dominant mode of strain accommodation. These basins are magmatically segmented, where axial dike intrusions accommodate the majority of upper crustal extension, and faulting localizes away from border faults into zones of magmatism in the rift center.

the width of the half-graben itself, is largely dependent on the rigidity of the plate (e.g., Bott, 1996; Ebinger et al., 1999; Muirhead et al., 2016), with half grabens in the EARS typically 30–60 km wide (Ebinger, 1989; Ebinger et al., 1999; Morley, 1988; Muirhead et al., 2016; Rosendahl et al., 1986). Within these basins, abundant intrarift faults can be observed dissecting the flexing border fault hanging wall, with faults tens of kilometers long typically exhibiting throws of up to hundreds of meters (e.g., Agostini et al., 2011; Mohr, 1974; Muirhead et al., 2016; Shillington et al., 2020).

The development of structural relief during rifting exerts first-order control on sedimentation and basin-filling patterns (Ebinger & Scholz, 2011; Morley, 1989). Across-strike, the locus of subsidence and rift sedimentation in these half-graben basins occurs adjacent to the border fault (Figure 2.3), with along-strike thickening of the rift fill observed toward the center of the border fault at the location of maximum fault displacement (Ebinger & Scholz, 2011; Morley, 1989; Rosendahl et al., 1986; Wolfenden et al., 2004; Wright et al., 2020). Across-strike stratal thickening toward major faults supports the structural development of accommodation space during the deposition of these syn-rift packages and provides constraints on the timing of border fault movements during different phases of rift extension (e.g., Morley et al., 1999; Wolfenden et al., 2004; Wright et al., 2020).

2.2.3. Rift Basin Kinematics

Earlier studies of rift basin kinematics generally favored a NW-SE extension direction across East Africa since the Late Pleistocene, and thus, structures trending oblique to this extension direction were interpreted to exhibit oblique and strike-slip motions (Chorowicz et al., 1987; Delvaux et al., 1992; Wheeler & Karson, 1994). Hypotheses of dominantly strike-slip motions within these basins are hard to justify given the amount of dip-slip recorded on their border faults, which, based on the depth of syn-rift fill, reaches 7 km in North Malawi (Accardo et al., 2018; Shillington et al., 2020) and ≥13 km in the Kigoma Basin of Lake Tanganyika (cumulative throw on western and eastern border faults; Wright et al., 2020).

Geodetic data indicate a clockwise rotation of the Somalian plate with respect to the Nubian plate about a pole located offshore of South Africa, supporting E-W to WNW-ESE regional extension across the EARS (e.g., Saria et al., 2013). These geodetic data are consistent with average spreading rates and transform-fault azimuths along the Southwest Indian Ridge, suggesting that this broad extension has persisted since at least 3.2 Ma (Stamps et al., 2008). Focal mechanism data also support dominantly dip-slip motions and orthogonal basin opening across the EARS, regardless of the strike of rift border faults relative to the regional-scale plate motion

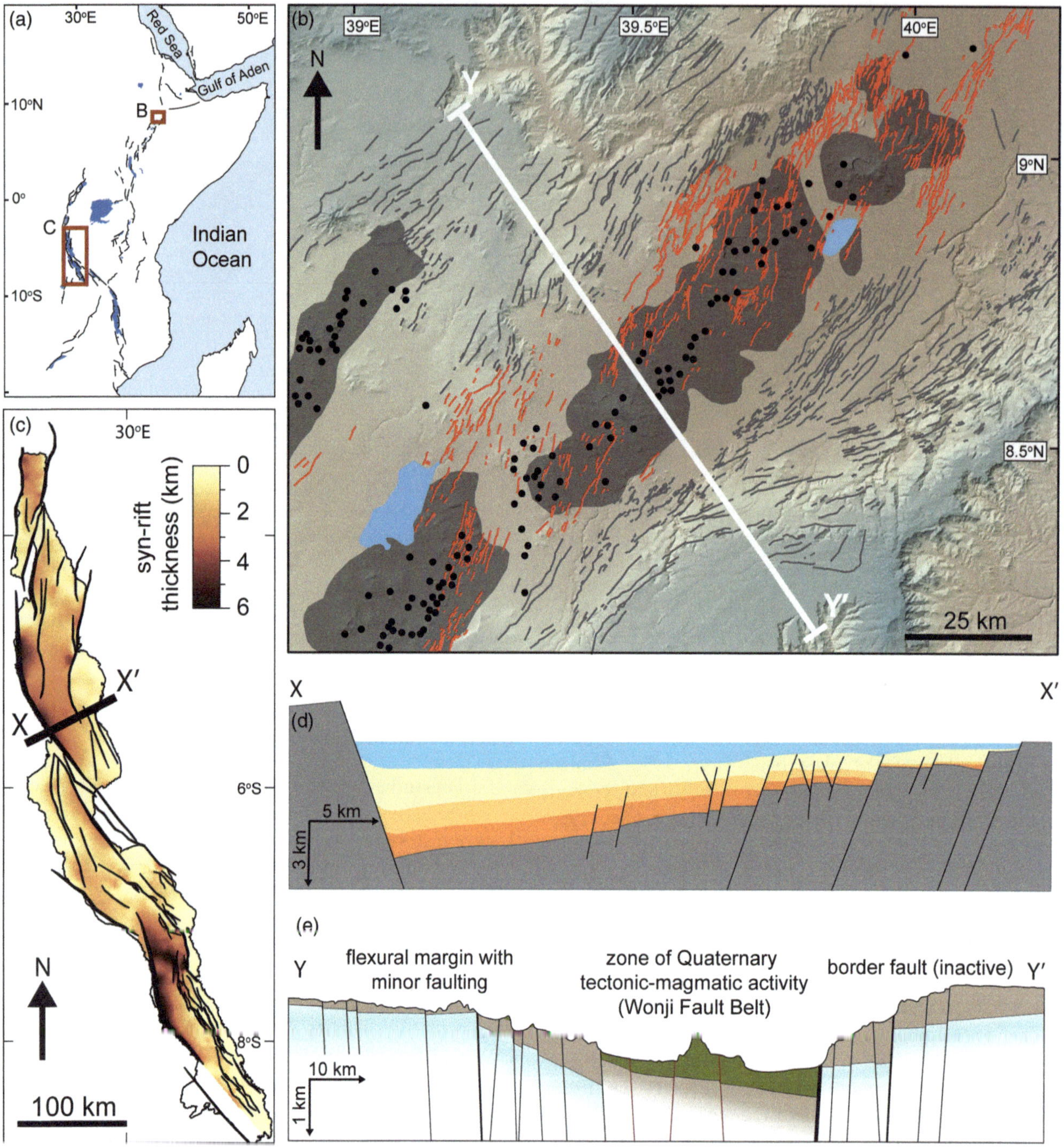

Figure 2.3 (a) Inset map showing the locations of panels B and C in the East African Rift System. (b) Annotated Shuttle Radar Topography Mission digital elevation model showing the distribution of intrarift faults (Wonji Fault Belt; red lines), border faults (gray lines), volcanic vents (black circles), and Quaternary volcanics (gray shading) in the Northern Main Ethiopian Rift. All mapped features are adapted from Figure 1 of Corti et al. (2018). (c) Thickness distribution of syn-rift sediments in the Lake Tanganyika Rift (slightly modified from Figure 2 of Wright et al., 2020 / John Wiley & Sons). The location of this multisegment rift basin is shown in (c), and the line X-X′ shows the location of the cross-section presented in (d). (d) Cross-section through the Lake Tanganyika Rift slightly modified from Figure 13 of Wright et al. (2020). Faults are shown as black lines, the gray fill represents crystalline basement, and yellow and orange colored fill shows the different syn-rift sequences. Location of the cross-section is shown in (a). (e) Cross-section through the Northern Main Ethiopian Rift, slightly modified from Figure 2 of Corti et al. (2018). Faults are shown as black and red lines, dark green fill represents Quaternary volcanics, Miocene to Pliocene syn-rift volcanics are brownish gray, and prerift volcanics are in light blue. Location of the cross-section is shown in (b).

(Delvaux & Barth, 2010). These observations support localized stress deflections within basins (Delvaux & Barth, 2010), which have been interpreted to be driven by lithospheric strength anisotropy associated with the Precambrian shear zones (e.g., Morley, 2010) and horizontal gradients in gravitational potential energy (e.g., Stamps et al., 2010; Rajaonarison et al., 2021). Stress rotations are also documented within interaction zones between rift segments and result in rift-oblique faults and fractures that exhibit dominantly dip-slip motions (Koehn et al., 2010; Muirhead & Kattenhorn, 2018).

2.2.4. Structural Evolution of EARS Basins

Rift basins in the EARS show characteristic and evolving patterns in extensional strain partitioning between border and intrarift faults (e.g., Hayward & Ebinger, 1996; Muirhead et al., 2016). During early-stage rifting (<5 Ma) in the Eastern Branch (e.g., Natron Basin, Tanzania), extension localizes rapidly onto border faults, which are estimated to accommodate 69–98% of the total basin extension (Muirhead et al., 2016). Sediment accumulation and across-strike stratal thickening are controlled by border fault displacements (Corti et al., 2015; Wolfenden et al., 2004), and intrarift faulting is driven primarily by extensional stresses associated with flexure of the border fault hanging wall (e.g., Muirhead et al., 2016). As rifting and lithospheric thinning proceeds, however, extensional strain localizes into the center of the rift basin (Bilham et al., 1999; Ebinger & Casey, 2001; Muirhead et al., 2016), where it is accommodated through a combination of axial diking and related shallow faulting (Keir et al., 2006; Rowland et al., 2007). This strain focusing is expressed at the surface as closely spaced networks of intrarift faults, aligned chains of volcanic cones, and inward dipping lavas and interbedded sediments that support basin subsidence and stratal thickening along the rift axis (Agostini et al., 2011; Corti et al., 2015; Keir et al., 2015; Muirhead et al., 2015; Rooney et al., 2011). Thus, as rifting proceeds to break up, the border faults that initially controlled the structural expression of rifting are rendered inactive, and the subsequent development of rift architecture in these more developed systems is largely the byproduct of magma-induced deformation along the rift axis (e.g., Ebinger & Casey, 2001; Keir et al., 2006, 2011; Rowland et al., 2007).

Structural investigations of basins in the Western Branch support a contrasting model of rift development in these magma-poor systems (Morley, 1988; Shillington et al., 2020; Wedmore et al., 2020). Although border fault movements still play a key role in controlling rift basin architecture and the development of accommodation space, intrarift faults appear to be more active at the earliest rift stages (first few million years). In southern Malawi, ~50% of Late Quaternary extension (faulting since 75 ka; Scholz et al., 2007) is estimated to have been accommodated on the intrarift faults (Wedmore et al., 2020). Similarly, in the Okavango Rift, extensional strain is widely distributed across a number of faults, with total throws no greater than a few hundred meters (Modisi et al., 2000). In line with observations from these incipient rift basins, structural restorations and syn-depositional fault analyses from seismic reflection data support increased intrarift fault activity during incipient rifting and a gradual reduction in intrarift fault activity over basin development in the Lake Tanganyika Rift (Wright et al., 2020). Overall, these reflection seismic data suggest that border faulting remains the primary mode of extensional strain accommodation in the first 10 Myrs of rifting in the Western Branch (Ebinger and Scholz, 2011; Ebinger, 1989), and provide no evidence for upper crustal strain focusing on the rift center in magma-poor systems of the EARS (Muirhead et al., 2019).

2.3. PLATE STRUCTURE AND PATTERN OF EARTHQUAKES

2.3.1. Broad-Scale Plate Structure in Africa

The thickness and internal structure of the African lithosphere, considered here as plate thickness, have been inferred from regional surface-wave tomographic images (Figure 2.4), which provide a broad picture of both the thickness of the crust and lithosphere (e.g., Fishwick, 2010; Priestley et al., 2018). The Archean cratonic cores of East Africa generally comprise lithosphere that is ~180–250 km thick; below younger, flanking terranes, the African plate is <100 km thick (Figure 2.4: e.g., Fishwick, 2010; Priestley et al., 2018). S-to-P receiver functions add further constraint on lithosphere-asthenosphere boundary depths in magma-poor regions (e.g., Hopper et al., 2020) but are generally insensitive to the base of the plate in magmatic rift sectors (e.g., Lavayssière et al., 2018). Joint inversion of P-to-S receiver functions with surface-wave dispersion data also constrain plate thickness (e.g., Dugda et al., 2007).

Our best constraints on crustal thickness and internal structure come from wide-angle seismic experiments such as KRISP94 (e.g., Prodehl et al., 1997) and EAGLE (e.g., Maguire et al., 2006) in Kenya and Ethiopia (Figure 2.5a), respectively. Additional constraints come from P-to-S receiver functions (e.g., Dugda et al., 2005; Hodgson et al., 2017; Plasman et al., 2017; Tugume et al., 2013) and the joint inversion of receiver functions and surface waves (e.g., Julià, 2007), but these are limited to areas where broadband seismograph stations have been deployed (Figure 2.4). These studies show that the

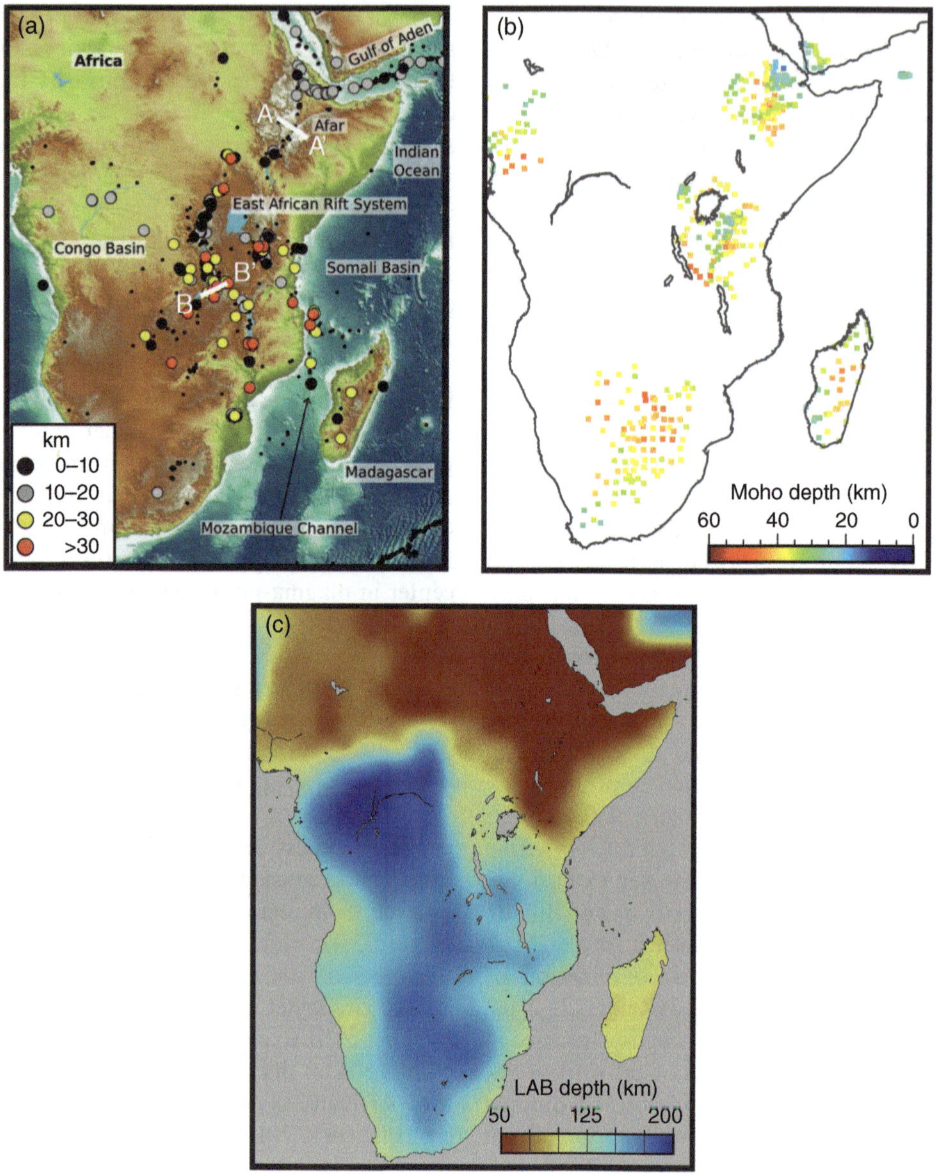

Figure 2.4 (a) Distribution of earthquakes in East Africa from the EHB global earthquake catalog with earthquakes colored by depth, adapted from Figure 1 of Craig et al. (2011). (b) Crustal thickness variation across East Africa (Boyce et al., 2021) using receiver function data presented in Ebinger et al. (2017) and Ogden et al. (2019). (c) Lithosphere thickness variations across East Africa inferred from surface waves. *Source:* Adapted from Priestley et al. (2018).

Archean and Proterozoic crust of East Africa is generally 40 ± 5 km thick (e.g., Tugume et al., 2013). The crust is, however, modified beneath the EARS.

2.3.2. Modification of the Plate Due to Rifting

In the relatively magma-poor sections of the EARS (the Western Branch and southern EARS), crustal P-wavespeeds (V_p) (e.g., Prodehl et al., 1997) and bulk crustal V_p/V_s ratios are consistent with average continental crust, with minimal evidence for mafic intrusion (e.g., Hodgson et al., 2017). Beneath these parts of the EARS, there is variably thinned crust (30–40 km thick) (Figures 2.4 and 2.5). Loci of maximum thinning (up to 10 km), equivalent to a stretching (β) factor of ~0.25, are commonly beneath the major fault-controlled rift sectors, and the magnitude of thinning is similar to that expected from the history of extension (Hodgson et al., 2017;

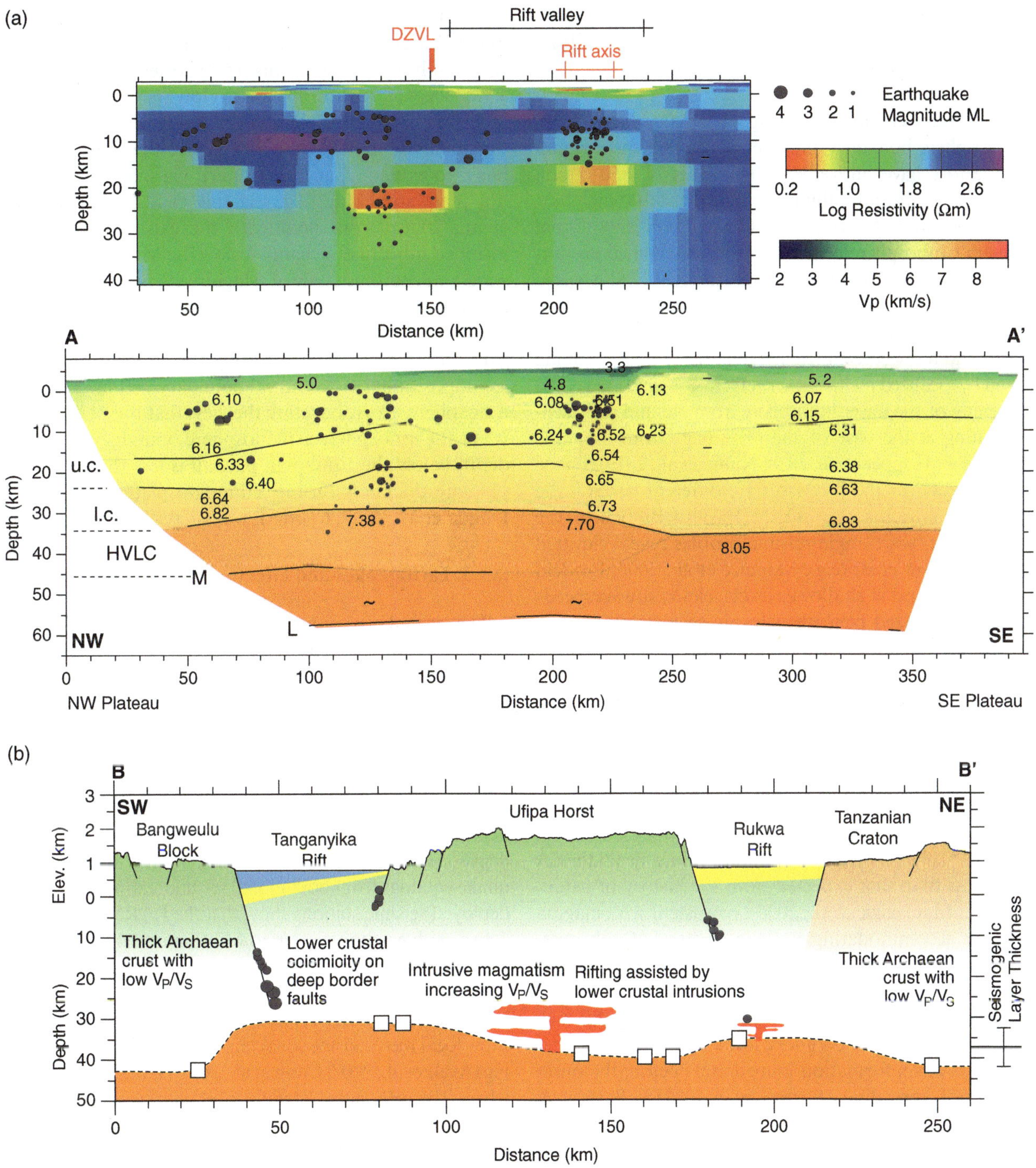

Figure 2.5 (a) Crustal scale cross-section of the magma-rich Main Ethiopian rift. The top panel shows a resistivity model from magnetotellurics (Whaler & Hautot, 2006), and the lower panel shows a V_p model from wide-angle controlled source imaging (Mackenzie et al., 2005). Both models have the same scaling and are aligned vertically such that the distance scaling is correlative. Both cross-sections are shown with seismicity overlain (Keir et al., 2009). DZVL, Debre Zeyit Volcanic Line (now Bishoftu); u.c., upper crust; l.c., lower crust; HVLC, high velocity lower crust. M and L are named reflectors. The cross-section alignment is NW-SW just above Tullu Moye and is labeled as A-A′ in panel (a) of Figure 2.4. (b) Crustal scale cross-section of the magma-poor Tanganyika and Rukwa rifts with crustal thickness constrained from receiver functions (Hodgson et al., 2017) and seismicity (Lavayssière et al., 2018) shown as white squares. The alignment of this cross-section is B-B′ and is shown and labeled as B-B′ in panel (a) of Figure 2.4.

Maguire et al., 1994; Muirhead et al., 2019) (Figure 2.5). Both of these observations are consistent with a rifting model for the crust dominated by stretching and associated thinning (McKenzie, 1978).

S-to-P receiver function imaging of the lithosphere-asthenosphere boundary shows evidence that the magma-poor sectors have lithosphere thicknesses of ~50–60 km, which is significantly thinner than the >100 km thick Proterozoic and Archean lithosphere. This shows that the mantle lithosphere beneath the rift has thinned proportionally more than the crust (e.g., Hopper et al., 2020; Njinju et al., 2019). The preferential thinning of the mantle lithosphere is potentially explained by it being compositionally weaker and more easily stretched than the crust or thinning caused by localized melting, thermal erosion, and associated weakening of the lower lithosphere from pre-and early syn-rift heating at the base of the plate (e.g., from a mantle plume) (e.g., Hopper et al., 2020; Njinju et al., 2019).

Plate structure and processes are different beneath the more magma-rich sectors in the Eastern Branch of the EARS. Here, wide-angle seismic profiles reveal elevated V_p of the lower crust (e.g., Maguire et al., 2006; Prodehl et al., 1997), and at shallower levels, elevated wavespeeds are more localized beneath volcanic sectors both inside the rift and near the flanks, strongly suggesting that mafic intrusions of sills and dikes modify the lithosphere and contribute to extension (e.g., Cornwell et al., 2006; Keranen et al., 2004; Mackenzie et al., 2005; Maguire et al., 2006; Oliva et al., 2019), consistent with petrographic observations (Rooney et al., 2005) and seismic observations of plate-scale melt-intrusion-related seismic anisotropy (e.g., Kendall et al., 2005; Bastow et al., 2010) (Figure 2.5). The crust does commonly thin beneath the rift but by less than that expected from the history of extension (e.g., Mackenzie et al., 2005), consistent with magma intrusion accommodating extension instead of only crustal stretching and extension. Only in northernmost Afar in the Danakil Depression does the crust exhibit marked thinning (to <15 km: Makris & Ginzburg, 1987). This may imply that the plate undergoes a late stage of thinning and associated basin formation immediately before the onset of seafloor spreading (Bastow et al., 2018; Bastow & Keir, 2011; Keir et al., 2013), perhaps as a result of heating and weakening of the plate due to protracted extension by magma intrusion (Daniels et al., 2014). This model is consistent with the pulsed nature of magmatism within the Eastern Branch of the EARS (Rooney, 2020a).

Constraints on plate thickness below magmatic sectors of the EARS are less corroborative than for the Moho. Below the MER and Afar, for example, S-to-P receiver function studies provide no constraint on plate thickness (Lavayssière et al., 2018), precluding the possibility of a sharp transition from slow wavespeed asthenosphere to fast wavespeed lithosphere. However, surface-wave studies that use wavespeed proxies for plate thickness (e.g., Dugda et al., 2007) have suggested a ~50 km-thick plate below the MER. Larger values come from joint surface and body-wave analysis of seismic anisotropy: oriented melt pocket-style seismic anisotropy with low aspect ratio melt inclusions is required to ~75 km depth, with the implication that the mantle lithosphere, like the crust above it, has also extended, at least in part, via magma intrusion, not plate stretching (e.g., Bastow et al., 2010; Kendall et al., 2006). Consistent with the existence of heavily intruded lithospheric mantle below the MER, the EAGLE wide-angle seismic study of Maguire et al. (2006) identified a mantle reflector at ~55 km below the MER (the "L" reflector in Figure 2.5), shallowing to ~40 km below Afar. This reflector was interpreted by Maguire et al. (2006) as a compositional or structural boundary within preexisting lithosphere. Corroborating the view that the "L" reflector is not the lithosphere asthenosphere boundary (LAB), its depth below the Ethiopian plateau is less than estimates of plate thickness on the Ethiopian plateau (70–80 km: Dugda et al., 2007; 67 km: Lavayssière et al., 2018).

2.3.3. Earthquakes and Effective Elastic Thickness (Te)

The spatial variations in lithosphere thickness broadly correlate with variation in effective elastic thickness (T_e), constrained by inversion of gravity and topography data (e.g., Hayward and Ebinger, 1996; Perez-Gussinye et al., 2009). Both lithosphere thickness and T_e are influenced by the compositional and thermal state of the lithosphere. Because an old, cool, strong, and thick plate fails at greater yield stresses and will also be brittle to greater depths, these variations significantly affect the size and depth distribution of earthquakes. In the EARS, earthquake catalogs from both global seismic stations and locally deployed seismic stations show that the EARS is seismically active, with earthquakes up to M7 (Figure 2.4). Seismicity is most common along the border-rift and intrarift faults, near active volcanic centers, and also sparse in some regions away from the rift (Figures 2.4 and 2.5). Most of the earthquakes show focal mechanisms consistent with normal dip-slip faulting (Ayele et al., 2007; Craig et al., 2011), although in regions where the rift steps laterally and in volcanically active regions, oblique and strike-slip earthquakes are also observed (e.g., La Rosa et al., 2019; Pagli et al., 2019).

Moment tensor inversion of larger (>M4) earthquakes shows a clear south-to-north decrease in earthquake depths and maximum size that spatially correlates with the reduction in T_e. In the central and southern sectors of the EARS, larger (>M6) earthquakes are common along the major rift bounding border faults and also occur throughout the crust (to ~40 km depth: e.g., Craig et al., 2011), providing evidence that brittle behavior is present throughout the whole crust (Figure 2.5). In these regions, accumulation of the seismic energy (seismic moment release) accounts for a significant proportion of that expected from the observed rate of extension (e.g., Déprez et al., 2013; Ebinger et al., 2013), showing that fault slip is an important mode of overall extension. In contrast, in the more magma-rich northern EARS, the maximum magnitude of earthquakes is smaller (<M5.5) and mainly restricted to the upper half of the crust (<20 km deep) (Figure 2.4), likely due to the thinner plate and lower T_e. The seismic moment release accounts for a small proportion of that expected (e.g., Déprez et al., 2013; Ebinger et al., 2013), with seismicity generally more focused on intrarift faults and volcanic centers (Greenfield et al., 2019; Wright et al., 2006), a set of observations consistent with the greater importance of magma-assisted extension.

2.4. THE MANTLE AND MELT GENERATION

2.4.1. Deep Mantle Structure

Some considerable time after debate had begun concerning how many mantle plumes could best explain East Africa's Cenozoic rifting, uplift, and magmatism, global tomographic models provided our first images of mantle wavespeed from the surface to the core-mantle boundary (e.g., Ritsema et al., 1999; Ritsema & Allen, 2003). These models yielded an unexpected surprise: a broad (~500 km wide), inclined, slow-wavespeed anomaly that extended from the LLSVP at the core-mantle boundary below South Africa to the surface somewhere around Ethiopia and Kenya. The term *African Superplume* was quickly adopted to distinguish it from the narrower (100–200 km diameter) plume tails predicted by the original Morgan (1971) plume hypothesis.

In addition to its elevated temperature, the Superplume in the lower mantle is now almost ubiquitously interpreted to be chemically distinct from the ambient mantle due to its inclination (e.g., Simmons et al., 2007) and markedly sharp sides (e.g., Ni, 2002). The inclination of the Superplume may also be partly due to large-scale mantle convection (e.g., Steinberger & Torsvik, 2012) or the tomographic manifestation of vertically rising plumes at sequentially advanced stages of ascent (e.g., Davaille & Vatteville, 2005). More recent global tomographic models (Figure 2.6) corroborate the Superplume hypothesis (French & Romanowicz, 2015; Hosseini et al., 2020; Li

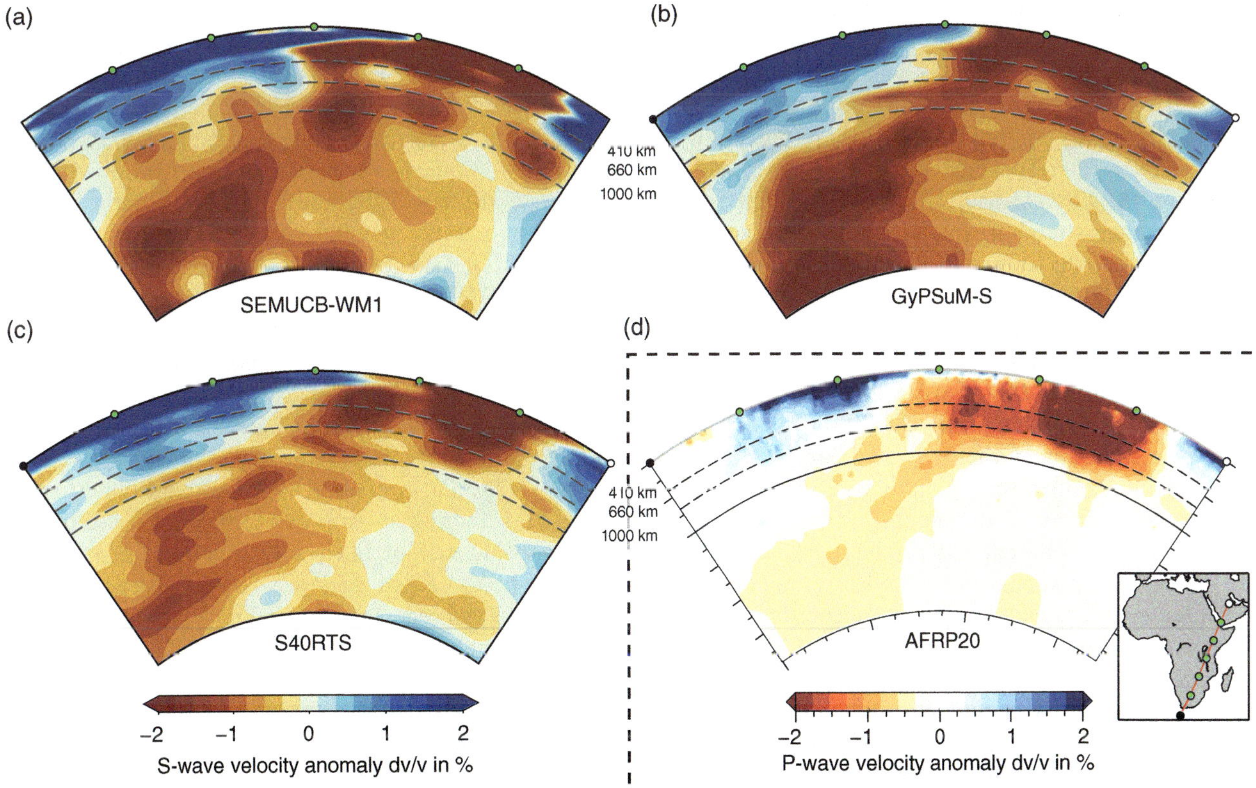

Figure 2.6 Tomographic cross-sections from the core-mantle boundary to the surface through four recent global tomographic models. Three of these are S-wave models: (a) SEMUCB-WM1 (adapted from French and Romanowicz, 2014), (b) GyPSuM-S (adapted from Simmons et al., 2010), and (c) S40-RTS (adapted from Ritsema et al., 2011). We also show a P-wave model, (d) AFRP20 (Boyce et al., 2021), separated from the other models by a dashed line. Tomographic cross-sections through S40-RTS, SEMUCB-WM1, and GyPSuM-S are produced using Submachine (Hosseini et al., 2018) with a profile that has a southerly end at (38°S, 18°E) and a northerly end at (25°N, 48°E).

et al., 2008; Montelli et al., 2006; Ritsema et al., 2011; Simmons et al., 2012); however, some of these models also provide tentative additional evidence for one or more narrower plume tails in the upper mantle below East Africa (e.g., Chang et al., 2020; Boyce et al., 2021).

2.4.2. The Mantle Transition Zone

The flow of Superplume material across the mantle transition zone is fundamental to our understanding of the link between the African Superplume in the lower mantle and Cenozoic hotspot tectonism that manifests at the surface. However, tomographic images, with their imprecise depth resolution, have been unable to address this issue unambiguously, so seismologists have appealed to alternate analytical methods aimed at high-resolution study of the mantle transition zone. A popular way to constrain mantle transition zone thickness, and thereby investigate the thermal and chemical state of the mantle, is via teleseismic receiver function analysis of the P-to-S conversions that emanate from the ~410 and ~660 km-deep (olivine-to-wadsleyite and ringwoodite-to-perovskite+ magnesiowustite, respectively) phase transitions in olivine (e.g., Helffrich, 2000). Numerous studies have been carried out in the region (e.g., Benoit et al., 2006; Cornwell et al., 2011; Huerta et al., 2009; Mulibo & Nyblade, 2013; Nyblade et al., 2000; Owens et al., 2000; Reed et al., 2016a, 2016b; Thompson et al., 2015), but surprisingly, these studies offer limited consensus on the number, geometry, and thermochemical nature of upwellings penetrating the East African upper mantle. In a broad-ranging, global study of the mantle transition zone, Tauzin et al. (2008) observed little correlation between mantle transition zone thickness and hotspot tectonism across Africa. Some EARS-focused studies (Benoit et al., 2006; Nyblade et al., 2000; Owens et al., 2000; Sun et al., 2017) favored thermal anomalies of 100–300 K as the best way to explain modest (typically 20–30 km) mantle transition zone thinning. Others have departed from the simple olivine-temperature paradigm to explain their observations in light of thermochemical heterogeneity. For example, Cornwell et al. (2011) interpreted a depressed "410" below Ethiopia/Afar as a ~250 K thermal anomaly, but heterogeneous "d660" depths as evidence for chemical heterogeneity; olivine-dominant below Afar, garnet-dominant below the MER. Subsequent work by Thompson et al. (2015) reported largely average mantle transition zone thicknesses below Afar and central Ethiopia, suggesting minimal thermal perturbations, with hydrous upwellings directly above the "410" below Afar. The transition from wadsleyite to ringwoodite at ~520 km is thought to be more gradational than the "410" and "660" phase transitions and is observed only intermittently across Africa (Julià & Nyblade, 2013).

2.4.3. The East African Upper Mantle

With the aim of imaging the upper mantle seismic structure below East Africa, numerous temporary broadband seismograph networks have been built since the 1990s, with apertures of typically hundreds of kilometers. Temporary networks generally record for up to 1–3 years and are often deployed in remote areas with sub-optimal burial methods that render their resulting seismograms noisy. Identifying the first breaking energy of body waves on such seismograms is often impossible without filtering the data, so relative arrival-time tomographic inversions have often been used instead to glean information about upper mantle seismic wavespeed (e.g., East Africa: Park & Nyblade, 2006; Mulibo & Nyblade, 2013; Ethiopia: Bastow et al., 2005; Benoit et al., 2006). These models have one vitally important distinction from the global models: their background mean (the $\delta V_{P,S} = 0$ contour) is equal to the regional, not global, zero mean. Thus, high-wavespeed anomalies (generally shaded blue) are not necessarily genuinely "fast" anomalies. In contrast, surface-wave methods generally provide absolute measures of shear wavespeed.

Ethiopia and Afar

Early regional body-wave tomographic inversions below Ethiopia presented by Bastow et al. (2005) cited evidence of markedly low wavespeeds and extremely delayed phase arrival times as evidence that the Ethiopian mantle is among the slowest worldwide. Markedly slow shear wavespeeds ($\delta VS \leq -11\%$) have also been observed in the uppermost mantle via surface-wave analysis (Bastow et al., 2010; Gallacher et al., 2016). Since ~8 Ma, the style of strain in the MER has shifted from faulting and stretching to extension by magma intrusion without marked crustal thinning (e.g., Mackenzie et al., 2005). The mantle lithosphere, too, is thought to be extending in this way (e.g., Bastow et al., 2010; Hammond, 2014; Kendall et al., 2006). When combined with evidence of the ultra-slow rate (~6 mm/yr: McClusky et al., 2010) of Nubia-Somalia separation in Ethiopia, it is evident that extension-related decompression melting likely adds little to asthenospheric low wavespeeds. These inferences are consistent with the lack of observed segmentation of low-wavespeed zones in the MER asthenosphere in terms of either position or length scale compared to zones of Quaternary magmatism observed at the surface. Instead, Bastow et al. (2005) noted that asthenospheric low-wavespeed anomalies mirror the location and length scales of earlier (Miocene) rifting, when first-order LAB topography likely formed in response to faulting and plate stretching. Surface-wave studies with lower lateral resolution than the body-wave studies also image segmented wavespeed patterns in the uppermost mantle (Gallacher et al., 2016).

More broadly below Ethiopia, body-wave relative arrival-time studies (e.g., Bastow et al., 2008; Benoit et al., 2006; Civiero et al., 2015; Hammond et al., 2013), reveal broad low-wavespeed anomalies within the upper mantle, consistent with the hypothesis that the African Superplume traverses the mantle transition zone, either as a broad anomaly (e.g., Benoit et al., 2006) or via smaller, discrete upwellings (e.g., Bastow et al., 2008; Civiero et al., 2015) rising from 410–660 km depth. Exploring causes of seismic heterogeneity below East Africa, Rooney et al. (2012c) examined mantle tomographic models below the region and concluded that petrologically determined mantle potential temperatures (100–140°C) alone are insufficient to explain the observed low-wavespeed anomalies. Instead, they concluded that CO_2-assisted melt production in the African Superplume likely contributes to the observations below depths where decompression melting of normal mantle peridotite alone can adequately explain the observations.

East Africa and the Tanzania Craton

Like Ethiopia, the East African mantle to the south has been probed extensively by surface (e.g., Adams et al., 2012; Celli et al., 2020; Emry et al., 2019; O'Donnell et al., 2013; Weeraratne et al., 2003) and body-wave tomography (e.g., Mulibo & Nyblade, 2013; Park & Nyblade, 2006). Wavespeeds below the Eastern Branch of the EARS are markedly slower than the magma-poor Western Branch, but not as slow as below Ethiopia.

Studies of the mantle below the EARS south of Ethiopia have placed constraints not only on Cenozoic rifting and hotspot tectonism but also on the structure and evolution of the Precambrian lithosphere. Surface wave models (e.g., Adams et al., 2012; Emry et al., 2019; O'Donnell et al., 2013; Weeraratne et al., 2003) have imaged asymmetrical fast wavespeeds below the Archean Tanzania Craton, consistent with an old, cold iron-depleted cratonic root.

Fast shear wavespeeds below the Tanzania Craton extend to much shallower depths than for most cratons worldwide (>200 km); they also demonstrate asymmetry, with fast-wavespeed anomalies penetrating to 135 km depth in the east compared to 175 km in the relatively magma-poor west. Accordingly, thermal erosion by underlying warm asthenospheric material has been proposed to explain the relatively thin Tanzania Craton lithosphere (e.g., Celli et al., 2020).

2.4.4. Melt Generation

At its most basic, the process of magma generation in a continental rift requires a perturbation of the pressure-temperature conditions of the ambient upper mantle such that partial melting may be initiated. In a continental rift, adiabatic decompression of the ambient upper mantle caused by lithospheric thinning is typically considered the dominant mechanism of melt formation; however, significant complexity arises when considering potential excess mantle temperature, heterogeneous mantle compositions, and the timescales of thinning (e.g., Bown & White, 1995)

Decompression

For anhydrous ambient mantle exhibiting average asthenospheric mantle potential temperature (T_P), thinning of the lithosphere by at least half ($b>2$) is required to facilitate melt creation, presuming an instantaneous decompression paradigm (McKenzie & Bickle, 1988). The initial thinning of the East African lithosphere is associated with Mesozoic rifting on the continent, which in East Africa most clearly manifests as the Anza graben (e.g., Bosworth & Morley, 1994). Although this rifting event is not associated with any significant magmatism in East Africa (Tiercelin et al., 2012), such modification of the continental lithosphere may have had an effect on the subsequent channeling of plume material and flood basalt eruptions during the Cenozoic (Ebinger & Sleep, 1998).

During the Cenozoic, a renewed period of lithospheric instability is associated with the interaction of the East African continental lithospheric mantle with the mantle thermochemical anomalies that created flood basalts in the region from ca. 45 to 27 Ma (Baker et al., 1996; Ebinger et al., 1993; George et al., 1998; Hofmann et al., 1997; Prave et al., 2016; Ukstins et al., 2002). Thermal-mechanical erosion of the continental lithosphere associated with plume-lithosphere interaction (e.g., Davies, 1994) is evident from the composition of basalts erupted during this initial phase of magmatic activity (e.g., Ayalew & Gibson, 2009; Rooney et al., 2013; Stewart & Rogers, 1996). Destabilization of the continental lithospheric mantle during this period may have also resulted in the delamination of the lower lithosphere into the convecting upper mantle, facilitating melt creation due to both decompression of the upper mantle and fusion of the delaminated material (e.g., Celli et al., 2020; Furman et al., 2016; Wallner & Schmeling, 2016).

Collectively, these studies show that the East African lithosphere has undergone protracted thinning, permitting conductive cooling of the underlying asthenosphere (e.g., Smye et al., 2017), demonstrating that a melting model based on instantaneous thinning is inappropriate when describing melt generation in East Africa. Despite these difficulties in predicting decompression melt generation processes, there is some evidence that in regions where extension is more pronounced, the geochemical signatures of rift basalts sometimes exhibit characteristics requiring that the aggregated melting column extend to shallower levels, consistent with a thinner lithosphere (Rooney et al., 2011). However, using a conceptual model that describes the petrogenesis of rift lavas solely through decompression of a homogeneous ambient upper mantle can result in apparent paradoxes. For example, in the Dabbahu–Manda Hararo rift zone (Afar) (Figure 2.1), the axial lavas exhibit an origin from deeper in the mantle than off-axis lavas (Ferguson et al., 2013). Such inconsistencies reflect the complexity of melt generation and the role of variable mantle temperature (Ferguson et al., 2013) and composition (Nelson et al., 2019), which are discussed next.

Thermal Perturbation

Despite the ambiguity in establishing plate thickness at any particular stage during the rifting process, the initial creation of some magmatism (in particular, large volumes) requires a thermal or chemical anomaly. Excess heat energy (i.e., that beyond the convecting upper mantle) is commonly denoted as excess mantle potential temperature (T_P). Such mantle has been thought to derive from a thermal boundary layer and is most often associated with a rising mantle plume, although the blanketing effect of super-continents may also create heterogeneous mantle T_P (e.g., O'Neill et al., 2009). Increased mantle temperatures may facilitate melting beneath thick lithospheric keels and contribute to melt creation in the continental flood basalt provinces that often precede rifting (e.g., Courtillot et al., 1999).

Evidence of heat anomalies in the East African upper mantle is first manifested with initial flood basalt pulses (45–27 Ma), where T_P values are up to 1,600°C—approximately 250 K above ambient upper mantle values (Beccaluva et al., 2009; Feyissa et al., 2019; Rooney et al., 2012c). Subsequent magmatism, which became centered on the nascent rift, also exhibits elevated temperatures that are beyond those possible in the ambient upper mantle. Modeling of geochemical data from modern basalts in Afar and coupled geochemical and seismic techniques both show T_P values that are elevated ~100 K above ambient mantle in the Afar Depression (Rooney et al., 2012c; Ferguson et al., 2013). In the Western Branch of the EARS, a thermal perturbation of the upper mantle of ~80 K above ambient was suggested (Rogers et al., 1998). Throughout the modern East African Rift, parallel observations have shown that the upper mantle beneath the East African Rift exhibits consistently elevated T_P (100 K above ambient mantle) that can be up to 140 K above ambient mantle (Rooney et al., 2012c). The broad thermal anomaly detected along the entirety of the EARS is suggestive of upper mantle that is contaminated with hot mantle rising from at least the mantle transition zone.

Isotopic anomalies in rift basalts require a source with time-integrated isolation from the asthenosphere (e.g., Deniel et al., 1994; Furman et al., 2004, 2006; Hart et al., 1989; Meshesha & and Shinjo, 2008; Rogers et al., 2000). Elevated values of $^3He/^4He$ detected in lavas from throughout the EARS require that the thermochemically anomalous material have at least some mass contribution from the lower mantle (Halldórsson et al., 2014; Hilton et al., 2011; Marty et al., 1993; Pik et al., 2006; Rooney et al., 2012a; Scarsi & Craig, 1996), consistent with seismic evidence presented earlier (e.g., Bastow et al., 2010; Boyce et al., 2021; Ritsema et al., 1999, 2011). The specific geometry of how this material is introduced into the upper mantle is ambiguous. However, converging arrays in $^{87}Sr/^{86}Sr$, $^{143}Nd/^{144}Nd$, $^{206}Pb/^{204}Pb$, $^{207}Pb/^{204}Pb$, $^{208}Pb/^{204}Pb$ of rift basalts observed throughout the EARS requires a commonality in the composition of the thermochemically anomalous material being introduced to the upper mantle (Rooney, 2020b). This data supports a model wherein material derived from the African Superplume has contaminated the East African upper mantle.

Compositional Heterogeneity

When considering melt generation and magma production in a rift environment, a common initial assumption is that the source lithology is a nominally anhydrous peridotite. However, the geochemical composition of many primitive rift lavas requires the presence of volatile-bearing phases in their source (e.g., Rogers et al., 1992). Unstable in the convecting upper mantle, the presence of such phases requires a mass contribution from the relatively cooler continental lithospheric mantle. The enrichment of the continental lithospheric mantle with these volatile-bearing phases likely originated by infiltration and chromatographic metasomatism of the lithospheric mantle by sub-lithosphere-derived melts and fluids, which may have occurred either during the initial stabilization of the lithosphere or the subsequent evolution of the plate (e.g., Dawson & Smith, 1988; Hudgins et al., 2015; Rogers et al., 1998).

Metasomes may play an important role in melt production during the early stages of rifting as they destabilize and melt under much less significant thermo-baric perturbations in comparison to the surrounding peridotite (Figure 2.7): for example, the perturbation imposed by the increased thermal gradient in the continental lithospheric mantle associated with a hotter ambient upper mantle (Rogers et al., 1998; Rooney et al., 2017). The two dominant metasomatic phases contributing to melt generation in the East African lithospheric mantle are phlogopite and amphibole (e.g., Foley et al., 2012). There is a distinct spatial variation in the effect of these phases on rift lavas such that the phlogopite only appears to affect magma compositions derived from deeper lithospheric mantle near the Tanzania Craton (Foley et al., 2012). In contrast, the influence of amphibole is evident throughout the EARS (Rooney, 2020b).

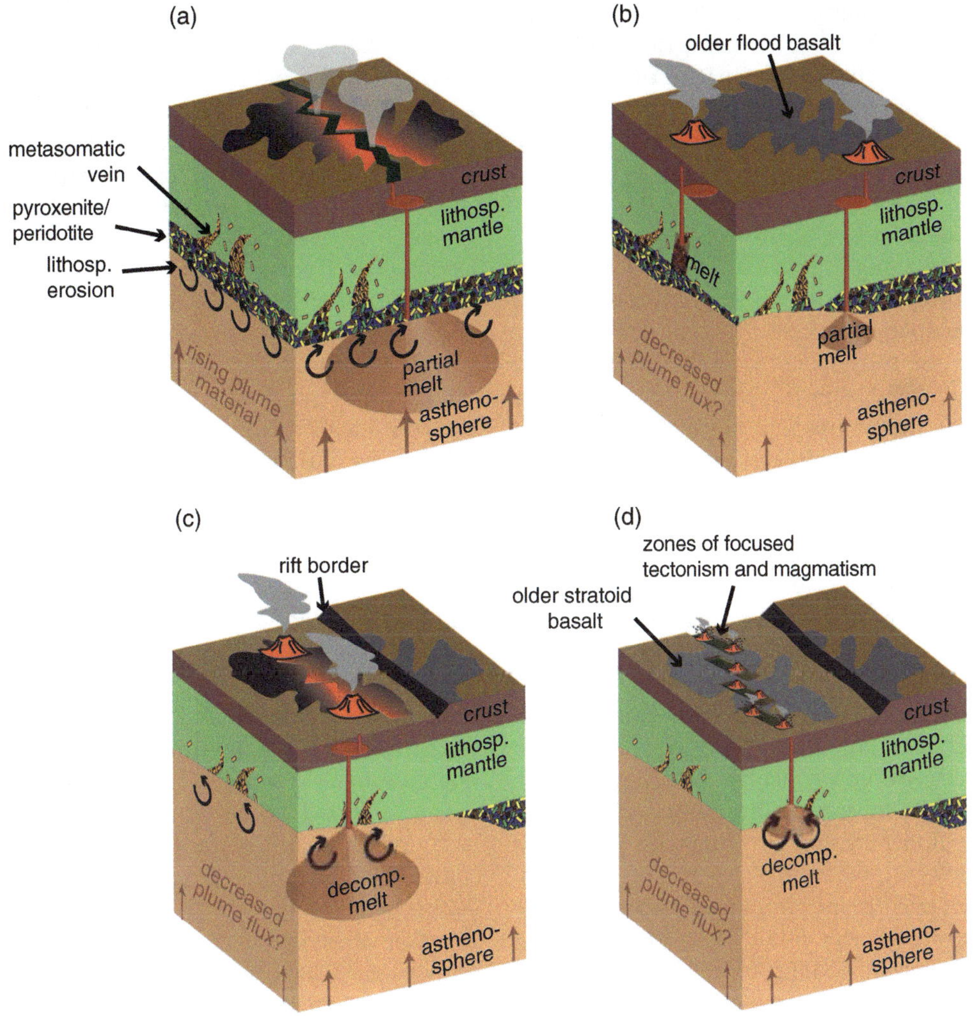

Figure 2.7 Cartoons illustrating different conceptual models of magma generation during the evolution of a mobile belt region of East Africa from initial flood basalt eruption to the modern zones of focused magmatic-tectonic intrusion. These cartoons are not a temporal sequence because repetition of extensional pulses and periods of quiescence is known to occur within the EARS. Accordingly, the history of rift development at any location within the EARS could be constructed (to a first order) by using these cartoons as building blocks. It should be noted that these figures do not address variability in the magmatic plumbing system within the crust during rift development. The four primary types of melt generation in East Africa: (a) Generalized model of melt generation during the Eocene and Oligocene flood basalt pulses. Melt is derived from an upper mantle that has been contaminated with thermochemically anomalous material rising from the African LLSVP and detached/eroded lithospheric materials. The continental lithospheric mantle comprises depleted peridotite (pale green) and metasomatically enriched lithosphere of anhydrous (patterned fill of yellow and blues) and hydrous (patterned fill of orange and black) components. The metasomes could date to as old as the formation of the plate or any subsequent event where magma may have intruded the continental lithosphere. Melt is generated and transmitted to the surface to erupt as fissural flood basalts. (b) Periods of relative magmatic quiescence. These periods occur after large basaltic pulses and likely result from a hiatus in plate thinning and associated decompression melting. The sources of melt during these periods may be hydrous mantle metasomes that experience thermobaric destabilization or thermochemically anomalous domains within the asthenosphere. During some of these quiescent phases, more silicic volcanism may be observed due to a decrease in the flux of magma into the continental lithosphere. (c) Period of widespread lithospheric thinning. These events are associated with pulses of rift basin development at the surface and the eruption of large volumes of basalt. Melt is generated dominantly through decompression of the asthenosphere during plate thinning. This event may occur multiple times during rift development. (d) Zones of focused tectonism and magmatism. During these periods, volcanism and extension are limited to small grabens within the rift valley. The limited thinning of the plate results in small-volume magmatism caused by localized upwelling of the asthenosphere and decompression melting. This period is typified by the eruption of cinder cones in between larger silicic central volcanoes.

The melting of metasomes, the mixing of such melts with magmas derived from sub-lithospheric depths, and the reaction of these hybrid magmas with the continental lithospheric mantle yields a complex array of magma compositions in East Africa (e.g., Nelson et al., 2019; Rooney et al., 2017).

Temporal Considerations

Lithospheric thinning associated with rifting is a fundamentally progressive process—as a rift evolves, the continental lithosphere necessarily thins in response (Figure 2.7). If one considers a paradigm where rifting results in continuous lithospheric thinning, then melt generation caused by decompression of the ambient upper mantle should become progressively more significant as the rift evolves. Such a progression is broadly evident in the EARS. In regions of incipient or limited rift development, such as Toro Ankole in the Western Branch of the EARS or the North Tanzania Divergence, magmas derived from destabilization of lithospheric mantle metasomes dominate the erupted products (Mana et al., 2015; Rosenthal et al., 2009). In regions where the rifting is more established (e.g., Kenya Rift), lava compositions can reflect melts of the metasomatized lithospheric mantle and ambient upper mantle (Rooney, 2020c). Finally, in regions of advanced rifting (e.g., Afar), melts of the ambient upper mantle are dominant (Ferguson et al., 2013). However, whether plate thinning is a continuous or punctuated event profoundly affects the timing and volume of melt generation. Slow thinning of the lithosphere may permit cooling of the decompressing mantle to the surface, suppressing melt generation over timescales typically associated with rift development (Bown & White, 1995). Thus, although the continental lithosphere must thin in response to rifting, discrete (pulsed) extensional events during the larger rifting process may be a necessary precondition for melt creation. Accordingly, linking modern lithospheric structure and melt formation requires knowledge of how and when the lithosphere may have reached its current state.

Where extension and associated lithospheric thinning are punctuated, pulses of magma generation can result in the eruption of large volumes of basalt over a short interval (Bastow & Keir, 2011), which is then followed by either periods of relative magmatic quiescence or the eruption of evolved magmatic products that formed due to the extended residence time of magmas within the crust (Hutchison et al., 2016a). After the Eocene-Oligocene flood basalts, three pulses of large-scale basaltic activity have been identified within the EARS:

1. Early Miocene Resurgence/Samburu event. This is the first event that manifests in both the Eastern and Western Branches of the EARS and heralds the first widespread magmatic episode associated with destabilization of the continental lithospheric mantle (Macgregor, 2015; Purcell, 2018). The event first occurs in the northern EARS as the Early Miocene Resurgence Phase (ca. 26.9–22 Ma: Rooney, 2017) and then expands southward as the Samburu Event (ca. 20–16 Ma: Rooney, 2020a).

2. Mid-Miocene Resurgence. This phase of activity is constrained to occur between ca. 12 and 9 Ma and is centered on Turkana and the margins of the Afar Depression (Rooney, 2020a). This event is linked with a significant period of extension in East Africa that originated in Afar and Turkana (Macgregor, 2015).

3. The Stratoid Phase. This phase is also centered on Turkana and Afar (Rooney, 2020a, 2020d) and is likewise linked with a period of rift development (Macgregor, 2015; Purcell, 2018). In Afar, this phase is constrained by the extensive sedimentary stratigraphic record to occur between 3.9 and 1.6 Ma (Campisano & Feibel, 2008; Deino et al., 2010; DiMaggio et al., 2015) and is followed by a focusing of strain and magmatism in the modern linear zones of focused magmatic-tectonic activity (Ebinger & Casey, 2001). South of Turkana, along the Eastern Branch of the EARS, the Stratoid Phase occurs over a more extended interval (4–0.5 Ma), likely linked with the asynchronous distribution of strain over this wide area (Rooney, 2020a).

The temporal relationship between these distinctive basaltic events and periods of extension highlights the control that lithospheric thinning exerts over magmatism in the EARS (Figure 2.7). Thus, any interpretation of the magmatism based on progressive thinning of the lithosphere must have superimposed on it the punctuated nature of extensional events in East Africa.

2.4.5. Melt and Rift Development

The traditional hypothesis that far-field extensional forces can be attributed primarily to plate subduction (Forsyth & Uyeda, 1975) is challenging in East Africa because the Nubian and Somalian plates are flanked to the east and west by mid-ocean ridges. Thus, "slab-pull" models may contribute only up to 25 MPa of the ~100 MPa stress required to rupture thick, cold, dry continental lithosphere (e.g., Kendall & Lithgow-Bertelloni, 2016). The presence of melt in the lithosphere is thought to reduce the yield strength of the lithosphere by up to an order of magnitude (Buck, 2006), but the question remains about precisely what the primary plate driving forces are along the EARS. On the one hand, some authors consider that horizontal tractions at the base of the lithosphere contribute significantly to observed large-scale EAR deformation patterns (e.g., Kendall & Lithgow-Bertelloni, 2016). Others favor the view that lithospheric buoyancy forces primarily drive present-day E-W extension across the EARS (e.g., Rajaonarison et al., 2021).

Although some theoretical models may support the view that one mechanism or another is sufficient to drive rifting, plate-driving forces will, in reality, never act in isolation. Uncertainties in fundamental parameters such as plate strength and thickness, including their variability in space and time, need to be considered carefully in future models when calculating the stresses required to initiate and maintain rifting. Calculating the viscous coupling between the convecting mantle and the overlying plate to determine the horizontal shear tractions generated by mantle flow is also challenging because of uncertainties in the thermochemical structure of the African Superplume. Whatever the balance between contributing plate driving forces, the geoscientific observations from East Africa confirm that melt must play an important role in lowering the yield strength of the lithosphere. Rifting is easier to start and maintain than has sometimes previously been presumed (e.g., Buck, 2006).

2.5. VOLCANOLOGY AND GEOTHERMAL RESOURCES OF THE EAST AFRICAN RIFT

The Eastern Branch of the EARS comprises a near-continuous magmatic rift with evidence of both on-axis and off-axis volcanism. The Western Branch, however, is marked by more discrete volcanic provinces largely concentrated in tectonic accommodation zones (Figure 2.1). Despite the dissimilarities in the rift-magmatic architecture, many parallels can be drawn on the rift volcanoes throughout the EARS in terms of their magma composition and thus rheology and eruptive style. This section focuses on the styles, products, and geothermal resources of rift volcanism that are most common across the present-day EARS. It so largely focuses on the volcanoes considered dormant-active in the Late Quaternary and up to the present day.

2.5.1. Rift Volcanism

Across the EARS, major element compositions of erupted magmas are largely bimodal, with predominantly (alkali) basalt or basanite/tephrite and sometimes melilite-nephelinite magmas on the mafic end, and phonolite, trachyte, and/or peralkaline rhyolite magmas on the silicic end, with volumetrically minor proportions of magmas of intermediate composition (e.g., Barette et al., 2017; Field et al., 2013; Fontijn et al., 2013; Hutchison et al., 2018; Macdonald, 2012; Mana et al., 2015; Peccerillo et al., 2007; Rooney, 2020b). Broadly speaking, magmas erupted from volcanoes in the Western Branch and the North Tanzanian Divergence, and to some extent in southern Central Kenya Rift (i.e., surrounding the Tanzanian Craton; Foley et al., 2012), tend to be silica-undersaturated and highly alkaline (e.g., Barette et al., 2017; Mana et al., 2015; Rooney, 2020c; Rosenthal et al., 2009). Whereas the Virunga and Toro Ankole magmas are highly potassic to ultrapotassic, the others are relatively more sodium-rich (e.g., Furman & Graham, 1999; Hudgins et al., 2015; Muravyeva & Senin, 2018; Rogers et al., 1998; Rosenthal et al., 2009). In most of these same areas, there is evidence for carbonatitic melts that are either still part of the present-day magmatic systems (i.e., at Oldoinyo Lengai volcano and in the Toro Ankole province; e.g., Bell & Simonetti, 1996; Guzmics et al., 2019; Keller et al., 2006; Muravyeva & Senin, 2018; Rosenthal et al., 2009) or have erupted from centers now considered extinct (Harkin, 1962; Le Bas & Rubie, 1977; Simonetti & Bell, 1994).

In the Central and Northern Kenya Rift, some predominantly trachytic volcanoes like Suswa and Paka straddle the silica-undersaturated to -saturated divide (e.g., Mibei et al., 2021; White et al., 2012), whereas others, like Olkaria and Oldoinyo Eburru, are silica-saturated and have erupted some of the most evolved magmas known in the EARS: that is, peralkaline rhyolites (both comendites and pantellerites; e.g., MacDonald et al., 2008; Ren et al., 2006). These latter compositions also dominate the silicic complexes of the MER (e.g., Fontijn et al., 2018; Hutchison et al., 2016a; Siegburg et al., 2018; Tadesse et al., 2018). Most peralkaline rhyolite complexes are thought to have undergone a trachytic (shield-building) phase followed by a caldera-forming eruption and post-caldera rhyolitic activity (e.g., Hutchison et al., 2016a; MacDonald & Scaillet, 2006). In Afar, basaltic magmas dominate the eruptive products; however, many volcanic centers also erupt intermediate compositions and peralkaline rhyolites (Ferguson et al., 2010; Field et al., 2013; Hutchison et al., 2018; Watts et al., 2020).

Whereas the basalt (trace element) geochemical compositions largely reflect melt generation and demonstrate heterogeneous source regions (cf. section 2.4.4), the differentiation paths of the more evolved erupted products along the EARS mainly reflect crustal processes. In most places, there is only limited or even no evidence for crustal assimilation (e.g., Mana et al., 2015; Rooney, 2020b). Magmatic differentiation paths that have been explored in some level of detail suggest a predominant process of protracted fractional crystallization from a basaltic parental melt toward phonolite or trachyte compositions and from the latter eventually toward peralkaline rhyolite compositions. (e.g., Chiasera et al., 2018; Giordano et al., 2014; Gleeson et al., 2017; Hutchison et al., 2018; Iddon et al., 2019; MacDonald et al., 2008; MacDonald & Scaillet, 2006; Mibei et al., 2021; Peccerillo et al., 2007; Rooney et al., 2012b; Tadesse et al., 2018; White et al., 2012).

The highly alkaline nature of most continental rift magmas has important implications for the magma rheology and thus eruptive processes. For a given temperature and volatile content, alkali-rich magmas can be one to several orders of magnitude less viscous than their calc-alkaline counterparts (e.g., Di Genova et al., 2013; Giordano et al., 2004; Stabile et al., 2016). The oxidation state of iron in the melt may have an additional effect, with lower oxygen fugacity conditions lowering viscosity (Di Genova et al., 2017; Stabile et al., 2016).

Due to reduced polymerization, alkalinity of magmas also influences (increases) volatile solubility, hence rheology (e.g., Carroll & Blank, 1997; Di Matteo et al., 2004; Lesne et al., 2011; Vetere et al., 2014). Preeruptive volatile budgets for magmas erupting from rift volcanoes are poorly constrained. Melt inclusion studies at Erta' Ale, with a permanent lava lake, suggest relatively volatile-poor magmas (<0.15 wt% H_2O; <200 ppm CO_2) (Field et al., 2012). For other mafic systems or eruptions (e.g., Nyamuragira, Nabro, Butajira), water contents up to ca. 1.5 wt%, CO_2 contents up to 5,000 ppm, and sulfur contents up to ca. 3,000 ppm have been measured (Donovan et al., 2018; Head et al., 2011; Iddon & Edmonds, 2020). Melt inclusion data on peralkaline rhyolite eruptive products from different MER volcanoes generally suggests relatively H_2O-rich melts (ca. 4 wt% H_2O, but with occasional measured values up to 8 wt%), with a few hundreds of ppm CO_2 and up to 600 ppm sulfur (Iddon & Edmonds, 2020). Finally, high halogen contents (F and Cl) have been reported in melt inclusions of both mafic and silicic eruptive products, often in the thousands of ppm (Donovan et al., 2018; Iddon & Edmonds, 2020).

2.5.2. Architecture of (Shallow) Plumbing Systems

Thermodynamic modeling of magmatic differentiation paths at caldera-hosted peralkaline trachyte and rhyolite volcanoes is generally consistent with a relatively dry (<1 wt% H_2O) (alkali) basalt parent melt fractionating under low oxygen fugacity conditions, around the FMQ buffer (e.g., Chiasera et al., 2018; Gleeson et al., 2017; Peccerillo et al., 2003; Rooney et al., 2007; Rooney, 2010; Trua et al., 1999; White et al., 2012). At Aluto, Gleeson et al. (2017) suggest that a strongly evolved (>90% crystallization) crystalline mush with peralkaline rhyolite melt is stored at pressures of ca. 150 MPa. Such conditions, sometimes with even lower storage pressures, are broadly consistent with those inferred at other peralkaline rhyolite volcanoes in the EARS (Peccerillo et al., 2003; Ronga et al., 2010; Rooney et al., 2012b; Scaillet & MacDonald, 2001).

Rather than constructing large central edifices, many trachyte and peralkaline rhyolite systems in the Eastern Branch are complexes with a caldera and multiple post-caldera vents that each may have formed during monogenetic eruptions of pumice cones (e.g., Clarke et al., 2019; Hutchison et al., 2016a; MacDonald et al., 2008). Together with the general absence of basaltic melts erupting at the caldera complexes (at best, only at the periphery), this draws up an image of a partial melt of basaltic composition stored at the base of the crust and which fractionates across a transcrustal plumbing system to trachytic and eventually peralkaline rhyolitic composition in shallow crustal reservoirs that are no more than a few kilometers deep. Small melt pockets near the top of a laterally extensive mush zone may thus feed individual post-caldera eruptions (Iddon et al., 2019; Iddon & Edmonds, 2020; MacDonald et al., 2008; Samrock et al., 2018).

However, other volcanoes, spanning the entire compositional spectrum—for example, from the largely basaltic Nyamuragira and Nyiragongo to the natrocarbonatitic-phonolitic Oldoynio Lengai, trachytic Longonot, and rhyolitic Gudda-Baricha—form steep-sided conical edifices with a prominent central crater. These volcanoes may have a more "classic" polybaric plumbing system whereby a shallow and relatively small reservoir is connected to a larger and deeper one (e.g., Wauthier et al., 2013).

For the 2005–2010 Dabbahu-rifting episode, at least three discrete shallow (~5 km deep) sub-volcanic crustal reservoirs (beneath Gabho, Dabbahu, and the segment-centered Ado'Ale Volcanic Complex) directly fed episodic upper crustal diking along the segment (Yirgu et al., 2006; Grandin et al., 2009). Combined evidence from petrology, magnetotellurics, geodesy, and seismology suggests these in turn are replenished from mid to base of crustal reservoirs, potentially offset laterally and along-rift from the surface expression of the rift segment (Desissa et al., 2013; Grandin et al., 2009).

2.5.3. Holocene/Historically Active Volcanoes in the EARS

Across the continental EARS (excluding the Arabian Peninsula and the Red Sea), a total of 98 volcanic centers are currently recognized as Holocene by the Smithsonian Institution's Global Volcanism Program (Global Volcanism Program, 2013) (Figure 2.1). About a quarter of these have been historically active, and several more show active signs of unrest.

Signs of Volcanic Unrest

Since the early 2000s, shallow crustal dike intrusions have been interpreted based on observations of volcano surface deformation by Interferometry Satellite Aperture Radar (InSAR), including between 2005 and 2010 at Dabbahu-Manda Harraro (e.g., Hamling et al., 2009), in 2015 northeast of Fentale (Temtime et al., 2020) in Ethiopia, and in 2007 in the Natron basin in northern Tanzania (Calais et al., 2008) (Figure 2.1). Interestingly, this latter intrusion occurred under the southern flank of the Pleistocene Gelai shield volcano (Figure 2.1), which is not known to have had any Holocene eruptions and also did not erupt on this occasion (Calais et al., 2008). However, a magmatic explosive eruption did follow at Oldoinyo Lengai ca. 20 km further southwest about 2 months later, after ca. 25 years of low-intensity effusive natrocarbonatite activity (Kervyn et al., 2010). At both Nyamuragira and Nyiragongo volcanoes, InSAR and Global Navigation Satellite System (GNSS) data have been extremely valuable in a post-eruptive assessment of the processes that led to recent eruptions and the associated pre- and post-eruptive deformation (e.g., Samsonov & d'Oreye, 2012; Wauthier et al., 2012, 2013)

Even without eruptive activity or clear indications of discrete dike intrusions, several volcanoes in the EARS, particularly in the MER and the Kenyan Rift, show ground deformation as detected by InSAR observations (Biggs et al., 2009, 2011; Albino & Biggs, 2021). The source of this deformation, whether magmatic or hydrothermal or a combination of both, is not always straightforward to constrain, as several of these volcanoes host mature geothermal reservoirs that are continuously active. However, in some cases, the deformation is episodic or of such a magnitude that hydrothermal sources alone are insufficient to explain the signal. For example, an inflation episode at Paka (Northern Kenyan Rift; Figure 2.1) showed a total of 21 cm of uplift over the course of 9 months between 2006 and 2007, which was attributed to a largely magmatic source at 2.8 km depth (Biggs et al., 2009).

In other, better-studied places, a combined magmatic-hydrothermal system is thought to generate the observed inflation and/or deflation patterns. At the caldera complex of Aluto (Figure 2.1), Ethiopia, an alternation of uplift and subsidence between 2004 and 2011 is best explained by a source at ca. 5 km depth roughly below the center of the caldera (Biggs et al., 2011; Hutchison et al., 2016b). Hutchison et al. (2016b) interpret this source as a magmatic reservoir fed by regular magmatic intrusions causing uplift, followed by leakage of magmatic gas and fluids into the overlying geothermal reservoir, causing subsidence. Further south along the MER, Corbetti is currently one of the fastest uplifting volcanic complexes in the world, with rates constrained to ca. 7 cm uplift per year since 2009 (Albino & Biggs, 2021; Biggs et al., 2011; Lloyd et al., 2018). This deformation is thought to have a largely magmatic origin and is attributed to a growing reservoir at ca. 6.6 km depth, receiving a volumetric input on the order of $10^7\,m^3$ per year (Lloyd et al., 2018). Based on reconstructions of past eruptions, Aluto and Corbetti both represent the most frequently active volcanoes of the MER, so their deformation patterns and their inferred magmatic origin are important to understand in the context of potential future eruptions.

Archives of Past Eruptive Activity

The historical record of eruptive activity in East Africa is short (ca. 100–200 years) and particularly poorly detailed. The largest historical eruption in East Africa is thought to have been the May 1861 CE eruption of Dubbi in Eritrea (Figure 2.1), which possibly resulted in more than 100 fatalities (e.g., Wiart & Oppenheimer, 2000). Other examples of notable recent eruptions with relatively significant effects on the local communities in otherwise remote areas have occurred at Oldoinyo Lengai (Tanzania) in 2007–2008 (e.g., Bosshard-Stadlin et al., 2014), Nabro (Eritrea) in 2011 (e.g., Goitom et al., 2015), and the 2005–2010 Dabbahu-Manda Hararo rifting episode (e.g., Ferguson et al., 2010). Erta' Ale's regular lava lake overflows and/or effusive eruptions from flank vents (the most recent of which started in January 2017 and lasted at least 18 months; Moore et al. (2019)) have largely remained effusive, typically causing little to no damage or disturbance to the local populations. In stark contrast, the most disastrous volcanic eruptions to have happened on the African continent in recent history are the 1977, 2002, and 2021 Nyiragongo flank fissure eruptions (Figure 2.1), which drained the lava lake (Hamaguchi et al., 1992; Komorowski, 2002). In 2002, the extremely low-viscous and highly mobile lava flows inundated the city of Goma, leading to up to 400,000 displaced people and ca. 150 fatalities (Komorowski & Karume, 2015).

These recent historical eruptions have nearly all been volumetrically small but have nonetheless had widespread (local) effects. They are, however, dwarfed by the magnitude of some of the geologically recent eruptions that occurred at many EARS volcanoes, for which more and more detailed studies exist. Both terrestrial and lacustrine sequences have been critical in reconstructing past eruptive activity throughout the EARS. Terrestrial sequences highlight the complexity of eruptive styles and emplacement mechanisms, especially at proximal-medial distances (i.e., roughly within 20 km of the source vent), and are therefore essential to understand past eruptive behavior and potential hazards from similar future eruptions (Clarke et al., 2019, 2020; Fontijn et al., 2010, 2012, 2018). At the same time, lacustrine sequences often capture fine ash layers that might not be preserved in terrestrial sequences because they are distal to the source or represent smaller-scale eruptions that have little chance of being preserved in the terrestrial record. These lacustrine sequences can thus help provide a more detailed view of past eruptions, and especially their timing (e.g., Balashova et al., 2018; Martin-Jones et al., 2017; McNamara et al., 2018), even if ash layers cannot always be correlated to a specific source volcano or eruption (Lane et al., 2018; Martin-Jones et al., 2020; Martin-Jones et al., 2017; Ross et al., 2015; Williams et al., 1993).

Many volcanoes along the EARS that have undergone one or more caldera-forming eruptions, often associated with widespread ash-rich tuff or (welded) ignimbrite deposits, are now considered to be in their "post-caldera" stage of evolution (Fontijn et al., 2018; Hutchison et al.,

2016b; Macdonald, 2012; Mibei et al., 2021), during which they typically display a mix of both effusive and explosive activity.

Volcanoes with confirmed or suspected recent large Plinian-style explosive eruptions (>1 km^3 of ash, dispersed over >100 km^2) that are typically preserved as thick tephra fall deposits, potentially associated with widespread deposits of pyroclastic density currents, occur in the Rungwe Volcanic Province (Fontijn et al., 2010, 2011, 2012), the North Tanzanian Divergence (Wilkinson et al., 1986), the Southern and Central Kenya Rift (Blegen et al., 2016; Clarke, 1990; Leat, 1984, 1985; Mibei et al., 2021; Scott, 1980; Scott & Skilling, 1999), the MER (Fontijn et al., 2018; Martin-Jones et al., 2017; Rapprich et al., 2016), and the Afar Triple Junction (Wiart et al., 2000). For the most frequently active volcanoes (and for which data exist: e.g., Aluto, Corbetti, or Rungwe), eruption frequencies throughout the Holocene are constrained to be one to several per millennium (Fontijn et al., 2010, 2018; Martin-Jones et al., 2017; McNamara et al., 2018).

2.5.4. Geothermal Resources

The abundant activity and presence of shallow active magmatic reservoirs in caldera-hosting complexes noted previously makes the rift particularly appropriate in providing sites of hydrothermal activity at which geothermal energy can be harvested at relatively shallow depths, resulting in both big and small-sized development initiatives. Exploration for geothermal energy resources is ongoing in almost all the countries traversed by the EARS. Some of these countries are being evaluated for high- or low- to medium-temperature resource development for power generation and/or direct use. In the assessments on the type of geothermal system, use is made of geophysical, geochemical, remote sensing, and GPS data, among other techniques (Abiye & Haile, 2008; Armadillo et al., 2020; Awaleh et al., 2020; Dávalos-Elizondo et al., 2021; Kanda et al., 2019). Investigations since the early 1950s (McNitt, 1982) have resulted in the identification of high-enthalpy geothermal systems associated with young volcanic centers and low-enthalpy systems associated with extremely weak surface manifestations. The total (energy) power potential of the entire region has been variously estimated but needs further refinement. The current energy estimates have only focused on electrical power potential that mainly relies on high-temperature resources (>200 °C), leaving out direct use, which can use geothermal resources at intermediate temperatures of <100 °C (Teklemariam, 2020; Kebede & Ambebrhan, 2020; Omenda, 2020). Kenya is already generating a sizable amount of its electrical power supply from geothermal resources. Other countries are still in the surface exploration stage and/or early stages of subsurface investigations through drilling.

Geothermal Systems in the Eastern Branch

The hydrothermal systems in the Eastern Branch are dominated by hot springs, fumaroles, and hot and altered grounds along the axis of the rift, and which are associated with central volcanoes that exhibit calderas and shallow magma chambers. Case studies from three sites in the Eastern Branch of the EARS—Olkaria (Kenya Rift), Tullu Moye (MER), and Era Boru (Afar) (Figure 2.1)—provide inferences about the type of controls that would suggest ideal situations for potential geothermal sites (Varet et al., 2020). Olkaria is a well-developed geothermal venture (with electricity production of over 800 MWe) and has been extensively studied since the 1960s (e.g., Clarke, 1990). Tullu Moye is now in the exploration drilling stages for geothermal fluids. Research on the Era Boru site in Afar that has involved volcanology, geophysics, and geochemistry since the early 1970s suggests that the hydrothermal surface manifestations have developed along crustal structures, especially where they intersect. Varet et al. (2020) observed that in these three cases, despite the different extensional rates involved, similar volcanological and structural processes control their associated hydrothermal systems, arising from prerift lines of weakness in the Proterozoic basement (Omenda, 1998).

Geothermal Systems in the Western Branch

Lower abundances of Holocene volcanic systems in the Western Branch of the EARS have resulted in geothermal resource research being less well-developed than in the Eastern Branch. Geothermal systems in the EARS typically require magma reservoir development within the shallow continental crust to provide sufficient heat. However, for the Western Rift, it is postulated, from geological and geophysical investigations, that magma reservoir development of sufficient volume is unlikely at the shallow depths necessary for volcano-related surficial hydrothermal systems (e.g., Jakovlev et al., 2013; Muirhead & Kattenhorn, 2018). Furman et al. (2016) suggested that geothermal systems in the Western Branch of the EARS probably occur within fault zones, with heat sources in the form of transient dikes. Known geothermal manifestations in the region are closely associated with structures that mainly occur at the toe of large fault systems, such as Gisenyi, Kapisya, and Buranga. Some hot springs also occur at contacts between different rock types, such as Chinyunyu in Zambia (Omenda et al., 2016). Rift faulting therefore promotes deep circulation of meteoric water to produce geothermal systems. The heat sources can be attributed to a combination of elevated crustal temperatures due to rifting processes, shallow magmatic intrusions, and the possible contribution of radioactive heat from elements within the Precambrian basement rocks.

2.6. INTO THE FUTURE

Advances over the past decade in recognizing the broad interconnectedness of the EARS have changed our understanding of the relationship between features in the deep mantle and surface processes. The recognition that material rising from the African LLSVP profoundly influences the upper mantle beneath East Africa, with consequential effects on melt generation and plate strength, requires a wide-scale view of the rifting process. Indeed, it is now evident that the EARS cannot be interpreted strictly within a surface plate tectonic paradigm and must instead incorporate insights from both deep-mantle and crustal processes. The fundamental scientific architecture that is now emerging from these efforts provides an important framework upon which to arrange existing work and deduce new avenues for exploration. The challenges ahead lie in crafting focused studies that provide enhanced details of the often-equivocal process operating within the EARS but with an eye to how such insights illuminate the interconnection of the larger system.

ACKNOWLEDGMENTS

This work is supported by funding from the following agencies to the authors: Tyrone Rooney—United States National Science Foundation GeoPRISMS awards: 1551872 and 1850606; Ian Bastow—UK Natural Environment Research Council (NERC) award number NE/S014136/1; Derek Keir—Ministero Università e Ricerca (MiUR) through PRIN grant 2017P9AT72 and NERC grant NE/L013932/1; Karen Fontijn—F.R.S.-FNRS MIS grant F.4515.20; James Muirhead—United States National Science Foundation GeoPRISMS award 1654518. We thank three anonymous peer reviewers for comments that helped improve the manuscript and the careful editorial handling of Elizabeth Catlos. This manuscript was accepted for publication in July 2021.

REFERENCES

Abiye, T. A., & Haile, T. (2008). Geophysical exploration of the Boku geothermal area, Central Ethiopian Rift. *Geothermics*, *37*, 586–596.

Accardo, N. J., Shillington, D. J., Gaherty, J. B., Scholz, C. A., Nyblade, A. A., Chindandali, P. R. N., et al. (2018). Constraints on rift basin structure and border fault growth in the northern Malawi rift from 3-D seismic refraction imaging. *Journal of Geophysical Research - Solid Earth*, *123*, 10,003–10,025.

Adams, A., Nyblade, A., & Weeraratne, D. (2012). Upper mantle shear wave velocity structure beneath the East African plateau: evidence for a deep, plateauwide low velocity anomaly. *Geophysical Journal International*, *189*, 123–142.

Agostini, A., Bonini, M., Corti, G., Sani, F., & Manetti, P. (2011). Distribution of quaternary deformation in the central Main Ethiopian Rift, East Africa. *Tectonics*, *30*, TC4010. doi: 10.1029/2010tc002833

Albino, F., & Biggs, J. (2021). Magmatic processes in the East African Rift system: insights from a 2015–2020 Sentinel-1 InSAR survey. *Geochemistry, Geophysics, Geosystems*, *22*, e2020GC009488. doi: 10.1029/2020GC009488

Armadillo, E., Rizzello, D., Pasqua, C., Pisani, P., Ghirotto, A., Kabaka, K., et al. (2020). Geophysical constraints on the Luhoi (Tanzania) geothermal conceptual model. *Geothermics*, *87*, 101875. doi: 10.1016/j.geothermics.2020.101875

Awaleh, M. O., Boschetti, T., Adaneh, A. E., Daoud, M. A., Ahmed, M. M., Dabar, O. A., et al. (2020). Hydrochemistry and multi-isotope study of the waters from Hanlé-Gaggadé grabens (Republic of Djibouti, East African Rift system): A low-enthalpy geothermal resource from a transboundary aquifer. *Geothermics*, *86*, 101805. doi: 10.1016/j.geothermics.2020.101805

Ayalew, D., & Gibson, S. A. (2009). Head-to-tail transition of the Afar mantle plume: geochemical evidence from a Miocene bimodal basalt-rhyolite succession in the Ethiopian Large Igneous Province. *Lithos*, *112*, 461–476. doi: 10.1016/j.lithos.2009.04.005

Ayele, A., Stuart, G., Bastow, I., & Keir, D. (2007). The August 2002 earthquake sequence in north Afar: insights into the neotectonics of the Danakil microplate. *Journal of African Earth Sciences*, *48*, 70–79. doi: 10.1016/j.jafrearsci.2006.06.011

Baker, J., Snee, L., & Menzies, M. (1996). A brief Oligocene period of flood volcanism in Yemen: implications for the duration and rate of continental flood volcanism at the Afro-Arabian triple junction. *Earth and Planetary Science Letters*, *138*, 39–55.

Balashova, A., Mattsson, H. B., & Hirt, A. M. (2018). New tephrostratigraphic data from Lake Emakat (northern Tanzania): implications for the eruptive history of the Oldoinyo Lengai volcano. *Journal of African Earth Sciences*, *147*, 374–382.

Banks, N. L., Bardwell, K. A., & Musiwa, S. (1995). Karoo rift basins of the Luangwa Valley, Zambia. *Geological Society, London, Special Publications*, *80*, 285–295.

Barette, F., Poppe, S., Smets, B., Benbakkar, M., & Kervyn, M. (2017). Spatial variation of volcanic rock geochemistry in the Virunga Volcanic Province: statistical analysis of an integrated database. *Journal of African Earth Sciences*, *134*, 888–903.

Bastow, I. D., Booth, A. D., Corti, G., Keir, D., Magee, C., Jackson, C. A.-L., et al. (2018). The development of late-stage Continental breakup: seismic reflection and borehole evidence from the Danakil depression, Ethiopia. *Tectonics*, *37*, 2848–2862.

Bastow, I. D., & Keir, D. (2011). The protracted development of the continent-ocean transition in Afar. *Nature Geoscience*, *4*, 248–250. doi: 10.1038/Ngeo1095

Bastow, I. D., Nyblade, A. A., Stuart, G. W., Rooney, T. O., & Benoit, M. H. (2008). Rifting at the edge of the African low velocity anomaly. *Geochemistry, Geophysics, Geosystems*, *9*(12), Q12022. doi: 10.1029/2008GC002107

Bastow, I. D., Pilidou, S., Kendall, J. M., & Stuart, G. W. (2010). Melt-Induced seismic anisotropy and magma assisted rifting in Ethiopia: evidence from surface waves. *Geochemistry, Geophysics, Geosystems*, *11*(6), Q0AB05. doi: 10.1029/2010GC003036

Bastow, I. D., Stuart, G. W., Kendall, J. M., & Ebinger, C. J. (2005). Upper-mantle seismic structure in a region of incipient continental breakup: Northern Ethiopian rift. *Geophysical Journal International*, *162*, 479–493.

Beccaluva, L., Bianchini, G., Natali, C., & Siena, F. (2009). Continental flood basalts and mantle plumes: a case study of the Northern Ethiopian Plateau. *Journal of Petrology*, *50*, 1377–1403.

Bell, K., & Simonetti, A. (1996). Carbonatite magmatism and plume activity: implications from the Nd, Pb and Sr isotope systematics of Oldoinyo Lengai. *Journal of Petrology*, *37*, 1321–1339.

Benoit, M. H., Nyblade, A. A., & VanDecar, J. C. (2006). Upper mantle P-wave speed variations beneath Ethiopia and the origin of the Afar hotspot. *Geology*, *34*, 329–332.

Biggs, J., Anthony, E. Y., & Ebinger, C. J. (2009). Multiple inflation and deflation events at Kenyan volcanoes, East African Rift. *Geology*, *37*, 979–982.

Biggs, J., Bastow, I. D., Keir, D., & Lewi, E. (2011). Pulses of deformation reveal frequently recurring shallow magmatic activity beneath the Main Ethiopian Rift. *Geochemistry, Geophysics, Geosystems*, *12*, Q0AB10. doi: 10.1029/2011GC003662

Bilham, R., Bendick, R., Larson, K., Mohr, P., Braun, J., Tesfaye, S., & Asfaw, L. (1999). Secular and tidal strain across the main Ethiopian rift. *Geophysical Research Letters*, *26*, 2789–2792.

Blegen, N., Brown, F. H., Jicha, B. R., Binetti, K. M., Faith, J. T., Ferraro, J. V., et al. (2016). The Menengai Tuff: a 36 ka widespread tephra and its chronological relevance to Late Pleistocene human evolution in East Africa. *Quaternary Science Reviews*, *152*, 152–168. doi: 10.1016/j.quascirev.2016.09.020

Boone, S. C., Kohn, B. P., Gleadow, A. J., Morley, C. K., Seiler, C., & Foster, D. A. (2019). Birth of the East African Rift system: nucleation of magmatism and strain in the Turkana Depression. *Geology*, *47*, 886–890.

Bosshard-Stadlin, S. A., Mattsson, H. B., & Keller, J. (2014). Magma mixing and forced exsolution of CO_2 during the explosive 2007–2008 eruption of Oldoinyo Lengai (Tanzania). *Journal of Volcanology and Geothermal Research*, *285*, 229–246.

Bosworth, W., Lambiase, J., & Keisler, R. (1986). A new look at Gregory's Rift: the structural style of continental rifting. *EOS. Transactions of the American Geophysical Union*, *67*, 577–583.

Bosworth, W., & Morley, C. K. (1994). Structural and stratigraphic evolution of the Anza rift, Kenya. *Tectonophysics*, *236*, 93–115.

Bott, M. H. P. (1996). Flexure associated with planar faulting. *Geophysical Journal International*, *126*, F21–F24. doi: 10.1111/j.1365-246X.1996.tb04692.x

Bown, J. W., & White, R. S. (1995). Effect of finite extension rate on melt generation at rifted continental margins. *Journal of Geophysical Research - Solid Earth*, *100*, 18011–18029.

Boyce, A., Bastow, I., Cottaar, S., Kounoudis, R., Guilloud De Courbeville, J., Caunt, E., & Desai, S. (2021). AFRP20: New P-wavespeed Model for the African mantle reveals two whole-mantle plumes below East Africa and Neoproterozoic modification of the Tanzania Craton. *Geochemistry, Geophysics, Geosystems*, *22*, e2020GC009302. doi: 10.1029/2020GC009302

Buck, W. R. (2006). The role of magma in the development of the Afro-Arabian Rift System. *Geological Society of London, Special Publication*, *259*, 43–54. doi: 10.1144/gsl.sp.2006.259.01.05

Calais, E., d'Oreye, N., Albaric, J., Deschamps, A., Delvaux, D., Deverchere, J., et al. (2008). Strain accommodation by slow slip and dyking in a youthful continental rift, East Africa. *Nature*, *456*, 783–787. doi: 10.1038/Nature07478

Calais, E., Ebinger, C., Hartnady, C., & Nocquet, J. M. (2006). Kinematics of the East African Rift from GPS and earthquake slip vector data. *Geological Society of London, Special Publication*, *259*, 9–22.

Campisano, C., & Feibel, C. S. (2008). Tephrostratigraphy of the Hadar and Busidima Formations at Hadar, Afar Depression, Ethiopia. In J. Quade & J. G. Wynn (Eds.), *The geology of early humans in the horn of Africa, special paper 446* (pp. 135–162). Boulder, Co: The Geological Society of America.

Carroll, M. R., & Blank, J. G. (1997). The solubility of H_2O in phonolitic melts. *American Mineralogist*, *82*, 549–556.

Celli, N. L., Lebedev, S., Schaeffer, A. J., & Gaina, C. (2020). African cratonic lithosphere carved by mantle plumes. *Nature Communications*, *11*, 92. doi: 10.1038/s41467-019-13871-2

Chang, S.-J., Kendall, E., Davaille, A., & Ferreira, A. M. (2020). The evolution of mantle plumes in East Africa. *Journal of Geophysical Research - Solid Earth*, *125*, e2020JB019929. doi: 10.1029/2020JB019929

Chiasera, B., Rooney, T. O., Girard, G., Yirgu, G., Grosfils, E. B., Ayalew, D., et al. (2018). Magmatically assisted off-rift extension—the case for broadly distributed strain accommodation. *Geosphere*, *14*, 1544–1563. doi: 10.1130/ges01615.1

Chorowicz, J., Fournier, J. L., & Vidal, G. (1987). A model for rift development in Eastern Africa. *Geological Journal*, *22*, 495–513.

Civiero, C., Hammond, J. O., Goes, S., Fishwick, S., Ahmed, A., Ayele, A., et al. (2015). Multiple mantle upwellings in the transition zone beneath the northern East-African Rift system from relative P-wave travel-time tomography. *Geochemistry, Geophysics, Geosystems*, *16*, 2949–2968.

Clarke, B., Calder, E. S., Dessalegn, F., Fontijn, K., Cortés, J. A., Naylor, M., et al. (2019). Fluidal pyroclasts reveal the intensity of peralkaline rhyolite pumice cone eruptions. *Nature Communications*, *10*, 1–10.

Clarke, B., Tierz, P., Calder, E., & Yirgu, G. (2020). Probabilistic volcanic hazard assessment for pyroclastic density currents from pumice cone eruptions at Aluto volcano, Ethiopia. *Frontiers in Earth Science*, *8*, 348. doi: 10.3389/feart.2020.00348

Clarke, M. C. (1990). *Geological, volcanological and hydrogeological controls on the occurrence of geothermal activity in the area surrounding Lake Naivasha, Kenya: with coloured 1: 250 000 geological maps*. Ministry of Energy.

Cornwell, D. G., Hetényi, G., & Blanchard, T. D. (2011). Mantle transition zone variations beneath the Ethiopian Rift and Afar: chemical heterogeneity within a hot mantle? *Geophysical Research Letters*, *38*, L16308. doi: 10.1029/2011gl047575

Cornwell, D. G., Mackenzie, G. D., England, R. W., Maguire, P. K. H., Asfaw, L. M., & Oluma, B. (2006). Northern Main Ethiopian Rift crustal structure from new high-precision gravity data. In G. Yirgu, C. Ebinger, & P. Maguire (Eds.), *The afar volcanic province within the East African Rift System* (pp. 307–322). London: Special Publication of the Geological Society.

Corti, G., Agostini, A., Keir, D., Van Wijk, J., Bastow, I. D., & Ranalli, G. (2015). Magma-induced axial subsidence during final-stage rifting: implications for the development of seaward-dipping reflectors. *Geosphere*, *11*, 563–571.

Corti, G., Molin, P., Sembroni, A., Bastow, I. D., & Keir, D. (2018). Control of pre-rift lithospheric structure on the architecture and evolution of continental rifts: insights from the Main Ethiopian Rift, East Africa. *Tectonics*, *37*, 477–496. doi: 10.1002/2017TC004799

Courtillot, V., Jaupart, C., Manighetti, I., Tapponnier, P., & Besse, J. (1999). On causal links between flood basalts and continental breakup. *Earth and Planetary Science Letters*, *166*, 177–195.

Craig, T. J., Jackson, J. A., Priestley, K., & McKenzie, D. (2011). Earthquake distribution patterns in Africa: their relationship to variations in lithospheric and geological structure, and their rheological implications. *Geophysical Journal International*, *185*, 403–434.

Daly, M. C., Chorowicz, J., & Fairhead, J. D. (1989). Rift basin evolution in Africa: the influence of reactivated steep basement shear zones. *Geological Society of London, Special Publication*, *44*, 309–334.

Daly, M. C., Green, P., Watts, A. B., Davies, O., Chibesakunda, F., & Walker, R. (2020). Tectonics and landscape of the Central African Plateau and their implications for a propagating Southwestern Rift in Africa. *Geochemistry, Geophysics, Geosystems*, *21*, e2019GC008746.

Daniels, K. A., Bastow, I. D., Keir, D., Sparks, R. S. J., & Menand, T. (2014). Thermal models of dyke intrusion during development of continent–ocean transition. *Earth and Planetary Science Letters*, *385*, 145–153.

Davaille, A., & Vatteville, J. (2005). On the transient nature of mantle plumes. *Geophysical Research Letters*, *32*, L14309. doi: 10.1029/2005GL023029

Dávalos-Elizondo, E., Atekwana, E. A., Atekwana, E. A., Tsokonombwe, G., & Laó-Dávila, D. A. (2021). Medium to low enthalpy geothermal reservoirs estimated from geothermometry and mixing models of hot springs along the Malawi Rift Zone. *Geothermics*, *89*, 101963. doi: 10.1016/j.geothermics.2020.101963

Davidson, A., & Rex, D. C. (1980). Age of volcanism and rifting in Southwestern Ethiopia. *Nature*, *283*, 657–658.

Davies, G. F. (1994). Thermomechanical erosion of the lithosphere by mantle plumes. *Journal of Geophysical Research - Solid Earth*, *99*, 15709–15722.

Dawson, J. B., & Smith, J. V. (1988). Metasomatized and veined upper-mantle xenoliths from Pello Hill, Tanzania - evidence for anomalously-light mantle beneath the Tanzanian Sector of the East-African rift-valley. *Contributions to Mineralogy and Petrology*, *100*, 510–527.

Deino, A. L., Scott, G. R., Saylor, B., Alene, M., Angelini, J. D., & Haile-Selassie, Y. (2010). 40Ar/39Ar dating, paleomagnetism, and tephrochemistry of Pliocene strata of the hominid-bearing Woranso-Mille area, west-central Afar Rift, Ethiopia. *Journal of Human Evolution*, *58*, 111–126.

Delvaux, D., & Barth, A. (2010). African stress pattern from formal inversion of focal mechanism data. *Tectonophysics*, *482*, 105–128.

Delvaux, D., Levi, K., Kajara, R., & Sarota, J. (1992). Cenozoic paleostress and kinematic evolution of the Rukwa-North Malawi rift valley (East African Rift System). *Bulletin des Centres de Recherches Exploration-Production Elf-Aquitaine*, *16*, 383–406.

Deniel, C., Vidal, P., Coulon, C., & Vellutini, P. J. (1994). Temporal evolution of mantle sources during continental rifting - the volcanism of Djibouti (Afar). *Journal of Geophysical Research - Solid Earth*, *99*, 2853–2869.

Déprez, A., Doubre, C., Masson, F., & Ulrich, P. (2013). Seismic and aseismic deformation along the East African Rift System from a reanalysis of the GPS velocity field of Africa. *Geophysical Journal International*, *193*, 1353–1369.

Desissa, M., Johnson, N. E., Whaler, K. A., Hautot, S., Fisseha, S., & Dawes, G. J. K. (2013). A mantle magma reservoir beneath an incipient mid-ocean ridge in Afar, Ethiopia. *Nature Geoscience*, *6*, 861–865.

Di Genova, D., Romano, C., Hess, K.-U., Vona, A., Poe, B. T., Giordano, D., et al. (2013). The rheology of peralkaline rhyolites from Pantelleria Island. *Journal of Volcanology and Geothermal Research*, *249*, 201–216.

Di Genova, D., Vasseur, J., Hess, K.-U., Neuville, D. R., & Dingwell, D. B. (2017). Effect of oxygen fugacity on the glass transition, viscosity and structure of silica-and iron-rich magmatic melts. *Journal of Non-Crystalline Solids*, *470*, 78–85.

Di Matteo, V., Carroll, M. R., Behrens, H., Vetere, F., & Brooker, R. A. (2004). Water solubility in trachytic melts. *Chemical Geology*, *213*, 187–196.

DiMaggio, E. N., Arrowsmith, J. R., Campisano, C. J., Johnson, R., Deino, A. L., Warren, M., et al. (2015). Tephrostratigraphy and depositional environment of young (<2.94 Ma) Hadar formation deposits at Ledi-Geraru, Afar, Ethiopia. *Journal of African Earth Sciences*, *112*, 234–250. doi: 10.1016/j.jafrearsci.2015.09.018

Donovan, A., Blundy, J., Oppenheimer, C., & Buisman, I. (2018). The 2011 eruption of Nabro volcano, Eritrea: perspectives on magmatic processes from melt inclusions. *Contributions to Mineralogy and Petrology*, *173*, 1.

Dugda, M. T., Nyblade, A. A., & Julia, J. (2007). Thin lithosphere beneath the Ethiopian plateau revealed by a joint inversion of Rayleigh wave group velocities and receiver functions. *Journal of Geophysical Research - Solid Earth*, *112*, B08305. doi: 10.1029/2006JB004918

Dugda, M. T., Nyblade, A. A., Julia, J., Langston, C. A., Ammon, C. J., & Simiyu, S. (2005). Crustal structure in Ethiopia and Kenya from receiver function analysis: implications for rift development in eastern Africa. *Journal of Geophysical Research - Solid Earth*, *110*, B01303. doi: 10.1029/2004JB003065

Ebinger, C., & Scholz, C. A. (2011). Continental rift basins: the East African perspective. *Tectonics of Sedimentary Basins: Recent Advances*, 183–208.

Ebinger, C. J. (1989). Tectonic development of the western branch of the East-African Rift System. *Geological Society of America Bulletin*, *101*, 885–903.

Ebinger, C. J., & Casey, M. (2001). Continental breakup in magmatic provinces: an Ethiopian example. *Geology*, *29*, 527–530.

Ebinger, C. J., Jackson, J. A., Foster, A. N., & Hayward, N. J. (1999). Extensional basin geometry and the elastic lithosphere. *Philosophical Transactions of the Royal Society of London. Series A: Mathematical, Physical and Engineering Sciences*, *357*, 741–762.

Ebinger, C. J., Keir, D., Bastow, I. D., Whaler, K., Hammond, J. O., Ayele, A., et al. (2017). Crustal structure of active deformation zones in Africa: implications for global crustal processes. *Tectonics*, *36*, 3298–3332.

Ebinger, C. J., Rosendahl, B. R., & Reynolds, D. J. (1987). Tectonic model of the Malaŵi rift, Africa. *Tectonophysics*, *141*, 215–235.

Ebinger, C. J., & Sleep, N. H. (1998). Cenozoic magmatism throughout East Africa resulting from impact of a single plume. *Nature*, *395*, 788–791.

Ebinger, C. J., van Wijk, J., & Keir, D. (2013). The time scales of continental rifting: implications for global processes. *Geological Society of America Special Papers*, *500*, 371–396.

Ebinger, C. J., Yemane, T., Harding, D. J., Tesfaye, S., Kelley, S., & Rex, D. C. (2000). Rift deflection, migration, and propagation: linkage of the Ethiopian and Eastern rifts. *Geological Society of America Bulletin*, *112*, 163–176.

Ebinger, C. J., Yemane, T., WoldeGabriel, G., Aronson, J. L., & Walter, R. C. (1993). Late Eocene-recent volcanism and faulting in the southern main Ethiopian Rift. *Journal of the Geological Society of London*, *150*, 99–108.

Emry, E. L., Shen, Y., Nyblade, A. A., Flinders, A., & Bao, X. (2019). Upper mantle earth structure in Africa from full-wave ambient noise tomography. *Geochemistry, Geophysics, Geosystems*, *20*, 120–147. doi: 10.1029/2018GC007804

Ferguson, D. J., Barnie, T. D., Pyle, D. M., Oppenheimer, C., Yirgu, G., Lewi, E., et al. (2010). Recent rift-related volcanism in Afar, Ethiopia. *Earth and Planetary Science Letters*, *292*, 409–418. doi: 10.1016/j.epsl.2010.02.010

Ferguson, D. J., Maclennan, J., Bastow, I. D., Pyle, D. M., Jones, S. M., Keir, D., et al. (2013). Melting during late-stage rifting in Afar is hot and deep. *Nature*, *499*, 70–73. doi: 10.1038/nature12292

Feyissa, D. H., Kitagawa, H., Bizuneh, T. D., Tanaka, R., Kabeto, K., & Nakamura, E. (2019). Transition from plume-driven to plate-driven magmatism in the evolution of the main Ethiopian Rift. *Journal of Petrology*, *60*, 1681–1715. doi: 10.1093/petrology/egz043

Field, L., Barnie, T., Blundy, J., Brooker, R. A., Keir, D., Lewi, E., & Saunders, K. (2012). Integrated field, satellite and petrological observations of the November 2010 eruption of Erta Ale. *Bulletin of Volcanology*, *74*, 2251–2271.

Field, L., Blundy, J., Calvert, A., & Yirgu, G. (2013). Magmatic history of Dabbahu, a composite volcano in the Afar Rift, Ethiopia. *Geological Society of America Bulletin*, *125*, 128–147. doi: 10.1130/b30560.1

Fishwick, S. (2010). Surface wave tomography: imaging of the lithosphere–asthenosphere boundary beneath central and southern Africa? *Lithos*, *120*, 63–73.

Foley, S. F., Link, K., Tiberindwa, J. V., & Barifaijo, E. (2012). Patterns and origin of igneous activity around the Tanzanian craton. *Journal of African Earth Sciences*, *62*, 1–18.

Fontijn, K., Delvaux, D., Ernst, G. G., Kervyn, M., Mbede, E., & Jacobs, P. (2010). Tectonic control over active volcanism at a range of scales: case of the Rungwe Volcanic Province, SW Tanzania; and hazard implications. *Journal of African Earth Sciences*, *58*, 764–777.

Fontijn, K., Elburg, M. A., Nikogosian, I. K., van Bergen, M. J., & Ernst, G. G. (2013). Petrology and geochemistry of Late Holocene felsic magmas from Rungwe volcano (Tanzania), with implications for trachytic Rungwe Pumice eruption dynamics. *Lithos*, *177*, 34–53.

Fontijn, K., Ernst, G. G., Bonadonna, C., Elburg, M. A., Mbede, E., & Jacobs, P. (2011). The~ 4-ka Rungwe Pumice (South-Western Tanzania): a wind-still Plinian eruption. *Bulletin of Volcanology*, *73*, 1353–1368.

Fontijn, K., McNamara, K., Tadesse, A. Z., Pyle, D. M., Dessalegn, F., Hutchison, W., et al. (2018). Contrasting styles of post-caldera volcanism along the Main Ethiopian Rift: implications for contemporary volcanic hazards. *Journal of Volcanology and Geothermal Research*, *356*, 90–113.

Fontijn, K., Williamson, D., Mbede, E., & Ernst, G. G. (2012). The Rungwe volcanic province, Tanzania–a volcanological review. *Journal of African Earth Sciences*, *63*, 12–31.

Forsyth, D., & Uyeda, S. (1975). On the relative importance of the driving forces of plate motion. *Geophysical Journal International*, *43*, 163–200.

Franke, D., Jokat, W., Ladage, S., Stollhofen, H., Klimke, J., Lutz, R., et al. (2015). The offshore East African rift system: structural framework at the toe of a juvenile rift. *Tectonics*, *34*, 2086–2104.

French, S. W., & Romanowicz, B. (2015). Broad plumes rooted at the base of the Earth's mantle beneath major hotspots. *Nature*, *525*, 95–99.

French, S. W., & Romanowicz, B. A. (2014). Whole-mantle radially anisotropic shear velocity structure from spectral-element waveform tomography. *Geophysical Journal International*, *199*, 1303–1327.

Furman, T., Bryce, J. G., Karson, J., & Iotti, A. (2004). East African Rift System (EARS) plume structure: insights from quaternary mafic lavas of Turkana, Kenya. *Journal of Petrology*, *45*, 1069–1088.

Furman, T., Bryce, J. G., Rooney, T., Hanan, B. B., Yirgu, G., & Ayalew, D. (2006). Heads and tails: 30 million years of the Afar plume. In G. Yirgu, C. Ebinger, & P. Maguire (Eds.), *The afar volcanic province within the East African Rift system* (pp. 95–120). London: Special Publication of the Geological Society.

Furman, T., & Graham, D. (1999). Erosion of lithospheric mantle beneath the East African Rift system; geochemical evidence from the Kivu volcanic province. *Lithos*, *48*, 237–262.

Furman, T., Nelson, W. R., & Elkins-Tanton, L. T. (2016). Evolution of the East African rift: drip magmatism, lithospheric thinning and mafic volcanism. *Geochimica et Cosmochimica Acta*, *185*, 418–434.

Gallacher, R. J., Keir, D., Harmon, N., Stuart, G., Leroy, S., Hammond, J. O., et al. (2016). The initiation of segmented buoyancy-driven melting during continental breakup. *Nature Communications*, *7*, 13110. doi: 10.1038/ncomms13110

Garnero, E. J., McNamara, A. K., & Shim, S.-H. (2016). Continent-sized anomalous zones with low seismic velocity at the base of Earth's mantle. *Nature Geoscience*, *9*, 481–489. doi: 10.1038/NGEO2733

George, R., Rogers, N., & Kelley, S. (1998). Earliest magmatism in Ethiopia: evidence for two mantle plumes in one flood basalt province. *Geology*, *26*, 923–926.

Giordano, D., Romano, C., Papale, P., & Dingwell, D. B. (2004). The viscosity of trachytes, and comparison with basalts, phonolites, and rhyolites. *Chemical Geology*, *213*, 49–61.

Giordano, F., D'Antonio, M., Civetta, L., Tonarini, S., Orsi, G., Ayalew, D., et al. (2014). Genesis and evolution of mafic and felsic magmas at Quaternary volcanoes within the Main Ethiopian Rift: insights from Gedemsa and Fanta 'Ale complexes. *Lithos*, *188*, 130–144. doi: 10.1016/j.lithos.2013.08.008

Gleeson, M. L. M., Stock, M. J., Pyle, D. M., Mather, T. A., Hutchison, W., Yirgu, G., & Wade, J. (2017). Constraining magma storage conditions at a restless volcano in the Main Ethiopian Rift using phase equilibria models. *Journal of Volcanology and Geothermal Research*, *337*, 44–61. doi: 10.1016/j.jvolgeores.2017.02.026

Global Volcanism Program. (2013). Volcanoes of the World (v. 4.11.2) [Database]. Distributed by Smithsonian Institution; compiled by Venzke, E. doi: 10.5479/si.GVP.VOTW4-2013

Goitom, B., Oppenheimer, C., Hammond, J. O., Grandin, R., Barnie, T., Donovan, A., et al. (2015). First recorded eruption of Nabro volcano, Eritrea, 2011. *Bulletin of Volcanology*, *77*, 85.

Grandin, R., Socquet, A., Binet, R., Klinger, Y., Jacques, E., de Chabalier, J.-B., et al. (2009). September 2005 Manda Hararo-Dabbahu rifting event, Afar (Ethiopia): constraints provided by geodetic data. *Journal of Geophysical Research - Solid Earth*, *114*, B08404. doi: 10.1029/2008JB005843

Greenfield, T., Keir, D., Kendall, J.-M., & Ayele, A. (2019). Low-frequency earthquakes beneath Tullu Moye volcano, Ethiopia, reveal fluid pulses from shallow magma chamber. *Earth and Planetary Science Letters*, *526*, 115782. doi: 10.1016/j.epsl.2019.115782

Gregory, J. (1896). *The Great Rift Valley*. London: John Murray.

Guzmics, T., Berkesi, M., Bodnar, R. J., Fall, A., Bali, E., Milke, R., et al. (2019). Natrocarbonatites: a hidden product of three-phase immiscibility. *Geology*, *47*, 527–530. doi: 10.1130/G46125.1

Halldórsson, S. A., Hilton, D. R., Scarsi, P., Abebe, T., & Hopp, J. (2014). A common mantle plume source beneath the entire East African Rift System revealed by coupled helium-neon systematics. *Geophysical Research Letters*, *41*, 2304–2311.

Hamaguchi, H., Nishimura, T., & Zana, N. (1992). Process of the 1977 Nyiragongo eruption inferred from the analysis of long-period earthquakes and volcanic tremors. *Tectonophysics*, *209*, 241–254.

Hamling, I. J., Ayele, A., Bennati, L., Calais, E., Ebinger, C. J., Keir, D., et al. (2009). Geodetic observations of the ongoing Dabbahu rifting episode: new dyke intrusions in 2006 and 2007. *Geophysical Journal International*, *178*, 989–1003. doi: 10.1111/j.1365-246X.2009.04163.x

Hammond, J. O. (2014). Constraining melt geometries beneath the Afar Depression, Ethiopia from teleseismic receiver functions: the anisotropic H-κ stacking technique. *Geochemistry, Geophysics, Geosystems*, *15*, 1316–1332.

Hammond, J. O. S., Kendall, J.-M., Stuart, G. W., Ebinger, C. J., Bastow, I. D., Keir, D., et al. (2013). Mantle upwelling and initiation of rift segmentation beneath the Afar Depression. *Geology*, *41*, 635–638.

Harkin, D.A. (1962). *The Rungwe volcanics at the northern end of Lake Nyasa* (PhD dissertation). Ph.D. Dissertation. University of Glasgow.

Hart, W. K., Woldegabriel, G., Walter, R. C., & Mertzman, S. A. (1989). Basaltic volcanism in Ethiopia - Constraints on continental rifting and mantle interactions. *Journal of Geophysical Research - Solid Earth and Planets*, *94*, 7731–7748.

Hayward, N. J., & Ebinger, C. J. (1996). Variations in the along-axis segmentation of the Afar Rift system. *Tectonics*, *15*, 244–257.

Head, E. M., Shaw, A. M., Wallace, P. J., Sims, K. W., & Carn, S. A. (2011). Insight into volatile behavior at Nyamuragira volcano (DR Congo, Africa) through olivine-hosted melt inclusions. *Geochemistry, Geophysics, Geosystems*, *12*, Q0AB11. doi: 10.1029/2011GC003699

Helffrich, G. (2000). Topography of the transition zone seismic discontinuities. *Reviews of Geophysics*, *38*, 141–158.

Hilton, D. R., Halldórsson, S. A., Barry, P. H., Fischer, T. P., de Moor, J. M., Ramirez, C. J., et al. (2011). Helium isotopes at Rungwe Volcanic Province, Tanzania, and the origin of East African Plateaux. *Geophysical Research Letters*, *38*, L21304. doi: 10.1029/2011GL049589

Hodgson, I., Illsley-Kemp, F., Gallacher, R. J., Keir, D., Ebinger, C. J., & Mtelela, K. (2017). Crustal structure at a young continental rift: a receiver function study from the Tanganyika Rift. *Tectonics*, *36*, 2806–2822.

Hofmann, C., Courtillot, V., Feraud, G., Rouchett, P., Yirgu, G., Ketefo, E., & Pik, R. (1997). Timing of the Ethiopian flood basalt event and implications for plume birth and global change. *Nature*, *389*, 838–841.

Hopper, E., Gaherty, J. B., Shillington, D. J., Accardo, N. J., Nyblade, A. A., Holtzman, B. K., et al. (2020). Preferential localized thinning of lithospheric mantle in the melt-poor Malawi Rift. *Nature Geoscience*, *13*, 584–589.

Hosseini, K., Matthews, K. J., Sigloch, K., Shephard, G. E., Domeier, M., & Tsekhmistrenko, M. (2018). SubMachine: Web-based tools for exploring seismic tomography and other models of Earth's deep interior. *Geochemistry, Geophysics, Geosystems*, *19*, 1464–1483.

Hosseini, K., Sigloch, K., Tsekhmistrenko, M., Zaheri, A., Nissen-Meyer, T., & Igel, H. (2020). Global mantle structure from multifrequency tomography using P, PP and P-diffracted waves. *Geophysical Journal International*, *220*, 96–141.

Hudgins, T. R., Mukasa, S. B., Simon, A. C., Moore, G., & Barifaijo, E. (2015). Melt inclusion evidence for CO_2-rich melts beneath the western branch of the East African Rift: implications for long-term storage of volatiles in the deep lithospheric mantle. *Contributions to Mineralogy and Petrology*, *169*, 1–18. doi: 10.1007/s00410-015-1140-9

Huerta, A. D., Nyblade, A. A., & Reusch, A. M. (2009). Mantle transition zone structure beneath Kenya and Tanzania: more evidence for a deep-seated thermal upwelling in the mantle. *Geophysical Journal International*, *177*, 1249–1255. doi: 10.1111/j.1365-246X.2009.04092.x

Hunt, J. A., Zafu, A., Mather, T. A., Pyle, D. M., & Barry, P. H. (2017). Spatially variable CO_2 degassing in the main Ethiopian Rift: implications for Magma storage, volatile transport, and Rift-related emissions. *Geochemistry, Geophysics, Geosystems*, *18*, 3714–3737.

Hutchison, W., Fusillo, R., Pyle, D. M., Mather, T. A., Blundy, J. D., Biggs, J., et al. (2016a). A pulse of mid-Pleistocene rift volcanism in Ethiopia at the dawn of modern humans. *Nature Communications*, *7*, 13192. doi: 10.1038/ncomms13192

Hutchison, W., Mather, T. A., Pyle, D. M., Boyce, A. J., Gleeson, M. L., Yirgu, G., et al. (2018). The evolution of magma during continental rifting: New constraints from the isotopic and trace element signatures of silicic magmas from Ethiopian volcanoes. *Earth and Planetary Science Letters*, *489*, 203–218.

Hutchison, W., Pyle, D. M., Mather, T. A., Yirgu, G., Biggs, J., Cohen, B. E., et al. (2016b). The eruptive history and magmatic evolution of Aluto volcano: new insights into silicic peralkaline volcanism in the Ethiopian rift. *Journal of Volcanology and Geothermal Research*, *328*, 9–33.

Iddon, F., & Edmonds, M. (2020). Volatile-rich magmas distributed through the upper crust in the Main Ethiopian Rift. *Geochemistry, Geophysics, Geosystems*, *21*, e2019GC008904. doi: 10.1029/2019GC008904

Iddon, F., Jackson, C., Hutchison, W., Fontijn, K., Pyle, D. M., Mather, T. A., et al. (2019). Mixing and crystal scavenging in the Main Ethiopian Rift revealed by trace element systematics in feldspars and glasses. *Geochemistry, Geophysics, Geosystems*, *20*, 230–259. doi: 10.1029/2018GC007836

Jakovlev, A., Rümpker, G., Schmeling, H., Koulakov, I., Lindenfeld, M., & Wallner, H. (2013). Seismic images of magmatic rifting beneath the western branch of the East African rift. *Geochemistry, Geophysics, Geosystems*, *14*, 4906–4920.

Jestin, F., Huchon, P., & Gaulier, J. M. (1994). The Somalia plate and the East African Rift System: present-day kinematics. *Geophysical Journal International*, *116*, 637–654.

Julià, J. (2007). Constraining velocity and density contrasts across the crust—mantle boundary with receiver function amplitudes. *Geophysical Journal International*, *171*, 286–301.

Julià, J., & Nyblade, A. A. (2013). Probing the upper mantle transition zone under Africa with P520s conversions: implications for temperature and composition. *Earth and Planetary Science Letters*, *368*, 151–162. doi: 10.1016/j.epsl.2013.02.021

Kanda, I., Fujimitsu, Y., & Nishijima, J. (2019). Geological structures controlling the placement and geometry of heat sources within the Menengai geothermal field, Kenya as evidenced by gravity study. *Geothermics*, *79*, 67–81.

Kebede, S., & Ambebrhan, Y. (2020). *Opportunities for direct geothermal energy resource utilization in Ethiopia*. In *Paper presented at the 8th African Rift Geothermal Conference*. Nairobi, Kenya.

Keir, D., Bastow, I., Whaler, K., Daly, E., Cornwell, D. G., & Hautot, S. (2009). Lower-crustal earthquakes near the Ethiopian rift induced by magma injection. *Geochemistry, Geophysics, Geosystems*, *10*, Q0AB02. doi: 10.1029/2009GC002382

Keir, D., Bastow, I. D., Corti, G., Mazzarini, F., & Rooney, T. O. (2015). The origin of along-rift variations in faulting and magmatism in the Ethiopian Rift. *Tectonics*, *34*, 464–477.

Keir, D., Bastow, I. D., Pagli, C., & Chambers, E. L. (2013). The development of extension and magmatism in the Red Sea rift of Afar. *Tectonophysics*, *607*, 98–114. doi: 10.1016/j.tecto.2012.10.015

Keir, D., Belachew, M., Ebinger, C. J., Kendall, J.-M., Hammond, J. O., Stuart, G. W., et al. (2011). Mapping the evolving strain field during continental breakup from crustal anisotropy in the Afar Depression. *Nature Communications*, *2*, 1–7.

Keir, D., Ebinger, C. J., Stuart, G. W., Daly, E., & Ayele, A. (2006). Strain accommodation by magmatism and faulting as rifting proceeds to breakup: seismicity of the northern Ethiopian rift. *Journal of Geophysical Research - Solid Earth*, *111*, B05314. doi: 10.1029/2005jb003748

Keller, J., Zaitsev, A. N., & Wiedenmann, D. (2006). Primary magmas at Oldoinyo Lengai: the role of olivine melilitites. *Lithos*, *91*, 150–172.

Kendall, J.-M., & Lithgow-Bertelloni, C. (2016). Why is Africa rifting? *Geological Society of London, Special Publication*, *420*, 11–30.

Kendall, J. M., Pilidou, S., Keir, D., Bastow, I., Stuart, G. W., & Ayele, A. (2006). Mantle upwellings, melt migration and the rifting of Africa: insights from seismic anisotropy. In G. Yirgu, C. Ebinger, & P. Maguire (Eds.), *The Afar volcanic province within the East African Rift system* (pp. 55–72). London: Geological Society of London Special Publication.

Kendall, J.-M., Stuart, G. W., Ebinger, C., Bastow, I. D., & Keir, D. (2005). Magma-assisted rifting in Ethiopia. *Nature*, *433*, 146–148.

Keranen, K., Klemper, S. L., Gloaguen, R., & Working Grp, E. A. G. L. E. (2004). Three-dimensional seismic imaging of a protoridge axis in the Main Ethiopian rift. *Geology*, *32*, 949–952.

Kervyn, M., Ernst, G. G., Keller, J., Vaughan, R. G., Klaudius, J., Pradal, E., et al. (2010). Fundamental changes in the activity of the natrocarbonatite volcano Oldoinyo Lengai, Tanzania. *Bulletin of Volcanology*, *72*, 913–931.

Kilembe, E. A., & Rosendahl, B. R. (1992). Structure and stratigraphy of the Rukwa rift. *Tectonophysics*, *209*, 143–158.

Klerkx, J., Theunissen, K., & Delvaux, D. (1998). Persistent fault controlled basin formation since the Proterozoic along the western branch of the East African Rift. *Journal of African Earth Sciences*, *26*, 347–361.

Koehn, D., Lindenfeld, M., Rümpker, G., Aanyu, K., Haines, S., Passchier, C. W., & Sachau, T. (2010). Active transsection faults in rift transfer zones: evidence for complex stress fields and implications for crustal fragmentation processes in the western branch of the East African Rift. *International Journal of Earth Sciences*, *99*, 1633–1642.

Kolawole, F., Atekwana, E. A., Laó-Dávila, D. A., Abdelsalam, M. G., Chindandali, P. R., Salima, J., & Kalindekafe, L. (2018). Active deformation of Malawi rift's north basin Hinge zone modulated by reactivation of preexisting Precambrian Shear zone fabric. *Tectonics*, *37*, 683–704.

Kolawole, F., Atekwana, E. A., Malloy, S., Stamps, D. S., Grandin, R., Abdelsalam, M. G., et al. (2017). Aeromagnetic, gravity, and differential interferometric synthetic aperture radar analyses reveal the causative fault of the 3 April 2017 Mw 6.5 Moiyabana, Botswana, earthquake. *Geophysical Research Letters*, *44*, 8837–8846.

Komorowski, J. C. (2002). The January 2002 flank eruption of Nyiragongo Volcano (Democratic Republic of Congo): chronology, evidence for a tectonic rift trigger, and impact of lava flows on the city of Goma. *Acta Vulcanologica*, 1000–1035.

Komorowski, J. C., & Karume, K. (2015). *Nyiragongo (Democratic Republic of Congo), January 2002: a major eruption in the midst of a complex humanitarian emergency* (p. 273). Hazards Risk: Glob. Volcan.

Koptev, A., Calais, E., Burov, E., Leroy, S., & Gerya, T. (2018). Along-axis variations of rift width in a coupled lithosphere-mantle system, application to East Africa. *Geophysical Research Letters*, *45*, 5362–5370.

La Rosa, A., Pagli, C., Keir, D., Sani, F., Corti, G., Wang, H., & Possee, D. (2019). Observing oblique slip during rift linkage in northern Afar. *Geophysical Research Letters*, *46*, 10782–10790.

Lane, C. S., Martin-Jones, C. M., & Johnson, T. C. (2018). A cryptotephra record from the Lake Victoria sediment core record of Holocene palaeoenvironmental change. *The Holocene*, *28*, 1909–1917.

Lavayssière, A., Rychert, C., Harmon, N., Keir, D., Hammond, J. O. S., Kendall, J.-M., et al. (2018). Imaging lithospheric discontinuities beneath the Northern East African Rift using S-to-P receiver functions. *Geochemistry, Geophysics, Geosystems*, *19*, 4048–4062. doi: 10.1029/2018GC007463

Le Bas, M. J., & Rubie, D. C. (1977). Kisingiri I: an ijolite–carbonatite–nephelinite volcanic complex in western Kenya. In M. J. Le Bas (Ed.), *Carbonatite-Nephelinite volcanism: an African case history* (pp. 39–46). John Wiley & Sons London.

Leat, P. T. (1984). Geological evolution of the trachytic caldera volcano Menengai, Kenya Rift Valley. *Journal of the Geological Society*, *141*, 1057–1069. doi: 10.1144/gsjgs.141.6.1057

Leat, P. T. (1985). Facies variations in peralkaline ash-flow tuffs from the Kenya Rift Valley. *Geological Magazine*, *122*, 139–150.

Lesne, P., Scaillet, B., Pichavant, M., Iacono-Marziano, G., & Beny, J.-M. (2011). The H_2O solubility of alkali basaltic melts: an experimental study. *Contributions to Mineralogy and Petrology*, *162*, 133–151.

Li, C., van der Hilst, R. D., Engdahl, E. R., & Burdick, S. (2008). A new global model for P wave speed variations in Earth's mantle. *Geochemistry, Geophysics, Geosystems*, *9*, Q05018. doi: 10.1029/2007gc001806

Lloyd, R., Biggs, J., Birhanu, Y., Wilks, M., Gottsmann, J., Kendall, J.-M., et al. (2018). Sustained uplift at a continental rift caldera. *Journal of Geophysical Research - Solid Earth*, *123*, 5209–5226.

Macdonald, R. (2012). Evolution of peralkaline silicic complexes: lessons from the extrusive rocks. *Lithos*, *152*, 11–22. doi: 10.1016/j.lithos.2012.01.014

MacDonald, R., Belkin, H. E., Fitton, J. G., Rogers, N. W., Nejbert, K., Tindle, A. G., & Marshall, A. S. (2008). The roles of fractional crystallization, magma mixing, crystal mush remobilization and volatile-melt interactions in the genesis of a young basalt-peralkaline rhyolite suite, the Greater Olkaria Volcanic Complex, Kenya rift valley. *Journal of Petrology*, *49*, 1515–1547.

MacDonald, R., & Scaillet, B. (2006). The central Kenya peralkaline province: insights into the evolution of peralkaline salic magmas. *Lithos*, *91*, 59–73.

Macgregor, D. (2015). History of the development of the East African Rift system: a series of interpreted maps through time. *Journal of African Earth Sciences*, *101*, 232–252.

Mackenzie, G., Thybo, H., & Maguire, P. (2005). Crustal velocity structure across the Main Ethiopian Rift: results from 2-dimensional wide-angle seismic modelling. *Geophysical Journal International*, *162*, 994–1006.

Maguire, P., Keller, G. R., Klemper, S. L., Mackenzie, G., Keranen, K., Harder, S., et al. (2006). Crustal structure of the Northern Main Ethiopian Rift from the EAGLE controlled source survey; a snapshot of incipient lithospheric break-up. In G. Yirgu, C. Ebinger, & P. Maguire (Eds.), *The Afar volcanic province within the East African Rift system* (pp. 269–292). London: Special Publication of the Geological Society.

Maguire, P. K. H., Swain, C. J., Masotti, R., & Khan, M. A. (1994). A crustal and uppermost mantle cross-sectional model of the Kenya Rift derived from seismic and gravity data. *Tectonophysics*, *236*, 217–249.

Makris, J., & Ginzburg, A. (1987). The Afar Depression - transition between continental rifting and sea-floor spreading. *Tectonophysics*, *141*, 199–214.

Mana, S., Furman, T., Turrin, B. D., Feigenson, M. D., & Swisher, C. C., III (2015). Magmatic activity across the East African North Tanzanian Divergence Zone. *Journal of the Geological Society*, *172*, 368–389.

Martin-Jones, C., Lane, C., Van Daele, M., Meeren, T. V. D., Wolff, C., Moorhouse, H., et al. (2020). History of scoria-cone eruptions on the eastern shoulder of the Kenya–Tanzania Rift revealed in the 250-ka sediment record of Lake Chala near Mount Kilimanjaro. *Journal of Quaternary Science*, *35*, 245–255.

Martin-Jones, C. M., Lane, C. S., Pearce, N. J., Smith, V. C., Lamb, H. F., Schaebitz, F., et al. (2017). Recurrent explosive eruptions from a high-risk Main Ethiopian Rift volcano throughout the Holocene. *Geology*, *45*, 1127–1130.

Marty, B., Appora, I., Barrat, J.-A. A., Deniel, C., Vellutini, P., Vidal, P., et al. (1993). He, Ar, Sr, Nd and Pb isotopes in volcanic rocks from Afar; evidence for a primitive mantle component and constraints on magmatic sources. *Geochemical Journal*, *27*, 219–228.

McClusky, S., Reilinger, R., Ogubazghi, G., Amleson, A., Healeb, B., Vernant, P., et al. (2010). Kinematics of the southern Red Sea–Afar Triple Junction and implications for plate dynamics. *Geophysical Research Letters*, *37*. doi: 10.1029/2009GL041127

McKenzie, D., & Bickle, M. J. (1988). The volume and composition of melt generated by extension of the lithosphere. *Journal of Petrology*, *29*, 625–679

McKenzie, D. P. (1978). Some remarks on the development of sedimentary basins. *Earth and Planetary Science Letters*, *40*, 25–32.

McNamara, K., Cashman, K. V., Rust, A. C., Fontijn, K., Chalié, F., Tomlinson, E. L., & Yirgu, G. (2018). Using lake sediment cores to improve records of volcanism at Aluto volcano in the Main Ethiopian Rift. *Geochemistry, Geophysics, Geosystems*, *19*, 3164–3188.

McNitt, J. R. (1982). *The geothermal potential of East Africa*. Paper presented at the Regional Seminar on Geothermal Energy in Eastern and Western Africa, Nairobi, Kenya.

Meshesha, D., & Shinjo, R. (2008). Rethinking geochemical feature of the Afar and Kenya mantle plumes and geodynamic implications. *Journal of Geophysical Research - Solid Earth*, *113*. doi: 10.1029/2007JB005549

Mibei, G., Harðrson, B. S., Franzson, H., Bali, E., Geirsson, H., & Guðfinnsson, G. H. (2021). Eruptive history and volcano-tectonic evolution of Paka volcanic complex in the northern Kenya rift: insights into the geothermal heat source. *Journal of African Earth Sciences*, *173*, 103951. doi: 10.1016/j.jafrearsci.2020.103951

Modisi, M. P., Atekwana, E. A., Kampunzu, A. B., & Ngwisanyi, T. H. (2000). Rift kinematics during the incipient stages of continental extension: evidence from the nascent Okavango rift basin, northwest Botswana. *Geology*, *28*, 939–942.

Mohr, P., & Zanettin, B. (1988). The Ethiopian flood basalt province. In J. D. MacDougall (Ed.), *Continental flood basalts* (pp. 63–110). Dordrecht: Kluwer Academic.

Mohr, P. A. (1974). *Mapping of the major structures of the African Rift system*. SAO Special Report #361).

Montagner, J.-P., Marty, B., Stutzmann, E., Sicilia, D., Cara, M., Pik, R., et al. (2007). Mantle upwellings and convective instabilities revealed by seismic tomography and helium isotope geochemistry beneath eastern Africa. *Geophysical Research Letters*, *34*. doi: 10.1029/2007gl031098

Montelli, R., Nolet, G., Dahlen, F. A., & Masters, G. (2006). A catalogue of deep mantle plumes: new results from finite-frequency tomography. *Geochemistry, Geophysics, Geosystems*, *7*, Q11007. doi: 10.1029/2006GC001248

Moore, C., Wright, T., Hooper, A., & Biggs, J. (2019). The 2017 Eruption of Erta'Ale Volcano, Ethiopia: insights into the shallow axial plumbing system of an incipient Mid-Ocean Ridge. *Geochemistry, Geophysics, Geosystems*, *20*, 5727–5743.

Morgan, J. P. (1971). Convection plumes in the lower mantle. *Nature*, *230*, 42–43.

Morley, C. K. (1988). Variable extension in lake Tanganyika. *Tectonics*, *7*, 785–801.

Morley, C. K. (1989). Extension, detachments, and sedimentation in continental rifts (with particular reference to East Africa). *Tectonics*, *8*, 1175–1192.

Morley, C. K. (2010). Stress re-orientation along zones of weak fabrics in rifts: an explanation for pure extension in 'oblique'rift segments? Earth Planet. *Science Letters*, *297*, 667–673.

Morley, C. K., Karanja, F. M., Wescott, W. A., Stone, D. M., Harper, R. M., Wigger, S. T., & Day, R. A. (1999). Geology and geophysics of the Western Turkana Basins. In C. K. Morley (Ed.), *Geoscience of rift systems—evolution of East Africa, AAPG Studies in Geology* (Vol. *44*, Ch. 2). American Association of Petroleum Geologists.

Moucha, R., & Forte, A. M. (2011). Changes in African topography driven by mantle convection. *Nature Geoscience*, *4*, 707–712.

Muirhead, J. D., & Kattenhorn, S. A. (2018). Activation of preexisting transverse structures in an evolving magmatic rift in East Africa. *Journal of Structural Geology*, *106*, 1–18. doi: 10.1016/j.jsg.2017.11.004

Muirhead, J. D., Kattenhorn, S. A., & Le Corvec, N. (2015). Varying styles of magmatic strain accommodation across the East African Rift. *Geochemistry, Geophysics, Geosystems*, *16*, 2775–2795.

Muirhead, J. D., Kattenhorn, S. A., Lee, H., Mana, S., Turrin, B. D., Fischer, T. P., et al. (2016). Evolution of upper crustal faulting assisted by magmatic volatile release during early-stage continental rift development in the East African Rift. *Geosphere*, *12*, 1670–1700.

Muirhead, J. D., Wright, L. J., & Scholz, C. A. (2019). Rift evolution in regions of low magma input in East Africa. *Earth and Planetary Science Letters*, *506*, 332–346.

Mulibo, G. D., & Nyblade, A. A. (2013). The P and S wave velocity structure of the mantle beneath eastern Africa and the African superplume anomaly. *Geochemistry, Geophysics, Geosystems*, *14*, 2696–2715.

Mulibo, G. D., & Nyblade, A. A. (2016). The seismotectonics of southeastern Tanzania: implications for the propagation of the eastern branch of the East African Rift. *Tectonophysics*, *674*, 20–30.

Muravyeva, N. S., & Senin, V. G. (2018). Xenoliths from Bunyaruguru volcanic field: some insights into lithology of East African Rift upper mantle. *Lithos*, *296*, 17–36.

Nelson, W. R., Hanan, B., Graham, D. W., Shirey, S. B., Yirgu, G., Ayalew, D., & Furman, T. (2019). Distinguishing plume and metasomatized lithospheric mantle contributions to post-flood basalt volcanism on the southeastern Ethiopian Plateau. *Journal of Petrology*, *60*, 1063–1094.

Ni, S. (2002). Sharp sides to the African Superplume. *Science*, *296*, 1850–1852.

Njinju, E. A., Atekwana, E. A., Stamps, D. S., Abdelsalam, M. G., Atekwana, E. A., Mickus, K. L., et al. (2019). Lithospheric structure of the Malawi Rift: implications for magma-poor rifting processes. *Tectonics*, *38*, 3835–3853.

Nyblade, A. A., Knox, R. P., & Gurrola, H. (2000). Mantle transition zone thickness beneath Afar: implications for the origin of the Afar hotspot. *Geophysical Journal International*, *142*, 615–619.

O'Donnell, J. P., Adams, A., Nyblade, A. A., Mulibo, G. D., & Tugume, F. (2013). The uppermost mantle shear wave velocity structure of eastern Africa from Rayleigh wave tomography: constraints on rift evolution. *Geophysical Journal International*, *194*, 961–978.

O'Neill, C., Lenardic, A., Jellinek, A. M., & Moresi, L. (2009). Influence of supercontinents on deep mantle flow. *Gondwana Research*, *15*, 276–287. doi: 10.1016/j.gr.2008.11.005

Ogden, C. S., Bastow, I. D., Gilligan, A., & Rondenay, S. (2019). A reappraisal of the H–κ stacking technique: implications for global crustal structure. *Geophysical Journal International*, *219*, 1491–1513.

Oliva, S. J., Ebinger, C. J., Wauthier, C., Muirhead, J. D., Roecker, S. W., Rivalta, E., & Heimann, S. (2019). Insights into fault-magma interactions in an early-stage continental rift from source mechanisms and correlated volcano-tectonic earthquakes. *Geophysical Research Letters*, *46*, 2065–2074.

Omenda, P. (2020). *Country update report for Kenya 2015-2019*. Paper presented at the World Geothermal Congress 2020, International Geothermal Association, Reykjavik.

Omenda, P., Ebinger, C., Nelson, W., Delvaux, D., Cumming, W., Marini, L., et al. (2016). *Characteristics and important factors that influence the development of geothermal systems in the western branch of East African Rift System*. Paper presented at the 6th African Rift Geothermal Conference.

Omenda, P. A. (1998). The geology and structural controls of the Olkaria geothermal system, Kenya. *Geothermics*, *27*, 55–74.

Owens, T. J., Nyblade, A. A., Gurrola, H., & Langston, C. A. (2000). Mantle transition zone structure beneath Tanzania, East Africa. *Geophysical Research Letters*, *27*, 827–830.

Pagli, C., Yun, S.-H., Ebinger, C., Keir, D., & Wang, H. (2019). Strike-slip tectonics during rift linkage. *Geology*, *47*, 31–34.

Park, Y., & Nyblade, A. A. (2006). P-Wave tomography reveals a westward dipping low velocity zone beneath the Kenya rift. *Geophysical Research Letters*, *33*, L07311. doi: 10.1029/2005GL025605

Pastier, A.-M., Dauteuil, O., Murray-Hudson, M., Moreau, F., Walpersdorf, A., & Makati, K. (2017). Is the Okavango Delta the terminus of the East African Rift System? Towards a new geodynamic model: geodetic study and geophysical review. *Tectonophysics*, *712*, 469–481.

Peccerillo, A., Barberio, M. R., Yirgu, G., Ayalew, D., Barbieri, M., & Wu, T. W. (2003). Relationships between mafic and peralkaline silicic magmatism in continental rift settings: a petrological, geochemical and isotopic study of the Gedemsa volcano, central Ethiopian rift. *Journal of Petrology*, *44*, 2003–2032.

Peccerillo, A., Donati, C., Santo, A. P., Orlando, A., Yirgu, G., & Ayalew, D. (2007). Petrogenesis of silicic peralkaline rocks in the Ethiopian rift: geochemical evidence and volcanological implications. *Journal of African Earth Sciences*, *48*, 161–173. doi: 10.1016/j.jafrearsci.2006.06.010

Perez-Gussinye, M., Metois, M., Fernandez, M., Verges, J., Fullea, J., & Lowry, A. R. (2009). Effective elastic thickness of Africa and its relationship to other proxies for lithospheric structure and surface tectonics. *Earth and Planetary Science Letters*, *287*, 152–167. doi: 10.1016/j.epsl.2009.08.004

Pik, R., Marty, B., & Hilton, D. R. (2006). How many mantle plumes in Africa? The geochemical point of view. *Chemical Geology*, *226*, 100–114.

Plasman, M., Tiberi, C., Ebinger, C., Gautier, S., Albaric, J., Peyrat, S., et al. (2017). Lithospheric low-velocity zones associated with a magmatic segment of the Tanzanian Rift, East Africa. *Geophysical Journal International*, *210*, 465–481. doi: 10.1093/gji/ggx177

Prave, A. R., Bates, C. R., Donaldson, C. H., Toland, H., Condon, D. J., Mark, D., & Raub, T. D. (2016). Geology and geochronology of the Tana Basin, Ethiopia: LIP volcanism, super eruptions and Eocene–Oligocene environmental change. *Earth and Planetary Science Letters*, *443*, 1–8.

Priestley, K., McKenzie, D., & Ho, T. (2018). A lithosphere–asthenosphere boundary—a global model derived from multimode surface-wave tomography and petrology. *Lithospheric Discontinuities*, 111–123.

Prodehl, C., Ritter, J. R. R., Mechie, J., Keller, G. R., Khan, M. A., Jacob, B., et al. (1997). The KRISP 94 lithospheric investigation of southern Kenya—the experiments and their main results. *Tectonophysics*, *278*, 121–147.

Purcell, P. G. (2018). Re-imagining and re-imaging the development of the East African Rift. *Petroleum Geoscience*, *24*, 21–40.

Rajaonarison, T. A., Stamps, D. S., & Naliboff, J. (2021). Role of lithospheric buoyancy forces in driving deformation in East Africa from 3D geodynamic modeling. *Geophysical Research Letters*, *48*, e2020GL090483. doi: 10.1029/2020GL090483

Rapprich, V., Žáček, V., Verner, K., Erban, V., Goslar, T., Bekele, Y., et al. (2016). Wendo Koshe Pumice: the latest Holocene silicic explosive eruption product of the Corbetti volcanic system (southern Ethiopia). *Journal of Volcanology and Geothermal Research*, *310*, 159–171.

Reed, C. A., Gao, S. S., Liu, K. H., & Yu, Y. (2016a). The mantle transition zone beneath the Afar Depression and adjacent regions: implications for mantle plumes and hydration. *Geophysical Journal International*, *205*, 1756–1766.

Reed, C. A., Liu, K. H., Chindandali, P. R., Massingue, B., Mdala, H., Mutamina, D., et al. (2016b). Passive rifting of thick lithosphere in the southern East African Rift: evidence from mantle transition zone discontinuity topography. *Journal of Geophysical Research - Solid Earth*, *121*, 8068–8079.

Ren, M., Omenda, P. A., Anthony, E. Y., White, J. C., Macdonald, R., & Bailey, D. K. (2006). Application of the QUILF thermobarometer to the peralkaline trachytes and pantellerites of the Eburru volcanic complex, East African Rift, Kenya. *Lithos*, *91*, 109–124.

Ritsema, J., & Allen, R. M. (2003). The elusive mantle plume. *Earth and Planetary Science Letters*, *207*, 1–12.

Ritsema, J., Deuss, A., van Heijst, H. J., & Woodhouse, J. H. (2011). S40RTS: a degree-40 shear-velocity model for the mantle from new Rayleigh wave dispersion, teleseismic traveltime and normal-mode splitting function measurements. *Geophysical Journal International*, *184*, 1223–1236. doi: 10.1111/j.1365-246X.2010.04884.x

Ritsema, J., van Heijst, H. J., & Woodhouse, J. H. (1999). Complex shear wave velocity structure imaged beneath Africa and Iceland. *Science*, *286*, 1925–1928.

Rogers, N., Macdonald, R., Fitton, J. G., George, R., Smith, M., & Barreiro, B. (2000). Two mantle plumes beneath the East African Rift system; Sr, Nd and Pb isotope evidence from Kenya Rift basalts. *Earth and Planetary Science Letters*, *176*, 387–400.

Rogers, N. W., Demulder, M., & Hawkesworth, C. J. (1992). An enriched mantle source for potassic basanites - evidence from Karisimbi Volcano, Virunga volcanic province, Rwanda. *Contributions to Mineralogy and Petrology*, *111*, 543–556.

Rogers, N. W., James, D., Kelley, S. P., & de Mulder, M. (1998). The generation of potassic lavas from the eastern Virunga Province, Rwanda. *Journal of Petrology*, *39*, 1223–1247.

Ronga, F., Lustrino, M., Marzoli, A., & Melluso, L. (2010). Petrogenesis of a basalt-comendite-pantellerite rock suite: the Boseti Volcanic Complex (Main Ethiopian Rift). *Mineralogy and Petrology*, *98*, 227–243. doi: 10.1007/s00710-009-0064-3

Rooney, T., Furman, T., Bastow, I. D., Ayalew, D., & Gezahegn, Y. (2007). Lithospheric modification during crustal extension in the Main Ethiopian Rift. *Journal of Geophysical Research: Solid Earth*, *112*, B10201. doi: 10.1029/2006JB004916

Rooney, T., Furman, T., Yirgu, G., & Ayalew, D. (2005). Structure of the Ethiopian lithosphere: xenolith evidence in the Main Ethiopian Rift. *Geochimica et Cosmochimica Acta*, *69*, 3889–3910.

Rooney, T. O. (2010). Geochemical evidence of lithospheric thinning in the southern Main Ethiopian Rift. *Lithos*, *117*, 33–48. doi: 10.1016/j.lithos.2010.02.002

Rooney, T. O. (2017). The Cenozoic magmatism of East-Africa: part I—flood basalts and pulsed magmatism. *Lithos*, *286*, 264–301.

Rooney, T. O. (2020a). The Cenozoic magmatism of East Africa: part II–rifting of the mobile belt. *Lithos*, *360*, 105291. doi: 10.1016/j.lithos.2019.105291

Rooney, T. O. (2020b). The Cenozoic magmatism of East Africa: part V–magma sources and processes in the East African Rift. *Lithos*, *360*, 105296, 105296. doi: 10.1016/j.lithos.2019

Rooney, T. O. (2020c). The Cenozoic magmatism of East Africa: part III–rifting of the craton. *Lithos*, *360*, 105390. doi: 10.1016/j.lithos.2019.105390

Rooney, T. O. (2020d). The Cenozoic magmatism of East Africa: part IV–the terminal stages of rifting preserved in the Northern East African Rift System. *Lithos*, *360*, 105381. doi: 10.1016/j.lithos.2019.105381

Rooney, T. O., Bastow, I. D., & Keir, D. (2011). Insights into extensional processes during magma assisted rifting: evidence from aligned scoria cones. *Journal of Volcanology and Geothermal Research*, *201*, 83–96.

Rooney, T. O., Hanan, B. B., Graham, D. W., Furman, T., Blichert-Toft, J., & Schilling, J.-G. (2012a). Upper mantle pollution during Afar plume–continental Rift Interaction. *Journal of Petrology*, *53*, 365–389. doi: 10.1093/petrology/egr065

Rooney, T. O., Hart, W. K., Hall, C. M., Ayalew, D., Ghiorso, M. S., Hidalgo, P., & Yirgu, G. (2012b). Peralkaline magma evolution and the tephra record in the Ethiopian Rift. *Contributions to Mineralogy and Petrology*, *164*, 407–426.

Rooney, T. O., Herzberg, C., & Bastow, I. D. (2012c). Elevated mantle temperature beneath East Africa. *Geology*, *40*, 27–30. doi: 10.1130/g32382.1

Rooney, T. O., Mohr, P., Dosso, L., & Hall, C. (2013). Geochemical evidence of mantle reservoir evolution during progressive rifting along the western Afar margin. *Geochimica et Cosmochimica Acta*, *102*, 65–88. doi: 10.1016/j.gca.2012.08.019

Rooney, T. O., Nelson, W. R., Ayalew, D., Hanan, B., Yirgu, G., & Kappelman, J. (2017). Melting the lithosphere: metasomes as a source for mantle-derived magmas. *Earth and Planetary Science Letters*, *461*, 105–118. doi: 10.1016/j.epsl.2016.12.010

Rosendahl, B. R., Reynolds, D. J., Lorber, P. M., Burgess, C. F., McGill, J., Scott, D., et al. (1986). Structural expressions of rifting: lessons from Lake Tanganyika, Africa. *Geological Society, London, Special Publications*, *25*, 29–43.

Rosenthal, A., Foley, S. F., Pearson, D. G., Nowell, G. M., & Tappe, S. (2009). Petrogenesis of strongly alkaline primitive volcanic rocks at the propagating tip of the western branch of the East African Rift. *Earth and Planetary Science Letters*, *284*, 236–248. doi: 10.1016/j.epsl.2009.04.036

Ross, K. A., Schmid, M., Ogorka, S., Muvundja, F. A., & Anselmetti, F. S. (2015). The history of subaquatic volcanism recorded in the sediments of Lake Kivu; East Africa. *Journal of Paleolimnology*, *54*, 137–152.

Rowland, J. V., Baker, E., Ebinger, C. J., Keir, D., Kidane, T., Biggs, J., et al. (2007). Fault growth at a nascent slow-spreading ridge: 2005 Dabbahu rifting episode, Afar. *Geophysical Journal International*, *171*, 1226–1246. doi: 10.1111/j.1365-246X.2007.03584.x

Rowland, J. V., & Sibson, R. H. (2001). Extensional fault kinematics within the Taupo Volcanic Zone, New Zealand: soft-linked segmentation of a continental rift system. *New Zealand Journal of Geology and Geophysics*, *44*, 271–283. doi: 10.1080/00288306.2001.9514938

Samrock, F., Grayver, A. V., Eysteinsson, H., & Saar, M. O. (2018). Magnetotelluric image of transcrustal magmatic system beneath the Tulu Moye geothermal prospect in the Ethiopian Rift. *Geophysical Research Letters*, *45*, 12,847–12,855.

Samsonov, S., & d'Oreye, N. (2012). Multidimensional time-series analysis of ground deformation from multiple InSAR data sets applied to Virunga Volcanic Province. *Geophysical Journal International*, *191*, 1095–1108.

Saria, E., Calais, E., Altamimi, Z., Willis, P., & Farah, H. (2013). A new velocity field for Africa from combined GPS and DORIS space geodetic solutions: contribution to the definition of the African reference frame (AFREF). *Journal of Geophysical Research - Solid Earth*, *118*, 1677–1697.

Scaillet, B., & MacDonald, R. (2001). Phase relations of peralkaline silicic magmas and petrogenetic implications. *Journal of Petrology*, *42*, 825–845.

Scarsi, P., & Craig, H. (1996). Helium isotope ratios in Ethiopian rift basalts. *Earth and Planetary Science Letters*, *144*, 505–516.

Scholz, C. A., Johnson, T. C., Cohen, A. S., King, J. W., Peck, J. A., Overpeck, J. T., et al. (2007). East African megadroughts between 135 and 75 thousand years ago and bearing on early-modern human origins. *Proceedings of the National Academy of Sciences*, *104*, 16416–16421.

Scholz, C. H., Koczynski, T. A., & Hutchins, D. G. (1976). Evidence for incipient rifting in southern Africa. *Geophysical Journal International*, *44*, 135–144.

Scott, S. C. (1980). The geology of Longonot volcano, Central Kenya: a question of volumes. *Philosophical Transactions of the Royal Society of London. Series A, Mathematical and Physical Sciences*, *296*, 437–465.

Scott, S. C., & Skilling, I. P. (1999). The role of tephrachronology in recognizing synchronous caldera-forming events at the Quaternary volcanoes Longonot and Suswa, south Kenya Rift. *Geological Society of London, Special Publication*, *161*, 47–67.

Shillington, D. J., Scholz, C. A., Chindandali, P. R., Gaherty, J. B., Accardo, N. J., Onyango, E., et al. (2020). Controls on rift faulting in the north basin of the Malawi (Nyasa) Rift, East Africa. *Tectonics*, *39*, e2019TC005633. doi: 10.1029/2019TC005633

Siegburg, M., Gernon, T. M., Bull, J. M., Keir, D., Barfod, D. N., Taylor, R. N., et al. (2018). Geological evolution of the Boset-Bericha volcanic complex, Main Ethiopian Rift: 40Ar/39Ar evidence for episodic Pleistocene to Holocene volcanism. *Journal of Volcanology and Geothermal Research*, *351*, 115–133. doi: 10.1016/j.jvolgeores.2017.12.014

Simmons, N. A., Forte, A. M., Boschi, L., & Grand, S. P. (2010). GyPSuM: a joint tomographic model of mantle density and seismic wave speeds. *Journal of Geophysical Research - Solid Earth*, *115*, B12310. doi: 10.1029/2010JB007631

Simmons, N. A., Forte, A. M., & Grand, S. P. (2007). Thermochemical structure and dynamics of the African superplume. *Geophysical Research Letters*, *34*, L02301. doi: 10.1029/2006GL028009

Simmons, N. A., Myers, S. C., Johannesson, G., & Matzel, E. (2012). LLNL-G3Dv3: global P wave tomography model for improved regional and teleseismic travel time prediction. *Journal of Geophysical Research - Solid Earth*, 117.

Simonetti, A., & Bell, K. (1994). Nd, Pb and Sr isotopic data from the Napak carbonatite-nephelinite centre, eastern Uganda: an example of open-system crystal fractionation. *Contributions to Mineralogy and Petrology*, *115*, 356–366.

Smith, M., & Mosley, P. (1993). Crustal heterogeneity and basement influence on the development of the Kenya Rift, East Africa. *Tectonics*, *12*, 591–606.

Smye, A., Seman, S., Hudak, M., & Crispin, K. (2017). Rates of mantle cooling and exhumation during rifting constrained by REE-in-pyroxene speedometry. *Geochemistry, Geophysics, Geosystems*, *18*, 2510–2525.

Stabile, P., Webb, S., Knipping, J. L., Behrens, H., Paris, E., & Giuli, G. (2016). Viscosity of pantelleritic and alkali-silicate melts: effect of Fe redox state and Na/(Na+ K) ratio. *Chemical Geology*, *442*, 73–82.

Stamps, D. S., Calais, E., Saria, E., Hartnady, C., Nocquet, J.-M., Ebinger, C. J., & Fernandes, R. M. (2008). A kinematic model for the East African Rift. *Geophysical Research Letters*, *35*, L05304. doi: 10.1029/2007GL032781

Stamps, D. S., Flesch, L. M., & Calais, E. (2010). Lithospheric buoyancy forces in Africa from a thin sheet approach. *International Journal of Earth Sciences*, *99*, 1525–1533.

Steinberger, B., & Torsvik, T. H. (2012). A geodynamic model of plumes from the margins of large low shear velocity provinces. *Geochemistry, Geophysics, Geosystems*, *13*, Q01W09. doi: 10.1029/2011GC003808

Steinbruch, F. (2010). Geology and geomorphology of the Urema Graben with emphasis on the evolution of Lake Urema. *Journal of African Earth Sciences*, *58*, 272–284.

Stewart, K., & Rogers, N. (1996). Mantle plume and lithosphere contributions to basalts from southern Ethiopia. *Earth and Planetary Science Letters*, *139*, 195–211.

Sueß, E. (1891). *Die Brüche des östlichen Afrika*. Tempsky.

Sun, M., Liu, K. H., Fu, X., & Gao, S. S. (2017). Receiver function imaging of mantle transition zone discontinuities beneath the Tanzania craton and adjacent segments of the East African Rift system. *Geophysical Research Letters*, *44*, 12,116–12,124.

Tadesse, A. Z., Ayalew, D., Pik, R., Yirgu, G., & Fontijn, K. (2018). Magmatic evolution of the Boku volcanic complex, main Ethiopian rift. *Journal of African Earth Sciences*, *149*, 109–130. doi: 10.1016/j.jafrearsci.2018.08.003

Tauzin, B., Debayle, E., & Wittlinger, G. (2008). The mantle transition zone as seen by global Pds phases: no clear evidence for a thin transition zone beneath hotspots. *Journal of Geophysical Research - Solid Earth*, *113*, B08309. doi: 10.1029/2007JB005364

Teklemariam, M. (2020). *Overview of geothermal resource utilization and potential in East African Rift System*. Paper presented at the 8th African Rift Geothermal Conference, Nairobi, Kenya.

Temtime, T., Biggs, J., Lewi, E., & Ayele, A. (2020). Evidence for active rhyolitic dike intrusion in the northern Main Ethiopian Rift from the 2015 Fentale seismic swarm. *Geochemistry, Geophysics, Geosystems*, *21*, e2019GC008550. doi: 10.1029/2019GC008550

Thompson, D. A., Hammond, J. O. S., Kendall, J.-M., Stuart, G. W., Helffrich, G. R., Keir, D., et al. (2015). Hydrous upwelling across the mantle transition zone beneath the Afar Triple Junction. *Geochemistry, Geophysics, Geosystems*, *16*, 834–846.

Tiercelin, J.-J., Potdevin, J.-L., Thuo, P. K., Abdelfettah, Y., Schuster, M., Bourquin, S., et al. (2012). Stratigraphy, sedimentology and diagenetic evolution of the Lapur Sandstone in northern Kenya: implications for oil exploration of the Meso-Cenozoic Turkana depression. *Journal of African Earth Sciences*, *71*, 43–79.

Trua, T., Deniel, C., & Mazzuoli, R. (1999). Crustal control in the genesis of Plio-Quaternary bimodal magmatism of the main ethiopian rift (MER); geochemical and isotopic (Sr, Nd, Pb) evidence. *Chemical Geology*, *155*, 201–231.

Tugume, F., Nyblade, A., Julià, J., & van der Meijde, M. (2013). Precambrian crustal structure in Africa and Arabia: evidence lacking for secular variation. *Tectonophysics*, *609*, 250–266.

Ukstins, I. A., Renne, P. R., Wolfenden, E., Baker, J., Ayalew, D., & Menzies, M. (2002). Matching conjugate volcanic rifted margins; 40Ar/39Ar chrono-stratigraphy of pre- and syn-rift bimodal flood volcanism in Ethiopia and Yemen. *Earth and Planetary Science Letters*, *198*, 289–306.

Varet, J., Omenda, P., & Birba, E. (2020). *Giant geothermal sites along the East Africa Rift System (EARS): The determinant volcano-structural setting*. Paper presented at the 8th African Rift Geothermal Conference, Nairobi, Kenya.

Vetere, F., Holtz, F., Behrens, H., Botcharnikov, R. E., & Fanara, S. (2014). The effect of alkalis and polymerization on the solubility of H_2O and CO_2 in alkali-rich silicate melts. *Contributions to Mineralogy and Petrology*, *167*, 1014. doi: 10.1007/s00410-014-1014-6

Vormann, M., & Jokat, W. (2021). The crustal structure of the Kerimbas basin across the offshore branch of the East African rift system. *Geophysical Journal International*, *226*, 2073–2102.

Wallner, H., & Schmeling, H. (2016). Numerical models of mantle lithosphere weakening, erosion and delamination induced by melt extraction and emplacement. *International Journal of Earth Sciences*, *105*, 1741–1760.

Watts, E. J., Gernon, T. M., Taylor, R. N., Keir, D., Siegburg, M., Jarman, J., et al. (2020). Evolution of the Alu-Dalafilla and Borale volcanoes, Afar, Ethiopia. *Journal of Volcanology and Geothermal Research*, *408*, 107094. doi: 10.1016/j.jvolgeores.2020.107094

Wauthier, C., Cayol, V., Kervyn, F., & d'Oreye, N. (2012). Magma sources involved in the 2002 Nyiragongo eruption, as inferred from an InSAR analysis. *Journal of Geophysical Research - Solid Earth*, *117*, B05411. doi: 10.1029/2011JB008257

Wauthier, C., Cayol, V., Poland, M., Kervyn, F., d'Oreye, N., Hooper, A., et al. (2013). Nyamulagira's magma plumbing system inferred from 15 years of InSAR. *Geological Society of London, Special Publication*, *380*, 39–65.

Wedmore, L. N., Biggs, J., Williams, J. N., Fagereng, Å., Dulanya, Z., Mphepo, F., & Mdala, H. (2020). Active fault scarps in southern Malawi and their implications for the distribution of strain in incipient continental rifts. *Tectonics*, *39*, e2019TC005834.

Weeraratne, D. S., Forsyth, D. W., Fischer, K. M., & Nyblade, A. A. (2003). Evidence for an upper mantle plume beneath the Tanzanian craton from Rayleigh wave tomography. *Journal of Geophysical Research - Solid Earth*, *108*. doi: 10.1029/2002JB002273

Weinstein, A., Oliva, S. J., Ebinger, C. J., Roecker, S., Tiberi, C., Aman, M., et al. (2017). Fault-magma interactions during early continental rifting: Seismicity of the Magadi-Natron-Manyara basins, Africa. *Geochemistry, Geophysics, Geosystems*, *18*, 3662–3686.

Whaler, K., & Hautot, S. (2006). Magnetotelluric studies of the northern Ethiopian rift. In G. Yirgu, C. Ebinger, & P. Maguire (Eds.), *The Afar volcanic province within the East African Rift System* (pp. 293–306). London: Special Publication of the Geological Society.

Wheeler, W. H., & Karson, J. A. (1994). Extension and subsidence adjacent to a "weak" continental transform: an example from the Rukwa rift, East Africa. *Geology*, *22*, 625–628.

White, J. C., Espejel-García, V. V., Anthony, E. Y., & Omenda, P. (2012). Open system evolution of peralkaline trachyte and phonolite from the Suswa volcano, Kenya rift. *Lithos*, *152*, 84–104. doi: 10.1016/j.lithos.2012.01.023

Wiart, P., & Oppenheimer, C. (2000). Largest known historical eruption in Africa: Dubbi volcano, Eritrea, 1861. *Geology*, *28*, 291–294.

Wiart, P. A. M., Oppenheimer, C., & Francis, P. (2000). Eruptive history of Dubbi volcano, northeast Afar (Eritrea), revealed by optical and SAR image interpretation. *International Journal of Remote Sensing*, *21*, 911–936.

Wilkinson, P., Mitchell, J. G., Cattermole, P. J., & Downie, C. (1986). Volcanic chronology of the Men–Kilimanjaro region, Northern Tanzania. *Journal of the Geological Society*, *143*, 601–605.

Williams, T. M., Henney, P. J., & Owen, R. B. (1993). Recent eruptive episodes of the Rungwe Volcanic Field (Tanzania) recorded in lacustrine sediments of the northern Malawi rift. *Journal of African Earth Sciences (and the Middle East)*, *17*, 33–39.

Wolfenden, E., Ebinger, C., Yirgu, G., Deino, A., & Ayalew, D. (2004). Evolution of the northern Main Ethiopian rift: birth of a triple junction. *Earth and Planetary Science Letters*, *224*, 213–228.

Wright, L. J., Muirhead, J. D., & Scholz, C. A. (2020). Spatiotemporal variations in upper crustal extension across the different basement terranes of the Lake Tanganyika rift, East Africa. *Tectonics*, *39*, e2019TC006019. doi: 10.1029/2019TC006019

Wright, T. J., Ebinger, C., Biggs, J., Ayele, A., Yirgu, G., Keir, D., & Stork, A. (2006). Magma-maintained rift segmentation at continental rupture in the 2005 Afar dyking episode. *Nature*, *442*, 291–294.

Yirgu, G., Ababa, A., & Ayele, A. (2006). Recent seismovolcanic crisis in northern Afar, Ethiopia. *Eos, Transactions American Geophysical Union*, *87*, 325–329.

3

Geological and Geophysical Constraints Guide New Tectonic Reconstruction of the Gulf of Mexico

Irina Filina[1] and Erin Beutel[2]

ABSTRACT

Tectonic models proposed for the Gulf of Mexico (GoM) vary dramatically in many aspects, ranging from the pre-rift locations of crustal blocks, to the timing of the break-up, to even the order of tectonic events. In this study, we integrated seismic data with gravity and magnetic fields to determine the crustal architecture in different parts of the basin and to map key tectonic features. As a result, a set of new geological constraints was derived, including the Triassic rifts, regions of Seaward Dipping Reflectors, and Jurassic pre-salt sedimentary basins in the eastern GoM and along the Yucatan margin and two distinct crustal zones in the oceanic domain. Our tectonic reconstruction takes into account an apparent temporal variability of the magmatic regime during basin formation that ranged from the Central Atlantic Magmatic Province (~200 Ma) to an ultra-slow amagmatic spreading during the initial stage of GoM opening (~ 165 Ma) followed by a major ridge reorganization (~ 152 Ma) associated with increased magmatic supply. This second phase of oceanic spreading ceased at early Cretaceous (~ 135 Ma) based on published correlation analysis of seismic and well data. Overall, our tectonic reconstruction takes into account previously known and newly derived geological constraints and integrates various geophysical datasets.

3.1. INTRODUCTION

The Gulf of Mexico (GoM) is a petroleum-rich basin located on the southeastern edge of North America (Figure 3.1). The basin has a more than a century-long exploration history (Galloway, 2008) that resulted in multiple wells, primarily targeting oil and gas deposits within a sedimentary section. The thickness of sediments exceeds 14 km (Laske et al., 2013), so the pre-GoM Paleozoic rocks and/or the basement are sampled only by a small number of wells around the rim of the basin (Figure 3.1a; Scott et al., 1961; Ramos, 1975; Ball et al., 1988; Dobson & Buffler, 1991; Woods et al., 1991; MacRae & Watkins, 1995; Coombs et al., 2019; Erlich & Pindell, 2020). The vast majority of geophysical data acquired in the basin is proprietary, although several seismic sections, such as the ones shown in Figure 3.2 (Snedden et al., 2014; Lin et al., 2019; Filina et al., 2022), are recently published and are available for qualitative analysis for a general geoscience community.

The tectonic history of the basin remains poorly understood even though new high-quality data are constantly being collected in the basin (e.g., Saunders et al., 2016; Horn et al., 2017; O'Reilly et al., 2017; Deighton et al., 2017) and new wells are being drilled in deepwater in both U.S. and Mexican sectors. Many tectonic models of the GoM opening are proposed in the literature, but they contradict each other in some respects, as summarized in Filina et al. (2022). The overall consensus implies that the formation of the basin relates to the disassembly of the supercontinent Pangaea that started in the

[1] Department of Earth and Atmospheric Sciences, University of Nebraska–Lincoln, Lincoln, Nebraska, USA

[2] Department of Geology and Environmental Geosciences, College of Charleston, Charleston, South Carolina, USA

Extensional Tectonics: Rifting and Continental Extension, Geophysical Monograph 278, First Edition.
Edited by İbrahim Çemen and Elizabeth J. Catlos.

DOI: 10.1002/9781119773757.ch03

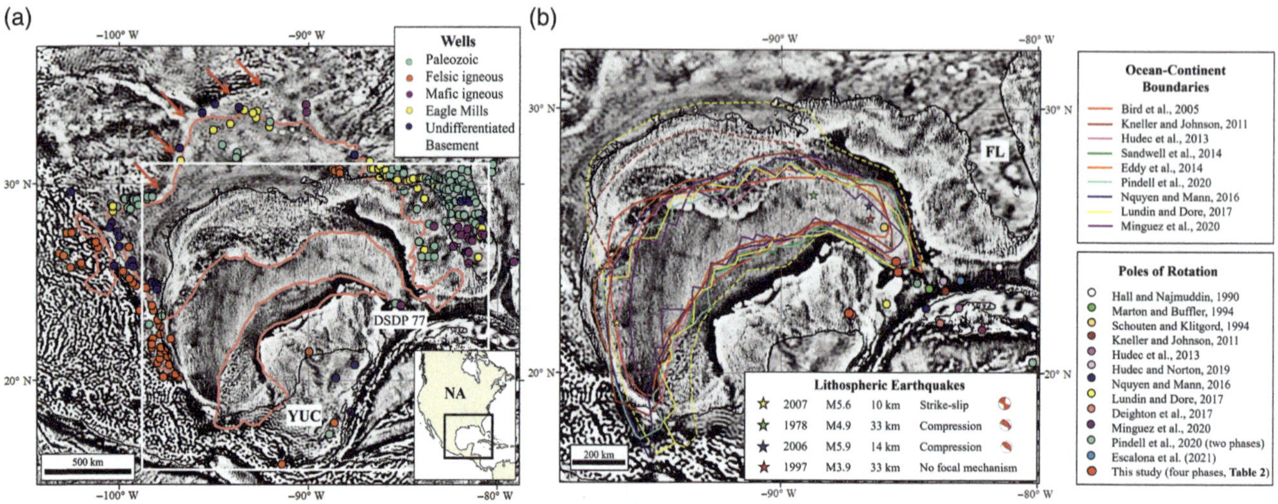

Figure 3.1 (a) The first vertical derivative of the gravity field from Sandwell et al. (2014); the greyscale color scheme is as in the original publication and means to highlight geological structures. Of particular interest are the apparent NW-SE trending lineaments in the middle of the Gulf of Mexico basin that are interpreted as segments of extinct spreading centers offset by transform faults. Pink polygons outline the extent of salt basins from Snedden and Galloway (2019). Red arrows point to the Paleozoic Ouachita-Marathon orogenic belt related to the assembly of Pangea. The white rectangle shows the extent of the zoomed section shown in panel (b). NA, North America; YUC, Yucatan crustal block; FL, Florida crustal block. (b) Locations of published ocean-continent boundaries (OCB) for the Gulf of Mexico plotted with different colors. The circles are poles of rotation from different publications. The two published tectonic models are shown with solid and dashed lines as they include tectonic boundaries for two stages: the initial breakup (dashed line) and the Yucatan rotational phase (solid line of the same color).

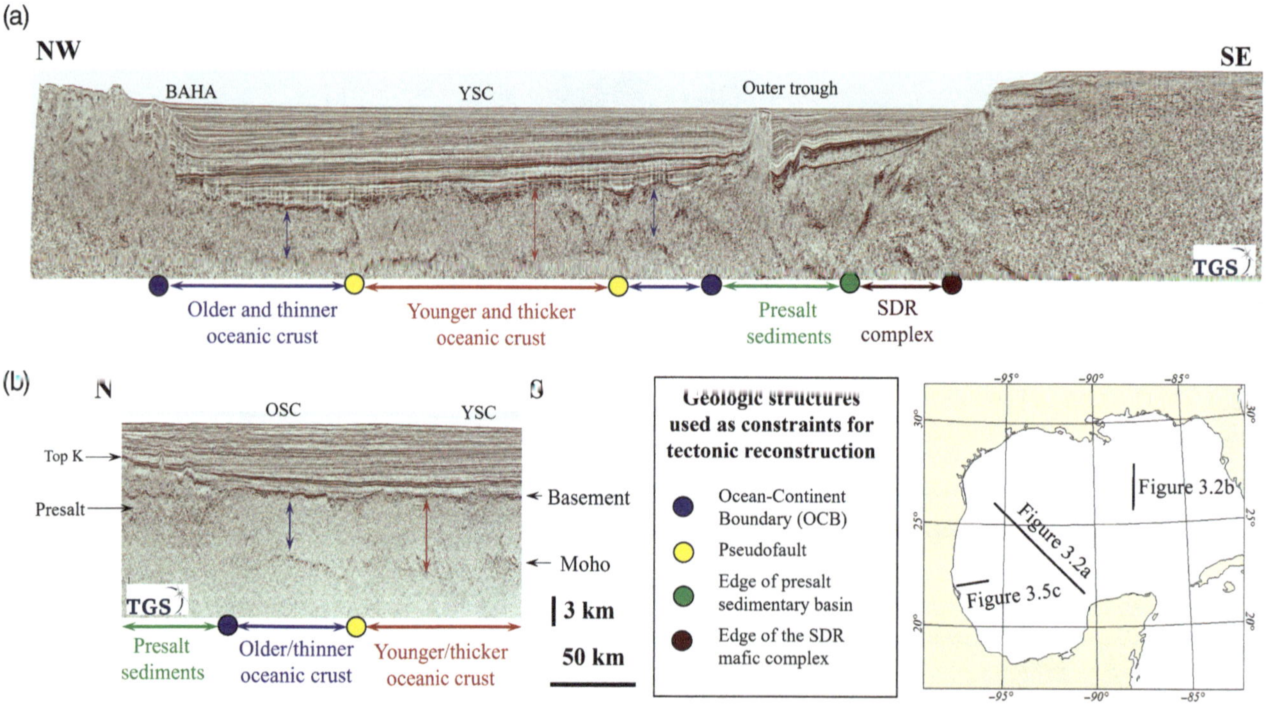

Figure 3.2 Seismic cross-sections (data courtesy of TGS) illustrating various tectonic zones and geological structures in the Gulf of Mexico. (a) A regional seismic line from the northwestern GoM to the Yucatan margin from the recent GIGANTE survey (Filina et al., 2022). Blue (5 km) and red (7.5 km) vertical arrows mark interpreted oceanic crust produced in Phase 1 (~165–152 Ma, ultraslow) and Phase 2 (~152–135 Ma, slow) oceanic spreading, respectively. YSC, Young spreading center; BAHA refers to the BAHA high outlined by Hudec et al. (2020) shown in Figure 3.5a. (b) a regional cross-section through the eastern GoM (Snedden et al., 2014 and Lin et al., 2019). Note apparent variations in the crustal thickness in the oceanic domain and challenges of the subsalt seismic image in the continental regions. OSC, Old Spreading Center (cf. Figure 3.5).

Late Triassic. The oceanic spreading initiated in the Late Jurassic and ceased in the Early Cretaceous as the Yucatan crustal block rotated away from North America. However, several controversies in tectonic history have still not been resolved. These include the tectonic affiliation of the Triassic redbeds (earlier syn-rift or postcollisional sedimentary basin), the nature of rifting and breakup (magma-rich vs. magma-poor), the timing of salt deposition with respect to the formation of oceanic crust (i.e., what was first), and the prerift location of the crustal blocks. The primary reason for these disputes is the poor seismic image of crustal structures, especially in the continental domain and over the ocean-continent transition zone, which allows for multiple, sometimes contradictory, interpretations. To overcome that, integration with other datasets, particularly potential fields (Figure 3.3), is necessary. In this chapter, we describe how the joint analysis of publicly available seismic data with potential fields (gravity and magnetic) and observed lithospheric seismicity (Figure 3.1b) helps delineate subsurface structures that honor all the data. We review published integrated models and summarize the key tectonic features mapped by those models, such as the ocean-continent boundary (OCB), various magmatic complexes, presalt sedimentary basins, and two distinct crustal zones in the oceanic domain. We then perform a spatial analysis of potential fields to trace these structures outside the models' coverage. We use these mapped geological features as constraints for the new tectonic reconstruction of the basin. Our integrated approach allows for a more confident tectonic reconstruction as geological structures at the conjugate margins end up being aligned when restored to their prebreakup locations.

3.2. GEOLOGIC HISTORY

Many tectonic models have been proposed to describe the opening of the Gulf of Mexico (such as Salvador, 1991; Hall and Najmuddin 1994; Marton and Buffler 1994; Schouten and Klitgord, 1994; Bird et al., 2005, Kneller and Johnson, 2011; Eddy et al., 2014; Nguyen and Mann, 2016; Lundin and Dore, 2017; Escalona et al., 2021 among many others). Major tectonic events leading to the formation of the GoM are illustrated in Figure 3.4. The basin opened to the south of the Ouachita-Marathon orogenic belt that marks the assembly of the supercontinent Pangaea that was completed by middle Permian (Soto-Kerans et al., 2020). The orogenic belt is easily traceable in potential fields to the north of the GoM (marked with red arrows in Figure 3.1a). Pre-GoM Paleozoic rocks were penetrated by some wells, mostly at the northern edge of the basin (Figure 3.1a).

The Pangea disassemble started in the Triassic, with the South Georgia Rift system (Figure 3.5a; McBride, 1991) that failed and is known to be capped by the flood basalts of the Central Atlantic Magmatic Province (CAMP) that occurred ~200 Ma (e.g., Marzoli et al., 1999; McHone, 2003). The overlying Triassic redbeds, such as the Eagle Mills Formation (the name comes from the well in southern Arkansas), are penetrated by multiple wells in the northern part of the basin (Figure 3.1a). The palynology analysis allowed these redbeds to be dated as Carnian (~237 Ma, Wood & Benson, 2000). In the southern GoM, the Eagle Mills equivalents include several formations of the Huizachal Group (Barboza-Gudiño et al., 2010; Martini & Ortega-Gutiérrez, 2016), although some of them are overprinted by the tectonics of the East Mexico Permo-Triassic Nazas arc (Stern & Dickinson, 2010).

Deposition of salt (shown with pink outlines in Figure 3.1a) plays a significant role in the majority of published tectonic reconstructions. Until 2019, the age of salt was thought to be Callovian (~166–163 Ma) based on the stratigraphy (Salvador, 1991), and recently, the earlier Bajocian (~170–168 Ma) salt age has been confirmed with Sr isotopes, so some models now have to readjust the timing of major tectonic phases (Pindell et al., 2020). Most authors suggest that the salt was deposited at the last stage of continental rifting (e.g., Eddy et al., 2014; Nguyen & Mann, 2016), whereas Hudec et al. (2013) argue for 7–12 Myr of postsalt continental stretching in the central part of the basin prior the oceanic spreading. Recently, several authors proposed that salt deposition was coincident with an oceanic spreading (Kneller & Johnson, 2011; Lundin & Doré, 2017; Pindell et al., 2020), and some even allow salt deposition in two separated basins after the oceanic crust was formed (Padilla y Sánchez, 2016).

Although the published models agree that the GoM opened via counterclockwise rotation of the Yucatan crustal block, details of this rotation vary dramatically between them. Some models utilize paleomagnetic data for Mexican outcrops to constrain the total degree of rotation of the Yucatan crustal block that varies in the literature from >70° (Molina-Garza et al., 1992) to ~40° (Godínez-Urban et al., 2011a, 2011b), and consequently does not provide a solid constraint. The proposed location of conjugate margins also varies between the models. As such, the study of Van Avendonk et al. (2015) concludes that the Texas and western Yucatan margins are conjugates based on the tectonic reconstruction that results in a tight fit of the crustal blocks. Alternatively, a companion paper of Eddy et al. (2014) suggests that the Louisiana margin is conjugate for the western Yucatan. In contrast, Lundin and Doré (2017) reconstruct the western Yucatan to northeastern GoM (Alabama margin), and Hudec and Norton (2019) report that the outer troughs mapped in northern Yucatan and Florida are conjugate features. In our study, we map matching geological features on both United States and Mexican margins based on the integration of multiple geophysical data to constrain the tectonic reconstruction of the basin.

The timing of the drifting phase ranges drastically in the literature because of the uncertainty in geologic interpretations (the start of drifting varies from 190 to 154 Ma

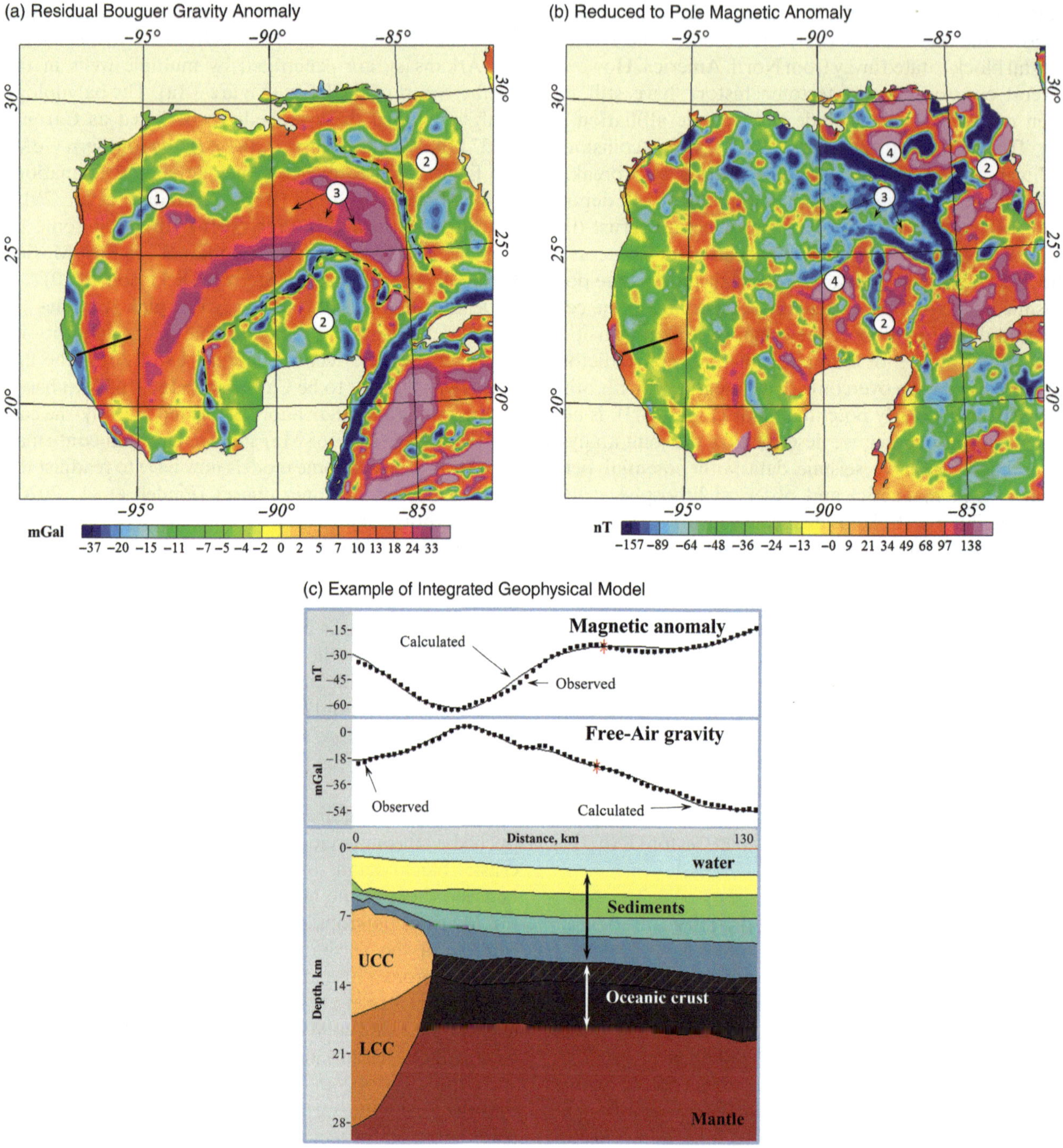

Figure 3.3 (a) Residual gravity field (original data from Sandwell et al., 2014; corrected for water and the upper mantle gravity signals; see text for details). The dashed black line follows the sharp edge of the carbonate platforms, which is apparent in the gravity data. 1: Gravity low over the thick salt structure known as the "salt wall" in the northwestern GoM. 2: Characteristic gravity and magnetic lows and highs interpreted as Triassic rifts in the northeastern GoM and the Yucatan margin; these structures are aligned in the tectonic restoration for 220 Ma in Figure 3.7a. 3: Pronounced gravity lows and magnetic highs associated with the Phase 2 extinct spreading center (mapped with black segments in Figure 3.5a). (b) Reduced to pole magnetics Minguez et al. (2020) in the center and Meyer et al. (2017) near the shoreline. 4: Magnetic anomalies responsible for interpreted SDR complexes. (c) Example of integrated geophysical modeling for the profile through the western GoM; the location is shown as a black line in panels (a) and (b). The top panels show observed and computed gravity and magnetic fields. The model shown in the bottom panel uses the seismic reflection data from Pindell et al. (2020) to constrain multiple subsurface layers. See the physical properties of each subsurface layer in the legend for Figure 3.5. UCC, upper continental crust; LCC, lower continental crust.

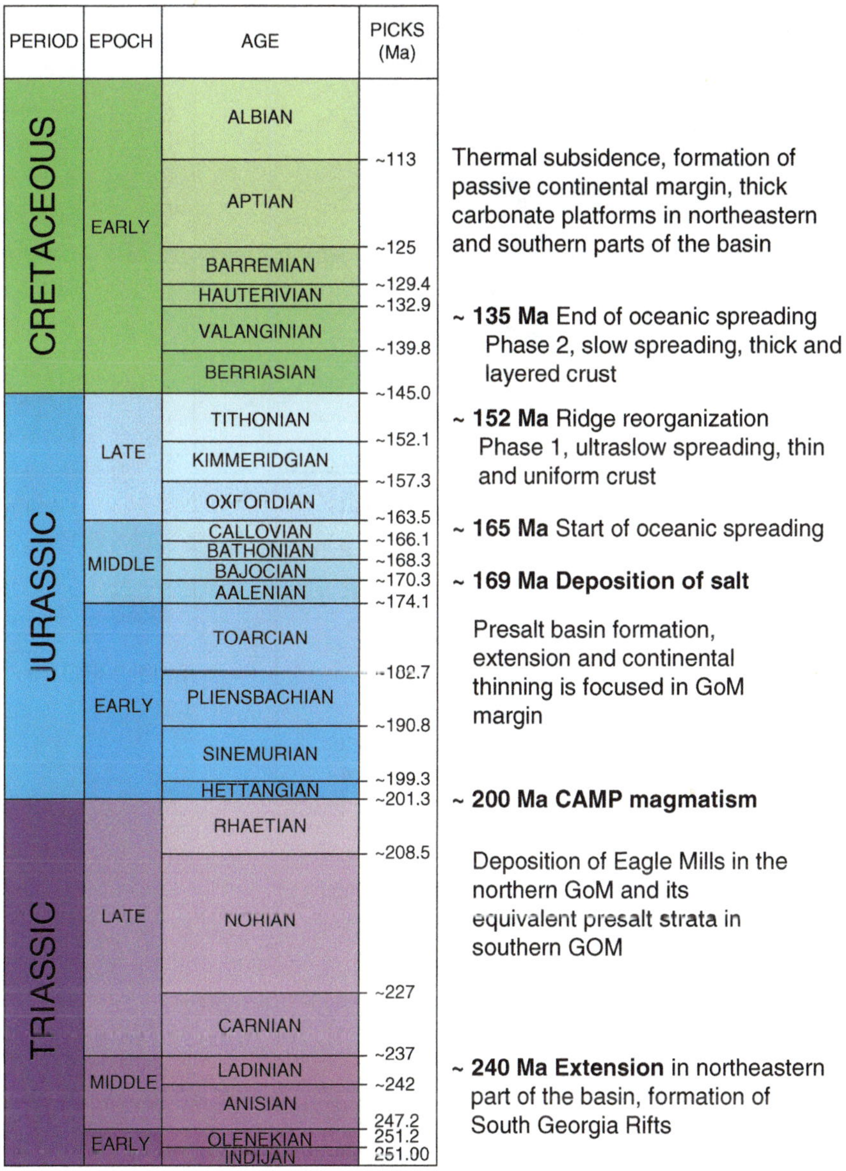

Figure 3.4 Simplified geologic time chart listing the major tectonic events related to the opening of the GoM.

among different papers, and the end of oceanic spreading ranges between 154 and 128 Ma). For example, Lundin and Doré (2017) interpret strong magnetic anomalies aligned with the basinward dipping reflectors in the northeastern GoM as evidence of subaerial basalt flows (referred to as seaward dipping reflectors [SDRs]) based on a similar appearance of these complexes to drilled SDRs in Norway and Greenland (e.g., Hinz, 1981; Mjelde et al., 2008). They further expand this interpretation to other pronounced magnetic anomalies around the GoM (Figure 3.3b) despite the fact that seismic imaging in that part of the basin is very challenging and no basinward dipping reflections are observed in seismic data. Lundin and Doré (2017) thus conclude a magma-rich nature of the GoM breakup and interpret all strong positive magnetic anomalies surrounding the basin as SDR complexes that guide their tectonic restoration (yellow polygons in Figure 3.1b). Multiple authors interpret the basinward dipping reflectors observed in the northeastern and southern parts of the basin as evidence of magma-rich basin breakup (Imbert & Post, 2005; Imbert et al., 2005; Eddy et al., 2014; Liu et al., 2019; Filina & Hartford, 2021). In contrast, the model of Minguez et al. (2020) suggests that the strong magnetic signature of the Louisiana magnetic anomaly (labeled 4 in Figure 3.3b) can be modeled without any magmatic SDR complex, thus arguing for a magma-poor breakup. In this study, we support the evidence of the magmatic addition during the rifting stage (not at the breakup, as in Lundin and Doré [2017]) based on modeling of magnetic data in various parts of the basin that requires the presence of highly magnetic rocks in the subsurface to explain the observed magnetic

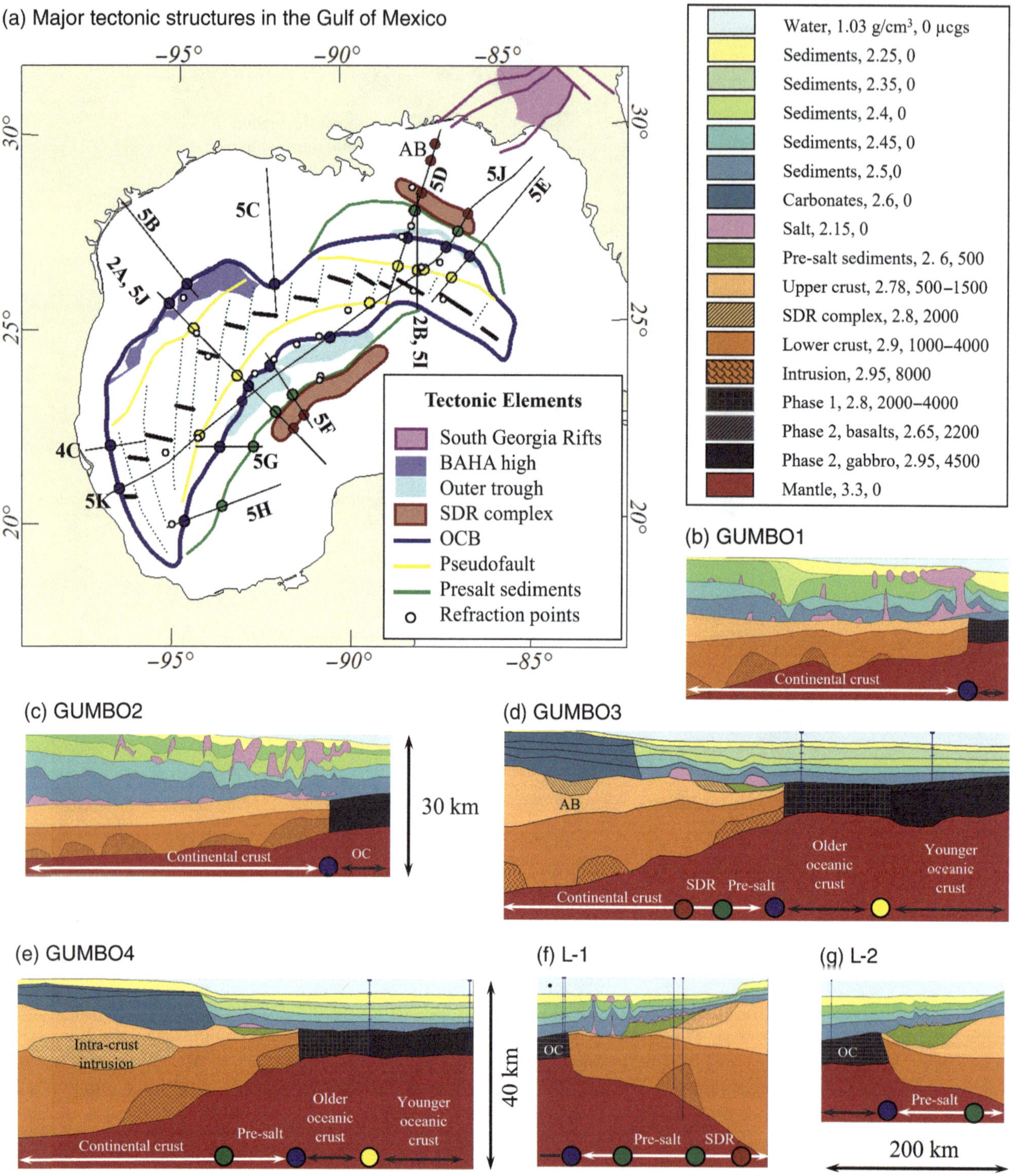

Figure 3.5 (a) Key tectonic structures mentioned in the literature and interpreted in our study (see text for details and references). The thin black lines show the cross-sections developed from the integration of seismic reflection, refraction, gravity, and magnetic data. (b) The model along GUMBO1 (Van Avendonk et al., 2015; Filina, 2019), (c) GUMBO2 (Eddy et al., 2018; Filina, 2019), (d) GUMBO3 (Eddy et al., 2014; Liu et al., 2019), (e) GUMBO4 (Christeson et al., 2014; Liu & Filina, 2018), (f–h) lines L1–L3 from William-Rojas et al. (2012) modeled by Filina and Hartford (2021), (i) modified model along a seismic line shown in Figure 3.2b (Adapted from Filina, 2020), (j) interpreted seismic profile from TGS GIGANTE survey shown in Figure 3.2a, and (k) an interpretation of the seismic line of Horn et al., 2017 / GeoExpro; note that the line is split into two segments to keep the same scale as other lines. The same markers as in Figure 3.2 are used to show locations of interpreted tectonic features: OCB (blue), presalt basin (green), magmatic complexes (referred to as SDR in literature) are brown, and pseudofaults marking the boundary between old and new oceanic crustal zones are marked with yellow circles. OC, oceanic crust.

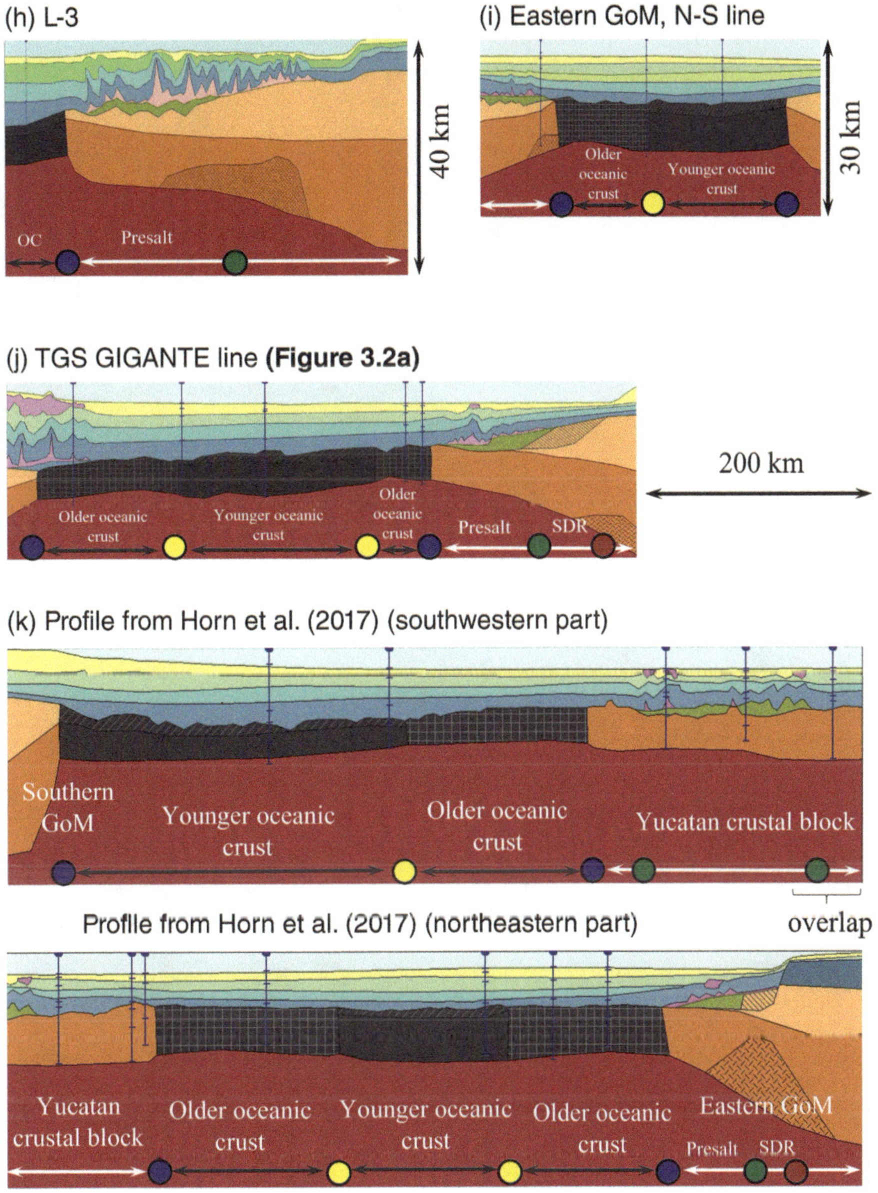

Figure 3.5 (Continued)

anomalies. We also acknowledge the time constraints on the tectonic events from Snedden et al. (2014) that are based on seismic reflection data tied to industry wells with biostratigraphy markers: namely, the initiation of spreading at ~160 Ma and the end of the oceanic crust formation at ~137 Ma.

In addition to the pre-CAMP South Georgia Rift system (Figure 3.5a), several major tectonic structures are mentioned in the literature, such as the outer trough (Hudec & Norton, 2019) and BAHA high (Hudec et al., 2020; the acronym comes from the exploration prospects along the high). The outer troughs represent the regions of a sudden drop up to 2 km deep in the basement boundary with evidence of unconfined salt flow together with an overlying Jurassic sedimentary section that are mapped by Hudec and Norton (2019) from proprietary seismic data on the Yucatan and Florida margins (Figures 3.2a and 3.5a). Filina and Hartford (2021; Figure 3.5f) interpreted this feature to be related to the exhumation of the lower crust that occurred at the final, postsalt stage of continental rifting. Another intriguing crustal feature in the GoM is the BAHA high in the west (Figure 3.5a), as the nature of the crust there remains highly debated. Fiduk et al. (1999) suggest that the crust in that region is extended continental based on multiple linear basement highs that were interpreted as horsts marking the tilted blocks. In contrast, Minguez et al. (2020) describe this region as exhumed upper mantle, based on the magnetic

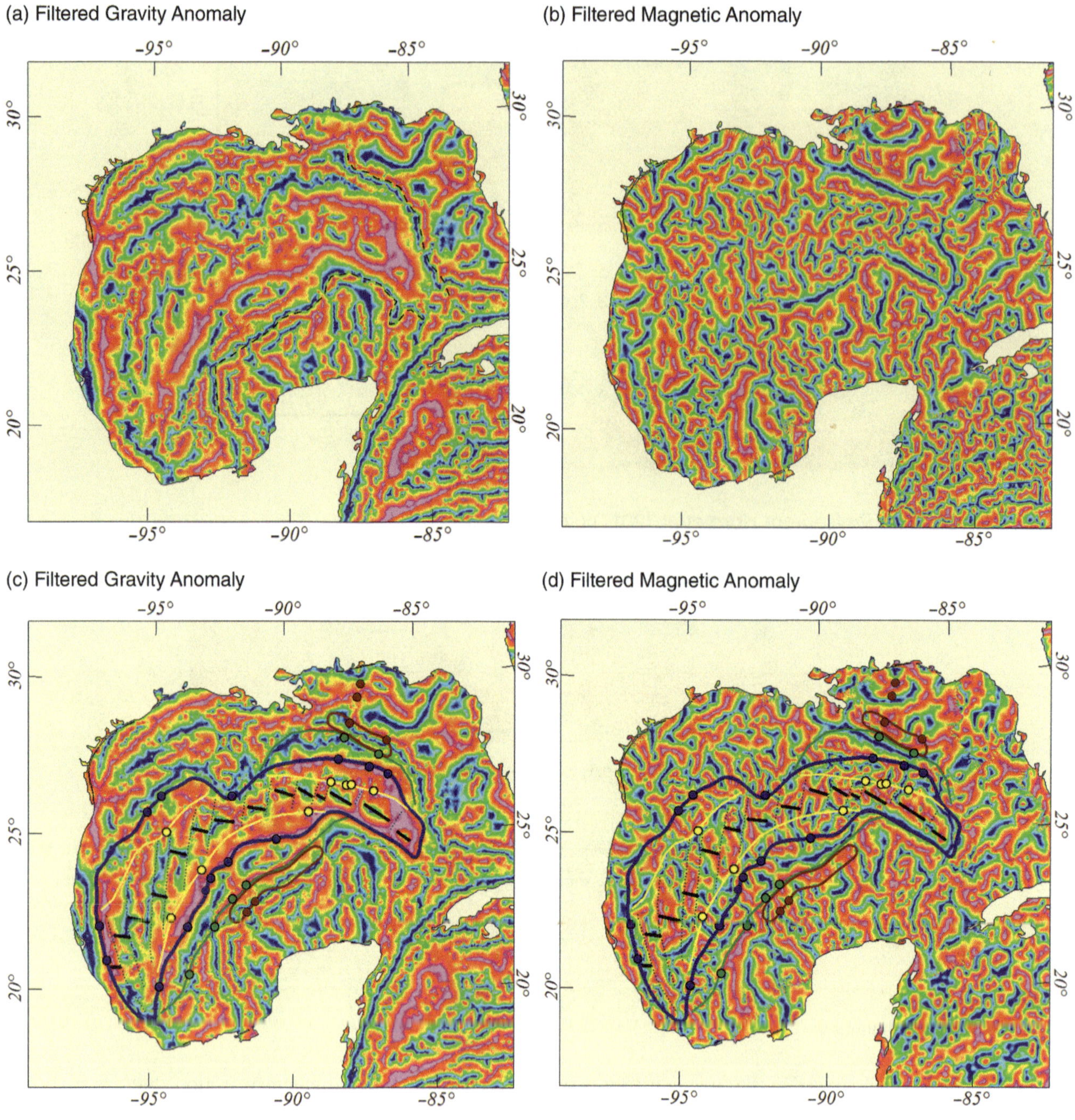

Figure 3.6 Filtered residual potential fields data. (a) Residual Bouguer gravity; the dashed black line marks the shelf break that is associated with a sharp change in bathymetry that is evident in filtered gravity data. (b) Residual RTP magnetic. (c, d) Interpreted tectonic zones. The same symbology is used as in Figures 3.2 and 3.5.

signature. Alternatively, Hudec et al. (2020) argue that the crust of the BAHA high is volcanic (i.e., oceanic) in origin and was emplaced at the time of salt deposition. In our study, we interpret this region as an oceanic crust produced in the initial spreading phase (Figures 3.4, 3.5, and 3.6), which is consistent with the crustal thickness determined by the refraction data of Nakamura et al. (1988). Another structure that is worth mentioning is the basement ramp (Hudec et al., 2013): the step-up of the oceanic basement with respect to the continental one that was interpreted by Hudec et al. (2013) from proprietary seismic data as a marker for the limit of oceanic crust (shown in pink in Figure 3.1b). The later interpretations of Hudec's team conclude that the nature of the crust around the ramp remains poorly understood (Curry et al., 2018; Hudec & Norton, 2019). Moreover, Curry et al. (2018) outline a region of "uncertain crust" generally coincident with the ocean-continental boundary. Many of the integrated models shown in Figure 3.5 acknowledge this step-up of oceanic crust with respect to the continental basement.

Multiple authors mapped the segments of the extinct spreading centers and corresponding transform faults in the center of the basin based on gravity data (Figures 3.1 and 3.3a). Despite these interpretations being done on the same dataset, the mapped tectonic features still show discrepancies exceeding 60 km. The drastic asymmetry of the basin with respect to the spreading center should also be noted (Figure 3.1a), which is especially apparent in the eastern part of the basin, where the crustal domain to the north of the spreading center is noticeably wider than the one to the south. Some models suggest that this results from an asymmetric spreading (Hudec et al., 2013; Nguyen & Mann, 2016). In our model, we explain this apparent asymmetry with two distinct spreading episodes: the initial one from ~165 Ma that produced thin crust adjacent to the OCB in many parts of the basin, including the crust of the BAHA high in the northwestern GoM, and the northmost zone of oceanic crust in the northeastern GoM (Figures 3.5 and 3.6). The second spreading episode follows the major ridge reorganization ~152 Ma and produces thicker and layered crust until ~135 Ma (Figure 3.4). We utilize pronounced gravity lows to map the two sets of spreading centers that further guide our tectonic reconstruction.

3.3. INTEGRATED GEOPHYSICAL MODELING

The concept of integrated geophysical modeling assumes multiple geophysical and geological observations jointly analyzed to derive the subsurface model that honors all of them. In this study, we utilize 2D seismic reflection sections (such as the ones shown in Figure 3.2), seismic refraction data (Ewing et al., 1960; Ibrahim et al., 1981; Nakamura et al., 1988; Eddy et al., 2014, 2018; Christeson et al., 2014; Van Avendonk et al., 2015), gravity (Sandwell et al., 2014; Figures 3.1 and 3.3a), and magnetic fields (Bankey et al., 2002; Meyer et al., 2017; Minguez et al., 2020; Figure 3.3), as well as limited core data to constrain physical properties of subsurface rocks (i.e., Schlager et al., 1984; Hilterman, 1998). An example of an integrated geophysical model is shown in Figure 3.3c.

3.3.1. Northwestern GoM (GUMBO1)

The model in the northwestern GoM (Figure 3.5b) follows the seismic refraction line GUMBO1 (Van Avendonk et al., 2015) that images a complex subsurface consisting of a highly variable sedimentary cover with multiple salt structures over a crustal layer with several high-velocity zones (V_p > 7.5 km/s). The structures imaged in this profile are interpreted differently by various authors. Lundin and Doré (2017) position the entire profile over the oceanic domain. Van Avendonk et al. (2015) suggest thinned and hyperextended continental crust with a ~40 km wide zone of interpreted exhumed mantle adjacent to oceanic crust (the last 80 km of the profile) overlain by salt and thins basinward from 9 to 5 km. Filina (2019) noticed that the seismic refraction results of Van Avendonk et al. (2015) lack a thick salt structure, known as "salt wall," associated with the Perdido fold belt (Filina et al., 2015; this feature is evident in gravity data and is labeled 1 in Figure 3.3a). The absence of this thick salt in the velocity model results in distorted crustal seismic velocities. Reinterpreted refraction data, further integrated with potential fields, revealed thinned and intruded continental crust (Filina, 2019; Figure 3.5b) that varies in thickness from ~15 km at the northwestern end of the line to ~10 km near the OCB located basinward of the "salt wall" near the Sigsbee Escarpment. The crust at the southwestern edge of the line, over the BAHA high, is interpreted to be oceanic with a total thickness of 6 km, which is consistent with the refraction data of Nakamura et al. (1988) and with the conclusion of Van Avendonk et al. (2015). The high-velocity zones are interpreted as magmatic intrusions, presumably emplaced during continental rifting as they require high magnetic susceptibilities to explain the observed magnetic signatures.

Our interpretation (Figure 3.5b) suggests no presalt sediments deposited in the northwestern GoM (see more details in Filina, 2019). This conclusion is consistent with the results of the 3D seismic survey described by Filina et al. (2015) in the northwestern GoM that was reprocessed with guidance from 3D gravity modeling. The integration with gravity resulted in significantly improved seismic imaging and ultimately allowed mapping of the acoustic basement, revealing no presalt sediments in the area.

3.3.2. Central GoM (GUMBO2)

The north-central GoM is sampled by a seismic refraction line GUMBO2 (Eddy et al., 2018) that is generally aligned with the seismic reflection line GULFSPAN 2000 (Radovich et al., 2011). All but two published tectonic models agree that the northern end of the line is located over the stretched continental crust (Figure 3.1b), although the location of the OCB varies between the models up to 149 km (87 mi) in that region. Kneller and Johnson (2011) and Lundin and Doré (2017) are the only authors proposing that the entire line is over the oceanic domain (Figure 3.1b). The OCB suggested by Eddy et al. (2018) is located near the Sigsbee Escarpment, which was validated by an integrated geophysical analysis by Filina (2019), concluding that the OCB near the Sigsbee (Figure 3.5c) is preferable to explain observed gravity and magnetic anomalies. However, this OCB challenges many published tectonic models that assume a tight fit between the western Yucatan and the northern GoM margin, demanding the OCB to be much more landward than is suggested by the GUMBO2 results (compare OCBs in Figures 3.1b and 3.5a).

Similar to other GUMBO lines, several high-velocity zones in the lower crust that are aligned with strong magnetic anomalies are interpreted as magmatic complexes that were presumably emplaced during the continental rifting phase. No presalt sediments are concluded for the GUMBO2 line as well, they are not evident either in seismic reflection data or in seismic refraction data, and they are not demanded by potential fields. The absence of presalt deposits in the northwestern and central GoM is consistent with the observations of Ewing and Galloway (2019) about a general lack of Early Jurassic deposits in the northwestern GoM as well as with the regional uplift in Texas in the Late Triassic–Early Jurassic that was documented by Dickinson et al. (2010).

3.3.3. Northeastern GoM (GUMBO3, GUMBO4)

There is much better agreement between published OCBs in the northeastern GoM (Figure 3.1b). However, the nature of the margin (magma-rich vs. magma-poor) is highly debated in the literature. Both landward and basinward dipping reflectors are imaged in seismic in this area that are interpreted by various authors as either magmatic complex (SDR) associated with magma-rich margin (e.g., Imbert & Post, 2005; Imbert et al., 2005; Eddy et al., 2014, Liu et al., 2019) or growth faults filled with sediments related to magma-poor basin formation (Rowan, 2014, Curry et al., 2018, Minguez et al., 2020). Liu et al. (2019) performed an integrated geophysical analysis of several seismic reflection and refraction lines in the northeastern part of the basin and established two regions of the seismic dipping reflectors that require high densities and high magnetic susceptibilities to explain the observed gravity and magnetic fields (see the example in Figure 3.5d).

The northern complex (referred to by Liu et al. [2019] as the Northern SDR) is aligned with the Apalachicola Basin (labeled AB in Figures 3.5a and 3.5d). We interpret this structure as a buried continuation of the Triassic South Georgia Rift System, known to be capped by the CAMP volcanics (McBride, 1991). Therefore, we interpret the highly magnetic rocks in that region demanded from the observed magnetic signature as evidence of CAMP-related magmatism. The second magmatic complex to the south (Figure 3.5d, referred to by Liu et al. [2019] as the Southern SDR) is also aligned with the basinward dipping reflectors in seismic data and requires high densities and magnetic susceptibilities to match potential fields. As already mentioned, this is interpreted as an SDR complex by many authors, although the timing of this structure remains discussed in the literature (Eddy et al., 2014; Lundin & Doré, 2017). In our reconstruction, we associate these complexes with the CAMP magmatic event (~200 Ma, Figure 3.4); that is, they were emplaced during the continental rifting phase (not at the breakup phase, as suggested by other authors). This interpretation agrees with several high-velocity zones in the lower crust imaged by several lines GUMBO lines (Figures 3.5b–e) representing igneous intrusive bodies emplaced at the same time during a magma-assisted continental rifting phase. This conclusion is also consistent with multiple basaltic igneous rocks encountered from the wells in the shallow waters and onshore in this region (Figure 3.1a) as well as with the findings of the DSDP Leg 77 wells offshore Yucatan (Schlager et al., 1984) that encountered diabase intrusions with Jurassic crystallization age in Paleozoic metasediments. The mapped magmatic complexes, particularly SDRs, are further used as constraints in our tectonic reconstruction as they are aligned with a similar structure at the conjugate margin (Figures 3.5 and 3.6).

3.3.4. Yucatan Margin

Filina and Hartford (2021) developed three integrated models (Figures 3.5f–h) along the seismic lines published by Williams-Rojas et al. (2012). Multiple seismic sections at the Yucatan margin show basinward dipping reflectors (Williams-Rojas et al., 2012; Saunders et al., 2016; O'Reilly et al., 2017; Steier & Mann, 2019), similar to the ones seen in Figure 3.2b, that are further aligned with pronounced magnetic anomalies (labeled 4 in Figure 3.3b). The magnetic signature for the profile shown in Figure 3.5f requests the presence of two anomalous bodies: one shallow (i.e., of a shorter wavelength), consistent with an SDR complex; and one deeper, located slightly basinward, interpreted as a "feeding" intrusive body (Filina & Hartford, 2021). In addition to the magmatic SDR complexes that are interpreted at both Yucatan and eastern GoM margins (Figures 3.5d–f), several other matching geological structures are mapped at both locations. Hudec and Norton (2019) outlined outer troughs up to 2 km deep from proprietary seismic data on both margins (Figures 3.2b and 3.5a); this feature is wider on the Yucatan side (~40 km) and is less than 20 km wide in the northeastern GoM (Hudec and Norton [2019] refer to that as the Florida margin). Integrated analysis of seismic and potential field data (Filina & Hartford, 2021) for the profile shown in Figure 3.5f suggests that the outer trough is underlain by exhumed lower continental crust. The exhumation likely occurred at the last (postsalt) phase of continental rifting (as in Hudec et al., 2013). The removal of the low-density upper crust immediately before the oceanic spreading triggered the rapid subsidence of that part of the margin that, in turn, triggered the unconfined salt flow documented by Hudec and Norton (2019). A similar region of nearly exhumed lower continental crust may be interpreted in the northeastern GoM (Figure 3.5e;

~10 km-thick region of V_p~7 km/s adjacent to OCB; Christeson et al., 2014). This interpreted exhumation of lower continental crust is consistent with an estimated ultraslow spreading rate (0.9 cm/yr; Filina et al., 2020), as well as with the geochemical analysis of the volcanic rocks encountered in salt diapirs, which have a postsalt crystallization age (~160 Ma) and originate from the deprived magmatic source (Stern et al., 2011). The rapid marginal collapse is also documented in the northeastern GoM from proprietary seismic data by Pindell et al. (2014), although it is attributed to the exhumation of the upper mantle rather than the lower continental crust. However, the integration of seismic data with potential fields does not support mantle exhumation in the GoM (Figure 3.5). In addition, seismic data (Eddy et al., 2014; Christeson et al., 2014) shows a strong reflection and an abrupt velocity contrast associated with the Moho boundary that is inconsistent with a mantle exhumation.

The thick presalt sedimentary basins represent other matching tectonic features on the Yucatan and northeastern GoM margins (Figure 3.5). These sediments were first documented at the Yucatan margin by Williams-Rojas et al. (2012), who presented a convincing seismic section with up to 5 km-thick sediments imaged beneath the salt, although the base of these presalt deposits was not confidently mapped in seismic reflection data. Filina and Harford (2021) combined seismic and potential fields to outline up to 100 km-wide presalt sedimentary basins in the Yucatan margin (Figure 3.5f–h) that can also be seen in seismic data in Figure 3.2b as well as in the regional seismic reflection profile of Horn et al. (2017) and in the corresponding models (Figures 3.5j,k).

3.3.5. Oceanic Domain

The GUMBO experiment (Eddy et al., 2014, 2018; Christeson et al., 2014) revealed that oceanic crust in the GoM shows significant variations both in thickness (ranges between 5 and 9 km) and physical properties (V_p ranges from 5.5–6 to >7 km/s). The changes in the GoM oceanic crust are evident in seismic reflection data in Figure 3.2 and interpreted in many models in Figure 3.5. The thickness of the crust in those models is constrained not only by seismic reflection data (as shown in Figure 3.2) but also by vintage refraction data (Ewing et al., 1960; Ibrahim et al., 1981; Nakamura et al., 1988) shown as blue columns in the cross-sections in Figure 3.5 (see the locations of seismic refraction constraints in Figure 3.5a). Overall, two distinct crustal zones in the eastern GoM are identified (Figure 3.5a): thinner and uniform crust near the OCB and thicker layered crust in the center of the basin. The thicknesses and physical properties are guided by seismic refraction data of Eddy et al. (2014, 2018) and Christeson et al. (2014), and the observed crustal variations are validated with observed gravity and magnetic fields. We refer to the contact between the two oceanic crustal zones as a *pseudofault*. We map that structure in five profiles in Figure 3.5 (marked with yellow circles).

3.4. SPATIAL ANALYSIS OF GRAVITY AND MAGNETIC DATA AND MAPPED TECTONIC FEATURES

Potential fields—gravity and magnetic—are sensitive to variations in physical properties of subsurface rocks, namely density and magnetic susceptibility. The major geological structures, such as the boundary between continental and oceanic crusts, are associated with significant contrasts in these properties, thus allowing us to map these structures from the filtered gravity and magnetic data.

In gravity, we computed the Bouguer correction based on the densities of water and the top sediments (Figure 3.5). We then computed the regional trend via an upward continuation of the Bouguer gravity anomaly to an elevation of 40 km. We chose this value by trial and error as the one removing high-frequency components in gravity due to shallower subsurface structures, so the resultant regional trend represents the broad and smooth signal from deep regional sources. In magnetics, we performed differential reduction to pole (RTP) to correct for the skewness of observed anomalies due to a nonvertical ambient field. We then found a regional magnetic trend via upward continuation to an elevation of 25 km. This lower elevation was used because the magnetic field decreases faster with distance than gravity (Telford et al., 1990). Once we removed the regional trends from both gravity and magnetic fields, we applied the tilt derivative filter (Salem et al., 2008) to highlight the regions where the residual fields change the most (Figure 3.6a,b).

3.4.1. Ocean-Continent Boundary

The OCB was determined in all models (blue markers in Figures 3.2, 3.5, and 3.6). As this boundary is associated with a significant contrast in the physical properties of continental and oceanic rocks, it corresponds to signals in both potential fields: a pronounced high in gravity and a high-trough pattern in magnetics (Figure 3.3a,b). This correlation allowed us to trace the OCB confidently for the entire basin (Figures 3.5a and 3.6). The fact that our analysis integrates multiple geophysical datasets gives us confidence in the derived OCB location (to be compared to multiple interpretations in Figure 3.1b).

3.4.2. Presalt Basins

Presalt basins are identified in the northeastern and southern parts of the GoM (Figures 3.2, 3.5, and 3.6). The presence of thick presalt deposits is evident in multiple seismic cross-sections (Williams-Rojas et al., 2012; Saunders et al., 2016; Horn et al., 2017; Liu et al., 2019; Lin et al., 2019), although these sediments have not been drilled and their age remains unconstrained to date. We model this section as dense and slightly magnetic sediments (Figure 3.5) to account for their inferred volcaniclastic nature, which is further validated by a good match between observed and computed potential fields (Liu et al., 2019; Filina & Hartford, 2021). Noticeably, a thick presalt section is only interpreted in the northeastern and southern parts of the Gulf of Mexico (Figures 3.5 and 3.6); none of these sediments are observed in the northwestern and central GoM (Figure 3.5b,c), which potentially relates to the Late Triassic–Early Jurassic uplift in Texas (Dickinson et al., 2010) that prevents these sediments from being deposited in those regions. We utilize the mapped presalt basins in the northeastern GoM and at the Yucatan margin as conjugate features formed at the stretched continental crust after the CAMP event (i.e., in the Early Jurassic) and use them as constraints for our tectonic reconstruction at that time.

3.4.3. Oceanic Domain: Two Spreading Episodes and Pseudofaults Marking the Contact Between Them

Our modeling (Figure 3.5) suggests the presence of two distinct spreading episodes in the GoM that produced drastically different oceanic crusts. The one immediately adjacent to the OCB is thin (~5 km) and uniform (i.e., modeled as a single layer) based on seismic velocities V_p~7 km/s determined by refraction data of Christeson et al. (2014). We interpret this crust to be produced by an initial ultraslow spreading episode (~165–152 Ma) with an estimated full spreading rate of 0.9 cm/yr (Filina et al., 2020). At 152 Ma, the ridge reorganization occurred that resulted in the NE-SW trending spreading centers (thick black lines in Figure 3.5a) that are evident in gravity data (Figures 3.1 and 3.3a). The oceanic crust produced in the second spreading episode (~152–135 Ma) is thicker (up to 9 km) and has two layers (Eddy et al., 2014), which are interpreted as basalts (V_p~5.5–6 km/s) over gabbro (V_p~7 km/s). Despite an apparent increase in the availability of magmatic material during the younger spreading episode, the spreading rate remained slow (estimated full rate of 1.1 cm/yr; Filina et al., 2020). The pseudofault—the boundary between the two crustal domains—can be traced in both filtered potential fields as an apparent change in character (Figure 3.6). This boundary is further aligned with the hypocenters of several recent earthquakes (Filina et al., 2020), suggesting that the pseudofaults in the eastern GoM represent zones of weakness that are currently being reactivated under compressional stress.

Our tectonic reconstruction (Figure 3.7) accounts for this two-phased formation of oceanic crust. The ridge reorganization during the formation of oceanic crust at ~152 Ma explains multiple puzzling geological observations: (a) the severe asymmetry in the oceanic domain (Figure 3.1a), (b) the presence of two distinct crustal zones with dramatically different thickness and physical properties (Figure 3.5), and (c) the observed seismicity within the oceanic domain that is not aligned with any known tectonic structure (Filina et al., 2020).

3.4.4. Magmatic Addition

Many of our models (Figure 3.5) suggest the presence of igneous magmatic complexes within the crust, such as the SDRs in the northeastern and southern margins, and multiple intrusive bodies in the lower crust all over the basin. The latter are aligned with high seismic velocities imaged from refraction profiles in the northern GoM (Eddy et al., 2014, 2018; Christeson et al., 2014; Van Avendonk et al., 2015) and with strong magnetic anomalies (Figure 3.3b). As already mentioned, many authors interpret those as rift-related magmatic addition (Eddy et al., 2014, 2018; Christeson et al., 2014; Filina, 2019; Liu et al., 2019). These features must be made of dense and highly magnetic rocks to explain the observed gravity and magnetic anomalies. Moreover, the shallow magmatic complexes (SDRs) are aligned with the basinward dipping reflectors observed in seismic data (such as those in Figure 3.2a). Their interpretation as rift-related magmatic addition is also consistent with the findings of the DSDP Leg 77 well (Figure 3.1a; Schlager et al., 1984) that sampled Paleozoic basement intruded by diabase dike with Early Jurassic crystallization age.

The extents of the SDR provinces were established from our spatial analysis (Figure 3.6) as up to 400 km long and up to 50 km wide in the Yucatan margin, whereas this complex appears shorter (~270 km) in the northeastern part of the basin. We used the mapped SDR complexes as conjugate features on both margins to constrain the tectonic reconstruction of the Early Jurassic time. In general, we conclude that the availability of magmatic material varied throughout the history of the basin from magma-rich CAMP (presumably responsible for the observed SDRs and igneous intrusions) to amagmatic postsalt continental stretching that resulted in the exhumation of the upper crust in Yucatan (Figure 3.5f) and northeastern GoM (Figure 3.5e), followed by an ultraslow spreading producing thin oceanic crust ~165–152 Ma. The magmatic input apparently increased after ridge reorganization at ~152 Ma, as thicker than normal oceanic crust (Figure 3.5d,e,i,k) was produced in the second spreading episode (~152–137 Ma). These temporal variations in the magmatic input are consistent with observations of Planke et al. (2000), who showed that the periods of significant magmatic addition

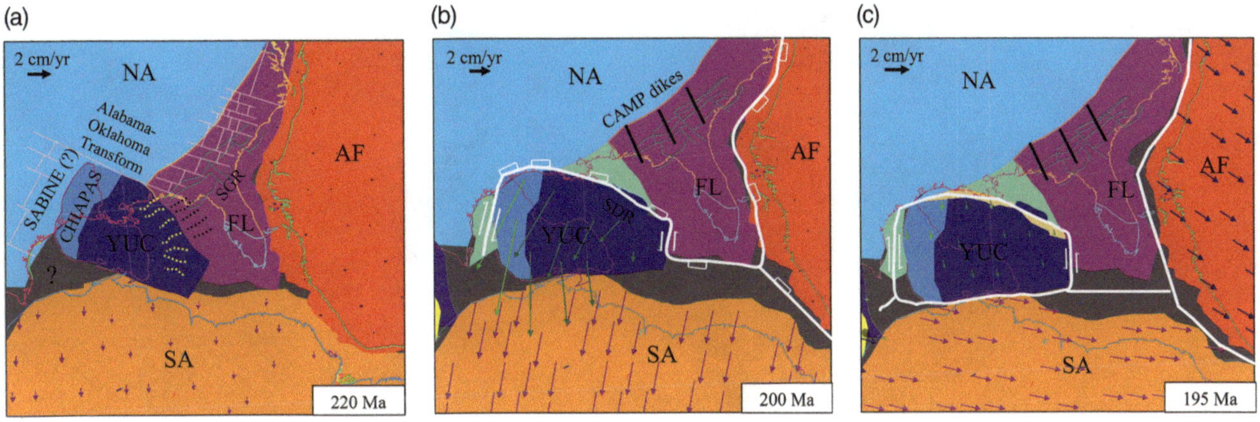

Figure 3.7 Tectonic reconstructions of the GoM using GPlates version 2.2 and modifying global plate motions adapted from Müller et al. (2019). Each plate, microplate, or geologic structure is assigned its own color of GPlates arrows generated by GPlates, which indicate the relative velocity of that point on a plate to a fixed North American plate. If no arrows are present, that plate is not moving relative to North America. Each reconstruction has a slightly different scale for the arrows, as indicated in each figure. Dark grey is unassigned lithosphere and could be oceanic or arc, but generally it is not a known continent. (a) 220 Ma reconstruction of the GoM. Reconstruction is based on the alignment of the Triassic rifts in the Yucatan and Florida (Pindell & Kennan, 2009), the deformation in continental rocks, and the preexisting structures created during the rifting of Rodinia (Thomas, 2006) (Figure 3.3). Reconstruction shows a snapshot of the plate positions and motions during the 230–205 Ma time period. To account for the continental stretching of the Yucatan and Florida between 230 and 205 Ma, they were emplaced in North America and moved south the approximate amount they were stretched. The South Georgia Rift was also formed during this time, and the approximate location of the faults is taken from Heffner (2013) (as seen in the arrows). The area between Chiapas and east Texas indicated with a question mark (?) is unknown and could actually be a slightly shifted Chiapas or accretionary wedge or Mexican terranes or backarc (there is no definitive data at this time). (b) 200 Ma reconstruction of the GoM based on the inner SDRs identified in Figure 3.5a shown as dark blue regions, and the structural features on land, including the northwest trending dikes and the left-lateral strike-slip faults in the piedmont of Georgia and South Carolina (Bartholomew pers. comm). CAMP volcanism from this time period is indicated via the NW-trending CAMP dikes seen throughout the SE United States. White lines show postulated plate boundaries; boxes indicate normal faulting/extension. Note that no plate boundary is assumed between South America and the Yucatan. (c) 195 Ma reconstructions based on the location of the seaward dipping reflectors (SDRs) and the development of transitional crust. When compared to the motion at 200 Ma, South America has changed its relative motion by almost 90°. Along with the intrusion of the CAMP mafics, this relative motion change may actually have facilitated the rotation of the Yucatan between 205 and 195 Ma. (d) 170 Ma reconstruction of the incipient GoM. Transitional crust has been almost entirely created at this time and has two different signatures. The butter yellow areas near the Yucatan and in the northeastern GoM are transitional crust that has sunk considerably as the magmatism from CAMP cools, resulting in the deposition of thick sedimentary sequences before 170 Ma (Figures 3.5 and 3.6). The transitional crust in the northwestern GoM (mint green) does not have any thick sedimentary sequences and appears to have been relatively high standing before this time period. Note the beginning of contact between the Maracaibo (MA) block and the Yucatan, which has been determined to be a subduction zone. (e) 163 Ma reconstruction of the incipient GoM. Due to collision with the Maracaibo block and the initiation of subduction, the Yucatan block is now rotating rapidly; all transitional crust is now formed, and ultraslow spreading will commence. Thick salt deposits now cover all the basin (Pindell et al., 2020). (f) 153 Ma reconstruction of the incipient GoM basin. New features include the thin, ultraslow spreading crust identified in Figure 3.5, shown as the present-day filtered residual satellite gravity of those regions (from Figure 3.6a). Note that the pole of rotation relative to North America for the Yucatan is the same as at 163 Ma and that the Yucatan is still attached to North America via transitional crust between the two ultraslow spreading centers. Possible ridges (thick black lines) and transforms (dashed lines) have been noted. The white line indicates plate boundaries, and the black double-headed arrows indicate the extension direction across the plate boundary. (g) 143 Ma reconstruction of the forming GoM basin. The Yucatan has now completely broken free from North America, which has caused an increase in the rate of rotation, resulting in the thicker and layered crust seen in Figure 3.5 and the formation of a new spreading ridge (all oceanic crust is shown via the residual gravity from Figure 3.6a). The pseudofaults generated during propagation of the new ridge are shown as yellow dashed lines. B, Bahamas. (h) 134 Ma reconstruction of the GoM basin. The Yucatan has now stopped moving after docking with southern Mexico. The oceanic domain consists of two different crustal zones separated by a pseudofault (yellow dotted line). The older oceanic crust is not layered and is thin, whereas the younger crust is the result of faster spreading and is thicker and has at least two layers, as shown in Figures 3.2 and 3.5.

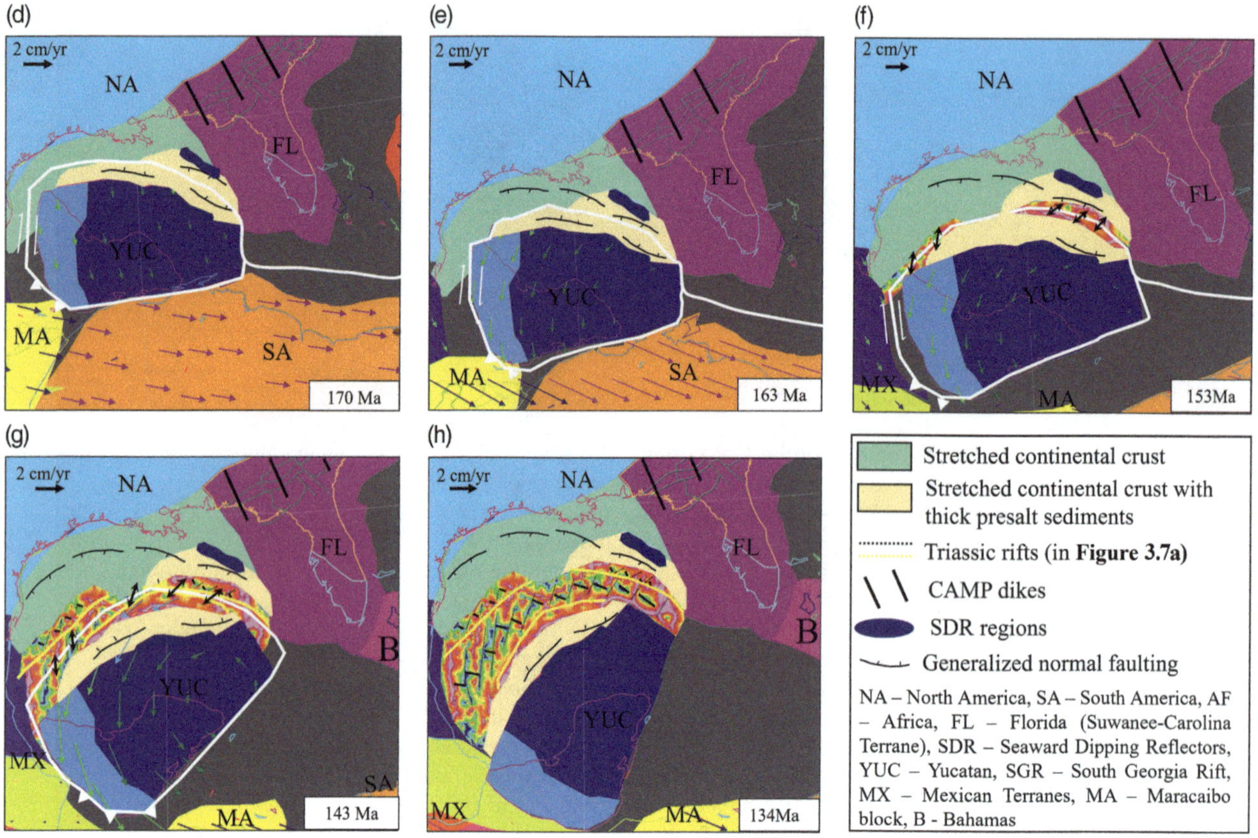

Figure 3.7 (Continued)

may alternate with episodes of very limited volcanism in passive continental margins.

3.5. TECTONIC RECONSTRUCTION OF THE BASIN

The extensive interpretation and identification of geologic features in the previous sections are used as a framework for the tectonic reconstruction of the GoM (Table 3.1). Particularly, we utilize the matching tectonic features at both margins, namely the Triassic basins (labeled 2 in Figure 3.3) and SDR complexes for the Late Triassic reconstruction (Figure 3.7a,b), the presalt sedimentary basins for the Early and Middle Jurassic ones (Figure 3.7c,d), the evidence of the exhumation of the lower continental crust at the final (postsalt) stage of continental rifting (Figure 3.7e), and the two phases of oceanic spreading (Figure 3.7f,g) for Late Jurassic–Early Cretaceous times. Although our reconstructions are guided by these tectonic constraints, they have also relied heavily on our understanding of the deformation seen in the crust itself.

Each rift basin evolves differently depending on the magma supply, the initial thickness of the crust, preexisting weaknesses, and the orthogonality of the extension to preexisting weaknesses, but overall, continental rifts follow a pattern during their evolution. Continental rifts generally start with widespread deformation, which is followed by the localization of deformation to a narrow zone, leading to the formation of thinned transitional crust and, eventually, to the formation of oceanic crust (e.g., Guan et al., 2019; Corti, 2012). No matter the angle of interaction or the presence or absence of magma, the ability of the two sides of the rift to influence each other's stress state is greatly reduced once the transitional stage is reached. However, there must still be some connection between the two sides of the rift because the continental crust is still extending.

We used the program Gplates2.2.0 along with the Müller et al. (2019) tectonic reconstruction data as a starting point for our reconstruction of the opening of the GoM. We then modified the motion of the Yucatan/Chiapas (henceforth referred to as Yucatan) and the Suwanee-Carolina Terrane (henceforth referred to as the Florida block) based on the geophysical and tectonic structures listed in Table 3.1. Each time period had its own unique set of variables used in the reconstruction, but the basic principle of the reconstruction is that the similarly aged unique features likely formed adjacent to each other and that the relative motion of the plates must

Table 3.1 The list of tectonic constraints used for each reconstruction period.

Observation	Interpretation	Timing	Implication	Constraint	Reconstruction
Triassic rifts evident in potential field data (Figure 3.3)	Stretching of Florida and Yucatan terrains attached to North America	~230–205 Ma based on dated Eagle Mills and South Georgia Rifts	Both Florida and Yucatan are welded to North America.	The rifts must be aligned when reconstructed.	Figure 3.7a
Magmatic addition (SDRs, igneous intrusions in the lower and middle crust, Figures 3.2, 3.5, and 3.6)	Evidence of magmatic activity during continental rifting stage	SDRs ~190 Ma (per DSDP well 535A), but interpreted as CAMP-related event (~200 Ma)	Magmatic burst during continental rifting	The SDRs on the Yucatan block and the North American margin should be near each other once reconstructed.	Figure 3.7b
Presalt sedimentary basins on transitional crust in the eastern and the southern GoM (Figures 3.5 and 3.6)	Syn-rift sediments before the deposition of salt at ~169 Ma (Figure 3.4)	From ~190 to ~169 Ma, most likely deposited in the sag formed due to cooling at the end of the SDR magmatism	Local subsidence in the east to accumulate thick sediments. Local uplift in the west as no presalt section developed.	Presalt basins should be proximal once restored.	Figure 3.7c,d
Outer Trough (Figures 3.2 and 3.5a) and zone of significant thickening of Jurassic strata in the Northern Yucatan	Evidence of uncontrolled flow of salt with the Jurassic overburden (Hudec & Norton, 2019)	Postsalt continental stretching (~169–165 Ma)	Interpreted as exhumation of lower continental crust, suggesting no magmatism immediately before oceanic spreading	Initial oceanic spreading likely to be ultraslow	Figure 3.7e
Phase 1 of oceanic spreading producing thin (~5 km) and uniform oceanic crust (Figures 3.2, 3.5, and 3.6)	Ultraslow spreading (estimated 0.9 cm/yr; Filina et al., 2020)	~165–~152 Ma Ridge reorganization at ~152 Ma	Rotation of the Yucatan around the nearby pole produces Phase 1 spreading centers, low availability of magmatic material	The Phase 1 crust from Yucatan and North America should be adjacent during reconstruction.	Figure 3.7f
Phase 2 of oceanic spreading resulting in thicker (up to 9 km) and layered crust (Figures 3.2, 3.5, and 3.6)	Slow spreading (estimated 1.1 cm/yr; Filina et al., 2020)	~152–~135 Ma Timing is loosely constrained by Snedden et al. (2014).	Final separation of Yucatan from North America allows southward migration of the Yucatan block Increase in magma supply	All oceanic crust should be formed by ~135 Ma.	Figure 3.7g,h

Table 3.2 Kinematic parameters determined from tectonic reconstruction in Figure 3.7. All poles are versus North America.

Time (Ma)	Latitude	Longitude	Angle
220	24.3332	−85.7972	−94.2209
205	24.0279	−85.7491	−93.7897
200	23.8349	−85.4426	−68.5622
162	22.1944	−87.271	−60.8887
151	22.2691	−87.3824	−44.8

create a stress field that matches the observed deformations. However, large-scale regional features rarely create straight lines or match perfectly during tectonic deformation as each individual basin, rift, fault, magmatic complex, and stress responds to the local conditions in the context of the larger tectonic picture. The poles of rotation derived from our reconstruction are listed in Table 3.2.

3.5.1. Late Triassic (230–205 Ma, Pre-CAMP)

Our model begins 230 million years ago after the supercontinent Pangea began to slowly separate (e.g., Traverse, 1987; Labails et al., 2010; Frizon de Lamotte et al., 2015; Leleu et al., 2016; Peace et al., 2020). This resulted in continental rifting in southeastern North America, including Florida and the Yucatan blocks, two of the several terranes accreted during the assembly of Pangea (Soreghan & Soreghan, 2013). The associated Triassic redbeds are documented on both sides and are penetrated in many wells in the northern margin (Figure 3.1a), as well as studied in outcrops in Mexico (Barboza-Gudiño et al., 2010; Martini & Ortega-Gutiérrez, 2016). Although small regional variations in rift orientation and structure are common in continental rifting, the rifts of approximately the same age in the same region should have approximately the same orientation (e.g., Peace et al., 2016). Therefore, we lined up the Triassic rifts from the northern margin of the Yucatan block with the rifts of the same age along the eastern shore of Florida that are apparent in potential fields (Figure 3.3) in our Late Triassic reconstruction (Figure 3.7a).

We have also used the known motions of the larger plates at this time to determine the location and motion of the Yucatan (Müller et al., 2019). Specifically, any reconstruction must result in plate motions that would produce the observed deformation from that time period. The NE-SW trending Triassic rifts in Florida that match the Yucatan rifts are also oriented similarly to the South Georgia Rift (Figure 3.5a) and other rifts in Georgia and South Carolina (Figure 3.7a; Heffner et al., 2012; Heffner 2013; Marzen et al., 2020; Pindell et al., 2020). These rifts all contain sediments indicating an age of origin ~230 Ma (e.g., Labails et al., 2010). Therefore, they all must be related to the same tectonic deformational event. As shown in Figure 3.7a, the relative motion of the southeastern corner of North America to South America and Africa would likely produce basins of this orientation. Therefore, both the Yucatan block and the Florida block were (a) located at the southeastern corner of North America, (b) connected enough to North America to be affected by its motion, and (c) connected to South America/Africa. However, to account for the stretching of the continental crust, both the Yucatan and Florida terranes are modeled as moving very slightly to the southeast (Figure 3.7a). This statement is based on the assumption that the stress (motion) is not transmitted across weak boundaries. Therefore, to create the rift basins in Yucatan and Florida, both needed to be strongly connected to the surrounding continents. Thus, they will continue to be affected on a large scale by the motions of the larger continents until there is a weak zone between them.

Based on the South Georgia Rift and fracture data (Bartholomew et al., 2012), the NE-SW oriented rifting due to NW-SE tensional stresses continued until around 205 Ma, at which time the South Georgia Rift appears to have shut off (e.g., McBride, 1991; Marzen et al., 2020). This timing agrees with the observation that the CAMP magmatic complexes (~200 Ma) in the South Georgia Rift are not faulted (McBride, 1991; Marzen et al., 2020). Therefore, the plate motion between North America, the Yucatan, the Florida block, and Gondwana was likely fairly steady between 230 and 205 Ma, and both the Yucatan and Florida blocks were likely relatively stationary. There are very few Triassic continental rifts similar to what is seen in Florida and the South Georgia Rift to the west of the Mississippi (Snedden & Galloway, 2019), suggesting that the tie between South America and North America was either very weak and/or the deformation was accommodated within the suture zone north of the Sabine terrane (Huerta & Harry, 2012; Keller et al., 2016).

Reconstruction of the Chiapas block is fraught with numerous interpretations overprinted by multiple collisions (e.g., Pindell et al., 2020). In our model, we reconstruct it as being attached to the Yucatan and embedded in southern North America (Figure 3.7a). This area was identified by Thomas (2006) as a significant embayment into the North American continent because of the rifting in the late Precambrian. Therefore, we have put the Chiapas terrane into the embayment; it could be in the open area between South America and North America below what is now Texas (indicated by a question mark (?) in Figure 3.7a); however, there is no data on land suggesting a different stress field than that generated by the rotation of the Yucatan.

3.5.2. Late Triassic–Early Jurassic (205–195 Ma)

The CAMP massive igneous intrusion marks the end of the Triassic with injected dikes, sills, and basaltic flows from the northern parts of South America and Africa to Iberia.

Modern dating suggests that it mostly occurred within 1 million years, around 200 Ma (e.g., Nomade et al., 2007; Marzoli et al., 2018), but there may have been later pulses. The older dating of what is now known as the CAMP complexes came up with the earlier age ~180–190 Ma that was later amended to around 200 Ma (Nomade et al., 2007; Lanphere and Gohn, 1983; Gottfried et al., 1983). We therefore hypothesize that the ~190 Ma dates of the diabase intrusions from the DSDP wells (Schlager et al., 1984) may be off, and the encountered magmatism indeed relates to CAMP (i.e., ~200 Ma). The South Georgia Rifts are known to be capped by the magmatic complexes from this event (e.g., McBride, 1991; Marzen et al., 2020). The potential continuation of the South Georgia Rift system—the Apalachicola Basin (labeled AB in Figure 3.5a,d)—requires dense and highly magnetic rocks to explain the observed magnetic anomalies (Liu et al., 2019) that we interpret as CAMP, thus being emplaced around 200 Ma. We expand this interpretation to other mapped magmatic complexes mapped from seismic and potential fields, particularly to SDRs (Figures 3.5 and 3.6), as being added during the CAMP magmatic event along both the eastern margin of Florida and the western Yucatan. None of the other seismic lines show this volume of volcanic activity; therefore, it is unlikely that these two inner SDR basalt basins formed adjacent to any of the other portions of the rifted basin. Therefore, these areas were adjacent to each other around 200 Ma (Figure 3.7b).

Paleogeographically, to have the two inner SDRs adjacent to each other, the Yucatan block needs to rotate dramatically counterclockwise in approximately 5 million years, and the rifting needs to progress from scattered continental rifting to focused rifting, resulting in the stretched and intruded continental crust as we have modeled in Figure 3.7b. Although this may appear to be too dramatic for these slow-moving continental blocks, there is other evidence for a major tectonic upheaval, including a dramatic change in the orientation of the stress field as identified by structural features such as the CAMP dikes and the fractures seen on land (Beutel, 2009; Bartholomew et al., 2012). The CAMP dikes in southeastern North America were initially injected oriented in an NW-SE direction parallel to the inner SDRs around 200 Ma (e.g., Nomade et al., 2007; Beutel, 2009; Marzen et al., 2020). Both the ~200 Ma dikes and SDRs indicate a change in the extension direction from NW-SE tensional stress that formed the NE-SW trending rifts to NNE-SSW tensional stress that formed the NNW-SSE trending dikes and inner SDRs, all of which make a major plate motion change reasonable. Further, the change in dominant fracture orientations at this time also indicates a major change in the plate motion (Bartholomew et al., 2012). Because the South Georgia Rift ceases to extend (i.e., McBride, 1991), all southward motion of Florida is modeled as ceasing at ~205 Ma. Tectonically, the rapid motion change of both the stress field and the motion of the Yucatan is likely caused by the separation of Florida from South America and the Yucatan. Prior to ~205 Ma, Florida and South America were still connected, as were Florida and Africa; this combined motion created the NE-SW trending extensional basins. However, when South America was separated from Florida by oceanic crust (Figure 3.7b), it ceased having any influence on Florida, thus changing the internal stress field of the Florida Block. Because South America was still attached to the Yucatan, it continued to influence the motion of the Yucatan, perhaps resulting in the rotation of the Yucatan, which created a stress field almost 90 degrees from the original.

3.5.3. Early–Middle Jurassic (195–165 Ma)

Between ~195 and ~165 Ma, our plate reconstruction (Figure 3.7c,d) is guided by the existence of presalt basins that developed in the east (what is now the northeastern GoM and the western Yucatan margin) but were not formed in the west (what is now the Texas margin and the southern Yucatan/Chiapas block). We explain this with the postmagmatic subsidence of the crust in the east that did not occur in the west. The western thickened crust formed during the Ouachita orogeny was not extended in the Triassic, in contrast to the crust in the east (Figure 3.7a). In the west, only the lower crust is intruded (Figure 3.5b,c), whereas in the east, there are upper, middle, and lower crustal mafic intrusions (Figure 3.5d,e,i,k). The volume of intrusions, thinner crust, and proximity to excessive mafic volcanism in South Florida around 200 Ma (Figure 3.1a) suggest that the eastern continental crust was significantly hotter than the western one. As the magmatism in the east ceased by around 195 Ma, the eastern continental crust began subsiding and accumulated a large body of sediments before the deposition of salt at 169 Ma. In contrast, the thicker and more buoyant western GoM crust remains elevated and failed to accumulate a significant presalt sedimentary section (Figure 3.7d,e). This is consistent with observations of Dickinson et al. (2010) in central Texas suggesting a local uplift of that region at Late Triassic–Early Jurassic, as well as with observations of Ewing and Galloway (2019) noting a general lack of Early Jurassic deposits in the northwestern GoM.

Our reconstruction suggests that the rate of southward motion of the Yucatan relative to North America increases as the rotation of the Yucatan, seen between 205 and 195 Ma, ceased. This is likely due to the development of a thinned transitional crust and the increased rate of motion of South America. North America's effect on the Yucatan is hampered by a transitional crust that doesn't transmit stress well but likely helps to guide the relative motion that is more strongly affected by South America (Figure 3.7c). When oceanic crust formation begins, it is perhaps best identified by the sudden increase in the rate of relative motion of continents to North America. Around 200 Ma, it is clear that South America is no longer completely tied to North America and that oceanic crust has likely developed between the Florida/Suwanee block and South America (compare Figure 3.7a to Figure 3.7b). We also see a dramatic increase in the rate of motion of Africa relative to North America at 195 Ma, suggesting that the two are now separated by a weak zone and some oceanic crust (compare Figure 3.7b,c) These relative motions and relationships between the plates continue until around 165 Ma. By that time, the salt is deposited throughout the entire GoM (Pindell et al., 2020). In the east, the salt is deposited on thick sedimentary sequences (butter yellow in Figure 3.7d,e), whereas in the west, deposition occurs on the transitional crust that is now subsiding as it stretches (Figure 3.7e, lime green). Africa no longer has any influence on the GoM area.

3.5.4. Late Jurassic Oceanic Phase 1 (165–152 Ma)

Between 165 and 152 million years ago, our reconstruction is based on the interaction of the Yucatan with the plates around it and the seafloor spreading, which initiates at ~165 Ma in our model (loosely constrained by the age of ~160 Ma determined by Snedden et al., 2014). Specifically, the interaction of the southwestern corner of the Yucatan and the Maracaibo block (Figure 3.7e), along with other subduction zones in that area, causes the Yucatan block to begin rotating. The rate of rotation is limited by the ongoing connection between the Yucatan and the North American continental crust in the center of the basin (the still-connected stretched continental crust with evidence of postsalt exhumation of lower continental crust; Figure 3.5f) and the thin, ultraslow spreading oceanic crust in the east and the west (as seen in Figure 3.5). The sudden acceleration of South America during this interval suggests that the connection between the Yucatan and South America has weakened to the point when South America and the Yucatan no longer influence each other. Instead, the Yucatan appears to be guided by the motion of the Mexican terrains that will make up Mexico and parts of Central America. Around 165 Ma, the transitional crust along the northeast and northwest corners of the Yucatan finally thins enough to start oceanic spreading (Figure 3.7f,g). However, the relative motion to North America remains very slow, so the produced oceanic crust, as seen in Figure 3.5, is magma-starved and very thin. The rate of spreading is likely slowed by the oblique spreading orientation (Figure 3.7f) and continued continental connection between North America and the Yucatan. Although highly oblique back-arc basins have been proposed to account for the motion of the Yucatan (Stern & Dickinson, 2010), the rotation and motion can be explained by the interaction of the Yucatan with the Southeast moving terranes of Mexico (Figure 3.7f,g).

3.5.5. Late Jurassic–Early Cretaceous Oceanic Phase 2 (152–135 Ma)

Between 152 and 135 million years ago, the reconstruction of the motion of the Yucatan was constrained mostly by the Phase 2 oceanic spreading centers and transform faults (Figure 3.6). However, the different rate of rotation is modeled with respect to Phase 1 because of the change in the connection between the Yucatan and North America (compare the northern boundary of the Yucatan in Figure 3.7e,f). The last connection of continental crust between North America and the Yucatan is broken by the onset of a slow-spreading ridge in the center of the basin around 152 Ma. An apparent increase in the rate of rotation (Figure 3.7g, Table 3.2) corresponds to an increase in the spreading rate and increased magmatic supply; as seen in Figure 3.5, the oceanic crust is now thicker and two-layered. By approximately 135 Ma (Figure 3.7h), the subduction zone between the Yucatan and the Mexican terranes has closed, and the Yucatan now moves as part of the North American plate. This causes the mid-ocean ridge in the GoM to cease spreading, and all the relative motion between North America and South America rearranges to the spreading ridge in the Caribbean.

3.6. CONCLUSIONS

Integration of multiple geophysical datasets with geological observations allowed us to derive strong constraints for tectonic reconstruction of the GoM, such as Triassic rifts, CAMP-related magmatic complexes, Early to Middle Jurassic sedimentary basins developed before deposition of salt, and two distinct oceanic spreading episodes in the Late Jurassic through Early Cretaceous. Our tectonic reconstruction allows us to draw the following conclusions:

1. The magmatic regime varied during the basin formation from the CAMP magmatic event to amagmatic postsalt rifting that resulted in the exhumation of the lower continental crust.
2. The number of matching geological observations—SDRs, presalt basin, and outer troughs underlined by an exhumed lower crust—suggest that the northeastern GoM and the western Yucatan are conjugate margins.
3. The rotation of the Yucatan between 205 and 195 Ma explains the termination of the South Georgia Rift system and the emplacement of NW-SE trending dikes on top of NE-SW trending rift basins.
4. Our reconstruction suggests that the ~170 Ma rotation of the Yucatan initiated with the formation of a subduction zone between the Yucatan and the Maracaibo block.

Tectonic Reconstruction Summary

~230–~205 Ma: The Yucatan is emplaced into North America and rotated so the present-day NW coast is parallel to the western edge of Florida. Slow continental stretching as North America moves to the NW away from the rest of Pangea results in NE-trending rift basins (Figure 3.7a).

~205–195 Ma: Yucatan and Florida terranes separate; Yucatan rotates ~30° counterclockwise, changing the stress field and continental deformation, and inner SDRs form as part of the CAMP. Yucatan is still tectonically connected to North America and South America; Florida is now permanently attached to North America (Figure 3.7b,c).

~195–~165 Ma: General south-southeast motion of the Yucatan block continues to be guided by the motion of North America and South America. Transitional crust forms around the GoM, but thick sedimentary sequences only form where excess magmatism during CAMP cooled and resulted in their subsidence. A thick layer of salt is deposited between 170 and 165 Ma (Pindell et al., 2020) (Figure 3.7d).

~165–~150 Ma: A subduction zone forms along the southwest margin of the Yucatan, dipping under the Maracaibo block, causing the Yucatan to begin to rotate. Ultraslow spreading crust is formed along the northwestern and northeastern margins of the Yucatan, but transitional crust still connects the Yucatan and North America in the center of the basin (Figure 3.7e,f).

~150 Ma: Southwest subduction continues under what will become Mexico. Transitional crust no longer connects North America and the Yucatan, resulting in an increase in rotation speed, a new ridge, and the formation of slow-spreading oceanic crust (Figure 3.7g).

~135 Ma: Yucatan docks with Mexico, and all tectonic motion in the GoM ceases (Figure 3.7h).

5. Ridge reorganization occurred during the formation of oceanic crust that resulted in two sets of extinct spreading centers. The older ultraslow spreading (~165–152 Ma) produced a thin and uniform crust, whereas the younger slow-spreading episode (~152–135 Ma) resulted in a thicker and layered crust with NW-SE trending spreading centers.

The use of geologic features identified using geophysical methods, combined with the analysis of how the Yucatan is interacting with all surrounding plates at any given time, lends robustness to our model.

ACKNOWLEDGMENTS

The authors are extremely grateful to TGS and particularly to Jim Howell for allowing us to use proprietary seismic data shown in Figure 3.2 for our analysis. Many thanks to Alex Peace and M. J. Bartholomew for very helpful reviews.

REFERENCES

Ball, M. M., Martin, R. G., Foote, R. Q., & Applegate, A. V. (1988). *Structure and stratigraphy of the western Florida shelf; Part I, Multichannel reflection seismic data* (Map No. 88-439). US Geological Survey.

Bankey, V., Cuevas, A., Daniels, D., Finn, C. A., Hernandez, I., Hill, P., et al. (2002). *Digital data grids for the magnetic anomaly map of North America* (Map No. 2002-414). U.S. Geological Survey.

Barboza-Gudiño, J. R., Zavala-Monsiváis, A., Venegas-Rodríguez, G., & Barajas-Nigoche, L. D. (2010). Late Triassic stratigraphy and facies from northeastern Mexico: tectonic setting and provenance. *Geosphere*, *6*(5), 621–640.

Bartholomew, M. J., Van Arsdale, R. B., Cox, R. T., Tuttle, M. P., Boyd, O. S., & Locat, J. (2012). Structural controls on intraplate earthquakes in the eastern United States. In R. T. Cox, et al. (Eds.), *Recent advances in North American paleoseismology and neotectonics east of the Rockies* (pp. 165–190). Geological Society of America.

Beutel, E. K. (2009). Magmatic rifting of Pangaea linked to onset of South American plate motion. *Tectonophysics*, *468*(1–4), 149–157.

Bird, D. E., Burke, K., Hall, S. A., & Casey, J. F. (2005). Gulf of Mexico tectonic history: Hotspot tracks, crustal boundaries, and early salt distribution. *AAPG bulletin*, *89*(3), 311–328.

Christeson, G. L., Van Avendonk, H. J. A., Norton, I. O., Snedden, J. W., Eddy, D. R., Karner, G. D., & Johnson, C. A. (2014). Deep crustal structure in the eastern Gulf of Mexico. *Journal of Geophysical Research: Solid Earth*, *119*(9), 6782–6801.

Coombs, H., Kerr, A., Pindell, J., Buchs, D., Weber, B., & Solari, L. (2019). *Petrogenesis of the crystalline basement along the western Gulf of Mexico: Postcollisional magmatism during the formation of Pangea. Southern and central Mexico: basement framework, tectonic evolution, and provenance of Mesozoic-Cenozoic basins* (Special Paper). Geological Society of America.

Corti, G. (2012). Evolution and characteristics of continental rifting: analog modeling-inspired view and comparison with examples from the East African Rift System. *Tectonophysics*, *522*, 1–33.

Curry, M. A., Peel, F. J., Hudec, M. R., & Norton, I. O. (2018). Extensional models for the development of passive-margin salt basins, with application to the Gulf of Mexico. *Basin Research*, *30*(6), 1180–1199.

Deighton, I. C., Winter, F., & Chisari, D. (2017). *Recent high-resolution seismic, magnetic and gravity data throws new light on the early development of the Gulf of Mexico*. Paper presented at the AAPG Annual Meeting, Houston.

Dickinson, W. R., Gehrels, G. E., & Stern, R. J. (2010). Late Triassic Texas uplift preceding Jurassic opening of the Gulf of Mexico: evidence from U-Pb ages of detrital zircons. *Geosphere*, *6*(5), 641–662.

Dobson, L. M. & Buffler, R. T. (1991). *Basement rocks and structure, northeast Gulf of Mexico*. Paper presented at the GCAGS and GC-SEPM Meeting, Houston.

Eddy, D. R., Van Avendonk, H. J., Christeson, G. L., & Norton, I. O. (2018). Structure and origin of the rifted margin of the northern Gulf of Mexico. *Geosphere*, *14*(4), 1804–1817.

Eddy, D. R., Van Avendonk, H. J., Christeson, G. L., Norton, I. O., Karner, G. D., Johnson, C. A., & Snedden, J. W. (2014). Deep crustal structure of the northeastern Gulf of Mexico: implications for rift evolution and seafloor spreading. *Journal of Geophysical Research: Solid Earth*, *119*(9), 6802–6822.

Erlich, R. N., & Pindell, J. (2020). Crustal origin of the West Florida Terrane, and detrital zircon provenance and development of accommodation during initial rifting of the southeastern Gulf of Mexico and western Bahamas. *Geological Society, London, Special Publications*, *504*, 77–118.

Escalona, A., Norton, I. O., Lawver, L. A., & Gahagan, L. (2021). Quantitative plate tectonic reconstructions of the Caribbean region from Jurassic to present. In C. Bartolini (Ed.), *Eastern Caribbean–northeastern South American boundary: tectonic evolution, basin architecture, and petroleum systems* (Vol. *123*, pp. 239–263). AAPG Memoir, American Association of Petroleum Geologists.

Ewing, J., Antoine, J., & Ewing, M. (1960). Geophysical measurements in the western Caribbean Sea and in the Gulf of Mexico. *Journal of Geophysical Research*, *65*(12), 4087–4126.

Ewing, T. E., & Galloway, W. E. (2019). Evolution of the northern Gulf of Mexico sedimentary basin. In A. D. Miall (Ed.), *The sedimentary basins of the United States and Canada* (pp. 627–694). Elsevier.

Fiduk, J. C., Welmer, P., Trudgill, B. D., Rowan, M. G., Gale, P. E., Phair, R. L., et al. (1999). The Perdido fold belt, northwestern deep Gulf of Mexico, part 2: Seismic stratigraphy and petroleum systems. *AAPG Bulletin*, *83*(4), 578–612.

Filina, I. (2019). Crustal architecture of the northwestern and central Gulf of Mexico from integrated geophysical analysis. *Interpretation*, *7*(4), T899–T910.

Filina, I., Austin, J. A., Doré, A. G., Johnson, E. A., Minguez, D. A., Norton, I. O., et al. (2022). Opening of the Gulf of Mexico: what we know, what questions remain, and how we might answer them. *Tectonophysics*, *822*, 229150.

Filina, I., Delebo, N., Mohapatra, G., Coble, C., Harris, G., Layman, J., et al. (2015). Integration of seismic and gravity data—A case study from the western Gulf of Mexico. *Interpretation*, *3*(4), SAC99–SAC106.

Filina, I., & Hartford, L. (2021). Subsurface structures along western Yucatan from integrated geophysical analysis. *Marine and Petroleum Geology*, *127*, 104964.

Filina, I., Liu, M., & Beutel, E. (2020). Evidence of ridge propagation in the eastern Gulf of Mexico from integrated analysis of potential fields and seismic data. *Tectonophysics*, *775*, 228307.

Frizon de Lamotte, D., Fourdan, B., Leleu, S., Leparmentier, F., & de Clarens, P. (2015). Style of rifting and the stages of Pangea breakup. *Tectonics*, *34*(5), 1009–1029.

Galloway, W. E. (2008). Depositional evolution of the Gulf of Mexico sedimentary basin. In A. D. Miall (Ed.), *Sedimentary basins of the world* (Vol. *5*, pp. 505–549). Elsevier Science.

Godínez-Urban, A., Lawton, T. F., Molina-Garza, R. S., Iriondo, A., Weber, B., & López-Martínez, M. (2011a). Jurassic volcanic and sedimentary rocks of the La Silla and Todos Santos Formations, Chiapas: record of Nazas arc magmatism and rift-basin formation prior to opening of the Gulf of Mexico. *Geosphere*, *7*(1), 121–144.

Godínez-Urban, A., Molina-Garza, R. S., Geissman, J. W., & Wawrzyniec, T. (2011b). Paleomagnetism of the Todos Santos and La Silla Formations, Chiapas: implications for the opening of the Gulf of Mexico. *Geosphere*, *7*(1), 145–158.

Gottfried, D., Annell, C. S., & Byerly, G. S. (1983). *Geochemistry and tectonic significance of subsurface basalts from Charleston, South Carolina: Clubhouse Crossroads test holes# 2 and# 3* (Professional Paper 1313, pp. A1–A10). U.S. Geological Survey.

Guan, H., Geoffroy, L., Gernigon, L., Chauvet, F., Grigne, C., & Werner, P. (2019). Magmatic ocean-continent transitions. *Marine and Petroleum Geology*, *104*, 438–450.

Hall, S. A., & Najmuddin, I. J. (1994). Constraints on the tectonic development of the eastern Gulf of Mexico provided by magnetic anomaly data. *Journal of Geophysical Research: Solid Earth*, *99*(B4), 7161–7175.

Heffner, D. M. (2013). *Tectonics of the South Georgia Rift* (PhD thesis). University of South Carolina.

Heffner, D. M., Knapp, J. H., Akintunde, O. M., & Knapp, C. C. (2012). Preserved extent of Jurassic flood basalt in the South Georgia Rift: a new interpretation of the J horizon. *Geology*, *40*(2), 167–170.

Hilterman, F. (1998). *Rock property framework for comprehending deep-water seismic response*. Paper presented at the 14th Annual Gulf Coast Technical Meeting.

Hinz, K. (1981). A hypothesis on terrestrial catastrophes: Wedges of very thick oceanward dipping layers beneath passive margins. *Geologisches Jahrbuch Reihe E, Geophysik*, *22*, 5–28.

Horn, B., Hartwig, A., Faw, J., Novianti, I., Goswami, A., & McGrail, A. (2017). Refining exploration opportunities in Mexico. *GEOExPro*, *14*, 64–69.

Hudec, M. R., Dooley, T. P., Peel, F. J., & Soto, J. I. (2020). Controls on the evolution of passive-margin salt basins: structure and evolution of the Salina del Bravo region, northeastern Mexico. *AAPG Bulletin*, *132*(5–6), 997–1012.

Hudec, M. R., & Norton, I. O. (2019). Upper Jurassic structure and evolution of the Yucatán and Campeche subbasins, southern Gulf of Mexico. *AAPG Bulletin*, *103*(5), 1133–1151.

Hudec, M. R., Norton, I. O., Jackson, M. P., & Peel, F. J. (2013). Jurassic evolution of the Gulf of Mexico salt basin. *AAPG Bulletin*, *97*(10), 1683–1710.

Huerta, A. D., & Harry, D. L. (2012). Wilson cycles, tectonic inheritance, and rifting of the North American Gulf of Mexico continental margin. *Geosphere*, *8*(2), 374–385.

Ibrahim, A. K., Carye, J., Latham, G., & Buffler, R. T. (1981). Crustal structure in Gulf of Mexico from OBS refraction and multichannel reflection data. *AAPG Bulletin*, *65*(7), 1207–1229.

Imbert, P., Philippe, Y., Post, P. J., Rosen, N. C., Olson, D. L., Palmes, S. L., et al. (2005). *The Mesozoic opening of the Gulf of Mexico: Part 2. Integrating seismic and magnetic data into a general opening model.* Paper presented at the 25th Annual GCSSEPM Research Conference: Petroleum systems of divergent continental margins.

Imbert, P. & Post, P. J. (2005). *The Mesozoic opening of the Gulf of Mexico: Part 1, Evidence for oceanic accretion during and after salt deposition.* Paper presented at the 25th Annual GCSSEPM Research Conference: petroleum systems of divergent continental margins.

Keller, G. R., Pulliam, J., Gurrola, H., & Mickus, K. (2016). Extended abstract: an overview of the structure and tectonic evolution of the Gulf Coast region of Texas and Louisiana. *GCAGS Transactions*, *66*, 785–788.

Kneller, E. A. & Johnson, C. A. (2011). *Plate kinematics of the Gulf of Mexico based on integrated observations from the Central and South Atlantic.* Paper presented at the GCAGS and GC-SEPM Annual Convention, Austin.

Labails, C., Olivet, J. L., Aslanian, D., & Roest, W. R. (2010). An alternative early opening scenario for the Central Atlantic Ocean. *Earth and Planetary Science Letters*, *297*(3–4), 355–368.

Lanphere, M. A., & Gohn, G. S. (1983). *40Ar/39Ar ages of basalt from Clubhouse Crossroads test hole# 2, near Charleston, South Carolina.* U.S. Department of the Interior, Geological Survey.

Laske, G., Masters, G., Ma, Z., & Pasyanos, M. (2013). Update on CRUST1. 0—A 1- degree global model of Earth's crust. *Geophysical Research Abstracts*, *15*, 2658.

Leleu, S., Hartley, A. J., van Oosterhout, C., Kennan, L., Ruckwied, K., & Gerdes, K. (2016). Structural, stratigraphic and sedimentological characterisation of a wide rift system: the Triassic rift system of the Central Atlantic Domain. *Earth-Science Reviews*, *158*, 89–124.

Lin, P., Bird, D. E., & Mann, P. (2019). Crustal structure of an extinct, late Jurassic-to-earliest Cretaceous spreading center and its adjacent oceanic crust in the eastern Gulf of Mexico. *Marine Geophysical Research*, *40*(3), 395–418.

Liu, M. & Filina, I. (2018). *Potential fields modelling in the northeastern Gulf of Mexico.* Paper presented at the AAPG Annual Convention & Exhibition.

Liu, M., Filina, I., & Mann, P. (2019). Crustal structure of Mesozoic rifting in the northeastern Gulf of Mexico from integration of seismic and potential fields data. *Interpretation*, *7*(4), T857–T867.

Lundin, E. R., & Doré, A. G. (2017). The Gulf of Mexico and Canada Basin: genetic siblings on either side of North America. *GSA Today*, *27*(1), 4–11.

MacRae, G., & Watkins, J. S. (1995). Early Mesozoic rift stage half graben formation beneath the DeSoto Canyon salt basin, northeastern Gulf of Mexico. *Journal of Geophysical Research: Solid Earth*, *100*(B9), 17795–17812.

Martini, M., & Ortega-Gutiérrez, F. (2016). Tectono-stratigraphic evolution of eastern Mexico during the break-up of Pangea. *Earth-Science Reviews*, *183*, 38–55. doi: 10.1016/j.earscirev.2016.06.013

Marton, G., & Buffler, R. T. (1994). Jurassic reconstruction of the Gulf of Mexico Basin. *International Geology Review*, *36*(6), 545–586.

Marzen, R. E., Shillington, D. J., Lizarralde, D., Knapp, J. H., Heffner, D. M., Davis, J. K., & Harder, S. H. (2020). Limited and localized magmatism in the Central Atlantic Magmatic Province. *Nature Communications*, *11*(1), 1–8.

Marzoli, A., Callegaro, S., Dal Corso, J., Davies, J. H., Chiaradia, M., Youbi, N., et al. (2018). The Central Atlantic magmatic province (CAMP): a review. In L. Tanner (Ed.), *The Late Triassic World* (pp. 91–125). Springer.

Marzoli, A., Renne, P. R., Piccirillo, E. M., Ernesto, M., Bellieni, G., & De Min, A. (1999). Extensive 200-million-year-old continental flood basalts of the Central Atlantic Magmatic Province. *Science*, *284*(5414), 616–618.

McBride, J. H. (1991). Constraints on the structure and tectonic development of the early Mesozoic South Georgia rift, southeastern United States; seismic reflection data processing and interpretation. *Tectonics*, *10*(5), 1065–1083.

McHone, J. G. (2003). Volatile emissions from Central Atlantic Magmatic Province basalts: mass assumptions and environmental consequences. *Geophysical Monograph-American Geophysical Union*, *136*, 241–254.

Meyer, B., Saltus, R., & Chulliat, A. (2017). EMAG2: earth magnetic anomaly grid (2-arc-minute resolution) version 3. *National Centers for Environmental Information.* 10.7289/V5H70CVX

Minguez, D., Gerald Hensel, E., & Johnson, E. A. (2020). A fresh look at Gulf of Mexico tectonics: testing rotations and breakup mechanisms from the perspective of seismically constrained potential-fields modeling and plate kinematics. *Interpretation*, *8*(4), SS31–SS45.

Mjelde, R., Breivik, A. J., Raum, T., Mittelstaedt, E., Ito, G., & Faleide, J. I. (2008). Magmatic and tectonic evolution of the North Atlantic. *Journal of the Geological Society*, *165*(1), 31–42.

Molina-Garza, R. S., Van Der Voo, R. O. B., & Urrutia-Fucugauchi, J. (1992). Paleomagnetism of the Chiapas Massif, southern Mexico: evidence for rotation of the Maya Block and implications for the opening of the Gulf of Mexico. *Geological Society of America Bulletin*, *104*(9), 1156–1168.

Müller, R. D., Zahirovic, S., Williams, S. E., Cannon, J., Seton, M., Bower, D. J., et al. (2019). A global plate model including lithospheric deformation along major rifts and orogens since the Triassic. *Tectonics*, *38*(6), 1884–1907.

Nakamura, Y., Sawyer, D. S., Shaub, F. J., MacKenzie, K., & Oberst, J. (1988). Deep crustal structure of the northwestern Gulf of Mexico. *Gulf Coast Association of Geological Societies Transactions*, *38*, 207–215.

Nguyen, L. C., & Mann, P. (2016). Gravity and magnetic constraints on the Jurassic opening of the oceanic Gulf of Mexico and the location and tectonic history of the Western Main transform fault along the eastern continental margin of Mexico. *Interpretation*, *4*(1), SC23–SC33.

Nomade, S., Knight, K. B., Beutel, E., Renne, P. R., Verati, C., Féraud, G., et al. (2007). Chronology of the Central Atlantic Magmatic Province: implications for the Central Atlantic rifting processes and the Triassic–Jurassic biotic crisis. *Palaeogeography, Palaeoclimatology, Palaeoecology*, *244*(1–4), 326–344.

O'Reilly, C., Keay, J., Birch-Hawkins, A., Bate, D., & Halliday, J. (2017). Regional play types in the Mexican offshore. *GEOExPro*, *14*, 36–45.

Padilla y Sánchez, R. J. (2016). *Late Triassic-Late Cretaceous paleogeography of Mexico and the Gulf of México. Mesozoic of the gulf rim and beyond: new progress in science and exploration of the Gulf of Mexico Basin*. Paper presented at the 35th Annual GCSSEPM Foundation Perkins-Rosen Research Conference.

Peace, A., McCaffrey, K., Imber, J., Phethean, J., Nowell, G., Gerdes, K., & Dempsey, E. (2016). An evaluation of Mesozoic rift-related magmatism on the margins of the Labrador Sea: implications for rifting and passive margin asymmetry. *Geosphere*, *12*(6), 1701–1724.

Peace, A. L., Phethean, J. J., Franke, D., Foulger, G. R., Schiffer, C., Welford, J. K., et al. (2020). A review of Pangaea dispersal and Large Igneous Provinces–In search of a causative mechanism. *Earth-Science Reviews*, *206*, 102902.

Pindell, J., Graham, R., & Horn, B. (2014). Rapid outer marginal collapse at the rift to drift transition of passive margin evolution, with a Gulf of Mexico case study. *Basin Research*, *26*(6), 701–725.

Pindell, J., Villagómez, D., Molina-Garza, R., Graham, R., & Weber, B. (2020). A revised synthesis of the rift and drift history of the Gulf of Mexico and surrounding regions in the light of improved age dating of the Middle Jurassic salt. *Geological Society, London, Special Publications*, *504*, 29–76.

Pindell, J. L., & Kennan, L. (2009). Tectonic evolution of the Gulf of Mexico, Caribbean and northern South America in the mantle reference frame: an update. *Geological Society, London, Special Publications*, *328*(1), 1–55.

Planke, S., Symonds, P. A., Alvestad, E., & Skogseid, J. (2000). Seismic volcanostratigraphy of large-volume basaltic extrusive complexes on rifted margins. *Journal of Geophysical Research: Solid Earth*, *105*(B8), 19335–19351.

Radovich, B., Horn, B., Nuttall, P., & McGrail, A. (2011). The only complete regional perspective. *GeoExPro*, *8*, 36.

Ramos, E. L. (1975). Geological summary of the Yucatan Peninsula. In A. E. M. Nairn & F. G. Stehli (Eds.), *The Gulf of Mexico and the Caribbean* (pp. 257–282). Boston, MA: Springer.

Rowan, M. G. (2014). Passive-margin salt basins: hyperextension, evaporite deposition, and salt tectonics. *Basin Research*, *26*(1), 154–182.

Salem, A., Williams, S., Fairhead, D., Smith, R., & Ravat, D. (2008). Interpretation of magnetic data using tilt-angle derivatives. *Geophysics*, *73*(1), L1–L10.

Salvador, A. (1991). Origin and development of the Gulf of Mexico basin. In A. Salvador (Ed.), *The Gulf of Mexico basin* (The Geology of North America, Vol. 1, pp. 389–444). Geological Society of America.

Sandwell, D. T., Müller, R. D., Smith, W. H., Garcia, E., & Francis, R. (2014). New global marine gravity model from CryoSat-2 and Jason-1 reveals buried tectonic structure. *Science*, *346*(6205), 65–67.

Saunders, M., Geiger, L., Rodriguez, K., & Hargreaves, P. (2016). *The delineation of pre-salt license blocks in the deep offshore Campeche-Yucatan Basin*. Paper presented at the AAPG Annual Convention and Exhibition, Calgary.

Schlager, W., Buffler, R. T., Angstadt, D., Bowdler, J. L., Cotillon, P. H., Dallmeyer, R. D., et al. (1984). Deep sea drilling project, leg 77, southeastern Gulf of Mexico. *Geological Society of America Bulletin*, *95*(2), 226–236.

Schouten, H., & Klitgord, K. D. (1994). Mechanistic solutions to the opening of the Gulf of Mexico. *Geology*, *22*(6), 507–510.

Scott, K. R., Hayes, W. E., & Fietz, R. P. (1961). Geology of the Eagle Mills formation. *Gulf Coast Association of Geological Societies* Transactions, *11*, 1–14.

Snedden, J. W., & Galloway, W. E. (2019). *The Gulf of Mexico sedimentary basin: depositional evolution and petroleum applications*. Cambridge University Press.

Snedden, J. W., Norton, I. O., Christeson, G. L., & Sanford, J. C. (2014). Interaction of deepwater deposition and a mid-ocean spreading center, Eastern Gulf of Mexico Basin, USA. *Gulf Coast Association of Geological Societies Transactions*, *64*, 371–383.

Soreghan, G. S., & Soreghan, M. J. (2013). Tracing clastic delivery to the Permian Delaware Basin, USA: implications for paleogeography and circulation in westernmost equatorial Pangea. *Journal of Sedimentary Research*, *83*(9), 786–802.

Soto-Kerans, G. M., Stockli, D. F., Janson, X., Lawton, T. F., & Covault, J. A. (2020). Orogen proximal sedimentation in the Permian foreland basin. *Geosphere*, *16*(2), 567–593.

Steier, A., & Mann, P. (2019). Late Mesozoic gravity sliding and Oxfordian hydrocarbon reservoir potential of the northern Yucatan margin. *Marine and Petroleum Geology*, *103*, 681–701.

Stern, R. J., Anthony, E. Y., Ren, M., Lock, B. E., Norton, I., Kimura, J. I., et al. (2011). Southern Louisiana salt dome xenoliths: First glimpse of Jurassic (ca. 160 Ma) Gulf of Mexico crust. *Geology*, *39*(4), 315–318.

Stern, R. J., & Dickinson, W. R. (2010). The Gulf of Mexico is a Jurassic backarc basin. *Geosphere*, *6*(6), 739–754.

Telford, W. M., Telford, W. M., Geldart, L. P., Sheriff, R. E., & Sheriff, R. E. (1990). *Applied geophysics*. Cambridge University Press.

Thomas, W. A. (2006). Tectonic inheritance at a continental margin. *GSA Today*, *16*(2), 4–11.

Traverse, A. (1987). Pollen and spores date origin of rift basins from Texas to Nova Scotia as early Late Triassic. *Science*, *236*(4807), 1469–1472.

Van Avendonk, H. J., Christeson, G. L., Norton, I. O., & Eddy, D. R. (2015). Continental rifting and sediment infill in the northwestern Gulf of Mexico. *Geology*, *43*(7), 631–634.

Williams-Rojas, C. T., Reyes-Tovar, E., Miranda-Peralta, L., Reyna-Martinez, G., Cardenas-Alvarado, A., Maldonado-Villalon, R., et al. (2012). Hydrocarbon potential of the deepwater portion of the "Salina del Istmo" Province, southeastern Gulf of Mexico, Mexico. *Gulf Coast Association of Geological Societies Transactions*, *62*, 641–644.

Wood, G. D., & Benson, D. G., Jr. (2000). The North American occurrence of the algal coenobium plaesiodictyon. Paleogeographic, paleoecologic, and biostratigraphic importance in the Triassic. *Palynology*, *24*(1), 9–20.

Woods, R. D., Salvador, A., & Miles, A. E. (1991). Pre-Triassic. In A. Salvador (Ed.), *The Gulf of Mexico Basin* (pp. 109–129). Boulder, CO: Geological Society of America.

4

Introduction to an Active Oblique Rift: Walker Lane–Gulf of California

Cathy J. Busby[1] and Keith Putirka[2]

ABSTRACT

Oblique/transtensional rifts dominate the geologic record, yet most studies have focused on orthogonal rifts. In this chapter, we provide a brief introduction to an active, superbly well-exposed, and highly accessible oblique rift: the Walker Lane–Gulf of California oblique rift. Because it is so young (12 Ma to present), the Walker Lane–Gulf of California oblique rift can be placed in the context of oceanic plate tectonics. The Walker Lane rift is an archetype of the first stage of oblique continental rifting, thinning of continental lithosphere, while its southern extension, the Gulf of California, is in the second stage, where the continent has ruptured and seafloor spreading is underway. Both segments were born in the axis of pre-existing extensional continental arcs, where the lithosphere was thinned and thermally weakened. However, the Gulf of California opened more quickly, because the entire Baja California Peninsula (1220 km long) was transferred to the Pacific plate all at once, while the Walker Lane is opening by progressive south-to-north unzipping. We present case studies from each rift segment to illustrate the structural and magmatic evolution of oblique rifts and, in the Gulf of California case study, relate the onshore record to the offshore record of oblique rifting.

4.1. INTRODUCTION TO THE WALKER LANE–GULF OF CALIFORNIA OBLIQUE RIFT

Oblique rifting (also known as transtensional rifting) has been called "the rule, not the exception," because 70% of all rifts since the Permian have exceeded an obliquity of 20° or greater (Brune et al., 2018). Examples of oblique continental rifts include the West Antarctic Rift System (Rossetti et al., 2006), the Tangra Yumco rift in Tibet (Reynolds & Laskowski, 2023), the Reykjanes Peninsula of southwest Iceland (Clifton & Kattenhorn, 2006), the Red Sea–Afar-Gulf of Aden system (Zwaan et al., 2020), the West Somali Basin of the East Africa rift (Ruocco et al., 2021), and the Bay of Biscay on the north margin of the Iberian Peninsula (Marsaglia, 2012). Rift obliquity was also important during the opening of the North Atlantic (Brune et al., 2018). However, most publications to date have focused on orthogonal rifts, perhaps because they can be modeled in two dimensions, whereas oblique rifts require more complicated three-dimensional models. Like orthogonal rifts, oblique rifts can be divided into three stages (Ingersoll & Busby, 1995). In the first stage, continental lithosphere is thinned. The Walker Lane oblique rift is in this stage (Figure 4.1). In the second stage, the continent has ruptured, and seafloor spreading has begun. The Gulf of California oblique rift is in this stage (Figure 4.1). In the third stage, the continents drift far apart and subside and are generally tectonically inactive at their passive margins. The Walker Lane–Gulf of California oblique rift has not reached this stage. Stage three continental margins are commonly buried by sedimentary

[1] *Department of Earth and Planetary Sciences, University of California at Davis, Davis, California, USA*

[2] *Department of Earth and Environmental Sciences, California State University, Fresno, California, USA*

Extensional Tectonics: Rifting and Continental Extension, Geophysical Monograph 278, First Edition.
Edited by İbrahim Çemen and Elizabeth J. Catlos.

DOI: 10.1002/9781119773757.ch04

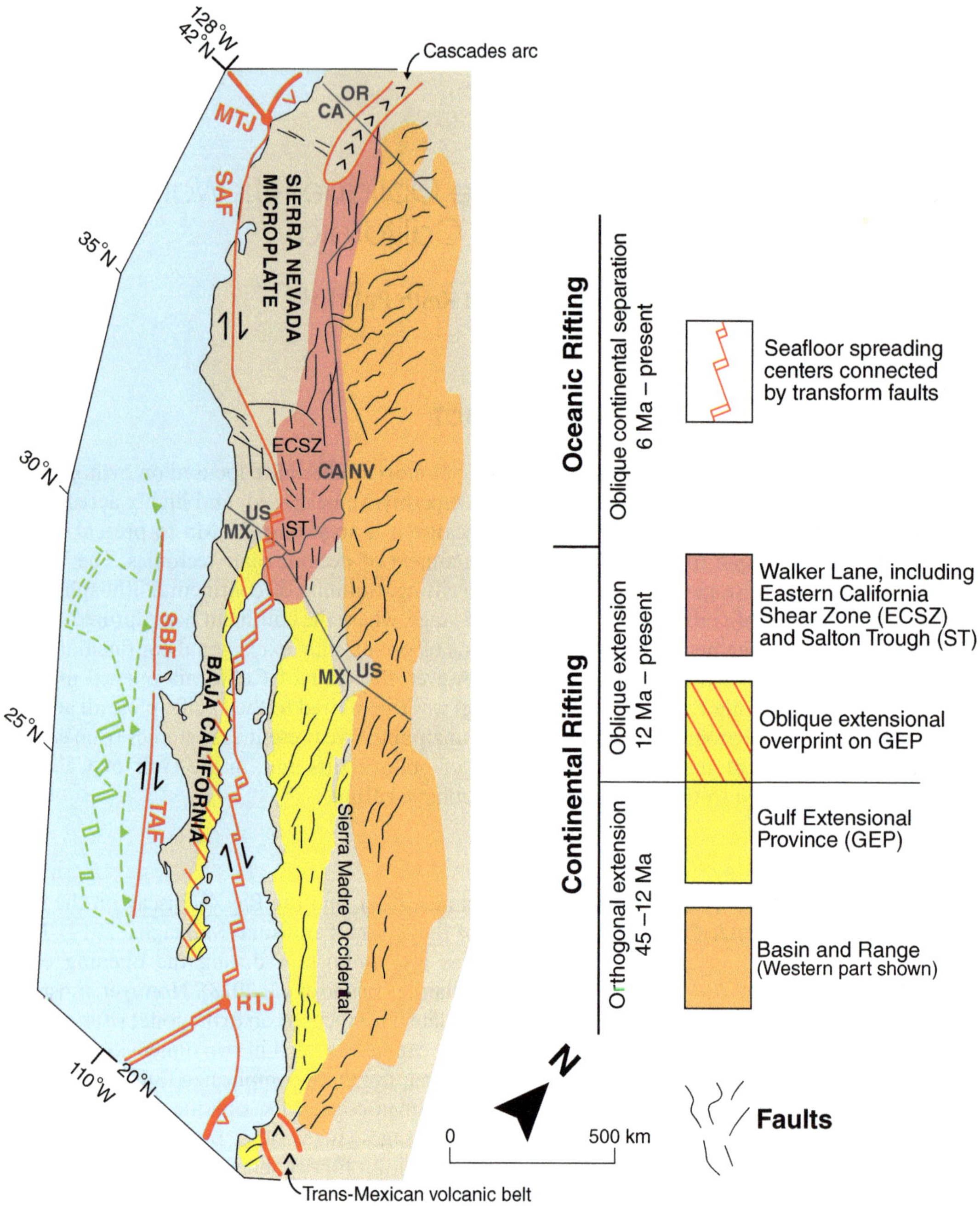

Figure 4.1 Map of the Walker Lane–Gulf of California oblique rift. The first stage of rifting, thinning of continental lithosphere, is underway in the north in the Walker Lane, including the Eastern California Shear Zone (ECSZ) and the Salton Trough (ST). The second stage of rifting, continental rupture, and seafloor spreading is underway in the south, in the Gulf of California. Continental rifting began at ca. 45 Ma in a broad zone of orthogonal extension that continued until 12 Ma (Basin and Range and Gulf Extensional provinces). Oblique extension began at 12 Ma, focused on the western edge of the Basin and Range–Gulf Extensional Province in what was the active arc axis, as shown in Figure 4.3. Seafloor spreading began at 6 Ma and continues today. MTJ, Mendocino Triple Junction; RTJ, Rivera Triple Junction; SAF, San Andreas fault; TAF, Tosco-Abreojos fault; SBF, San Benito fault. Extinct spreading centers, transforms, and trench shown in green. Adapted from Faulds and Henry (2008).

rocks, so the products of the earlier stages can only be imaged with geophysical surveys. This chapter will focus first on stage 1 (Walker Lane) and then move to stage 2 (Gulf of California).

In this chapter, we provide a brief overview of an active, superbly well-exposed, and highly accessible oblique rift: the Late Miocene to Recent Gulf of California-Walker Lane oblique rift, which extends about 3,000 km from the Mendocino triple junction of northern California to the Rivera triple junction off the south tip of Baja California (Figures 4.1 and 4.2). First, we present our interpretation of its origin and evolution during the interaction between the North American plate and the Farallon and Pacific plates. We highlight the differences between the Gulf of California oblique rift segment, which formed all at once, and the Walker Lane oblique segment, which is opening

Figure 4.2 Examples of the superb exposures in the Walker Lane–Gulf of California oblique rift. (a) Walker Lane: oblique aerial photomosaic of the syndepositionally faulted Cataract paleochannel fill (location in Figure 4.6) showing superior exposure of stratigraphy and faults, from Busby et al. (2016) (view looking north; length of the ridge shown is 5 km). Tvs, Valley Springs Formation (sequence 1, Figure 4.6); Trpdf, Relief Peak Formation (sequence 2, Figure 4.6); Ts, Stanislaus Group high-K_2O volcanic rocks (sequence 3, Figure 4.6), divided here into Table Mountain Latite lava flows (Tstml), Eureka Valley Tuff Tollhouse Flat Member welded ignimbrite (Tvset), Eureka Valley Tuff Lava Flow Member (Tsel) that includes basalt vent facies (Tselbvf), Eureka Valley Tuff By-Day Member welded ignimbrite (Tseb), and Eureka Valley Tuff Upper Member nonwelded ignimbrite, (Tseu); Td, Dardanelles Formation shoshonite, including lava (Tsdl), intrusions (Tsdi), vent facies deposits (Tsdvf), and a debris avalanche deposit (Tdda); Tdpi, Disaster Peak Formation (sequence 4, Figure 4.6) andesitic intrusions. Photo by Cathy J. Busby (Chapter Author). (b) Gulf of California: oblique aerial photomosaic of the Concepción Peninsula, viewed looking toward the east. The peninsula is ca. 28 km long and up to 10 km wide. It consists of an east-tilted block bounded on the west side by a system of west-dipping normal faults that drop the hanging wall down below sea level (Concepción Bay). Footwall uplift along these faults exposes a stratigraphic sequence that dips east and includes (from base to top): Cretaceous granitic basement, Oligocene forearc ignimbrites and eolian sandstones, Early to Middle Miocene arc volcanic and volcaniclastic rocks and intrusions, and Late Miocene continental rift ignimbrites, lavas, and alluvial fan conglomerates (Graettinger et al., 2022; Busby et al., 2024). With permission of Rebecca Dorsey.

by progressive south-to-north "unzipping" (Figure 4.3). Second, we summarize the major tectonic and magmatic features of the Walker Lane oblique rift segment, which is in the first stage of rifting, and provide a case study from the central Walker Lane as a more detailed example. Third, we summarize the major tectonic and magmatic features of the Gulf of California oblique rift segment, which is in the second stage of rifting, and provide a case study from the central Baja California margin of the Gulf of California oblique rift.

The goal of this chapter is to describe the Walker Lane–Gulf of California oblique rift in a succinct and highly accessible way, not to present an exhaustive description of it (for an exhaustive review of Walker Lane, see Busby, 2013). In this spirit, we refer to recent publications with exhaustive reference lists rather than cluttering this chapter with hundreds of primary citations. This chapter is thus presented more as a textbook chapter and less as a research paper for the purpose of reaching as many students and researchers as possible with the message that this is an oblique/transtensional rift system, one deserving of global interest and further work.

The Walker Lane has only been recognized as an oblique rift in the past two decades (see full summary of evidence in Busby, 2013), and to this day, is commonly misunderstood as solely a strike-slip fault system. Meanwhile, the Gulf of California has long been recognized as an oblique rift because it has seafloor spreading centers connected by transform faults that take up the oblique motion (Figure 4.1). We emphasize the fact that the Walker Lane oblique rift segment represents an earlier stage of oblique rift evolution, and the Gulf of California represents a later stage. They are part of the same oblique rift.

The Gulf of California (Figure 4.1) consists of short seafloor spreading centers connected by long, right-stepping dextral oceanic transform faults and forms the boundary between the North American and Pacific plates. This system passes northward into the Salton Trough (ST in Figure 4.1), where rapid sedimentation from the Colorado River has created new continental crust as continental crust is removed by extension, delaying continental rupture (van Wijk et al., 2019). The San Andreas fault emanates from the north end of the Salton Trough and consists of a relatively narrow continental transform fault. However, just north of the Salton Trough, the San Andreas fault takes a major left jog, producing the transpressional "big bend" of southern California, whereas the oblique rift continues straight through the Eastern California Shear Zone (ECSZ) and onward to Walker Lane (WL in Figure 4.1). Seventy-five percent of the motion between the Pacific plate and the North American plate is taken up along the San Andreas fault, and the other 25% is taken up along the ECSZ and Walker Lane (Bennett et al., 1999; see references in Faulds & Henry, 2008; Busby, 2013). The region between the San Andreas fault and the Walker Lane oblique rift is referred to as the Sierra Nevada microplate (Unruh et al., 2003), described further later in the chapter. Walker Lane is widely referred to as "the future plate boundary" (Unruh et al., 2003) because it is inferred that the San Andreas fault will eventually be abandoned and California will be transferred to the Pacific plate, becoming separated from Nevada by new oceanic crust like that in the Gulf of California.

The Walker Lane–Gulf of California oblique rift is an outstanding natural laboratory for studying oblique rifting. Exposures are truly superb (Figure 4.2) because these rocks lie in the Sonora-Mojave deserts and the high Sierra, and it is a very young system with active uplifts. Furthermore, the youth of the Walker Lane–Gulf of California oblique rift can be placed in the context of oceanic plate tectonics because the relevant oceanic plates have largely not yet been subducted. In the following section, we describe the plate tectonic evolution (which is not simple) in the simplest way possible.

4.2. PLATE TECTONIC SETTING OF THE WALKER LANE–GULF OF CALIFORNIA OBLIQUE RIFT

The tectonic history of the western United States and Mexico can best be described by their interactions with the Farallon and Pacific oceanic plates (Figure 4.3). The Farallon plate began subducting under western North America in Permian time (Rains et al., 2012; Cecil et al., 2019; Dobbs et al., 2021; Busby et al., 2023; Busby, 2023). By 35 Ma (Figure 4.3a), that subduction brought the ridge system between the Farallon and Pacific plates closer to the North American plate, but it still lay far offshore. The subduction-related arc was segmented by an abrupt left jog at the Nevada-Oregon border, inferred to record a slab tear, with the subducting slab dipping more steeply in the north and more shallowly in the south (Figure 4.3a; Colgan et al., 2011). South of the slab tear, the arc migrated westward with time (Figure 4.3a–c) due to progressive slab steepening (see references in Busby, 2013).

By 24 Ma (Figure 4.3b), the position of the subduction-related arc swept westward to occupy the future position of the Walker Lane–Gulf of California oblique rift (rift position shown in Figure 4.3c). These arcs (Ancestral Cascades arc in California and Comondú arc in Mexico) thermally weakened the continental lithosphere, priming it to become the future site of continental rifting (Busby, 2013; Busby et al., 2020). Meanwhile, the Pacific-Farallon ridge system was subducted beneath the North American continental margin in southern California, putting the Pacific plate directly in contact with the North American plate.

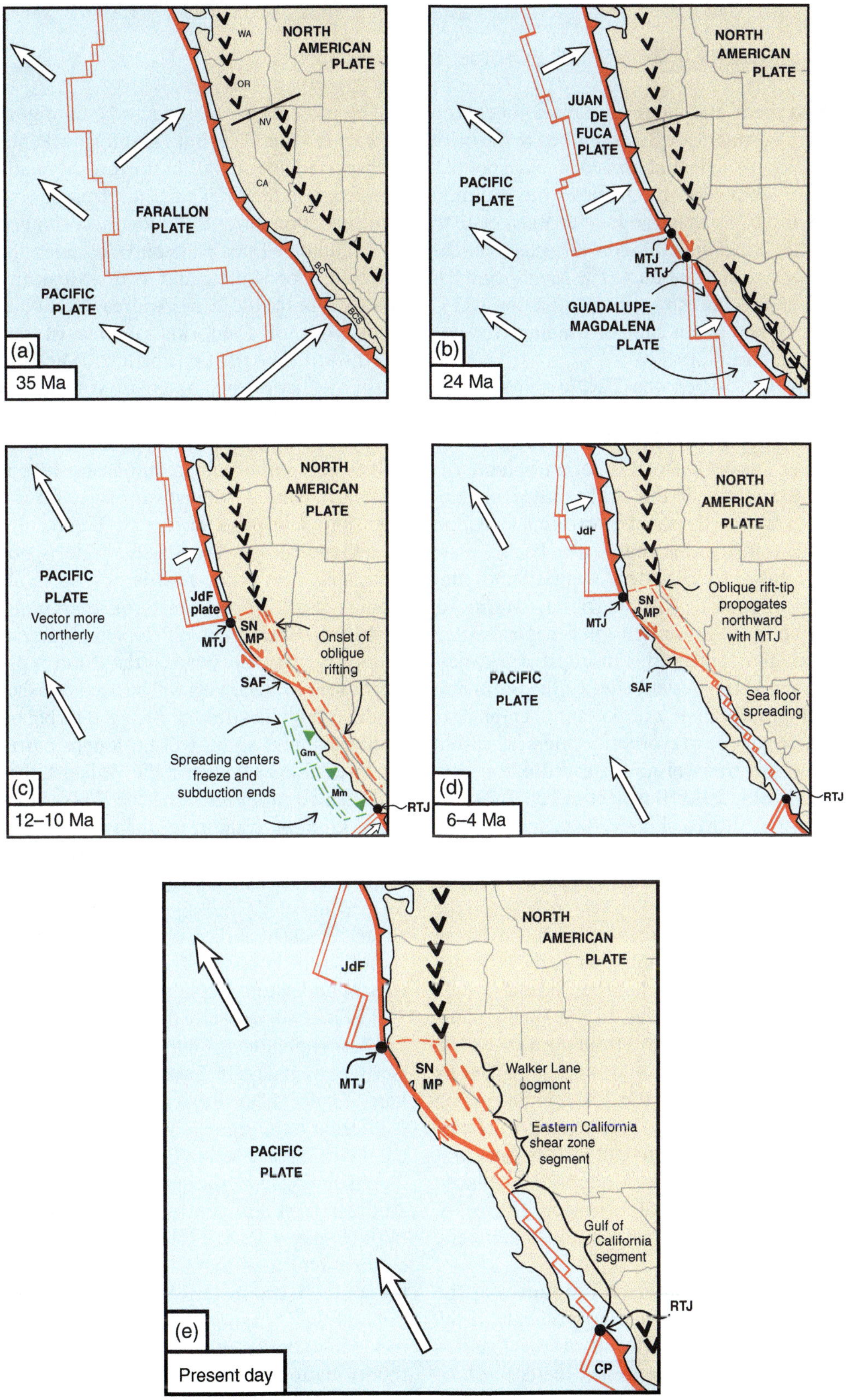

Figure 4.3 Plate tectonic setting for the Walker Lane–Gulf of California oblique rift. Active trenches, transform faults, and rifts/spreading centers shown in red; green dashed lines indicate inactive trench, transform faults, and spreading centers. Details of the tectonic evolution are provided in the text. White arrows represent oceanic plate motion vectors. Black V's represent the position of the subduction-related arc. In time frames (a) and (b), the arc takes an abrupt left jog at the Nevada-Oregon border, inferred to represent a slab tear (see text). Rift segments shown in (e) are described in the text (Walker Lane, Eastern California Shear Zone, and Gulf of California). CP, Cocos plate; Gm, Guadalupe microplate; JdF, Juan de Fuca microplate; MTJ, Mendocino triple junction; SNMP, Sierra Nevada microplate, RTJ, Rivera triple junction. States are labeled on (a): in United States: AZ, Arizona; Ca, California; OR, Oregon; NV, Nevada; WA, Washington; in Mexico: BC, Baja California; BCS, Baja California Sur; Mm, Magdalena microplate.

Subduction ceased there, and the relative motion between the Pacific and North American plates resulted in the birth of the San Andreas fault, a dextral transform plate boundary. By this ridge subduction mechanism, the Farallon plate was broken into two plates: the Juan de Fuca plate to the north and the Guadalupe-Magdalena plate to the south, and both continued to subduct. The Mendocino triple junction (MTJ) and the Rivera triple junction (RTJ) formed at the north and south ends of the San Andreas fault, respectively (Figure 4.3b).

At 12–10 Ma (Figure 4.3c), the Pacific plate vector shifted to a more northerly direction (Atwater & Stock, 1998). We propose that this shift resulted in the birth of the Walker Lane–Gulf of California oblique rift (Busby, 2013). In the continent, this shift caused deformation to switch from E-W orthogonal extension (Basin and Range) to NW-SE transtension; that is, the Pacific plate began to drag the entire southwestern United States and western Mexico northward (McQuarrie & Wernicke, 2005). However, the NW-SE transtension became focused within the thermally weakened Ancestral Cascades-Comondú arc system at the western edge of the Basin and Range province (Figure 4.3c). It is important to emphasize here that the earliest oblique rift volcanic centers lay within the arc system, where transtension replaced Basin and Range extension (Busby, 2013; Busby et al., 2018, 2020). Also at this time (12–10 Ma, Figure 4.3c), subduction ended abruptly along the length of the western Baja California margin, perhaps due to slab breakoff (Michaud et al., 2006). This caused the spreading centers between the Guadalupe-Magdalena plate and the Pacific plate to freeze off the west coast of Baja California; that is, they did not subduct (Figure 4.3c). By this mechanism, the entire Baja California Peninsula was transferred to the Pacific plate, and the plate margin jumped inboard from the dead trench to the Comondú arc axis. This jump was accompanied by a 1,000-km-southward jump of the RTJ (Figure 4.3c). In contrast, in California, the strain became partitioned between the San Andreas transform fault and the Ancestral Cascades arc, where Walker Lane oblique rifting replaced the Basin and Range orthogonal extension (Busby & Putirka, 2009). Thus, the Sierra Nevada microplate (SNMP) was born (Figure 4.3c).

At 6–4 Ma (Figure 4.3d), continental rupture and the onset of seafloor spreading occurred very quickly in the Gulf of California because nearly all of the Pacific–North American strain was concentrated there. Seafloor spreading began first in the central part of the Gulf of California at 6 Ma (Guaymas Basin, described further later in this chapter) and then in the northern and southern parts by 4.5 Ma (Lizarralde et al., 2007). Meanwhile, in California, the San Andreas transform fault continued to lengthen northward, progressively shutting off subduction northward, and the SNMP lengthened due to northward propagation of the Walker Lane rift tip.

The present-day plate tectonic configuration is shown in Figure 4.3e. The Baja California Peninsula has been transported about 280 km northward relative to mainland Mexico, and the Gulf of California has been widened by seafloor spreading. Subduction continues north of the MTJ (Juan de Fuca plate and Cascades arc) and south of the RTJ (Cocos plate and Trans-Mexican volcanic belt), whereas both the San Andreas fault and the Walker Lane–Gulf of California oblique rift have lengthened northward. The rift tip (north end) now lies beneath and in the vicinity of the Lassen arc volcanic center (see references in Busby, 2013; Kattenhorn et al., 2016; Waldien et al., 2019; Figure 4.4).

It is very important to emphasize here that the Pacific plate is moving northwestward away from North America (see the white plate vectors in Figures 4.3c–e), resulting in oblique rifting. The Pacific plate is not merely being translated northward relative to North America, which would result in a transform margin. In future, the SNMP will be rifted off North America, just as Baja California was at 6 Ma ("the future plate boundary" mentioned earlier). This will be accomplished when the San Andreas fault dies due to locking at its big bend in southern California, and strain will no longer partitioned between the San Andreas fault and the Walker Lane. The continent will then be ruptured along the Walker Lane. The Walker Lane–Gulf of California oblique rift can in no way be regarded as a type of transform margin (e.g., leaky transform or transtensional transform). It is a true rift, but one formed under dextral shear rather than pure orthogonal shear. As stated earlier, oblique rifts are much more common than orthogonal rifts in the geologic record, but they remain understudied and poorly recognized.

Previous workers have divided the Walker Lane–Gulf of California oblique rift into segments using varying criteria (see the discussion in Busby, 2013), so we define our usage here (Figure 4.3e), from south to north: (a) the Gulf of California is the segment with seafloor spreading (stage 1); (b) the ECSZ lies within the continent north of that (stage 2), emanating from the east end of the big bend of the San Andreas fault and continuing northward to the Garlock fault (Figure 4.4); and (c) the Walker Lane lies within the continent north of that (stage 2), from the Garlock fault to the southern end of the Cascades arc (Figure 4.4).

The ECSZ (Figures 4.3e, 4.4) is structurally very complex due to deformation events that immediately preceded the formation of the oblique rift, as well as the reactivation of much older structures, and it has been described in detail in papers by Dorsey et al. (2021) and Mavor et al. (2023). For these reasons, the ECSZ will not be described further in this chapter. Suffice it to say that the onset of dextral transtension in the ECSZ is estimated at 13–9 Ma (Mavor et al., 2023), which is supported by U-Pb geochronology on synkinematic opal (Nuriel et al., 2019) and synkinematic calcite (Mavor et al., 2023),

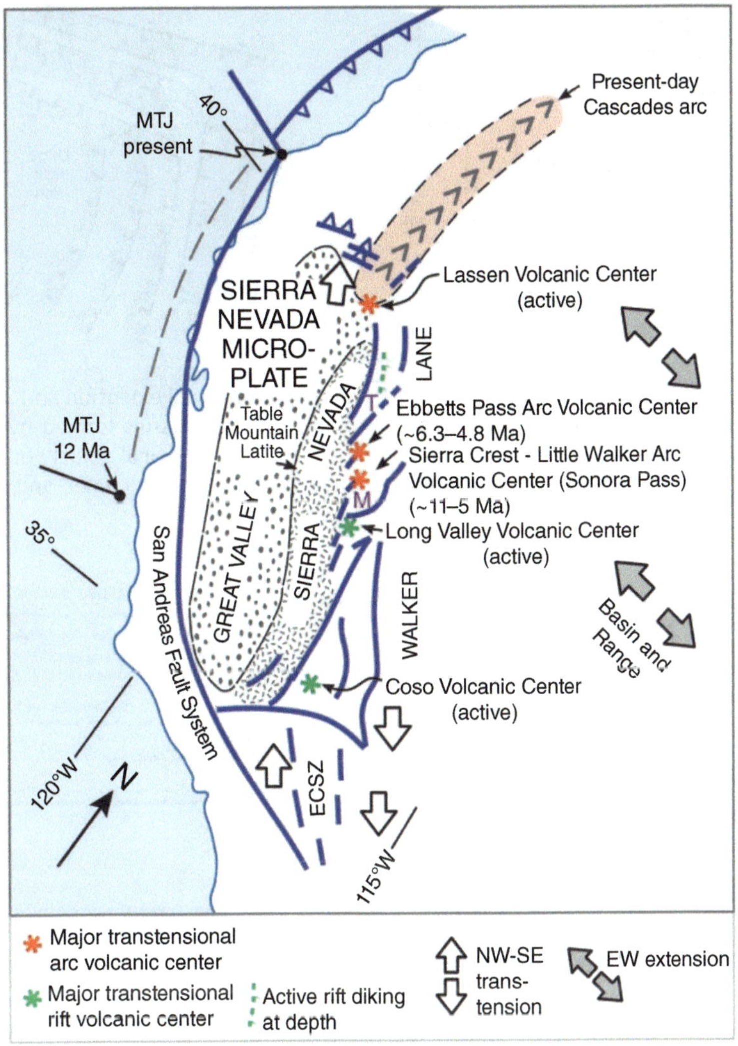

Figure 4.4 Late Miocene to Holocene transtensional arc and rift volcanism (12–0 Ma) along the trailing edge of the SNMP, referred to as the Walker Lane (from Busby, 2013 © The Geological Society of America, used with permission). Shown in oblique projection of the western Cordillera about the preferred Sierra Nevada–North American Euler pole (following the perspective of Unruh et al., 2003). For references to seafloor reconstructions, see Busby (2013). ECSZ, eastern California shear zone; MTJ, Mendocino triple junction. Purple letters: M, Mina deflection, T, Lake Tahoe. The Sierra Crest–Little Walker arc volcanic center and the Ebbetts Pass arc volcanic center are shown in map view in Figure 4.6.

so it is widely regarded as a segment of the oblique rift. In this chapter, we focus on the Walker Lane as the archetype of the first stage of oblique rifting.

4.3. WALKER LANE—THE FIRST STAGE: OBLIQUE CONTINENTAL RIFTING

The Walker Lane (Figure 4.4) offers an exceptional opportunity to study an oblique continental rift in its prebreakup stage, with superior exposures and very easy access. The Walker Lane is a ~100 km-wide fault zone that lies between the Sierra Nevada and the Basin and Range, and it parallels the San Andreas fault to the west. Walker Lane faults differ from Basin and Range faults in several fundamental ways. Walker Lane faults are dominantly dextral strike-slip and dextral-oblique normal faults that trend NW to NNW (Faulds et al., 2005; Kreemer et al., 2009; Jayko & Bursik, 2012); these lie at a high angle to the N-S to NNE-SSW faults of the Basin and Range (Figure 4.4). The Walker Lane also contrasts markedly with the Basin and Range by having NE- to ENE-striking sinistral-oblique faults that step right from the NW to NNW dextral-oblique faults; the most prominent example of this is the Mina deflection in the central Walker Lane (M in Figure 4.4). These NE- to ENE-striking sinistral-oblique normal faults occur at all scales (e.g., see Figures 4.5 and 4.6), forming right-releasing stepovers.

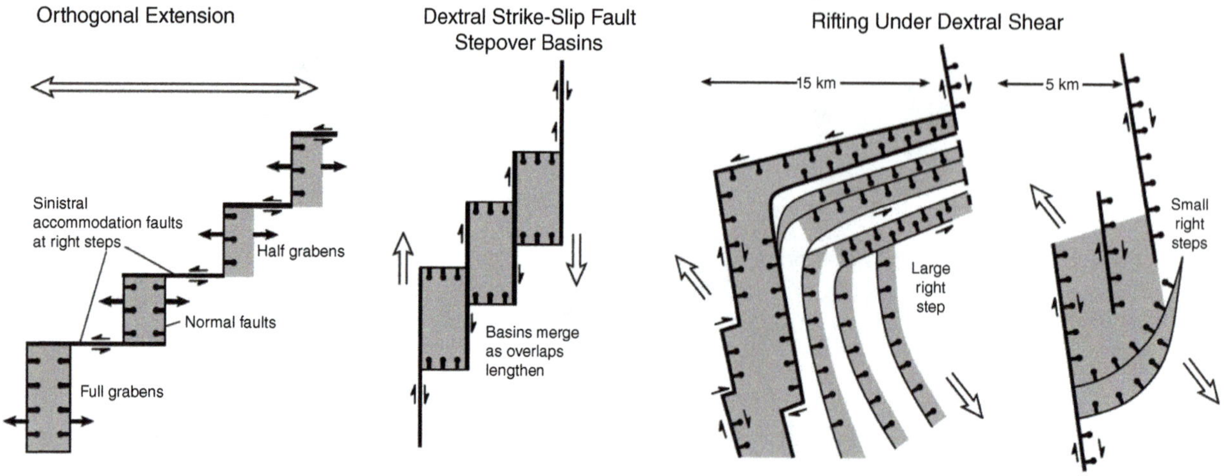

Figure 4.5 Simplified depiction of faults and basins formed under orthogonal extension and under right-stepping dextral strike-slip faults (left two panels) for comparison with faults and basins formed by transtension/rifting under dextral shear (right two panels), from Busby et al. (2018); © The Geological Society of America, used with permission. A large right step produced the Sierra Crest–Little Walker volcanic center, and a smaller right step produced the Ebbetts Pass volcanic center, both shown on Figure 4.6.

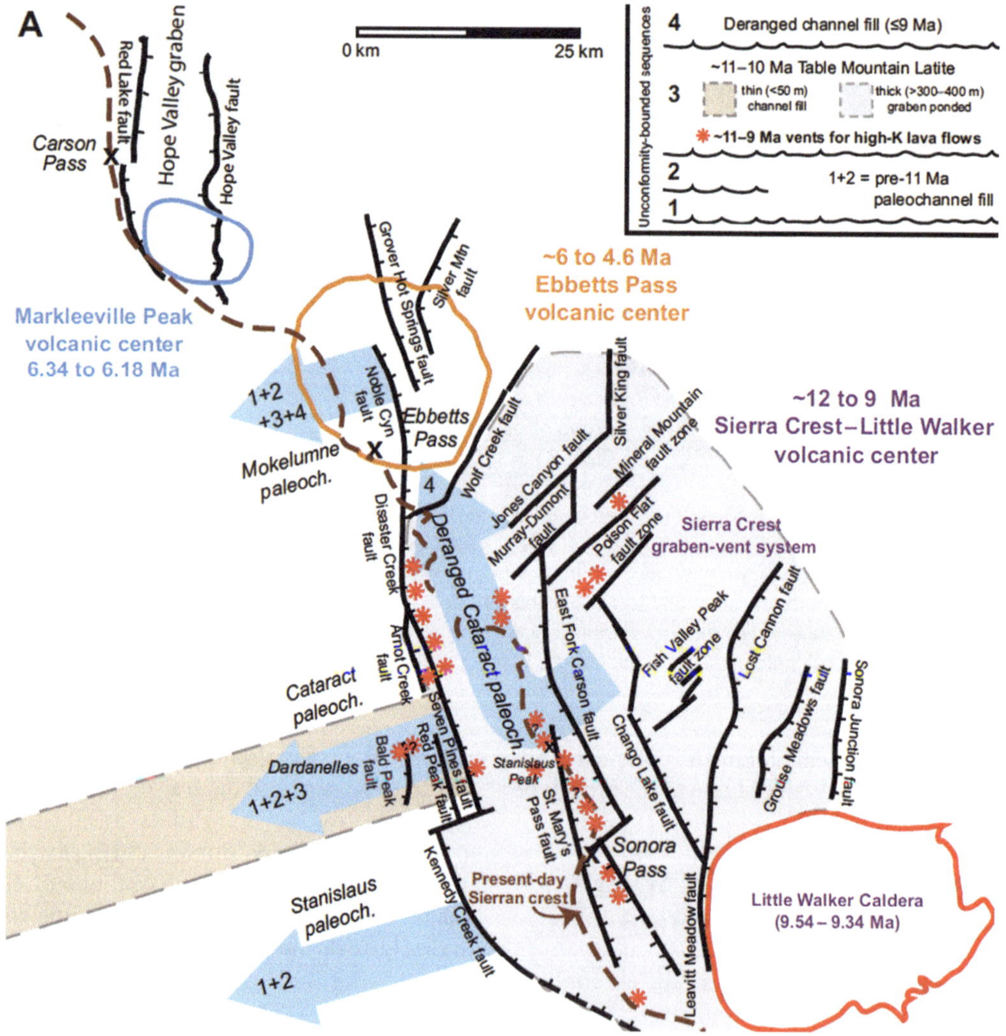

Figure 4.6 Case study of oblique rift basins in the Walker Lane (Busby et al., 2018). Tectonic and volcanic setting of the Sonora–Ebbetts–Carson Pass region (Figure 4.4), from Busby (2013) and Busby et al. (2013a, 2013b, 2016, 2018). Locations of Ebbetts Pass and Sierra Crest–Little Walker arc volcanic centers are in Figure 4.4. All of the NNW faults are dextral-normal oblique faults, and all of the NE faults are sinistral-normal oblique faults, as shown in Figure 4.5 for rifting under dextral shear. The Sierra Crest–Little Walker volcanic center formed at a large (releasing) right step of 15 km, and the Ebbetts Pass volcanic center formed at a smaller right step of 5 km, as shown in Figure 4.5. Transtension culminated in the formation of the Little Walker caldera in a structural setting similar to the Pleistocene Long Valley caldera (see text).

Busby (2013) proposed that all of the large volcanic centers in the Walker Lane, both active and extinct, are sited on major right-releasing stepovers or bends (Figure 4.4). Today, and for the past 12 Ma, the northern tip of the Walker Lane oblique rift has exploited the southern end of the Cascades arc, propagating northward with time. Today, the rift tip lies beneath the active Lassen arc volcanic center, and active volcanic centers south of Lassen along Walker Lane consist of continental rift centers that have formed in the wake of the migrating triple junction (Figure 4.4). Thus, rift magmas may erupt at any point along the Walker Lane south of Lassen. In contrast, the extinct Walker Lane arc volcanic centers south of Lassen are progressively older, moving southward in the Walker Lane oblique rift (Figure 4.4), and they have progressively greater structural levels of exposure due to ongoing rifting. This provides the opportunity to study the structural controls on their plumbing systems, vents, and basin fills on a timescale of millions of years (Busby & Putirka, 2009; Busby et al., 2018).

The largest rift volcanic centers in the Walker Lane oblique rift lie at major right-releasing stepovers (Busby, 2013; green stars in Figure 4.4): the Long Valley volcanic field (Bursik, 2009; Riley et al., 2012) and the Coso volcanic field (Pluhar et al., 2006). As noted by Busby et al. (2013a, 2013b), the structure, size, and volcanology of the active Long Valley rift volcanic center is very similar to that of the 12–9 Ma Sierra Crest–Little Walker arc volcanic center (Figure 4.6), described further later in the chapter, with both centers culminating in the formation of giant continental calderas. Both occur at major right-releasing stepovers. The Coso rift volcanic center also lies at a right-releasing stepover in a dextral fault zone, atop sinistral-oblique normal faults that bound rotating blocks (Pluhar et al., 2006). The Coso volcanic center is still undergoing tectonic subsidence (Hammond et al., 2012). In addition, active emplacement of rift dikes (green dashes in Figure 4.4) is currently taking place in the region between Lake Tahoe and the Lassen volcanic center (Smith et al., 2016).

The largest extinct arc volcanic centers in the Walker Lane oblique rift also lie at major right-releasing stepovers (Busby, 2013; red stars in Figure 4.4). These include the Sierra Crest–Little Walker arc volcanic center and the Ebbetts Pass arc volcanic center in the central Walker Lane. These are described further later in the chapter (Figures 4.4, 4.5).

We have focused our studies on the central Walker Lane because it has more abundant volcanic rocks than the southern or northern Walker Lane; volcanic rocks are more abundant because rifting is more oblique there, so there is greater extension (Busby, 2013). The northern and southern segments of the Walker Lane are dominated by translation along long NNW dextral strike-slip faults (Figure 4.4). In contrast, the central segment shows evidence of distributed dextral shear accommodated by shorter, NNW dextral and ESE sinistral strike-slip fault systems, with more abundant oblique-normal faults, including NE-striking faults, as well as vertical-axis block rotations (Figures 4.1, 4.2a; Unruh et al., 2003; Wesnousky, 2005; Surpless, 2008; Wesnousky et al., 2012; Bormann et al., 2016). The abundant volcanic rocks in the central Walker Lane make it ideal for detailed tectonic reconstructions because (a) the timing of fault slip can be constrained by dating volcanic strata using the $^{40}Ar/^{39}Ar$ method (e.g., Putirka & Busby, 2007; Busby et al., 2008a, 2008b, 2013a, 2013b, 2016, 2018; Hagan et al., 2009; Nagorsen-Rinke et al., 2013; Fleck et al., 2015); and (b) paleomagnetic studies can be used to determine the importance of clockwise vertical-axis block rotations due to dextral shear, and the relationships of such rotations to well-exposed faults (Carlson et al., 2013; Carlson, 2017).

We emphasize that magmatism in the oblique Walker Lane rift is focused into large transtensional stepovers but is not the direct product of localized transtension. The magmatism is a regional manifestation of subduction (for the arc volcanic centers) and asthenospheric upwelling (for the rift volcanic centers). However, we infer that structures of the large stepovers must penetrate the lithosphere to focus regional magmatism into localized basins. This carries the implication that the large Walker Lane volcanic centers are controlled by translithospheric faults, not faults limited to the upper crust, and that translithospheric faults formed by 12 Ma in the central Walker Lane, at the Sierra Crest–Little Walker arc volcanic center (described later in the chapter). This contrasts with the Death Valley region in the southern Walker Lane, where it is inferred that dextral transtensive faulting was restricted to the upper crust from 13–8 Ma and did not become translithospheric until 8 Ma (Lutz et al., 2022). In the Death Valley case, the upper crust was initially decoupled from the lower crust due to previous localized, extreme crustal thickening caused by the subduction of an oceanic plateau in Late Cretaceous time, which led to the development of a weak middle crust. Then the weak middle crust became strengthened between 13 and 8 Ma by thinning, cooling, and exhumation, achieved by transtensional detachment faulting. Only then could transtension became translithospheric, from 8 Ma to the present (Lutz et al., 2022). This model does not apply to the central Walker Lane, where there is no evidence for extreme crustal thickening.

4.3.1. Structural Styles of Oblique/Trantensional Rifting Under Dextral Shear

A simple set of diagrams highlights the ways in which oblique/transtensional rift basins may differ from orthogonal rift basins or from pull-apart basins in a purely strike-slip system (Figure 4.5). Orthogonal extension (e.g., Basin and Range) produces normal faults that bound grabens, with localized strike-slip faults accommodating jogs in the rift system. A purely strike-slip system may produce local extension at a releasing stepover with localized normal faults. Rifting under dextral shear is illustrated for two possibilities: a large right step and a smaller right step (Figure 4.5). In both, obliquity between the Sierra Nevada and North American microplate motions (large open arrows) and the orientation of faults resulted in NNW normal-dextral-oblique faults. In the large Sierra Crest–Little Walker oblique rift basin (Figure 4.5), a major right step (15 km) was accommodated by a system of approximately NE normal-sinistral oblique faults. In contrast, in the smaller Ebbetts Pass oblique rift basin, a series of small right steps along approximately NNW normal-dextral faults produced a narrower pull-apart structure (5 km wide), with normal faults along its trailing edge (Figure 4.5). These central Walker Lane oblique rift basins exploited the south end of the thermally weakened Ancestral Cascades arc and propagated northward with time as the rift tip grew northward in concert with the northward migration of the MTJ from the Sierra Crest–Little Walker volcanic center to the Ebbetts Pass volcanic center (Figure 4.6).

4.3.2. Case Study: Central Walker Lane

The onset of plate-margin-scale Walker Lane transtension was signaled by the development of an unusually large and voluminous transtensional arc volcanic center, the ca. 12–9 Ma Sierra Crest–Little Walker arc volcanic center (Figure 4.6). This Miocene arc volcanic center is as aerially extensive (ca. 4,000 km^3) as the Pliocene to Holocene Long Valley transtensional rift volcanic field and, as mentioned earlier, has a very similar structure. Extreme transtension in the very large Sierra Crest–Little Walker pull-apart triggered rapid ascent of low-degree partial melts, causing an outpouring of high-K_2O lava "flood andesites" (Table Mountain Latite; Figure 4.6). These were erupted from fissure vents that resemble flood basalt vents (lines of red stars in Figure 4.6), the melts being transported up NNW dextral-oblique normal faults that bound the largest graben, which now lies along the Sierra Nevada range crest (brown dashes in Figure 4.6). Smaller, point-source vents (single red stars in Figure 4.6) were plumbed through faults bounding smaller NNW half grabens, as well as a series of NE-SW sinistral-oblique normal faults that define a major transfer zone. Reconstruction of fault-slip histories shows that the area of maximum transtension occurred in the southeast and ultimately accommodated the accumulation of high-K_2O silicic magmas at shallow crustal levels, producing the Little Walker caldera, in a structural position similar to that of the Long Valley caldera (Busby et al., 2013b). Three high-K_2O ignimbrites erupted from the Little Walker caldera (Eureka Valley Tuff, 9.54–9.34 Ma) and are laterally extensive, making excellent strain markers (Busby et al., 2018). Also shown are three ancient Sierran paleochannels (blue arrows in Figure 4.6) that were progressively disrupted, from south to north, as the rift tip propagated northward with time (Busby et al., 2013a). Sierran paleochannels are west-flowing ancestral river channels that were carved on the shoulder of a broad Cretaceous-Paleocene uplift, termed the Nevadaplano (DeCelles, 2004), with its crest in central Nevada. These paleochannels funneled volcanic and volcaniclastic rocks westward from Nevada across the Sierra Nevada in Oligocene time (sequence 1, Figure 4.6) and early Miocene time (sequence 2, Figure 4.6). These included ignimbrites, lavas, block-and-ash-flows, volcanic debris flows, and fluvial volcaniclastics. The Stanislaus paleochannel in the south was beheaded first (Busby et al., 2016; Figure 4.6). This was followed by the derangement of the Cataract paleochannel into the newly developed NNW structural grain ("deranged Cataract paleochannel," Figure 4.6), which then fed into the Mokelumne paleochannel; the latter remained connected to sources in Nevada because transtension had not yet begun at Ebbetts Pass (sequence 4, Figure 4.6). Magmatism ceased at 9.1 Ma in the southern half of the Sierra Crest–Little Walker volcanic center, but small volumes continued to erupt in the northern half until 5 Ma, overlapping with the onset of magmatism and transtensional faulting in the Ebbetts Pass arc volcanic center to the north at 6.4 Ma (Busby et al., 2018).

The Ebbetts Pass arc volcanic center and oblique rift basin is smaller than the Sierra Crest–Little Walker (Figure 4.6) but nonetheless has an estimated original volume (~270 km^3) comparable to the estimated volume for the Lassen transtensional arc volcanic center. The Lassen Volcanic Center formed in the last 825 k.y., and its volume is estimated at 200 km^3 by Hildreth (2007). The smaller size of the Ebbetts Pass transtensional rift basin, relative to the Sierra Crest–Little Walker transtensional rift basin, is consistent with the observations that it has only one basalt, few high-K_2O rocks, and a much greater proportion of dacite; it also erupted rhyolites, which are not present in the Sierra Crest–Little Walker transtensional rift basin. We attribute this to a higher degree of crustal contamination due to the slower ascent of magmas (Manzo et al., 2023; Manzo, 2023).

The Ebbetts Pass volcanic center began as small-volume lavas that began to leak from incipient pull-apart basin faults at 6.4 Ma and was then overfilled by the growth of a stratovolcano from ca. 5–4.5 Ma, which buried all the basin-margin faults (Busby et al., 2018). The Ebbetts Pass stratovolcano is young enough to partially retain the features of a large central cone despite erosion by Quaternary glaciers (see Figure 10a in Busby et al., 2018). The growth of the stratovolcano beheaded the Mokelumne paleochannel, whose fill is capped by volcanic debris flow deposits (sequence 4, Figure 4.6) dated at 6 Ma by U-Pb detrital zircon geochronology (Hodges et al., 2025).

Around the same time that the Ebbetts Pass oblique rift basin formed, a small full graben formed nearby to the northwest (Hope Valley graben), and a small volcanic center was focused into the graben (Markleeville Peak volcanic center, Figure 4.6). The graben-bounding faults do not have an oblique component of slip, as shown by pure normal offset of a paleochannel system that crosses them (Hagan et al., 2009). However, strain is partitioned between normal faults and strike-slip faults at many localities in the Walker Lane (Surpless, 2008). Examples include the Tahoe Basin (Kent et al., 2005; Schweickert et al., 2019), the Gardnerville Basin (Cashman et al., 2009), and the Owens Valley strike-slip fault and adjacent southern Sierra Nevada range front normal fault (Jayko & Bursik, 2012; Haddon et al., 2016). However, we wish to emphasize that all of the large volcanic centers in the Walker Lane that have been identified so far, both active and extinct, are sited on major right-releasing stepovers or bends.

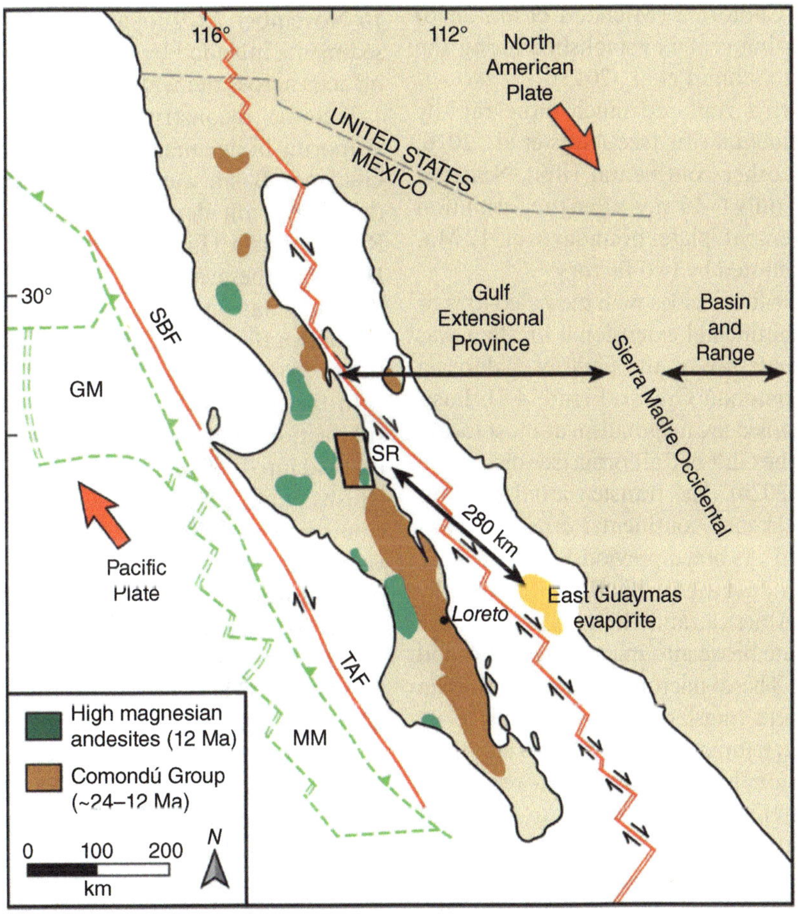

Figure 4.7 Simplified map of Late Cenozoic rocks of the Baja California Peninsula and seafloor structures in the Gulf of California and Pacific margins to the east and west, respectively (data sources given in Busby et al., 2020). Red single line indicates active fault; red double line indicates active spreading center; green dashed lines indicate inactive spreading centers, transforms, and trench. Large red arrows indicate present-day relative motion between Pacific and North American plates; oceanic spreading centers are present along the entire length of the rift (Schmitt et al., 2013). The Santa Rosalia Basin (SR) represents the on-land part of the Guaymas Basin, offset by 280 km. Evaporites exposed on land in the Santa Rosalía Basin (not shown separately) were correlated with evaporites in the subsurface of the Guaymas Basin (shown in yellow) by Miller and Lizarralde (2013). A schematic summary of the Santa Rosalía Basin is shown in Figure 4.8. The Comondú Group (shown in brown) includes 24–12 Ma arc volcanic rocks as well as younger rift volcanic rocks (Figure 4.8). The high-magnesian andesites (shown in green) are post-subduction volcanic rocks interpreted to represent slab window magmatism (see text). GM, Guadalupe microplate; MM, Magdalena microplate; TAF, Tosco-Abreojos fault; SBF, San Benito fault. From Busby et al. (2020) and Alves et al. (2020).

4.4. GULF OF CALIFORNIA—THE SECOND STAGE: SEAFLOOR SPREADING AND OBLIQUE CONTINENTAL MARGIN

The Gulf of California (Figure 4.7) is an archetype of continental rupture through transtensional/oblique rifting. It forms a narrow sea that lies between the Baja California Peninsula and the western margin of mainland Mexico, consisting of several short spreading segments connected by transform faults that together represent the northern extent of the East Pacific Rise plate boundary. Capture of the Baja California microplate by the Pacific plate is essentially complete, with ~90% of Pacific–North American plate motion in that region now accommodated by the Gulf of California (Michaud et al., 2006) and the oceanic spreading centers established along the entire length of the rift (Schmitt et al., 2013).

The Gulf of California ruptured much more rapidly than is typical of continental rifts (see Alves et al., 2020, for comparisons with other continental rifts). Seafloor spreading commenced only 6–10 m.y. after the formation of the oblique- convergent plate boundary at 12 Ma. Rapid rupture was promoted by two factors:

1. The transtensional rift coincides with the paleo-axis of a subduction-related continental extensional arc that was centered over the ultimate site of the Gulf of California from ca. 24–12 Ma (Comondú Group; Figure 4.7). East-west extension accompanied arc magmatism at most localities along the length of the Gulf of California (see discussion in Busby et al., 2020, 2024). The transtensional rift thus exploited thermally weakened continental crust that was already partially thinned. As noted previously (Figure 4.3), subduction ended when the East Pacific Rise was partially overridden by North America, and the remnants of the convergent Farallon plate broke into microplates (GM and MM in Figure 4.7). These microplates and the Baja California Peninsula were transferred to the Pacific plate when the plate boundary jumped eastward into the thermally weakened extensional arc axis, and east-west intra-arc extension was replaced by northwest-southeast extension at the trailing edge of the Pacific plate.

2. Pull-apart basins are the most efficient tectonic setting for localizing rapid thinning of the crust (Christie-Blick & Biddle, 1985; Pitman & Andrews, 1985; Nilsen & Sylvester, 1995; Allen & Allen, 2005; Xie & Heller, 2009). In the Gulf of California, this has produced several short spreading segments separated by transform faults (Figure 4.7).

4.4.1. Case Study: Guaymas–Santa Rosalía

We focus here on the earliest ruptured segment of the Gulf of California transtensional rift, the Guaymas–Santa Rosalía segment in the center of the Gulf (Figures 4.7 and 4.8) because that is the segment with the most combined offshore and onshore data. The Guaymas Rift is a magmatically robust rift that achieved continental breakup 3–4 m.y. before transtensional pull-apart basins to the north and south ruptured the continental crust (Lizarralde et al., 2007). It produced about an 8-km-thick igneous crust, compared to mid-ocean ridge production or basins in the rest of the Gulf of California, which are 6 km in thickness (Lizarralde et al., 2007). The Guaymas Basin has been the focus of a Deep Sea Drilling Project (DSDP) expedition, as well as wide-angle and multichannel seismic surveys (Curray & Moore, 1982; Lizarralde et al., 2007; Miller & Lizarralde, 2013). The International Ocean Discovery (IODP) expedition from 16 September to 16 November 2019 focused on Late Pleistocene organic sediments intruded by magmatic sills that lie up to 40 km off axis across the width of the basin (Teske et al., 2018).

Tectonic reconstructions show that the 2-km-thick evaporite body imaged by multichannel seismic in the Guaymas Basin was fringed by thinner (up to ca. 70 m thick) gypsum deposits exposed onshore in the Santa Rosalía Basin (Figure 4.7; Miller & Lizarralde, 2013). However, the evaporite and underlying crustal rocks in the Guaymas Basin have not been sampled or dated. The Santa Rosalía area (Figures 4.7 and 4.8) offers the opportunity to do this.

In the Santa Rosalía area (Figure 4.8), ca. 13–10 Ma subduction-related volcanic rocks pass conformably upward into 9.69–8.84 Ma rift transitional volcanic rocks (Busby et al., 2020). Clastic sedimentary rocks are absent from the early rift stage because explosive and effusive volcanism flooded the basin. Like other parts of the Gulf of California, the onset of oblique continental rifting at Santa Rosalía is marked by deposits of large-volume silicic ignimbrite eruptions (rift transitional ignimbrites; Figure 4.8). These are overlain by rift transitional lavas, which were ponded in grabens. The rift transitional volcanic rocks were then rotated eastward, presumably toward a north-south fault that now lies below sea level in the Gulf of California to the east. This is consistent with the fact that the shape of the east Guaymas evaporite indicates that the basin it formed in was bounded by N-S normal faults that were active before seafloor spreading began (Miller & Lizarralde, 2013). Tectonic reconstructions show that the 2-km-thick evaporite body imaged by multichannel seismic in the Guaymas Basin was fringed by thinner (up to ca. 70 m thick) gypsum deposits exposed onshore in the Santa Rosalía Basin (Miller and Lizarralde, 2013). However, the evaporite and underlying crustal rocks in the Guaymas Basin have not been sampled or dated. The Santa Rosalía area offers the opportunity to do this.

The breakup sequence of the Santa Rosalía Basin (Boleo Formation; Figure 4.8) consists of a basal marine

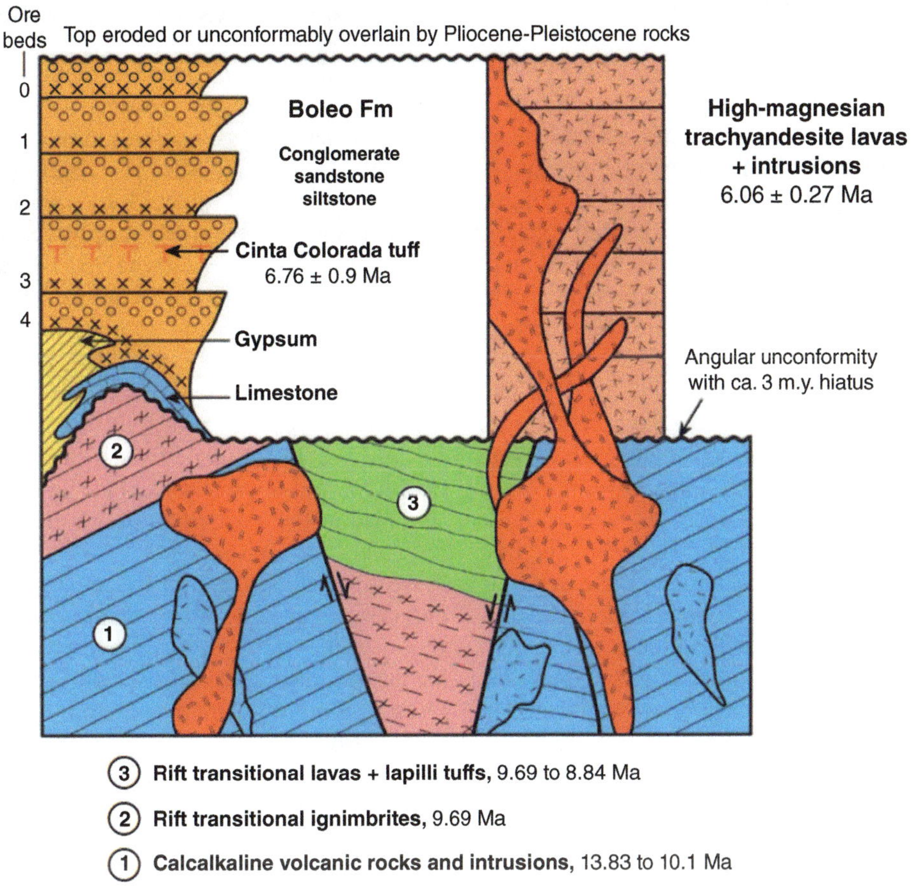

Figure 4.8 Schematic representation of stratigraphic, structural, intrusive, and geochronological relations in the Santa Rosalía area (location in Figure 4.7). This is based on detailed mapping presented in Figures 3 and 4 of Busby et al. (2020). Cinta Colorada tuff age from Holt et al. (2000); all other ages are from Busby et al. (2020). Calcalkaline (arc) volcanic rocks (1) and rift transitional ignimbrite (2) and lavas (3) ranging from ca. 14–9 Ma were tilted about normal faults and eroded before deposition of the 6 Ma Boleo Formation continental breakup sequence in a transtensional rift basin. Also shown are high-magnesian andesites that erupted and intruded at the south margin of the Boleo Basin at 6 Ma, attributed to slab window magmatism (see text). From Busby et al. (2020).

limestone overlain by evaporites. This in turn is overlain by a series of clastic fan-delta progradational sequences sourced locally by erosion of andesitic volcanic rocks, concurrent with the deposition of evaporites (Niemi & Salgado Muñoz, 2019; Dorsey et al., 2022). Copper-Co-Zn ore beds occur at the base of each cycle and are inferred to have formed from brine pools in a marginal marine environment, analogous to the Red Sea (Conly et al., 2006, 2011). High-magnesian andesite intrusions in the Santa Rosalía area are approximately the same age as the Boleo Formation and may have provided some of the heat engine and fluids for Boleo mineralization; more likely, however, the onset of seafloor spreading at ca. 6 Ma in the Guaymas Basin provided the source of the hydrothermal brines (Busby et al., 2020). The seismic structure beneath Baja California shows a distinct low-velocity (i.e., hot) zone crossing the Baja California Peninsula (Di Luccio et al., 2014) at the latitude of the Santa Rosalía–Guaymas rift. This lies above the boundary between the Guadalupe and Magdalena slab remnants (Figure 4.7), and the slab appears to bow out toward the paleotrench at this locality, suggesting that there is a long slab window there. The high-magnesian andesites in the Santa Rosalía area are further inboard from the trench than any others found so far in Baja California (Figure 4.7), perhaps due to the presence of a large slab tear in this area. Perhaps this also contributed to anomalous high magma production demonstrated by Lizarralde et al. (2007) for the Guaymas Basin and its early rifting relative to other basins in the Gulf of California (Busby et al., 2020).

In summary, the Gulf of California offers a view of a young "Intraplate Continental Margin" (Alves et al., 2020), where plate reconstructions are very well constrained.

4.5. CONCLUSIONS

We have focused our studies on segments of the Walker Lane–Gulf of California oblique rift that have abundant volcanic rocks because $^{40}Ar/^{39}Ar$ dates on them, in conjunction with detailed mapping and geochemical work, can be used to construct tectonic models for oblique rifting. Furthermore, extinct uplifted and eroded volcanic centers provide a view of the structural controls on magmatic plumbing systems and eruptive products that cannot be discerned at active volcanoes, where these features are deeply buried.

The Walker Lane–Gulf of California oblique rift offers the opportunity to compare first-stage oblique continental rifting and the processes that thin continental lithosphere (Walker Lane) with second-stage oblique continental rifting, where the continent has ruptured and seafloor spreading is underway (Gulf of California). For the first stage, we have shown that large Walker Lane volcanic centers are located on major releasing fault stepovers. The Walker Lane oblique rift plate boundary was born at a large stepover, with an outpouring of large-volume "flood andesites" erupted from fissure vents up to 12 km long. For the second stage, we compare the onshore record of oblique rift initiation (Boleo Basin, Santa Rosalía) with the offshore record of the onset of seafloor spreading (the adjacent Guaymas Basin, where the seafloor spreading in the Gulf of California began). Although the Guaymas Basin has been imaged by wide-angle and multichannel seismic surveys and targeted by ODP and IODP drilling expeditions, its deeper levels have not been sampled, thus making its onshore equivalent (Boleo Basin) valuable for understanding the process of continental breakup. The Boleo Basin is floored by Late Miocene continental arc and rift transitional volcanic rocks deposited in extensional basins. These record thinning and thermal weakening of the continental lithosphere prior to the onset of oblique rifting. The overlying Boleo Basin has a breakup sequence, with basal marine limestone overlain by gypsum and delta progradational sequences with Cu-Co-Zn ore beds. The ore beds formed from brine pools that formed during the onset of seafloor spreading in the Guaymas Basin at 6 Ma.

Further studies are needed in the Gulf of California oblique rift segment to compare it with the Walker Lane oblique rift segment, which is much better studied. In Baja California Sur, there remain vast tracts of undivided volcanic rock that have not been mapped in detail or dated or studied geochemically. The exposures are outstanding and can be accessed on foot, by four-wheel drive vehicle, or by boat, and are nearly all on public land. We hope this summary chapter will introduce and promote further research on this globally significant example of an accessible, spectacularly well-exposed oblique rift.

ACKNOWLEDGMENTS

We thank formal reviewers Kathleen Marsaglia and Tony Doré and an anonymous reviewer for their very helpful reviews. This contribution was supported by NSF-EAR-1917361 and NSF EAR 1921182 (to Busby and Putirka) for the Walker Lane work. The Gulf of California work was largely supported by the Baja Basins National Science Foundation (NSF) Research Experience for Undergraduates (REU) grant 151120 to principal investigator Tina Niemi and co-principal investigator Cathy Busby and a 2017–2018 UC Mexus-Conacyt Collaborative Research Grant to Cathy Busby and Margarita López-Martínez. Partial support for the Gulf of California work came from NSF grant EAR-1545676 (to Busby and Susan DeBari) for geochronology and for supporting the participation of Busby's postdoctoral researcher, Sarah Medynski.

We would like to thank Elizabeth Catlos for inviting this overview chapter. Busby would also like to thank the convenors of the 2021 NSF Rift2Ridge workshop for facilitating the exchange of ideas, especially Scott Bennett, Rebecca Dorsey, Simon Klattenhorn, and Kathleen Marsaglia. Busby also wishes to thank Tiago Alves for involving her in a major review paper on continental breakup (Alves et al., 2020), from which she learned a great deal.

Last but not least, Busby would like to thank Tina Niemi for organizing all of the considerable logistics for the field and lab components of the NSF Baja Basins Research Experience for Undergraduates (2014–2018) and the NSF International Research Experience for Students (2019–2022). Busby also thanks Alison Graettinger for co-supervising the undergraduate students involved in the volcanology part of the Baja Basins project, and she thanks the following undergraduate students who published GSA and AGU abstracts with us (in alphabetical order): N. Acuña, J. Allard, C. Andrews, A. Bearden, E. Bowman, E.P. Gutierrez, S. Hutchinson, J. Ibarra Mendoza, E. Lodes, T. McEnaney, Q. Mindrup, J. Ojeda, S. Rack and J.J. Rice.

AVAILABILITY STATEMENT

No new data were generated, analyzed, or presented in the manuscript.

REFERENCES

Allen, P. A., & Allen, J. R. (2005). *Basin analysis: principles and applications* (p. 549). Oxford: Blackwell Scientific Publications.

Alves, T., Fetter, M., Busby, C., Gontijo, R., Cunha, T. A., & Mattos, N. H. (2020). A tectonostratigraphic review of continental breakup on intraplate continental margins and its impact on resultant hydrocarbon systems. *Marine and Petroleum Geology*, *117*, 1–39. doi: 10.1016/j.marpetgeo.2020.104341

Atwater, T., & Stock, J. (1998). Pacific-North America plate tectonics of the Neogene southwestern United States: an update. *International Geology Review*, *40*(5), 375–402. doi: 10.1080/00206819809465216

Bennett, R. A., Davis, J. L., & Wernicke, B. P. (1999). Present-day pattern of Cordilleran deformation in the western United States. *Geology*, *27*, 371–374.

Bormann, J., Hammond, W. C., Kreemer, C., & Blewitt, G. (2016). Accommodation of missing shear in the Central Walker Lane, western North America: Constraints from dense GPS measurements. *Earth and Planetary Science Letters, 440,* 169–177. doi: 10.1016/j.epsl.2016.01.015

Brune, S., Williams, S. E., & Müller, R. D. (2018). Oblique rifting: the rule, not the exception. *Solid Earth*, *9*(5), 1187–1206. doi: 10.5194/se-9-1187-2018

Bursik, M. (2009). A general model for tectonic control of magmatism: examples from Long Valley Caldera (USA) and El Chichon (Mexico). *Geofisica Internacional*, *48*, 171–183. doi: 10.22201/igeof.00167169

Busby, C. J. (2013). Birth of a plate boundary at ca. 12 Ma in the ancestral Cascades arc, Walker Lane belt of California and Nevada. *Geosphere*, *9*, 1147–1160. doi: 10.1130/ges00928.1

Busby, C. J. (2023). The Guerrero-Alisitos-Vizcaino superterrane of Western Mexico and its ties to the Mexican continental margin (Gondwana and SW Laurentia). *Geological Society of America Memoirs*, *220*, 773–805. doi: 10.1130/2022.1220(34)

Busby, C. J., Andrews, G. D. M., Koerner, A. K., Brown, S. R., Melosh, B. L., & Hagan, J. C. (2016). Progressive derangement of ancient (Mesozoic) E-W Nevadaplano paleochannels into modern (Miocene-Recent) NNW trends in the Walker Lane belt, Central Sierra Nevada. *Geosphere*, *12*(1), 135–175. doi: 10.1130/ges01182.1

Busby, C. J., DeOreo, S. B., Skilling, I., Hagan, J. C., & Gans, P. B. (2008b). Carson Pass-Kirkwood paleocanyon system: implications for the Tertiary evolution of the Sierra Nevada, California. *Geological Society of America Bulletin*, *120*(3/4), 274–299. doi: 10.1130/b25849.1

Busby, C. J., Graettinger, A., López Martínez, M., Medynski, S., Andrews, C., Bowman, E., et al. (2020). Volcanic record of the arc-to-rift transition onshore of the Guaymas basin in the Santa Rosalía area, Gulf of California, Baja California. *Geosphere*, *16*, 30. doi: 10.1130/GES02094.1

Busby, C. J., Graettinger, A. H., López-Martínez, M., Orozco-Esquivel, T., Putirka, K., Acuña Henry, N., Allard, J., Hutchinson, S. J., & Niemi, T. M. (2024). Early Miocene to Pliocene magmatic and structural evolution of the Gulf of California margin in central Baja California: The Mulegé-La Trinidad region, Baja California Sur. *The virtue of fieldwork in volcanology, sedimentology, structural geology, and tectonics, celebrating the career of Cathy Busby*. Geological Society of America Special Paper 573. https://doi.org/10.1130/2025.2563(12)

Busby, C. J., Hagan, J. C., Putirka, K., Pluhar, C. J., Gans, P. B., Wagner, D. L., et al. (2008a). The ancestral Cascades arc: Cenozoic evolution of the central Sierra Nevada (California) and the birth of the new plate boundary. In J. E. Wright & J. W. Shervais (Eds.), *Ophiolites, arcs, and batholiths: a tribute to Cliff Hopson* (pp. 331–378). Geological Society of America. doi: 10.1130/2008.2438(12)

Busby, C. J., Hagan, J. C., & Renne, P. (2013b). Initiation of Sierra Nevada range front - Walker Lane faulting at ~12 Ma in the Ancestral Cascades arc. *Geosphere*, *9*(5), 1–22. doi: 10.1130/GES00927.1

Busby, C. J., Koerner, A. A., Melosh, B. L., Andrews, G. D. M., & Hagan, J. C. (2013a). The Sierra Crest graben-vent system: A Walker Lane pull-apart in the Ancestral Cascades arc. *Geosphere*, *9*(4), 736–780. doi: 10.1130/GES00670.1

Busby, C. J., Pavlis, T., Roeske, S., & Tikoff, B. (2023). The North American Cordillera during the Mesozoic to Paleogene: selected questions and controversies. *Geological Society of America Memoirs*, *220*, 635–658. doi: 10.1130/2022.1220(31)

Busby, C. J., & Putirka, K. (2009). Miocene evolution of the western edge of the Nevadaplano in the central and northern Sierra Nevada: paleocanyons, magmatism and structure: In Gary Ernst, editor, The rise and fall of the Nevadaplano. *International Geology Review*, *51*(7–8), 671–701. doi: 10.1080/00206810902978265

Busby, C. J., Putirka, K., Renne, P. R., Hagan, J. C., Melosh, B., & Gambs, M. (2018). A tale of two Walker Lane pull-aparts in the Ancestral Cascades Arc, central Sierra Nevada, California. *Geosphere*, *14*(5), 2068–2117. doi: 10.1130/ges01398.1

Carlson, C. W. (2017). *Kinematics and transfer mechanisms of strain accommodation at the transition between the Northern and Central Walker Lane, Western Nevada* (PhD thesis). Reno, Nevada: University of Nevada.

Carlson, C. W., Pluhar, C. J., Glen, J. M. G., & Farner, M. J. (2013). Kinematics of the west-central Walker Lane: spatially and temporally variable rotations evident in the late Miocene Stanislaus Group. *Geosphere*, *9*(6), 1530–1551. doi: 10.1130/GES00955.1

Cashman, P. H., Trexler, J. H., Muntean, T. W., Faulds, J. E., Louie, J. N., & Oppliger, G. L. (2009). Neogene tectonic evolution of the Sierra Nevada - Basin and Range transition zone at the latitude of Carson City, Nevada. *Geological Society of America Special Paper*, *447*, 171–188. doi: 10.1130/2009.2447(10)

Cecil, M. R., Ferrer, M. A., Riggs, N. R., Marsaglia, K., Kylander-Clark, A., Ducea, M. N., & Stone, P. (2019). Early arc development recorded in Permian-Triassic plutons of the northern Mojave Desert region, California: *USA. Bulletin*, *131*, 749–765. doi: 10.1130/B31963.1

Christie-Blick, N., & Biddle, K. T. (1985). Deformation and basin development along strike-slip faults. *Society of Economic Paleontologists and Mineralogists. Special Publication*, *37*, 1–34. doi: 10.2110/pec.85.37.0001

Clifton, A. E., & Kattenhorn, S. A. (2006). Structural architecture of a highly oblique divergent plate boundary segment. *Tectonophysics*, *419*, 27–40.

Colgan, J. P., Egger, A. E., John, D. A., Cousens, B., Fleck, R. J., & Henry, D. (2011). Oligocene and Miocene arc volcanism in northeastern California: evidence for post-Eocene segmentation of the subducting Farallon plate. *Geosphere*, *7*(3), 733–755. doi: 10.1130/GES00650.1

Conly, A. G., Beaudoin, G., & Scott, S. D. (2006). Isotopic constraints on fluid evolution and precipitation mechanisms for the Boléo Cu-Co-Zn deposit, Mexico. *Mineralium Deposita*, *41*, 127–151. doi: 10.1007/s00126-005-0045-3

Conly, A. G., Scott, S. D., & Bellon, H. (2011). Metalliferous manganese oxide mineralization associated with the Boléo Cu-Co-Zn district. *Economic Geology and the Bulletin of the Society of Economic Geologists*, *106*, 1173–1196. doi: 10.2113/econgeo.106.7.1173

Curray, J. R., & Moore, D. G. & Shipboard Scientific Party(1982). Guaymas basin: sites 477, 478 and 481. *Initial Reports of the Deep Sea Drilling Project*, *64*, 211–415. doi: 10.2973/dsdp.proc.64.104.1982

DeCelles, P. G. (2004). Late Jurassic to Eocene evolution of the Cordilleran thrust belt and foreland basin system, western USA. *American Journal of Science*, *304*, 105, 105–168. doi: 10.2475/ajs.304.2

Di Luccio, F., Persaud, P., & Clayton, R. W. (2014). Seismic structure beneath the Gulf of California: A contribution from group velocity measurements. *Geophysical Journal International*, *199*(3), 1861–1877. doi: 10.1093/gji/ggu338

Dobbs, S. C., Riggs, N. R., Marsaglia, K. M., Gonzalez-Leon, C. M., Cecil, R. M., & Smith, M. (2021). The Permian Monos Formation: Stratigraphic and detrital zircon evidence for Permian Cordilleran arc development along the southwest margin of Laurentia northwest Sonora Mexico. *Geosphere*, *7*(2), 520–537. doi: 10.1130/GES02320.1

Dorsey, R. J., O'Connell, B., Gardner, K. K., Homan, M. B., Bennett, S. E. K., Thacker, J. O., & Darin, M. H. (2021). Tectonostratigraphic record of late Miocene– early Pliocene transtensional faulting in the Eastern California shear zone, southwestern USA. *Geosphere*, *17*(4), 1101–1125. doi: 10.1130/GES02337.1

Dorsey, R., Niemi, T. M., Darin, M., Salgado Munoz, V. O., Bennett, S., Pecha, M., et al. (2022). Stratigraphy, sedimentology, age, and tectonic significance of the Upper Miocene Boleo formation in the Santa Rosalia Basin Baja California Sur, Mexico. *Geological Society of America Abstracts with Programs*, *54*(5). doi: 10.1130/abs/2022AM-381103

Faulds, J. E., & Henry, C. D. (2008). Tectonic influences and temporal evolution of the Walker Lane: an incipient transform along the evolving Pacific-North American plate boundary. *Arizona Geological Society Digest*, *22*, 437–470. doi: 10.1130/g21274.1

Faulds, J. E., Henry, C. D., & Hinz, N. H. (2005). Kinematics of the northern Walker Lane: an incipient transform fault along the Pacific-North American plate boundary. *Geology*, *33*(6), 505–508. doi: 10.1130/g21274.1

Fleck, R. J., duBray, E. A., John, D. A., Vikre, P. G., Cosca, M. A., Snee, L. W., & Box, S. E. (2015). *Geochronology of Cenozoic rocks in the Bodie Hills, California and Nevada* (Data Series 916), p. 26. U.S. Geological Survey. doi: 10.3133/ds916

Graettinger, A., Busby, C., Mindrup, Q., & McEnaney, T. (2022). *Volcanology and structure of the Mulege-Bahia Concepcion region, Gulf of California, Baja California Sur, Mexico: results from the Baja basin research experience for undergraduates (REU) and International Research Experience for Students (IRES)*. Paper presented at GSA Connects.

Haddon, E. K., Amos, C. B., Zielke, O., Jayko, A. S., & Burgmann, R. (2016). Surface slip during large Owens Valley earthquakes. *Geochemistry, Geophysics, Geosystems*, *17*, 2239–2269. doi: 10.1002/2015GC006033

Hagan, J. C., Busby, C. J., Putirka, K., & Renne, P. (2009). Cenozoic paleocanyon evolution, Ancestral Cascades arc volcanism, and structure of the Hope Valley–Carson Pass region, Sierra Nevada, California. *International Geology Review*, *51*, 777–823. doi: 10.1080/00206810903028102

Hammond, W. C., Blewitt, G., Li, Z., Plag, H.-P., & Kreemer, C. (2012). Contemporary uplift of the Sierra Nevada, western United States, from GPS snd InSAR measurements. *Geology*, *40*(7), 667–670. doi: 10.1130/G32968.1

Hildreth, W. (2007). *Quaternary magmatism in the Cascades—geologic perspective* (U.S. Geological Survey Professional Paper 1744). doi: 10.3133/pp1744

Hodges, M., Newby, T., Swarner, H., Busby, C., Shapiro, R., Francek, G., & Hodges, C. (2025). Detrital zircon geochronology of a fossiliferous Miocene paleochannel. In N. Riggs et al. (Eds.), *Special Paper: The virtue of fieldwork in volcanology, sedimentology, structural geology, and tectonics, celebrating the career of Cathy Busby* (vol. 563), Geological Society of America, in press.

Holt, J. W., Holt, E. W., & Stock, J. M. (2000). An age constraint on Gulf of California rifting from the Santa Rosalía basin, Baja California Sur, Mexico. *Geological Society of America Bulletin*, *112*, 540–549. doi: 10.1130/0016-7606(2000)112<540:AACOGO>2.0.CO;2

Ingersoll, R. V., & Busby, C. J. (1995). Tectonics of sedimentary basins. In C. J. Busby & R. V. Ingersoll (Eds.), *Tectonics of sedimentary basins* (pp. 1–52). Blackwell Science.

Jayko, A. S., & Bursik, M. (2012). Active transtensional intracontinental basins: Walker Lane belt in the western Great Basin. In C. Busby & A. Azor (Eds.), *Recent advances in tectonics of sedimentary basins* (pp. 226–248). Wiley-Blackwell. doi: 10.1002/9781444347166.ch11

Kattenhorn, S. A., Krantz, B., Walker, E. L., & Blakeslee, M. W. (2016). Evolution of the Hat Creek fault system, Northern California. In: 3-D structural interpretation. *AAPG Memoir*, *111*, 121–154. doi: 10.1306/13561990m1113674

Kent, G. M., Babcock, J. A., Driscoll, N. W., Harding, A. J., Seitz, G. G., Dingler, J. A., et al. (2005). A 60 k.y. record of extension across the western boundary of the Basin and Range Province: estimate of slip rates from offset shoreline terraces and a catastrophic slide beneath Lake Tahoe. *Geology*, *33*, 365–368. doi: 10.1130/g21230.1

Kreemer, C., Blewitt, G., & Hammond, W. C. (2009). Geodetic constraints on contemporary deformation in the northern Walker Lane: 2. Velocity and strain rate tensor analysis. *Geological Society of America Special Paper*, *447*, 17–31. doi: 10.1130/2009.2447(02)

Lizarralde, D., Axen, G. J., Brown, H. E., Fletcher, J., Gonzáles-Fernández, A., Harding, A. J., et al. (2007). Variation in styles of rifting in the Gulf of California. *Nature*, *448*, 466–469. doi: 10.1038/nature06035

Lutz, B. M., Axen, G. J., van Wijk, J. W., & Phillips, F. M. (2022). Whole-lithosphere shear during oblique rifting. *Geology*, *50*, 412–416. doi: 10.1130/G49603.1

Manzo, A. (2023). *Not all large silicic volcanic systems are transcrustal: evidence from the Sierra Crest - Little Walker and*

Ebbetts Pass Volcanic Centers, central Sierra Nevada, California (Master's thesis). California State University, Fresno.

Manzo, A., Busby, C., & Putirka, K. (2023). Not all large silicic volcanic systems are transcrustal: evidence from the Sierra Crest-Little Walker and Ebbetts Pass volcanic centers, central Sierra Nevada, California. *Geological Society of America Abstracts with Programs*, *55*(4). doi: 10.1130/abs/2023CD-387261

Marsaglia, K. M. (2012). Sedimentation at plate boundaries. In C. Busby & A. Pérez (Eds.), *transition: recent advances in the tectonics of sedimentary basins* (pp. 291–309). Blackwell.

Mavor, S. P., Bennett, S. E. K., Crow, R. S., Singleton, J. S., Langenheim, V., Stockli, D., et al. (2023). Evolution of Miocene normal and dextral faulting in the lower Colorado River region near Blythe, California, USA. *Geosphere*, *19*(5), 1180–1209. doi: 10.1130/GES02608.1

McQuarrie, N., & Wernicke, B. P. (2005). An animated tectonic reconstruction of southwestern North America since 36 Ma. *Geosphere*, *1*, 147–172. doi: 10.1130/GES00016.1

Michaud, F., Royer, J. Y., Bourgois, J., Dyment, J., Calmus, T., Bandy, W., et al. (2006). Oceanic-ridge subduction vs. slab break off: Plate tectonic evolution along the Baja California Sur continental margin since 15 Ma. *Geology*, *34*(1), 13–16. doi: 10.1130/g22050.1

Miller, N. C., & Lizarralde, D. (2013). Thick evaporates and early rifting in the Guaymas basin, Gulf of California. *Geology*, *41*, 283–286. doi: 10.1130/G33747.1

Nagorsen-Rinke, S., Lee, J., & Calvert, A. (2013). Pliocene sinistral slip across the Adobe Hills, eastern California–western Nevada: kinematics of fault slip transfer across the Mina deflection. *Geosphere*, *9*(1), 37–53. doi: 10.1130/GES00825.S3

Niemi, T. M. & Salgado Muñoz, V. O. (2019). *Gypsum facies analysis of the copper-bearing Upper Miocene Boleo Formation in the transtensional Santa Rosalia Basin, Baja California Sur, México*. Paper presented at the AAPG Europe Meeting on "Eurasian Mature Salt Basins," Krakow, Poland.

Nilsen, T. H., & Sylvester, A. G. (1995). Strike slip basins. In C. J. Busby & R. V. Ingersoll (Eds.), *Tectonics of Sedimentary Basins* (pp. 425–456). Blackwell Science.

Nuriel, P., Miller, D. M., Schmidt, K. M., Coble, M. A., & Maher, K. (2019). Ten-million years of activity within the Eastern California Shear Zone from U-Pb dating of fault-zone opal. *Earth and Planetary Science Letters*, *521*, 37–45. doi: 10.1016/j.epsl.2019.05.047

Pitman, W. C., III, & Andrews, J. A. (1985). Subsidence and thermal history of small pull-apart basins. *Special Publication, Society of Economic Paleontologists and Mineralogists*, *37*, 45–119. doi: 10.2110/pec.85.37.0045

Pluhar, C. J., Coe, R. S., Lewis, J. C., Monastero, F. C., & Glen, J. M. G. (2006). Fault block kinematics at a releasing stepover of the Eastern California shear zone: partitioning of rotation style in and around the Coso geothermal area and nascent metamorphic core complex. *Earth and Planetary Science Letters*, *250*, 134–163. doi: 10.1016/j.epsl.2006.07.034

Putirka, K., & Busby, C. J. (2007). The tectonic significance of high K_2O volcanism in the Sierra Nevada, California. *Geology*, *35*(10), 923–926. doi: 10.1130/g23914a.1

Rains, J., Marsaglia, K. M., & Dunne, G. C. (2012). The stratigraphic and petrologic record of Permian subduction inception along the North American margin. *Lithosphere*, *4*, 533–552.

Reynolds, A. & Lakowski, A.K. (2023). *Dynamics of continental rifting in southern Tibet: Interpretations from spatiotemporal trends in thermochronology data*. Paper presented at AGU23.

Riley, P., Tikoff, B., & Hildreth, W. (2012). Transtensional deformation and structural control of contiguous but independent magmatic systems: Mono-Inyo Craters, Mammoth Mountain, and Long Valley Caldera, California. *Geosphere*, *8*, 740–751. doi: 10.1130/ges00662.1

Rossetti, F., Storti, F., Busetti, M., Lisker, F., Vicenzo, G. D., Laufer, A. L., et al. (2006). Eocene initiation of Ross Sea dextral faulting and implications for East Antarctic neotectonics. *Journal of the Geological Society, London*, *163*, 119–126.

Ruocco, S., Iacopini, D., Tavani, S., Ebinger, C., Dottore Stagna, M., & Reynolds, D. (2021). *Neogene rift tectonic activity in the West Somali Basin, offshore Tanzania: example of a segmented oblique rift structure*. Paper presented at the EGU General Assembly, EGU21-9066.

Schmitt, A. K., Martín-Barajas, A., Weber, B., Stockli, D. F., Zou, H., & Shen, C. C. (2013). Oceanic magmatism in sedimentary basins of the northern Gulf of California rift. *Geological Society of America Bulletin*, *125*, 1833–1850. doi: 10.1130/B30787.1

Schweickert, R. A., Moore, J. G., Lahren, M. M., Kortemeier, W., Kitts, C., & Ademek, T. (2019). The Tahoe-Sierra frontal fault zone, Emerald Bay area, Lake Tahoe, California: history, displacements and rates. *Geosphere*, *15*(3), 783–819. doi: 10.1130/GES02022.1

Smith, K. D., Keny, G. M., von Seggern, D. P., Driscoll, N. W., & Eisses, A. (2016). Evidence for Moho-lower crustal transition depth diking and rifting of the Sierra Nevada microplate. *Geophysical Research Letters*, *43*, 10738–10744. doi: 10.1002/2016GL070283

Surpless, B. (2008). Modern strain localization in the central Walker Lane, western United States: implications for the evolution of intraplate deformation in transtensional settings. *Tectonophysics*, *457*, 239–253. doi: 10.1016/j.tecto.2008.07.001

Teske, A., Lizarralde, D., & Höfig, T. W. (2018). *International ocean discovery program Expedition 385 scientific prospectus: Guaymas basin tectonics and biosphere*. doi: 10.14379/iodp.sp.385.2018

Unruh, J. R., Humphrey, J., & Barron, A. (2003). Transtensional model for the Sierra Nevada frontal fault system, eastern California. *Geology*, *31*, 327–330. doi: 10.1130/0091-7613(2003)031<0327:TMFTSN>2.0CO;2

Van Wijk, J. W., Heyman, S. P., Axen, G. J., & Persaud, P. (2019). Nature of the crust in the northern Gulf of California and Salton Trough. *Geosphere*, *15*(5), 1598–1616. doi: 10.1130/GES02082.1

Waldien, T. S., Meigs, A. J., & Madin, I. P. (2019). Active dextral strike-slip faulting record termination of the Walker Lane belt at the southern Cascade arc in the Klamath graben, Oregon, USA. *Geosphere*, *15*, 882–900. doi: 10.1130/GES02043.1

Wesnousky, S. G. (2005). The San Andreas and Walker Lane fault systems, western North America: transpression, transtension, cumulative slip and the structural evolution of a major transform plate boundary. *Journal of Structural Geology*, *27*(8), 1505–1512. doi: 10.1016/j.jsg.2005.01.015

Wesnousky, S. G., Bormann, J. M., Kreemer, C., Hammond, W. C., & Brune, J. N. (2012). Neotectonics, geodesy, and seismic hazards in the Northern Walker Lane of Western North America: thirty kilometers of crustal shear and no strike-slip? *Earth and Planetary Science Letters*, *329–330*, 133–140. doi: 10.1016/j.epsl.2012.02.018

Xie, X., & Heller, P. L. (2009). Plate tectonics and basin subsidence history. *Geological Society of America Bulletin*, *121*(1/2), 55–64. doi: 10.1130/B26398.1

Zwaan, F., Corti, G., Sani, F., Keir, D., Muluneh, A., & Papini, M. (2020). Structural analysis of the Western Afar Margin, East Africa: evidence for multiphase rotational rifting. *Tectonics*, *39*, e2019TC006043. doi: 10.1029/2019TC006043

5

Extensional Tectonics in Western Anatolia, Turkey: Eastward Continuation of the Aegean Extension

Elizabeth J. Catlos[1], Thomas M. Etzel[1,2], and İbrahim Çemen[3]

ABSTRACT

Western Anatolia, at the boundary between the Aegean and Anatolian microplates, is considered a type-location for the significant transition between compressional and extensional tectonics. The region is marked by diverse tectonic events from the Paleoproterozoic through the Cambrian, Devonian, and Late Cretaceous, recorded by its suture zones, metamorphic rocks, and igneous intrusions. Extension initiated in a complex collage of multiple sutured crustal fragments. Today, the region is transitioning or has transitioned to a stress regime dominated by strike-slip tectonics. Control for extension in Western Anatolia is widely accepted as the rollback of the African (Nubian) slab along the Hellenic arc. Outstanding questions include the timing and geometry of the Hellenic arc and connections to other subduction systems. Slab tear is proposed for many regions across the Anatolian and Aegean microplates. This process impacts lithospheric dynamics, generates economic and energy deposits, facilitates lithospheric thinning, and influences the onset of transfer zones that accommodate deformation and provides magma conduits. The role of magma in facilitating extension and where and why switches in stress regimes occurred are still under consideration. The correlation between Aegean and Anatolian tectonic events also requires a better understanding of the metamorphic history recorded in Western Anatolia rocks.

5.1. INTRODUCTION

The Aegean and eastern Mediterranean are considered the most rapidly deforming regions across the Alpine-Himalayan chain (Figure 5.1) (Papazachos & Delibasis 1969; Papazachos & Comninakis, 1971; McKenzie, 1972; Şengör et al., 1985; Taymaz et al., 1991; Jackson, 1994; Reilinger et al., 1997; Nyst & Thatcher, 2004; Le Pichon et al., 2019; Meng et al., 2021). The Aegean and Anatolia microplates, sometimes classified as the single Aegean-Anatolian microplate, are a complex amalgamation of a series of terranes that today experience seismicity (Şengör & Yılmaz, 1981; Okay et al., 1996; Reilinger et al., 1997; Nyst & Thatcher, 2004; Tan, 2013). The Anatolian microplate is a large peninsula that coincides with over two-thirds of the country of Turkey (Figure 5.1) (Le Pichon et al., 1995; Oral et al., 1995; Reilinger et al., 1997; Papazachos, 1999). It is the westernmost protrusion of the Asian continent, with a pole of rotation located in the northern Sinai Peninsula (Reilinger et al., 2010). The Black Sea bounds it to the north and the Mediterranean Sea to the south. The Aegean microplate has mainly continental crust and sediments obscured by the Aegean Sea (Le Pichon & Angelier, 1981; Jolivet & Patriat, 1999; Makris et al., 2013). The Sea of Marmara connects

[1]*Jackson School of Geosciences, Department of Earth and Planetary Sciences, University of Texas at Austin, Austin, Texas, USA*

[2]*ExxonMobil, Parkway Spring, Texas, USA*

[3]*Department of Geological Sciences, University of Alabama, Tuscaloosa, Alabama, USA*

Extensional Tectonics: Rifting and Continental Extension, Geophysical Monograph 278, First Edition.
Edited by İbrahim Çemen and Elizabeth J. Catlos.

DOI: 10.1002/9781119773757.ch05

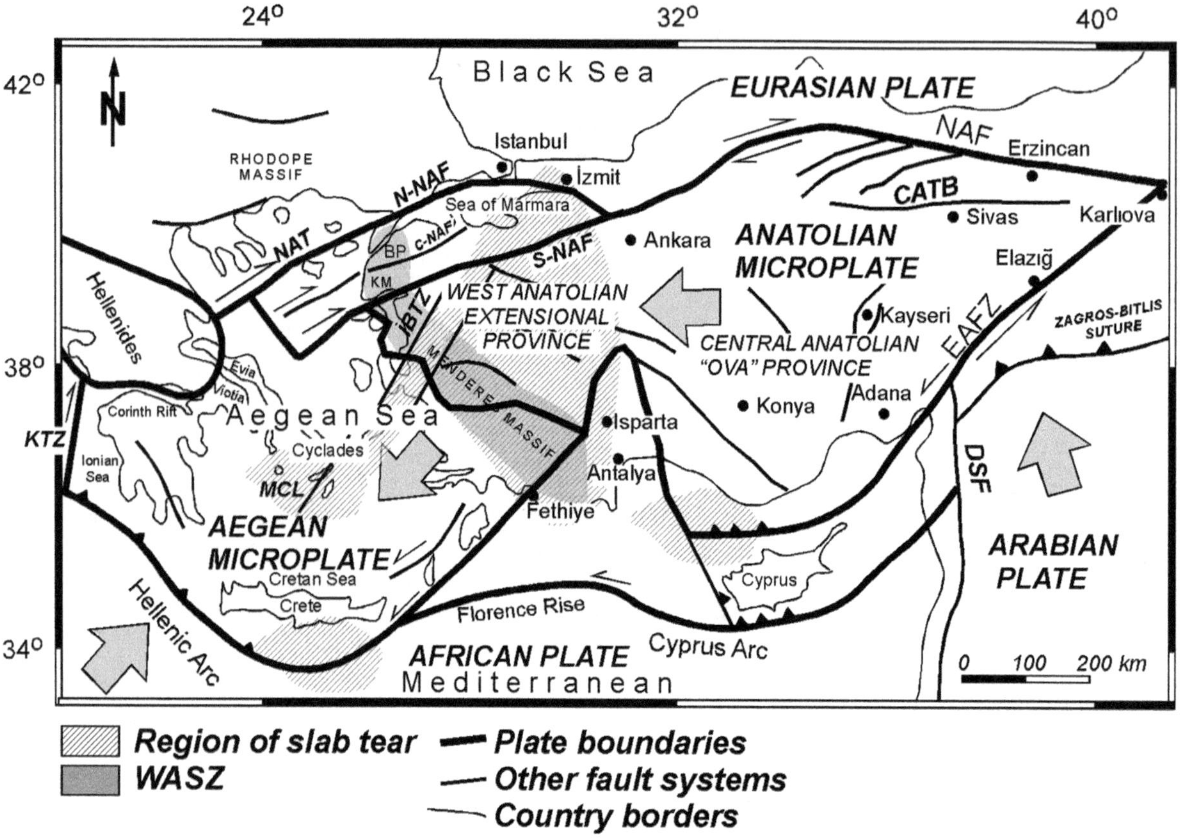

Figure 5.1 Tectonic map of the Aegean and Anatolian microplates. Plate boundaries after McClusky et al. (2000), Nyst and Thatcher (2004), Piper et al. (2010), Harrison et al. (2013), and Tan (2013). Only some major fault systems are labeled. NAF, North Anatolian Fault; EAFZ, East Anatolian Fault Zone; CATB, Central Anatolian Thrust Belt; DSF, Dead Sea Fault; KTZ, Kephalonia Transform Zone; MCL, Mid-Cycladic lineament; İBTZ, İzmir–Balıkesir transfer zone; NAT, North Aegean Trough; NAF, North Anatolian Fault (N- northern, C- central, and S- southern segments); KM, Kazdağ Massif. Region of slab tear in western Turkey and the Aegean after Jolivet et al. (2015), near Crete (Özbakır et al., 2013), Cyprus (Woodside et al., 2002), and between the Aegean domain and the Menderes Massif (Roche et al., 2019). Boundaries between Central and Western Anatolia after Şengör et al. (1985).

the Black and Aegean Seas through the Bosphorus and Dardanelles straits, separating a fragment of Eurasia's microplate (Nyst & Thatcher, 2004).

Deciphering the assembly of the Aegean and Anatolian microplates and their past and present-day deformation drivers impacts our understanding of continental tectonics, subduction zone processes, lithospheric deformation, ore generation process, and hazards (Jackson, 1994; Meng et al., 2021; Rabayrol & Hart, 2021). The borders of the Aegean and Anatolian microplates coincide with fault systems that played vital roles in triggering changes in their tectonic nature (McKenzie, 1972, 1978). The microplates share some borders, including the right-lateral strike-slip North Anatolian Fault Zone (NAFZ) and the Western Anatolian Extensional Province (WAEP) (Figure 5.1) (Ketin, 1948; Şengör et al., 1985; Barka, 1992; Armijo et al., 1999; McClusky et al., 2000; Çemen et al., 2006; Chousianitis et al., 2015). The subducting Hellenic and Cyprus arcs and the complex dynamics coinciding with the Florence Rise makes up their southern borders (Le Pichon & Angelier, 1979; Angelier et al., 1982; Anastasakis & Kelling 1991; Papazachos et al., 2000; Ergün et al., 2005; Suckale et al., 2009; Royden & Papanikolaou, 2011). Global positioning system (GPS) constraints show that the principal northern boundaries of the southwestern Aegean plate are the North Aegean Trough (NAT) and Kephalonia (also Cephalonia and Kefalonia) Transform Zone (KTZ) (McKenzie 1972; Le Pichon et al. 1995; Kahle et al. 2000; Pearce et al., 2012; Chousianitis et al. 2015; Haddad et al. 2020). The southern boundary is separated from the Anatolia plate by the WAEP, a zone of N-S extension (Figure 5.1) (McClusky et al., 2000; Chousianitis et al., 2015). Although many of its bounding fault systems are presently active, both the Anatolian and Aegean microplates contain internal structures, including transfer zones (Figures 5.1 and 5.2) (Nyst & Thatcher, 2004; Çemen et al., 2006; Oner et al., 2010; Aktuğ et al., 2013; Özkaymak et al., 2013; Uzel et al., 2013; Seghedi et al., 2015; Barbot & Weiss, 2021).

Several tectonic models applied to the Aegean and Anatolian microplates have transformed our ideas about the lithosphere's response to extensional, strike-slip, and compressional forces (see the review in Aktuğ et al., 2013). Advances in tomography and GPS technology have contributed to our understanding of its present-day dynamics (Barka & Reilinger, 1997; McClusky et al., 2000; Ganas & Parsons, 2009; Komut et al., 2012; Aktuğ et al., 2013; Jolivet et al., 2015; Ventouzi et al., 2018). The deformation, metamorphism, and igneous activity exposed in the upper portions of the microplate's lithosphere constrain processes that operated in its lower lithosphere over long periods of geological time (Jackson, 1994; Komut et al., 2012).

This review chapter is divided into two primary parts. The first section reviews some of the chronology and

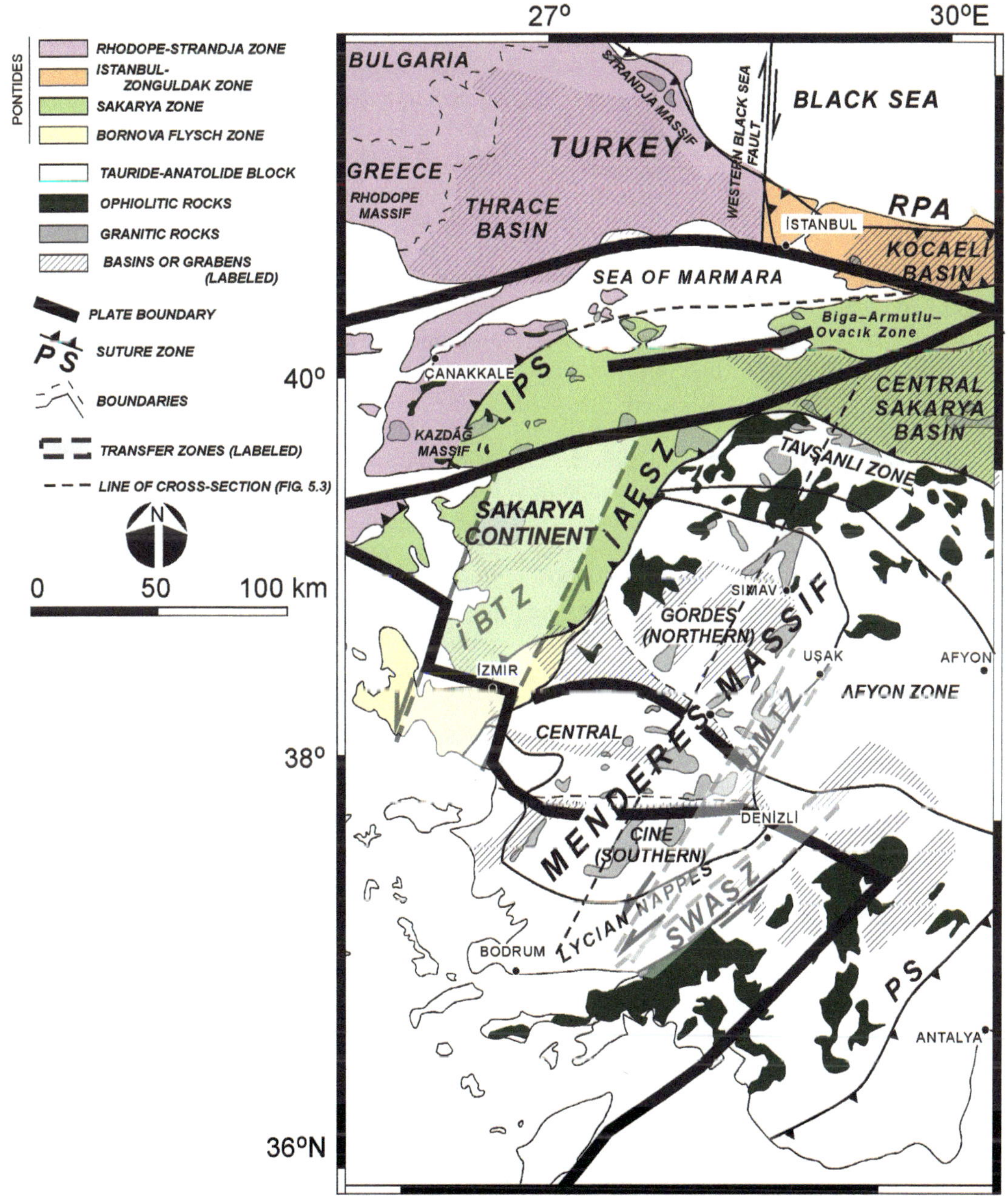

Figure 5.2 Geological map of Western Anatolia focusing on the ophiolite and granite assemblages along the boundary between the Aegean and Anatolia microplates. Plate boundary after Nyst and Thatcher (2004). Terrane boundaries, major fault systems, and transfer zones after Okay (2008), Akbayram et al. (2017), Oner et al. (2010), and Karaoğlu and Helvacı (2014). RPA, Rhodope-Pontide Arc; İBTZ, İzmir–Balıkesir Transfer Zone (also sometimes referred to as the Western Anatolia Transfer Zone, Gessner et al., 2013); SWASZ, South West Anatolian Shear Zone; IPS, Intra-Pontide suture zone; İAESZ, İzmir-Ankara-Erzincan suture zone; PS, Pamphylian suture zone; UMTZ, Uşak-Mugla Transfer Zone.

tectonic history of the juncture between the Aegean and Anatolian microplates from data available in Western Anatolia (Figures 5.1 and 5.2). The goal is to outline how the boundary results from an accumulation of a series of tectonic processes that record stress transitions in the geological past. The second part of the chapter aims to present outstanding questions that remain in unraveling its complex dynamics. This particular area of the Anatolian microplate has been studied for almost 50 years (McKenzie, 1972) and has become the type-locality for understanding subduction zone dynamics, a focus of diverse and multidisciplinary studies.

5.2. GEOLOGICAL BACKGROUND

5.2.1. Assembly of Key Components (Paleoproterozoic-Eocene)

The Anatolian microplate comprises multiple continental fragments separated by oceans that collided and ultimately combined by the Late Cretaceous-Eocene, with exposures of ophiolitic and high-pressure/low-temperature rock assemblages that mark the suture zones (Figures 5.2, 5.3, 5.4) (Şengör & Yılmaz, 1981; Okay, 2008; Moix et al., 2008; Okay & Tüysüz, 1999; Pourteau et al., 2016; Okay et al., 2020). Western Anatolia is explicitly defined by the amalgamation of two terranes: the Pontides to the north and the Anatolides-Taurides to the south (Şengör & Yılmaz, 1981; Yilmaz et al., 1997; Okay & Tüysüz, 1999; Pourteau et al., 2016). The Pontides extends across northern Turkey and comprises mainly Pan-African basement blocks and Phanerozoic sedimentary cover units that may have originated from the southern Eurasia margin before back-arc extension initiated and created the Black Sea (Yilmaz et al., 1997; Moix et al., 2008; Pourteau et al., 2010; Okay et al., 2013).

The Intra-Pontide suture zone (IPS) is mapped within the Pontide zone between the Sakarya continental zones and the İstanbul-Zonguldak Unit (also İstanbul-Zonguldak Zone, İstanbul Nappe, or İstanbul Zone, see Yiğitbaş et al., 2004) (Figures 5.2, 5.3, 5.4). The İstanbul portion of the unit exists in the west (İstanbul, Gebze, south Çamdağ regions) and the Zonguldak to the east (north Çamdağ, Zonguldak, Safranbolu regions), both being Gondwanan fragments (Okay et al., 2006; Bozkaya et al., 2012). The IPS has varying interpretations, including an accretionary complex, a suprasubduction zone, and a remnant of a former ocean basin that may have extended into eastern Europe (Okay et al., 1996; Robertson & Ustaömer, 2004; Göncüoğlu et al., 2012, 2014; Marroni et al., 2014; Akbayram et al., 2017; Sayit et al., 2016; Frassi et al., 2018). Geological units within the IPS may also be from components from the İstanbul-Zonguldak or Sakarya zone, which has led to a debate about its presence and utility of the IPS in paleogeographic reconstructions (Moix et al., 2008).

Magmatic assemblages help us understand the tectonic processes involved in Western Anatolia, so we summarize some available time constraints for several key granite bodies dispersed throughout this region in Tables 5.1–5.8. and Figure 5.4. Zircon ages extracted from metagranites and quartzite units indicate that the İstanbul-Zonguldak Unit has a Precambrian basement with Gondwanan units (Chen et al., 2002; Yiğitbaş et al., 2004; Ustaömer et al., 2005, 2011) and stratigraphic similarities with Paleozoic rocks from the southern margin of Laurasia (Görür et al., 1997; Kaldova et al., 2003). Some of the oldest Neoproterozoic granites in Western Anatolia are found in the İstanbul Zone (Table 5.1, Karadere or Karabük metagranite, Figure 5.4; Chen et al., 2002; Ustaömer et al., 2016; Di Rosa et al. 2019), although zircons from the Karacabey (Taşmalı) and Karaburun plutons in the Western Sakarya Zone and the Çine Massif in the southern portion of the Menderes Massif also yield Paleoproterozoic and Neoproterozoic dates (Tables 5.5 and 5.8; Loos & Reischmann 1999; Aysal et al., 2012; Ustaömer et al., 2012). The Triassic ages from granites that intrude the İstanbul-Zonguldak Unit are thought to time partial closure of the Paleotethyan Ocean (Table 5.1, Ustaömer et al., 2016). Some of the youngest mineral ages from İstanbul-Zonguldak granites are Late Cretaceous ($^{40}Ar/^{39}Ar$ ages, 93.3±2.0 Ma, 86.1±2.0 Ma; Delaloye &

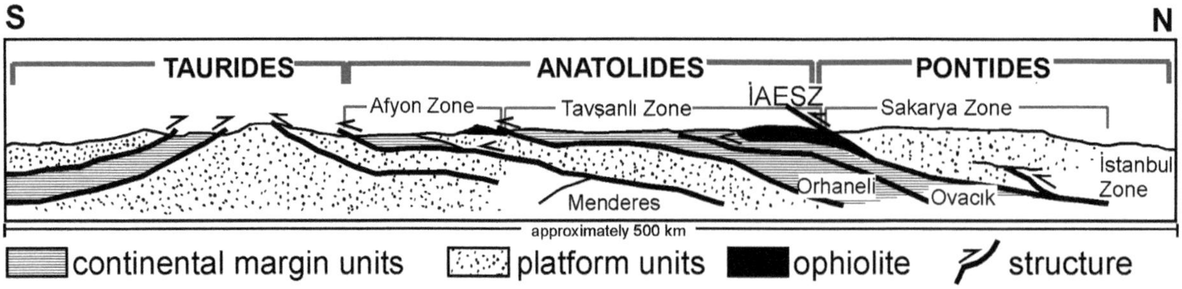

Figure 5.3 North-south generalized cross-section across western Turkey after Okay (1986) and Shin et al. (2013). İAESZ, İzmir-Ankara-Erzincan Suture Zone. See Figure 5.2 for the approximate line of section on the geological map.

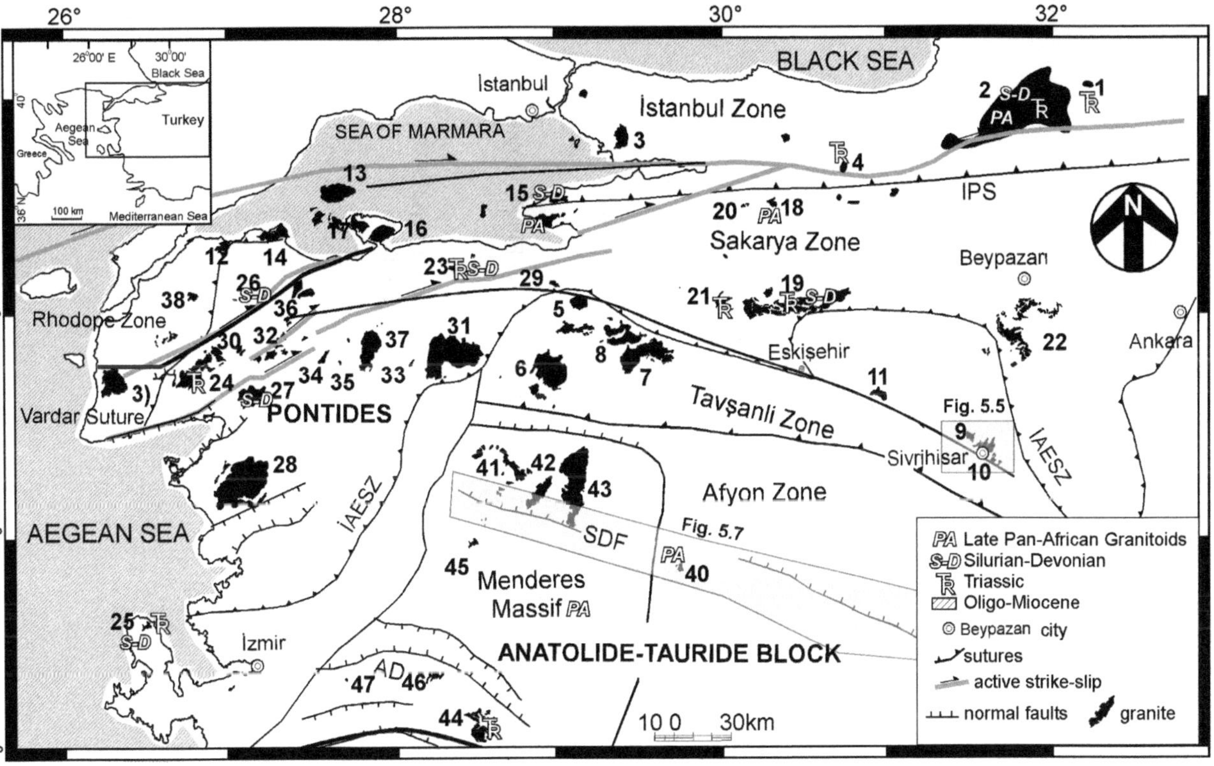

Figure 5.4 Geological map showing structures and locations of Western Anatolia granite bodies. Base map after Delaloye and Bingöl (2000), Senel and Aydal (2002), and Okay (2008). See Tables 5.1–5.8 for the granite names corresponding to this figure's numbers. IPS, Intra-Pontide Suture Zone, İAESZ, İzmir-Ankara-Erzincan Suture Zone, SDF, Simav Detachment Fault, AD, Alaşehir Detachment. Locations of Figures 5.5 and 5.7 are indicated.

Bingöl, 2000), which are similar to estimates for the activity within the subduction-accretion complex associated with the İzmir-Ankara-Erzincan suture zone (İAESZ) (Figures 5.2, 5.3, 5.4) (Okay et al., 2020).

The İAESZ separates the Pontide's Sakarya Composite Terrane in the north from the Anatolide-Tauride block to the south (Figures 5.2 and 5.4) (Şengör & Yılmaz, 1981; Okay & Tüysüz, 1999; Tekin et al., 2002; Göncüoğlu, 2010). The IPS and İAESZ mark the Late Cretaceous–earliest Tertiary closure of Neo-Tethyan ocean basins (Pourteau et al., 2010; Akbayram et al., 2017). In the Aegean microplate, the İAESZ is thought to record the closure of the Vardar ocean and link with the Vardar ophiolite (or Axios-Vardar suture zone) (Channell & Kozur, 1997; Okay & Tüysüz, 1999; Tekin et al., 2002; Moix et al., 2008), but its exposure beneath the Aegean Sea is masked (Burtman, 1994; Stampfli, 2000; Yılmaz et al., 2001; Burchfiel et al., 2008). The Vardar suture may also connect to the IPS that separates the Sakarya Zone from the İstanbul Zone (Şengör & Yılmaz, 1981; Okay & Satir, 2000; Okay et al., 2001, 2010; Beccaletto & Jenny, 2004; d'Atri et al., 2012; Di Rosa et al. 2019) and may connect to the Meliata-Balkan suture of Greece (Stampfli, 2000). The IPS and Vardar connection may be evidenced in the Biga Peninsula by an isolated ophiolite-bearing accretionary complex that was active until the Late Cretaceous (Figures 5.2 and 5.4) (Okay et al., 1991). Some disagree and do not map any suture within the Biga Peninsula (Altunkaynak & Genç, 2008; Burchfiel et al., 2008; Şengün et al., 2011). Because of the uncertain link between the sutures, the relationship of the basement of the Biga Peninsula to that in the Rhodope-Thrace Massif is debated (Bonev & Beccaletto, 2007; Elmas, 2012). In Western Anatolia, the Pamphylian Suture (Figure 5.2) may connect to the Alanya and Bitlis suture zones further to the east (Centikaplan et al., 2016) and beneath the Lycian nappes to the Cycladic domain to the west (Stampfli & Kozur, 2006).

In Western Anatolia, blueschist assemblages exposed along the İAESZ are intruded by suture zone granitoids (SZGs; Topuk, Orhaneli, Tepeldağ [Gürgenyayla and Gürgenyayla]; Table 5.2; Figure 5.4). These granitoids have Paleocene (63.5 ± 2.8 Ma) to Oligocene (31.4 ± 0.6 Ma) dates but are thought to have crystallized in the early Eocene (~45–47 Ma; Okay & Satir, 2006; Altunkaynak, 2007). The SGZs intrude into the western portion of the Tavşanlı Zone, a blueschist sequence overlain by a Cretaceous accretionary complex and ophiolitic sheet. The zone formed as a result of northward-dipping

Table 5.1 A summary of some dates of granitic assemblages that intrude the İstanbul-Zonguldak Zone.

Granite	Location[a]	Approach[b]	Age	Reference
Late Pan-African granitoids or Cadomian granitoids				
Karadere	1	U-Pb zrn	924 ± 4	Chen et al. (2002)
(Karabük metagranite)			620 ± 2	
Karadere	1	U-Pb zrn	668 ± 7	Chen et al. (2002)
(Karabük metatonalite)			589 ± 4	
Bolu (Tüllükırış)	2	U-Pb zrn	576 ± 6	Ustaömer et al. (2005)
Bolu (Kapıkaya)	2	U-Pb zrn	565.3 ± 1.9	Ustaömer et al. (2005)
Karadere (Karabük)	1	Sm-Nd grt + wr	559 ± 8	Chen et al. (2002)
Devonian				
Bolu	2		389, 200	Ustaömer et al. (2012)
			273–255	
			229.6 ± 4.2/2.3	
Bolu	2	$^{40}Ar/^{39}Ar$ or + hbl	381.1 ± 7.1	Delaloye and Bingöl (2000)
			93.3 ± 2.0	
Permo-Triassic				
Bolu (Sünnice group)	2	$^{207}Pb/^{206}Pb$ zrn	262 ± 19	Ustaömer et al. (2005)
Sancaktepe	3	U-Pb zrn	257.3 ± 1.5	Aysal et al. (2018)
			253.7 ± 1.8	
Akyazı	4	$^{40}Ar/^{39}Ar$ or + chl	240.4 ± 4.9	Delaloye and Bingöl (2000)
			86.1 ± 2.0	

[a] See Figure 5.4 for the locations of these granite bodies.
[b] Abbreviations after Whitney and Evans (2010); wr, whole rock.

Table 5.2 Summary of some dates of granitic assemblages that intrude the Tavşanlı Zone.

Granite	Location[a]	Approach[b]	Age	Reference
Western Tavşanli Zone: Suture zone granitoids				
Topuk	5	$^{40}Ar/^{39}Ar$ bt+kfs	63.5 ± 2.8	Delaloye and Bingöl (2000)
			43.0 ± 2.7	
Orhaneli	6	$^{40}Ar/^{39}Ar$ bt+hbl	57.9 ± 1.2	Delaloye and Bingöl (2000)
			31.4 ± 0.6	
Tepeldağ	7	U-Pb zrn	54.56 ± 0.21	Kocatürk et al. (2024)
Orhaneli	6	$^{40}Ar/^{39}Ar$ bt+hbl	52.6 ± 0.4	Harris et al. (1994)
			52.4 ± 1.4	
Topuk	5	$^{40}Ar/^{39}Ar$ bt+hbl	47.8 ± 0.4	Harris et al. (1994)
Topuk	5	U-Pb zrn	47.35 ± 0.21	Kocatürk et al. (2024)
			45.90 ± 0.29	
Tepeldağ (Gürgenyayla)	7	U-Pb zrn	44.9 ± 0.2	Okay and Satir (2006)
Tepeldağ (Gürgenyayla)	7	Rb-Sr bt	44.7 ± 0.4	Okay and Satir (2006)
Eastern Tavşanli granitoids				
Kaymaz	9	U-Pb zrn	84.98 ± 6.27	Gautier (1984)
Sivrihisar	10	U-Pb zrn	79.9 ± 8.6	Shin et al. (2013)
			42.4 ± 2.4	
Sarıkavak (Topkaya)	11	U-Pb zrn	65.9 ± 3.8	Gautier (1984)
Sivrihisar	10	$^{40}Ar/^{39}Ar$ bt+hbl	62.9 ± 1.3	Delaloye and Bingöl (2000)
			56.3 ± 0.2	
Karacaören (Günyüzü)	10	$^{40}Ar/^{39}Ar$ hbl+bt	59.3 ± 3.0	Demirbilek et al. (2018)
			46.7 ± 2.3	
Tekören granodiorite (Günyüzü)	10	U-Pb zrn	57.5 ± 0.8	Okay et al. (2022)
Tekören granodiorite (Günyüzü)	10	$^{40}Ar/^{39}Ar$ hbl+bt	57.3 ± 2.3	Demirbilek et al. (2018)
			23.4 ± 1.1	
Dinek granodiorite (Günyüzü)	10	$^{40}Ar/^{39}Ar$ hbl+kfs	55.9 ± 2.7	Demirbilek et al. (2018)
			45.3 ± 1.8	
Kadıncık (Günyüzü)	10	U-Pb zrn	56.5 ± 0.4	Okay et al. (2022)
Sivrihisar	10	U-Pb zrn	56.5 ± 0.6	Okay et al. (2022)
Sivrihisar	10	U-Pb zrn	55.52 ± 0.61	Ersoy et al. (2023)
Sivrihisar	10	U-Pb zrn	55.3 ± 0.3	Okay et al. (2022)
Kaymaz	9	U-Pb zrn	55.3 ± 1.2	Okay et al. (2022)
Sivrihisar (Karacaören/Günyüzü)	10	U-Pb zrn	54.7 ± 3.1	Okay et al. (2022)
Kaymaz	9	$^{40}Ar/^{39}Ar$ kfs	54.0 ± 2.1	Demirbilek et al. (2018)
			52.1 ± 2.0	
Kaymaz	9	U-Pb zrn	53.73 ± 0.54	Ersoy et al. (2023)
Sivrihisar	10	$^{40}Ar/^{39}Ar$ hbl	53.2 ± 2.1	Demirbilek et al. (2018)
			44.7 ± 1.7	

(*Continued*)

Table 5.2 (Continued)

Granite	Location[a]	Approach[b]	Age	Reference
Kadıncık (Günyüzü)	10	$^{40}Ar/^{39}Ar$ hbl+wr	52.8 ± 2.4	Demirbilek et al. (2018)
			45.7 ± 1.7	
Kadıncık (Günyüzü)	10	U-Pb zrn	51.30 ± 0.61	Ersoy et al. (2023)
			40.85 ± 0.34	
Sivrihisar (Kadnıcık/Günyüzü)	10	Rb-Sr kfs+bt	47.0 ± 1.6	Bağcı (2012)
Sivrihisar	10	$^{40}Ar/^{39}Ar$ kfs	46.02 ± 0.21	This study
Sarıkavak (Topkaya)	11	U-Pb zrn	45.3 ± 0.2	Okay et al. (2022)
Kaymaz	9	U-Pb zrn	44.3 ± 4.9	Shin et al. (2013)
			19.4 ± 4.5	
Sarıkavak (Topkaya)	11	U-Pb zrn	44.30 ± 0.69	Ersoy et al. (2023)
Sivrihisar (Karacaören /Günyüzü)	10	Rb-Sr kfs+bt	40.8 ± 3.0	Bağcı (2012)

[a] See Figure 5.4 for the locations of these granite bodies.

[b] Abbreviations after Whitney and Evans (2010); wr, whole rock.

Table 5.3 A summary of some dates of granitic assemblages associated with rocks between the Sakarya and İstanbul Zones.

Granite	Location[a]	Approach[b]	Age	Reference
Middle Eocene magmatic rocks (South Marmara granitoids)				
Şevketiye	12	$^{40}Ar/^{39}Ar$ ms	71.9 ± 1.8	Delaloye and Bingöl (2000)
İlyasdağ tonalite (Marmara Island)	13	U-Pb zrn	56.7 ± 0.8 46.1 ± 0.7	Ustaömer et al. (2009)
Karabiga (Lapseki)	14	U-Pb xtm	52.7 ± 1.9	Beccaletto et al. (2007)
Karabiga (Lapseki)	14	U-Pb zrn	48.27 ± 0.21 47.06 ± 0.32	Özdamar et al. (2023)
Fıstıklı (Armutlu–Yalova)	15	$^{40}Ar/^{39}Ar$ bt+ms	48.2 ± 1.0 34.3 ± 0.9	Delaloye and Bingöl (2000)
Fıstıklı (Armutlu–Yalova)	15	U-Pb zrn	48.0 ± 0.4	Okay et al. (2022)
Karabiga (Lapseki)	14	$^{40}Ar/^{39}Ar$ bt	47.34 ± 0.43 46.30 ± 0.52	Özdamar et al. (2023)
Karabiga (Lapseki)	14	$^{40}Ar/^{39}Ar$ bt	45.3 ± 0.9	Delaloye and Bingöl (2000)
Kapidağ	16	$^{40}Ar/^{39}Ar$ hbl+bt	42.2 ± 1.0 38.2 ± 0.8	Delaloye and Bingöl (2000)
Avşa Island	17	K-Ar bt	40.9 ± 1.1	Karacık et al. (2008)

[a] See Figure 5.4 for the locations of these granite bodies.
[b] Abbreviations after Whitney and Evans (2010).

subduction and represents the Mesozoic to Eocene closing of the northern branch of the Neo-Tethyan Ocean (Okay, 1986, 2008; Okay & Kelley, 1994; Sherlock et al., 1999; Moix et al., 2008, Shin et al., 2013; Plunder et al., 2013; Fornash & Whitney, 2020). The Tavşanlı Zone is narrow (~50 km) and trends E-W for approximately 250–350 km (Okay & Whitney, 2010; Plunder et al., 2013). The western and central portions contain blueschist facies metavolcanic and metasedimentary rocks with rare metabasalts (Okay, 1980, 1981, 1982; Okay & Kelley, 1994, see Seaton et al., 2009).

The Sivrihisar Massif further to the east is the only portion of the Tavşanlı Zone to contain eclogite, blueschist and Barrovian sequences (Figures 5.4 and 5.5) (Gautier, 1984; Seaton et al., 2009). Rb-Sr and $^{40}Ar/^{39}Ar$ phengite ages from the Sivrihisar Massif constrain high-pressure/low-temperature (HP/LT) metamorphism to ~88–80 Ma (Sherlock et al., 1999; Seaton et al., 2009; Whitney et al., 2011; Pourteau et al., 2013; Shin et al., 2013). Older dates from the HP/LT assemblages reported from the western portion of the Tavşanlı Zone may suffer from excess argon (see the review in Shin et al., 2013). Barrovian-metamorphosed marble from the Sivrihisar Massif contains ~59 Ma muscovite ($^{40}Ar/^{39}Ar$), timing their exhumation (Seaton et al., 2009). Late Cretaceous and early Paleocene ages are also reported from eastern Tavşanlı Zone granitoids, which are medium to high K, calc-alkaline, metaluminous, I-type, and post-collisional (Kaymaz, Sivrihisar, Sarıkavak [Topkaya], Günyüzü [Karacaören, Tekören, Dinek, Kadıncık bodies]; Figures 5.4 and 5.5; Table 5.2) (Shin et al., 2013; Demirbilek et al., 2018; Ersoy et al., 2023). However, these results are interpreted as inheritance (Shin et al., 2013; Demirbilek et al., 2018). The Sivrihisar granite's age is often cited as 53 ± 3 Ma based on a hornblende $^{40}Ar/^{39}Ar$ date affected by excess argon (Sherlock et al., 1999) (Figure 5.5b,c). However, the Sivrihisar granite contains zircon that is 78.4 ± 8.5 Ma (likely inherited) to 41.9 ± 2.3 Ma (U-Pb, ±1σ, Shin et al., 2013). Figure 5.5b,c shows the K-feldspar $^{40}Ar/^{39}Ar$ age from the same sample, which yields a plateau of 46.02 ± 0.21 Ma (MSWD 4.21), similar to those reported for the Sivrihisar and nearby Kaymaz granite and SZGs (Table 5.2). The flat $^{40}Ar/^{39}Ar$ age spectrum is consistent with rapid cooling during exhumation (Figure 5.5d). Paleocene-Eocene dates from the Tavşanlı Zone granites mark the timing of the closure of the İAESZ (Okay et al., 2020).

The Tavşanlı zone is one component of the larger Anatolide-Tauride block, a microcontinent that rifted away from the northern margin of Gondwana beginning in the early Permian (Figures 5.2, 5.3, and 5.4) (Stampfli & Kozur, 2006) or Triassic (Şengör & Yılmaz, 1981; Şengör et al., 1984; Okay & Tüysüz, 1999; Robertson & Ustaömer, 2009a, 2009b). The Taurides comprise the southern portion of the Anatolide-Tauride block and are Neoproterozoic-Early Cambrian (Infracambrian) basement overlain by Cambrian to Eocene marine sediments (Gutnic et al., 1979; Özgül, 1997; Candan et al., 2016). The Anatolide terrane is the metamorphic equivalent of the Taurides and is subdivided into zones based on lithologies and the type and age of metamorphism (see the review in Bozkurt & Oberhänsli, 2001; Candan et al., 2016; Moix et al., 2008). These include the Tavşanlı Zone, Afyon Zone, Menderes Massif, and Lycian nappes

Table 5.4 Summary of some dates of granitic assemblages that intrude the Central Sakarya Zone.

Granite	Location[a]	Approach[b]	Age	Reference
Late Pan-African granitoids or Cadomian granitoids				
Pamukova	18	U-Pb zrn	582.0 ± 9.1 446.0 ± 3.8	Okay et al. (2008)
Gemlik	15	U-Pb zrn	575.5 ± 3.6 438.9 ± 4.5	Okay et al. (2008)
Söğüt granite	19	U-Pb zrn	588.0 ± 4.3 575.2 ± 5.1	Karsli et al. (2023)
Ordovician				
Söğüt granite	19	U-Pb zrn	482.0 ± 6.5 461.0 ± 7.7	Karsli et al. (2023)
Silurian-Devonian				
Sarıcakaya	19	U-Pb zrn	419 ± 6 434 ± 7 319 ± 5 Ma	Topuz et al. (2020)
Carboniferous				
İnhisar	18	$^{40}Ar/^{39}Ar$ ms+chl	348.5 ± 6.6 213.5 ± 4.4	Delaloye and Bingöl (2000)
Geyve	20	U-Pb zrn	327 ± 12	Ustaömer et al. (2016)
Söğüt granite (Sarıcakaya, Çaltı)	19	U-Pb zrn	327.2 ± 1.9	Ustaömer et al. (2012)
Söğüt granite (Sarıcakaya, Küplü)	19	U-Pb zrn	324.3 ± 1.3	Ustaömer et al. (2012)
Söğüt granite (Sarıcakaya, Borçak)	19	U-Pb zrn	319.5 ± 1.1	Ustaömer et al. (2012)
Bilecik	21	$^{40}Ar/^{39}Ar$ bt+or	312.1 ± 6.0 233.5 ± 4.8	Delaloye and Bingöl (2000)
Permian				
Söğüt granite	19	$^{40}Ar/^{39}Ar$ bt	290 ± 4.8	Okay et al. (2002)
Jurassic to Late Cretaceous				
Pamukova	18	$^{40}Ar/^{39}Ar$ or +chl	168.2 ± 3.5 123.0 ± 2.8	Delaloye and Bingöl (2000)
Beypazarı	22	U-Pb zrn	95.4 ± 4.2 70.5 ± 3.4	Speciale et al. (2012)
Beypazarı	22	$^{40}Ar/^{39}Ar$ bt	80.1 ± 1.4 79.2 ± 0.9	Okay et al. (2020)
Beypazarı	22	U-Pb zrn	74.8 ± 0.4 73.2 ± 1.4	Okay et al. (2020)
Beypazarı	22	$^{40}Ar/^{39}Ar$ hbl	82.9 ± 1.8 77.7 ± 4.5	Delaloye and Bingöl (2000)

[a] See Figure 5.4 for the locations of these granite bodies.
[b] Abbreviations after Whitney and Evans (2010); wr, whole rock.

Table 5.5 Summary of some dates of granitic assemblages that intrude into the Western Pontide Zone.

Granite	Location[a]	Approach[b]	Age	Reference
Proterozoic				
Karacabey (Taşmalı)	23	U-Pb zrn (inherited cores)	1961.9 ± 16.4	Aysal et al. (2012)
			804 ± 10.5	
Evciler (Kazdağ)	24	U-Pb zrn	805, 286	Ustaömer et al. (2012)
Karaburun	25	U-Pb zrn	1800, 960, 380, 297	Ustaömer et al. (2012)
Devonian				
Güveylerobası (Çamlık-related)	26	U-Pb zrn	401.5 ± 4.8	Aysal et al. (2012)
Karacabey (Taşmalı)	23	U-Pb zrn	400.3 ± 1.4	Aysal et al. (2012)
Eybek (Çamlık)	27	U-Pb zrn	397.5 ± 1.4	Okay et al. (2006)
Karacabey (Taşmalı)	23	Pb-Pb zrn	395.9 ± 4.1	Sunal (2012)
			393.8 ± 2.7	
Güveylerobası	25	U-Pb zrn	371.2 ± 2.3	Ustaömer et al. (2016)
Permo-Triassic				
Karacabey (Taşmalı)	23	$^{40}Ar/^{39}Ar$ bt	304.5 ± 3.7	Sunal (2012)
			223.0 ± 7.5	
Karacabey (Taşmalı)	23	$^{40}Ar/^{39}Ar$ bt	298.3 ± 5.8	Delaloye and Bingöl (2000)
			199.4 ± 4.0	
Kozak	28	U-Pb zrn	280.2 ± 18.2	Black et al. (2013)
			259.1 ± 13.8	
Karaburun	25	U-Pb zrn	244.4 ± 1.5	Ustaömer et al. (2012)
Evciler	24	U-Pb zrn	229.6 ± 0.60	Ustaömer et al. (2012)
Karacabey (Taşmalı)	23	(U/Th)-He zrn	93.0 ± 6.9	Sunal (2012)
Late Eocene-Oligocene-Miocene				
Kozak	28	$^{40}Ar/^{39}Ar$ or +bt	37.6 ± 3.3	Delaloye and Bingöl (2000)
			19.5 ± 0.4	
Kozak	28	U-Pb zrn	36.5 ± 6.6	Black et al. (2013)
			17.1 ± 0.7	
Evciler (Kazdağ)	24	$^{40}Ar/^{39}Ar$ chl+bt	36.0 ± 1.4	Delaloye and Bingöl (2000)
			26.4 ± 0.6	
Uludağ	29	U-Pb zrn	34.71 ± 0.34	Topuz and Okay (2017)
			28.24 ± 0.39	
Eybek	27	U-Pb zrn	32.5 ± 3.0	Black et al. (2013)
			21.0 ± 1.2	
Uludağ	29	U-Pb zrn	29.41 ± 0.24	Özdamar et al. (2024)
Evciler (Kazdağ)	24	$^{40}Ar/^{39}Ar$ hbl	28.32 ± 0.09	Özdamar et al. (2024)
Evciler (Kazdağ)	24	^{207}Pb-^{206}Pb zrn	28.2 ± 4.1	Erdoğan et al. (2013)
			26.0 ± 5.6	
Evciler (Kazdağ)	24	U-Pb zrn	28.21 ± 0.18	Özdamar et al. (2024)

(Continued)

Table 5.5 (Continued)

Granite	Location[a]	Approach[b]	Age	Reference
Katrandağ	30	$^{40}Ar/^{39}Ar$ hbl+chl	27.6 ± 0.6	Delaloye and Bingöl (2000)
			24.7 ± 0.6	
Uludağ	29	$^{40}Ar/^{39}Ar$ bt	26.8 ± 0.8	Delaloye and Bingöl (2000)
			24.7 ± 0.7	
Uludağ	29	$^{40}Ar/^{39}Ar$ bt	26.73 ± 0.05	Özdamar et al. (2024)
Eybek	27	$^{40}Ar/^{39}Ar$ bt	22.63 ± 0.07	Özdamar et al. (2024)
Eybek	27	$^{40}Ar/^{39}Ar$ bt	26.6 ± 0.8	Delaloye and Bingöl (2000)
			21.1 ± 0.4	
Eybek	27	$^{40}Ar/^{39}Ar$ hbl	26.57 ± 0.15	Özdamar et al. (2024)
Eybek	27	U-Pb zrn	26.31 ± 0.19	Özdamar et al. (2024)
Çataldağ (Bozenköy)	31	K-Ar bt+hbl	25.9 ± 0.5	Boztuğ et al. (2009)
			21.27 ± 0.44	
Evciler (Kazdağ)	24	$^{40}Ar/^{39}Ar$ bt	25.08 ± 0.11	Özdamar et al. (2024)
Evciler (Kazdağ)	24	Rb-Sr	25.0 ± 0.3	Birkle (1992)
				Genç (1998)
Evciler (Kazdağ)	24	U-Pb zrn	24.8 ± 4.6	Erdoğan et al. (2013)
Kozak	28	K-Ar bt+hbl	23.0 ± 3.8	Boztuğ et al. (2009)
			14.6 ± 1.0	
Çataldağ (Çataltepe)	31	K-Ar bt	22.0 ± 0.3	Boztuğ et al. (2009)
			21.7 ± 0.1	
Çataldağ (Turfandağ)	31	K-Ar bt	21.9 ± 0.6	Boztuğ et al. (2009)
			21.2 ± 0.6	
Çataldağ (Balıçıkhisar)	31	$^{40}Ar/^{39}Ar$ bt	20.8 ± 0.4	Delaloye and Bingöl (2000)
Kozak	28	$^{40}Ar/^{39}Ar$ hbl	20.78 ± 0.06	Özdamar et al. (2024)
Evciler (Kazdağ)	24	Rb-Sr	20.7 ± 0.2	Okay and Satir (2000)
			20.5 ± 0.2	
Kozak	28	$^{40}Ar/^{39}Ar$ bt	19.87 ± 0.05	Özdamar et al. (2024)
Kozak	28	U-Pb zrn	19.80 ± 0.20	Özdamar et al. (2024)
Younger South Marmara granitoid bodies				
Yenice	32	$^{40}Ar/^{39}Ar$ hbl	47.6 ± 1.4	Delaloye and Bingöl (2000)
			20.1 ± 1.1	
Ilıca	33	K-Ar hbl	37.9 ± 0.1	Boztuğ et al. (2009)
			25.6 ± 1.9	
Yenice	32	U-Pb zrn	24.39 ± 0.18	Özdamar et al. (2024)
Yenice	32	$^{40}Ar/^{39}Ar$ hbl	24.07 ± 0.08	Özdamar et al. (2024)
Kızıldam	34	K-Ar wr+bt	23.9 ± 0.6	Karacık et al. (2008)
			20.7 ± 0.8	

Yenice	32	$^{40}Ar/^{39}Ar$ bt	23.22 ± 0.09	Özdamar et al. (2024)
Danişment	35	K-Ar wr+bt	23.2 ± 1.1	Karacık et al. (2008)
			22.1 ± 0.6	
Ilıca	33	K-Ar wr+bt	22.8 ± 0.5	Karacık et al. (2008)
			18.4 ± 2.2	
Ilıca	33	$^{40}Ar/^{39}Ar$ hbl	22.31 ± 0.18	Özdamar et al. (2024)
Ilıca	33	U-Pb zrn	22.26 ± 0.13	Özdamar et al. (2024)
Sarılıoluk	36	K-Ar hbl	22.6 ± 0.8	Karacık et al. (2008)
Yenice	32	K-Ar wr+bt	21.9 ± 1.1	Karacık et al. (2008)
			18.8 ± 1.3	
Davutlar	37	K-Ar wr+bt	21.6 ± 0.6	Karacık et al. (2008)
			18.4 ± 1.1	
Ilıca	33	$^{40}Ar/^{39}Ar$ bt	21.55 ± 0.10	Özdamar et al. (2024)
Yeniköy	36	K-Ar wr	20.1 ± 1.0	Karacık et al. (2008)

[a] See Figure 5.4 for locations of these granite bodies.
[b] Abbreviations after Whitney and Evans (2010); wr, whole rock.

Table 5.6 Summary of some dates of granitic assemblages that intrude the Rhodope-Strandja Zone (Biga Peninsula only).

Granite	Location[a]	Approach[b]	Age	Reference
Kuşçayır	38	$^{40}Ar/^{39}Ar$ hbl	39.4 ± 0.8 35.7 ± 0.8	Delaloye and Bingöl (2000)
Kestanbol (Ezine)	39	U-Pb zrn	26.2 ± 2.0 18.8 ± 1.0	Black et al. (2013)
Kestanbol (Ezine)	39	$^{40}Ar/^{39}Ar$ hbl	22.70 ± 0.99	Doner et al. (2024)
Kestanbol (Ezine)	39	$^{40}Ar/^{39}Ar$ lct & bt	22.21 ± 0.07 21.22 ± 0.09	Akal (2013)
Kestanbol (Ezine)	39	U-Pb zrn	21.91 -21.52	Doner et al. (2024)
Kestanbol (Ezine)	39	$^{40}Ar/^{39}Ar$ bt	21.41 ± 0.40	Doner et al. (2024)
Kestanbol (Ezine)	39	$^{40}Ar/^{39}Ar$ hbl	20.5 ± 0.6	Delaloye and Bingöl (2000)

[a] See Figure 5.4 for locations of these granite bodies.
[b] Abbreviations after Whitney and Evans (2010).

(Figures 5.2, 5.3, and 5.4). The Tavşanlı and Afyon Zones are sometimes considered part of a single Kütahya-Bolkardağ Belt (Özcan et al., 1988; Göncüoğlu et al., 1997, 2012).

A series of granite bodies intrude into the IPS between the Sakarya and İstanbul Zones with ages that resemble the SZGs and eastern portions of the Tavşanlı Zone. These Middle Eocene magmatic rocks (MEMR), also known as the South Marmara granitoids (Şevketiye, İlyasdağ tonalite [Marmara Island], Karabiga [Lapseki], Fıstıklı [Armutlu–Yalova], Kapidağ, and Avşa Island; Figure 5.4; Table 5.3) are located in close association with the IPS and range from the Late Cretaceous (71.9 ± 1.8 Ma) to Late Eocene (34.3 ± 0.9 Ma). The MEMR are unique in these ages, as further east, along strike of the IPS and into the central portion of the Sakarya Zone, some of the oldest plutons in Western Anatolia are exposed (Pamukova, Gemlık, İnhısar, Geyve, Bilecik, Söğüt; Figure 5.4; Table 5.4.). Some intrusions are associated with economically important kaolinite deposits (Kadir & Kart, 2009). The Cambrian Gemlik granite body is located in the vicinity of the MEMR granites (Figure 5.4). Its age is more consistent with Cadomian Orogeny (650–550 Ma) granites further north in the İstanbul-Zonguldak and Strandja zones (Şahin et al., 2014) and similar-age rocks from the basement or core of the Afyon Zone and Menderes Massif (Dannat, 1997; Loos & Reischmann, 1999; Şahin et al., 2014; Hetzel & Reischmann, 1996). Western Anatolian granites with Cambrian ages are termed Late Pan-African granitoids or Cadomian granitoids and are associated with tectonic events along the northern margin of Gondwana (Gürsu & Göncüoğlu, 2006; Şahin et al., 2014). We identify some of these granites in their particular zones in Figure 5.4 and distinct sections of Tables 5.1, 5.4, and 5.8. Note that the entire core of the Menderes Massif is considered Pan-African (primarily late Neoproterozoic to Cambrian) basement (see the review in Oberhäensli et al., 2010).

Proterozoic zircon dates are found in the Pontides zone, but some of its central and western granite assemblages also record Silurian-Devonian dates (Sarıcakaya, Table 5.4; Karaburun, Güveylerobası [Çamlık-related], Karacabey [Taşmalı], Eybek [Çamlık], Güveylerobası, Table 5.5; Figure 5.4). These ages are linked to the amalgamation of a fragment of Avalonia terrane in a subduction-zone-type setting (Aysal et al., 2012; Sunal, 2012; Topuz et al., 2020). Variscan-age (Carboniferous) granites are also reported for granites in the Central and Western Sakarya Zone and Afyon Zone (Tables 5.4. and 5.6; Figure 5.4). Some of these results could represent inherited cores or xenocrystic grains from the surrounding metamorphic assemblages. For example, the Miocene-age Alaçam granite in the Afyon Zone has reported Carboniferous ages, but the older dates were likely entrained from its basement units (Hasözbek et al. 2010; Candan et al. 2016).

The Afyon Zone is considered the southward palaeogeographic extension of the Tavşanlı Zone (Candan et al., 2005; Pourteau et al., 2010; Akal, 2013; Özdamar et al., 2013). Although it is often mapped as closely and narrowly paralleling the Tavşanlı Zone, the southern extent of the Afyon Zone is unclear, and a portion may also be exposed between the southern Menderes Massif and Lycian Nappes (Okay, 1986; Candan et al., 2005; Pourteau et al., 2013; Ustaömer et al., 2020). The zone has also been termed the Afyon-Bolkardağ Zone (Okay, 1986; Özdamar et al., 2013) and Ören-Afyon Zone (Pourteau et al., 2013). The zone consists of Pan-African-related basement underlying shelf-type Paleozoic-Mesozoic sequence of the Taurides and metasedimentary and metavolcanic rocks, portions of which have undergone regional greenschist to blueschist facies (Fe-Mg carpholite and glaucophane) metamorphism (Figure 5.3)

Table 5.7 Brief summary of some dates from granitic assemblages that intrude the Afyon Zone.

Granite	Location[a]	Approach[b]	Age	Reference
Paleozoic granitoids				
Sandıklı	39	U-Pb zrn	541 ± 9	Gürsu et al. (2004)
Alaçam (basement)	41	U-Pb zrn	331.3 ± 1.7 314.3 ± 4.8	Candan et al. (2016)
Alaçam (basement)	41	U-Pb zrn	314.9 ± 2.7	Hasözbek et al. (2010)
Late Eocene-Oligocene-Miocene				
Balkan (Muratdağ)	40	$^{40}Ar/^{39}Ar$ or	35.5 ± 3.0	Delaloye and Bingöl (2000)
Koyunoba	42	U-Pb zrn	30.0 ± 3.9 14.7 ± 2.6	Catlos et al. (2012)
Alaçam	41	$^{40}Ar/^{39}Ar$ or	27.1 ± 1.0 18.5 ± 1.8	Delaloye and Bingöl (2000)
Alaçam	41	U-Pb zrn	25.3 ± 1.5 17.5 ± 0.9	Catlos et al. (2012)
Eğrigöz	43	$^{40}Ar/^{39}Ar$ bt+or	24.6 ± 1.4 20.0 ± 0.7	Delaloye and Bingöl (2000)
Eğrigöz	43	U-Pb zrn	24.1 ± 1.3 5.7 ± 0.6	Catlos et al. (2012)
Eğrigöz	43	U-Pb zrn	20.7 ± 0.6	Ring and Collins (2005)
Koyunoba	42	$^{40}Ar/^{39}Ar$ kfs	20.37 ± 0.03	Etzel et al. (2020)
Alaçam	41	Rb-Sr bt	20.17 ± 0.20 20.01 ± 0.20	Hasözbek et al. (2010)
Eğrigöz	43	$^{40}Ar/^{39}Ar$ ms	20.2 ± 0.3	Işık et al. (2004)
Eğrigöz	43	$^{40}Ar/^{39}Ar$ kfs	20.02 ± 0.03	Etzel et al. (2020)
Alaçam	41	U-Pb zrn	20.0 ± 1.4 20.3 ± 3.3	Hasözbek et al. (2010)
Baklan	40	U-Pb zrn	19.5 ± 0.2	Özdamar et al. (2024)
Baklan	40	$^{40}Ar/^{39}Ar$ hbl	19.387 ± 0.068	Özdamar et al. (2024)
Baklan	40	$^{40}Ar/^{39}Ar$ wr	19.3 ± 0.9 17.8 ± 0.7	Aydoğan et al. (2008)
Baklan	40	$^{40}Ar/^{39}Ar$ bt	18.861 ± 0.065	Özdamar et al. (2024)

[a] See Figure 5.4 for locations of these granite bodies.
[b] Abbreviations after Whitney and Evans (2010); wr, whole rock.

Table 5.8 Brief summary of some dates of granitic assemblages that intrude the Menderes Massif.

Granite	Location[a]	Approach[b]	Age	Reference
Late Pan-African granitoids or Cadomian granitoids				
Çine Massif metagranites	North of Milas	U-Pb zrn	662 ± 3	Loos and Reischmann (1999)
			517 ± 6	
Demirci–Gördes		$^{207}Pb/^{206}Pb$ zrn	537.2 ± 2.4	Dannat (1997)
			544.1 ± 4.3	
Ödemiş–Kiraz		$^{207}Pb/^{206}Pb$ zrn	528.0 ± 4.3	Dannat (1997)
			570 ± 5	
Ödemiş–Kiraz		$^{207}Pb/^{206}Pb$ zrn	546.0 ± 1.6	Hetzel and Reischmann (1996)
			546.4 ± 0.8	
Çine Massif		$^{207}Pb/^{206}Pb$ zrn	521 ± 5	Loos and Reischmann (1999)
			572 ± 7	
Bafa Lake-Çine Massif		U-Pb zrn	541 ± 14	Gessner et al. (2004)
			566 ± 9	
Yatağan		$^{207}Pb/^{206}Pb$ zrn	555.5 ± 6.2	Dora et al. (2005)
North of Yatağan		U/Pb zrn	549 ± 26	Dora et al. (2005)
Triassic				
Alaşehir	44	U-Pb zrn	222.9 ± 1.1	Ustaömer et al. (2016)
Late Eocene-Oligocene-Miocene				
Alaşehir	44	$^{40}Ar/^{39}Ar$ bt	36.4 ± 2.2	Delaloye and Bingöl (2000)
			16.6 ± 0.3	
Gördes	45	$^{40}Ar/^{39}Ar$ ms	28.8 ± 0.6	Delaloye and Bingöl (2000)
			19.4 ± 0.7	
Salihli	46	Th-Pb mnz	21.7 ± 4.5	Catlos et al. (2010)
			9.6 ± 1.6	
Turgutlu	47	Th-Pb mnz	19.2 ± 5.1	Catlos et al. (2010)
			11.5 ± 0.8	
Salihli	46	U-Pb ttn	17.07 ± 0.2	Rossetti et al. (2017)
			14.36 ± 0.3	
Turgutlu	47	U-Pb mnz	16.1 ± 0.2	Glodny and Hetzel (2007)
Salihli	46	U-Pb aln	15.0 ± 0.3	Glodny and Hetzel (2007)
Turgutlu	47	$^{40}Ar/^{39}Ar$ kfs	14.06 ± 0.03	Etzel et al. (2020)
Salihli	46	$^{40}Ar/^{39}Ar$ kfs	5.05 ± 0.02	Etzel et al. (2020)

[a] See Figure 5.4 for locations of these granite bodies.
[b] Abbreviations after Whitney and Evans (2010).

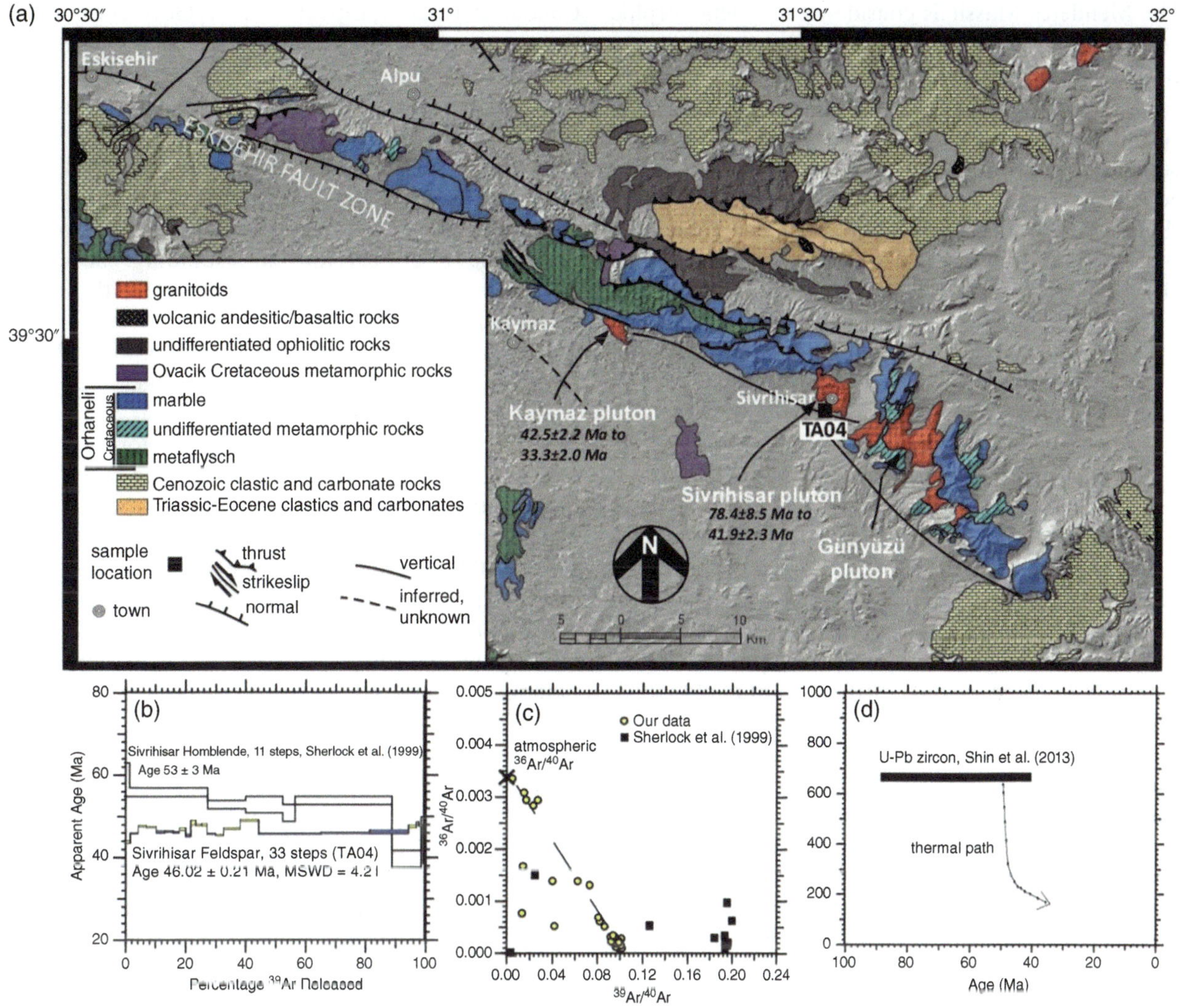

Figure 5.5 (a) A simplified geologic map of the Sivrihisar Massif (eastern Tavşanlı Zone) overlain on a hillshade raster. Map after Senel and Aydal (2002), Özsayin and Dirik (2007), and Shin et al. (2013). See the data repository for the color figure. (b) Sivrihisar granite K-feldspar age spectra for sample TA04. The upper profile by Sherlock et al. (1999); the lower are our results. (c) $^{36}Ar/^{40}Ar$ versus $^{39}Ar/^{40}Ar$ plot comparing our data to Sherlock et al. (1999). Our results show mixing between a radiogenic and atmospheric component of argon with four lower points from initial isothermal steps. Sherlock et al. (1999) data are affected by excess argon (ArE). (d) One possible thermal history path for the Sivrihisar granite is based on the rapidly cooled K-feldspar ages, zircon ages, and zircon saturation temperature from Shin et al. (2013).

(Okay, 1984; Candan et al., 2005, Pourteau et al., 2010; Özdamar et al., 2013). In this sense, its stratigraphy resembles that of the Tavşanlı Zone (Candan et al., 2005). Rhyolitic volcanic assemblages contain zircon that crystallized in the Late Triassic time extension along the northern margin of Gondwana as the Neo-Tethyan Ocean developed (230 ± 2 Ma and 229 ± 2 Ma; Özdamar et al., 2013). Triassic ages reported for granitic assemblages found within the İstanbul-Zonguldak Zone, central and western Sakarya, and Menderes Massif are also attributed to this event (Figure 5.4; Okay et al., 2020). LT/HP metamorphism in the Afyon Zone is thought to have occurred at 70–65 Ma coincident with the closure of the Neo-Tethyan Ocean (Pourteau et al., 2010, 2013; Özdamar et al., 2013; Plunder et al., 2013). Based on zircon ages from granites intruding Tavşanlı Zone blueschist and altered ophiolitic assemblages, portions of the Afyon Zone may have subducted beneath the Tavşanlı Zone during the Late Cretaceous (Speciale et al., 2012; Shin et al., 2013). Upper Paleocene-Lower Eocene sedimentary rocks overlie the metamorphic rocks of the Afyon Zone (Candan et al., 2005).

The Menderes Massif is considered the metamorphic basement on which the rocks of the Afyon Zone were deposited before regional metamorphism (Okay, 1984). The Menderes Massif exposes ~40,000 km^2 of metamorphic and igneous rocks, and its stratigraphy was initially described as a gneiss "core" and Paleozoic schist envelope with overlying Mesozoic-Cenozoic marble "cover" (Schuiling, 1962; Dürr, 1975; Şengör et al., 1984). The massif has also been mapped as a large-scale recumbent fold (Okay, 2001; Gessner et al., 2002), a series of nappes stacked during south-directed thrusting (Ring et al., 1999, 2001; Gessner et al., 2001), or north-directed thrusting (Hetzel et al., 1995a, 1995b) (see Gessner et al., 2013). In the nappe model, the core is represented by the Çine and Bozdağ nappes, whereas the cover would be the Bayındır and Selimiye nappes (Ring et al., 2001). However, all nappes may be part of the Menderes Massif core series stacked during Eocene out-of-sequence thrusting (Régnier et al., 2007). Timeframes recorded by the massif begin in the Archean and Neoproterozoic based on zircons extracted from metagranites and orthogneisses with geochemical signatures dominated by reworking of old crust (Oberhäensli et al., 2010; Zlatkin et al., 2013). During this time, the Menderes Massif was part of a collage of terranes associated with NE Africa and Arabia (Şengör et al., 1984; von Raumer et al., 2015). Some Neoproterozoic zircons (ca. 570 Ma) have an older crust signature, but others suggest a proximal juvenile source resembling the Arabian-Nubian shield (Zlatkin et al., 2013).

Cambrian metagranites, orthogneisses, granulites, ecologites, and micaschists are exposed throughout the massif (Hetzel & Reischmann, 1996; Dannat, 1997; Loos & Reischmann, 1999; Neubauer, 2002; Oberhäensli et al., 2010; Zlatkin et al., 2013; Koralay, 2015). Cambrian-Ordovician monazite and zircon inclusions are found in Menderes Massif garnets (Catlos & Çemen, 2005; Etzel et al., 2019). During this time, the Menderes Massif was affected by events related to the Cadomian Orogeny, and its core units were affected by Pan-African S- and I-type granites and metamorphosed (Neubauer, 2002). Other terranes within Western Anatolia likewise have a Cadomian signature (Figure 5.4, Kozur & Göncüoğlu, 1998). Granulite-facies metamorphism in the Menderes Massif is suggested to have occurred at 580.0 ± 5.7 Ma to 660 ± 61 Ma (U-Pb monazite dates, Oelsner et al., 1997; U-Pb zircon ages, Koralay et al., 2006). Middle-Triassic zircons in metagranites are found in its central region (Figure 5.4; Dannat, 1997; Koralay et al., 2001).

The timing of Menderes Massif nappe stacking is thought to have occurred during the Eocene-Oligocene or sometime after the Late Cretaceous (main Menderes metamorphism, [MMM]; Satir & Friedrichsen, 1986; Konak et al., 1987; Dora et al., 1995; Bozkurt & Park, 1999; Bozkurt & Satir, 2000; Bozkurt & Oberhänsli, 2001; Candan et al., 2001; Lips et al., 2001; Gessner et al., 2011). Gessner et al. (2001) report that the Bayındır nappe deformed once during the Eocene related to MMM, whereas the Bozdağ, Çine, and Selimiye nappes record pre-MMM and MMM events. Figure 5.6 shows a paleogeographic reconstruction of the possible setting of the fragments forming Western Anatolia during the closure of the İAESZ during the Eocene. This paleographic timeframe is critical for understanding the complex tectonic scenario that set the scene before the onset of extension.

The Aegean Orogeny (Searle & Lamont, 2020) is proposed for the tectonic history further west of the Menderes Massif, including the Cycladic Metamorphic Core Complexes, but may mirror its development. In this scenario, subduction and a continent-continent collision occur between the Eurasian and Adria-Apulia/Cyclades plates as marked by ophiolite obduction at 74 Ma (Lamont et al., 2020a) and HP eclogite and blueschist facies metamorphism at 57–46.5 Ma (Tomaschek 2003; Lagos et al., 2007; Bulle et al., 2010; Dragovic et al., 2012). The HP metamorphism (P = 11–12 kbar) is documented by ophiolitic mélanges that may record a cycle of Alpine collisional thickening followed by extension and overprinting via extension (Papanikolaou, 1987; Okrusch & Bröcker, 1990; Avigad & Garfunkel, 1991; Katzir et al., 2000; Parra et al., 2002; Laurent et al., 2018; Lamont et al., 2020b). HP metamorphism is recognized as part of an NE-trending subduction-exhumation channel (Xypolias & Alsop, 2014; Laurent et al., 2018; Gerogiannis et al., 2019). Crustal thickening and regional kyanite-sillimanite grade Barrovian-type metamorphism occur from 22–14 Ma, followed by orogenic collapse. The island of Naxos exemplifies the process with structural data that suggest it is the result of the gravitational collapse of the Aegean orogenic wedge (Vanderhaeghe, 2004). This model emphasizes the role of compression in forming Aegean metamorphic core complexes (Coney and Harms, 1984; Searle and Lamont, 2020, 2022), which is an alternative to the perspective of solely extensionally driven core complexes discussed in the next section.

5.2.2. Extensional History (Oligocene-Miocene)

Following the final amalgamation of the various terranes as described in the previous section, Western Anatolia experienced a switch from the dynamics of collision to extension and extrusion (Berckhemer, 1977; Le Pichon & Angelier, 1979, 1981; Şengör & Yılmaz, 1981; Şengör et al., 1985; Meulenkamp et al., 1988; Buick, 1991; Jolivet et al., 1994; Seyitoğlu & Scott, 1996; Okay & Satir, 2000; Bozkurt, 2001; Çemen et al., 2006). A sequence of partial melting, Barrovian metamorphism, and granitoid emplacement has been cited

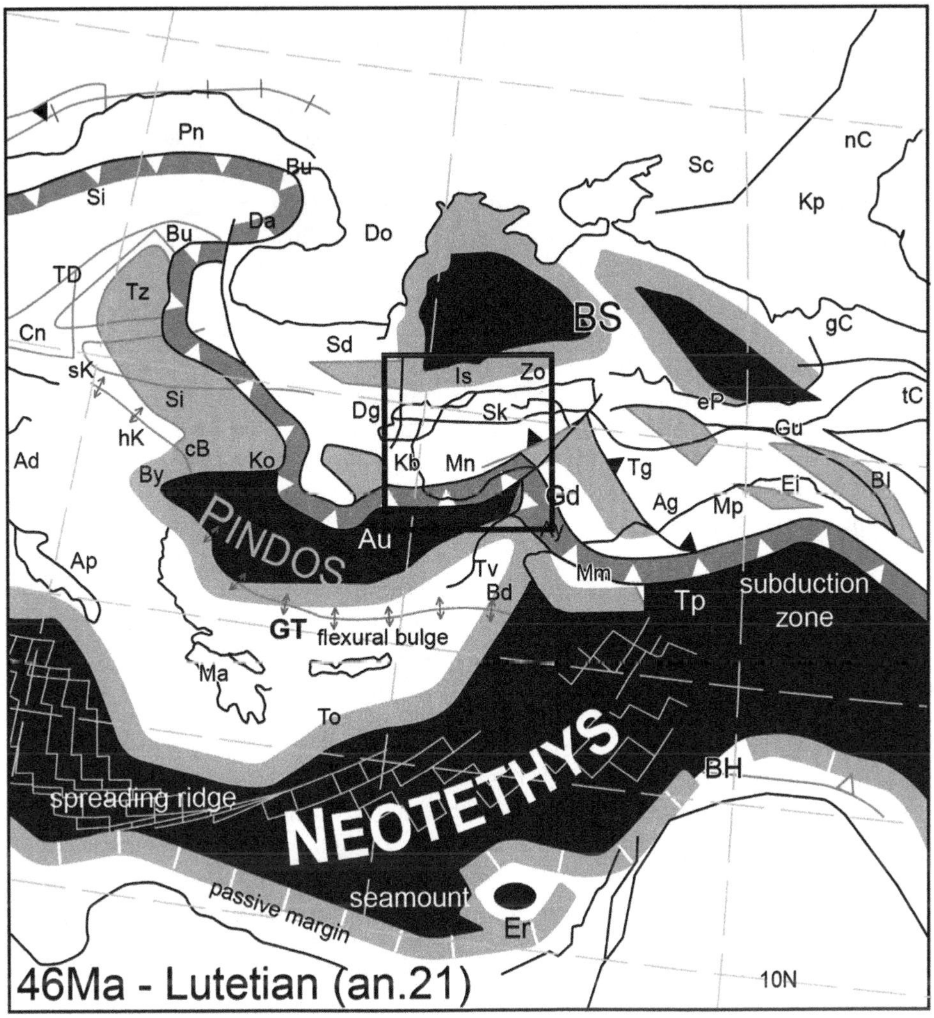

Figure 5.6 Paleogeographic reconstruction of Western Anatolia (center box) and the surrounding region at 46 Ma before the onset of extension (Adapted from Stampfli et al., 2006). Abbreviations relevant to Western Anatolia are Mn, Menderes Massif; Kb, Karaburun; Dg, Denizgören ophiolite; Sk, Sakarya; Is, İstanbul; Zo, Zonguldak; BS, Black Sea; Er, Eratosthenes seamount. For other abbreviations, see Stampfli and Kozur (2006).

for providing evidence of a change from crustal shortening to extensional tectonism (Keay et al., 2001; Altunkaynak, 2007; Dilek & Altunkaynak, 2007; Altunkaynak et al., 2012; Rossetti et al., 2017). The process may be recorded by numerous Oligocene to Miocene-age granites (Figure 5.4, Tables 5.5–5.8.) and linked to the development of metamorphic core complexes located from northeastern Greece and southern Bulgaria through the Aegean Sea and western Turkey.

In continental orogenic domains, metamorphic core complexes are deep crustal domes exhumed and deformed during extension. They are commonly surrounded by sedimentary and volcanic rocks, which may be partly deposited during exhumation (Tirel et al., 2008). Core complexes in western Turkey and the Aegean region include the Rhodope, Kazdağ, Uludağ, Cyclades, Menderes, and Crete massifs (Figures 5.1, 5.2, and 5.4) (Sokoutis et al., 1993; Hetzel et al., 1995a, 1995b; Burg et al., 1996; Lips et al., 1999; Bozkurt & Oberhänsli, 2001; Candan et al., 2001; Lips et al., 2001; Ring et al., 2003; Bozkurt & Sözbilir, 2004; Duru et al., 2004; Vanderhaeghe, 2004; Catlos & Çemen, 2005; Brun & Sokoutis, 2007; Okay et al., 2008; Cavazza et al., 2009; Kruckenberg et al., 2011; Gessner et al., 2013; Baran et al., 2017).

In Western Anatolia specifically, the Menderes, Kazdağ, and Uludağ massifs are central locations for studying postcollision extensional tectonics (Figures 5.1, 5.2, and 5.4) (Şengör et al., 1984, Bozkurt & Park, 1994; Hetzel et al., 1995a, 1995b; Yılmaz et al., 2001; Işik & Tekeli, 2001; Çemen et al., 2006; Topuz & Okay, 2017). The Menderes Massif has global importance due to its role as the largest zone of active continental extension (Jolivet & Faccenna, 2000; Çemen et al., 2006). The region has long attracted the attention of those seeking to understand the driving forces of extension from a variety of perspectives (Lister et al., 1984; Thomson & Ring, 2006; Régnier et al., 2007; Gessner et al., 2013; Uzel et al., 2015). Both low-angle detachment faults and high-angle normal faults bound sedimentary basins and separate the Menderes Massif into northern (Gördes), central (Ödemiş), and southern (Çine) submassifs (Figure 5.2). In the central Menderes Massif, Miocene-age granites are cut by the low-angle Alaşehir detachment, helping to constrain the timing of extension (Alaşehir, Salihli, Turgutlu; Table 5.8.). The Kazdağ Massif is smaller in scale compared to the Menderes Massif and is a NE-SW oriented structural dome or tectonic window flanked by detachment structures (Figures 5.2 and 5.4; Okay et al., 1991; Okay & Satir, 2000; Duru et al., 2004; Bonev et al., 2009; Cavazza et al., 2009). This massif's Evciler (Kazdağ) pluton routinely yields Oligo-Miocene crystallization ages from a range of chronometers (Table 5.5). The Uludağ Massif is NW-SE trending and has high-grade metamorphic and intrusive Eocene-Miocene granitic rocks (Figure 5.4; Table 5.5; Okay et al., 2008). Large Neogene basins bind the northern and southern sections of the Uludağ Massif, and late-stage exhumation is largely thought to have occurred during the Early Miocene (Topuz & Okay, 2017).

Besides these localities, Miocene ages have been reported for granites in the eastern Tavşanlı zone (Table 5.2) (Kaymaz and Tekören granodiorite [Günyüzü]; Shin et al., 2013; Demirbilek et al., 2018). These ages likely represent metamorphism and subsequent alteration associated with the large-scale extension/exhumation affecting Western Anatolia during this time. Early Miocene ages also characterize granites closely associated with the Menderes, Kazdağ, and Uludağ metamorphic core complexes. For example, Miocene ages are reported for a group of granites near the Kazdağ Massif, extensively exposed in the Biga Peninsula and western Pontides (Kozak, Eybek, Katrandağ, Çataldağ [Bozenköy, Çataltepe, Turfandağ, Balıçıkhisar], Kuşçayır, and Kestanbol [Ezine]; Figure 5.4; Table 5.5) and from a series of plutons grouped as the Younger South Marmara Granitoids (Yenice, Ilıca, Kızıldam, Danişment, Sarılıoluk, Davutlar, and Yeniköy; Figure 5.4; Table 5.5; Karacık et al. 2008). North of the Menderes Massif, Miocene-age plutons also intrude the Afyon Zone, in close association with the Simav Fault system, which includes the lower-angle Simav Detachment Fault (SDF) and higher-angle Simav Fault further south (Koyunoba, Alaçam, and Eğrigöz; Figures 5.2, 5.4, and 5.7; Table 5.7; Işik et al., 2003).

The Simav structures are at the boundary between two dynamically distinct regions in western Turkey: a northern component dominated by the NAFZ that accommodates the lateral extrusion of the Anatolian block and a southern zone of large-scale crustal extension (Seyitoğlu, 1997; Ersoy et al., 2010). The Simav Fault is a distinct, high-angle (~45–60°) system that extends ~150 km between the towns of Banaz in the east and Sındırgı in the west (Figure 5.7) (Ambraseys & Tchalenko, 1972; Seyitoğlu, 1997; Ersoy et al., 2010; Hetzel et al., 2013). The structure near the town of Simav has >200 m of relief between the top of the hanging wall and footwall and dips steeply to the north, roughly perpendicular to the current extension direction (Işık et al., 2003). This fault is thought to have formed during the Pliocene and is currently active (Seyitoğlu, 1997; Ring & Collins, 2005). Deciphering the sense of motion of the Simav Fault has implications for understanding Turkey's neotectonic regime and is discussed in the section regarding outstanding questions in Aegean tectonics.

Estimates of timing core complex exhumation and extension in Western Anatolia have relied on calc-alkaline magmatism, widespread continental sedimentation, and mineral chronometers (Sokoutis et al., 1993; Gautier et al., 1999; Catlos & Çemen, 2005; Altunkaynak & Genç, 2008; Brun & Sokoutis 2010; Brun et al., 2016; Özdamar et al., 2024). In some locations, the complexes record progression of magmatism from earlier Eocene-age mantle melts and input from asthenosphere upwelling to later Oligocene to Late Miocene crustal contamination and subduction signatures, with emplacement ages that young to the south (Delaloye & Bingöl, 2000; Altunkaynak & Dilek, 2006; Dilek & Altunkaynak, 2007; Altunkaynak, 2007; Altunkaynak & Genç, 2008; Altunkaynak et al., 2012; Karaoğlu & Helvacı, 2014). However, this simple scenario of melt origin and emplacement can be complicated, as the melts are influenced by varied protoliths of varying sources, ages, and degrees of crustal anatexis (Pe-piper, 2000; Stouraiti et al., 2010, 2018).

Late Cenozoic (since ~32 Ma) plutonic rocks are also widespread in the Aegean (Altherr et al., 1988; Henjes-Kunst et al., 1988; Pe-piper, 2000; Keay et al., 2001; Brichau et al., 2007, 2008). The origin of the granites is linked to subduction migration along the Hellenic arc (Fytikas et al., 1984; Schaarschmidt et al., 2021) or regional, widespread extensional deformation (Boztuğ et al., 2009). Barrovian metamorphism on Naxos is thought to have influenced the development of fluid-fluxed melts at ca. 8–10 kbar between 18.5 and 17 Ma (Lamont et al., 2020a; Searle & Lamont, 2022). Peak metamorphism is thought to have occurred at 20.7–16.7 Ma (Keay et al., 2001). In some locations, coeval mafic and felsic melts were emplaced (Seyitoğlu & Scott, 1996; Aldanmaz et al., 2000; Okay & Satir, 2000; Pe-Piper & Piper, 2001; Ozgenç & Ilbeyli, 2008). Magma compositions were influenced by a range of factors, including inflowing mantle at the site of melting, the nature of the subduction component and the degree of

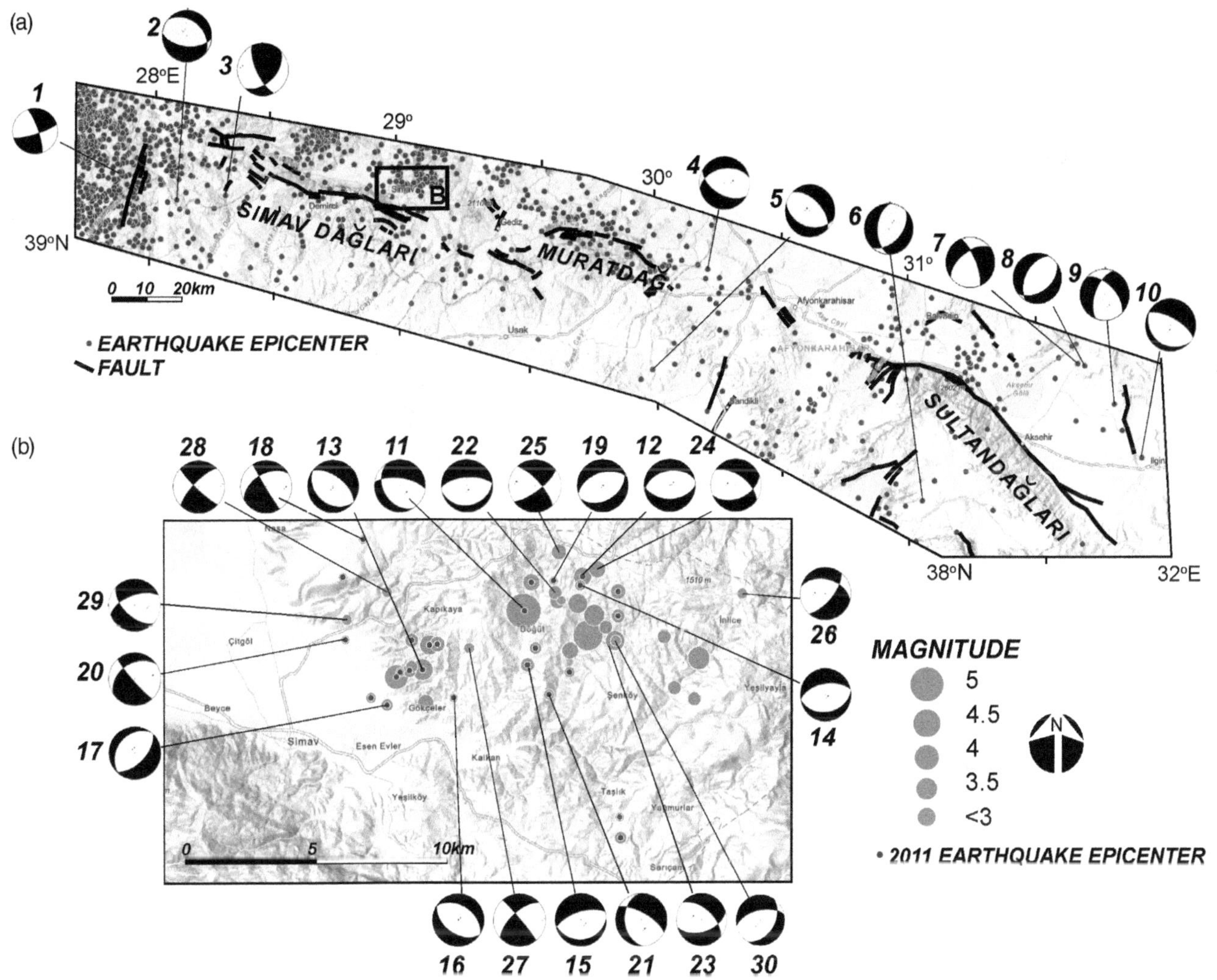

Figure 5.7 (a) Map of the Simav Fault and associated structures. Small dots are extracted from the USGS Earthquake Catalog magnitude 2.5+ (http://earthquake.usgs.gov/earthquakes/search) of events from 1952 to 2021. Location of fault strands after Konak (2002). The inset shows the location near the town on Simav in panel (b). (b) Map of the surrounding area of Simav with earthquakes plotted. In this map, events were extracted from the Turkish Ministry of the Interior, Disaster and Emergency Management Presidency, Earthquake Department Earthquake Catalog (M> = 4.0), 1900–2021 (https://deprem.afad.gov.tr/depremkatalogu). The size of the circle represents magnitude. The figure highlights 2011 earthquakes with additional solid dots. Base maps in both panels are from ESRI. Focal mechanism solutions in both panels were extracted from the Turkish catalog. See Table 5.9. for details of the events. For locations of faults in panel (b), see Mutlu (2020).

interaction between mantle and subduction components, as well as the melting of fluid-rich mantle and the assimilation/crystallization history of the resulting hydrous magma (Pearce & Stern, 2006). Extensive geochemical and isotopic studies of Miocene I-type granitoid plutons of the central Aegean Sea show little evidence of a significant contribution of mantle-derived magmas (Altherr & Siebel, 2002).

Cenozoic magmatism in the Anatolian microplate consists of three distinct, continuous geochemical phases (Innocenti et al., 2005; Dilek & Altunkaynak, 2007; Altunkaynak & Genç, 2008; Akay, 2009; Altunkaynak et al., 2012). Magmatic rocks represent a Late Eocene-Middle Miocene phase with orogenic character and a petrological affinity ranging from calc-alkaline to dominant high-K calc-alkaline to shoshonitic. During the Late Miocene-Early Pliocene, alkaline volcanic rocks appear. The third phase is characterized by Pliocene-Quaternary Na-enriched alkali basalts with an oceanic island basalt (OIB) signature (Aldanmaz, 2012). The first volcanic activity in the South Aegean Active Volcanic Arc occurred between 5 and 2 Ma (Müller et al., 1979; Fytikas et al., 1984; Matsuda et al., 1999; Elburg & Smet, 2020). The driver of extension is widely thought

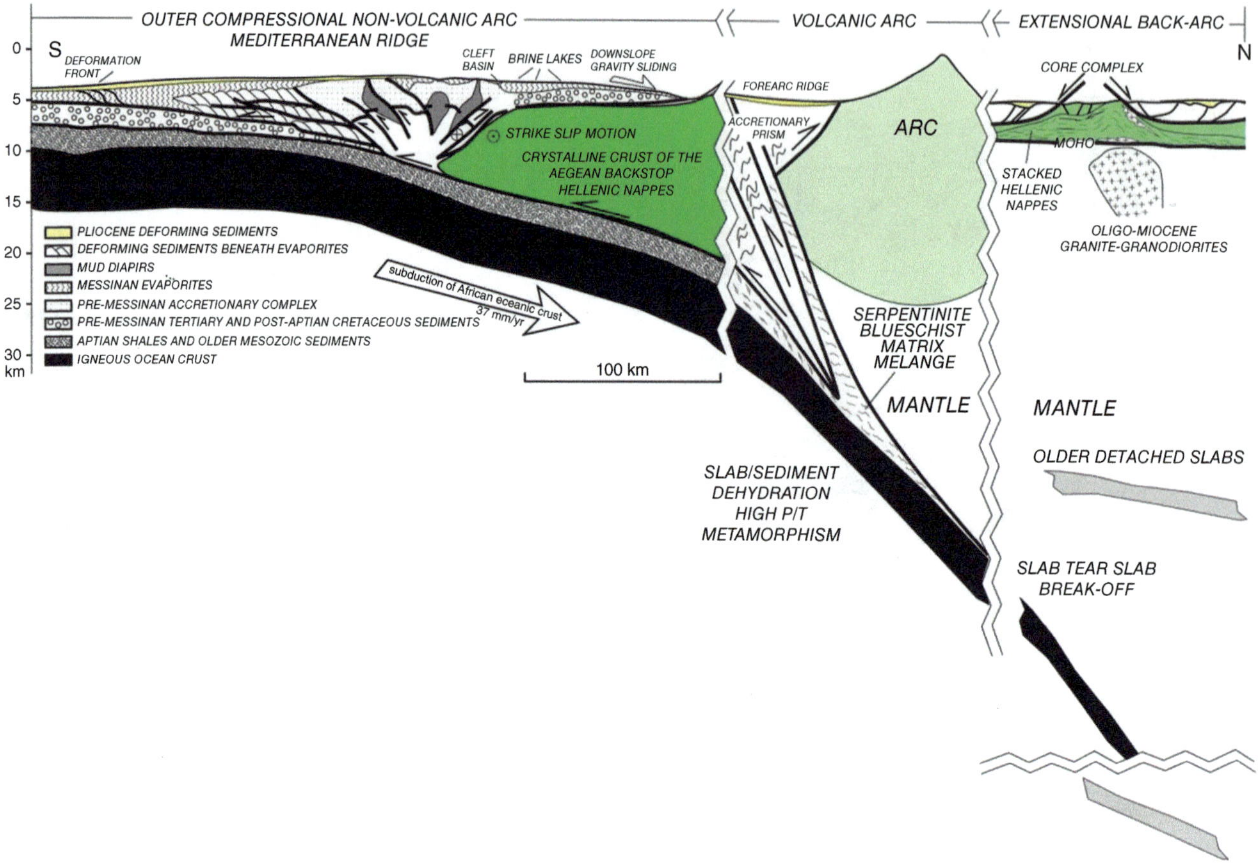

Figure 5.8 North-south generalized cross-section through the Hellenic arc system showing the key structural elements. Map of the Mediterranean Ridge adapted from Westbrook et al. (2002).

to be the rollback of a subducting African slab (Figures 5.8, 5.9, and 5.10) (Jolivet & Faccenna, 2000; Çemen et al., 2006; van Hinsbergen et al., 2010; Royden, 1993; Faccenna et al. 2003, 2014; Brun & Faccenna 2008). We discuss the slab and arc dynamics, geometry, and initiation age in the section regarding outstanding questions in Aegean tectonics.

5.2.3. Strike-Slip History (Late Miocene, Pliocene to Present)

The Aegean and Anatolian microplates have emerged to be type-localities for the model of tectonic escape based on GPS vectors (Reilinger et al., 2006). In this scenario, the Anatolian plate moves westward in response to the collision of Arabia and Eurasia (Şengör & Yılmaz, 1981; Şengör et al., 1985; Bozkurt, 2001). The North and East Anatolian transform fault systems accommodate extrusion, and rollback along the Hellenic arc is suggested to provide space to accommodate the escaping plate (McKenzie, 1972; Dewey & Şengör, 1979; Le Pichon & Angelier, 1979; Jackson & McKenzie, 1984; Barka & Kadinsky-Cade, 1988; Taymaz et al., 1991; Reilinger et al., 1997; McClusky et al., 2000; Tatar et al., 2013). Philippon et al. (2014) suggest a two-stage evolution of the arc. At 30 Ma, extension was only driven by the southward retreat of the Hellenic trench at a rate lower than 1 cm/yr, but since the last 13 Ma, the interaction of trench retreat with Anatolia escape accelerated the rate of trench retreat in the southwest direction at a rate of up to 3 cm/yr.

In western Turkey, extrusion tectonics is dominated by the active right-lateral NAFZ and North Anatolian Shear Zone (NASZ), which extends for ~1,200 km from the Karlıova triple junction through the Sea of Marmara and Biga Peninsula (Figure 5.1) (Ketin, 1948; Barka, 1992; Armijo et al., 1999; Şengör & Zabcı, 2019). The NASZ contains the NAF and is speculated to have accommodated from 25 to 110 km of displacement, depending on location, since the Late Miocene (Westaway 1994; Yoshioka 1996; Armijo et al., 1999; Hubert-Ferrari et al. 2002; Şengör & Zabcı, 2019). The structure accommodates ~24 mm/year of slip along northern Turkey (McClusky et al., 2000; Bulut et al., 2018). The geometries of its western and eastern terminations are poorly defined (Barbot & Weiss, 2021).

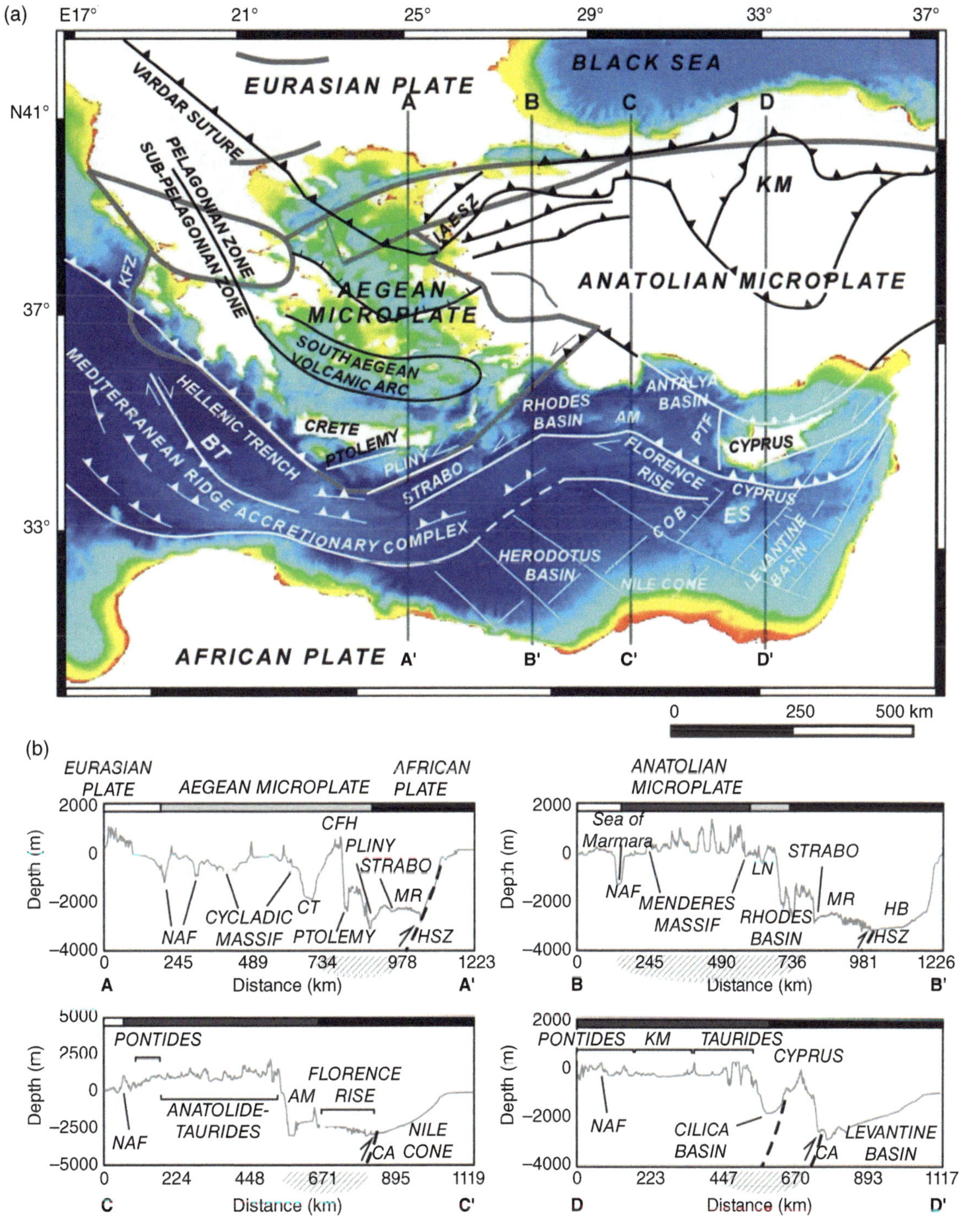

Figure 5.9 (a) EMODnet Digital Bathymetry map with some structures overlain. The Aegean and Anatolian microplate boundaries are shown in gray after Nyst and Thatcher (2004). Other structures after Hall et al. (1984, 2009), Woodside et al. (2002), Peterek and Schwarze (2004), Meier et al. (2007), Kinnaird and Robertson (2013), and Symeou et al. (2018). BT, Backthrust; KFZ, Kephalonia Fault Zone; İAESZ, İzmir-Ankara-Erzincan Suture Zone; KM, Kirşehir Massif; AM, Anaximander Mountains; PTF, Paphos Transform Fault; ES, Eratosthenes Seamount. (b) Profiles along the lines of the section shown in panel (a). CT, CFH; LN, Lycian nappes; MR, Mediterranean Ridge Accretionary Complex; HB, Herodotus Basin; HSZ, Hellenic Shear Zone; NAF, North Anatolian Fault; AM, Anaximander Mountains; CA, Cyprus arc. Hashed regions in panel (b) indicate areas speculated to be affected by slab tear (Woodside et al., 2002; Özbakır et al., 2013; Jolivet et al., 2015).

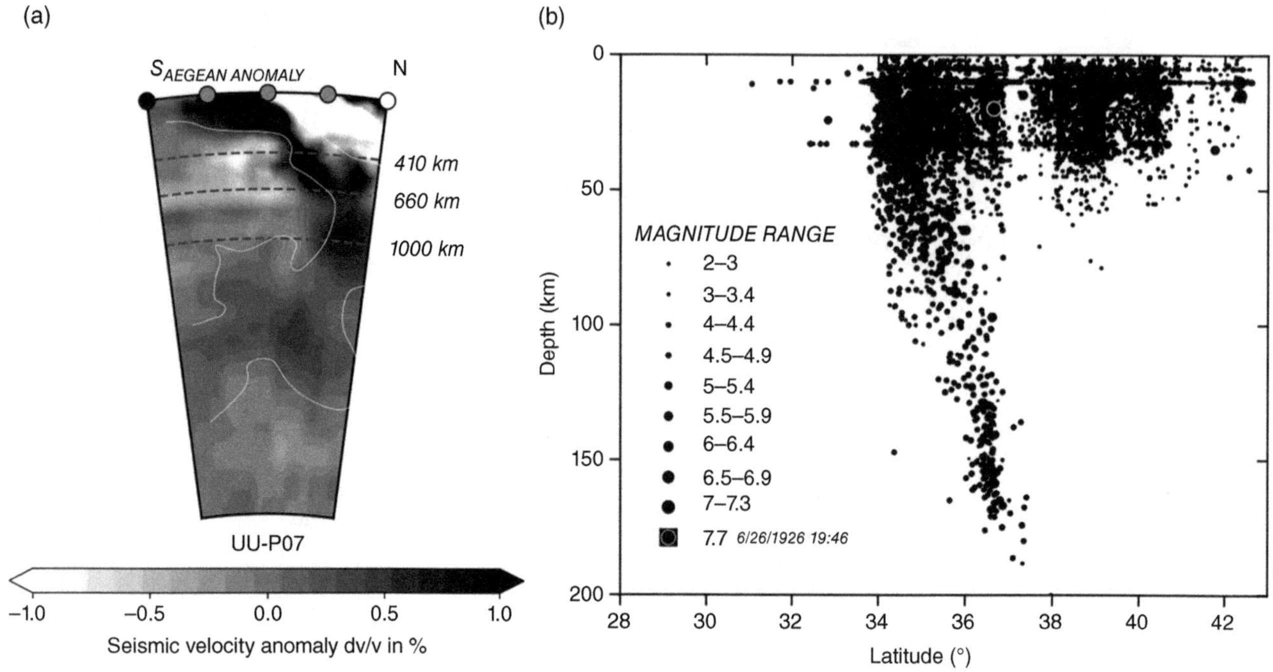

Figure 5.10 (a) Cross-section of the Aegean anomaly interpreted as the African slab using the UUP07 P-wave model (Amaru, 2007). The line of section used latitudes of 28–43° and longitudes of 24–28°. For more detailed views of the anomaly, see van der Meer et al. (2018), Wei et al. (2019), Blom et al. (2020), and El-Sharkawy et al. (2020). The depths of the dashed lines are 410, 660, and 1,000 km from the surface. Interpretations of the geology below 1,000 are debated and discussed in the text. Image created using Hosseini et al. (2018). (b) Depth versus estimated earthquake depth for the same latitude and longitude as in panel (a). In this map, events were adapted from https://deprem.afad.gov.tr/last-earthquakes. Events are from 01/24/1900 to 6/17/2021. We indicate the largest event (6/26/1926, 19:46).

The NAFZ splits into three strands as it trends westward into Western Anatolia and the Aegean Sea (Figure 5.1) (Emre et al., 1998; Kürçer et al., 2008; Beniest et al., 2016; Şengör & Zabcı, 2019). Each segment comprises several en échelon fragments (Emre et al., 1998; Kürçer et al., 2008). The northernmost E-W striking segments within the Sea of Marmara change strike in the northern Aegean Sea toward a NE-SW orientation in the NAT, maintaining its right-lateral strike-slip character, but splits across three basins and two transpressional ridges (Bulut et al., 2018). A branch between the northern and central segments originates southeast of Sapanca Lake (Kürçer et al., 2008) and terminates at the western end of the NAT (Ferentinos et al., 2018). This structure enters the Aegean Sea and trends into the Northern Skyros Basin. Strands of the NAFZ have also been linked to the KTZ through the transtensional Central Hellenic Shear Zone (Royden & Papanikolaou, 2011; Evangelidis, 2017). The NAFZ may have been active in the Western Anatolia–Marmara region since the Pliocene (Ünay et al., 2001).

Sakellariou et al. (2017) suggest that the southwestward expansion and stretching of the Aegean microplate during the Plio-Quaternary is accommodated by a northern right-lateral tectonic boundary marked by the KTZ and NASZ and a southern left-lateral tectonic boundary marked by the Pliny and Strabo trenches (Figure 5.9). Papanikolaou and Royden (2007) note that regional extension has a much-reduced role in the dynamics of the Aegean microplate and that no active extensional strain is present except for a small southeastern domain (Figure 5.1) (Corinth rift, south Viotia, south of Evia, and across the Sperchios–Kamena Vourla rift; Brooks & Ferentinos 1984; Chousianitis et al., 2013, 2015). Maggini and Caputo (2020) report that seismogenic faults in the internal Aegean domain associated with the Hellenic subduction arc are characterized by pure normal and strike-slip kinematics or by a combination and that active thrusting is limited to the central and western sectors of the Hellenic subduction zone and the offshore regions external to it.

Figure 5.11 shows the focal mechanisms for selected earthquakes (2010–2020) that appear along the Aegean-Anatolian microplate boundary. Earthquakes with focal mechanisms consistent with reverse faulting have occurred south of Crete, including those associated with an Mw 6.4 earthquake on 2 May 2020. These earthquakes occurred at relatively shallow depths (6.5–9.6 km; Table 5.10) and

Table 5.9 List of selected earthquake events along the Simav Fault and associated fault systems.

No.[a]	Event-ID[b]	Time (UTC)	Latitude	Longitude	Depth (km)	Rms[c]	Mag[d]
1	465625	2/18/2020 16:09	39.1015	27.8453	14.68	0.45	5.2
2	150860	12/10/2011 5:15	38.8625	30.1883	13.44	0.96	4.2
3	319040	12/2/2015 15:52	39.1495	28.154	10.85	0.4	4.0
4	132605	6/10/2011 22:47	39.0975	28.3405	34.38	0.85	4.7
5	160143	3/29/2012 10:13	38.6035	30.004	12.77	0.73	4.2
6	367059	3/29/2017 18:10	38.2003	31.0575	14.87	0.36	4.0
7	495401	2/9/2021 15:51	38.5965	31.6318	7.01	0.49	4.7
8	495403	2/9/2021 15:53	38.59	31.6495	4.61	0.41	4.1
9	367501	4/3/2017 9:05	38.4801	31.7975	13.84	0.25	4.0
10	136512	7/27/2011 9:53	38.3278	31.8802	17.79	0.33	4.8
11	128573	5/19/2011 20:15	39.1328	29.082	24.46	0.49	5.7
12	128577	5/19/2011 20:25	39.1442	29.1078	7.00	0.44	4.6
13	128603	5/19/2011 21:12	39.113	29.0377	7.74	0.57	4.8
14	128672	5/20/2011 0:13	39.1413	29.1065	16.92	0.62	4.1
15	128701	5/20/2011 0:58	39.1147	29.0837	17.38	0.78	4.3
16	129252	5/21/2011 21:43	39.1037	29.0513	7.00	0.11	4.0
17	129791	5/24/2011 2:55	39.1013	29.0217	16.80	0.45	4.2
18	131192	5/30/2011 22:03	39.1567	29.0112	15.29	0.85	4.0
19	132022	6/5/2011 21:29	39.143	29.095	6.98	0.55	4.0
20	134386	6/29/2011 11:40	39.1232	29.0032	9.28	0.75	4.0
21	135896	7/19/2011 21:16	39.1048	29.093	17.66	0.67	4.1
22	138300	8/25/2011 4:19	39.139	29.0957	22.54	0.77	4.3
23	161414	4/16/2012 10:10	39.1227	29.1222	6.90	0.5	4.7
24	161595	4/17/2012 20:45	39.1468	29.1142	6.99	0.58	4.5
25	161902	4/20/2012 16:39	39.1525	29.0975	20.59	0.81	4.4
26	177315	10/30/2012 0:12	39.1385	29.1787	21.35	0.76	4.1
27	188611	3/12/2013 20:47	39.1203	29.0583	12.81	0.52	4.1
28	197002	6/9/2013 14:18	39.1392	29.022	15.61	0.68	4.1
29	234353	7/15/2014 12:25	39.13	29.0041	9.92	0.32	4.1
30	309933	9/3/2015 8:23	39.1226	29.1225	10.24	0.49	4.1

[a] See Figure 5.7a for events 1–10 and Figure 5.7b for events 11–30.

[b] Parameters were extracted from https://deprem.afad.gov.tr/depremkatalogu 1900–20XX Earthquake Catalog (M> = 4.0), Turkish Ministry of the Interior, Disaster and Emergency Management Presidency, Earthquake Department (AFAD).

[c] Rms, root-mean-square travel time residual in seconds.

[d] All magnitudes are ML (original magnitude relationship defined for local earthquakes) except events 1, 3, 6, 7, 9, 29, and 30, which are moment magnitudes (Mw).

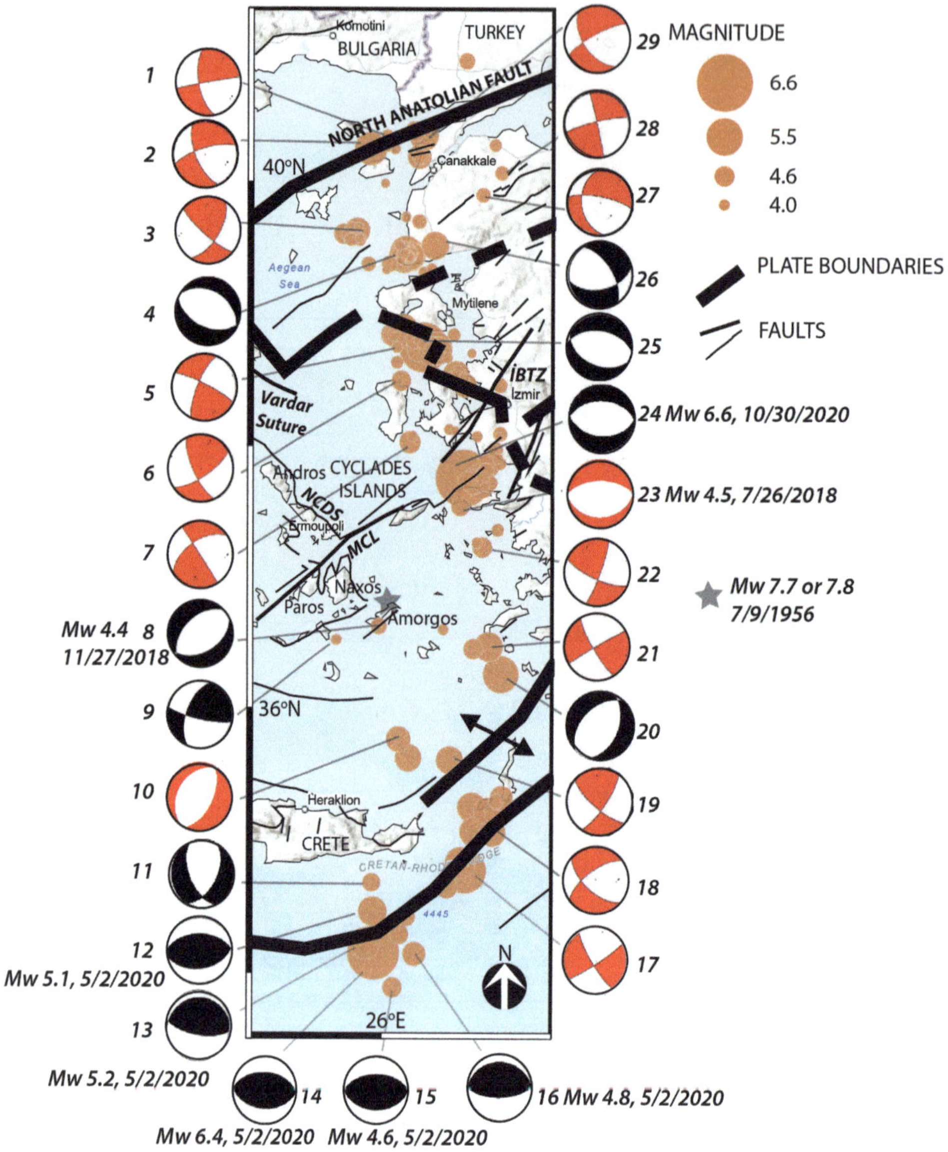

Figure 5.11 Map of plate boundaries between the Aegean and Anatolian microplates with some faults indicated (after Nyst & Thatcher, 2004; Uzel et al., 2013; Pe-Piper et al., 2002; Menant et al., 2016). Focal mechanisms are from the Turkish Ministry of the Interior, Disaster and Emergency Management Presidency, Earthquake Department Earthquake Catalog (M> = 4.0), 1900–2021 (https://deprem.afad.gov.tr/depremkatalogu). Events are only from 2010 to 2020; details are presented in Table 5.10. The size of the circle represents magnitude. The 9 July 1956 Amorgos earthquake epicenter is also indicated after Alatza et al. (2020). See Okal et al. (2009) for discussions regarding the focal mechanism of this event. The base map is from ESRI. İBTZ, İzmir–Balıkesir transfer zone; NCSD, North Cyclades Detachment System; MCL, Mid-Cycladic Lineament.

may be associated with a plate-interface zone defined by the upper plate and splay-thrust faults (Saltogianni et al., 2020). Observations and modeling of historical and recent earthquakes have shown that uplift along the Hellenic arc margin offshore of Crete is controlled by reverse fault motion with little contribution from plate-interface slip (Mouslopoulou et al., 2015).

Extrusion and deformation in Western Anatolia are also accommodated by transfer zones, where strain is transferred from one structural element to another and displacement changes between individual fault and basin segments (Gawthorpe & Hurst, 1993; Barbot & Weiss, 2021). Some examples of these zones include the NE-SW trending strike-slip dominated İzmir–Balıkesir Transfer Zone (İBTZ), Uşak-Mugla Transfer Zone (UMTZ), and Southwestern Anatolian Shear Zone (SWASZ) (Figures 5.1 and 5.2) (Çemen et al., 2006; Oner et al., 2010; Sözbilir et al., 2011; Gessner et al., 2013; Özkaymak et al., 2013; Uzel et al., 2013; Karaoğlu & Helvacı, 2014; Seghedi et al., 2015). These transfer zones have been considered

Table 5.10 List of selected earthquake events along the Aegean-Anatolian plate boundary.

No.[a]	Event-ID[b]	Time (UTC)	Latitude	Longitude	Depth (km)	Rms[c]	Mag[d]
1	199626	7/12/2013 0:36	40.3738	25.946	27.85	0.45	4.3
2	201060	7/30/2013 5:33	40.3028	25.7902	20.01	0.52	5.3
3	184151	1/19/2013 19:26	39.6382	25.6795	20.91	0.5	4.2
4	360268	2/6/2017 10:58	39.5275	26.1373	9.83	0.21	5.3
5	183497	1/12/2013 13:47	39.6447	25.6733	6.89	0.8	4.0
6	155511	1/29/2012 21:03	38.7387	26.0447	32.39	0.68	4.2
7	101387	3/26/2010 18:35	38.1457	26.177	24.26	0.2	4.7
8	426091	11/27/2018 23:16	36.7565	25.877	16.15	0.62	4.4
9	426096	11/27/2018 23:46	36.6493	25.4535	5.95	0.75	4.1
10	418888	8/19/2018 5:46	35.8861	26.0695	28.49	0.77	4.9
11	309516	8/27/2015 0:25	34.7751	25.8068	7.06	0.52	4.5
12	472843	5/2/2020 16:44	34.5521	25.8181	6.76	0.56	5.1
13	472824	5/2/2020 13:45	34.2973	25.7371	9.63	0.98	5.2
14	472819	5/2/2020 12:51	34.2226	25.8253	6.65	0.98	6.4
15	472825	5/2/2020 13:33	33.9548	26.0141	6.5	0.96	4.6
16	472827	5/2/2020 14:21	34.2123	26.232	5.86	0.93	4.8
17	294406	4/16/2015 18:07	34.8643	26.7275	12.34	0.62	5.9
18	169403	7/4/2012 23:46	35.1613	26.9993	34.09	0.35	5.0
19	293183	3/27/2015 23:34	35.7295	26.576	56.13	0.47	5.0
20	507881	8/1/2021 4:31	36.3843	27.0805	10.86	0.17	5.5
21	187555	2/27/2013 22:05	36.7298	26.5115	140.27	0.43	4.1
22	417483	7/26/2018 8:17	37.6546	26.6698	4.5	0.52	4.5
23	483762	10/30/2020 11:51	37.879	26.703	14.9	1	6.6
24	375576	6/12/2017 12:28	38.8436	26.313	15.96	0.28	6.2
25	431610	2/20/2019 18:23	39.6011	26.4261	5.8	0.37	5.0
26	411695	5/3/2018 2:04	39.967	26.8993	10.39	0.35	4.3
27	284923	12/16/2014 9:02	40.1298	27.0845	17.35	0.29	4.3
28	115792	11/3/2010 2:51	40.3997	26.3147	28.9	0.59	5.1
29	199626	7/12/2013 0:36	40.3738	25.946	27.85	0.45	4.3

[a] See Figure 5.11 for events.

[b] Parameters were extracted from https://deprem.afad.gov.tr/depremkatalogu 1900–20XX Earthquake Catalog (M> = 4.0), Turkish Ministry of the Interior, Disaster and Emergency Management Presidency, Earthquake Department (AFAD).

[c] Rms, root-mean-square travel time residual in seconds.

[d] All magnitudes are ML (original magnitude relationship defined for local earthquakes) except events 4, 8–14, 16, 17, 19, 20, 23–28, which are moment magnitudes (Mw).

significant portions of the larger Western Anatolian Shear Zone (WASZ) or WAEP (Figure 5.1) and may have developed due to mantle processes related to the subduction of the Aegean slab (Gessner et al., 2013; Uzel et al. 2020)—some transfer zones trend into other fault systems. For example, the İBTZ is speculated to connect to the Mid-Cycladic Lineament (MCL) in central Greece and the NASZ in northern Turkey (Figures 5.1 and 5.11; Uzel et al., 2013; Seghedi et al., 2015; Westerweel et al., 2020). The MCL is a strike-slip structure that may be the result of the reactivation of the Vardar suture zone, evidenced by the North Cycladic Detachment System (Figure 5.11), to accommodate the westward extrusion of Anatolia in the Late Miocene (Philippon et al., 2014). These transfer zones have been used to illustrate that the Aegean and Anatolian microplates experienced or are currently transitioning from a stress regime dominated by extension to transform tectonics (Papanikolaou & Royden, 2007; Cavazza et al., 2009).

Presently, normal fault motion exists within the İBTZ as illustrated by focal mechanisms from a 2020 Mw6.6 earthquake and a 2018 Mw4.5 earthquake within the zone. An Mw4.4 earthquake with normal motion occurred off the coast of Amorgos near the 1956 Mw7.7 (or 7.8) earthquake, one of the strongest earthquakes of the 20th century in the area of the South Aegean (Okal et al., 2009; Alatza et al., 2020). The 1956 event has debated focal mechanisms as strike-slip or normal faulting geometries (see Okal et al., 2009). A normal sense of motion is also found with earthquakes near the NASZ, including the 2017 Mw6.2 and 2017 Mw5.3 earthquakes (Figure 5.11, Table 5.10). These events are likely associated with transtensional motion.

5.3. OUTSTANDING QUESTIONS IN AEGEAN TECTONICS

As outlined in the previous section, significant contributions have been made regarding the fundamental tectonics and geological history recorded by rocks throughout the Western Anatolian microplate. However, outstanding questions remain regarding the boundary between the Aegean and Anatolian microplates that affect our understanding of the mechanisms that drive extension in the Earth's lithosphere. Most of these questions center on how upper lithospheric and crustal deformation are linked to lower lithosphere and mantle processes.

5.3.1. Slab Dynamics

African Slab Geometry and Connections to Other Subduction Systems

Based on several geophysical, tectonic, and geochemical developments, the subducting African (Nubian or Aegean) slab has emerged as the primary driver for extension in the Aegean and Anatolian microplates and the development of their metamorphic core complexes (Figures 5.1, 5.2, 5.8, 5.9, and 5.10) (Jolivet et al. 2013; Jolivet & Faccenna, 2000; Çemen et al., 2006; Dilek & Sandvol, 2009; van Hinsbergen et al., 2010; van Hinsbergen & Schmid 2012; Salaün et al., 2012; Faccenna et al., 2014; El-Sharkawy et al., 2020; Barbot & Weiss, 2021). The Hellenic and Cyprus arcs are the surface expression of the subducting African plate and eastern Mediterranean lithosphere beneath the Anatolian and Aegean microplates (Le Pichon & Angelier, 1979; Angelier et al., 1982; Anastasakis & Kelling, 1991; Papazachos et al., 2000; Ergün et al., 2005; Ganas & Parsons, 2009; Hall et al., 2009; Royden & Papanikolaou, 2011; Hall et al., 2014; Symeou et al., 2018; Ventouzi et al. 2018).

Although it has a well-developed Wadati-Benioff zone dipping ~30° from 20–100 km depth and ~45° from 100–150 km depth (Figure 5.10b) (Papazachos & Comninakis, 1971; Papazachos et al., 2000; Suckale et al., 2009; Hayes, 2018), it has a debated slab geometry at intermediate depths (150–250 km, Suckale et al., 2009; Agostini et al., 2010; see the review in Hansen et al., 2019; El-Sharkawy et al., 2020). Seismic body wave tomography shows it extends into the upper and lower mantle to 1,400 ± 100 km depth (Figure 5.10a; Spakman et al., 1988; Bijwaard et al., 1998; van der Meer et al., 2018; see the review in Bocchini et al., 2018). However, the slab may be a single folded body that overturned in the lower mantle (Faccenna et al., 2003) or two slabs located between 2,000 and 1,500 km and from 1,500 km to the surface (van Hinsbergen et al., 2005; van der Meer et al., 2018). Mantle tomography has shown multiple subducted slabs beneath the Aegean and Anatolian microplates (Spakman et al., 1988; Spakman, 1990; 1991; Wortel & Spakman, 2000; Govers & Fichtner, 2016; van der Meer et al., 2018; Wei et al., 2019). Blom et al. (2020) show the Hellenic slab, visible in both S and P velocity, extending from the surface to the transition zone in a bent, arcuate shape. A high-velocity structure exists beneath the Hellenic arc and the Aegean Sea that flattens from the 410 km discontinuity and is not seen at deeper levels. Wei et al. (2019) show a gap in the subducting slab at depths 60–100 km west of the South Hellenides. In the South Hellenides, slab tear may be visible at the 660 km discontinuity, whereas four slabs are imaged beneath the North Hellenides.

Interpretations of these tomographic images have indicated that more slab is imaged than is reflected by seismicity (Spakman et al., 1988; Papadopoulos, 1997; Bijwaard et al., 1998) and that a variation of slab thickness exists across the Aegean Sea (Karagianni et al., 2002). Mantle tomography has also shown that not all slabs in the Mediterranean region are connected to the lithosphere at the surface, consistent with past delamination (Spakman et al., 1988; Dilek & Sandvol, 2009; Wortel & Spakman, 2000). Challenges in imaging the subduction zone includes its small size, its spatially highly variable nature, and the uneven distribution of its seismic stations (El-Sharkawy et al., 2020).

The Hellenic subduction system comprises three regions: an outer compressional nonvolcanic arc, a volcanic arc, and an extensional back-arc region that makes up the broader Aegean Sea region (Figure 5.8; McKenzie 1972; Papazachos, 2019). Although the Western Hellenic Arc (also termed the North and Southern Hellenic Arc; Royden & Papanikolaou, 2011) has a well-defined topography, trench, sedimentation, and strain pattern (Stanley et al., 1978; Papadopoulos et al., 1988; Hatzfeld et al., 1990; Cocard et al., 1999), the central and eastern portions of the Hellenic arc are more difficult to characterize as the boundary becomes diffuse (Beißer et al., 1990; Shaw & Jackson, 2010; Özbakır et al., 2013). The Hellenic arc's connection with the Cyprus arc and even the nature of plate motion along strike of the Cyprus arc has been debated (Anastasakis & Kelling, 1991; Woodside et al., 2002; Ergün et al., 2005; Hall et al., 2009; Harrison et al., 2013; Kinnaird & Robertson, 2013; Symeou et al., 2018). The surface morphology of the southern and eastern portions of the Hellenic arc and its connection to the Cyprus arc are obstructed by a section of sediments up to 300 km wide and 6–10 km thick that form the Mediterranean Ridge (Figures 5.8 and 5.9; Heezen & Ewing, 1963; Emery et al., 1966; Le Pichon et al., 1982; Kenyon et al. 1982; Kastens et al., 1992; Foucher et al., 1993; Westbrook & Reston, 2002; Kopf et al., 2003). The ridge is a giant accretionary complex, extending ~2,000 km from the Calabrian Rise east of Greece to the Florence Rise, and is the largest structural unit of the Eastern Mediterranean Sea (Limonov et al., 1996; Cita et al., 1996). The front of the subduction of the Hellenic arc is located south of the Mediterranean Ridge (Jost et al., 2002; Westbrook & Reston, 2002; Jolivet et al., 2013). Most of the subducting African plate beneath the ridge is oceanic, except along the central sector of its southern margin, where the accretionary complex collides with the African continental margin (Chaumillon & Mascle, 1997; Westbrook & Reston, 2002). The ridge may be the fastest outward-growing wedge in most recent Earth history, with a rate of up to 10 km/m.y. (Kastens, 1991; Kopf et al., 2003). It has been speculated to grow by off-scraping against a backstop formed by the Alpine nappes of the Hellenic Arc (Kastens, 1991).

The intensively folded and faulted rocks of the Mediterranean Ridge vary in geometry along strike (Cita et al., 1996; Chaumillon & Mascle, 1997; Westbrook & Reston, 2002; Kopf et al., 2003). The wedge accumulates sediments in its western and eastern portions, but in its central portion between Libya and Crete, the ridge behaves unlike a typical accretionary complex. In this area, a trench system (the Hellenic trenches; Ptolemy, Pliny, and Strabo; Figure 5.9) developed in between the accretionary complex and volcanic arc, likely as a result of back-thrusting beneath the northern edge of the complex (Galindo-Zaldivar et al., 1996; Westbrook & Reston, 2002). The accretionary complex is unusual compared to others worldwide, not only because of these back thrusts but also because it appears to have formed in a continent-continent collisional setting and contains shallow, Messinian-age evaporites (Cita et al., 1996; Chaumillon & Mascle, 1997). These evaporites influence its deformation and fast growth rate due to their mechanical properties and effect on fluid flow and pressure (Kastens, 1991; Westbrook & Reston, 2002; Kopf et al., 2003). Understanding the development of the Mediterranean Ridge is critical to determining the initiation age of the Hellenic arc, as described in the next section.

The Age of Subduction of the African Slab

The subduction zone initiation (SZI) age is defined as the onset of downward plate motion forming a new slab, which later evolves into a self-sustaining subduction zone (Crameri et al., 2020). Constraints regarding SZI age of the present-day expression of the Hellenic arc developed from several independent approaches, including timing sedimentation within the Mediterranean Ridge (Kastens, 1991; Kopf et al., 2003), analysis of topography combined with estimates of slab age and depth (McKenzie, 1978; Le Pichon et al., 2019), reconstructions of subducted slabs using tomography (Spakman et al., 1988), paleomagnetism (Savostin et al., 1986; Marsellos et al., 2010), and the timing of metamorphism and volcanic activity (Fytikas et al., 1984; see the review in Catlos & Çemen, 2023). Early estimates for the initiation of Hellenic arc subduction are 13 ± 3–5 Ma (Le Pichon & Angelier, 1979) to 5–10 Ma (McKenzie, 1978; Mercier, 1981) based on interpretations of seismic activity coupled with assumptions regarding the age of subducted lithosphere and subduction depths. These ages are similar to the onset of the KTZ based on geodynamic modeling and GPS data (Figure 5.1) (6–8 Ma; Royden & Papanikolaou, 2011) and the timing of the earliest volcanic activity in the South Aegean arc (Pliocene; Pe-Piper & Piper, 2005). Reconstructions of fault systems in the northern margin of the eastern Mediterranean Sea are consistent with estimates of 15 Ma (Le Pichon et al., 2019).

However, interpretations of the Aegean seismic velocity structure, tomography, and seismicity data in the Aegean area suggest older estimates (26–40 Ma; Meulenkamp et al., 1988; Spakman et al., 1988; Papadopoulos, 1997; Brun & Sokoutis, 2010). These ages are more consistent with the ages of granitic intrusions found throughout Western Anatolia (Figure 5.4, Tables 5.5–5.8) and the timing of the onset of sedimentation associated with the Mediterranean Ridge at 23.6–33 Ma (Fytikas et al., 1984; Kastens, 1991). Younger estimates from the ridge are also reported (~19 Ma; Kopf et al., 2003). Plate reconstructions suggest that the Northern Hellenic trench experienced the onset of subduction from 27 to 34 Ma, whereas the southern Hellenic segment was active at 34 Ma (Royden & Papanikolaou, 2011).

If the incoming lithosphere is heterogeneous in terms of thickness and composition, subduction zones may behave chaotically, retreating, advancing, or remaining stationary at different stages (Royden & Husson, 2009; Husson et al., 2009). The progressive deceleration in motion of Africa with respect to Europe in the Mediterranean region is observed to have occurred since 35 Ma and in the eastern Mediterranean from 35 Ma to 10 Ma to a convergence rate of a few mm/yr (Savostin et al., 1986; Marsellos et al., 2010). The rate of trench retreat is estimated to have accelerated from ~0.6 cm/y during the first 30 m.y. of subduction to 3.2 cm/yr during the past 15 m.y., perhaps due to Middle Miocene–Pliocene slab tear (Brun et al., 2017). Differences in the timing of initiation and rate of subduction exist between segments along the Western Hellenic Arc and should also be expected to occur along other portions of the Hellenic and Cyprus arcs (Royden & Papanikolaou, 2011; Pearce et al., 2012). The timing of interpreted ductile "extensional" shear fabrics in metamorphic rocks can also be complicated as these may record extrusion instead of processes associated with slab rollback (Searle & Lamont, 2022).

These Late Cenozoic estimates are difficult to reconcile with the model in which the Hellenic arc is a single, evolving subduction zone system that was initiated in the Mesozoic (Jurassic) (Faccenna et al., 2003; van Hinsbergen et al., 2005; Royden & Papanikolaou, 2011; Jolivet et al., 2013; Malandri et al., 2017). In this scenario, the Vardar suture in Greece (equivalent to the İAESZ; Channell & Kozur, 1997; Okay & Tüysüz, 1999; Moix et al., 2008) and Pindos suture zone (equivalent to units within the Antalya domain and Dilek peninsula; Stampfli & Kozur, 2006) had buoyant microcontinents that entered and locked subduction, triggering southward slab rollback and migration of the volcanic arc (van Hinsbergen et al. 2005; Brun & Faccenna 2008; Jolivet & Brun 2010; Jolivet et al., 2013; Cornée et al., 2019). The model eliminates the need for multiple sutures and subducted slabs to be present beneath western Turkey and the Aegean and simplifies the evolution of the Aegean microplate to a single evolving, long-lived subduction system. The present-day curvature of the Hellenic forearc thus represents oblique subduction and a plate-boundary expression that grew systematically over long periods of geological time (Huchon et al., 1982; Le Pichon et al., 1995; ten Veen & Kleinspehn, 2002; Gautier et al. 1999; Le Pichon et al., 2002; Wallace et al., 2005, 2008; van Hinsbergen & Schmid, 2012; Philippon et al., 2014; Cornée et al., 2019).

The single-subduction system requires all the lower plate continental crust to be accreted into the upper plate while subducting continental lithosphere and requires that the entire Aegean Crust from the Vardar suture to the Mediterranean Ridge was derived from the lower plate (van Hinsbergen et al. 2005). Oceans between the accreted domains were of significant size (500 km in some cases), and the process would lead to significant elevation changes, crustal thicknesses, and critical changes in the zone of subduction transitions occurred from oceanic to continental shear zones (see the discussion in Le Pichon et al., 2019). Not all units record blueschist facies conditions, and some experienced Barrovian prograde (burial) P-T paths, such as on the island of Naxos (Lamont et al., 2020a).

Currently, the Hellenic arc is migrating SW faster than the counterclockwise rotation of Anatolia (ten Veen & Kleinspehn, 2002), and the rate of convergence between Africa and Eurasia is 4 cm/yr (Reilinger et al., 1997; Kahle et al., 2000; McClusky et al., 2000; Hollenstein et al., 2008). Timing constraints on Aegean forearc curvature due to opposite rotations, clockwise in the west and counterclockwise in the east, are Eocene and Middle Miocene (Morris & Robertson, 1993; Cornée et al., 2019). Trench bending and rollback increased subduction obliquity over time, which has been accommodated by strain partitioning within the upper Eurasian plate (Philippon et al., 2014; Brun et al., 2016; Cornée et al., 2019). Subduction zones with limited trench-parallel lengths on the order of the Hellenic arc (600–800 km) and narrow slabs (<1,500 km) typically have rapid retreat rates (Schellart et al., 2007; Bolhar et al., 2010).

The Number, Location, and Impacts of Slab Detachments and Tears

An additional key focus has been identifying the location, depth, and relationship of ancient and present-day active subducting slabs and their detachment mechanisms beneath the Aegean and Anatolian microplates (see the review in Hansen et al., 2019; El-Sharkawy et al., 2020). Several locations across the Aegean and Anatolian microplates have been suggested to be affected by slab tears, either trench parallel or perpendicular (Figure 5.1). The tearing process in the near term can lead to intermediate-depth seismicity (Meighan et al., 2013) and explain earthquakes that appear inconsistent with a coherent subducting slab (Clark et al., 2008). Tears can lead to large-volume magmatism (Cocchi et al., 2017), changes in igneous geochemistry, and facilitate the ore-forming process and mineral deposits (de Boorder et al., 1998; Rabayrol et al., 2019; Rabayrol & Hart, 2021). The process leads to asthenosphere upwelling and changes in thermal and fluid regimes (Roche et al., 2018; Gessner et al., 2018). Slab tear has been related to present and past geothermal activity in Western Anatolia and the generation of a late Eocene-Miocene metallogenic period

(Pb-Zn- followed by Au-rich) (Menant et al., 2018; Gessner et al., 2018; Rabayrol & Hart, 2021). Their presence significantly affects plate dynamics, including subduction rates, plate motion, and mantle dynamics (Gianni et al., 2019).

These sites vary in scale from regional to local and include the boundary between the Hellenic and Cyprus arcs (Wortel & Spakman, 1992; Biryol et al., 2011), at the Anaximander Mountains (Woodside et al., 2002), south of Crete at the Pliny-Strabo Shear Zone (Özbakır et al., 2013), the İBTZ transfer zone (Kaya, 1981; Gessner et al., 2013), and beneath the Menderes Massif itself (Biryol et al., 2011; Rabayrol & Hart, 2021). A tear is speculated to generate a ~200-km-deep, low-velocity anomaly below western Turkey (Roche et al., 2019). Slab tear has been used to interpret the deep Rhodes Basin (Faccenna et al., 2014; Woodside et al., 2000) and tectonic activity within southwest Anatolia (Biryol et al., 2011; Roche et al., 2019).

Trench-parallel tear affects the subducting African lithosphere between northern Greece and the Gulf of Corinth along the Western Hellenic Arc (Hansen et al., 2019). Trench-perpendicular tear may accommodate the region between the Hellenic and Cyprian arcs, which differ in subduction steepness and material subducted (Dilek & Sandvol, 2009). The Cyprian arc involves shallower subduction dynamics, with the Eratosthenes seamount and Anixamander Mountains (mud volcanoes; Lykousis et al., 2009) impinging on the trench (Figure 5.9; Kempler & Ben-Avraham 1987; Zitter et al. 2003). The back thrusts and tectonic geometry of the Mediterranean Ridge have led to speculation that the African slab detached in the region between Libya and Crete (Kopf et al., 2003). Alternatively, this region may have a subduction transform edge propagator (STEP; tear or hinge fault; Govers & Wortel, 2005; Özbakır et al., 2013). Nine STEP structures have been proposed to exist beneath southern Greece, segmenting the subducting African slab and contributing to seismicity and deformation (Sachpazi et al., 2016). A STEP is also proposed for the transition between the Cyprus and Hellenic arcs (Salaün et al., 2012; Elitez et al., 2016; Portner et al., 2018).

The KTZ (Figures 5.1 and 5.9) has been a particular subject of the debate regarding slab tear (see Bocchini et al., 2018; Hansen et al., 2019). The structure is part of the Western Hellenic Subduction Zone, considered one of Europe's most seismically active areas (Pearce et al., 2012; Halpaap et al., 2018). The KTZ may represent a vertical tear along the oceanic and continental lithosphere (Suckale et al., 2009), forming the KTZ as a STEP fault (Govers & Wortel, 2005). The STEP fault may be in its initial stages of forming (Evangelidis, 2017; Özbakır et al., 2020), or the slab may have entirely detached (Wortel & Spakman, 2000). A smooth transition has also been proposed between two segments, without a tear between, at least at depths shallower than 100 km (Pearce et al., 2012; Halpaap et al., 2018).

Despite the fragmentation of the subducting African lithosphere, the thickness of the Aegean and Anatolian crust is remarkably similar (Zhu et al., 2006; Sodoudi et al., 2006; Karabulut et al., 2019). Estimates from the central Menderes Massif are 28–30 km (Zhu et al., 2006), whereas the thickness beneath the Aegean Sea averages ~25 km (Zhu et al., 2006; Tirel et al., 2004; Kind et al., 2015). The crustal thickness in the southern and central parts of the Aegean is reported to be thinner (20–22 km), whereas the northern Aegean Sea shows a relatively thicker crust (25–28 km) (Karagianni et al., 2005; Sodoudi et al., 2006). Depending on the model used, the crustal thickness beneath western Crete could be 32.5–35 km or up to 45 km (Snopek et al., 2007). Karabulut et al. (2019) demonstrate significant crustal thickness variations (20–47 km) from western Greece to eastern Anatolia but show that these are relatively uniform within specific regions. In Western Anatolia, the crustal thicknesses are 25–30 km, increasing slightly to the north. In contrast, in southern Anatolia, crustal thicknesses decrease from 35 to 25 km in the Mediterranean Sea, except north of Antalya Bay, where it locally reaches 40 km. A thickness of 40 km is in line with estimates of eastern Anatolia (Kind et al., 2015), western Greece (Karagianni et al., 2005), and the Anatolian plateau (Saunders et al., 1998).

These thickness estimates seem at odds with large-scale back-arc thinning typically seen in subduction zone settings (Saunders & Tarney 1984). The Aegean Sea is not a typical back-arc basin (Agostini et al., 2010; Doglioni et al., 2002) because a thick layer of continental crust underlies it, and it lacks an ocean floor (Makris, 1978). The Aegean Sea is disrupted by the active NASZ in its northern portion (Brooks & Ferentinos, 1984; Gürer et al., 2006; Kokkalas et al., 2006; Kreemer et al., 2004; Lyberis, 1984). The region displays a complex tensional regime where crustal stretching is inconsistent with the geometry and direction of the subducting Hellenic slab (Mantovani et al., 1997; Agostini et al., 2010). The premise of extrusion tectonics driven by convergence in the west requires a free lateral boundary in the east. However, the Aegean plate is constructed mainly of continental lithosphere and has a similar thickness as the Anatolian plate, as seen in both bathymetry (Figure 5.9) and seismic reflection (Zhu et al., 2006; Sodoudi et al., 2006). But slab ruptures associated with the differential retreat, inherited lower plate lithospheric heterogeneities, and mantle upwelling would provide accommodation for the microplates to extrude (Agostini et al., 2010; Govers & Fichtner, 2016; Karabulut et al., 2019). The onset of the NASZ may be the result of slab deformation and detachment beneath the Bitlis-Hellenic subduction zone, which accelerated slab retreat in the west and indentation of the continent along the Bitlis-Zagros suture zone (Figure 5.1) (Faccenna et al., 2006; Schildgen et al., 2014).

5.3.2. Timing, Number, and Geometry of Transfer Zones

Transfer zones play a significant role in accommodating tectonic escape (Barbot & Weiss, 2021), and despite their importance in accommodating the present-day subduction dynamics, when, how, and why specific transfer zones occur across Western Anatolia is debated. For example, the İBTZ is a deep crustal transform fault zone consisting of NE-trending active strike-slip dominated faults and accommodates differential deformation between the Cycladic and Menderes core complexes (Uzel et al., 2013, 2020). The İBTZ is also mapped as the Western Anatolian Transfer Zone (WATZ; Gessner et al., 2013, 2018). The zone may be the surface expression of a tear in the subducting African slab (Gessner et al., 2013; Uzel et al., 2015; Sümer et al., 2018) or a transition between extensional and strike-slip dynamics due to the southward rotation rollback of the subduction zone (Ersoy & Palmer, 2013; Özkaymak et al., 2013; Ersoy et al., 2014; Ersoy et al., 2017; Uzel et al., 2020). Based on a compilation of data from igneous rocks throughout Western Anatolia, Uzel et al. (2020) suggest that volcanic activity in the region is always associated with the İBTZ as recorded by the positions of the eruption centers that follow the trend of the transfer zone. A lack of $^{40}Ar/^{39}Ar$ ages from igneous assemblages between 15.97 and 13.82 Ma is attributed to a pulse of core complex exhumation and a change in partitioning extension between the Cyclades and Menderes Massif. Geochemical compositions of Miocene-age (17.48–14.94 Ma) volcanoes within the transfer zone indicate their origins are decompression melting of the upper mantle/lower crust, consistent with the outcome of regional transtensional movements in a post-collisional setting (Seghedi et al., 2015). Slab tear typically results in asthenosphere-derived (ocean-island basalt, OIB-like) Na-alkaline basalts, which are only exposed in the region within the Northern Menderes Massif (Kula volcanic rocks; Holness & Bunbury, 2006; Ersoy et al., 2017).

The İBTZ may trend further south into the MCL, an extensional fault exposed near or on the island of Paros that records orogen-parallel extension or transform fault motion (Figure 5.11) (Morris & Anderson, 1996; Avigad et al., 1998; Walcott & White, 1998; Pe-Piper et al., 2002; Gessner et al., 2013; Philippon et al., 2014; Beniest et al., 2016; Malandri et al., 2017). Besides the İBTZ, the SWASZ and UMTZ are located near each other on the border of the Menderes Massif, but their influence on each other is presently unclear (Figure 5.2).

5.3.3. Magmatic Influence in Driving Extension

Throughout Western Anatolia, magmatic pulses are exposed as geochemically variable extrusive and intrusive igneous rocks (Tables 5.1–5.8; Figure 5.4; Rossetti et al., 2017). Metamorphic core complexes with their associated postcollisional magmatic suites offer insights into the tectonic processes controlling crustal extension (Perkins et al., 2018). Extensional systems cut igneous intrusions in Western Anatolia metamorphic core complexes, and their ages are critical for timing events that facilitated their emplacement. Geochemical data regarding the depths of granite formation lends additional insight into how the mantle processes operated in the past. The picture, however, is complicated by the influence of the collisional dynamics that characterized the earlier assembly of the microplate (see the Assembly section). Granite crystallization ages provide information regarding how extension during the Eocene to Miocene migrated through Western Anatolia and the Aegean region in the past (Delaloye & Bingöl, 2000; Pe-Piper, 2000; Altunkaynak & Dilek, 2006; Altunkaynak et al., 2012).

Magma bodies can drive extension through the conductive transfer of heat from the upwelling of hot, asthenospheric mantle beneath significantly extended crust, and small-volume partial melts can exploit crustal pathways developed during extensional deformation (McKenzie & Bickle 1988; von Blanckenburg & Davies 1995; Perkins et al., 2018). Volatiles facilitate additional crustal deformation and metamorphism, resulting in feedback between decompression and mantle upwelling and driving additional lithospheric melting (Teyssier & Whitney, 2002; Kendall et al., 2005; Whitney et al., 2013; Platt et al., 2015; Perkins et al., 2018). The Menderes Massif of western Turkey is suggested to be a key area to study feedback relationships between magma generation and emplacement, rheological weakening, and activation of extensional detachment tectonics (Rossetti et al., 2017). The island of Naxos likewise illustrates the interplays between lower crustal flow and upper crustal extension and between buoyancy- and isostasy-driven controls in developing migmatite domes (Kruckenberg et al., 2011). The connections between detachment faulting and magma emplacement have also been explored in the Cyclades (Rabillard et al., 2018).

Determining the role of magma generation and extension requires understanding intrusive rock relationships to fault structures. In Western Anatolia, maps of the same pluton are commonly inconsistent regarding the locations of structures that may have affected or resulted from exhumation. For example, the northern boundary of the Kozak pluton (Figure 5.4) is shown by some as an intrusive contact (Akal & Helvacı, 1999) but by others as fault-bounded (Altunkaynak & Yilmaz, 1998, 1999; Yilmaz et al., 2001). The Eğrigöz, Koyunoba, and Alaçam plutons (Figure 5.4) have been the focus of many field-based, geochemical, and geochronological studies, but conflicting ideas exist regarding their relationship to the SDF (Figure 5.7) (see Catlos et al., 2012). For example, Işık and Tekeli (2001) map the SDF only along the northern portion of the Eğrigöz pluton, whereas Ring and Collins (2005) and Işık et al. (2004) indicate the SDF is exposed along the western edge of both the northern Eğrigöz and Koyonba plutons. Seyitoğlu et al. (2004) place the SDF within the central portion of the Eğrigöz pluton, whereas Ersoy et al. (2010) mark the structure as following the outer boundaries of the Eğrigöz and Koyunoba bodies. Thomson and Ring (2006) place the detachment prominently along the northern edge and central portion of the Eğrigöz granite and the eastern edge of the Koyunoba body. Recent gravity measurements suggest an igneous intrusion at depth near the Simav Fault (Toker et al., 2018, 2019). The 12–15-km-thick intrusion is located in the NE margin of the Simav graben at 2.5–3 km below the surface and has been suggested to be a primary driver of recent-day seismicity. Developing links between magmatism and extensional dynamics requires a critical structural understanding of the granite petrology structures and a clear delineation of how it appears affected by fault systems (Kruckenberg et al., 2011; Rabillard et al., 2018).

In Western Anatolia, many published maps also do not distinguish different granite types or textural orientations (Karacik & Yılmaz, 1998; Akal & Helvacı, 1999; Şahin et al., 2010). Mineral lineations and solid-state or magmatic fabrics associated with faulting or shearing are rarely reported. Besides the standard structural and petrographic analyses, cathodoluminescence (CL) images of extensional-related Western Anatolia granites (Salihli and Turgutlu; Catlos et al., 2010; Eğrigöz, Koyunoba, and Alaçam, Catlos et al., 2012; Figure 5.4) help document mineral zoning, deformation, and fluid alteration (Ramseyer et al., 1992).

Western Anatolia granites share similar microtextural characteristics in CL, including evidence for fluid interactions and multiple generations of microcracks. The samples show secondary alteration textures, mineral growth generations, and evidence for fluid migration. The generations of microfractures, microcracks, and microfaults in CL document that these granites experienced brittle deformation multiple times, both at depth and lower temperatures near the surface (Catlos et al., 2010, 2012). CL imagery is a powerful tool for identifying mineral textural relationships, growth histories, and deformation structures of Western Anatolia granite assemblages.

5.3.4. Timing the Switches in the Stress Regimes in Western Anatolia

The Simav Fault system illustrates another outstanding question regarding deciphering stress regimes within Western Anatolia (Figure 5.7). On 19 May 2011, a magnitude 5.7 (Mww, USGS and Turkish Ministry of the Interior, Disaster and Emergency Management Presidency, Earthquake Department, AFAD) earthquake occurred near Simav. The epicenter was located ~53 km NNW of Uşak and ~82 km WSW from Kütahya in western Turkey at 20:15:23.4 GMT. The estimated depth of the earthquake varies (Doğangün et al., 2013). Table 5.9 reports the 24.46 km result from AFAD, although the USGS Earthquake Catalog suggests a shallower 7.0 km depth. Görgün (2014) estimated a best-fit hypocenter depth of 10 km and 6.0 magnitude (Mw). Karasözen et al. (2016) indicate that the centroid depth was 7–9 km, but the hypocenters of the mainshock and largest aftershocks were located systematically deeper at depths of 10–22 km. In approximately the same location, an Mw ~5.1 event preceded the mainshock on 17 February 2009, and an Mw 4.4 foreshock occurred 15 min before the mainshock (Karasözen et al., 2016).

The Simav region is considered to be one of the most seismically active portions of Western Anatolia (Inel et al., 2013; Görgün, 2014), and the 19 May 2011 Simav (Kütahya) earthquake was the largest felt in the region since the destructive 1969 Demirci and 1970 Gediz earthquake sequences (Ilhan, 1971; Ambraseys & Tchalenko, 1972; Eyidoğan & Jackson, 1985). All of these earthquakes involved dominant normal faulting with nucleation zones from 6 to 10 km depth and dips of 30–50° (Eyidoğan & Jackson, 1985; Emre & Duman, 2011; Görgün, 2014; Karasözen et al., 2016). However, a strike-slip component is recorded by some of the aftershocks of the 1969 and 1970 earthquakes and the 2011 Simav event (Figure 5.7b) (Ambraseys & Tchalenko, 1972; Eyidoğan & Jackson, 1985; Emre & Duman, 2011). In addition, Figure 5.7b shows that some earthquakes in the Simav region after the 2011 event also yielded fault plane solutions that include some or a significant strike-slip component.

The epicenters of these earthquakes occurred near the Simav Fault (Figure 5.7) (Seyitoğlu, 1997; Ersoy et al., 2010). The fault extends ~150 km between the towns of Banaz in the east and Sındırgı in the west (Ambraseys & Tchalenko, 1972; Seyitoğlu, 1997; Ersoy et al., 2010). It may be part of a larger extensional Akşehir-Simav Fault System (Koçyiğit & Deveci, 2007), which extends >250 km from the town of Akşehir in south-central Turkey and includes the Sultandağ Fault in the east (Aksarı et al., 2010). Or, it may be part of the Sındırgı-Sincanlı Fault Zone (SSFZ) between the towns of Soma and Afyon (Doğan & Emre, 2006). The Simav Fault may also connect to the Muratdağ Fault near Gediz in an en-échelon pattern, which supports a right-lateral system (Ambraseys & Tchalenko, 1972). The location of the Akşehir-Simav Fault System between Uşak and Afyon is unclear (Karasözen et al. 2016).

The Simav Fault is assigned as an active right-lateral strike-slip fault in active tectonic maps of Turkey (Şaroğlu et al., 1992; Emre et al., 2011). This sense of motion is based on offsets of metamorphic zones east of Simav (Konak, 1982; Seyitoğlu, 1997) and its relationship to the formation of the NAFZ (Konak, 1982; Doğan & Emre, 2006; Emre & Duman, 2011). The strike-slip motion is also consistent with uniform (magnitude and orientation) GPS plate velocity vectors that show the region is extruding through an SW motion from 30–40 mm/yr (McClusky et al., 2000; Reilinger et al., 2006, 2010). However, the detailed analysis of the Simav Fault mechanisms consistently indicates a normal mechanism (Görgün, 2014; Yolsal-Çevikbilen et al., 2014; Demirci et al., 2015; Karasözen et al., 2016; Bello et al., 2017; Mutlu, 2020). This origin is linked to subduction-related extension along the Hellenic and Cyprus arcs (Seyitoğlu, 1997; Işik et al., 2003; Ersoy et al., 2010; Görgün, 2014; Yolsal-Çevikbilen et al., 2014; Demirci et al., 2015; Karasözen et al. 2016; Bello et al., 2017).

The Simav Fault may have been initiated as a strike-slip system but switched to extension sometime after the Late Miocene (Oygür & Erler, 2000). Strike-slip motion has also been speculated to predate subsidence currently experienced by Western Anatolia and may be related to Eocene to Oligocene compression (Oygür & Erler, 2000). Based on an analysis of the available data from the 19 May 2011 event, Görgün (2014) indicates that the hypocenter distribution is consistent with the activation of two nearly parallel faults: one northern one with a fault plane trending mainly E W and dipping toward SE and a southern fault plane trending NW-SE and dipping toward SE. The strike-slip mechanisms are delegated to smaller fault segments that experience a stress change after the mainshock and more minor secondary faults in the region with different mechanisms. Karasözen et al. (2016) suggest the potential involvement of structures inherited from earlier deformation phases of shortening and extension in evaluating the nature of motion along the structure.

The Simav E-W-trending graben hosts one of Turkey's most important geothermal systems (Bello et al., 2017). Based on a study of geothermal activity, soil radon gas release, and regional seismicity patterns, İnan et al. (2012) suggest that the epicentral area of the 19 May 2011 Simav earthquake is located within a block that is tectonically separated from the Aegean Extensional Province and the Marmara Region. The observation is also supported by geodetic data that show a region surrounding the event behaves distinctly from the Aegean Extensional Province (Tiryakioğlu, 2015). Yolsal-Çevikbilen et al. (2014) suggest the magnitude of the stress drop associated with the 19 May 2011 event (62 bars) is more consistent with an intraplate earthquake compared to those associated with Aegean plate boundaries (3–11 bars).

5.3.5. Relating Geological Units and Events Across Boundaries

As noted in the Geological Background section, several units and structures can be correlated from Western Anatolia to the Aegean region. For example, the Cyclades Blueschist Unit (CBU) from the southern portion of the Menderes Massif (Figure 5.12a) is often matched to outcrops exposed in the Cyclades (Ring et al., 1999; Roche et al., 2018; Çetinkaplan et al., 2020; Barbot & Weiss, 2021), but distinguishing structures developed during subduction-related burial and prograde metamorphism from those that formed due to decompression, and

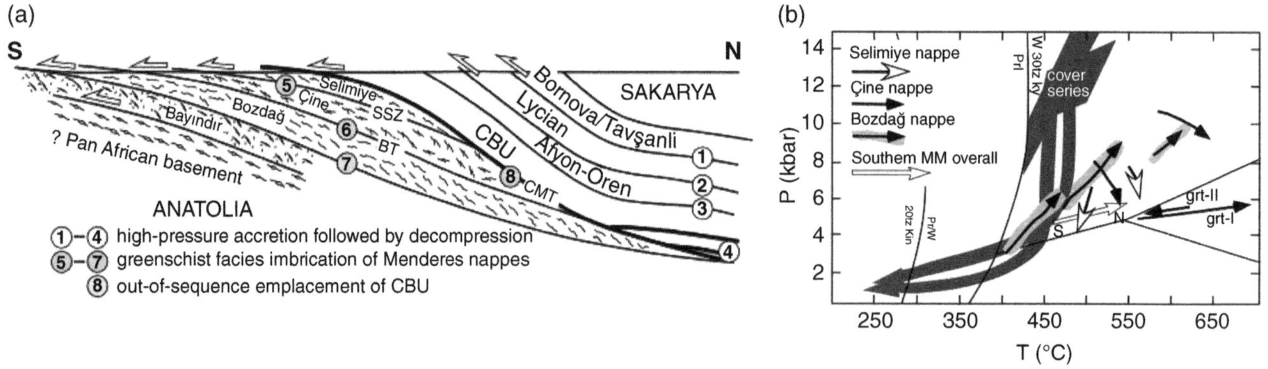

Figure 5.12 (a) Interpretative thrust sequence during the formation of Anatolide belt adapted from Gessner et al. (2013). CBU, Cyclades Blueschist Unit; CMT, Cyclades-Menderes Thrust; SSZ, Selimiye Shear Zone; BT, Bozdağ Thrust. (b) P-T paths from Menderes Massif nappes (Ring et al., 2001; Whitney & Bozkurt, 2002; Rimmelé et al., 2005; Régnier et al., 2007).

retrogression is problematic (Rosenbaum et al., 2002; Xypolias et al., 2012; Çetinkaplan et al., 2020). The CBU experienced multiple phases of deformation and mineralogical transformations (Seman et al., 2017; Gerogiannis et al., 2019). Discriminating between local internal structures and those correlating as significant deformation zones poses a challenge. Çetinkaplan et al. (2020) suggest that the contact between the Menderes Massif and the CBU, now defined by a ductile thrust fault, was originally a lithosphere-scale transform fault zone.

The timing of detachment systems in the Menderes Massif is similar to those estimated in the Cyclades. Three major Aegean microplate detachment systems include the North Cycladic Detachment System on Andros, Tinos, and Mykonos (Figure 5.11; Jolivet et al., 2010), the Naxos-Paros Detachment on Naxos and Paros (Buick, 1991; John & Howard, 1995; Cao et al., 2013), and the West Cycladic Detachment on Serifos (Grasemann et al., 2012). The North Cycladic Detachment System may have initiated activity in the Oligocene until the Late Miocene (Jolivet et al., 2010). The Naxos-Paros Detachment records retrogression associated with its latest activity in the Late Miocene (Cao et al., 2017). These time frames are similar to constraints estimated for the activity of detachment faulting in the central Menderes Massif (Hetzel et al., 1995a, 1995b; Işık et al., 2003; Glodny & Hetzel, 2007; Catlos et al., 2010). The Cyclades Detachments cross-cut blueschist-amphibolite facies fabrics and post-date HP metamorphism and peak Barrovian metamorphism (Searle & Lamont, 2020).

Another correlation links the lithologies, conditions, and metamorphic history of Menderes Massif nappes to those in the Cyclades (Robertson et al., 1991; Ring et al., 1999; Stampfli, 2000; Çetinkaplan et al., 2020). Menderes Massif nappes have zoned garnets useful for generating P-T conditions and paths (Figures 5.12.b and 5.13). These paths are often developed by connecting peak metamorphic conditions of individual rocks, inferences from mineral assemblages, pseudosections, or Gibbs method thermodynamic modeling (Ashworth & Evirgen, 1984, 1985a, 1985b; Ring et al., 2001; Whitney & Bozkurt, 2002; Cenki-Tok et al., 2016; Etzel et al., 2019, 2020). Despite these studies, the number and timing of garnet-growth events recorded in the rocks remain unclear. Some Çine nappe rocks experienced two stages of garnet growth (Ring et al., 2001), whereas other samples are consistent with one episode (Régnier et al., 2007). Pan-African garnet growth is recorded in the Menderes Massif, and conditions could reflect events unrelated to MMM (Ring et al., 2004; Catlos & Çemen, 2005). Gessner et al. (2001) report that the Bayındır nappe deformed once during the Eocene related to MMM, whereas the Bozdağ, Çine, and Selimiye nappes record pre-MMM and MMM events. This observation contradicts Oberhäensli et al. (1997), who suggest the cover sequence records deformation during the Eocene but structurally lower units record pre-MMM events. Studies of Bozdağ nappe rocks show prograde burial, but conditions decrease downward by ~40°C/kbar per km of structural section (inverted metamorphism, Ring et al., 2001). Selimiye nappe rocks record exhumation and retrogression (Régnier et al., 2007). Paths in Figure 5.12.b were generated by connecting peak metamorphic conditions of individual rocks, inferring from mineral assemblages, pseudosections, or Gibbs method thermodynamic modeling. P-T paths that decrease in pressure or temperature suggest the potential for tectonic switching as unloading and refrigeration occur when the thrust reverses and experiences extension.

Challenges for generating P-T conditions and paths include a prior garnet-producing history and retrograde fluid inducted alteration and overprinting as the core complex formed (Satir & Taubald, 2001; Régnier et al., 2003; Catlos & Çemen, 2005; Baker et al., 2008; Candan et al., 2011). Menderes Massif rocks are known to yield problematic P-T estimates based on evidence of disequilibrium among phases and the application of barometers to inappropriate (uncalibrated) mineral compositions (Ashworth & Evirgen, 1984, 1985a, 1985b). In some cases, calculated conditions appear at odds with observed mineral assemblages and structural data (Ring et al., 2001; Whitney & Bozkurt, 2002). Pressure estimates using conventional approaches are challenging to obtain due to the lack of appropriate mineral assemblages (Iredale et al., 2013). Problems may arise if the chosen mineral compositions for thermobarometric calculations are associated with retrogression instead of the desired prograde conditions. P-T paths that only rely on core and rim measurements are also limited in their ability to test models developed regarding lithospheric response to perturbations, including motion within fault zones.

One promising avenue to address this issue is the application of isochemical phase equilibria modeling. Figure 5.13 shows this approach applied to garnets from the Menderes Massif's Çine, Selimiye, and Bayindir nappe from Etzel et al. (2019) and Etzel et al. (2020) and a sample from the Northern Menderes Massif from Cenki-Tok et al. (2016). The researchers report petrological details, X-ray element maps, and geochemical data from the rocks. They compositionally analyzed micaschists with a mineral assemblage of garnet + biotite + plagioclase + muscovite + quartz + rutile ± ilmenite ± apatite ± pyrite ± zircon ± monazite. The sample from the Northern Menderes Massif contains kyanite and small porphyroblasts of staurolite. Geochemical data reported in the papers were used to construct isochemical phase diagrams were created using rock bulk compositions, the software package Theriak-Domino (de Capitani & Brown, 1987; de Capitani & Petrakakis, 2010) with the Holland and Powell (1998, 2011) thermodynamic

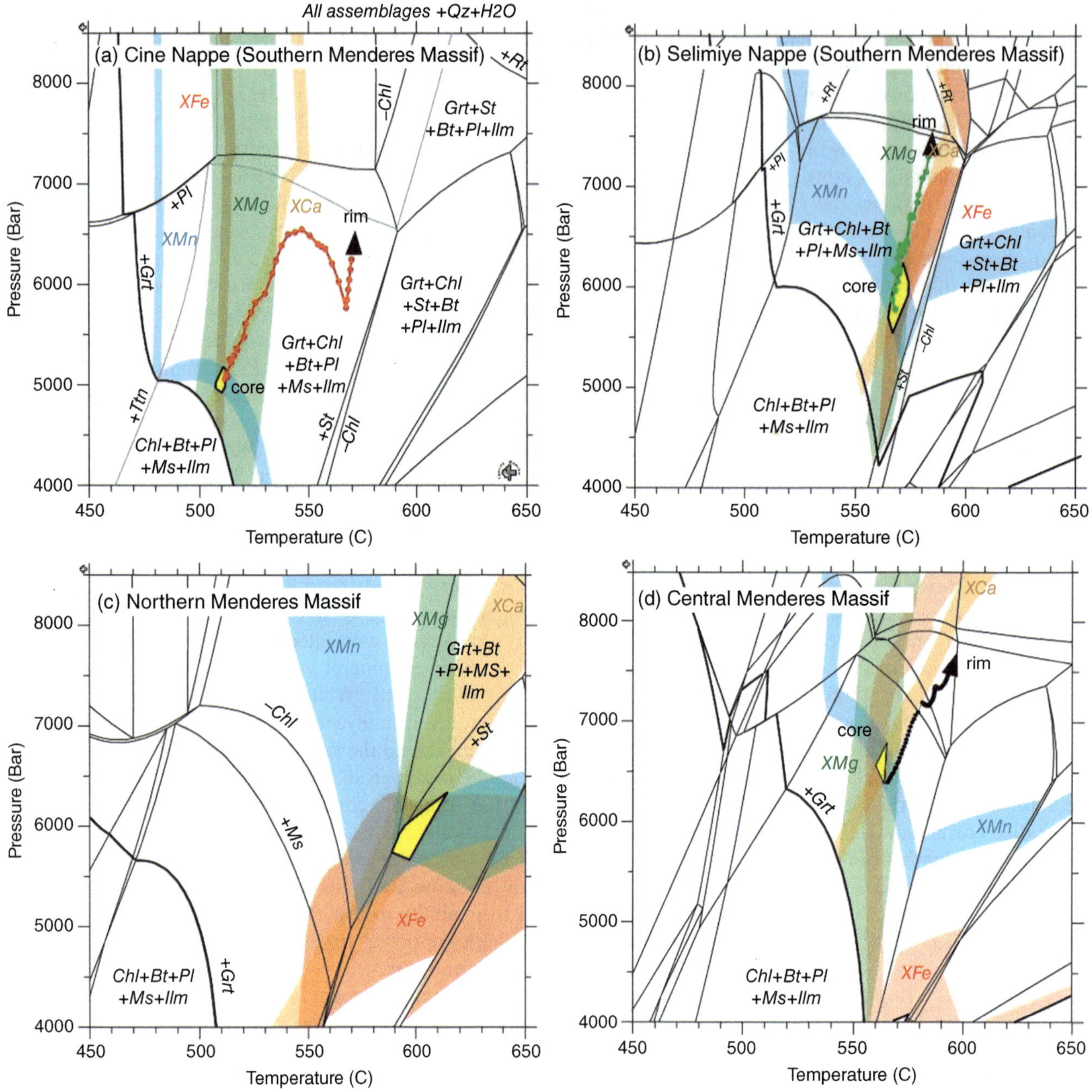

Figure 5.13 Isochemical phase diagrams with overlapping garnet core compositional isopleths for garnet-bearing samples from the (a) Çine nappe, (b) Selimiye nappe adapted from Etzel et al. (2019), (c) Northern Menderes Massif using data adapted from Cenki-Tok et al. (2016), and (d) the Central Menderes Massif (Adapted from Etzel et al., 2020). Mineral abbreviations after de Capitani & Brown (1987) and de Capitani & Petrakakis (2010). Labeled stripes are compositional isopleths of ±0.1 mole fraction for end-member garnet core compositional contents, except for panel (d), which overlies ±0.2 mole fraction and is for the reported representative composition for that garnet by Cenki-Tok et al. (2016). The gray polygon in each diagram represents the conditions estimated for garnet growth in the samples. High-resolution P-T paths for the samples are shown in panels (a), (b), and (d).

dataset, and appropriate mixing models in the system $MnO–Na_2O–CaO–K_2O–FeO–MgO–Al_2O_3–SiO_2–H_2O–TiO_2$. Isopleths of ±0.01-mole-fraction spessartine, almandine, pyrope, and grossular corresponding with the garnet core composition are plotted on the phase diagram. This portion of the diagram with intersecting isopleths approximates the chemical system when garnet began growth. This diagram also tests if the thermodynamic dataset and mixing models used in the modeling are appropriate for these particular samples, as expected mineral assemblages appeared in the phase diagrams with intersecting isopleths.

After the garnet core conditions were estimated, a Matlab script was applied to each step along a garnet compositional traverse from core to rim to yield both an estimate of the P-T conditions of incremental growth and a new effective bulk rock composition, ultimately culminating in a high-resolution P-T path. High-resolution P-T paths are defined as those derived from fractionated equilibrium phase diagram modeling, and the resolution is an outcome of the number of garnet fractionated steps. Garnets with complex zoning profiles, modified by diffusion, or rocks that experienced major changes in bulk composition over their growth history are not candidates (Catlos et al., 2018). However, even these types of samples may provide clues by exploring the reason for their failure (Catlos et al., 2018; Etzel et al., 2020). Ideal samples are those with garnets that preserve prograde, gradational core-to-rim zoning profiles. Garnets from the Selimiye and Bayindir nappes of the Southern and Central Menderes Massif, respectively, show similar trajectories. However, the Çine nappe garnet yields an N-shape path and a significantly different metamorphic history.

Either tectonically driven extension may have created the N-shaped P-T path during orogenesis or the result of erosional exhumation during pulses of thrust motion (Etzel et al., 2019). Etzel et al. (2019) developed two thermal models: erosional denudation followed by fault reactivation (Figure 5.14a) and tectonic switching (Figure 5.14b), which are briefly summarized here. Figure 5.14a,b show an upper equilibrium thermal grid (depth versus horizontal distance) before faulting with

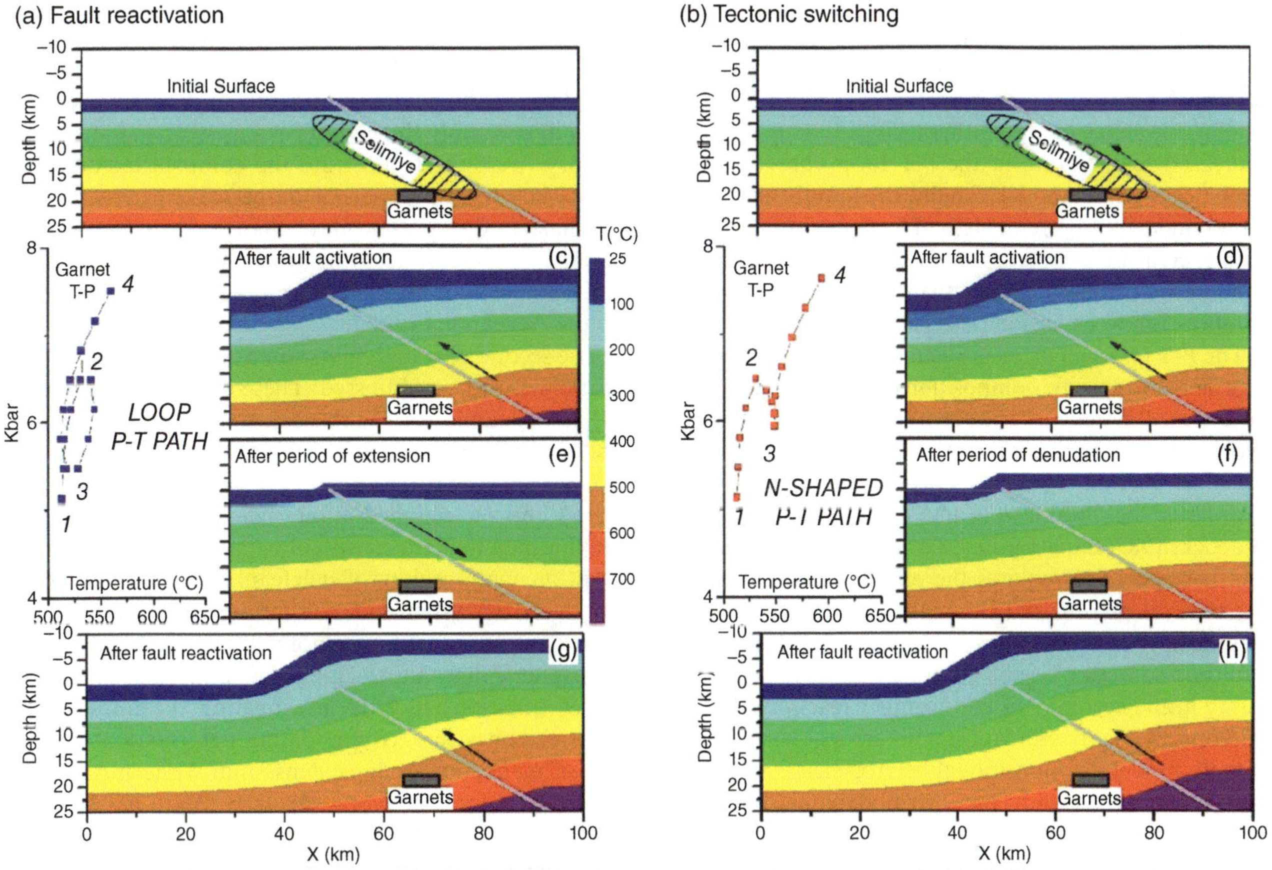

Figure 5.14 Snapshots of thermal models of the Çine nappe for the (a) tectonic switching and (b) fault reactivation model adapted from Etzel et al. (2019). (a) and (b) are the upper equilibrium thermal grid (depth versus horizontal distance) before faulting with the position of fault (gray line) arbitrarily selected at 30°. Fault displacement varies linearly. The grid includes reflecting side boundaries, top, and bottom maintained at 25 and 700 °C, and an initial geothermal gradient at 25 °C/km indicated by shaded bars. A hatch area infers the position of the Selimiye samples, and the gray bar represents the approximate initial location of the Çine nappe garnet with the N-shaped P-T path. This scenario is also represented by point 1 in P-T path insets. In panels (c) and (d), the fault is activated, and a finite-difference solution to the diffusion-advection equation is used to examine the P-T variations in the hanging wall and footwall as a result of motion. The rock sample experiences the path from 1 to 2 on the P-T path insets. In panels (e) and (f), motion stops. Extension occurs in panel (e), and denudation occurs in panel (f). This scenario is modeled based on the mid-rim lower pressure portion of the garnet P-T path and is represented by points 2 to 3 on the P-T path insets. The fault is reactivated in panels (g) and (h), represented by points 3 to 4 on the P-T path insets.

the position of fault (gray line) arbitrarily selected at 30°. Fault displacement varies linearly across shear zones. The grid includes reflecting side boundaries, top and bottom maintained at 25 and 700 °C, and an initial geothermal gradient at 25 °C/km indicated by shaded zones. A hatched area shows the position of the Selimiye samples, and the gray bar represents the approximate initial location of the Çine nappe garnet with the N-shaped P-T path. This position is also represented by point 1 in the P-T path insets. In Figure 5.14c,d, the fault is active. A finite-difference solution to the diffusion-advection equation examines the P-T variations in the hanging wall and footwall due to its motion. The rock sample experiences points 1 to 2 in the P-T path insets. Fault motion stops, denudation occurs in Figure 5.14e, and extension occurs in Figure 5.14f. This process is based on the mid-rim lower pressure portion of the garnet P-T path and is represented by points 2 to 3 on the P-T path insets. Although in the end, the surface geometry in the denudation phase (Figure 5.13e) and extensional phase (Figure 5.14f) are similar, the shape of the isotherms is different. It leads to decreased temperature in the P-T loop observed in the tectonic switching model. Finally, the fault is reactivated, represented by Figure 5.14g,h. The decrease in pressure with increasing temperature is related to an episode of denudation (model 1) rather than a tectonic switch from compression to extension (Etzel et al., 2019).

The P-T paths reported in Figure 5.13 approximate how a garnet with specific compositional zoning would behave in a closed system of a known bulk composition as it evolves during increasing T. A critical assumption is that the minerals in a sample experienced equilibrium, which can never be proven for any rock system (Spear & Peacock, 1989; Lanari & Duesterhoeft, 2019). Closed system behavior also requires the original compositions of the mineral phases, and the bulk rock has not changed significantly since metamorphism (Lanari & Engi, 2017). Multiple sources of error are inherent, including uncertainty in the accuracy of end-member reactions, electron microprobe analyses, calibration errors, variations in activity models, compositional heterogeneity, and uncertainty associated with the thermodynamic properties inherent in the choice of internally consistent database (Kohn & Spear, 1991; White et al., 2014; Palin et al., 2016; Lanari & Duesterhoeft, 2019). Garnets with significant changes in composition over short distances from core to rim and those affected by diffusion cannot be modeled. Garnets in samples that experienced significant changes in bulk composition or multiple deformation episodes resulting in modification of composition are also unsuitable.

A significant value of the high-resolution P-T path and isopleth approaches is that a user can detect when systems stray from the equilibrium and closed system assumptions. Confidence in paths and conditions increases when minerals assemblages agree with rock observations and if the P-T paths reproduce trends in garnet zoning. Samples collected from the same outcrop or nearby should yield similar P-T conditions and paths. In addition, a user can gauge the extent of overlapping mineral isopleths in P-T space, including if matrix mineral compositions overlap the garnet rim conditions. These paths are the first steps in developing critical insights into the metamorphic history of the assembly of the Menderes Massif and, combined with age information from the garnet itself or matrix or mineral inclusions, can be used to test models for the development of Western Anatolia.

5.4. CONCLUSIONS

This chapter is divided into two major sections. The first outlines, as much as possible, our present-day understanding of the geological history of Western Anatolia from its assembly through its extensional and strike-slip history. We aim to illustrate the complex tectonic scenario before the onset of large-scale extension and emphasize the present-day change in stress regime toward strike-slip tectonics. The transitions are also comparable in duration and timing to those experienced by the Aegean microplate.

The second part highlights some outstanding questions that remain to be addressed. These include issues regarding the dynamics of the African slab along the Hellenic arc, the arc's geometry and connections to other subduction systems, and reconciling the Jurassic SZI age with Late Cenozoic sedimentation, magmatic, and paleogeographic data that are consistent with younger initiation. In addition, many regions of slab tear are proposed throughout the African slab, and their influence on accommodating extrusion, creating economic resources, and driving lithospheric thinning and magmatism should be explored. Other questions include investigating the influence of transfer zones in accommodating deformation and the role of magma in driving extension in Western Anatolia.

The interface between Western Anatolia and the Aegean region exemplifies tectonic transitions and how the interplay between large-scale tectonics influences smaller-scale processes. The Aegean and western Turkey contain helpful assemblages that can be exploited to time these processes that shape the lithosphere and are critical in understanding the region's hazards and mineralizations. Extracting high-resolution P-T paths from Western Anatolia garnet-bearing rocks is a promising approach to evaluate tectonic models and correlate and compare metamorphic histories of nearby assemblages and from those across long distances.

ACKNOWLEDGMENTS

We appreciate constructive reviews of the original manuscript from Spyros Pavlides (Aristotle University of Thessaloniki, Greece), Thomas Lamont (University of Bristol), and three anonymous reviewers. We appreciate analytical assistance with the $^{40}Ar/^{39}Ar$ analysis Dr. Marnie Forster and Davood Vasegh from the Australia National University Argon Facility. Modeling of the P-T data was done with the assistance of Dr. Oscar Lovera (UCLA).

AVAILABILITY STATEMENT

Data supporting this chapter's conclusions are available from the Texas Data Repository Dataverse (Catlos, 2021).

REFERENCES

Agostini, S., Doglioni, C., Innocenti, F., Manetti, P., & Tonarini, S. (2010). On the geodynamics of the Aegean rift. *Tectonophysics*, *488*(1–4), 7–21. doi: 10.1016/j.tecto.2009.07.025

Akal, C., & Helvacı, C. (1999). Mafic microgranular enclaves in the Kozak granodiorite, western Anatolia. *Turkish Journal of Earth Sciences*, *8*, 1–17.

Akal, C. (2013). Coeval Shoshonitic-ultrapotassic dyke emplacements within the Kestanbol Pluton, Ezine – Biga Peninsula (NW Anatolia). *Turkish Journal of Earth Sciences*, *22*(2), 220–238. doi: 10.3906/yer-1202-1

Akay, E. (2009). Geology and petrology of the Simav Magmatic Complex (NW Anatolia) and its comparison with the Oligo-Miocene granitoids in NW Anatolia: implications on Tertiary tectonic evolution of the region. *International Journal of Earth Sciences*, *98*(7), 1655–1675. doi: 10.1007/s00531-008-0325-0

Akbayram, K., Şengör, A. M. C., & Özcan, E. (2017). The evolution of the Intra-Pontide suture: implications of the discovery of Late Cretaceous–early Tertiary mélanges. In R. Sorkhabi (Ed.), *Tectonic evolution, collision, and seismicity of Southwest Asia: in Honor of Manuel Berberian's forty-five years of research contributions*. Geological Society of America. doi: 10.1130/2016.2525(18)

Aksarı, D., Karabulut, H., & Özalaybey, S. (2010). Stress interactions of three moderate size earthquakes in Afyon, southwestern Turkey. *Tectonophysics*, *485*(1–4), 141–153. doi: 10.1016/j.tecto.2009.12.010

Aktuğ, B., Parmaksız, E., Kurt, M., Lenk, O., Kılıçoğlu, A., Gürdal, M. A., & Özdemir, S. (2013). Deformation of Central Anatolia: GPS implications. *Journal of Geodynamics*, *67*, 78–96. doi: 10.1016/j.jog.2012.05.008

Alatza, S., Papoutsis, I., Paradissis, D., Kontoes, C., & Papadopoulos, G. A. (2020). Multi-temporal InSAR analysis for monitoring ground deformation in Amorgos Island, Greece. *Sensors*, *20*(2), 338. doi: 10.3390/s20020338

Aldanmaz, E. (2012). Osmium isotope and highly siderophile element geochemistry of mantle xenoliths from NW Turkey: implications for melt depletion and metasomatic history of the sub-continental lithospheric mantle. *International Geology Review*, *54*(7), 799–815. doi: 10.1080/00206814.2011.581799

Aldanmaz, E., Pearce, J. A., Thirlwall, M. F., & Mitchell, J. G. (2000). Petrogenetic evolution of late Cenozoic, post-collision volcanism in western Anatolia, Turkey. *Journal of Volcanology and Geothermal Research*, *102*(1–2), 67–95. doi: 10.1016/S0377-0273(00)00182-7

Altherr, R., Henjes-Kunst, F., Matthews, A., Friedrichsen, H., & Hansen, B. T. (1988). O-Sr isotopic variations in Miocene granitoids from the Aegean: evidence for an origin by combined assimilation and fractional crystallization. *Contributions to Mineralogy and Petrology*, *100*(4), 528–541. doi: 10.1007/BF00371381

Altherr, R., & Siebel, W. (2002). I-type plutonism in a continental back-arc setting: Miocene granitoids and monzonites from the central Aegean Sea, Greece. *Contributions to Mineralogy and Petrology*, *143*(4), 397–415. doi: 10.1007/s00410-002-0352-y

Altunkaynak, Ş., & Yılmaz, Y. (1998). The Mount Kozak magmatic complex, Western Anatolia. *Journal of Volcanology and Geothermal Research*, *85*(1–4), 211–231. doi: 10.1016/S0377-0273(98)00056-0

Altunkaynak, Ş., & Yilmaz, Y. (1999). The Kozak Pluton and its emplacement. *Geological Journal*, *34*(3), 257–274. doi: 10.1002/(SICI)1099-1034(199907/09)34:33.0.CO;2-Q

Altunkaynak, Ş. (2007). Collision-driven slab breakoff magmatism in Northwestern Anatolia, Turkey. *The Journal of Geology*, *115*(1), 63–82. doi: 10.1086/509268

Altunkaynak, Ş., & Dilek, Y. (2006). Timing and nature of postcollisional volcanism in western Anatolia and geodynamic implications. In Y. Dilek & S. Pavlides (Eds.), *Postcollisional tectonics and magmatism in the Mediterranean Region and Asia*. Geological Society of America. doi: 10.1130/2006.2409(17)

Altunkaynak, Ş., Dilek, Y., Genç, C. Ş., Sunal, G., Gertisser, R., Furnes, H., et al. (2012). Spatial, temporal and geochemical evolution of Oligo–Miocene granitoid magmatism in western Anatolia, Turkey. *Gondwana Research*, *21*(4), 961–986. doi: 10.1016/j.gr.2011.10.010

Altunkaynak, Ş., & Genç, Ş. C. (2008). Petrogenesis and time-progressive evolution of the Cenozoic continental volcanism in the Biga Peninsula, NW Anatolia (Turkey). *Lithos*, *102*(1–2), 316–340. doi: 10.1016/j.lithos.2007.06.003

Amaru, M. L. (2007). *Global travel time tomography with 3-D reference models* (PhD thesis). Netherlands: Utrecht University. Retrieved from http://dspace.library.uu.nl/bitstream/handle/1874/19338/index.htm;jsessionid=88B6AA4941C5E76FD65034D89D4065E0?sequence=17

Ambraseys, N. N., & Tchalenko, J. S. (1972). Seismotectonic aspects of the Gediz, Turkey, earthquake of March 1970. *Geophysical Journal of the Royal Astronomical Society*, *30*(3), 229–252.

Anastasakis, G., & Kelling, G. (1991). Tectonic connection of the Hellenic and Cyprus arcs and related geotectonic elements. *Marine Geology*, *97*(3–4), 261–277. doi: 10.1016/0025-3227(91)90120-S

Angelier, J., Lybéris, N., Le Pichon, X., Barrier, E., & Huchon, P. (1982). The tectonic development of the Hellenic arc and the sea of Crete: a synthesis. *Tectonophysics*, *86*(1–3), 159–196. doi: 10.1016/0040-1951(82)90066-X

Armijo, R., Meyer, B., Hubert, A., & Barka, A. (1999). Westward propagation of the North Anatolian fault into the northern Aegean: timing and kinematics. *Geology*, *27*(3), 267. doi: 10.1130/0091-7613(1999)027<0267:WPOTNA >2.3.CO;2

Ashworth, J. R., & Evirgen, M. M. (1984). Garnet and associated minerals in the southern margin of the Menderes Massif, southwest Turkey. *Geological Magazine*, *121*(4), 323–337. doi: 10.1017/S0016756800029228

Ashworth, J. R., & Evirgen, M. M. (1985a). Plagioclase relations in pelites, central Menderes Massif, Turkey. II. Perturbation of garnet-plagioclase geobarometers. *Journal of Metamorphic Geology*, *3*(3), 219–229. doi: 10.1111/j.1525-1314.1985.tb00318.x

Ashworth, J. R., & Evirgen, M. M. (1985b). Plagioclase relations in pelites, central Menderes Massif, Turkey. I. The peristerite gap with coexisting kyanite. *Journal of Metamorphic Geology*, *3*(3), 207–218. doi: 10.1111/j.1525-1314.1985.tb00317.x

Avigad, D., Baer, G., & Heimann, A. (1998). Block rotations and continental extension in the central Aegean Sea: palaeomagnetic and structural evidence from Tinos and Mykonos (Cyclades, Greece). *Earth and Planetary Science Letters*, *157*(1–2), 23–40. doi: 10.1016/S0012-821X(98)00024-7

Avigad, D., & Garfunkel, Z. (1991). Uplift and exhumation of high-pressure metamorphic terrains: the example of the cycladic blueschist belt (Aegean Sea). *Tectonophysics*, *188*(3–4), 357–372. doi: 10.1016/0040-1951(91)90464-4

Aydoğan, M. S., Çoban, H., Bozcu, M., & Akıncı, Ö. (2008). Geochemical and mantle-like isotopic (Nd, Sr) composition of the Baklan Granite from the Muratdağı Region (Banaz, Uşak), western Turkey: implications for input of juvenile magmas in the source domains of Western Anatolia Eocene–Miocene granites. *Journal of Asian Earth Sciences*, *33*(3–4), 155–176. doi: 10.1016/j.jseaes.2006.10.007

Aysal, N., Öngen, S., Peytcheva, I., & Keskin, M. (2012). Origin and evolution of the Havran Unit, Western Sakarya basement (NW Turkey): new LA-ICP-MS U-Pb dating of the metasedimentary metagranitic rocks and possible affiliation to Avalonian microcontinent. *Geodinamica Acta*, *25*(3–4), 226–247. doi: 10.1080/09853111.2014.882536

Aysal, N., Yılmaz Şahin, S., Güngör, Y., Peytcheva, I., & Öngen, S. (2018). Middle Permian–early Triassic magmatism in the Western Pontides, NW Turkey: Geodynamic significance for the evolution of the Paleo-Tethys. *Journal of Asian Earth Sciences*, *164*, 83–103. doi: 10.1016/j.jseaes.2018.06.026

Bagci, M. (2012). *Geochemical and geochronological (Rb/Sr) properties of the Günyüzü granitoids (Sivrihisar, Eskişehir)* Paper presented at the SGEM2012 12th International Multidisciplinary Scientific GeoConference and EXPO, Sofia, Bulgaria. doi: 10.5593/sgem2012/s01.v1007

Baker, C. B., Catlos, E. J., Sorensen, S. S., Çemen, I., & Hancer, M. (2008). Evidence for polymetamorphic garnet growth in the Çine (southern Menderes) Massif, Western Turkey. *IOP Conference Series: Earth and Environmental Science*, *2*, 012020. doi: 10.1088/1755-1307/2/1/012020

Baran, Z. O., Dilek, Y., & Stockli, D. (2017). Diachronous uplift and cooling history of the Menderes core complex, western Anatolia (Turkey), based on new Zircon (U-Th)/He ages. *Tectonophysics*, *694*, 181–196. doi: 10.1016/j.tecto.2016.12.005

Barbot, S., & Weiss, J. R. (2021). Connecting subduction, extension and shear localization across the Aegean Sea and Anatolia. *Geophysical Journal International*, *226*(1), 422–445. doi: 10.1093/gji/ggab078

Barka, A. A. (1992). The North Anatolian fault zone. *Annales Tectonicae*, *6*, 64–195.

Barka, A. A., & Kadinsky-Cade, K. (1988). Strike-slip fault geometry in Turkey and its influence on earthquake activity. *Tectonics*, *7*(3), 663–684. doi: 10.1029/TC007i003p00663

Barka, A. A., & Reilinger, R. (1997). Active tectonics of the Eastern Mediterranean Region: deduced from GPS, neotectonic and seismicity data. *Annelis de Geofisica*, *40*(3), 587–610.

Beccaletto, L., & Jenny, C. (2004). Geology and correlation of the Ezine Zone: a rhodope fragment in NW Turkey? *Turkish Journal of Earth Sciences*, *13*(13), 45–176.

Beccaletto, L., Bonev, N., Bosch, D., & Bruguier, O. (2007). Record of a Palaeogene syn-collisional extension in the north Aegean region: evidence from the Kemer micaschists (NW Turkey). *Geological Magazine*, *144*(2), 393–400. 10.1017/S001675680700310X

Beißer, M., Wyss, M., & Kind, R. (1990). Inversion of source parameters for subcrustal earthquakes in the Hellenic Arc. *Geophysical Journal International*, *103*(2), 439–450. doi: 10.1111/j.1365-246X.1990.tb01783.x

Bello, O. A., Özgür, N., & Çalışkan, T. A. (2017). Hydrogeological, hydrogeochemical and isotope geochemical features of geothermal waters in Simav and environs, Western Anatolia, Turkey. *Procedia Earth and Planetary Science*, *17*, 29–32. doi: 10.1016/j.proeps.2016.12.014

Beniest, A., Brun, J. P., Gorini, C., Crombez, V., Deschamps, R., Hamon, Y., & Smit, J. (2016). Interaction between trench retreat and Anatolian escape as recorded by neogene basins in the northern Aegean Sea. *Marine and Petroleum Geology*, *77*, 30–42. doi: 10.1016/j.marpetgeo.2016.05.011

Berckhemer, H. (1977). Some aspects of the evolution of marginal seas deduced from observations in the Aegean region. In B. Biju-Duval & L. Montadert (Eds.), *Structural history of the Mediterranean basins* (pp. 303–311). Paris, France: Editions Technip.

Bijwaard, H., Spakman, W., & Engdahl, E. R. (1998). Closing the gap between regional and global travel time tomography. *Journal of Geophysical Research: Solid Earth*, *103*(B12), 30055–30078. doi: 10.1029/98JB02467

Birkle, P. (1992). *Petrologie-Geochemie und Geochronologie des miozänen Magmatismus auf der Biga-Halbinsel (Ezine, NW-Türkie)* (PhD thesis). Tübingen, Germany: Eberhard-Karls-Universität Tübingen.

Biryol, C. B., Beck, S. L., Zandt, G., & Özacar, A. A. (2011). Segmented African lithosphere beneath the Anatolian region inferred from teleseismic P-wave tomography: segmented lithosphere beneath Anatolia. *Geophysical Journal International*, *184*(3), 1037–1057. doi: 10.1111/j.1365-246X.2010.04910.x

Black, K. N., Catlos, E. J., Oyman, T., & Demirbilek, M. (2013). Timing Aegean extension: evidence from in situ U–Pb geochronology and cathodoluminescence imaging of granitoids from NW Turkey. *Lithos*, *180–181*, 92–108. doi: 10.1016/j.lithos.2013.09.001

Blom, N., Gokhberg, A., & Fichtner, A. (2020). Seismic waveform tomography of the central and eastern Mediterranean upper mantle. *Solid Earth*, *11*(2), 669–690. doi: 10.5194/se-11-669-2020

Bocchini, G. M., Brüstle, A., Becker, D., Meier, T., Van Keken, P. E., Ruscic, M., et al. (2018). Tearing, segmentation, and backstepping of subduction in the Aegean: new insights from seismicity. *Tectonophysics*, *734–735*, 96–118. doi: 10.1016/j.tecto.2018.04.002

Bolhar, R., Ring, U., & Allen, C. M. (2010). An integrated zircon geochronological and geochemical investigation into the Miocene plutonic evolution of the Cyclades, Aegean Sea, Greece: Part 1: Geochronology. *Contributions to Mineralogy and Petrology*, *160*(5), 719–742. doi: 10.1007/s00410-010-0504-4

Bonev, N., & Beccaletto, L. (2007). From syn- to post-orogenic Tertiary extension in the north Aegean region: constraints on the kinematics in the eastern Rhodope–Thrace, Bulgaria–Greece and the Biga Peninsula, NW Turkey. *Geological Society, London, Special Publications*, *291*(1), 113–142. doi: 10.1144/SP291.6

Bonev, N., Beccaletto, L., Robyr, M., & Monié, P. (2009). Metamorphic and age constraints on the Alakeçi shear zone: implications for the extensional exhumation history of the northern Kazdağ Massif, NW Turkey. *Lithos*, *113*(1–2), 331–345. doi: 10.1016/j.lithos.2009.02.010

Bozkaya, Ö., Yalçın, H., & Göncüoğlu, M. C. (2012). Diagenetic and very low-grade metamorphic characteristics of the Paleozoic series of the İstanbul Terrane (NW Turkey). *Swiss Journal of Geosciences*, *105*(2), 183–201. doi: 10.1007/s00015-012-0108-2

Bozkurt, E., & Park, L. R. G. (1994). Southern Menderes Massif: an incipient metamorphic core complex in western Anatolia, Turkey. *Journal of the Geological Society*, *151*(2), 213–216. doi: 10.1144/gsjgs.151.2.0213

Bozkurt, E., & Satir, M. (2000). The southern Menderes Massif (western Turkey): geochronology and exhumation history. *Geological Journal*, *35*(3–4), 285–296. doi: 10.1002/gj.849

Bozkurt, E., & Sözbilir, H. (2004). Tectonic evolution of the Gediz Graben: field evidence for an episodic, two-stage extension in western Turkey. *Geological Magazine*, *141*(1), 63–79. doi: 10.1017/S0016756803008379

Bozkurt, E. (2001). Neotectonics of Turkey – a synthesis. *Geodinamica Acta*, *14*(1–3), 3–30. doi: 10.1080/09853111.2001.11432432

Bozkurt, E., & Oberhänsli, R. (2001). Menderes Massif (Western Turkey): Structural, metamorphic and magmatic evolution - a synthesis. *International Journal of Earth Sciences*, *89*(4), 679–708. doi: 10.1007/s005310000173

Bozkurt, E., & Park, R. G. (1999). The structure of the Palaeozoic schists in the Southern Menderes Massif, western Turkey: a new approach to the origin of the Main Menderes Metamorphism and its relation to the Lycian Nappes. *Geodinamica Acta*, *12*(1), 25–42. doi: 10.1080/09853111.1999.11105329

Boztuğ, D., Harlavan, Y., Jonckheere, R., Can, İ., & Sari, R. (2009). Geochemistry and K-Ar cooling ages of the Ilıca, Çataldağ (Balıkesir) and Kozak (İzmir) granitoids, west Anatolia, Turkey. *Geological Journal*, *44*(1), 79–103. doi: 10.1002/gj.1132

Brichau, S., Ring, U., Carter, A., Bolhar, R., Monié, P., Stockli, D., & Brunel, M. (2008). Timing, slip rate, displacement and cooling history of the Mykonos detachment footwall, Cyclades, Greece, and implications for the opening of the Aegean Sea basin. *Journal of the Geological Society*, *165*(1), 263–277. doi: 10.1144/0016-76492006-145

Brichau, S., Ring, U., Carter, A., Monié, P., Bolhar, R., Stockli, D., & Brunel, M. (2007). Extensional faulting on Tinos Island, Aegean Sea, Greece: how many detachments? *Tectonics*, *26*(4), 2006TC001969. doi: 10.1029/2006TC001969

Brooks, M., & Ferentinos, G. (1984). Tectonics and sedimentation in the Gulf of Corinth and the Zakynthos and Kefallinia channels, Western Greece. *Tectonophysics*, *101*(1–2), 25–54. doi: 10.1016/0040-1951(84)90040-4

Brun, J.-P., & Faccenna, C. (2008). Exhumation of high-pressure rocks driven by slab rollback. *Earth and Planetary Science Letters*, *272*(1–2), 1–7. doi: 10.1016/j.epsl.2008.02.038

Brun, J.-P., Faccenna, C., Gueydan, F., Sokoutis, D., Philippon, M., Kydonakis, K., & Gorini, C. (2017). Effects of slab rollback acceleration on Aegean extension. *Bulletin of the Geological Society of Greece*, *50*(1), 5. doi: 10.12681/bgsg.11697

Brun, J.-P., & Sokoutis, D. (2010). 45 m.y. of Aegean crust and mantle flow driven by trench retreat. *Geology*, *38*(9), 815–818. doi: 10.1130/G30950.1

Brun, J.-P., Faccenna, C., Gueydan, F., Sokoutis, D., Philippon, M., Kydonakis, K., & Gorini, C. (2016). The two-stage Aegean extension, from localized to distributed, a result of slab rollback acceleration. *Canadian Journal of Earth Sciences*, *53*(11), 1142–1157. doi: 10.1139/cjes-2015-0203

Brun, J.-P., & Sokoutis, D. (2007). Kinematics of the Southern Rhodope Core Complex (North Greece). *International Journal of Earth Sciences*, *96*(6), 1079–1099. doi: 10.1007/s00531-007-0174-2

Buick, I. S. (1991). The late Alpine evolution of an extensional shear zone, Naxosv, Greece. *Journal of the Geological Society*, *148*(1), 93–103. doi: 10.1144/gsjgs.148.1.0093

Bulle, F., Bröcker, M., Gärtner, C., & Keasling, A. (2010). Geochemistry and geochronology of HP mélanges from Tinos and Andros, cycladic blueschist belt, Greece. *Lithos*, *117*(1–4), 61–81. doi: 10.1016/j.lithos.2010.02.004

Bulut, F., Özener, H., Doğru, A., Aktuğ, B., & Yaltırak, C. (2018). Structural setting along the Western North Anatolian Fault and its influence on the 2014 North Aegean earthquake (Mw 6.9). *Tectonophysics*, *745*, 382–394. doi: 10.1016/j.tecto.2018.07.006

Burchfiel, B. C., Nakov, R., Dumurdzanov, N., Papanikolaou, D., Tzankov, T., Serafimovski, T., et al. (2008). Evolution and dynamics of the Cenozoic tectonics of the South Balkan extensional system. *Geosphere*, *4*(6), 919. doi: 10.1130/GES00169.1

Burg, J., Ricou, L., Ivano, Z., Godfriaux, I., Dimov, D., & Klain, L. (1996). Syn-metamorphic nappe complex in the Rhodope Massif. Structure and kinematics. *Terra Nova*, *8*(1), 6–15. doi: 10.1111/j.1365-3121.1996.tb00720.x

Burtman, V. S. (1994). Meso-Tethyan oceanic sutures and their deformation. *Tectonophysics*, *234*(4), 305–327. doi: 10.1016/0040-1951(94)90230-5

Candan, O., Akal, C., Koralay, O. E., Okay, A. I., Oberhänsli, R., Prelević, D., & Mertz-Kraus, R. (2016). Carboniferous granites on the northern margin of Gondwana, Anatolide-Tauride Block, Turkey – evidence for southward subduction of Paleotethys. *Tectonophysics*, *683*, 349–366. doi: 10.1016/j.tecto.2016.06.030

Candan, O., Dora, O., Oberhänsli, R., Cetinkaplan, M., Partzsch, J., Warkus, F., & Durr, S. (2001). Pan-African high-pressure metamorphism in the Precambrian basement of the Menderes Massif, western Anatolia, Turkey. *International Journal of Earth Sciences*, *89*(4), 793–811. doi: 10.1007/s005310000097

Candan, O., Çetinkaplan, M., Oberhänsli, R., Rimmelé, G., & Akal, C. (2005). Alpine high-P/low-T metamorphism of the Afyon Zone and implications for the metamorphic evolution of Western Anatolia, Turkey. *Lithos*, *84*(1–2), 102–124. doi: 10.1016/j.lithos.2005.02.005

Candan, O., Oberhänsli, R., Dora, O. O., Cetinkaplan, M., Koralay, E., Rimmelé, G., et al. (2011). Polymetamorphic evolution of the Pan-African basement and Palaeozoic-early Tertiary cover series of the Menderes Massif. *Mineral Research and Exploration Bulletin*, *142*, 121–163.

Cao, S., Neubauer, F., Bernroider, M., Genser, J., Liu, J., & Friedl, G. (2017). Low-grade retrogression of a high-temperature metamorphic core complex: Naxos, Cyclades, Greece. *Geological Society of America Bulletin*, *129*(1–2), 93–117. doi: 10.1130/B31502.1

Cao, S., Neubauer, F., Bernroider, M., & Liu, J. (2013). The lateral boundary of a metamorphic core complex: the Moutsounas shear zone on Naxos, Cyclades, Greece. *Journal of Structural Geology*, *54*, 103–128. doi: 10.1016/j.jsg.2013.07.002

Catlos, E. (2021). Replication data for: extensional tectonics in Western Anatolia, Turkey: eastward continuation of the Aegean extension [Dataset]. Texas Data Repository. doi: 10.18738/T8/ER3WQV

Catlos, E., Jacob, L., Oyman, T., & Sorensen, S. (2012). Long-term exhumation of an Aegean metamorphic core complex granitoids in the Northern Menderes Massif, western Turkey. *American Journal of Science*, *312*(5), 534–571. doi: 10.2475/05.2012.03

Catlos, E. J., & Çemen, I. (2005). Monazite ages and the evolution of the Menderes Massif, western Turkey. *International Journal of Earth Sciences*, *94*(2), 204–217. doi: 10.1007/s00531-005-0470-7

Catlos, E. J., Lovera, O. M., Kelly, E. D., Ashley, K. T., Harrison, T. M., & Etzel, T. (2018). Modeling high-resolution pressure-temperature paths across the Himalayan Main Central Thrust (Central Nepal): implications for the dynamics of collision. *Tectonics*, *37*(8), 2363–2388. doi: 10.1029/2018TC005144

Catlos, E. J., Baker, C., Sorensen, S. S., Çemen, I., & Hançer, M. (2010). Geochemistry, geochronology, and cathodoluminescence imagery of the Salihli and Turgutlu granites (central Menderes Massif, western Turkey): implications for Aegean tectonics. *Tectonophysics*, *488*(1–4), 110–130. doi: 10.1016/j.tecto.2009.06.001

Catlos, E. J., & Çemen, İ. (2023). A review of the dynamics of subduction zone initiation in the Aegean region. In E. J. Catlos & İ. Çemen (Eds.), *Geophysical Monograph Series* (1st ed., Vol. *1*, pp. 87–117). Wiley. doi: 10.1002/9781119773856.ch4

Cavazza, W., Okay, A. I., & Zattin, M. (2009). Rapid early-middle Miocene exhumation of the Kazdağ Massif (western Anatolia). *International Journal of Earth Sciences*, *98*(8), 1935–1947. doi: 10.1007/s00531-008-0353-9

Çemen, I., Catlos, E. J., Göğüs, O., & Özerdem, C. (2006). Postcollisional extensional tectonics and exhumation of the Menderes massif in the Western Anatolia extended terrane, Turkey. In Y. Dilek & S. Pavlides (Eds.), *Postcollisional Tectonics and Magmatism in the Mediterranean Region and Asia*. Geological Society of America. doi: 10.1130/2006.2409(18)

Cenki-Tok, B., Expert, M., Işık, V., Candan, O., Monié, P., & Bruguier, O. (2016). Complete Alpine reworking of the northern Menderes Massif, western Turkey. *International Journal of Earth Sciences*, *105*(5), 1507–1524. doi: 10.1007/s00531-015-1271-2

Çetinkaplan, M., Candan, O., Oberhänsli, R., Sudo, M., & Cenki-Tok, B. (2020). *P–T–t* evolution of the Cycladic Blueschist Unit in Western Anatolia/Turkey: geodynamic implications for the Aegean region. *Journal of Metamorphic Geology*, *38*(4), 379–419. doi: 10.1111/jmg.12526

Çetinkaplan, M., Pourteau, A., Candan, O., Koralay, O. E., Oberhänsli, R., Okay, A. I., et al. (2016). P–T–t evolution of eclogite/blueschist facies metamorphism in Alanya Massif: time and space relations with HP event in Bitlis Massif, Turkey. *International Journal of Earth Sciences*, *105*(1), 247–281. doi: 10.1007/s00531-014-1092-8

Channell, J. E. T., & Kozur, H. W. (1997). How many oceans? Meliata, Vardar and Pindos oceans in Mesozoic Alpine paleogeography. *Geology*, *25*(2), 183. doi: 10.1130/0091-7613(1997)025<0183:HMOMVA>2.3.CO;2

Chaumillon, E., & Mascle, J. (1997). From foreland to forearc domains: new multichannel seismic reflection survey of the Mediterranean ridge accretionary complex (Eastern Mediterranean). *Marine Geology*, *138*(3–4), 237–259. doi: 10.1016/S0025-3227(97)00002-9

Chen, F., Siebel, W., Satir, M., Terzioğlu, M., & Saka, K. (2002). Geochronology of the Karadere basement (NW Turkey) and implications for the geological evolution of the İstanbul zone. *International Journal of Earth Sciences*, *91*(3), 469–481. doi: 10.1007/s00531-001-0239-6

Chousianitis, K., Ganas, A., & Evangelidis, C. P. (2015). Strain and rotation rate patterns of mainland Greece from continuous GPS data and comparison between seismic and geodetic moment release. *Journal of Geophysical Research: Solid Earth*, *120*(5), 3909–3931. doi: 10.1002/2014JB011762

Chousianitis, K., Ganas, A., & Gianniou, M. (2013). Kinematic interpretation of present-day crustal deformation in central Greece from continuous GPS measurements. *Journal of Geodynamics*, *71*, 1–13. doi: 10.1016/j.jog.2013.06.004

Cita, M. B., Erba, E., Lucchi, R., Pott, M., Van Der Meer, R., & Nieto, L. (1996). Stratigraphy and sedimentation in the Mediterranean Ridge diapiric belt. *Marine Geology*, *132*(1–4), 131–150. doi: 10.1016/0025-3227(96)00157-0

Clark, S. A., Sobiesiak, M., Zelt, C. A., Magnani, M. B., Miller, M. S., Bezada, M. J., & Levander, A. (2008). Identification and tectonic implications of a tear in the South American plate at the southern end of the Lesser Antilles. *Geochemistry, Geophysics, Geosystems*, *9*(11), 2008GC002084. doi: 10.1029/2008GC002084

Cocard, M., Kahle, H.-G., Peter, Y., Geiger, A., Veis, G., Felekis, S., et al. (1999). New constraints on the rapid crustal motion of the Aegean region: recent results inferred from GPS measurements (1993–1998) across the West Hellenic Arc, Greece. *Earth and Planetary Science Letters*, *172*(1–2), 39–47. doi: 10.1016/S0012-821X(99)00185-5

Cocchi, L., Passaro, S., Tontini, F. C., & Ventura, G. (2017). Volcanism in slab tear faults is larger than in island-arcs and back-arcs. *Nature Communications*, *8*(1), 1451. doi: 10.1038/s41467-017-01626-w

Coney, P. J., & Harms, T. A. (1984). Cordilleran metamorphic core complexes: Cenozoic extensional relics of Mesozoic compression. *Geology*, *12*(9), 550. doi: 10.1130/0091-7613(1984)12<550:CMCCCE>2.0.CO;2

Cornée, J.-J., Quillévéré, F., Moissette, P., Fietzke, J., López-Otálvaro, G. E., Melinte-Dobrinescu, M., et al. (2019). Tectonic motion in oblique subduction forearcs: insights from the revisited middle and upper Pleistocene deposits of Rhodes, Greece. *Journal of the Geological Society*, *176*(1), 78–96. doi: 10.1144/jgs2018-090

Crameri, F., Magni, V., Domeier, M., Shephard, G. E., Chotalia, K., Cooper, G., et al. (2020). A transdisciplinary and community-driven database to unravel subduction zone initiation. *Nature Communications*, *11*(1), 3750. doi: 10.1038/s41467-020-17522-9

d'Atri, A., Zuffa, G. G., Cavazza, W., Okay, A. I., & Di Vincenzo, G. (2012). Detrital supply from subduction/accretion complexes to the Eocene–Oligocene post-collisional southern Thrace Basin (NW Turkey and NE Greece). *Sedimentary Geology*, *243–244*, 117–129. doi: 10.1016/j.sedgeo.2011.10.008

Dannat, C. (1997). *Geochemie, Geochronologie und Nd- und Sr-Isotopie der granitoiden Kerngneise des Menderes Massivs, SW-Türkei* (PhD thesis). Mainz, Germany: Johannes Gutenberg University.

De Boorder, H., Spakman, W., White, S. H., & Wortel, M. J. R. (1998). Late Cenozoic mineralization, orogenic collapse and slab detachment in the European Alpine Belt. *Earth and Planetary Science Letters*, *164*(3–4), 569–575. doi: 10.1016/S0012-821X(98)00247-7

De Capitani, C., & Brown, T. H. (1987). The computation of chemical equilibrium in complex systems containing non-ideal solutions. *Geochimica et Cosmochimica Acta*, *51*(10), 2639–2652. doi: 10.1016/0016-7037(87)90145-1

De Capitani, C., & Petrakakis, K. (2010). The computation of equilibrium assemblage diagrams with Theriak/Domino software. *American Mineralogist*, *95*(7), 1006–1016. doi: 10.2138/am.2010.3354

Delaloye, M., & BINGOL, E. (2000). Granitoids from Western and Northwestern Anatolia: geochemistry and modeling of geodynamic evolution. *International Geology Review*, *42*(3), 241–268. doi: 10.1080/00206810009465081

Demirbilek, M., Mutlu, H., Fallick, A. E., Sarıiz, K., & Kibici, Y. (2018). Petrogenetic evolution of the Eocene granitoids in eastern part of the Tavşanlı Zone in northwestern Anatolia, Turkey. *Lithos*, *314–315*, 236–259. doi: 10.1016/j.lithos.2018.06.003

Demirci, A., Özden, S., Bekler, T., Kalafat, D., & Pınar, A. (2015). An active extensional deformation example: 19 May 2011 Simav earthquake (Mw = 5.8), Western Anatolia, Turkey. *Journal of Geophysics and Engineering*, *12*(4), 552–565. doi: 10.1088/1742-2132/12/4/552

Dewey, J. F., & Şengör, A. M. C. (1979). Aegean and surrounding regions: complex multiplate and continuum tectonics in a convergent zone. *Geological Society of America Bulletin*, *90*(1), 84. doi: 10.1130/0016-7606(1979)90<84:AASRCM>2.0.CO;2

Di Rosa, M., Farina, F., Marroni, M., Pandolfi, L., Göncüoğlu, M. C., Ellero, A., & Ottria, G. (2019). U-Pb zircon geochronology of intrusive rocks from an exotic block in the Late Cretaceous – Paleocene Taraklı Flysch (northern Turkey): constraints on the tectonics of the Intrapontide suture zone. *Journal of Asian Earth Sciences*, *171*, 277–288. doi: 10.1016/j.jseaes.2018.11.017

Dilek, Y., & Altunkaynak, Ş. (2007). Cenozoic crustal evolution and mantle dynamics of post-collisional magmatism in Western Anatolia. *International Geology Review*, *49*(5), 431–453. doi: 10.2747/0020-6814.49.5.431

Dilek, Y., & Sandvol, E. (2009). Seismic structure, crustal architecture and tectonic evolution of the Anatolian-African Plate Boundary and the Cenozoic Orogenic Belts in the Eastern Mediterranean Region. *Geological Society, London, Special Publications*, *327*(1), 127–160. doi: 10.1144/SP327.8

Doğan, A., & Emre, Ö. (2006). *Ege Graben Sisteminin Kuzey Sınırı: Sındırgı-Sincanlı Fay Zonu (northern boundary of Aegean Graben System: Sındırgı-Sincanlı Fault Zone)*. Paper presented at the 59th Türkiye Jeoloji Kurultayı, Ankara, Turkey: General Directorate of Mineral Research and Exploration (MTA). Retrieved from https://www.jmo.org.tr/resimler/ekler/afc8fe9c61d6760_ek.pdf

Doğangün, A., Ural, A., Sezen, H., Güney, Y., & Fırat, F. (2013). The 2011 earthquake in Simav, Turkey and seismic damage to reinforced concrete buildings. *Buildings*, *3*(1), 173–190. doi: 10.3390/buildings3010173

Doglioni, C., Agostini, S., Crespi, M., Innocenti, F., Manetti, P., Riguzzi, F., & Savasçin, Y. (2002). On the extension in western Anatolia and the Aegean sea. In G. S. Lister (Ed.), *Reconstruction of the evolution of the Alpine-Himalayan Orogen* (Vol. *8*, pp. 161–176). Journal of the Virtual Explorer.

Doner, Z., Unluer, A. T., Özdamar, Ş., Sarıkaya, O., Roden, M. F., Kaya, M., et al. (2024). The origin of alkali granites and Th-U ± REE enrichments in Kestanbol Magmatic complex (NW Anatolia) revisited: evidences from bulk-rock geochemistry and isotopic data, zircon U Pb, biotite Ar/Ar and apatite (U Th)/He geochronology. *Lithos*, *484–485*, 107751. doi: 10.1016/j.lithos.2024.107751

Dora, O. O., Candan, O., Durr, S., & Oberhansli, R. (1995). *New evidence on the geotectonic evolution of the Menderes Massif*. Paper presented at the International Earth Science Colloquium on the Aegean Region, Izmir, Turkey.

Dora, O. O., Candan, O., Kaya, E., & Akal, C. (2005). *Menderes Masifi Çine Asmasifi' ndeki Koçarlý - Bafa - Yataðan - Karacasu arasýnda uzanan gnays / þist dokanaðýnýn niteliði: Jeolojik, tektonik, petrografik ve jeokronolojik bir yaklaþým. (The nature of the gneiss/schist contact extending between Koçarlı - Bafa - Yataðan - Karacasu in the Menderes Massif Çine Submassif: a geological, tectonic, petrographic and geochronological approach.)* (DABÇAG—Turkish Scientific and Technological Research Council [TÜBÝTAK] Project No. 101 Y 132, p. 197). TÜBÝTAK.

Dragovic, B., Samanta, L. M., Baxter, E. F., & Selverstone, J. (2012). Using garnet to constrain the duration and rate of water-releasing metamorphic reactions during subduction: an example from Sifnos, Greece. *Chemical Geology*, *314–317*, 9–22. doi: 10.1016/j.chemgeo.2012.04.016

Dürr, S. (1975). *Über Alter und geotektonische Stellung des Menderes Kristallins/SW Anatolien und seine Äquivalente in der Mittleren Aegean* (PhD thesis). University of Marburg, Germany.

Duru, M., Pehlivan, S., Şenturk, Y., Yavaş, F., & Kar, H. (2004). New results on the lithostratigraphy of the Kazdağ Massif in Northwest Turkey. *Turkish Journal of Earth Sciences*, *13*, 177–186.

Elburg, M. A., & Smet, I. (2020). Geochemistry of lavas from Aegina and Poros (Aegean Arc, Greece): distinguishing upper crustal contamination and source contamination in the Saronic Gulf area. *Lithos*, *358–359*, 105416. doi: 10.1016/j.lithos.2020.105416

Elitez, İ., Yaltırak, C., & Aktuğ, B. (2016). Extensional and compressional regime driven left-lateral shear in southwestern Anatolia (eastern Mediterranean): the Burdur-Fethiye Shear Zone. *Tectonophysics*, *688*, 26–35. doi: 10.1016/j.tecto.2016.09.024

Elmas, A. (2012). Basement types of the Thrace Basin and a new approach to the pre-Eocene tectonic evolution of the northeastern Aegean and northwestern Anatolia: a review of data and concepts. *International Journal of Earth Sciences*, *101*(7), 1895–1911. doi: 10.1007/s00531-012-0756-5

El-Sharkawy, A., Meier, T., Lebedev, S., Behrmann, J. H., Hamada, M., Cristiano, L., et al. (2020). The slab puzzle of the Alpine-Mediterranean Region: insights from a new, high-resolution, shear wave velocity model of the upper mantle. *Geochemistry, Geophysics, Geosystems*, *21*(8), e2020GC008993. doi: 10.1029/2020GC008993

Emery, K. O., Heezen, B. C., & Allan, T. D. (1966). Bathymetry of the eastern Mediterranean Sea. *Deep Sea Research and Oceanographic Abstracts*, *13*(2), 173–192. doi: 10.1016/0011-7471(66)91098-9

Emre, O., & Duman, T. (2011). *19 May 2011 Simav-Kutahya earthquake (Mw:5.8) in Turkey pre-assessment, earthquake report*. Ankara, Turkey: General Directorate of Mineral Research and Exploration (MTA), Earth Dynamics Research Center.

Emre, Ö., Duman, T. Y., & Özalp, S. (2011). *Active fault map series of Turkey, Kutahya* (NJ 35-4; Quadrangle, Serial Number 10). Ankara, Turkey: General Directorate of Mineral Research and Exploration.

Emre, O., Erkal, T., Tchepalyga, A., Kazancı, N., Keçer, M., & Ünay, E. (1998). Neogene quaternary evolution of the Eastern Marmara Region, Northwest Turkey. *Bulletin of the Mineral Research Exploration Institute of Turkey*, *120*, 119–145.

Erdoğan, B., Akay, E., Hasözbek, A., Satır, M., & Siebel, W. (2013). Stratigraphy and tectonic evolution of the Kazdağı Massif (NW Anatolia) based on field studies and radiometric ages. *International Geology Review*, *55*(16), 2060–2082. doi: 10.1080/00206814.2013.818756

Ergün, M., Okay, S., Sari, C., Zafer Oral, E., Ash, M., Hall, J., & Miller, H. (2005). Gravity anomalies of the Cyprus Arc and their tectonic implications. *Marine Geology*, *221*(1–4), 349–358. doi: 10.1016/j.margeo.2005.03.004

Ersoy, E. Y., & Palmer, M. R. (2013). Eocene-Quaternary magmatic activity in the Aegean: implications for mantle metasomatism and magma genesis in an evolving orogeny. *Lithos*, *180–181*, 5–24. doi: 10.1016/j.lithos.2013.06.007

Ersoy, E. Y., Palmer, M. R., Genç, Ş. C., Prelević, D., Akal, C., & Uysal, İ. (2017). Chemo-probe into the mantle origin of the NW Anatolia Eocene to Miocene volcanic rocks: implications for the role of, crustal accretion, subduction, slab roll-back and slab break-off processes in genesis of post-collisional magmatism. *Lithos*, *288–289*, 55–71. doi: 10.1016/j.lithos.2017.07.006

Ersoy, E. Y., Çemen, İ., Helvacı, C., & Billor, Z. (2014). Tectono-stratigraphy of the Neogene basins in Western Turkey: implications for tectonic evolution of the Aegean Extended Region. *Tectonophysics*, *635*, 33–58. doi: 10.1016/j.tecto.2014.09.002

Ersoy, E. Y., Akal, C., Palmer, M. R., & Mertz-Kraus, R. (2023). U-Pb dating of arc to post-collisional magmatic events in northwestern Anatolia: the Eocene Granitoids in NW Anatolia revisited. *Journal of Asian Earth Sciences: X*, *9*, 100148. doi: 10.1016/j.jaesx.2023.100148

Ersoy, Y. E., Helvacı, C., & Sözbilir, H. (2010). Tectono-stratigraphic evolution of the NE–SW-trending superimposed Selendi basin: implications for late Cenozoic crustal extension in Western Anatolia, Turkey. *Tectonophysics*, *488*(1–4), 210–232. doi: 10.1016/j.tecto.2010.01.007

Etzel, T. M., Catlos, E. J., Ataktürk, K., Lovera, O. M., Kelly, E. D., Çemen, I., & Diniz, E. (2019). Implications for thrust-related shortening punctuated by extension from P-T paths and geochronology of garnet-bearing schists, southern (Çine) Menderes Massif, SW Turkey. *Tectonics*, *38*(6), 1974–1998. doi: 10.1029/2018TC005335

Etzel, T. M., Catlos, E. J., Cemen, I., Ozerdem, C., Oyman, T., & Miggins, D. (2020). Documenting exhumation in the central and northern Menderes Massif (western Turkey): new insights from garnet-based P-T estimates and K-Feldspar $^{40}Ar/^{39}Ar$ geochronology. *Lithosphere*, *2020*(1), 8818289. doi: 10.2113/2020/8818289

Evangelidis, C. P. (2017). Seismic anisotropy in the Hellenic subduction zone: effects of slab segmentation and subslab mantle flow. *Earth and Planetary Science Letters*, *480*, 97–106. doi: 10.1016/j.epsl.2017.10.003

Eyidoğan, H., & Jackson, J. (1985). A seismological study of normal faulting in the Demirci, Ala ehir and Gediz earthquakes of 1969-70 in western Turkey: implications for the nature and geometry of deformation in the continental crust. *Geophysical Journal International*, *81*(3), 569–607. doi: 10.1111/j.1365-246X.1985.tb06423.x

Faccenna, C., Becker, T. W., Auer, L., Billi, A., Boschi, L., Brun, J. P., et al. (2014). Mantle dynamics in the Mediterranean: Mediterranean dynamic. *Reviews of Geophysics*, *52*(3), 283–332. doi: 10.1002/2013RG000444

Faccenna, C., Bellier, O., Martinod, J., Piromallo, C., & Regard, V. (2006). Slab detachment beneath eastern Anatolia: a possible cause for the formation of the North Anatolian fault. *Earth and Planetary Science Letters*, *242*(1–2), 85–97. doi: 10.1016/j.epsl.2005.11.046

Faccenna, C., Jolivet, L., Piromallo, C., & Morelli, A. (2003). Subduction and the depth of convection in the Mediterranean mantle. *Journal of Geophysical Research: Solid Earth*, *108*(B2), 2001JB001690. doi: 10.1029/2001JB001690

Ferentinos, G., Georgiou, N., Christodoulou, D., Geraga, M., & Papatheodorou, G. (2018). Propagation and termination of a strike slip fault in an extensional domain: the westward growth of the North Anatolian Fault into the Aegean Sea. *Tectonophysics*, *745*, 183–195. doi: 10.1016/j.tecto.2018.08.003

Fornash, K. F., & Whitney, D. L. (2020). Lawsonite-rich layers as records of fluid and element mobility in subducted crust (Sivrihisar Massif, Turkey). *Chemical Geology*, *533*, 119356. doi: 10.1016/j.chemgeo.2019.119356

Foucher, J. P., Chamot-Roocke, N., Alexandry, S., Augustin, J. M., Monti, S., Pavlakis, P., & Voisset, M. (1993). *Multibeam bathymetry and seabed reflectivity maps of the MEDRIFT corridor across the eastern Mediterranean Ridge*. Paper presented at UEG VII, Strasbourg, France.

Frassi, C., Marroni, M., Pandolfi, L., Göncüoğlu, M. C., Ellero, A., Ottria, G., et al. (2018). Burial and exhumation history of the Daday Unit (Central Pontides, Turkey): implications for the closure of the Intra-Pontide oceanic basin. *Geological Magazine*, *155*(2), 356–376. doi: 10.1017/S0016756817000176

Fytikas, M., Innocenti, F., Manetti, P., Peccerillo, A., Mazzuoli, R., & Villari, L. (1984). Tertiary to Quaternary evolution of volcanism in the Aegean region. *Geological Society, London, Special Publications*, *17*(1), 687–699. doi: 10.1144/GSL.SP.1984.017.01.55

Galindo-Zaldivar, J., Nieto, L., & Woodside, J. (1996). Structural features of mud volcanoes and the fold system of the Mediterranean Ridge, south of Crete. *Marine Geology*, *132*(1–4), 95–112. doi: 10.1016/0025-3227(96)00155-7

Ganas, A., & Parsons, T. (2009). Three-dimensional model of Hellenic Arc deformation and origin of the Cretan uplift. *Journal of Geophysical Research: Solid Earth*, *114*(B6), 2008JB005599. doi: 10.1029/2008JB005599

Gautier, P., Brun, J.-P., Moriceau, R., Sokoutis, D., Martinod, J., & Jolivet, L. (1999). Timing, kinematics and cause of Aegean extension: a scenario based on a comparison with simple analogue experiments. *Tectonophysics*, *315*(1–4), 31–72. doi: 10.1016/S0040-1951(99)00281-4

Gautier, Y. (1984). *Déformations et métamorphismes associés à la fermeture téthysienne en Anatolie centrale (région de Sivrihisar, Turquie)* (PhD thesis). Université Paris-Sud, France.

Gawthorpe, R. L., & Hurst, J. M. (1993). Transfer zones in extensional basins: their structural style and influence on drainage development and stratigraphy. *Journal of the Geological Society*, *150*(6), 1137–1152. doi: 10.1144/gsjgs.150.6.1137

Genç, Ş. C. (1998). Evolution of the Bayramiç magmatic complex, northwestern Anatolia. *Journal of Volcanology and Geothermal Research*, *85*(1–4), 233–249. doi: 10.1016/S0377-0273(98)00057-2

Gerogiannis, N., Xypolias, P., Chatzaras, V., Aravadinou, E., & Papapavlou, K. (2019). Deformation within the Cycladic subduction–exhumation channel: new insights from the enigmatic Makrotantalo nappe (Andros, Aegean). *International Journal of Earth Sciences*, *108*(3), 817–843. doi: 10.1007/s00531-019-01680-3

Gessner, K., Ring, U., Passchier, C. W., Hetzel, R., & Okay, A. I. (2002). Stratigraphic and metamorphic inversions in the central Menderes Massif; a new structural model; discussion and reply. *International Journal of Earth Sciences*, *91*(1), 168–178.

Gessner, K., Collins, A. S., Ring, U., & GÜngÖr, T. (2004). Structural and thermal history of poly-orogenic basement: U–Pb geochronology of granitoid rocks in the southern Menderes Massif, Western Turkey. *Journal of the Geological Society*, *161*(1), 93–101. doi: 10.1144/0016-764902-166

Gessner, K., Gallardo, L. A., Markwitz, V., Ring, U., & Thomson, S. N. (2013). What caused the denudation of the Menderes Massif: review of crustal evolution, lithosphere structure, and dynamic topography in southwest Turkey. *Gondwana Research*, *24*(1), 243–274. doi: 10.1016/j.gr.2013.01.005

Gessner, K., Markwitz, V., & Güngör, T. (2018). Crustal fluid flow in hot continental extension: tectonic framework of geothermal areas and mineral deposits in western Anatolia. *Geological Society, London, Special Publications*, *453*(1), 289–311. doi: 10.1144/SP453.7

Gessner, K., Piazolo, S., Güngör, T., Ring, U., Kröner, A., & Passchier, C. (2001). Tectonic significance of deformation patterns in granitoid rocks of the Menderes nappes, Anatolide belt, southwest Turkey. *International Journal of Earth Sciences*, *89*(4), 766–780. doi: 10.1007/s005310000106

Gessner, K., Ring, U., & Güngör, T. (2011). *Field guide to Samos and the Menderes Massif.* Geological Society of America. doi: 10.1130/9780813700236

Gianni, G. M., Navarrete, C., & Spagnotto, S. (2019). Surface and mantle records reveal an ancient slab tear beneath Gondwana. *Scientific Reports*, *9*(1), 19774. doi: 10.1038/s41598-019-56335-9

Glodny, J., & Hetzel, R. (2007). Precise U–Pb ages of syn-extensional Miocene intrusions in the central Menderes Massif, western Turkey. *Geological Magazine*, *144*(2), 235–246. doi: 10.1017/S0016756806003025

Göncüoğlu, M. C. (2010). Introduction to the Geology of Turkey: geodynamic evolution of the pre-alpine and alpine terranes. *General Directorate of Mineral Research and Exploration Monography Series*, *5*, 1–66.

Göncüoğlu, M. C., Dirik, K., & Kozlu, H. (1997). General characteristics of pre-alpine and Alpine Terranes in Turkey: explanatory notes to the terrane map of Turkey. *Annales Géologique de Pays Héllenique*, *37*, 515–536.

Göncüoglu, M. C., Marroni, M., Pandolfi, L., Ellero, A., Ottria, G., Catanzariti, R., et al. (2014). The Arkot Dağ Mélange in Araç area, central Turkey: evidence of its origin within the geodynamic evolution of the Intra-Pontide suture zone. *Journal of Asian Earth Sciences*, *85*, 117–139. doi: 10.1016/j.jseaes.2014.01.013

Göncüoğlu, M. C., Marroni, M., Sayit, K., Tekin, U. K., Ottria, G., Pandolfi, L., & Ellero, A. (2012). The Ayli Dag ophiolite sequence (central-northern Turkey); a fragment of Middle Jurassic oceanic lithosphere within the Intra-Pontide suture zone. *Ofioliti*, *37*, 77–92.

Görgün, E. (2014). Source characteristics and Coulomb stress change of the 19 May 2011 Mw 6.0 Simav–Kütahya earthquake, Turkey. *Journal of Asian Earth Sciences*, *87*, 79–88. doi: 10.1016/j.jseaes.2014.02.016

Görür, N., Monod, O., Okay, A. I., Sengör, A. M. C., Tüysüz, O., Yigitbas, E., et al. (1997). Palaeogeographic and tectonic position of the Carboniferous rocks of the western Pontides (Turkey). In the frame of the Variscan belt. *Variscan Belt. Bulletin de La Société Géologique de France*, *168*, 197–205.

Govers, R., & Fichtner, A. (2016). Signature of slab fragmentation beneath Anatolia from full-waveform tomography. *Earth and Planetary Science Letters*, *450*, 10–19. doi: 10.1016/j.epsl.2016.06.014

Govers, R., & Wortel, M. J. R. (2005). Lithosphere tearing at STEP faults: response to edges of subduction zones. *Earth and Planetary Science Letters*, *236*(1–2), 505–523. doi: 10.1016/j.epsl.2005.03.022

Grasemann, B., Schneider, D. A., Stöckli, D. F., & Iglseder, C. (2012). Miocene bivergent crustal extension in the Aegean: evidence from the western Cyclades (Greece). *Lithosphere*, *4*(1), 23–39. doi: 10.1130/L164.1

Gürer, Ö. F., Sangu, E., & Özburan, M. (2006). Neotectonics of the SW Marmara region, NW Anatolia, Turkey. *Geological Magazine*, *143*(2), 229–241. doi: 10.1017/S0016756805001469

Gürsu, S., Cemal Göncüoglu, M., & Bayhan, H. (2004). Geology and Geochemistry of the Pre-early Cambrian Rocks in the Sandikli Area: implications for the Pan-African Evolution of NW Gondwanaland. *Gondwana Research*, *7*(4), 923–935. doi: 10.1016/S1342-937X(05)71075-3

Gürsu, S., & Göncüoglu, M. C. (2006). Petrogenesis and tectonic setting of Cadomian felsic igneous rocks, Sandıklı area of the western Taurides, Turkey. *International Journal of Earth Sciences*, *95*(5), 741–757. doi: 10.1007/s00531-005-0064-4

Gutnic, M., Bellier, O., Poisson, A., & Dumont, J. F. (1979). Geologie des Taurides Occidentales (Turquie). *Memoirs of the Geological Society of France*, *58*(137), 109.

Haddad, A., Ganas, A., Kassaras, I., & Lupi, M. (2020). Seismicity and geodynamics of western Peloponnese and central Ionian Islands: insights from a local seismic deployment. *Tectonophysics*, *778*, 228353. doi: 10.1016/j.tecto.2020.228353

Hall, J., Aksu, A. E., Elitez, I., Yaltırak, C., & Çifçi, G. (2014). The Fethiye–Burdur Fault Zone: a component of upper plate extension of the subduction transform edge propagator fault linking Hellenic and Cyprus Arcs, Eastern Mediterranean. *Tectonophysics*, *635*, 80–99. doi: 10.1016/j.tecto.2014.05.002

Hall, J., Aksu, A. E., Yaltırak, C., & Winsor, J. D. (2009). Structural architecture of the Rhodes Basin: a deep depocentre that evolved since the Pliocene at the junction of Hellenic and Cyprus Arcs, eastern Mediterranean. *Marine Geology*, *258*(1–4), 1–23. doi: 10.1016/j.margeo.2008.02.007

Hall, R., Audley-Charles, M. G., & Carter, D. J. (1984). The significance of Crete for the evolution of the Eastern Mediterranean. *Geological Society, London, Special Publications*, *17*(1), 499–516. doi: 10.1144/GSL.SP.1984.017.01.37

Halpaap, F., Rondenay, S., & Ottemöller, L. (2018). Seismicity, deformation, and metamorphism in the western Hellenic Subduction Zone: new constraints from tomography. *Journal of Geophysical Research: Solid Earth*, *123*(4), 3000–3026. doi: 10.1002/2017JB015154

Hansen, S. E., Evangelidis, C. P., & Papadopoulos, G. A. (2019). Imaging slab detachment within the western Hellenic subduction zone. *Geochemistry, Geophysics, Geosystems*, *20*(2), 895–912. doi: 10.1029/2018GC007810

Harris, N. B. W., Kelley, S., & Okay, A. I. (1994). Post-collision magmatism and tectonics in northwest Anatolia. *Contributions to Mineralogy and Petrology*, *117*(3), 241–252. doi: 10.1007/BF00310866

Harrison, R. W., Tsiolakis, E., Stone, B. D., Lord, A., McGeehin, J. P., Mahan, S. A., & Chirico, P. (2013). Late Pleistocene and Holocene uplift history of Cyprus: implications for active tectonics along the southern margin of the Anatolian microplate. *Geological Society, London, Special Publications*, *372*(1), 561–584. doi: 10.1144/SP372.3

Hasözbek, A., Akay, E., Erdoğan, B., Satır, M., & Siebel, W. (2010). Early Miocene granite formation by detachment tectonics or not? A case study from the northern Menderes Massif (Western Turkey). *Journal of Geodynamics*, *50*(2), 67–80. doi: 10.1016/j.jog.2010.03.002

Hatzfeld, D., Pedotti, G., Hatzidimitriou, P., & Makropoulos, K. (1990). The strain pattern in the western Hellenic arc deduced from a microearthquake survey. *Geophysical Journal International*, *101*(1), 181–202. doi: 10.1111/j.1365-246X.1990.tb00767.x

Hayes, G. (2018). Slab2—a comprehensive subduction zone geometry model [Dataset]. U.S. Geological Survey. doi: 10.5066/F7PV6JNV

Heezen, B. C., & Ewing. (1963). The mid oceanic ridge. In M. N. Hill (Ed.), *The seas* (Vol. *3*, pp. 388–410). New York, NY: Interscience.

Henjes-Kunst, F., Altherr, R., Kreuzer, H., & Hansen, B. T. (1988). Disturbed U-Th-Pb systematics of young zircons and uranothorites: the case of the Miocene Aegean Granitoids (Greece). *Chemical Geology: Isotope Geoscience Section*, *73*(2), 125–145. doi: 10.1016/0168-9622(88)90011-5

Hetzel, R., & Reischmann, T. (1996). Intrusion age of Pan-African augen gneisses in the southern Menderes Massif and the age of cooling after Alpine ductile extensional deformation. *Geological Magazine*, *133*(5), 565–572. doi: 10.1017/S0016756800007846

Hetzel, R., Ring, U., Akal, C., & Troesch, M. (1995b). Miocene NNE-directed extensional unroofing in the Menderes Massif, southwestern Turkey. *Journal of the Geological Society*, *152*(4), 639–654. doi: 10.1144/gsjgs.152.4.0639

Hetzel, R., Zwingmann, H., Mulch, A., Gessner, K., Akal, C., Hampel, A., et al. (2013). Spatiotemporal evolution of brittle normal faulting and fluid infiltration in detachment fault systems: a case study from the Menderes Massif, western Turkey. *Tectonics*, *32*(3), 364–376. doi: 10.1002/tect.20031

Hetzel, R., Passchier, C. W., Ring, U., & Dora, Ö. O. (1995a). Bivergent extension in orogenic belts: the Menderes massif (southwestern Turkey). *Geology*, *23*(5), 455. doi: 10.1130/0091-7613(1995)023<0455:BEIOBT>2.3.CO;2

Holland, T. J. B., & Powell, R. (1998). An internally consistent thermodynamic dataset for phases of petrological interest. *Journal of Metamorphic Geology*, *16*(3), 309–343. doi: 10.1111/j.1525-1314.1998.00140.x

Holland, T. J. B., & Powell, R. (2011). An improved and extended internally consistent thermodynamic dataset for phases of petrological interest, involving a new equation of state for solids. *Journal of Metamorphic Geology*, *29*(3), 333–383. doi: 10.1111/j.1525-1314.2010.00923.x

Hollenstein, C., Müller, M. D., Geiger, A., & Kahle, H.-G. (2008). Crustal motion and deformation in Greece from a decade of GPS measurements, 1993–2003. *Tectonophysics*, *449*(1–4), 17–40. doi: 10.1016/j.tecto.2007.12.006

Holness, M. B., & Bunbury, J. M. (2006). Insights into continental rift-related magma chambers: cognate nodules from the Kula Volcanic Province, Western Turkey. *Journal of Volcanology and Geothermal Research*, *153*(3–4), 241–261. doi: 10.1016/j.jvolgeores.2005.12.004

Hosseini, K., Matthews, K. J., Sigloch, K., Shephard, G. E., Domeier, M., & Tsekhmistrenko, M. (2018). SubMachine: web-based tools for exploring seismic tomography and other models of Earth's deep interior. *Geochemistry, Geophysics, Geosystems*, *19*(5), 1464–1483. doi: 10.1029/2018GC007431

Hubert-Ferrari, A., Armijo, R., King, G., Meyer, B., & Barka, A. (2002). Morphology, displacement, and slip rates along the North Anatolian Fault, Turkey. *Journal of Geophysical Research: Solid Earth*, *107*(B10) ETG-9. doi: 10.1029/2001JB000393

Huchon, P., Lybéris, N., Angelier, J., Le Pichon, X., & Renard, V. (1982). Tectonics of the hellenic trench: a synthesis of sea-beam and submersible observations. *Tectonophysics*, *86*(1–3), 69–112. doi: 10.1016/0040-1951(82)90062-2

Husson, L., Brun, J.-P., Yamato, P., & Faccenna, C. (2009). Episodic slab rollback fosters exhumation of HP-UHP rocks. *Geophysical Journal International*, *179*(3), 1292–1300. doi: 10.1111/j.1365-246X.2009.04372.x

Ilhan, E. (1971). Earthquakes in Turkey. In A. S. Campbell (Ed.), *Geology and history of Turkey* (pp. 431–442). Tripoli, Libya: Petroleum Exploration Society of Libya.

İnan, S., Pabuçcu, Z., Kulak, F., Ergintav, S., Tatar, O., Altunel, E., et al. (2012). Microplate boundaries as obstacles to pre-earthquake strain transfer in Western Turkey: inferences from continuous geochemical monitoring. *Journal of Asian Earth Sciences*, *48*, 56–71. doi: 10.1016/j.jseaes.2011.12.016

Inel, M., Ozmen, H. B., & Akyol, E. (2013). Observations on the building damages after 19 May 2011 Simav (Turkey) earthquake. *Bulletin of Earthquake Engineering*, *11*(1), 255–283. doi: 10.1007/s10518-012-9414-3

Innocenti, F., Agostini, S., Di Vincenzo, G., Doglioni, C., Manetti, P., Savaşçin, M. Y., & Tonarini, S. (2005). Neogene and Quaternary volcanism in Western Anatolia: magma sources and geodynamic evolution. *Marine Geology*, *221*(1–4), 397–421. doi: 10.1016/j.margeo.2005.03.016

Iredale, L. J., Teyssier, C., & Whitney, D. L. (2013). Cenozoic pure-shear collapse of the southern Menderes Massif, Turkey. *Geological Society, London, Special Publications*, *372*(1), 323–342. doi: 10.1144/SP372.15

Işık, V., Seyitoğlu, G., & Çemen, İ. (2003). Ductile–brittle transition along the Alaşehir detachment fault and its structural relationship with the Simav detachment fault, Menderes massif, western Turkey. *Tectonophysics*, *374*(1–2), 1–18. doi: 10.1016/S0040-1951(03)00275-0

Işik, V., & Tekeli, O. (2001). Late orogenic crustal extension in the northern Menderes massif (western Turkey): evidence for metamorphic core complex formation. *International Journal of Earth Sciences*, *89*(4), 757–765. doi: 10.1007/s005310000105

Isik, V., Tekeli, O., & Seyitoglu, G. (2004). The $^{40}Ar/^{39}Ar$ age of extensional ductile deformation and granitoid intrusion in the northern Menderes core complex: implications for the initiation of extensional tectonics in western Turkey. *Journal of Asian Earth Sciences*, *23*(4), 555–566. doi: 10.1016/j.jseaes.2003.09.001

Jackson, J., & McKenzie, D. (1984). Active tectonics of the Alpine–Himalayan Belt between western Turkey and Pakistan. *Geophysical Journal International*, *77*(1), 185–264. doi: 10.1111/j.1365-246X.1984.tb01931.x

Jackson, J. (1994). Active tectonics of the Aegean region. *Annual Review of Earth and Planetary Sciences*, *22*(1), 239–271. doi: 10.1146/annurev.ea.22.050194.001323

John, B. E., & Howard, K. A. (1995). Rapid extension recorded by cooling-age patterns and brittle deformation, Naxos, Greece. *Journal of Geophysical Research: Solid Earth*, *100*(B6), 9969–9979. doi: 10.1029/95JB00665

Jolivet, L., Brun, J. P., Gautier, P., Lallemant, S., & Patriat, M. (1994). 3D Kinematics of extension in the Aegean region from the early Miocene to the present, insights from the ductile crust. *Bulletin de la Societe Geologique de France*, *165*, 195–209.

Jolivet, L., & Patriat, M. (1999). Ductile extension and the formation of the Aegean Sea. *Geological Society, London, Special Publications*, *156*(1), 427–456. doi: 10.1144/GSL.SP.1999.156.01.20

Jolivet, L., & Brun, J.-P. (2010). Cenozoic geodynamic evolution of the Aegean. *International Journal of Earth Sciences*, *99*(1), 109–138. doi: 10.1007/s00531-008-0366-4

Jolivet, L., & Faccenna, C. (2000). Mediterranean extension and the Africa-Eurasia collision. *Tectonics*, *19*(6), 1095–1106. doi: 10.1029/2000TC900018

Jolivet, L., Faccenna, C., Huet, B., Labrousse, L., Le Pourhiet, L., Lacombe, O., et al. (2013). Aegean tectonics: strain localisation, slab tearing and trench retreat. *Tectonophysics*, *597–598*, 1–33. doi: 10.1016/j.tecto.2012.06.011

Jolivet, L., Lecomte, E., Huet, B., Denèle, Y., Lacombe, O., Labrousse, L., et al. (2010). The North Cycladic Detachment System. *Earth and Planetary Science Letters*, *289*(1–2), 87–104. doi: 10.1016/j.epsl.2009.10.032

Jolivet, L., Menant, A., Sternai, P., Rabillard, A., Arbaret, L., Augier, R., et al. (2015). The geological signature of a slab tear below the Aegean. *Tectonophysics*, *659*, 166–182. doi: 10.1016/j.tecto.2015.08.004

Jost, M. L., Knabenbauer, O., Cheng, J., & Harjes, H.-P. (2002). Fault plane solutions of microearthquakes and small events in the Hellenic arc. *Tectonophysics*, *356*(1–3), 87–114. doi: 10.1016/S0040-1951(02)00378-5

Kadir, S., & Kart, F. (2009). The occurrence and origin of the söğüt kaolinite deposits in the Paleozoic Saricakaya granite-granodiorite complexes and overlying Neogene sediments (Bilecik, northwestern Turkey). *Clays and Clay Minerals*, *57*(3), 311–329. doi: 10.1346/CCMN.2009.0570304

Kahle, H., Cocard, M., Peter, Y., Geiger, A., Reilinger, R., Barka, A., & Veis, G. (2000). GPS-derived strain rate field within the boundary zones of the Eurasian, African, and Arabian Plates. *Journal of Geophysical Research: Solid Earth*, *105*(B10), 23353–23370. doi: 10.1029/2000JB900238

Kaldova, J., Leichmann, J., Babek, O., & Melichar, R. (2003). Brunovistulian terrane (Central Europe) and İstanbul zone (NW Turkey): late Proterozoic and Paleozoic tectonostratigraphic development and paleogeography. *Geologica Carpathica*, *54*(3), 139–152.

Karabulut, H., Paul, A., Özbakır, A. D., Ergün, T., & Şentürk, S. (2019). A new crustal model of the Anatolia–Aegean domain: evidence for the dominant role of isostasy in the support of the Anatolian plateau. *Geophysical Journal International*, *218*(1), 57–73. doi: 10.1093/gji/ggz147

Karacık, Z., & Yılmaz, Y. (1998). Geology of the ignimbrites and the associated volcano–plutonic complex of the Ezine area, northwestern Anatolia. *Journal of Volcanology and Geothermal Research*, *85*(1–4), 251–264. doi: 10.1016/S0377-0273(98)00058-4

Karacık, Z., Yılmaz, Y., Pearce, J. A., & Ece, Ö. I. (2008). Petrochemistry of the south Marmara granitoids, northwest Anatolia, Turkey. *International Journal of Earth Sciences*, *97*(6), 1181–1200. doi: 10.1007/s00531-007-0222-y

Karagianni, E. E., Panagiotopoulos, D. G., Panza, G. F., Suhadolc, P., Papazachos, C. B., Papazachos, B. C., et al. (2002). Rayleigh wave group velocity tomography in the Aegean area. *Tectonophysics*, *358*(1–4), 187–209. doi: 10.1016/S0040-1951(02)00424-9

Karagianni, E. E., Papazachos, C. B., Panagiotopoulos, D. G., Suhadolc, P., Vuan, A., & Panza, G. F. (2005). Shear velocity structure in the Aegean area obtained by inversion of Rayleigh waves: shear velocity structure in the Aegean area. *Geophysical Journal International*, *160*(1), 127–143. doi: 10.1111/j.1365-246X.2005.02354.x

Karaoğlu, Ö., & Helvacı, C. (2014). Isotopic evidence for a transition from subduction to slab-tear related volcanism in western Anatolia, Turkey. *Lithos*, *192–195*, 226–239. doi: 10.1016/j.lithos.2014.02.006

Karasözen, E., Nissen, E., Bergman, E. A., Johnson, K. L., & Walters, R. J. (2016). Normal faulting in the Simav graben of western Turkey reassessed with calibrated earthquake relocations. *Journal of Geophysical Research: Solid Earth*, *121*(6), 4553–4574. doi: 10.1002/2016JB012828

Karsli, O., Şengün, F., Aydin, F., Moghadam, H. S., Dokuz, A., Griffin, W. L., et al. (2023). Ordovician S-type granites in the western Sakarya Zone, NW Turkey; linkage to a back-arc setting. *Lithos*, *442–443*, 107075. doi: 10.1016/j.lithos.2023.107075

Kastens, K. A. (1991). Rate of outward growth of the Mediterranean ridge accretionary complex. *Tectonophysics*, *199*(1), 25–50. doi: 10.1016/0040-1951(91)90117-B

Kastens, K. A., Breen, N. A., & Cita, M. B. (1992). Progressive deformation of an evaporite-bearing accretionary complex: SeaMARC I, SeaBeam and piston-core observations from the Mediterranean Ridge. *Marine Geophysical Researches*, *14*(4), 249–298. doi: 10.1007/BF01203620

Katzir, Y., Avigad, D., Matthews, A., Garfunkel, Z., & Evans, B. W. (2000). Origin, HP/LT metamorphism and cooling of ophiolitic mélanges in southern Evia (NW Cyclades), Greece. *Journal of Metamorphic Geology*, *18*(6), 699–718. doi: 10.1046/j.1525-1314.2000.00281.x

Kaya, O. (1981). Miocene reference section for the coastal parts of West Anatolia. *Newsletters on Stratigraphy*, *10*(3), 164–191. doi: 10.1127/nos/10/1981/164

Keay, S., Lister, G., & Buick, I. (2001). The timing of partial melting, Barrovian metamorphism and granite intrusion in the Naxos metamorphic core complex, Cyclades, Aegean Sea, Greece. *Tectonophysics*, *342*(3–4), 275–312. doi: 10.1016/S0040-1951(01)00168-8

Kempler, D., & Ben-Avraham, Z. (1987). The tectonic evolution of the Cyprean Arc. *Annales Tectonicae*, *1*, 58–71.

Kendall, J.-M., Stuart, G. W., Ebinger, C. J., Bastow, I. D., & Keir, D. (2005). Magma-assisted rifting in Ethiopia. *Nature*, *433*(7022), 146–148. doi: 10.1038/nature03161

Kenyon, N. H., Belderson, R. H., & Stride, A. H. (1982). Detailed tectonic trends on the central part of the Hellenic Outer Ridge and in the Hellenic Trench System. *Geological Society, London, Special Publications*, *10*(1), 335–343. doi: 10.1144/GSL.SP.1982.010.01.22

Ketin, I. (1948). Über die tektonisch-mechanischen Folgerungen aus den großen anatolischen Erdbeben des letzten Dezenniums. *Geologische Rundschau*, *36–36*(1), 77–83. doi: 10.1007/BF01791916

Kind, R., Eken, T., Tilmann, F., Sodoudi, F., Taymaz, T., Bulut, F., et al. (2015). Thickness of the lithosphere beneath Turkey and surroundings from S-receiver functions. *Solid Earth*, *6*(3), 971–984. doi: 10.5194/se-6-971-2015

Kinnaird, T., & Robertson, A. (2013). Tectonic and sedimentary response to subduction and incipient continental collision in southern Cyprus, easternmost Mediterranean region. *Geological Society, London, Special Publications*, *372*(1), 585–614. doi: 10.1144/SP372.10

Kocatürk, H., Kumral, M., Creaser, R. A., DuFrane, S. A., Ünlüer, A. T., Sendir, H., et al. (2024). Magma nature and tectono-magmatic context of the Eocene Uludağ granitoids (NW-Türkiye): insights into the Cenozoic geodynamics of the Tethyan Orogenic Belt. *Geochemistry*, 126170. doi: 10.1016/j.chemer.2024.126170

Koçyiğit, A., & Deveci, Ş. (2007). A N-S-trending active extensional structure, the Şuhut (Afyon) Graben: commencement age of the extensional Neotectonic period in the Isparta Angle, SW Turkey. *Turkish Journal of Earth Sciences*, *16*, 391–416.

Kohn, M. J., & Spear, F. S. (1991). Error propagation for barometers: 2. Application to rocks. *American Mineralogist*, *76*(1–2), 138–147.

Kokkalas, S., Xypolias, P., Koukouvelas, I., & Doutsos, T. (2006). Postcollisional contractional and extensional deformation in the Aegean region. In Y. Dilek & S. Pavlides (Eds.), *Postcollisional tectonics and magmatism in the Mediterranean Region and Asia*. Geological Society of America. doi: 10.1130/0-8137-2409-0.97

Komut, T., Gray, R., Pysklywec, R., & Göğüş, O. H. (2012). Mantle flow uplift of western Anatolia and the Aegean: interpretations from geophysical analyses and geodynamic modeling. *Journal of Geophysical Research: Solid Earth*, *117*(B11), 2012JB009306. doi: 10.1029/2012JB009306

Konak, N. (1982). *Geology of the Simav region*. İstanbul, Turkey: İstanbul University.

Konak, N. (2002). *The geological map of Turkey, Izmir area map*. Ankara, Turkey: General Directorate of Mineral Research and Exploration.

Konak, N., Akdeniz, N., & Öztürk, E. M. (1987). Geology of the south of Menderes Massif. Correlation of Variscan and Pre-Variscan events of the Alpine Mediterranean Mountain Belt. Field meeting, IGCP project. *Publications of the Mineral Research and Exploration Institute of Turkey*, *5*, 42–53.

Kopf, A., Mascle, J., & Klaeschen, D. (2003). The Mediterranean Ridge: a mass balance across the fastest growing accretionary complex on Earth. *Journal of Geophysical Research: Solid Earth*, *108*(B8), 2001JB000473. doi: 10.1029/2001JB000473

Koralay, O., Chen, F., Oberhansli, R., Wan, Y., & Candan, O. (2006). Age of granulite facies metamorphism in the Menderes Massif, Western Anatolia / Turkey. Paper presented at the 59th Geological Congress Turkey, Ankara, Turkey: General Directorate of Mineral Research and Exploration.

Koralay, O. E. (2015). Late Neoproterozoic granulite facies metamorphism in the Menderes Massif, Western Anatolia/ Turkey: implication for the assembly of Gondwana. *Geodinamica Acta*, *27*(4), 244–266. doi: 10.1080/09853111. 2015.1014987

Koralay, O. E., Satir, M., Dora, O., & #x000D6;. (2001). Geochemical and geochronological evidence for Early Triassic calc-alkaline magmatism in the Menderes Massif, western Turkey. *International Journal of Earth Sciences*, *89*(4), 822–835. doi: 10.1007/s005310000134

Kozur, H. W., & Göncüoğlu, M. C. (1998). Main features of the pre-Variscan development in Turkey. *Acta Universitatis Carolinae. Geologica*, *42*, 459–464.

Kreemer, C., Chamotrooke, N., & Le Pichon, X. (2004). Constraints on the evolution and vertical coherency of deformation in the Northern Aegean from a comparison of geodetic, geologic and seismologic data. *Earth and Planetary Science Letters*, *225*(3–4), 329–346. doi: 10.1016/ j.epsl.2004.06.018

Kruckenberg, S. C., Vanderhaeghe, O., Ferré, E. C., Teyssier, C., & Whitney, D. L. (2011). Flow of partially molten crust and the internal dynamics of a migmatite dome, Naxos, Greece. *Tectonics*, *30*(3), 2010TC002751. doi: 10.1029/2010TC002751

Kürçer, A., Chatzipetros, A., Tutkun, S. Z., Pavlides, S., Ateş, Ö., & Valkaniotis, S. (2008). The Yenice–Gönen active fault (NW Turkey): active tectonics and palaeoseismology. *Tectonophysics*, *453*(1–4), 263–275. doi: 10.1016/j.tecto. 2007.07.010

Lagos, M., Scherer, E. E., Tomaschek, F., Münker, C., Keiter, M., Berndt, J., & Ballhaus, C. (2007). High precision Lu–Hf geochronology of Eocene eclogite-facies rocks from Syros, Cyclades, Greece. *Chemical Geology*, *243*(1–2), 16–35. doi: 10.1016/j.chemgeo.2007.04.008

Lamont, T. N., Searle, M. P., Gopon, P., Roberts, N. M. W., Wade, J., Palin, R. M., & Waters, D. J. (2020b). The Cycladic Blueschist Unit on Tinos, Greece: cold NE subduction and SW directed extrusion of the Cycladic continental margin under the Tsiknias Ophiolite. *Tectonics*, *39*(9), e2019TC005890. doi: 10.1029/2019TC005890

Lamont, T. N., Searle, M. P., Waters, D. J., Roberts, N. M. W., Palin, R. M., Smye, A., et al. (2020a). Compressional origin of the Naxos metamorphic core complex, Greece: structure, petrography, and thermobarometry. *GSA Bulletin*, *132*(1–2), 149–197. doi: 10.1130/B31978.1

Lanari, P., & Duesterhoeft, E. (2019). Modeling metamorphic rocks using equilibrium thermodynamics and internally consistent databases: past achievements, problems and perspectives. *Journal of Petrology*, *60*(1), 19–56. doi: 10.1093/ petrology/egy105

Lanari, P., & Engi, M. (2017). Local bulk composition effects on metamorphic mineral assemblages. *Reviews in Mineralogy and Geochemistry*, *83*(1), 55–102. doi: 10.2138/ rmg.2017.83.3

Laurent, V., Lanari, P., Naïr, I., Augier, R., Lahfid, A., & Jolivet, L. (2018). Exhumation of eclogite and blueschist (Cyclades, Greece): pressure–temperature evolution determined by thermobarometry and garnet equilibrium modelling. *Journal of Metamorphic Geology*, *36*(6), 769–798. doi: 10.1111/ jmg.12309

Le Pichon, X., & Angelier, J. (1979). The Hellenic arc and trench system: a key to the neotectonic evolution of the eastern mediterranean area. *Tectonophysics*, *60*(1–2), 1–42. doi: 10.1016/0040-1951(79)90131-8

Le Pichon, X., & Angelier, J. (1981). The Aegean Sea. *Philosophical Transactions of the Royal Society of London. Series A, Mathematical and Physical Sciences*, *300*(1454), 357–372. doi: 10.1098/rsta.1981.0069

Le Pichon, X., Chamot-Rooke, N., Lallemant, S., Noomen, R., & Veis, G. (1995). Geodetic determination of the kinematics of central Greece with respect to Europe: implications for eastern Mediterranean tectonics. *Journal of Geophysical Research: Solid Earth*, *100*(B7), 12675–12690. doi: 10.1029/ 95JB00317

Le Pichon, X., Lallemant, S. J., Chamot-Rooke, N., Lemeur, D., & Pascal, G. (2002). The Mediterranean Ridge backstop and the Hellenic nappes. *Marine Geology*, *186*(1–2), 111–125. doi: 10.1016/S0025-3227(02)00175-5

Le Pichon, X., Lybéris, N., Angelier, J., & Renard, V. (1982). Strain distribution over the east Mediterranean ridge: a synthesis incorporating new Sea-Beam data. *Tectonophysics*, *86*(1–3), 243–274. doi: 10.1016/0040-1951(82)90069-5

Le Pichon, X., Şengör, A. M. C., & İmren, C. (2019). A new approach to the opening of the eastern Mediterranean Sea and the origin of the Hellenic subduction zone. Part 2: the Hellenic subduction zone. *Canadian Journal of Earth Sciences*, *56*(11), 1144–1162. doi: 10.1139/cjes-2018-0315

Limonov, A. F., Woodside, J. M., Cita, M. B., & Ivanov, M. K. (1996). The Mediterranean Ridge and related mud diapirism: a background. *Marine Geology*, *132*(1–4), 7–19. doi: 10.1016/0025-3227(96)00150-8

Lips, A. L. W., Wijbrans, J. R., & White, S. H. (1999). New insights from $^{40}Ar/^{39}Ar$ laserprobe dating of white mica fabrics from the Pelion Massif, Pelagonian Zone, Internal Hellenides, Greece: implications for the timing of metamorphic episodes and tectonic events in the Aegean region. *Geological Society, London, Special Publications*, *156*(1), 457–474. doi: 10.1144/GSL.SP.1999.156.01.21

Lips, A. L. W., Cassard, D., Sözbilir, H., Yilmaz, H., & Wijbrans, J. R. (2001). Multistage exhumation of the Menderes Massif, western Anatolia (Turkey). *International Journal of Earth Sciences*, *89*(4), 781–792. doi: 10.1007/s005310000101

Lister, G. S., Banga, G., & Feenstra, A. (1984). Metamorphic core complexes of Cordilleran type in the Cyclades, Aegean Sea, Greece. *Geology*, *12*(4), 221. doi: 10.1130/0091-7613(1984)12<221:MCCOCT>2.0.CO;2

Loos, S., & Reischmann, T. (1999). The evolution of the southern Menderes Massif in SW Turkey as revealed by zircon dating. *Journal of the Geological Society*, *156*(5), 1021–1030. doi: 10.1144/gsjgs.156.5.1021

Lybéris, N. (1984). Tectonic evolution of the North Aegean trough. *Geological Society, London, Special Publications*, *17*(1), 709–725. doi: 10.1144/GSL.SP.1984.017.01.57

Lykousis, V., Alexandri, S., Woodside, J., De Lange, G., Dählmann, A., Perissoratis, C., et al. (2009). Mud volcanoes and gas hydrates in the Anaximander mountains (Eastern Mediterranean Sea). *Marine and Petroleum Geology*, *26*(6), 854–872. doi: 10.1016/j.marpetgeo.2008.05.002

Maggini, M., & Caputo, R. (2020). Sensitivity analysis for crustal rheological profiles: examples from the Aegean Region. *Annals of Geophysics*, *63*(3), 2. doi: 10.4401/ag-8244

Makris, J. (1978). The crust and upper mantle of the Aegean region from deep seismic soundings. *Tectonophysics*, *46*(3–4), 269–284. doi: 10.1016/0040-1951(78)90207-X

Makris, J., Papoulia, J., & Yegorova, T. (2013). A 3-D density model of Greece constrained by gravity and seismic data. *Geophysical Journal International*, *194*(1), 1–17. doi: 10.1093/gji/ggt059

Malandri, C., Soukis, K., Maffione, M., Özkaptan, M., Vassilakis, E., Lozios, S., & Van Hinsbergen, D. J. J. (2017). Vertical-axis rotations accommodated along the Mid-Cycladic lineament on Paros Island in the extensional heart of the Aegean orocline (Greece). *Lithosphere*, *9*(1), 78–99. doi: 10.1130/L575.1

Mantovani, E., Albarello, D., Tamburelli, C., Babbucci, D., & Viti, M. (1997). Plate convergence, crustal delamination, extrusion tectonics and minimization of shortening work as main controlling factors of the recent Mediterranean deformation pattern. *Annals of Geophysics*, *40*(3), 4. doi: 10.4401/ag-3894

Marroni, M., Frassi, C., Göncüoğlu, M. C., Di Vincenzo, G., Pandolfi, L., Rebay, G., et al. (2014). Late Jurassic amphibolite-facies metamorphism in the Intra-Pontide Suture Zone (Turkey): an eastward extension of the Vardar Ocean from the Balkans into Anatolia? *Journal of the Geological Society*, *171*(5), 605–608. doi: 10.1144/jgs2013-104

Marsellos, A. E., Kidd, W. S. F., & Garver, J. I. (2010). Extension and exhumation of the HP/LT rocks in the Hellenic forearc ridge. *American Journal of Science*, *310*(1), 1–36. doi: 10.2475/01.2010.01

Matsuda, J., Senoh, K., Maruoka, T., Sato, H., & Mitropoulos, P. (1999). K-Ar ages of the Aegean volcanic rocks and their implication for the arc-trench system. *Geochemical Journal*, *33*(6), 369–377. doi: 10.2343/geochemj.33.369

McClusky, S., Balassanian, S., Barka, A., Demir, C., Ergintav, S., Georgiev, I., et al. (2000). Global positioning system constraints on plate kinematics and dynamics in the eastern Mediterranean and Caucasus. *Journal of Geophysical Research: Solid Earth*, *105*(B3), 5695–5719. doi: 10.1029/1999JB900351

McKenzie, D. (1972). Active tectonics of the Mediterranean Region. *Geophysical Journal International*, *30*(2), 109–185. doi: 10.1111/j.1365-246X.1972.tb02351.x

McKenzie, D. (1978). Active tectonics of the Alpine–Himalayan belt: the Aegean Sea and surrounding regions. *Geophysical Journal International*, *55*(1), 217–254. doi: 10.1111/j.1365-246X.1978.tb04759.x

McKenzie, D., & Bickle, M. J. (1988). The volume and composition of melt generated by extension of the lithosphere. *Journal of Petrology*, *29*(3), 625–679. doi: 10.1093/petrology/29.3.625

Meier, T., Becker, D., Endrun, B., Rische, M., Bohnhoff, M., Stöckhert, B., & Harjes, H.-P. (2007). A model for the Hellenic subduction zone in the area of Crete based on seismological investigations. *Geological Society, London, Special Publications*, *291*(1), 183–199. doi: 10.1144/SP291.9

Meighan, H. E., Ten Brink, U., & Pulliam, J. (2013). Slab tears and intermediate-depth seismicity. *Geophysical Research Letters*, *40*(16), 4244–4248. doi: 10.1002/grl.50830

Menant, A., Jolivet, L., Tuduri, J., Loiselet, C., Bertrand, G., & Guillou-Frottier, L. (2018). 3D subduction dynamics: a first-order parameter of the transition from copper- to gold-rich deposits in the eastern Mediterranean region. *Ore Geology Reviews*, *94*, 118–135. doi: 10.1016/j.oregeorev.2018.01.023

Menant, A., Jolivet, L., & Vrielynck, B. (2016). Kinematic reconstructions and magmatic evolution illuminating crustal and mantle dynamics of the eastern Mediterranean region since the late Cretaceous. *Tectonophysics*, *675*, 103–140. doi: 10.1016/j.tecto.2016.03.007

Meng, J., Sinoplu, O., Zhou, Z., Tokay, B., Kusky, T., Bozkurt, E., & Wang, L. (2021). Greece and Turkey shaken by African tectonic retreat. *Scientific Reports*, *11*(1), 6486. doi: 10.1038/s41598-021-86063-y

Mercier, J. L. (1981). Extensional-compressional tectonics associated with the Aegean Arc: comparison with the Andean Cordillera of south Peru - north Bolivia. *Philosophical Transactions of the Royal Society of London. Series A, Mathematical and Physical Sciences*, *300*(1454), 337–355. doi: 10.1098/rsta.1981.0068

Meulenkamp, J. E., Wortel, M. J. R., Van Wamel, W. A., Spakman, W., & Hoogerduyn Strating, E. (1988). On the Hellenic subduction zone and the geodynamic evolution of Crete since the late Middle Miocene. *Tectonophysics*, *146*(1–4), 203–215. doi: 10.1016/0040-1951(88)90091-1

Moix, P., Beccaletto, L., Kozur, H. W., Hochard, C., Rosselet, F., & Stampfli, G. M. (2008). A new classification of the Turkish terranes and sutures and its implication for the paleotectonic history of the region. *Tectonophysics*, *451*(1–4), 7–39. doi: 10.1016/j.tecto.2007.11.044

Morris, A., & Anderson, M. (1996). First palaeomagnetic results from the Cycladic Massif, Greece, and their implications for Miocene extension directions and tectonic models in the Aegean. *Earth and Planetary Science Letters*, *142*(3–4), 397–408. doi: 10.1016/0012-821X(96)00114-8

Morris, A., & Robertson, A. H. F. (1993). Miocene remagnetisation of carbonate platform and Antalya Complex units within the Isparta angle, SW Turkey. *Tectonophysics*, *220*(1–4), 243–266. doi: 10.1016/0040-1951(93)90234-B

Mouslopoulou, V., Nicol, A., Begg, J., Oncken, O., & Moreno, M. (2015). Clusters of megaearthquakes on upper plate faults control the Eastern Mediterranean hazard. *Geophysical Research Letters*, *42*(23). doi: 10.1002/2015GL066371

Müller, P., Kreuzer, H., Lenz, H., & Harre, W. (1979). Radiometric dating of two extrusives from a Lower Pliocene Marine Section on Aegina Island, Greece. *Newsletters on Stratigraphy*, *8*(1), 70–78. doi: 10.1127/nos/8/1979/70

Mutlu, A. K. (2020). Seismicity, focal mechanism, and stress tensor analysis of the Simav region, western Turkey. *Open Geosciences*, *12*(1), 479–490. doi: 10.1515/geo-2020-0010

Neubauer, F. (2002). Evolution of late Neoproterozoic to early Paleozoic tectonic elements in Central and Southeast European Alpine mountain belts: review and synthesis. *Tectonophysics*, *352*(1–2), 87–103. doi: 10.1016/S0040-1951(02)00190-7

Nyst, M., & Thatcher, W. (2004). New constraints on the active tectonic deformation of the Aegean. *Journal of Geophysical Research: Solid Earth*, *109*(B11), 2003JB002830. doi: 10.1029/2003JB002830

Oberhäensli, R., Osman, C., & Wilke, F. (2010). Geochronological evidence of Pan-African eclogites from the Central Menderes Massif, Turkey. *Turkish Journal of Earth Sciences*, *19*(4), 431–447. doi: 10.3906/yer-1001-20

Oberhänsli, R., Candan, O., Dora, O. Ö., Dürr, S., & H. (1997). Eclogites within the Menderes Massif/western Turkey. *Lithos*, *41*(1–3), 135–150. doi: 10.1016/S0024-4937(97)82009-9

Oelsner, F., Candan, O., & Oberhänsli, R. (1997). New evidence for the time of the high-grade metamorphism in the Menderes Massif, SW-Turkey. *Terra Nostra*, *97*, A16.

Okal, E. A., Synolakis, C. E., Uslu, B., Kalligeris, N., & Voukouvalas, E. (2009). The 1956 earthquake and tsunami in Amorgos, Greece. *Geophysical Journal International*, *178*(3), 1533–1554. doi: 10.1111/j.1365-246X.2009.04237.x

Okay, A., Satır, M., Tüysüz, O., Akyüz, S., & Chen, F. (2001). The tectonics of the Strandja Massif: late-Variscan and mid-Mesozoic deformation and metamorphism in the northern Aegean. *International Journal of Earth Sciences*, *90*(2), 217–233. doi: 10.1007/s005310000104

Okay, A. I. (1980). Mineralogy, petrology, and phase relations of glaucophane-lawsonite zone blueschists from the Tavşanli Region, Northwest Turkey. *Contributions to Mineralogy and Petrology*, *72*(3), 243–255. doi: 10.1007/BF00376143

Okay, A. I. (1981). Lawsonite zone blueschists and a sodic amphibole producing reaction in the Tavşanli Region, Northwest Turkey. *Contributions to Mineralogy and Petrology*, *75*(3), 179–186. doi: 10.1007/BF01166758

Okay, A. I. (1982). Incipient blueschist metamorphism and metasomatism in the Tavşanli region, Northwest Turkey. *Contributions to Mineralogy and Petrology*, *79*(4), 361–367. doi: 10.1007/BF01132065

Okay, A. I. (1984). Distribution and characteristics of the northwest Turkish blueschists. *Geological Society, London, Special Publications*, *17*(1), 455–466. doi: 10.1144/GSL.SP.1984.017.01.33

Okay, A. I. (1986). High-pressure/low-temperature metamorphic rocks of Turkey. In B. W. Evans & E. H. Brown (Eds.), *Blueschists and eclogites* (Vol. *164*, pp. 333–347). Memoir-Geological Society of America.

Okay, A. I. (2008). Geology of Turkey: a synopsis. *Der Anschnitt*, *21*, 19–42.

Okay, A. I., & Kelley, S. P. (1994). Tectonic setting, petrology and geochronology of jadeite + glaucophane and chloritoid + glaucophane schists from north-west Turkey. *Journal of Metamorphic Geology*, *12*(4), 455–466. doi: 10.1111/j.1525-1314.1994.tb00035.x

Okay, A. I., Monod, O., & Monié, P. (2002). Triassic blueschists and eclogites from northwest Turkey: vestiges of the Paleo-Tethyan subduction. *Lithos*, *64*(3–4), 155–178. doi: 10.1016/S0024-4937(02)00200-1

Okay, A. İ., Özcan, E., Cavazza, W., Okay, N., & Less, G. (2010). Basement types, lower Eocene series, upper Eocene olistostromes and the initiation of the Southern Thrace Basin, NW Turkey. *Turkish Journal of Earth Sciences*, *19*(1), 1–25. doi: 10.3906/yer-0902-10

Okay, A. I., Satir, M., Maluski, H., Siyako, M., Monié, P., Metzger, R., & Akyüz, S. (1996). Paleo- and Neo-Tethyan events in northwestern Turkey: geologic and geochronologic constraints. In A. Yin & T. M. Harrison (Eds.), *The tectonic evolution of Asia* (pp. 420–441). Cambridge: Cambridge University Press.

Okay, A. I., Satır, M., & Siebel, W. (2006). Pre-Alpide Palaeozoic and Mesozoic orogenic events in the Eastern Mediterranean region. *Geological Society, London, Memoirs*, *32*(1), 389–405. doi: 10.1144/GSL.MEM.2006.032.01.23

Okay, A. I., Siyako, M., & Bürkan, K. A. (1991). Geology and tectonic evolution of the Biga Peninsula, Northwest Turkey. *Bulletin of the Technical University of İstanbul*, *44*(1–2), 191–256.

Okay, A. I., & Whitney, D. L. (2010). Blueschists, eclogites, ophiolites and suture zones in northwest Turkey: a review and a field excursion guide. *Ofioliti*, *35*(2), 131–172.

Okay, A. I. (2001). Stratigraphic and metamorphic inversions in the central Menderes Massif: a new structural model. *International Journal of Earth Sciences*, *89*(4), 709–727. doi: 10.1007/s005310000098

Okay, A. I., Bozkurt, E., Satır, M., Yiğitbaş, E., Crowley, Q. G., & Shang, C. K. (2008). Defining the southern margin of Avalonia in the Pontides: geochronological data from the Late Proterozoic and Ordovician granitoids from NW Turkey. *Tectonophysics*, *461*(1–4), 252–264. doi: 10.1016/j.tecto.2008.02.004

Okay, A. I., & Satir, M. (2000). Coeval plutonism and metamorphism in a latest Oligocene metamorphic core complex in northwest Turkey. *Geological Magazine*, *137*(5), 495–516. doi: 10.1017/S0016756800004532

Okay, A. I., Sunal, G., Sherlock, S., Altner, D., Tüysüz, O., Kylander-Clark, A. R. C., & Aygül, M. (2013). Early Cretaceous sedimentation and orogeny on the active margin of Eurasia: Southern Central Pontides, Turkey. *Tectonics*, *32*(5), 1247–1271. doi: 10.1002/tect.20077

Okay, A. I., Sunal, G., Sherlock, S., Kylander-Clark, A. R. C., & Özcan, E. (2020). İzmir-Ankara suture as a Triassic to cretaceous plate boundary data from central Anatolia. *Tectonics*, *39*(5), e2019TC005849. doi: 10.1029/2019TC005849

Okay, A. I., Topuz, G., Kylander-Clark, A. R. C., Sherlock, S., & Zattin, M. (2022). Late Paleocene – Middle Eocene magmatic flare-up in Western Anatolia. *Lithos*, *428–429*, 106816. doi: 10.1016/j.lithos.2022.106816

Okay, A. I., & Tüysüz, O. (1999). Tethyan sutures of northern Turkey. *Geological Society, London, Special Publications*, *156*(1), 475–515. doi: 10.1144/GSL.SP.1999.156.01.22

Okay, A., & Satir, M. (2006). Geochronology of Eocene plutonism and metamorphism in northwest. *Geodinamica Acta*, *19*(5), 251–266. doi: 10.3166/ga.19.251-266

Okrusch, M., & Bröcker, M. (1990). Eclogites associated with high-grade blueschists in the Cyclades archipelago, Greece: a review. *European Journal of Mineralogy*, *2*(4), 451–478. doi: 10.1127/ejm/2/4/0451

Oner, Z., Dilek, Y., & Kadioglu, Y. K. (2010). Geology and geochemistry of the synextensional Salihli granitoid in the Menderes core complex, Western Anatolia, Turkey. *International Geology Review*, *52*(2–3), 336–368. doi: 10.1080/00206810902815871

Oral, M. B., Reilinger, R. E., Nafi Toksöz, M., King, R. W., Aykut Barka, A., Kinik, I., & Lenk, O. (1995). Global Positioning System offers evidence of plate motions in eastern Mediterranean. *Eos, Transactions American Geophysical Union*, *76*(2), 9–11. doi: 10.1029/EO076i002p00009-01

Oygür, V., & Erler, A. (2000). Metallogeny of Simav graben. *Geological Bulletin of Turkey*, *43*(1), 7–19.

Özbakır, A. D., Şengör, A. M. C., Wortel, M. J. R., & Govers, R. (2013). The Pliny–Strabo trench region: a large shear zone resulting from slab tearing. *Earth and Planetary Science Letters*, *375*, 188–195. doi: 10.1016/j.epsl.2013.05.025

Özbakır, A. D., Govers, R., & Fichtner, A. (2020). The Kefalonia Transform Fault: a STEP fault in the making. *Tectonophysics*, *787*, 228471. doi: 10.1016/j.tecto.2020.228471

Özcan, A., Göncüoğlu, M. C., Turhan, N., Uysal, S., Şentürk, K., & Işık, A. (1988). Late Paleozoic evolution of the Kütahya–Bolkardağ Belt. *METU Journal of Pure and Applied Science*, *21*(1/3), 211–220.

Özdamar, Ş., Billor, M. Z., Sunal, G., Esenli, F., & Roden, M. F. (2013). First U–Pb SHRIMP zircon and 40Ar/39Ar ages of metarhyolites from the Afyon–Bolkardağ Zone, SW Turkey: implications for the rifting and closure of the Neo-Tethys. *Gondwana Research*, *24*(1), 377–391. doi: 10.1016/j.gr.2012.10.006

Özdamar, Ş., Sarıkaya, O., Sütçü, N. M., Zou, H., & Sherlock, S. (2023). Geochronology and petrochemistry of Karabiga pluton in western Sakarya Zone (NW Turkey): implications from new zircon U-Pb and biotite Ar/Ar ages, Sr-Nd isotope data and bulk-rock geochemistry. *Geochemistry*, *126052*. doi: 10.1016/j.chemer.2023.126052

Özdamar, Ş., Zou, H., Billor, M. Z., Hames, W., Roden, M. F., Sarıkaya, O., & Georgiev, S. (2024). Petrogenesis, geochronology and thermochronology of Oligocene to Miocene Western Anatolia granitoid plutons in Turkey. *Lithos*, *464–465*, 107430. doi: 10.1016/j.lithos.2023.107430

Ozgenç, I., & Ilbeyli, N. (2008). Petrogenesis of the Late Cenozoic Eğrigöz Pluton in Western Anatolia, Turkey: implications for magma genesis and crustal processes. *International Geology Review*, *50*(4), 375–391. doi: 10.2747/0020-6814.50.4.375

Özgül, N. (1997). Stratigraphy of the tectono-stratigraphic units in the region Bozkır–Hadim–Taşkent (northern central Taurides). *Maden Tetkik ve Arama Dergisi*, *119*, 113–174.

Özkaymak, Ç., Sözbilir, H., & Uzel, B. (2013). Neogene–Quaternary evolution of the Manisa Basin: evidence for variation in the stress pattern of the İzmir-Balıkesir Transfer Zone, Western Anatolia. *Journal of Geodynamics*, *65*, 117–135. doi: 10.1016/j.jog.2012.06.004

Özsayin, E., & Dirik, K. (2007). Quaternary activity of the Cihanbeyli and Yeniceoba fault zones: Inönü-Eskiflehir fault system, central Anatolia. *Turkish Journal of Earth Sciences*, *16*, 471–492.

Palin, R. M., Weller, O. M., Waters, D. J., & Dyck, B. (2016). Quantifying geological uncertainty in metamorphic phase equilibria modelling; a Monte Carlo assessment and implications for tectonic interpretations. *Geoscience Frontiers*, *7*(4), 591–607. doi: 10.1016/j.gsf.2015.08.005

Papadopoulos, G. A. (1997). On the interpretation of large-scale seismic tomography images in the Aegean sea area. *Annals of Geophysics*, *40*(1), 3. doi: 10.4401/ag-3933

Papadopoulos, T., Wyss, M., & Schmerge, D. L. (1988). Earthquake locations in the western Hellenic arc relative to the plate boundary. *Bulletin of the Seismological Society of America*, *78*(3), 1222–1231. doi: 10.1785/BSSA0780031222

Papanikolaou, D. J. (1987). Tectonic Evolution of the Cycladic Blueschist Belt (Aegean Sea, Greece). In H. C. Helgeson (Ed.), *Chemical transport in metasomatic processes* (pp. 429–450). Dordrecht: Springer Netherlands. doi: 10.1007/978-94-009-4013-0_16

Papanikolaou, D. J., & Royden, L. H. (2007). Disruption of the Hellenic arc: Late Miocene extensional detachment faults and steep Pliocene-Quaternary normal faults—Or what happened at Corinth? *Tectonics*, *26*(5), 2006TC002007. doi: 10.1029/2006TC002007

Papazachos, B. C., & Comninakis, P. E. (1971). Geophysical and tectonic features of the Aegean Arc. *Journal of Geophysical Research*, *76*(35), 8517–8533. doi: 10.1029/JB076i035p08517

Papazachos, B. C., & Delibasis, N. D. (1969). Tectonic stress field and seismic faulting in the area of Greece. *Tectonophysics*, *7*(3), 231–255. doi: 10.1016/0040-1951(69)90069-9

Papazachos, B. C., Karakostas, V. G., Papazachos, C. B., & Scordilis, E. M. (2000). The geometry of the Wadati–Benioff zone and lithospheric kinematics in the Hellenic arc. *Tectonophysics*, *319*(4), 275–300. doi: 10.1016/S0040-1951(99)00299-1

Papazachos, C. B. (1999). Seismological and GPS evidence for the Aegean-Anatolia Interaction. *Geophysical Research Letters*, *26*(17), 2653–2656. doi: 10.1029/1999GL900411

Papazachos, C. B. (2019). Deep structure and active tectonics of the South Aegean volcanic arc. *Elements*, *15*(3), 153–158. doi: 10.2138/gselements.15.3.153

Parra, T., Vidal, O., & Jolivet, L. (2002). Relation between the intensity of deformation and retrogression in blueschist metapelites of Tinos Island (Greece) evidenced by chlorite–mica local equilibria. *Lithos*, *63*(1–2), 41–66. doi: 10.1016/S0024-4937(02)00115-9

Pearce, F. D., Rondenay, S., Sachpazi, M., Charalampakis, M., & Royden, L. H. (2012). Seismic investigation of the transition from continental to oceanic subduction along the western Hellenic Subduction Zone. *Journal of Geophysical Research: Solid Earth*, *117*(B7), 2011JB009023. doi: 10.1029/2011JB009023

Pearce, J. A., & Stern, R. J. (2006). Origin of back-arc basin magmas: trace element and isotope perspectives. In D. M. Christie, et al. (Eds.), *Geophysical Monograph Series* (Vol. *166*, pp. 63–86). American Geophysical Union. doi: 10.1029/166GM06

Pe-Piper, G., & Piper, D. J. W. (2005). The South Aegean active volcanic arc: relationships between magmatism and tectonics.

In *Developments in Volcanology* (Vol. 7, pp. 113–133). Elsevier. doi: 10.1016/S1871-644X(05)80034-8

Pe-Piper, G. (2000). Origin of S-type granites coeval with I-type granites in the Hellenic subduction system, Miocene of Naxos, Greece. *European Journal of Mineralogy*, *12*(4), 869–875.

Pe-Piper, G., & Piper, D. J. W. (2001). Late Cenozoic, post-collisional Aegean igneous rocks: Nd, Pb and Sr isotopic constraints on petrogenetic and tectonic models. *Geological Magazine*, *138*(6), 653–668. doi: 10.1017/S0016756801005957

Pe-Piper, G., Piper, D. J. W., & Matarangas, D. (2002). Regional implications of geochemistry and style of emplacement of Miocene I-type diorite and granite, Delos, Cyclades, Greece. *Lithos*, *60*(1–2), 47–66. doi: 10.1016/S0024-4937(01)00068-8

Perkins, R. J., Cooper, F. J., Condon, D. J., Tattitch, B., & Naden, J. (2018). Post-collisional Cenozoic extension in the northern Aegean: the high-K to shoshonitic intrusive rocks of the Maronia Magmatic Corridor, northeastern Greece. *Lithosphere*, *10*(5), 582–601. doi: 10.1130/L730.1

Peterek, A., & Schwarze, J. (2004). Architecture and Late Pliocene to recent evolution of outer-arc basins of the Hellenic subduction zone (south-central Crete, Greece). *Journal of Geodynamics*, *38*(1), 19–55. doi: 10.1016/j.jog.2004.03.002

Philippon, M., Brun, J.-P., Gueydan, F., & Sokoutis, D. (2014). The interaction between Aegean back-arc extension and Anatolia escape since Middle Miocene. *Tectonophysics*, *631*, 176–188. doi: 10.1016/j.tecto.2014.04.039

Piper, J. D. A., Gürsoy, H., Tatar, O., Beck, M. E., Rao, A., Koçbulut, F., & Mesci, B. L. (2010). Distributed neotectonic deformation in the Anatolides of Turkey: a palaeomagnetic analysis. *Tectonophysics*, *488*(1–4), 31–50. doi: 10.1016/j.tecto.2009.05.026

Platt, J. D., Brantut, N., & Rice, J. R. (2015). Strain localization driven by thermal decomposition during seismic shear. *Journal of Geophysical Research: Solid Earth*, *120*(6), 4405–4433. doi: 10.1002/2014JB011493

Plunder, A., Agard, P., Chopin, C., & Okay, A. I. (2013). Geodynamics of the Tavşanlı zone, western Turkey: insights into subduction/obduction processes. *Tectonophysics*, *608*, 884–903. doi: 10.1016/j.tecto.2013.07.028

Portner, D. E., Delph, J. R., Biryol, C. B., Beck, S. L., Zandt, G., Özacar, A. A., et al. (2018). Subduction termination through progressive slab deformation across Eastern Mediterranean subduction zones from updated P-wave tomography beneath Anatolia. *Geosphere*, *14*(3), 907–925. doi: 10.1130/GES01617.1

Pourteau, A., Candan, O., & Oberhänsli, R. (2010). High-pressure metasediments in central Turkey: constraints on the Neotethyan closure history. *Tectonics*, *29*(5), 1–18. doi: 10.1029/2009TC002650

Pourteau, A., Sudo, M., Candan, O., Lanari, P., Vidal, O., & Oberhänsli, R. (2013). Neotethys closure history of Anatolia: insights from ^{40}Ar–^{39}Ar geochronology and P–T estimation in high-pressure metasedimentary rocks. *Journal of Metamorphic Geology*, *31*(6), 585–606. doi: 10.1111/jmg.12034

Pourteau, A., Oberhänsli, R., Candan, O., Barrier, E., & Vrielynck, B. (2016). Neotethyan closure history of Western Anatolia: a geodynamic discussion. *International Journal of Earth Sciences*, *105*(1), 203–224. doi: 10.1007/s00531-015-1226-7

Rabayrol, F., & Hart, C. J. R. (2021). Petrogenetic and tectonic controls on magma fertility and the formation of post-subduction porphyry and epithermal mineralization along the late Cenozoic Anatolian Metallogenic Trend, Turkey. *Mineralium Deposita*, *56*(2), 279–306. doi: 10.1007/s00126-020-00967-9

Rabayrol, F., Hart, C. J. R., & Creaser, R. A. (2019). Tectonic triggers for postsubduction magmatic-hydrothermal gold metallogeny in the Late Cenozoic Anatolian metallogenic trend, Turkey. *Economic Geology*, *114*(7), 1339–1363. doi: 10.5382/econgeo.4682

Rabillard, A., Jolivet, L., Arbaret, L., Bessière, E., Laurent, V., Menant, A., et al. (2018). Synextensional granitoids and detachment systems within cycladic metamorphic core complexes (Aegean Sea, Greece): toward a regional tectonomagmatic model. *Tectonics*, *37*(8), 2328–2362. doi: 10.1029/2017TC004697

Ramseyer, K., Aldahan, A. A., Collini, B., & Landström, O. (1992). Petrological modifications in granitic rocks from the siljan impact structure: evidence from cathodoluminescence. *Tectonophysics*, *216*(1–2), 195–204. doi: 10.1016/0040-1951(92)90166-4

Régnier, J.-L., Mezger, J. E., & Passchier, C. W. (2007). Metamorphism of Precambrian–Palaeozoic schists of the Menderes core series and contact relationships with Proterozoic orthogneisses of the western Çine Massif, Anatolide belt, western Turkey. *Geological Magazine*, *144*(1), 67–104. doi: 10.1017/S0016756806002640

Régnier, J. L., Ring, U., Passchier, C. W., Gessner, K., & Güngör, T. (2003). Contrasting metamorphic evolution of metasedimentary rocks from the Çine and Selimiye nappes in the Anatolide belt, western Turkey. *Journal of Metamorphic Geology*, *21*(7), 699–721. doi: 10.1046/j.1525-1314.2003.00473.x

Reilinger, R., McClusky, S., Paradissis, D., Ergintav, S., & Vernant, P. (2010). Geodetic constraints on the tectonic evolution of the Aegean region and strain accumulation along the Hellenic subduction zone. *Tectonophysics*, *488*(1–4), 22–30. doi: 10.1016/j.tecto.2009.05.027

Reilinger, R., McClusky, S., Vernant, P., Lawrence, S., Ergintav, S., Cakmak, R., et al. (2006). GPS constraints on continental deformation in the Africa-Arabia-Eurasia continental collision zone and implications for the dynamics of plate interactions. *Journal of Geophysical Research: Solid Earth*, *111*(B5), 2005JB004051. doi: 10.1029/2005JB004051

Reilinger, R. E., McClusky, S. C., Oral, M. B., King, R. W., Toksöz, M. N., Barka, A. A., et al. (1997). Global positioning system measurements of present-day crustal movements in the Arabia-Africa-Eurasia plate collision zone. *Journal of Geophysical Research: Solid Earth*, *102*(B5), 9983–9999. doi: 10.1029/96JB03736

Rimmelé, G., Parra, T., Goffé, B., Oberhänsli, R., Jolivet, L., & Candan, O. (2005). Exhumation paths of high-pressure–low-temperature metamorphic rocks from the Lycian Nappes and the Menderes Massif (SW Turkey): a multi-equilibrium approach. *Journal of Petrology*, *46*(3), 641–669. doi: 10.1093/petrology/egh092

Ring, U., & Collins, A. S. (2005). U–Pb SIMS dating of synkinematic granites: timing of core-complex formation in the northern Anatolide belt of western Turkey. *Journal of the Geological Society*, *162*(2), 289–298. doi: 10.1144/0016-764904-016

Ring, U., Gessner, K., Güngör, T., & Passchier, C. W. (1999). The Menderes Massif of western Turkey and the Cycladic Massif in the Aegean—do they really correlate? *Journal of the Geological Society*, *156*(1), 3–6. doi: 10.1144/gsjgs.156.1.0003

Ring, U., Willner, A. P., & Lackmann, W. (2001). Stacking of nappes with different pressure-temperature paths: an example from the Menderes nappes of western Turkey. *American Journal of Science*, *301*(10), 912–944. doi: 10.2475/ajs.301.10.912

Ring, U., Buchwaldt, R., & Gessner, K. (2004). Pb/Pb dating of garnet from the Anatolide belt in western Turkey: regional implications and speculations on the role Anatolia played during the amalgamation of Gondwana. *Zeitschrift der Deutschen Geologischen Gesellschaft*, *154*(4), 537–555. doi: 10.1127/zdgg/154/2004/537

Ring, U., Johnson, C., Hetzel, R., & Gessner, K. (2003). Tectonic denudation of a Late Cretaceous–Tertiary collisional belt: regionally symmetric cooling patterns and their relation to extensional faults in the Anatolide belt of western Turkey. *Geological Magazine*, *140*(4), 421–441. doi: 10.1017/S0016756803007878

Robertson, A. H. F., Clift, P. D., Degnan, P. J., & Jones, G. (1991). Palaeogeographic and palaeotectonic evolution of the Eastern Mediterranean Neotethys. *Palaeogeography, Palaeoclimatology, Palaeoecology*, *87*(1–4), 289–343. doi: 10.1016/0031-0182(91)90140-M

Robertson, A. H. F., & Ustaömer, T. (2004). Tectonic evolution of the Intra-Pontide suture zone in the Armutlu Peninsula, NW Turkey. *Tectonophysics*, *381*(1–4), 175–209. doi: 10.1016/j.tecto.2002.06.002

Robertson, A. H. F., & Ustaömer, T. 7/(2009a). Formation of the Late Palaeozoic Konya Complex and comparable units in southern Turkey by subduction–accretion processes: implications for the tectonic development of Tethys in the Eastern Mediterranean region. *Tectonophysics*, *473*(1–2), 113–148. doi: 10.1016/j.tecto.2008.10.027

Robertson, A. H. F., & Ustaömer, T. 9/(2009b). Upper Palaeozoic subduction/accretion processes in the closure of Palaeotethys: evidence from the Chios Melange (E Greece), the Karaburun Melange (W Turkey) and the Teke Dere Unit (SW Turkey). *Sedimentary Geology*, *220*(1–2), 29–59. doi: 10.1016/j.sedgeo.2009.06.005

Roche, V., Jolivet, L., Papanikolaou, D., Bozkurt, E., Menant, A., & Rimmelé, G. (2019). Slab fragmentation beneath the Aegean/Anatolia transition zone: insights from the tectonic and metamorphic evolution of the Eastern Aegean region. *Tectonophysics*, *754*, 101–129. doi: 10.1016/j.tecto.2019.01.016

Roche, V., Sternai, P., Guillou-Frottier, L., Menant, A., Jolivet, L., Bouchot, V., & Gerya, T. (2018). Emplacement of metamorphic core complexes and associated geothermal systems controlled by slab dynamics. *Earth and Planetary Science Letters*, *498*, 322–333. doi: 10.1016/j.epsl.2018.06.043

Rosenbaum, G., Avigad, D., & Sánchez-Gómez, M. (2002). Coaxial flattening at deep levels of orogenic belts: evidence from blueschists and eclogites on Syros and Sifnos (Cyclades, Greece). *Journal of Structural Geology*, *24*(9), 1451–1462. doi: 10.1016/S0191-8141(01)00143-2

Rossetti, F., Asti, R., Faccenna, C., Gerdes, A., Lucci, F., & Theye, T. (2017). Magmatism and crustal extension: constraining activation of the ductile shearing along the Gediz detachment, Menderes Massif (western Turkey). *Lithos*, *282–283*, 145–162. doi: 10.1016/j.lithos.2017.03.003

Royden, L. H. (1993). The tectonic expression slab pull at continental convergent boundaries. *Tectonics*, *12*(2), 303–325. doi: 10.1029/92TC02248

Royden, L. H., & Husson, L. (2009). Subduction with variations in slab buoyancy: models and application to the Banda and Apennine Systems. In S. Lallemand & F. Funiciello (Eds.), *Subduction zone geodynamics* (pp. 35–45). Berlin, Heidelberg: Springer Berlin Heidelberg. doi: 10.1007/978-3-540-87974-9_2

Royden, L. H., & Papanikolaou, D. J. (2011). Slab segmentation and late Cenozoic disruption of the Hellenic arc. *Geochemistry, Geophysics, Geosystems*, *12*(3), 2010GC003280. doi: 10.1029/2010GC003280

Sachpazi, M., Laigle, M., Charalampakis, M., Diaz, J., Kissling, E., Gesret, A., et al. (2016). Segmented Hellenic slab rollback driving Aegean deformation and seismicity. *Geophysical Research Letters*, *43*(2), 651–658. doi: 10.1002/2015GL066818

Şahin, S. Y., Aysal, N., Güngör, Y., Peytcheva, I., & Neubauer, F. (2014). Geochemistry and U–Pb zircon geochronology of metagranites in Istranca (Strandja) Zone, NW Pontides, Turkey: implications for the geodynamic evolution of Cadomian orogeny. *Gondwana Research*, *26*(2), 755–771. doi: 10.1016/j.gr.2013.07.011

Şahin, S. Y., Örgün, Y., Güngör, Y., Göker, A., Gültekin, A. H., & Karacik, Z. (2010). Mineral and whole-rock geochemistry of the Kestanbol granitoid (Ezine-Çanakkale) and its mafic microgranular enclaves in Northwestern Anatolia: evidence of felsic and mafic magma interaction. *Turkish Journal of Earth Sciences*, *19*(1), 101–122. doi: 10.3906/yer-0809-3

Sakellariou, D., Mascle, J., & Lykousis, V. (2017). Strike slip tectonics and transtensional deformation in the Aegean region and the Hellenic arc: preliminary results. *Bulletin of the Geological Society of Greece*, *47*(2), 647. doi: 10.12681/bgsg.11098

Salaün, G., Pedersen, H. A., Paul, A., Farra, V., Karabulut, H., Hatzfeld, D., et al. (2012). High-resolution surface wave tomography beneath the Aegean-Anatolia region: constraints on upper-mantle structure: tomography of Aegea-Anatolia upper mantle. *Geophysical Journal International*, *190*(1), 406–420. doi: 10.1111/j.1365-246X.2012.05483.x

Saltogianni, V., Mouslopoulou, V., Oncken, O., Nicol, A., Gianniou, M., & Mertikas, S. (2020). Elastic fault interactions and earthquake rupture along the southern Hellenic subduction plate interface zone in Greece. *Geophysical Research Letters*, *47*(13), e2019GL086604. doi: 10.1029/2019GL086604

Şaroğlu, F., Emre, Ö., & Kuşçu, İ. (1992). *Active fault map of Turkey*. Ankara, Turkey: General Directorate of Mineral Research and Exploration.

Satir, M., & Taubald, H. (2001). Hydrogen and oxygen isotope evidence for fluid-rock interactions in the Menderes massif, western Turkey. *International Journal of Earth Sciences*, *89*(4), 812–821. doi: 10.1007/s005310000135

Satir, M., & Friedrichsen, H. (1986). The origin and evolution of the Menderes Massif, W-Turkey: a rubidium/strontium and oxygen isotope study. *Geologische Rundschau*, *75*(3), 703–714. doi: 10.1007/BF01820642

Saunders, A. D., & Tarney, J. (1984). Geochemical characteristics of basaltic volcanism within back-arc basins. *Geological*

Society, London, Special Publications, *16*(1), 59–76. doi: 10.1144/GSL.SP.1984.016.01.05

Saunders, P., Priestley, K., & Taymaz, T. (1998). Variations in the crustal structure beneath western Turkey: crustal structure beneath W. Turkey. *Geophysical Journal International*, *134*(2), 373–389. doi: 10.1046/j.1365-246x.1998.00571.x

Savostin, L. A., Sibuet, J.-C., Zonenshain, L. P., Le Pichon, X., & Roulet, M.-J. (1986). Kinematic evolution of the Tethys belt from the Atlantic ocean to the pamirs since the Triassic. *Tectonophysics*, *123*(1–4), 1–35. doi: 10.1016/0040-1951(86)90192-7

Sayit, K., Marroni, M., Göncüoğlu, M. C., Pandolfi, L., Ellero, A., Ottria, G., & Frassi, C. (2016). Geological setting and geochemical signatures of the mafic rocks from the Intra-Pontide Suture Zone: implications for the geodynamic reconstruction of the Mesozoic Neotethys. *International Journal of Earth Sciences*, *105*(1), 39–64. doi: 10.1007/s00531-015-1202-2

Schaarschmidt, A., Haase, K. M., Voudouris, P. C., Melfos, V., & Klemd, R. (2021). Migration of arc magmatism above mantle wedge diapirs with variable sediment contribution in the Aegean. *Geochemistry, Geophysics, Geosystems*, *22*(6), e2020GC009565. doi: 10.1029/2020GC009565

Schellart, W. P., Freeman, J., Stegman, D. R., Moresi, L., & May, D. (2007). Evolution and diversity of subduction zones controlled by slab width. *Nature*, *446*(7133), 308–311. doi: 10.1038/nature05615

Schildgen, T. F., Yıldırım, C., Cosentino, D., & Strecker, M. R. (2014). Linking slab break-off, Hellenic trench retreat, and uplift of the Central and Eastern Anatolian plateaus. *Earth-Science Reviews*, *128*, 147–168. doi: 10.1016/j.earscirev.2013.11.006

Schuiling, R. D. (1962). On petrology, age and structure of the Menderes migmatite complex (SW-Turkey). *Bulletin of the Mineral Research Exploration Institute of Turkey*, *58*, 71–84.

Searle, M. P., & Lamont, T. N. (2020). Compressional metamorphic core complexes, low-angle normal faults and extensional fabrics in compressional tectonic settings. *Geological Magazine*, *157*(1), 101–118. doi: 10.1017/S0016756819000207

Searle, M. P., & Lamont, T. N. (2022). Compressional origin of the Aegean Orogeny, Greece. *Geoscience Frontiers*, *13*(2), 101049. doi: 10.1016/j.gsf.2020.07.008

Seaton, N. C. A., Whitney, D. L., Teyssier, C., Toraman, E., & Heizler, M. T. (2009). Recrystallization of high-pressure marble (Sivrihisar, Turkey). *Tectonophysics*, *479*(3–4), 241–253. doi: 10.1016/j.tecto.2009.08.015

Seghedi, I., Helvacı, C., & Pécskay, Z. (2015). Composite volcanoes in the south-eastern part of İzmir–Balıkesir Transfer Zone, Western Anatolia, Turkey. *Journal of Volcanology and Geothermal Research*, *291*, 72–85. doi: 10.1016/j.jvolgeores.2014.12.019

Seman, S., Stockli, D. F., & Soukis, K. (2017). The provenance and internal structure of the Cycladic Blueschist Unit revealed by detrital zircon geochronology, Western Cyclades, Greece. *Tectonics*, *36*(7), 1407–1429. doi: 10.1002/2016TC004378

Senel, M., & Aydal, N. (2002). *Geological Map of Turkey (İzmir and İstanbul)*. Eskisehir Yolu, Turkey: Maden Tetkik ve Arama Genel Mudurlugu.

Şengör, A. M. C., Görür, N., & Şaroğlu, F. (1985). Strike-slip faulting and related basin formation in zones of tectonic escape: Turkey as a case study. In K. T. Biddle & N. Christie-Blick (Eds.), *Strike-slip deformation, basin formation, and sedimentation*. Society for Sedimentary Geology. doi: 10.2110/pec.85.37.0211

Şengör, A. M. C., Satir, M., & Akkök, R. (1984). Timing of tectonic events in the Menderes Massif, western Turkey: implications for tectonic evolution and evidence for pan-African basement in Turkey. *Tectonics*, *3*(7), 693–707. doi: 10.1029/TC003i007p00693

Şengör, A. M. C., & Yilmaz, Y. (1981). Tethyan evolution of Turkey: a plate tectonic approach. *Tectonophysics*, *75*(3–4), 181–241. doi: 10.1016/0040-1951(81)90275-4

Şengör, A. M. C., & Zabcı, C. (2019). The North Anatolian Fault and the North Anatolian Shear Zone. In C. Kuzucuoğlu, A. Çiner, & N. Kazancı (Eds.), *Landscapes and landforms of Turkey* (pp. 481–494). Cham: Springer International Publishing. doi: 10.1007/978-3-030-03515-0_27

Şengün, F., Yiğitbaş, E., & Tunç, İ. O. (2011). Geology and tectonic emplacement of eclogite and blueschists, Biga peninsula, northwest Turkey. *Turkish Journal of Earth Sciences*, *20*(3), 273–285. doi: 10.3906/yer-0912-75

Seyitoğlu, G. (1997). Late Cenozoic tectono-sedimentary development of the Selendi and Uşak-Güre basins: a contribution to the discussion on the development of east–west and north trending basins in western Turkey. *Geological Magazine*, *134*(2), 163–175. doi: 10.1017/S0016756897006705

Seyitoğlu, G., Işık, V., & Çemen, İ. (2004). Complete Tertiary exhumation history of the Menderes massif, western Turkey: an alternative working hypothesis. *Terra Nova*, *16*(6), 358–364. doi: 10.1111/j.1365-3121.2004.00574.x

Seyitoğlu, G., & Scott, B. C. (1996). The cause of N-S extensional tectonics in western Turkey: tectonic escape vs back-arc spreading vs orogenic collapse. *Journal of Geodynamics*, *22*(1–2), 145–153. doi: 10.1016/0264-3707(96)00004-X

Shaw, B., & Jackson, J. (2010). Earthquake mechanisms and active tectonics of the Hellenic subduction zone. *Geophysical Journal International*, *181*(2), 966–984. doi: 10.1111/j.1365-246X.2010.04551.x

Sherlock, S., Kelley, S., Inger, S., Harris, N., & Okay, A. (1999). ^{40}Ar-^{39}Ar and Rb-Sr geochronology of high-pressure metamorphism and exhumation history of the Tavşanli Zone, NW Turkey. *Contributions to Mineralogy and Petrology*, *137*(1–2), 46–58. doi: 10.1007/PL00013777

Shin, T. A., Catlos, E. J., Jacob, L., & Black, K. (2013). Relationships between very high pressure subduction complex assemblages and intrusive granitoids in the Tavşanlı Zone, Sivrihisar Massif, central Anatolia. *Tectonophysics*, *595–596*, 183–197. doi: 10.1016/j.tecto.2012.07.012

Snopek, K., Meier, T., Endrun, B., Bohnhoff, M., & Casten, U. (2007). Comparison of gravimetric and seismic constraints on the structure of the Aegean lithosphere in the forearc of the Hellenic subduction zone in the area of Crete. *Journal of Geodynamics*, *44*(3–5), 173–185. doi: 10.1016/j.jog.2007.03.005

Sodoudi, F., Kind, R., Hatzfeld, D., Priestley, K., Hanka, W., Wylegalla, K., et al. (2006). Lithospheric structure of the Aegean obtained from P and S receiver functions. *Journal of Geophysical Research: Solid Earth*, *111*(B12), 2005JB003932. doi: 10.1029/2005JB003932

Sokoutis, D., Brun, J. P., Van Den Driessche, J., & Pavlides, S. (1993). A major Oligo-Miocene detachment in southern Rhodope controlling north Aegean extension. *Journal of the Geological Society*, *150*(2), 243–246. doi: 10.1144/gsjgs.150.2.0243

Sözbilir, H., Sarı, B., Uzel, B., Sümer, Ö., & Akkiraz, S. (2011). Tectonic implications of transtensional supradetachment basin development in an extension-parallel transfer zone: the Kocaçay Basin, Western Anatolia, Turkey. *Basin Research*, *23*(4), 423–448. doi: 10.1111/j.1365-2117.2010.00496.x

Spakman, W., Wortel, M. J. R., & Vlaar, N. J. (1988). The Hellenic Subduction Zone: a tomographic image and its geodynamic implications. *Geophysical Research Letters*, *15*(1), 60–63. doi: 10.1029/GL015i001p00060

Spakman, W. (1990). Tomographic images of the upper mantle below central Europe and the Mediterranean. *Terra Nova*, *2*(6), 542–553. doi: 10.1111/j.1365-3121.1990.tb00119.x

Spear, F. S., & Peacock, S. M. (Eds.) (1989). *Metamorphic pressure-temperature-time paths*. Washington, DC: American Geophysical Union. doi: 10.1029/SC007

Speciale, P. A., Catlos, E. J., Yıldız, G. O., Shin, T. A., & Black, K. N. (2012). Zircon ages from the Beypazarı granitoid pluton (north central Turkey): tectonic implications. *Geodinamica Acta*, *25*(3–4), 162–182. doi: 10.1080/09853111.2013.858955

Stampfli, G. M., & Kozur, H. W. (2006). Europe from the Variscan to the Alpine cycles. *Geological Society, London, Memoirs*, *32*(1), 57–82. doi: 10.1144/GSL.MEM.2006.032.01.04

Stampfli, G. M. (2000). Tethyan oceans. In E. Bozkurt, J. A. Winchester, & J. D. A. Piper (Eds.), *Tectonics and magmatism in turkey and the surrounding area* (Vol. *173*, pp. 1–23). London: Geological Society of London. https://www.lyellcollection.org/doi/10.1144/GSL.SP.2000.173.01.01

Stanley, D. J., Knight, R. J., Stuckenrath, R., & Catani, G. (1978). High sedimentation rates and variable dispersal patterns in the western Hellenic Trench. *Nature*, *273*(5658), 110–113. doi: 10.1038/273110a0

Stouraiti, C., Baziotis, I., Asimow, P. D., & Downes, H. (2018). Geochemistry of the Serifos calc-alkaline granodiorite pluton, Greece: constraining the crust and mantle contributions to I-type granitoids. *International Journal of Earth Sciences*, *107*(5), 1657–1688. doi: 10.1007/s00531-017-1565-7

Stouraiti, C., Mitropoulos, P., Tarney, J., Barreiro, B., McGrath, A. M., & Baltatzis, E. (2010). Geochemistry and petrogenesis of Late Miocene granitoids, Cyclades, southern Aegean: nature of source components. *Lithos*, *114*(3–4), 337–352. doi: 10.1016/j.lithos.2009.09.010

Suckale, J., Rondenay, S., Sachpazi, M., Charalampakis, M., Hosa, A., & Royden, L. H. (2009). High-resolution seismic imaging of the western Hellenic subduction zone using teleseismic scattered waves. *Geophysical Journal International*, *178*(2), 775–791. doi: 10.1111/j.1365-246X.2009.04170.x

Sümer, Ö., Uzel, B., Özkaymak, Ç., & Sözbilir, H. (2018). Kinematics of the Havran-Balıkesir Fault Zone and its implication on geodynamic evolution of the Southern Marmara Region, NW Anatolia. *Geodinamica Acta*, *30*(1), 306–323. doi: 10.1080/09853111.2018.1540145

Sunal, G. (2012). Devonian magmatism in the western Sakarya Zone, Karacabey region, NW Turkey. *Geodinamica Acta*, *25*(3–4), 183–201. doi: 10.1080/09853111.2013.858947

Symeou, V., Homberg, C., Nader, F. H., Darnault, R., Lecomte, J., & Papadimitriou, N. (2018). Longitudinal and temporal evolution of the tectonic style along the Cyprus Arc System, assessed through 2-D reflection seismic interpretation. *Tectonics*, *37*(1), 30–47. doi: 10.1002/2017TC004667

Tan, O. (2013). The dense micro-earthquake activity at the boundary between the Anatolian and South Aegean microplates. *Journal of Geodynamics*, *65*, 199–217. doi: 10.1016/j.jog.2012.05.005

Tatar, O., Akpınar, Z., Gürsoy, H., Piper, J. D. A., Koçbulut, F., Mesci, B. L., et al. (2013). Palaeomagnetic evidence for the neotectonic evolution of the Erzincan Basin, North Anatolian Fault Zone, Turkey. *Journal of Geodynamics*, *65*, 244–258. doi: 10.1016/j.jog.2012.03.009

Taymaz, T., Jackson, J., & McKenzie, D. (1991). Active tectonics of the north and central Aegean Sea. *Geophysical Journal International*, *106*(2), 433–490. doi: 10.1111/j.1365-246X.1991.tb03906.x

Tekin, U. K., Göncüoğlu, M. C., & Turhan, N. (2002). First evidence of Late Carnian radiolarians from the Izmir–Ankara suture complex, central Sakarya, Turkey: implications for the opening age of the Izmir–Ankara branch of Neo-Tethys. *Geobios*, *35*(1), 127–135. doi: 10.1016/S0016-6995(02)00015-3

Ten Veen, J. H., & Kleinspehn, K. L. (2002). Geodynamics along an increasingly curved convergent plate margin: Late Miocene-Pleistocene Rhodes, Greece. *Tectonics*, *21*(3). doi: 10.1029/2001TC001287

Teyssier, C., & Whitney, D. L. (2002). Gneiss domes and orogeny. *Geology*, *30*(12), 1139. doi: 10.1130/0091-7613(2002)030<1139:GDAO>2.0.CO;2

Thomson, S. N., & Ring, U. (2006). Thermochronologic evaluation of postcollision extension in the Anatolide orogen, western Turkey. *Tectonics*, *25*(3), 2005TC001833. doi: 10.1029/2005TC001833

Tirel, C., Brun, J., & Burov, E. (2008). Dynamics and structural development of metamorphic core complexes. *Journal of Geophysical Research: Solid Earth*, *113*(B4), 2005JB003694. doi: 10.1029/2005JB003694

Tirel, C., Gueydan, F., Tiberi, C., & Brun, J.-P. (2004). Aegean crustal thickness inferred from gravity inversion. Geodynamical implications. *Earth and Planetary Science Letters*, *228*(3–4), 267–280. doi: 10.1016/j.epsl.2004.10.023

Tiryakioğlu, I. (2015). Geodetic aspects of the 19 May 2011 Simav earthquake in Turkey. *Geomatics, Natural Hazards and Risk*, *6*(1), 76–89. doi: 10.1080/19475705.2013.831379

Toker, C. E., Ulugergerli, E. U., & Kılıç, A. R. (2018). The Naşa intrusion (Western Anatolia) and its tectonic implication: a joint analyses of gravity and earthquake catalog data. *Bulletin of the Mineral Research and Exploration*, 247–258. doi: 10.19076/mta.346159

Toker, C.E., Cakir, R., & Ulugergerli, E. U. (2019). *Using non-derivative filters for the tectonic implications: a case study in Simav graben, western Turkey*. Paper presented at the Symposium on the Application of Geophysics to Engineering and Environmental Problems. doi: 10.4133/sageep.32-067

Tomaschek, F. (2003). Zircons from Syros, Cyclades, Greece—recrystallization and mobilization of zircon during high-pressure metamorphism. *Journal of Petrology*, *44*(11), 1977–2002. doi: 10.1093/petrology/egg067

Topuz, G., Candan, O., Okay, A. I., Von Quadt, A., Othman, M., Zack, T., & Wang, J. (2020). Silurian anorogenic basic and acidic magmatism in Northwest Turkey: implications for the opening of the Paleo-Tethys. *Lithos*, *356–357*, 105302. doi: 10.1016/j.lithos.2019.105302

Topuz, G., & Okay, A. I. (2017). Late Eocene–Early Oligocene two-mica granites in NW Turkey (the Uludağ Massif): water-fluxed melting products of a mafic metagreywacke. *Lithos*, *268–271*, 334–350. doi: 10.1016/j.lithos.2016.11.010

Ünay, E., Emre, Ö., Erkal, T., & Keçer, M. (2001). The rodent fauna from the Adapazarı pull-apart basin (NW Anatolia): its bearings on the age of the North Anatolian fault. *Geodinamica Acta*, *14*(1–3), 169–175. doi: 10.1080/09853111.2001.11432442

Ustaömer, P. A., Ustaömer, T., Collins, A. S., & Robertson, A. H. F. (2009). Cadomian (Ediacaran–Cambrian) arc magmatism in the Bitlis Massif, SE Turkey: Magmatism along the developing northern margin of Gondwana. *Tectonophysics*, *473*(1–2), 99–112. doi: 10.1016/j.tecto.2008.06.010

Ustaömer, P. A., Mundil, R., & Renne, P. R. (2005). U/Pb and Pb/Pb zircon ages for arc-related intrusions of the Bolu Massif (W Pontides, NW Turkey): evidence for Late Precambrian (Cadomian) age. *Terra Nova*, *17*(3), 215–223. doi: 10.1111/j.1365-3121.2005.00594.x

Ustaömer, P. A., Ustaömer, T., Gerdes, A., & Zulauf, G. (2011). Detrital zircon ages from a Lower Ordovician quartzite of the İstanbul exotic terrane (NW Turkey): evidence for Amazonian affinity. *International Journal of Earth Sciences*, *100*(1), 23–41. doi: 10.1007/s00531-009-0498-1

Ustaömer, P. A., Ustaömer, T., & Robertson, A. (2012). Ion probe U-Pb dating of the Central Sakarya basement: a peri-Gondwana terrane intruded by late lower carboniferous subduction/collision-related granitic rocks. *Turkish Journal of Earth Sciences*. doi: 10.3906/yer-1103-1

Ustaömer, T., Ustaömer, P. A., Robertson, A. H. F., & Gerdes, A. (2016). Implications of U–Pb and Lu–Hf isotopic analysis of detrital zircons for the depositional age, provenance and tectonic setting of the Permian–Triassic Palaeotethyan Karakaya Complex, NW Turkey. *International Journal of Earth Sciences*, *105*(1), 7–38. doi: 10.1007/s00531-015-1225-8

Ustaömer, T., Ustaömer, P. A., Robertson, A. H. F., & Gerdes, A. (2020). U-Pb-Hf isotopic data from detrital zircons in late Carboniferous and Mid-Late Triassic sandstones, and also Carboniferous granites from the Tauride and Anatolide continental units in S Turkey: implications for Tethyan palaeogeography. *International Geology Review*, *62*(9), 1159–1186. doi: 10.1080/00206814.2019.1636415

Uzel, B., Kuiper, K., Sözbilir, H., Kaymakci, N., Langereis, C. G., & Boehm, K. (2020). Miocene geochronology and stratigraphy of Western Anatolia: insights from new Ar/Ar dataset. *Lithos*, *352–353*, 105305. doi: 10.1016/j.lithos.2019.105305

Uzel, B., Langereis, C. G., Kaymakci, N., Sözbilir, H., Özkaymak, Ç., & Özkaptan, M. (2015). Paleomagnetic evidence for an inverse rotation history of Western Anatolia during the exhumation of Menderes core complex. *Earth and Planetary Science Letters*, *414*, 108–125. doi: 10.1016/j.epsl.2015.01.008

Uzel, B., Sözbilir, H., Özkaymak, Ç., Kaymakcı, N., & Langereis, C. G. (2013). Structural evidence for strike-slip deformation in the İzmir–Balıkesir transfer zone and consequences for late Cenozoic evolution of Western Anatolia (Turkey). *Journal of Geodynamics*, *65*, 94–116. doi: 10.1016/j.jog.2012.06.009

Van Der Meer, D. G., Van Hinsbergen, D. J. J., & Spakman, W. (2018). Atlas of the underworld: slab remnants in the mantle, their sinking history, and a new outlook on lower mantle viscosity. *Tectonophysics*, *723*, 309–448. doi: 10.1016/j.tecto.2017.10.004

Van Hinsbergen, D. J. J., Hafkenscheid, E., Spakman, W., Meulenkamp, J. E., & Wortel, R. (2005). Nappe stacking resulting from subduction of oceanic and continental lithosphere below Greece. *Geology*, *33*(4), 325. doi: 10.1130/G20878.1

Van Hinsbergen, D. J. J., Kaymakci, N., Spakman, W., & Torsvik, T. H. (2010). Reconciling the geological history of western Turkey with plate circuits and mantle tomography. *Earth and Planetary Science Letters*, *297*(3–4), 674–686. doi: 10.1016/j.epsl.2010.07.024

Van Hinsbergen, D. J. J., & Schmid, S. M. (2012). Map view restoration of Aegean–West Anatolian accretion and extension since the Eocene. *Tectonics*, *31*(5), 2012TC003132. doi: 10.1029/2012TC003132

Vanderhaeghe, O. (2004). Structural development of the Naxos migmatite dome. In D. L. Whitney, C. Teyssier, & C. S. Siddoway (Eds.), *Gneiss domes in orogeny* (Vol. *380*, pp. 211–228). Geological Society of America. doi: 10.1130/0-8137-2380-9.211

Ventouzi, C., Papazachos, C., Hatzidimitriou, P., Papaioannou, C., & EGELADOS Working Group. (2018). Anelastic P- and S- upper mantle attenuation tomography of the southern Aegean Sea subduction area (Hellenic Arc) using intermediate-depth earthquake data. *Geophysical Journal International*, *215*(1), 635–658. doi: 10.1093/gji/ggy292

Von Blanckenburg, F., & Davies, J. H. (1995). Slab breakoff: a model for syncollisional magmatism and tectonics in the Alps. *Tectonics*, *14*(1), 120–131. doi: 10.1029/94TC02051

Von Raumer, J. F., Stampfli, G. M., Arenas, R., & Sánchez Martínez, S. (2015). Ediacaran to Cambrian oceanic rocks of the Gondwana margin and their tectonic interpretation. *International Journal of Earth Sciences*, *104*(5), 1107–1121. doi: 10.1007/s00531-015-1142-x

Walcott, C. R., & White, S. H. (1998). Constraints on the kinematics of post-orogenic extension imposed by stretching lineations in the Aegean region. *Tectonophysics*, *298*(1–3), 155–175. doi: 10.1016/S0040-1951(98)00182-6

Wallace, L. M., Ellis, S., & Mann, P. (2008). Tectonic block rotation, arc curvature, and back-arc rifting: insights into these processes in the Mediterranean and the western Pacific. *IOP Conference Series: Earth and Environmental Science*, *2*, 012010. doi: 10.1088/1755-1307/2/1/012010

Wallace, L. M., McCaffrey, R., Beavan, J., & Ellis, S. (2005). Rapid microplate rotations and backarc rifting at the transition between collision and subduction. *Geology*, *33*(11), 857. doi: 10.1130/G21834.1

Wei, W., Zhao, D., Wei, F., Bai, X., & Xu, J. (2019). Mantle dynamics of the Eastern Mediterranean and Middle East: constraints from *P*-wave anisotropic tomography. *Geochemistry, Geophysics, Geosystems*, *20*(10), 4505–4530. doi: 10.1029/2019GC008512

Westaway, R. (1994). Present-day kinematics of the Middle East and eastern Mediterranean. *Journal of Geophysical Research: Solid Earth*, *99*(B6), 12071–12090. doi: 10.1029/94JB00335

Westbrook, G. K., & Reston, T. J. (2002). The accretionary complex of the Mediterranean Ridge: tectonics, fluid flow and the formation of brine lakes – an introduction to the special issue of Marine Geology. *Marine Geology*, *186*(1–2), 1–8. doi: 10.1016/S0025-3227(02)00169-X

Westerweel, J., Uzel, B., Langereis, C. G., Kaymakci, N., & Sözbilir, H. (2020). Paleomagnetism of the Miocene Soma basin and its structural implications on the central sector of a crustal-scale transfer zone in Western Anatolia (Turkey). *Journal of Asian Earth Sciences*, *193*, 104305. doi: 10.1016/j.jseaes.2020.104305

White, R. W., Powell, R., Holland, T. J. B., Johnson, T. E., & Green, E. C. R. (2014). New mineral activity–composition relations for thermodynamic calculations in metapelitic systems. *Journal of Metamorphic Geology*, *32*(3), 261–286. doi: 10.1111/jmg.12071

Whitney, D. L., & Evans, B. W. (2010). Abbreviations for names of rock-forming minerals. *American Mineralogist*, *95*(1), 185–187. doi: 10.2138/am.2010.3371

Whitney, D. L., Teyssier, C., Rey, P., & Buck, W. R. (2013). Continental and oceanic core complexes. *Geological Society of America Bulletin*, *125*(3–4), 273–298. doi: 10.1130/B30754.1

Whitney, D. L., Teyssier, C., Toraman, E., Seaton, N. C. A., & Fayon, A. K. (2011). Metamorphic and tectonic evolution of a structurally continuous blueschist-to-Barrovian terrane, Sivrihisar Massif, Turkey. *Journal of Metamorphic Geology*, *29*(2), 193–212. doi: 10.1111/j.1525-1314.2010.00915.x

Whitney, D. L., & Bozkurt, E. (2002). Metamorphic history of the southern Menderes massif, western Turkey. *Geological Society of America Bulletin*, *114*(7), 829–838. doi: 10.1130/0016-7606(2002)114<0829:MHOTSM>2.0.CO;2

Woodside, J., Mascle, J., Huguen, C., & Volkonskaia, A. (2000). The Rhodes Basin, a post-Miocene tectonic trough. *Marine Geology*, *165*(1–4), 1–12. doi: 10.1016/S0025-3227(99)00140-1

Woodside, J. M., Mascle, J., Zitter, T. A. C., Limonov, A. F., Ergün, M., & Volkonskaia, A. (2002). The Florence rise, the western bend of the Cyprus Arc. *Marine Geology*, *185*(3–4), 177–194. doi: 10.1016/S0025-3227(02)00191-9

Wortel, M. J. R., & Spakman, W. (1992). Structure and dynamics of subducted lithosphere in the Mediterranean region. *Proceedings of the Koninklijke Nederlandse Akademie van Wetenschappen*, *95*(3).

Wortel, M. J. R., & Spakman, W. (2000). Subduction and slab detachment in the Mediterranean-Carpathian region. *Science*, *290*(5498), 1910–1917. doi: 10.1126/science.290.5498.1910

Xypolias, P., & Alsop, G. I. (2014). Regional flow perturbation folding within an exhumation channel: a case study from the Cycladic Blueschists. *Journal of Structural Geology*, *62*, 141–155. doi: 10.1016/j.jsg.2014.02.001

Xypolias, P., Iliopoulos, I., Chatzaras, V., & Kokkalas, S. (2012). Subduction- and exhumation-related structures in the Cycladic Blueschists: insights from south Evia Island (Aegean region, Greece). *Tectonics*, *31*(2), 2011TC002946. doi: 10.1029/2011TC002946

Yiğitbaş, E., Kerrich, R., Yılmaz, Y., Elmas, A., & Xie, Q. (2004). Characteristics and geochemistry of Precambrian ophiolites and related volcanics from the İstanbul–Zonguldak Unit, Northwestern Anatolia, Turkey: following the missing chain of the Precambrian South European suture zone to the east. *Precambrian Research*, *132*(1–2), 179–206. doi: 10.1016/j.precamres.2004.03.003

Yılmaz, Y., Genç, Ş. C., Karacık, Z., & Altunkaynak, Ş. (2001). Two contrasting magmatic associations of NW Anatolia and their tectonic significance. *Journal of Geodynamics*, *31*(3), 243–271. doi: 10.1016/S0264-3707(01)00002-3

Yilmaz, Y., Tüysüz, O., Yiğitbaş, E., Genç, Ş. C., & Şengor, A. M. C. (1997). Geology and Tectonic evolution of the Pontides. In A. G. Robinson (Ed.), *Regional and petroleum geology of the Black Sea and surrounding region* (Vol. *68*, pp. 183–226). American Association of Petroleum Geologists. doi: 10.1306/M68612C11

Yolsal-Çevikbilen, S., Taymaz, T., & Helvacı, C. (2014). Earthquake mechanisms in the Gulfs of Gökova, Sığacık, Kuşadası, and the Simav Region (western Turkey): neotectonics, seismotectonics and geodynamic implications. *Tectonophysics*, *635*, 100–124. doi: 10.1016/j.tecto.2014.05.001

Yoshioka, T. (1996). Evolution of fault geometry and development of strike-slip basins: comparative studies on the transform zones in Turkey and Japan. *Island Arc*, *5*(4), 407–419. doi: 10.1111/j.1440-1738.1996.tb00162.x

Zhu, L., Mitchell, B. J., Akyol, N., Çemen, I., & Kekovali, K. (2006). Crustal thickness variations in the Aegean region and implications for the extension of continental crust. *Journal of Geophysical Research: Solid Earth*, *111*(B1), 2005JB003770. doi: 10.1029/2005JB003770

Zitter, T. A. C., Woodside, J. M., & Mascle, J. (2003). The Anaximander Mountains: a clue to the tectonics of southwest Anatolia. *Geological Journal*, *38*(3–4), 375–394. doi: 10.1002/gj.961

Zlatkin, O., Avigad, D., & Gerdes, A. (2013). Evolution and provenance of Neoproterozoic basement and Lower Paleozoic siliciclastic cover of the Menderes Massif (western Taurides): coupled U–Pb–Hf zircon isotope geochemistry. *Gondwana Research*, *23*(2), 682–700. doi: 10.1016/j.gr.2012.05.006

Part II
Extensional Tectonic Processes

6

Oceanic Isostasy: Self-Sustaining Seafloor Spreading and Rift Localization

James A. Conder

ABSTRACT

Why Earth is the only known planet with plate tectonics is a long-standing question. The primary driving forces of plate tectonics are widely understood to entail slab-pull, ridge-push, and mantle basal tractions. Mantle tractions are usually understood in terms of active upwellings within the mantle convective system, but such upwellings are fraught with difficulties in sustaining long term continuity with the overriding plates. Recognition that basal tractions can be modified by the isostatic response of thermal subsidence in the presence of an overlying ocean may reconcile the difficulties. In particular, greater thermal subsidence near the ridge axis, requires a systematic outward mass flux beneath the plate to maintain gravitational equilibrium. This flow biases basal tractions from resistance to plate push. This outward directed flow gives the plate a perpetual resistance-free base facilitating self-sustaining plate separation and boundary localization. Localization of the plate boundary can explain why oceanic boundaries tend to narrow in comparison to subaerial boundaries. Oceanic isostasy has likely impacted tectonics from Iceland to backarc spreading to the continental slivers dotting the oceans. That the ocean plays an integral part to perpetuating seafloor spreading may be an important puzzle piece as to why only Earth exhibits plate tectonics.

6.1. INTRODUCTION

To some degree, plate tectonics is a surface manifestation of underlying mantle dynamics (Bercovici et al., 2000; Coltice et al., 2017). That plate tectonics is only known to occur on Earth suggests something fundamentally different about the Earth's mantle dynamics compared to other terrestrial planets (Martin et al., 2008). An important key to the perpetuation of plate tectonics on a planet is the how and why of self-sustaining continuous generation of new seafloor at a spreading axis.

Seafloor spreading poses a bit of a paradox. Seafloor spreading seems to be a nearly ubiquitous process in the geological record. It has been active on Earth for perhaps 3 billion years or more (Cawood et al., 2018). It is responsible for the creation of the majority of Earth's surface and is an integral part of Earth's current dynamics (Crameri et al., 2019). Earth's supercontinents have come together and ruptured apart multiple times in Earth's history (Nance & Murphy, 2013). Rifted margins of widely varying structure evolve to spreading (Perez-Gussinye et al., 2023; Sapin et al., 2021), sometimes quickly (Umhoefer, 2011; Wang et al., 2019) and with rapid along-strike propagation (Taylor et al., 1995).

Simultaneously, seafloor spreading has been fairly difficult to account for. Despite rift structures on virtually all terrestrial planets, no other known planet exhibits seafloor spreading (Martin et al., 2008). The forces required to rupture continental lithosphere do not present a simple pathway to evolve to spreading (Brune et al., 2023; Buck, 2007). Failed rifts like the mid-continent rift in

School of Earth Systems and Sustainability, Southern Illinois University, Carbondale, Illinois, USA

Extensional Tectonics: Rifting and Continental Extension, Geophysical Monograph 278, First Edition.
Edited by İbrahim Çemen and Elizabeth J. Catlos.

DOI: 10.1002/9781119773757.ch06

North America (Ojakangas et al., 2001; Stein et al., 2018) and the West African Rift System (Ghomsi et al., 2022) are of a size and magmatic output comparable to or exceeding other rifts that did transition to spreading. Even some rifted margins that did transition, such as the north-central Atlantic, took a remarkably long time to fully rupture after initial rifting (Worthington et al., 2021).

Suggested answers to the question of why Earth has plate tectonics include large extraterrestrial impacts such as the one that created the moon (O'Neill et al., 2017), plume activity (Gerya et al., 2015), and the presence of water in the mantle (Bercovici, 1998; Dymkova & Gerya, 2013; Regenauer-Lieb et al., 2001; van der Lee et al., 2008). The importance of water in facilitating plate tectonics has been identified as affecting the rheology of the mantle and crust, likely both contributing to the development of a weak asthenosphere and enhancing deformation at plate boundaries (e.g., Bercovici, 1998). Additionally, water can enhance faulting by lubrication and increasing pore pressures (Dymkova & Gerya, 2013). Although each is potentially an important factor, here I look at another direct effect of water—specifically, the overlying ocean—in facilitating and localizing seafloor spreading by inducing basal tractions aligned with spreading.

6.2. PLATE DRIVING FORCES

The energy behind plates moving on the planet ultimately derives from heat and gravity. The first viable connection between mobile continents and planetary forces mobilizing them was made in the first half of the 20th century by Arthur Holmes (Holmes, 1931) in what he termed "substratum convection"—mantle convection, as we now know it. His real genius was demonstrating how a planetary heat engine could produce horizontal forces from essentially vertical dynamics. This remains the central condition of all plate tectonic forces where gravitational instabilities result in horizontal forces.

The primary forces driving plate tectonics are some combination of ridge-push, slab-pull, and mantle tractions (Becker & O'Connell, 2001; Coltice et al., 2019; Forsyth & Uyeda, 1975), again horizontal forces arising from gravitational instabilities. Considerable evidence points to slab pull as being a dominant force (Conrad & Lithgow-Bertelloni, 2002; Forsyth & Uyeda, 1975), although plate motions must already be developed to some degree prior to subduction initiation. The magnitude of ridge push is smaller than slab pull by roughly an order of magnitude (e.g., Fowler et al., 1990). It is a force that arises with the difference in elevation across a plate and varies by plate age as elevation differences increase. Young plates (<5 m.y.) have forces ~10^{11} N/m with values increasing to >10^{12} N/m for plates older than ~25 m.y. Of course, although ridge push increases with plate growth, basal resistance over a stationary mantle also increases correspondingly. There is broad consensus that mantle tractions are necessary for plate tectonic development (e.g., Bercovici et al., 2000), but our understanding of exactly how mantle tractions participate in plate motions still has large deficiencies. For example, it takes only a cursory look at popular explanations of seafloor spreading to find graphics of large convection cells beneath diverging plates with the upwelling cell limbs aligned with the spreading center between. Besides suggesting a likely erroneous Rayleigh number of the mantle (Weeraratne & Manga, 1998), such pictures fail when considering how plates grow, shrink, and migrate. To maintain upwelling limbs beneath a migrating mid-ocean ridge and growing plate, the aspect ratios of the cells will become inconsistent with each other, as well as the overlying plates. Subducting ridges would imply superimposed upwelling and downwelling limbs, etc. Still, the prevalence of such figures is likely to remain in the absence of a clear alternative relationship between underlying mantle flow and plate motion.

6.3. OCEANIC ISOSTASY

Oceanic isostasy is an only recently explored forcing function on oceanic mantle dynamics and hence mantle tractions (Conder, 2012, 2022a). One reason it has been long overlooked is that isostasy is typically viewed as a static or transient state and does not immediately lend itself as an obvious contributor to a steady-state process like seafloor spreading. However, under certain geological conditions, isostasy can act in nontransient ways. For example, one of the pioneers in continental isostasy suggested long-term isostatically induced horizontal compression in a system like the North American eastern seaboard: the compression arising from the isostatic return mass flux "undertow" as mass is continually eroded from the nearby mountains and deposited on the continental shelf (Hayford, 1911) (Figure 6.1). The undertow ceases only with cessation of the erosion-deposition mass redistribution.

Analogous to the process of constant return mass flux in Figure 6.1, differential thermal subsidence of the seafloor involves a continual mass exchange at the surface as water deepens at different rates across the plate. To maintain gravitational equilibrium, this surface water mass exchange must be balanced by a deeper counter-mass exchange (Figure 6.2). Our current geophysical understanding of thermal isostasy in the oceans largely follows the work of Parsons and Sclater (1977). The depth of the seafloor is widely understood to be controlled by Pratt isostasy, resulting in the well-known depth–square root of age relationship where the mid-ocean ridge system is topographically high and the seafloor subsides as it ages and moves away from the ridge due to cooling. In the

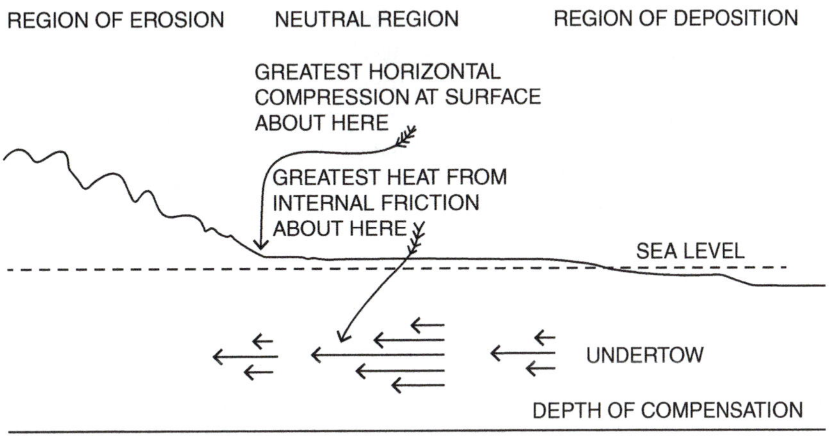

Figure 6.1 Figure from Hayford (1911) / American Association for the Advancement of Science / Public domain as out of copyright (1911) showing isostatic response of continual mass exchange at the surface being balanced by a continual counter mass flux at depth. The "undertow" mass flux will induce stresses on the crust above, in this case resulting in compression in the crust near the base of the mountain belt.

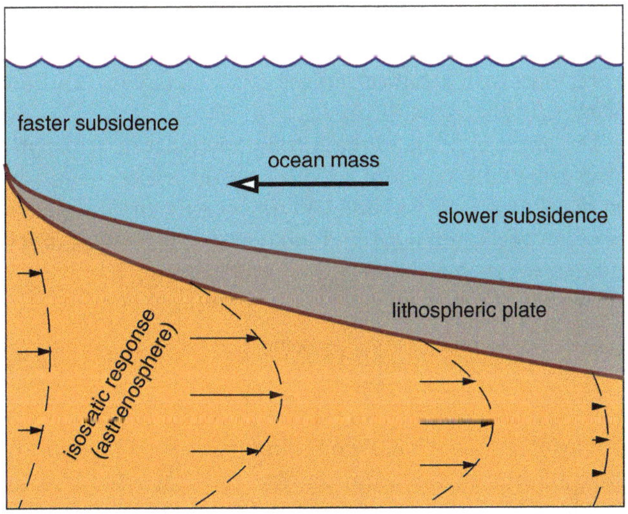

Figure 6.2 Asthenosphere isostatic response to surface mass exchange that occurs with differential thermal subsidence at the seafloor. Cartoon is in a plate-fixed reference frame. Faster subsidence regions (younger ages) gain mass more rapidly as water deepens than regions with slower subsidence (older ages), inducing a counter mass-flux that leads the plate in the plate spreading direction. In this plate-fixed reference frame, plate growth is to the left.

decades following this seminal work, the overlying ocean has invariably been treated as a static load correction, necessary to account for the excess mass effect on the shape of the seafloor (e.g., Turcotte & Schubert, 2002, section 4.22). Although substantively correct, with long-standing predictive power for understanding the depth of the seafloor, the Parsons and Sclater work has perhaps been a little too successful. In particular, the ocean and mantle dynamic interplay required to maintain the isostatic equilibrium that their work depends on remains almost entirely unconsidered.

In essence, to maintain gravitational equilibrium, the flux of ocean mass toward more rapidly thermally subsiding younger seafloor must be balanced by a corresponding flux of asthenosphere mass directed toward older lithosphere. As thermal subsidence is continual and systematic, this isostatic response must likewise be continual and systematic (Figure 6.2). Although isostatic in nature, this dynamically maintained equilibrium does not fit cleanly as either Pratt-type or Airy-type isostasy. Here I term it "oceanic isostasy." The 1D analytic solution of the physics of this oceanic isostasy for a constant velocity plate is derived in Conder (2012).

The outward-directed asthenosphere flow induced from oceanic isostasy only incrementally leads the overlying lithospheric plate with tractions less than those estimated for ridge push (Conder, 2012). Nonetheless, oceanic isostasy may have profound implications on the plate motions. Firstly, although the forces are small in absolute terms (10^8–10^{11} N/m) compared to ridge push or slab pull, asthenosphere leading the lithosphere alters the plate tectonic force balance by giving a systematically positive push to the plate and supplanting what would otherwise be basal resistance. At a minimum, this gives a resistance-free base for plates to glide on, allowing forces like ridge push greater efficiency. Further, because these tractions come from the isostatic response to the plate itself, the form of the self-sustaining asthenosphere flow stays aligned with the plate geometry through plate growth and migration. Second, the outward tractions can help localize the divergent plate boundary.

Although continental rifting precedes seafloor spreading, there are clear fundamental differences between the two processes. Continental rifting typically entails stretching continental crust, frequently but not always with significant magma emplacement (Bialas et al., 2010). Continental rifts may be wider or narrower (hundreds of kilometers wide versus 100 km or less) (Buck, 2007). Continental rifting involves thicker crust and lithosphere than that near spreading axes. In contrast, seafloor spreading generates entirely new igneous crust (as opposed to stretching) in a more continuous fashion reflected in regular and symmetric gravity and magnetic anomalies.

The ocean has always been treated as incidental to seafloor spreading. However, these isostatic considerations suggest that it could play an important role in facilitating plate separation during seafloor spreading. As noted, there are fundamental differences in structure and process between continental rifting and seafloor spreading, and to date, the tipping point for a system to move from rifting to spreading has been elusive. Arguably, the most fundamental difference is that seafloor spreading is more continuous and self-sustaining in its behavior, reflected in systematic morphology, magnetic anomalies, gravity anomalies, and ultimately the steady-state creation of new oceanic crust. Study of rifts and rifted margins to understand the transition to spreading has led to a wealth of information about crustal behavior and evolution during extension (e.g., Férot, 2021). However, the diversity of rifted margins in structure and strain history gives no overarching theory as to the successful tipping points of when a continental rift will transition to seafloor spreading (Brune et al., 2023).

The elimination of basal resistance that arises with thermal subsidence in the presence of an ocean may present this tipping point for the transition of continental rifting to seafloor spreading (Conder, 2022a). It has long been recognized that tractions on the base of the lithosphere are central to plate tectonics, but basal tractions may be either driver or drag to plate motions depending on whether the underlying viscous mantle leads or lags the motion of the overlying plate (e.g., Forsyth & Uyeda, 1975). In mature spreading systems, forces like ridge push and slab pull likely dominate, although their effectiveness is greater when basal drag is diminished. As continental rifts must be exerting forces sufficient for lithosphere extension over resistance from the lithosphere itself and from viscous basal tractions, an active rift system experiencing inundation can receive a positive push toward stable extension with the reduction in resistance that comes with the onset of isostatically driven undertow.

6.3.1. Quantitative Development of Oceanic Isostasy

Drawing largely on Conder (2012), a brief sketch of the underlying physics is presented here. A useful starting point for understanding oceanic isostasy is with Pratt isostasy and the square root of age–depth equation for seafloor

$$d = d_{\text{ridge}} + c\sqrt{t} \tag{6.1}$$

where d is ocean depth, d_{ridge} is ridge depth, and t is seafloor age. The subsidence constant, c, depends on several factors, including thermal expansion, mantle temperature, thermal diffusivity, mantle density at reference mantle temperature, ρ_m, and ocean density, ρ_w. Lumping properties other than densities into a constant f gives

$$c = f\frac{\rho_m}{\rho_m - \rho_w} \tag{6.2}$$

The factor $\rho_m/(\rho_m - \rho_w)$ accounts for the mass due to the overlying substrate. Subaerially, this factor goes to unity. Subaerial cooling is fully Pratt compensated as no mass enters or leaves with cooling and contraction. For mantle density of 3,330 kg/m^3 and ocean density of 1,030 kg/m^3, this implies submarine subsidence 45% greater than subaerial subsidence for crust of similar age (Figure 6.3). As this excess 45% comes from ocean mass being continually redistributed among columns during subsidence, a mechanism other than Pratt isostasy must accommodate nearly one-third of the gravitational disequilibrium. Reattaining gravitational equilibrium to account for the water mass redistribution requires a counter-redistribution of mass in the asthenosphere.

The amount of mass that ultimately must be redistributed from any column is equal to the excess mass, γ, increase arising from the overlying ocean, the difference between the submarine and subaerial subsidence curves (Figure 6.3a). Assuming $c_{\text{submarine}} \sim 0.35\,\text{km/m.y.}^{1/2}$,

$$\gamma = \left(0.11\frac{\text{km}}{\text{m.y.}^{\frac{1}{2}}}\right)\sqrt{t} \tag{6.3}$$

As this is a continually ongoing process with previous disequilibria already adjusted for, the relevant curve to consider is the rate of mass increase (Figure 6.3b). Differentiating equation (6.3) with respect to time gives the rate of mass accumulation in each column:

$$\dot{\gamma} = 0.055\frac{\text{km}}{\text{m.y.}^{\frac{1}{2}}} / \sqrt{t} \tag{6.4}$$

Because younger seafloor subsides more rapidly than older seafloor, excess is added more rapidly to young seafloor, requiring a counter asthenospheric flow away from young seafloor toward older seafloor.

The change in mass in a column due to lateral asthenosphere counter flow is the mass flux into the column minus the mass flux out of the column. In 1D, the governing equation becomes

$$\dot{\gamma}^* - \frac{dU}{dx} = 0 \tag{6.5}$$

where

$$\dot{\gamma}^* = \dot{\gamma} - \dot{\gamma}_{\text{mean}} \tag{6.6}$$

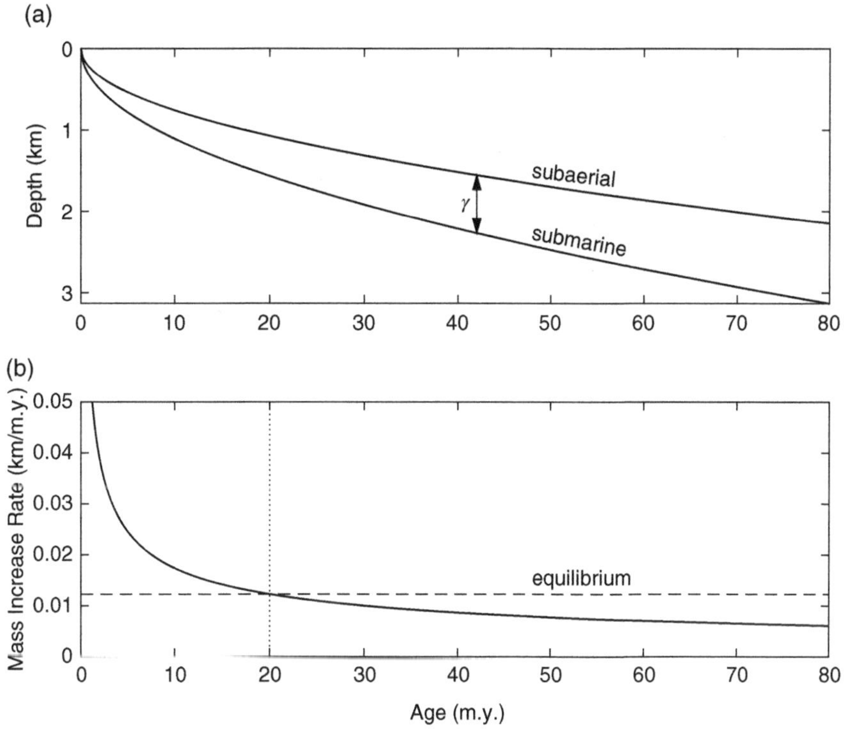

Figure 6.3 Subaerial and submarine square root of age (a) and mass gain (b) curves (Conder (2012) / with permission of GeoScienceWorld). γ corresponds to both the difference in the two curves and the height of the asthenosphere column equivalent to the mass of the overlying ocean.

and U is the lateral asthenosphere mass flux (in units of $km^2/m.y.$). The flux as a function of position, $U(x)$, can be found by

$$U(x) = \int_0^X \dot{\gamma}^* dx \tag{6.7}$$

As shown in the supplementary material of Conder (2022a), associated basal tractions, σ, may be estimated by assuming U is maintained through Poiseille flow in the asthenosphere. Given a channel width, H, and viscosity, η,

$$\sigma = 1.5\eta U / H^2 \tag{6.8}$$

To maintain consistency with the previous work, calculations of tractions in this chapter assume H to be 100 km and η to be 10^{20} Pa s. A numerical solution is given in Conder (2022a) and maintained as a MATLAB function (Conder, 2023).

6.4. LOCALIZATION OF SPREADING CENTERS

A clear difference of mid-ocean spreading centers relative to continental rifts is the narrow, stable, well-defined axis of extension giving rise to localized seismicity and a regular pattern of magnetic anomalies. Although the weakest part of the plate will preferentially rupture under extension, the systematic undertow induced by oceanic isostasy may directly factor in the tighter orderliness of spreading centers. The outward mass flux centers on the locus of greatest submarine subsidence (most rapid ocean deepening) (Figure 6.4). For most of the ridge system, this is the spreading axis, so the alignment of outward flow from the axis reinforces mantle tractions tending toward a single locus of extension. A quantification of localization forces can be found by integrating outward tractions as a function of location (Figure 6.4d). Positive values promote extension, and negative values promote compression.

Even in a messy thermal state, potentially like that of a continental rift, a preferential spreading locus may develop with submergence. Figure 6.5 shows localization preference for an instance with two loci of rifting. Even with two loci, the basal mantle tractions give a preference for localizing at one over the other. Going further, Figure 6.6 shows two examples of randomly distributed thermal ages (a), their associated mass fluxes arising with submergence (b), and the strength of localization across the system (c). Even with randomness in the thermal system, once submergence occurs, there are preferential loci for extension where subsequent new crust is likely to be emplaced, leading to a self-reinforcing feedback loop of raised temperatures and further localization in that place.

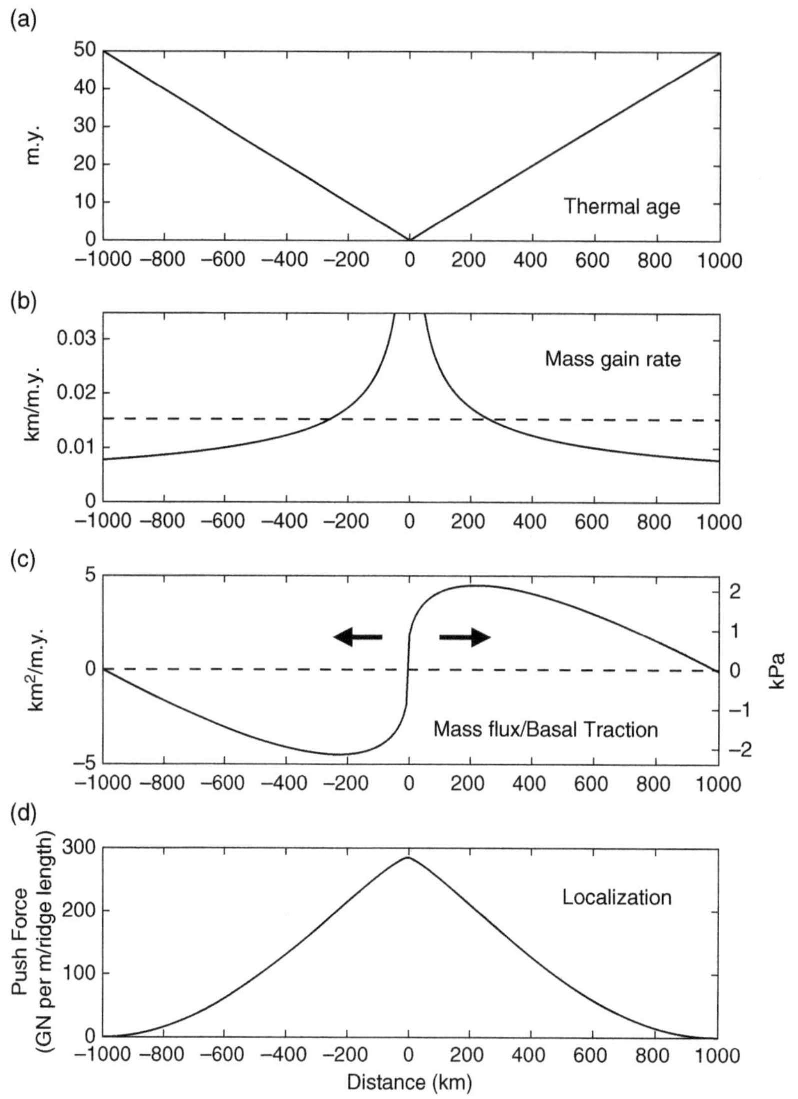

Figure 6.4 Calculations for 1D across-axis isostatic response of differential thermal subsidence at the seafloor. (a) Seafloor thermal age to either side of a constant velocity, symmetrically spreading ridge. (b) Mass gain that comes with subsidence and a deepening ocean. Dashed line is the mean gain line. To maintain gravitational equilibrium, the seafloor with ocean mass gains above the line must transfer asthenosphere mass to the seafloor with ocean mass gains below the line, which has relative mass deficits. (c) Resulting asthenosphere mass flux to maintain equilibrium (left axis) and associated basal tractions (right axis). Positive fluxes and tractions are directed to the right. Negative fluxes and tractions are directed to the left. Equilibrium (zero) line is dashed. Directions of flow are also denoted by arrows. (d) Integrated outward push force at the seafloor as a function of location. Positive values promote extension; negative values promote compression. The push force peaks and tends to localize extension at the axis.

6.5. GEOLOGICAL EXAMPLES

There are a number of areas across the globe worth examining their geophysical character and/or geological history in the context presented here. For instance, considerations could be made at rifted margins, including complex rifts like the Galicia Bank (Druet et al., 2018; Grevemeyer et al., 2022), the Gulf of Mexico (Kneller et al., 2012), and/or back-arc spreading centers.

6.5.1. Subaerial versus Submarine Spreading

With its subaerial spreading, Iceland presents a unique natural test and illustration of the process at work. Extensional tectonics on the island are known to differ in substantial ways from typical seafloor spreading. Extension takes place over a broader area, dikes and earthquake swarms are common, and ridge-parallel strike-slip faulting is documented (Einarsson, 2008;

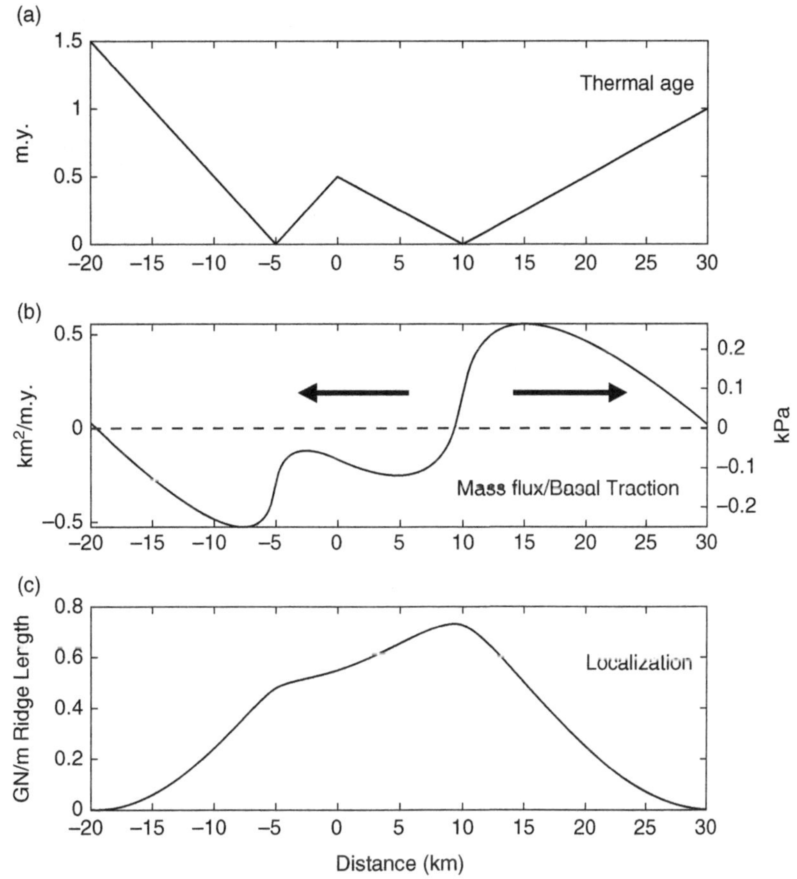

Figure 6.5 Example of two axes of extension. (a) Thermal age across the system, with two locations having zero age. (b) Isostatic mass flux response across the system after submergence (left axis) and associated basal tractions (right axis). The dashed line is equilibrium. (c) Outward mantle tractions associated with extension. When submergence occurs, the isostatic response likely has a preference for one of the axes to extend and reinforces localization at the expense of the other.

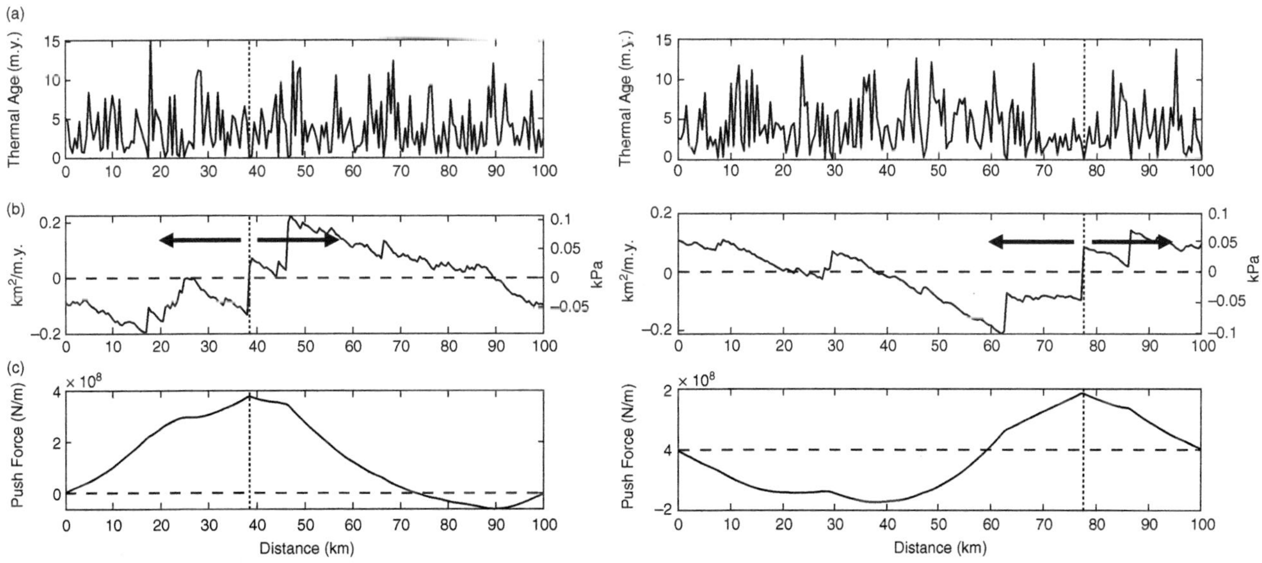

Figure 6.6 Examples of random thermal structure of a rift and isostatic response after submergence. (a) (Randomly distributed) thermal ages across the system. (b) Isostatic mass flux response across the system after submergence. (c) Mantle tractions associated with extension. When submergence occurs, the isostatic response will likely result in a preferred locus of extension that will reinforce itself with continued extension.

Karson, 2017; Sigmundsson et al., 2020). Thicker crust (Einarsson, 2008) and/or plume processes (Karson, 2017) likely contribute to these differences, but its subaerial nature may also be a contributing factor. To investigate the relative importance of the subaerial nature of extension on the island relative to thicker crust and/or hotspot processes, one can look at the portion of the spreading system that is both on-platform (thick crust) and submarine. Contrasting this section of the spreading system to adjacent submarine off-platform and subaerial on-platform sections offers a valuable examination of the relative importance of an overlying ocean on spreading center dynamics versus thicker crust and/or hotspot.

As a first-order observation, ridge morphology suggests that the coastline may be at least as important a factor in the spreading process as crustal thickness. Figure 6.7 shows a digital elevation model where the Reykjanes Ridge crosses the Icelandic platform and onto the island. The transition in morphology of the Reykjanes Ridge from a narrow band of *en echelon* spreading segments (Saemundsson et al., 2020) to the wider and more distributed volcanic zones of Iceland proper occurs across the Reykjanes Peninsula. In contrast, there is minimal response observed in the crossing from off-platform to on-platform. If the difference in morphology were primarily controlled by the structure of the platform, such as its thicker crust, the morphological differences should track with the edge of the platform rather than the coast as observed. Likewise, it would be a surprising happenstance for the change to correspond tightly with the coastline if the morphology change were strictly plume-controlled.

More than just ridge morphology, changes in the character of earthquake distribution, gravity, and magnetic anomalies, and even the eastward shift of the plate boundary, are all more tightly associated with the coastline than with the edge of the Icelandic Platform (Conder, 2022a). Together, these further the case that the presence of the ocean is an important factor in the observed differences. Figure 6.8 shows the isostatic gravity anomaly on and around the platform (Bonvalot et al., 2012). Figure 6.9 shows seismicity and slip vectors of extensional earthquakes from the Centroid-Moment-Tensor (CMT) catalog (globalcmt.org) on and around

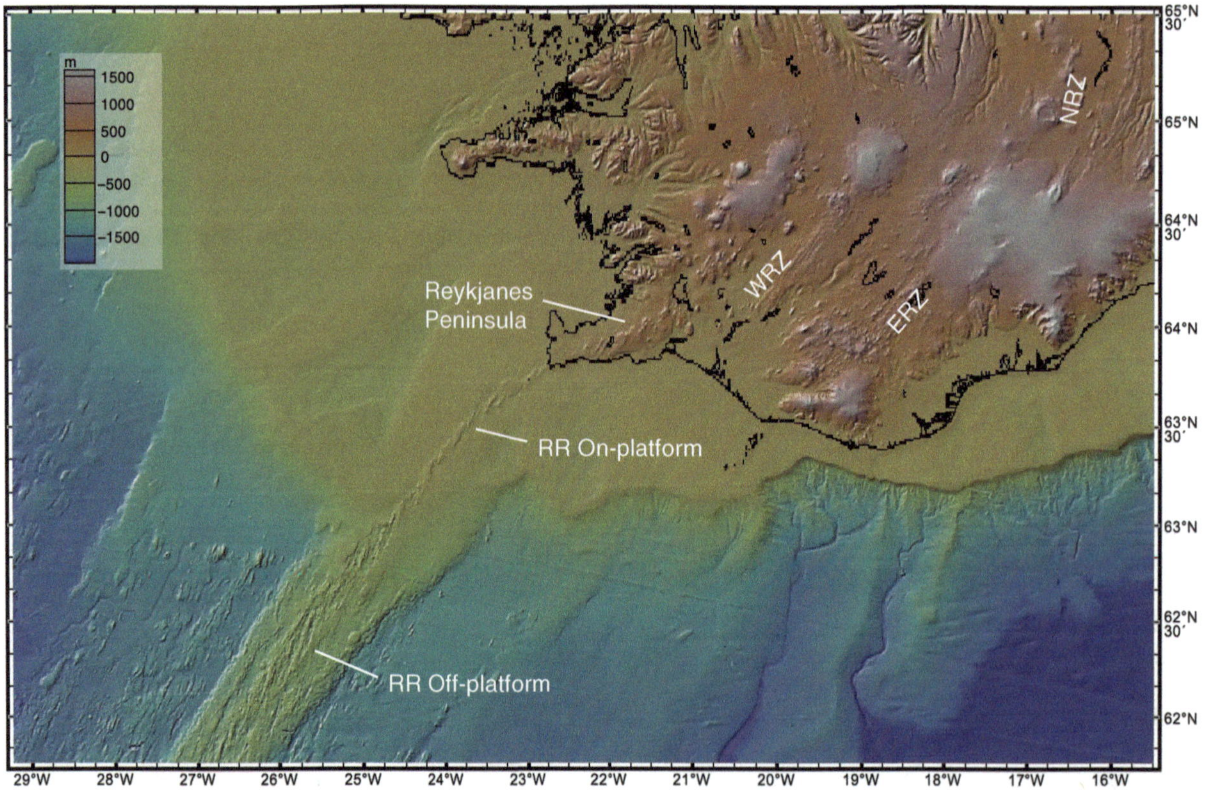

Figure 6.7 Morphology of the mid-ocean ridge system across the Icelandic platform and coastline. Ridge morphology changes are greatest across the Reykjanes Peninsula as opposed to across the off-platform/on-platform transition. RR, Reykjanes Ridge; WRZ, ERZ, NRZ, Western Rift Zone, Eastern Rift Zone, and Northern Rift Zone, respectively (following Karson [2017]). Colors denote depth/elevation. Figure made in GeoMapApp (geomapapp.org).

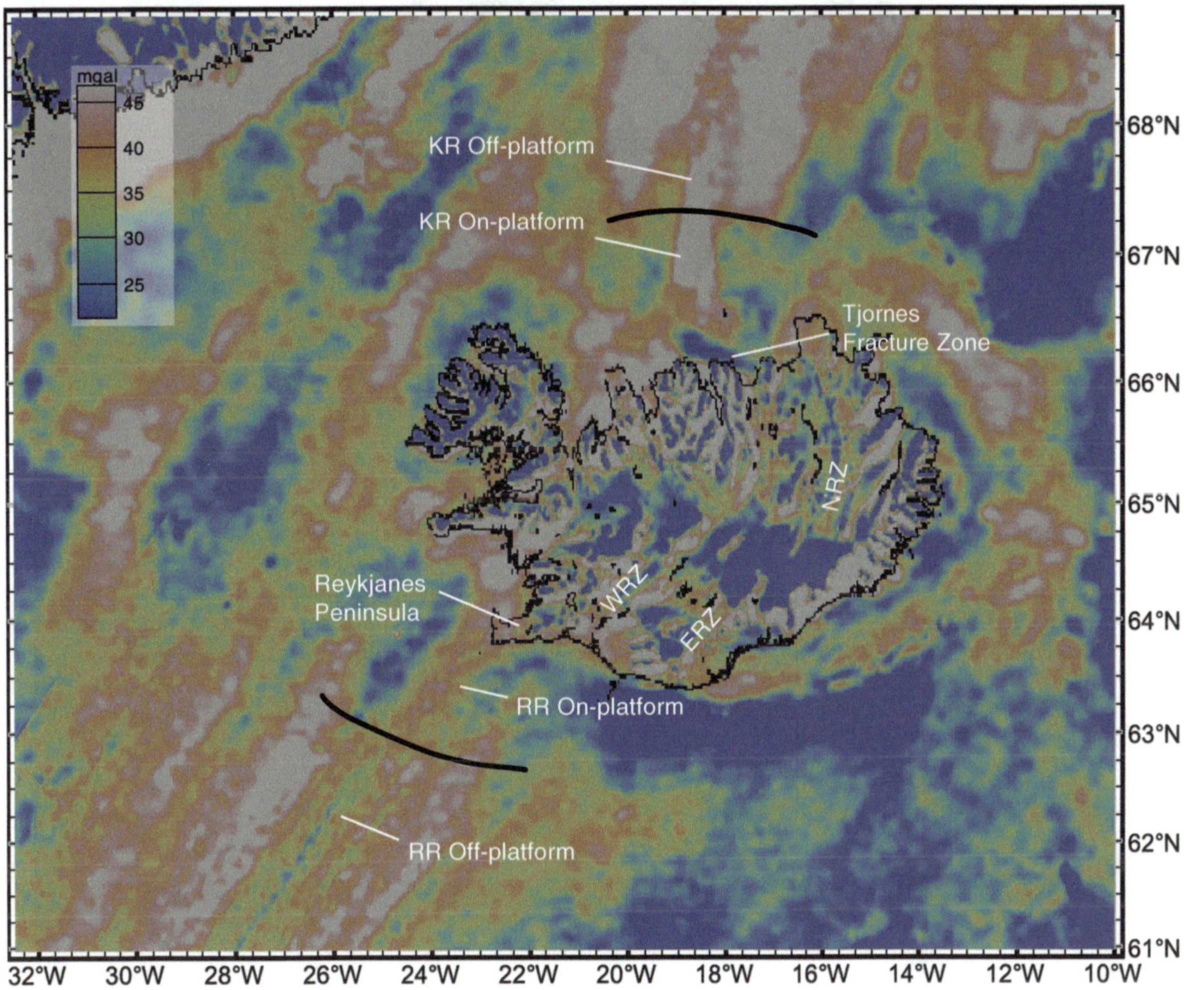

Figure 6.8 Isostatic gravity anomaly of Iceland and surrounding region (Bonvalot et al., 2012). The isostatic anomaly is the difference from what would be expected if the system were in Airy isostatic equilibrium. Thin black line denotes coastline. The edge of the platform, where intersected by the spreading system, is also marked. Labels are the same as in Figure 6.6. KR, Kolbeinsey Ridge. Figure made in GeoMapApp (geomapapp.org).

the platform. Events are determined to be extensional, compressional, or strike-slip by the most vertical stress axis of the moment tensor: P, T, or B. Earthquakes on the submarine platform occur along a localized spreading axis and are primarily associated with normal faulting more akin to earthquakes off-platform than on the island. Across the Reykjanes Peninsula, events are dominated by strike-slip bookshelf faulting rather than normal faulting (Einarsson et al., 2020). Further into the island, many events are associated with volcano-tectonic processes rather than simple plate separation (Einarsson, 2008), but nowhere on the island can the same coherency of extensional slip be found as on the submarine platform. Similarly, isostatic gravity anomalies have a greater change in character across the coastline than across the platform edge, with anomalies on the island being less spatially coherent and showing minimal definition with the local rift axes. In virtually every geophysical dataset observed, the coastline demarcates a greater change in geophysical character than delineations expected from associations with thick crust or plume processes. The tight correlation of tectonic character changes with the coastline is a strong indicator that the ocean is not passive but an active participant in the seafloor spreading process.

As Iceland is bounded by submarine regions to each side, its tectonic behavior may be subtly different than for a simple submerged/nonsubmerged dichotomy. Specifically, although the localization of the ridge axis for Iceland may be lacking relative to the totally submerged case in Figure 6.4, the mantle tractions arising around the island may be important to consider. Figure 6.10 shows the isostatic flow response and localization for a subaerial axis

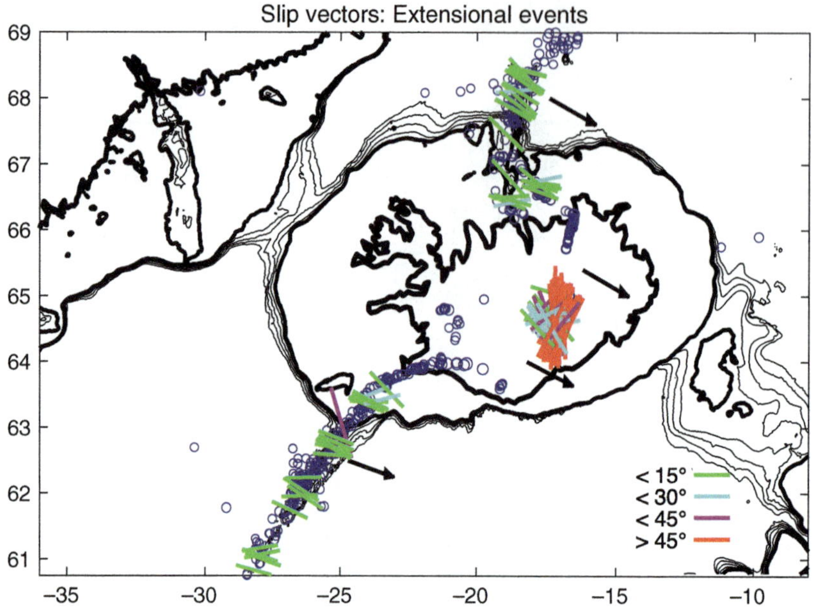

Figure 6.9 Seismicity and slip vectors for extensional earthquakes on and around Iceland. Blue circles are earthquakes of magnitude >4.5 1960–2022. Thick black line denotes coastlines. Medium thick black line is 400 m contour outlining the Iceland platform and thinner black lines show deeper 200 m contour intervals. Black arrows show predicted slip direction based on MORVEL Euler pole (DeMets et al., 2010). Green, cyan, magenta, and red lines show slip directions for extensional earthquakes from 1990 to 2023 in the CMT catalog (globalcmt.org). Extensional events are absent on the Reykjanes Peninsula as it is dominated by strike-slip bookshelf faulting. Colors show deviations from predicted directions: green within 15°, cyan within 30°, magenta within 45°, and red greater than 45° from predicted.

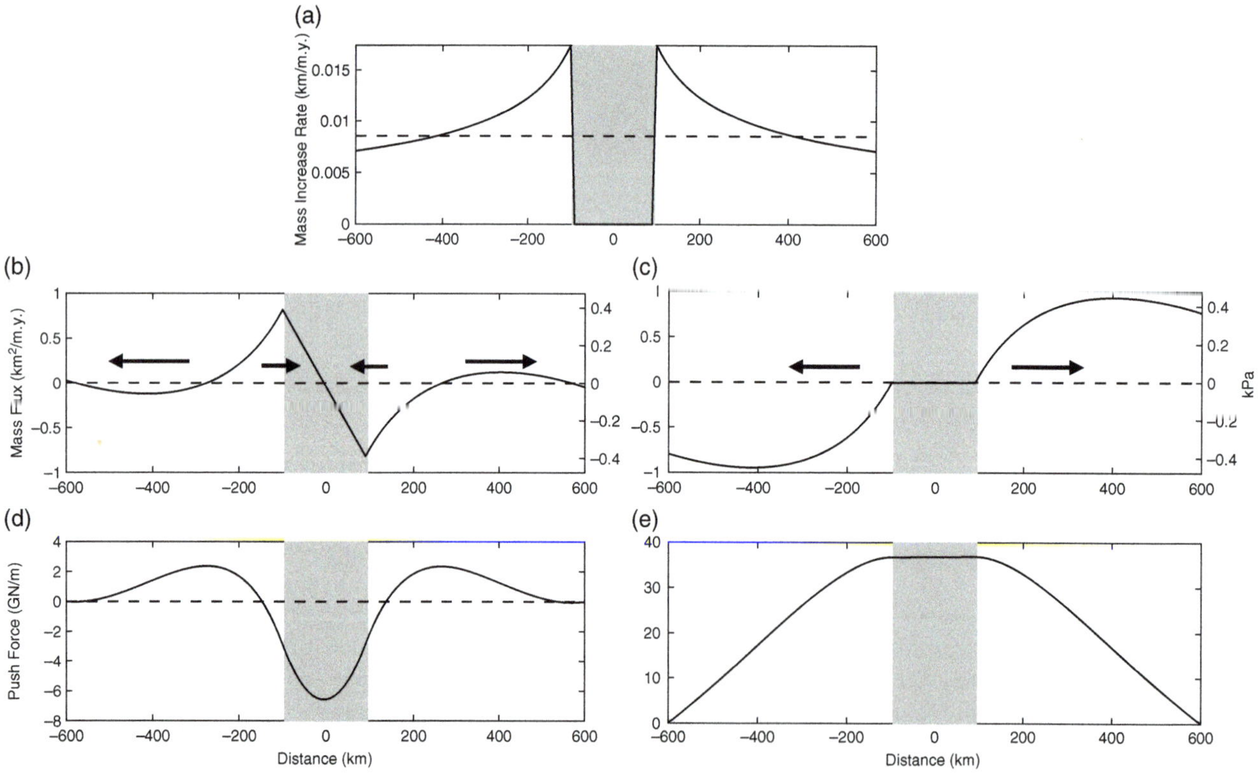

Figure 6.10 1D calculations of mantle tractions for a representative oceanic region with a subaerial portion (gray region) where the thermal age is youngest. The half-spreading rate is set to 10 km/m.y. (a) Mass adjustments arising from subsidence with an overlying ocean. The subaerial region gains zero mass with subsidence. The dashed line is a line of equilibration. (b, c) Corresponding mass exchange (left axis) and associated basal tractions (right axis), assuming that either the subaerial region participates in the mass exchange (b) or does not participate (c). (d, e) Corresponding associated outward push from induced mantle traction in (b) and (c), respectively. If the subaerial region fully participates in isostatic equilibration, preferred localization is just outside the edges of the island (d). If participation is limited, preferred localization is on the island itself, but no preference beyond that (e).

with submerged flanks. Importantly, the subaerial region does not gain additional mass with thermal subsidence. The location most rapidly accumulating mass via subsidence is the submarine region just beyond the shoreline (Figure 6.10a). As the outward mass flux aiding seafloor spreading is not present beneath the subaerial region, the flux beneath the plateau may be directed inward (Figure 6.10b). If the island participates fully in the gravitational equilibrium process with the surrounding oceanic regions, localization is preferred just outside the island's edges (Figure 6.10d). However, if the island does not efficiently participate in the gravitational equilibration process—as the isostatic anomalies suggest (Figure 6.8)–localization will preferentially occur within the subaerial region, but no strong preference beyond that (Figure 6.10c). This lack of strong localization may be why the on-island axes develop in multiple zones (Figure 6.7) and routinely shift around the platform (Foulger et al., 2020). Similarly, the lack of localization informs why extensional earthquake slip vectors lose consistency on the island while maintaining consistency on the submarine platform (Figure 6.9).

Progressing back in time, this effect of biasing extensional axes within subaerial regions flanked by ocean may be seen in the geological history of the Greenland-Iceland-Faroe ridge. Past rift axes associated with Iceland were not restricted to the current island. Multiple former rift axes jumped around nearly the entire length of the Greenland-Faroe ridge (Foulger et al., 2020). Importantly, these rift axes were mostly subaerial at the time of their existence as the Greenland-Iceland-Faroe ridge was a land bridge until 10–15 m.y. ago (Denk et al., 2011; Ellis & Stoker, 2014). As the submergence of the ridge progressed, localization, as seen in Figure 6.10e, likely continually constrained rift axis jumps to remaining subaerial regions—culminating in the current configuration.

6.5.2. Seaward Dipping Reflectors

Rifted margins are often described in terms of "magma-rich" and "magma-poor" end members (Franke, 2013; Peron-Pinvidic et al., 2019; Tugend et al., 2020). Magma-rich margins are typically characterized as having a package of seaward dipping reflectors (SDRs), frequently interpreted as thick packages of lava flows occurring during an increase of magma production associated with the creation of the first oceanic crust (e.g., Chauvet et al., 2021). It may be the case that SDRs in the distal portion of the rift system are indicators of subaerial spreading (Collier et al., 2017; Jackson et al., 2000; Karner et al., 2020; Mutter et al., 1982) with analogs today in Iceland (Karner et al., 2020) and the Afar subaerial section of the Red Sea Rift (Bastow & Keir, 2011; Corti et al., 2015). Given the prevalence of magma-rich margins (e.g., Geoffroy, 2005), a high frequency of possible subaerial spreading in the geological record appears to point away from the ocean having an important role in the development of seafloor spreading. On close inspection, though, these may be exceptions that prove the rule. Despite their prevalence along the Atlantic and other margins, Iceland and Afar are the only analogs we see today. Iceland and Afar are clearly unusual in the current spreading system in that (a) subaerial spreading accounts for <1% of the current spreading system and (b) as noted earlier, their behavior contrasts in fundamental ways with the submarine spreading system.

If distal SDRs are indicators of subaerial spreading, the intrinsic differences of seafloor spreading are important. It seems probable that without the facilitating mantle tractions, extension in Iceland and Afar is largely a kinematic response to actively facilitated spreading in the adjacent submerged plate boundary. Magma-rich, in this case, may indicate the development of a steady-state magma chamber with lithospheric thinning but without the localizing effects that come with submergence. Conspicuously, even a magma-rich region like Iceland with long-lived magma chambers appears to be largely driven by far-field tectonics as opposed to magma-driven opening (Kolzenburg et al., 2022). That magma-rich and magma-poor margins can vary over short spatial scales (Gouiza & Paton, 2017; Shillington et al., 2009) is consistent with a model of spreading tracking with inundation but with instances of Icelandic-type extension leading inundation. Spreading leading inundation may occur with adjacent inundated self-sustaining spreading centers forcing the subaerial neighbors in the system. Presumably most of these instances would evolve to stable and submerged seafloor spreading.

6.5.3. Rift to Drift

Although the number of locations where the mid-ocean ridge system crosses the coastline is limited, additional geological evidence for the ocean actively facilitating seafloor spreading can be found with the rift-to-drift transition. Places like the Afar triple junction with three extensional boundaries at different stages of the rift-to-drift transition, as well as Atlantic rifted margins, show the rift-to-drift transition tracking more closely with inundation than other factors such as obliquity, rifting style, and even rifting intensity (Conder, 2022a). The list of locations in considering the relationship between rupture and inundation is extensive, including rifted margins (Sapin et al., 2021; Unternehr et al., 2010), back-arc spreading centers, and rifts on other planets.

One example is rifting in the Gulf of California (GoC), a complex rifting and drifting system (Umhoefer et al., 2020) exhibiting along-strike variability in rifting structures (Lizarralde et al., 2007; van Wijk et al., 2017). The extension is primarily accommodated along multiple long transform faults connecting several short spreading segments (Figure 6.11). Although young, seafloor spreading and extension in the GoC has a complex evolution culminating in fully developed spreading in the south and pull-apart basin extension in the north (van Wijk et al., 2017). Prior to opening, subduction of the Farallon plate occurred off the west coast of Baja California. Rifting began in Baja after the subducting plate was consumed (~12 Ma). Subsequently, the GoC captured the primary plate boundary along the previously active continental arc, moving it inland (Michaud et al., 2006; Umhoefer, 2011). Rupture along the ~1500 km long system occurred rapidly, co-opting the plate boundary within 6–10 m.y. after cessation of subduction (Umhoefer, 2011). Although the rapid coalescence of the boundary must have had much to do with the weak arc lithosphere (Umhoefer, 2011), observed variations in along-strike extension style suggest other superposed tectonic effects at work as well, such as effects of sedimentation (Bialas & Buck, 2009), mantle fertility (Lizarralde et al., 2007), or structural styles (van Wijk et al., 2017). Of particular importance here, marine incursion in the GoC also appears to have a close relationship with the plate boundary coalescence and subsequent evolution and development of seafloor spreading (Umhoefer et al., 2018).

To the first order, it is easily observed that seafloor spreading is occurring in the southern gulf, which had more ready access to the sea, contrasting with earlier stages of rifting occurring in the northern section. Of course, the details are more nuanced. Marine incursion occurred in the southernmost gulf by 8–9 Ma, corresponding with spreading propagation of the East Pacific Rise toward the arc into the southernmost gulf (Umhoefer et al., 2018). Marine incursion reached all the

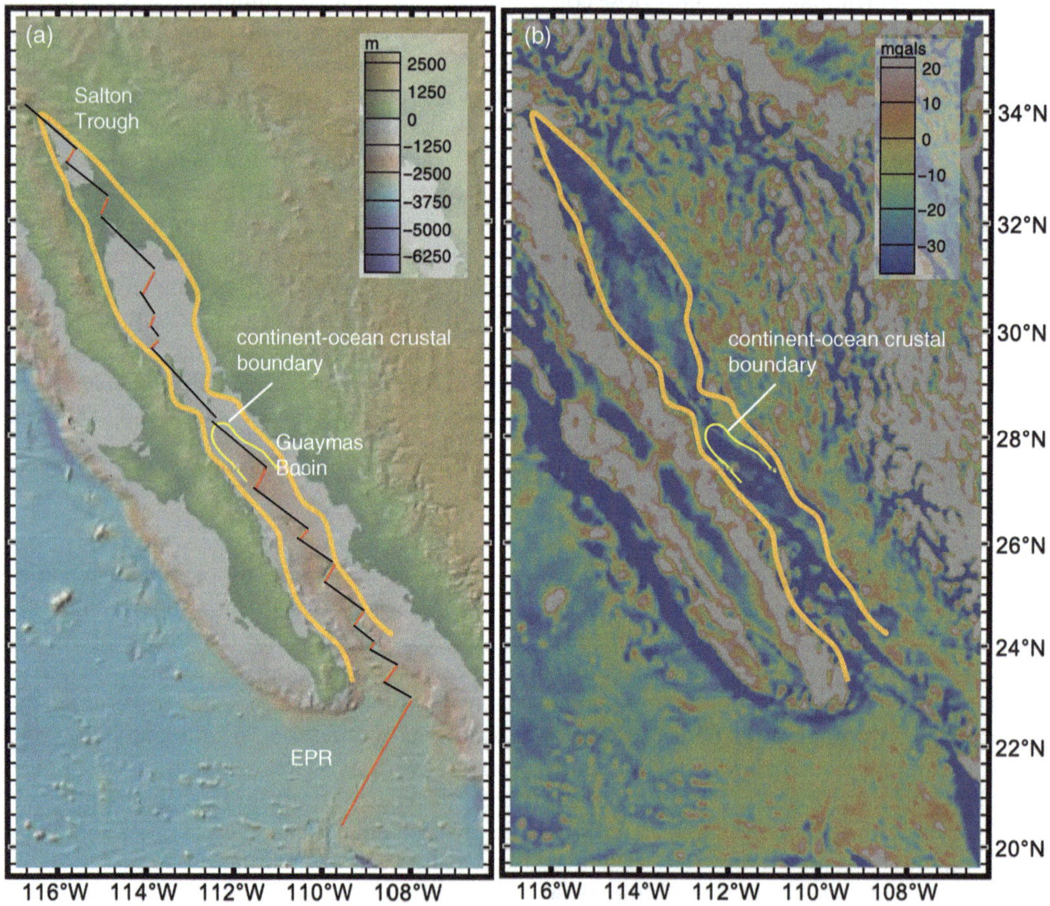

Figure 6.11 Gulf of California topography (a) and free-air gravity anomaly (Sandwell v30) (b). Plate boundary marked in red (extension) and black (transform). Boundary follows (Umhoefer et al., 2020) and (Martin-Barajas et al., 2013). EPR, East Pacific Rise. The exact boundary is more speculative in the northern gulf. The area of low free-air gravity seen along the gulf (thick orange line; roughly the 0 mGal contour) extends to and terminates at the Salton trough in the north, indicating the extent of the region where the isostatic accommodation process is correlated. The continental-oceanic crustal boundary is clear in the gravity across the gulf but is subdued along the gulf. Figure made in GeoMapApp (geomapapp.org).

way to the Salton Trough around 6.5–6.3 Ma (Bennett et al., 2015; Umhoefer et al., 2018), at which time the majority of the plate boundary localized in the GoC and the Guaymas Basin in the central gulf fully transitioned to spreading (Umhoefer et al., 2018). Although the Salton Trough is no longer inundated, marine incursion into the trough may have occurred as late as the Pleistocene (Ross et al., 2020). Free-air gravity anomalies suggest an isostatically coherent rifted domain bounded by subaerial flanks extending from the southern gulf to the Salton Trough (Figure 6.11). That the gravity coherency crosses the oceanic-continental crust boundary within the gulf but not across the coastline to either side (Figure 6.11b) suggests an ocean link to a similar isostatic behavior along the full GoC spreading system. Because of the narrowness of the GoC, it should be underscored that the viability of oceanic isostasy as a tectonic driver depends on load (incursion) width and strength of the lithosphere to support that load (Conder, 2022a), High-strength lithosphere and narrow loads are resistant to developing outward mantle tractions. Still, the gravity signal suggests that the entire rift system participates in the system's isostatic mass exchange.

Although the Salton Trough remains a pull-apart basin, it is not wholly clear whether currently submerged northern gulf basins have recently ruptured fully to seafloor spreading or remain in pull-apart mode (Martin-Barajas et al., 2013; van Wijk et al., 2019), as thick sediments obscure straightforward interpretation. In addition to obscuring the northern basins, the thick sediments of the northern gulf have been suggested to affect GoC tectonics. Thick sediments can reduce the buoyancy forces across the rift and tend to narrow rifting and aid the rift-to-drift transition (Bialas & Buck, 2009). However, the rift-to-drift correlation with the sedimentary package in the GoC runs counter to that, suggesting thermal effects may be more important (Martin-Barajas et al., 2013). Although sediments are not efficient as a thermal blanket to weaken the lithosphere (Bialas & Buck, 2009), it may be that sediments will tend to smooth the thermal variations, reducing the degree of differential thermal subsidence and thereby slowing the localization from oceanic isostasy and delaying the rift-to-drift transition.

6.5.4. Rafted Continental Slivers

Many rafted pieces of continental crust litter the world's ocean basins (Ashwal et al., 2017; Grevemeyer et al., 2022; King et al., 2020; Kumar et al., 2019; Nauret et al., 2019; Polteau et al., 2019; Santos et al., 2019; Scotchman et al., 2010). The fluid nature of the ocean can lead to rapid changes in its influence on underlying mantle tractions. In particular, rapid changes in the preferential locus of spreading can occur as the ocean recedes or advances in specific locations. Competing rifts may be inundated at different times. If a weaker rift is inundated first, it may begin to transition only to give way when a water path inundates a stronger competing second rift. For example, the Sao Paulo High presents an exception to the otherwise regular northward progression of the rift-to-drift transition in the South Atlantic (Heine et al., 2013; Moulin et al., 2010) (Figure 6.12). During this stall, the block comprising the rise was transferred from the African plate to the South American plate (Heine et al., 2013; Scotchman et al., 2010). It is plausible that a water path northward crossed an immature rift on the South American side of the block to link to the rift system northward. As spreading matured north and south, a second water path may have inundated a rift on the African side of the block, transferring the locus of extension to the other side (Conder, 2022b). It is not difficult to imagine similar scenarios for other stranded blocks of continental material following the rift-to-drift transition. Although speculative for any individual location, such a process would be an expected implication of inundation playing an active role in seafloor spreading development.

6.5.5. Back-Arc Spreading

Although the GoC is a former back arc that evolved to a spreading system, many current back arcs also have developed spreading (Martinez et al., 2007). Back-arc spreading presents a curious condition of extension in an overall compressional environment. In general, back arcs in subduction systems can be compressional or extensional, with extensional systems potentially developing into spreading systems. Back-arc spreading commonly results from the rifting of an oceanic volcanic arc that generates and migrates into a back-arc basin (Martinez & Taylor, 2003). The difference in systems with back-arc extension and compression has long been couched in terms of low-stress (extensional) "Marianas" type and high-stress (compressional) "Chilean" type subduction, with high-stress more typical for continental arcs and low-stress more frequent for oceanic arcs (e.g., Uyeda, 1987). The variability in subduction and back-arc systems has led to considerations of multiple variables to explain the differences, such as slab dip (Bott et al., 1989) or sediment thickness (Cloos & Shreve, 1996). More refined distinctions, such as accretionary versus erosional margins (e.g., Harris et al., 2014) or tsunamigenic versus non-tsunamigenic (e.g., Bilek & Lay, 2018), have been further advanced. The mechanisms driving the development of back-arc spreading are likewise varied in the literature (Goes et al., 2017; Heuret & Lallemand, 2005; Martinez & Taylor, 2006; Sdrolias & Muller, 2006; Wallace et al., 2005). Still, it is clear that extensional back-arc basins dominate the oceans, whereas compressional back-arcs are more prevalent in continental regions (Heuret &

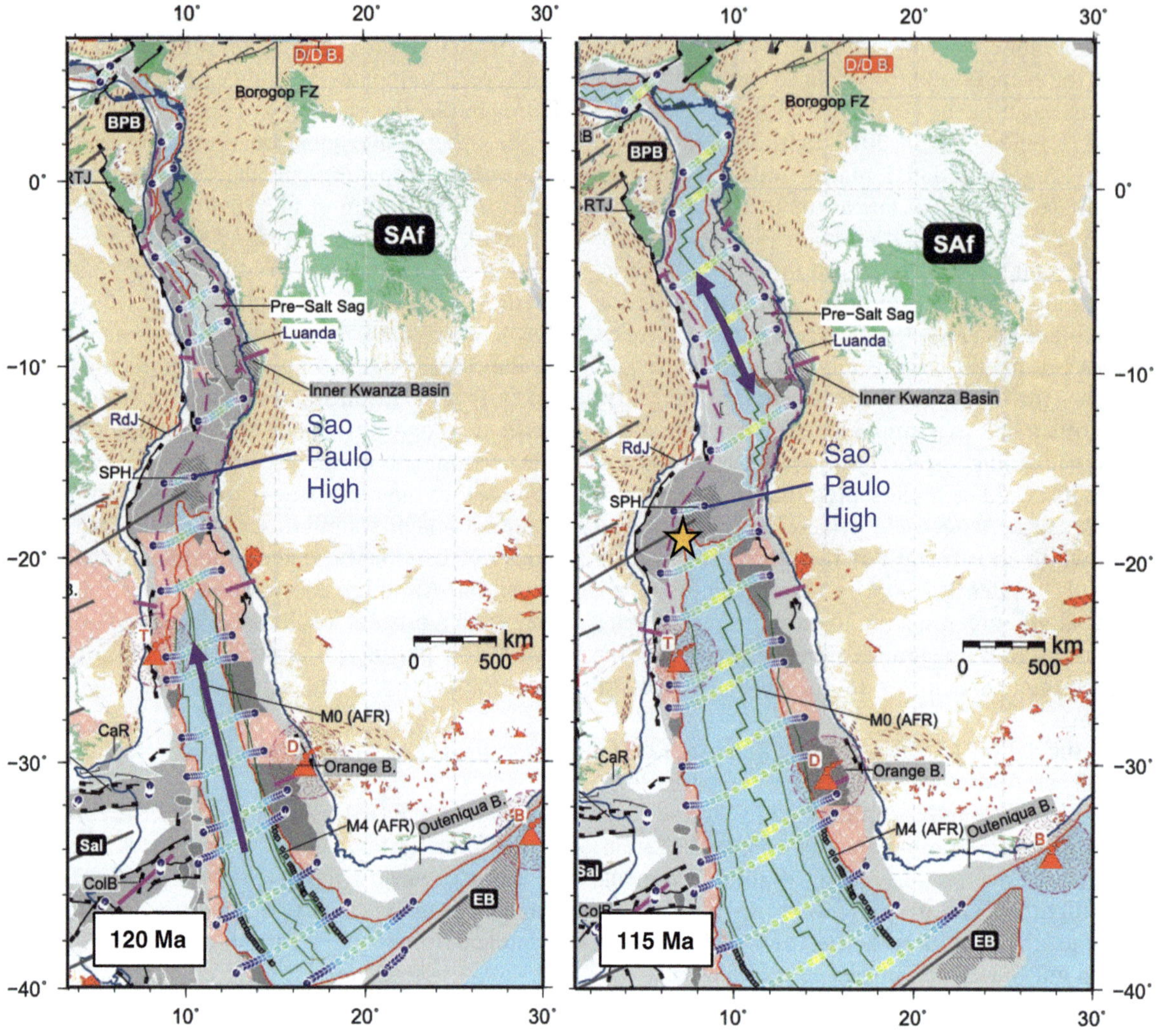

Figure 6.12 South Atlantic plate reconstructions at 120 and 115 Ma from Heine et al. (2013). Purple arrows show seafloor spreading propagation direction. A continuous northward progression stalled at ~120 Ma at the Sao Paulo High (SPH). Spreading picked up a few million years later to the north, again propagating northward but also southward back to the SPH. The connection between the two is made on the African side of the high. Orange star denotes the location of the extinct Abimael propagator (Sandwell et al., 2014) that propagated northward on the South American side of the SPH before succumbing to spreading on the other side of the high.

Lallemand, 2005). Although admittedly appearing somewhat tautological, the presence of ocean in oceanic back arcs may be a contributing factor to the development of back-arc spreading.

An ocean-ocean subduction zone in particular presents a geometrical layout favorable to active mantle tractions for facilitating spreading. The arc presents a strip of crust with thinned lithosphere and elevated temperatures (Ha et al., 2020) and high heat flow (e.g., Von Herzen et al., 2001) and therefore differential thermal subsidence. For largely submerged arc–back arc systems such as the Marianas (Figure 6.13a), Tonga (Taylor et al., 1996), or South Sandwich (Leat et al., 2016), this differential subsidence will tend to drive flow outward into the back arc away from the arc. Direct loading of volcanic deposits will add to the isostatic response, although likely in transient pulses. The resultant mantle tractions will augment existing tensional stresses in the overlying plate and tend to localize extension beneath the arc. If localization and full rupture do occur, the locus of differential subsidence will be strongest at the spreading axis (Figure 6.5), potentially leading to sustained spreading and migration into the back-arc basin with continued crustal accretion.

Pointedly, the rift-to-drift transition in back-arc basins can be fundamentally different than on continental rifted margins (Wang et al., 2019). Inherent differences may arise in that back-arc basins typically begin in a fully submerged state, although continental rifting is frequently

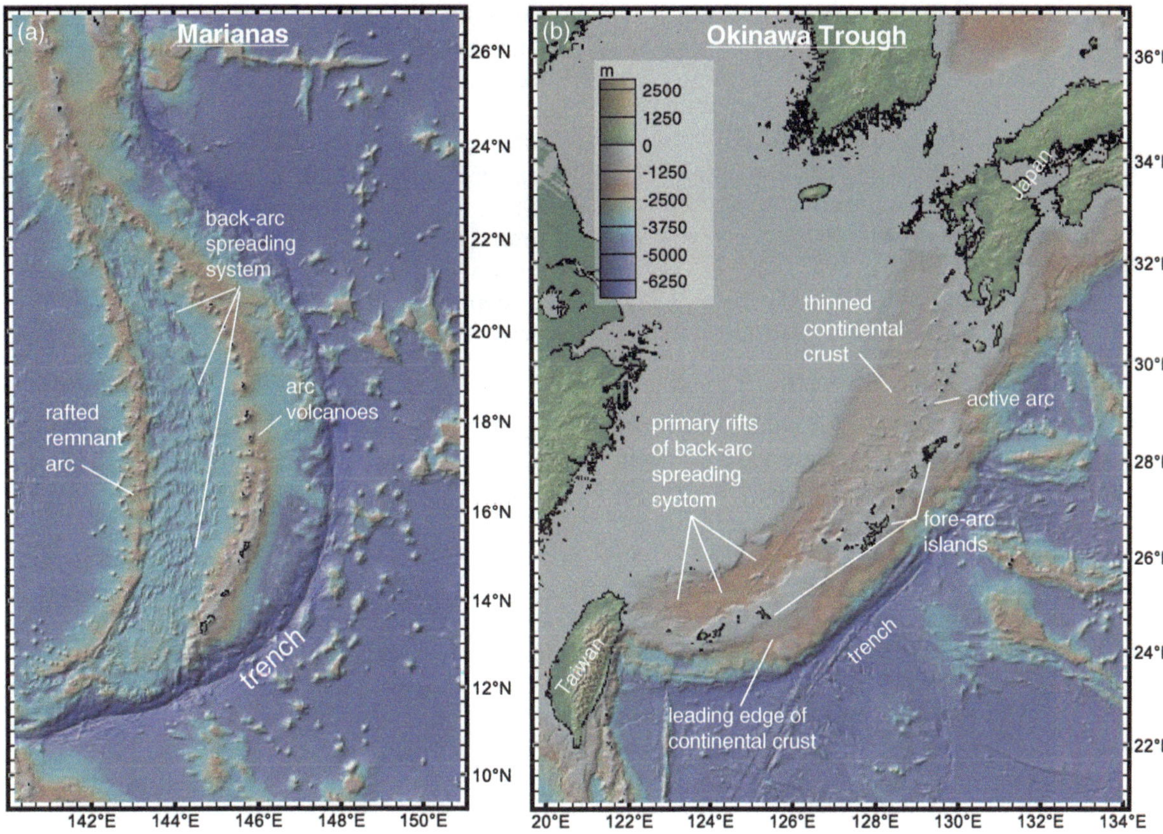

Figure 6.13 Bathymetry of the Marianas (a) and Okinawa Trough (b) back-arc spreading systems. Colorbar is the same for both. The Marianas is a typical ocean-ocean back-arc system. The mostly submerged arc system and remnant arc rafted off with spreading can be clearly observed. The Okinawa Trough is an ocean-continent back-arc system where extension is occurring much like that seen in ocean-ocean systems. Extension has thinned the continental crust to where distinct rift axes can be observed in the southern system. Figures made in GeoMapApp (geomapapp.org).

associated with advancing inundation. Advancing inundation can invoke a transient additional push for rift opening (Conder, 2022a). Such differences may lead to inherently different opening progressions for back arcs relative to continents. For instance, rather than progressive extension leading to rupture (e.g., Brune et al., 2016; Davis & Lavier, 2017; Lavier & Manatschal, 2006), back-arc spreading axes are shown to intrude laterally on diffusely extending regions through propagation or synchronous jumping (Dunn et al., 2013; Taylor et al., 1995; Wang et al., 2019). That an already extensively faulted and extending region preferentially localizes along a crack originating outside the area is peculiar; the implication is that the outside propagating crack has something intrinsic to localization that the already present extensional cracks within do not. That a localized and propagating axis may have outwardly directed mantle tractions that track with its growth in contrast to faults within the area can straightforwardly account for the co-option.

The Okinawa Trough (Figure 6.13b) is an unusual case of back-arc rifting within the continental crust (Arai et al., 2017). Extension in the ~1,000-km-long Okinawa Trough began in the Miocene (Fournier et al., 2001). Current extension rates vary considerably along strike, from ~1 cm/yr in the northern trough to ~5 cm/yr in the south (Nishimura et al., 2004). Extension is diffuse in the slower northern trough and localized along narrow axes in the central and southern troughs (Arai, 2021; Nishizawa et al., 2019). Such structure must be transient as continued spreading will undoubtedly create oceanic crust within the basin. Still, the current literature for the rift-to-drift transition makes little allowance for the possibility of localization within continental crust prior to the transition to spreading. As this is also an unusual case of a submerged continental arc and back arc, this joint occurrence may be a manifestation of the effectiveness of isostatically driven mantle tractions contributing to back-arc spreading development and evolution.

6.6. OCEANIC VERSUS CONTINENTAL TECTONICS

Notwithstanding the remarkable successes of plate tectonics in understanding the history and deformation of the Earth's crust, continental deformation only loosely adheres to the fundamental assumptions of nondeforming plate interiors and narrowly confined boundaries (England & Jackson, 1989; Molnar, 1988; Thatcher, 1995). Many plate boundaries on Earth can be classified as "diffuse" in that they do not fit the classical definition of deformation, being limited to a narrow boundary (Gordon, 1998; Stein & Sella, 2002) (Figure 6.14). For continents, plate boundaries frequently entail wide topographic and seismically active zones, these being the norm for both compressional (e.g., Tibetan plateau, Hindu Kush) and extensional regions (e.g., Basin and Range, portions of the East African Rift) rather than the exception (Gordon & Stein, 1992). Even "narrow" single-valley extensional boundaries like the Main Ethiopian Rift (Corti, 2009) and Rio Grande Rift (Hudson & Grauch, 2013) are several tens of kilometers wide, with 3 : 1 being a high length-to-width aspect ratio for individual basins. Continental strike-slip boundaries like the San Andreas can be 100 km wide or more (Bennett et al., 2002; Thatcher, 1995). In contrast, oceans tend to more closely adhere to the plate tectonic assumptions of rigid interiors and narrowly confined boundaries. This dichotomy presented a large obstacle to accepting the ideas leading to plate tectonics until after the oceans began to be systematically explored in the 20th century (Sandwell, 2001). The increasing quality and quantity of geodetic data, such as GPS, repeatedly expose more regional complexity of continental deformation. The complexity frequently requires additional microplates, tectonic blocks, or diffusely deforming regions to adequately explain the data (Hasterok et al., 2022; Stamps et al., 2021; Wang & Shen, 2020). Although diffuse boundaries are also recognized in the oceans (Gordon, 1998, 2023; Stein & Sella, 2002), they are of a decidedly different nature than those on continents (Figure 6.14). Diffuse oceanic plate boundaries are more subtle, being identified primarily by kinematic closure misfits and/or a moderate number of earthquakes (Stein & Sella, 2002) and characterized by smaller strain rates than other boundary zones (Zatman & Richards, 2002). Primary examples are found in the Indian Ocean (Conder & Forsyth, 2001; Royer & Gordon, 1997; Wiens et al., 1985) and in the Atlantic between the North and South American plates (DeMets & Merkouriev, 2019).

The dichotomy between continental and oceanic tectonics is typically explained in terms of contrasting

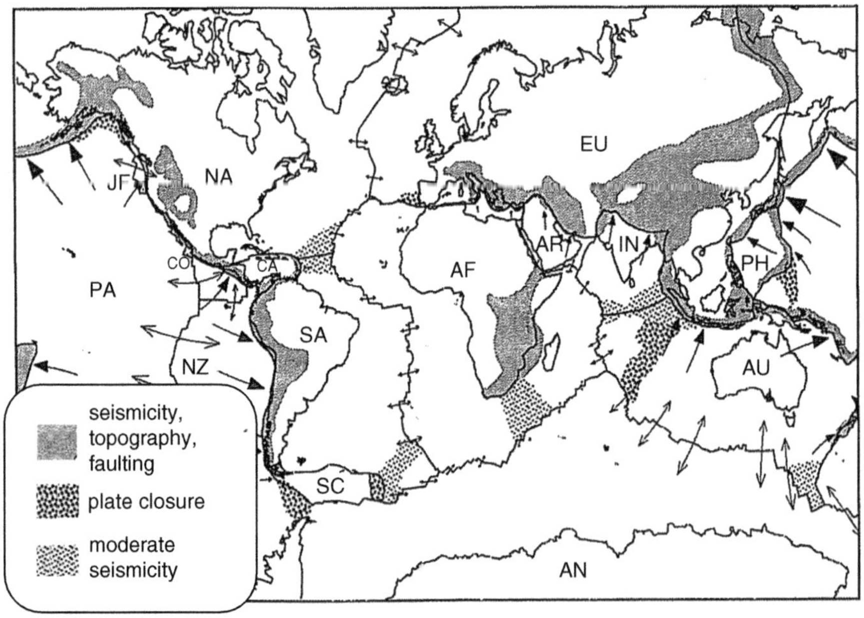

Figure 6.14 Diffuse plate boundaries from Stein and Sella (2002) / with permission of John Wiley & Sons. Stippling indicates diffuse tectonic boundaries as well as their overall character. Broad seismicity, topography, and faulting tend to characterize subaerial plate boundaries. Virtually all subaerial boundaries can be classified as diffuse. A smaller percentage of submarine boundaries can be classified as diffuse. Submarine diffuse plate boundaries tend to be more subtle than subaerial diffuse boundaries and are identifiable by plate closure and/or a moderate number of earthquakes. The mid-ocean ridge system tends to be narrow and nondiffuse.

crustal rheologies, with continental crust being thicker and often having a weak ductile layer in the lower crust (Molnar, 1988; Thatcher, 1995). Although the differences in rheological profiles are undoubtedly important controls on deformation, especially for enabling deformation in the third dimension like crustal thickening (Molnar, 1988), it is worth considering whether the presence of an ocean may also play a role in the dichotomy. As discussed earlier, Iceland's switch from a narrowly confined boundary to wider deformation occurs at the coastline rather than the edge of the submarine platform (Figure 6.7). For mantle tractions to localize spreading centers, there are two requirements: systematic differential subsidence (say, from thermal variations) and an overlying fluid layer to redistribute mass with that differential subsidence. As diffuse oceanic boundaries do not tend to reflect major topographical or thermal variability, there is little facilitation of localizing the boundaries until rifting or some other process imposes a systematically varying thermal structure across the region.

6.7. OUTSTANDING ISSUES

The idea of oceanic isostasy as a facilitator of seafloor spreading offers potential answers to many aspects of our understanding of seafloor spreading and plate tectonics. However, it does pose several questions of its own that are currently unanswered and listed here. This list should be considered representative rather than comprehensive.

Possibly most importantly, the generated mantle tractions need to be considered within the broader system of mantle upwellings and convective circulation. The induced flow adds to the overall flow pattern, biasing mantle tractions toward positive push, but does not require the integrated flow at any given location to be in the plate spreading direction. So, other flows in the asthenosphere driven by plumes or other mechanisms may still be dominant mantle tractions when present. There are numerous places across the globe that show evidence for asthenosphere flow in opposition to plate motion directions (Behn et al., 2004; Conder et al., 2002; Conrad et al., 2007; Ghosh et al., 2013; Naliboff et al., 2009). Perhaps biasing the "average" basal traction to positive is sufficient for the overall spreading system. Or perhaps the fact that the isostatic response will happen as shallowly as possible, whereas convective flows may dominate more deeply, keeps them from acting strongly in opposition. This will only be answered with a better understanding of the interaction of the various flows acting in the mantle.

The physics in Conder (2012) and used here for discussion assumes an axial high at the ridge axis, which is most appropriate for fast-spreading centers. Slow-spreading systems like the Mid-Atlantic Ridge exhibit median valleys (Small et al., 1998). Spreading from within a median valley requires dynamic uplift of young crust as it moves away from the axis prior to thermally subsiding (Phipps Morgan et al., 1987), following a different form of subsidence near the ridge axis than that seen in Figure 6.3. This different form will likely reduce the effectiveness of localization of the plate boundary. Neovolcanic zones within median valleys are typically wider than at fast-spreading centers, sometimes wandering laterally or broadening to 5–10 km within the median valley (Macdonald, 1982). These variations may be due to the lessened efficiency of localization, but to what degree that derives from having rift shoulders rather than monotonic subsidence remains to be determined.

It is clear that not every inundated or submerged rift transitions to seafloor spreading. In some cases, like the Gulf of Suez, it is straightforward as to why: strong lithosphere can support a narrow overlying load, precluding an isostatic response (Conder, 2022a, 2022b). However, locations like the submerged northern Okinawa Trough have wide loads of submergence and likely moderate to low elastic thicknesses and are extending diffusely. It seems possible that the slow rates of opening (Nishimura et al., 2004) may preclude full rupture, although slower spreading systems, such as the Gakkel Ridge in the Arctic (Cochran et al., 2003), have transitioned.

It is fair to say that our understanding of the interplay of oceanic isostasy as a facilitator of horizontal plate motions is in the early stages. There are numerous outstanding questions to be addressed, as well as undoubtedly many as yet unidentified research avenues for understanding the process's contributions and limitations to shaping our planet.

6.8. SUMMARY POINTS

Differential thermal subsidence of the crust in the presence of an overlying ocean can induce a flow in the asthenosphere to adjust for isostatic equilibration. Roughly one-third of oceanic plate gravitational equilibrium must be accommodated in this manner.

This self-sustained, isostatically driven asthenosphere flow is directed outward from the thermally youngest (and most rapidly subsiding) part of the system and may be a necessary component to the occurrence of seafloor spreading.

Mantle tractions induced by oceanic isostasy may lead to highly localized extension in contrast to less localized subaerial extension.

Subaerial spreading is different than submarine spreading in that it is neither self-perpetuating nor may have a strongly localized boundary. Instead, subaerial spreading is likely propelled along by the neighboring spreading system. Although less smooth than seafloor spreading, continual regular extension may be enough to sustain a steady-state magma chamber.

Extensional modes on Iceland contrast more strongly across the coastline than seen across the off-on platform transition or any other closed contour on the island. The coastline acting as the most prominent transition line illustrates a direct impact of the overlying ocean layer on facilitating self-perpetuating, localized seafloor spreading.

In addition to potentially facilitating seafloor spreading, oceanic isostasy may play important roles in the development of other seafloor processes, such as back-arc opening and the existence of stranded continental slivers in the oceans.

The dichotomy of continental versus oceanic tectonics may be viewed at least partially as a dichotomy of subaerial versus submarine tectonics.

Contrary to the long-held view of the ocean as incidental to plate tectonics, the ocean itself may well be a crucial element to the seafloor spreading process and potentially why Earth is the only known planet to exhibit plate tectonics.

ACKNOWLEDGMENTS

I thank the SIU Board of Trustees for granting a sabbatical leave for work on this project. Insightful comments and suggestions from Laurent Montesi and an anonymous reviewer greatly improved this manuscript. This work was partially supported by NSF grant EAR-1753637.

AVAILABILITY STATEMENT

No new data are presented in this chapter. Software for calculating mass fluxes as isostatic response is housed at Zenodo (Conder, 2023) and the SIU open software site: https://opensiuc.lib.siu.edu/geol_comp/.

REFERENCES

Arai, R. (2021). Characteristics of seismicity in the southern Okinawa Trough and their relation to back-arc rifting processes. *Earth, Planets and Space*, *73*. doi: 10.1186/s40623-021-01491-4

Arai, R., Kodaira, S., Yuka, K., Takahashi, T., Miura, S., & Kaneda, Y. (2017). Crustal structure of the southern Okinawa Trough: symmetrical rifting, submarine volcano, and potential mantle accretion in the continental back-arc basin. *Journal of Geophysical Research. Solid Earth*, *122*, 622–641. doi: 10.1002/2016jb013448

Ashwal, L. D., Wiedenbeck, M., & Torsvik, T. H. (2017). Archaean zircons in Miocene oceanic hotspot rocks establish ancient continental crust beneath Mauritius. *Nature Communications*, *8*, 14086. doi: 10.1038/ncomms14086

Bastow, I. D., & Keir, D. (2011). The protracted development of the continent-ocean transition in Afar. *Nature Geoscience*, *4*, 248–250. doi: 10.1038/Ngeo1095

Becker, T. W., & O'Connell, R. J. (2001). Predicting plate velocities with mantle circulation models. *Geochemistry, Geophysics, Geosystems*, *2*. doi: 10.1029/2001gc000171

Behn, M. D., Conrad, C. P., & Silver, P. G. (2004). Detection of upper mantle flow associated with the African Superplume. *Earth and Planetary Science Letters*, *224*, 259–274. doi: 10.1016/j.epsl.2004.05.026

Bennett, R. A., Davis, J. L., Normandeau, J. E., & Wernicke, B. P. (2002). Space geodetic measurements of plate boundary deformation in the western U.S. cordillera. In S. Stein & J. T. Freymueller (Eds.), *Plate boundary zones* (pp. 27–55). Washinton, DC: American Geophysical Union.

Bennett, S. E. K., Oskin, M. E., Dorsey, R. J., Iriondo, A., & Kunk, M. J. (2015). Stratigraphy and structural development of the southwest Isla Tiburon marine basin: implications for latest Miocene tectonic opening and flooding of the northern Gulf of California. *Geosphere*, *11*, 977–1007. doi: 10.1130/Ges01153.1

Bercovici, D. (1998). Generation of plate tectonics from lithosphere-mantle flow and void-volatile self-lubrication. *Earth and Planetary Science Letters*, *154*, 139–151. doi: 10.1016/S0012-821x(97)00182-9

Bercovici, D., Ricard, Y., & Richards, M. A. (2000). The relation between mantle dynamics and plate tectonics: A Primer. In A. Richards, R. G. Gordon, & R. D. van der Hilst (Eds.), *The history and dynamics of global plate motions*. Washington, DC: American Geophysical Union.

Bialas, R. W., & Buck, W. R. (2009). How sediment promotes narrow rifting: application to the Gulf of California. *Tectonics*, *28*. doi: 10.1029/2008tc002394

Bialas, R. W., Buck, W. R., & Qin, R. (2010). How much magma is required to rift a continent? *Earth and Planetary Science Letters*, *292*, 68–78. doi: 10.1016/j.epsl.2010.01.021

Bilek, S. L., & Lay, T. (2018). Subduction zone megathrust earthquakes. *Geosphere*, *14*, 1468–1500. doi: 10.1130/Ges01608.1

Bonvalot, S., Balmino, G., Briais, A., Kuhn, M., Peyrefitte, A., Vales, N., et al. (2012). *World gravity map*. Paris: Commission for the Geological Map of the World.

Bott, M. H. P., Waghorn, G. D., & Whittaker, A. (1989). Plate boundary forces at subduction zones and trench-arc compression. *Tectonophysics*, *170*, 1–15. doi: 10.1016/0040-1951(89)90099-1

Brune, S., Kolawole, F., Olive, J. A., Stamps, D. S., Buck, W. R., Buiter, S. J. H., et al. (2023). Geodynamics of continental rift initiation and evolution. *Nature Reviews Earth and Environment*, *4*(4), 235–253. doi: 10.1038/s43017-023-00391-3

Brune, S., Williams, S. E., Butterworth, N. P., & Muller, R. D. (2016). Abrupt plate accelerations shape rifted continental margins. *Nature*, *536*, 201–204.

Buck, W. (2007). Dynamic processes in extensional and compressional settings: the dynamics of continental breakup and extension. *Treatise Geophys*, *6*, 335–376.

Cawood, P. A., Hawkesworth, C. J., Pisarevsky, S. A., Dhuime, B., Capitanio, F. A., & Nebel, O. (2018). Geological archive of the onset of plate tectonics. *Philosophical Transactions of the Royal Society A: Mathematical, Physical and Engineering Sciences*, *376*, 20170405. doi: 10.1098/rsta.2017.0405

Chauvet, F., Sapin, F., Geoffroy, L., Ringenbach, J. C., & Ferry, J. N. (2021). Conjugate volcanic passive margins in the austral segment of the South Atlantic – Architecture and development. *Earth-Science Reviews*, *212*, 103461. doi: 10.1016/j.earscirev.2020.103461

Cloos, M., & Shreve, R. L. (1996). Shear-zone thickness and the seismicity of Chilean- and Marianas-type subduction zones. *Geology*, *24*, 107–110. doi: 10.1130/0091-7613(1996)024<0107:Sztats>2.3.Co;2

Cochran, J. R., Kurras, G. J., Edwards, M. H., & Coakley, B. J. (2003). The Gakkel Ridge: Bathymetry, gravity anomalies, and crustal accretion at extremely slow spreading rates. *Journal of Geophysical Research. Solid Earth*, *108*. doi: 10.1029/2002jb001830

Collier, J. S., McDermott, C., Warner, G., Gyori, N., Schnabel, M., McDermott, K., & Horn, B. W. (2017). New constraints on the age and style of continental breakup in the South Atlantic from magnetic anomaly data. *Earth and Planetary Science Letters*, *477*, 27–40. doi: 10.1016/j.epsl.2017.08.007

Coltice, N., Gerault, M., & Ulvrova, M. (2017). A mantle convection perspective on global tectonics. *Earth-Science Reviews*, *165*, 120–150. doi: 10.1016/j.earscirev.2016.11.006

Coltice, N., Husson, L., Faccenna, C., & Arnould, M. (2019). What drives tectonic plates? *Science Advances*, *5*. doi: 10.1126/sciadv.aax4295

Conder, J. A. (2012). Non-Pratt component of oceanic isostasy. *Lithosphere*, *4*, 430–434.

Conder, J. A. (2022a). Oceanic isostasy as a trigger for the rift-to-drift transition. *Geology*, *50*, 843–847. doi: 10.1130/G49914.1

Conder, J. A. (2022b). Oceanic isostasy as a trigger for the rift-to-drift transition: Reply. *Geology*, *50*, e555. doi: 10.1130/G50661Y.1

Conder, J. A. (2023). Isostaticmassflux1D. *Zenodo*. doi: 10.5281/zenodo.8360259

Conder, J. A., Forsyth, D. W., & Parmentier, E. M. (2002). Asthenospheric flow and asymmetry of the East Pacific Rise, MELT area. *Journal of Geophysical Research*, *107*, 2344.

Conder, J. A., & Forsyth, D. W. (2001). Seafloor spreading on the Southeast Indian Ridge over the last one million years: a test of the Capricorn plate hypothesis. *Earth and Planetary Science Letters*, *188*, 91–105. doi: 10.1016/S0012-821x(01)00326-0

Conrad, C. P., Behn, M. D., & Silver, P. G. (2007). Global mantle flow and the development of seismic anisotropy: differences between the oceanic and continental upper mantle. *Journal of Geophysical Research. Solid Earth*, *112*. doi: 10.1029/2006jb004608

Conrad, C. P., & Lithgow-Bertelloni, C. (2002). How mantle slabs drive plate tectonics. *Science*, *298*, 207–209. doi: 10.1126/science.1074161

Corti, G. (2009). Continental rift evolution: from rift initiation to incipient break-up in the Main Ethiopian Rift, East Africa. *Earth-Science Reviews*, *96*, 1–53. doi: 10.1016/j.earscirev.2009.06.005

Corti, G., Agostini, A., Keir, D., Van Wijk, J., Bastow, I. D., & Ranalli, G. (2015). Magma-induced axial subsidence during final-stage rifting: implications for the development of seaward-dipping reflectors. *Geosphere*, *11*, 563–571. doi: 10.1130/Ges01076.1

Crameri, F., Conrad, C. P., Montesi, L., & Lithgow-Bertelloni, C. R. (2019). The dynamic life of an oceanic plate. *Tectonophysics*, *760*, 107–135. doi: 10.1016/j.tecto.2018.03.016

Davis, J. K., & Lavier, L. L. (2017). Influences on the development of volcanic and magma-poor morphologies during passive continental rifting. *Geosphere*, *13*, 1524–1540. doi: 10.1130/Ges01538.1

DeMets, C., Gordon, R. G., & Argus, D. F. (2010). Geologically current plate motions. *Geophysical Journal International*, *181*, 1–80. doi: 10.1111/j.1365-246X.2009.04491.x

DeMets, C., & Merkouriev, S. (2019). High-resolution reconstructions of South America plate motion relative to Africa, Antarctica and North America: 34 Ma to present. *Geophysical Journal International*, *217*, 1821–1853. doi: 10.1093/gji/ggz087

Denk, T., Grimsson, F., Zetter, R., & Simonarson, L. A. (2011). The biogeographic history of Iceland—the North Atlantic land bridge revisited. *Topics in Geobiology*, *35*, 647–668. doi: 10.1007/978-94-007-0372-8_12

Druet, M., Munoz-Martin, A., Granja-Bruna, J. L., Carbo-Gorosabel, A., Acosta, J., Llanes, P., & Ercilla, G. (2018). crustal structure and continent-ocean boundary along the galicia continental margin (nw iberia): insights from combined gravity and seismic interpretation. *Tectonics*, *37*, 1576–1604. doi: 10.1029/2017tc004903

Dunn, R. A., Martinez, F., & Conder, J. A. (2013). Crustal construction and magma chamber properties along the Eastern Lau Spreading Center. *Earth and Planetary Science Letters*, *371*, 112–124. doi: 10.1016/j.epsl.2013.04.008

Dymkova, D., & Gerya, T. (2013). Porous fluid flow enables oceanic subduction initiation on Earth. *Geophysical Research Letters*, *40*, 5671–5676. doi: 10.1002/2013gl057798

Einarsson, P. (2008). Plate boundaries, rifts and transforms in Iceland. *Jökull*, *58*, 35–58.

Einarsson, P., Hjartardottir, A. R., Hreinsdottir, S., & Imsland, P. (2020). The structure of seismogenic strike-slip faults in the eastern part of the Reykjanes Peninsula Oblique Rift, SW Iceland. *Journal of Volcanology and Geothermal Research*, *391*, 106372. doi: 10.1016/j.jvolgeores.2018.04.029

Ellis, D., & Stoker, M. S. (2014). The Faroe-Shetland Basin: a regional perspective from the Paleocene to the present day and its relationship to the opening of the North Atlantic Ocean. *Geological Society, London, Special Publications*, *397*, 11–31. doi: 10.1144/Sp397.1

England, P., & Jackson, J. (1989). Active deformation of the continents. *Annual Review of Earth and Planetary Sciences*, *17*, 197–226. doi: 10.1146/annurev.ea.17.050189.001213

Férot, A. (2021). A successful model for interdisciplinary research. *EOS*, https://eos.org/opinions/geoprisms-a-successful-model-for-interdisciplinary-research

Forsyth, D., & Uyeda, S. (1975). On the relative importance of the driving forces of plate motion. *Geophysical Journal International*, *43*, 163–200.

Foulger, G. R., Dore, T., Emeleus, C. H., Franke, D., Geoffroy, L., Gernigon, L., et al. (2020). The Iceland Microcontinent and a continental Greenland-Iceland-Faroe Ridge. *Earth-Science Reviews*, *206*, 102926. doi: 10.1016/j.earscirev.2019.102926

Fournier, M., Fabbri, O., Angelier, J., & Cadet, J. P. (2001). Regional seismicity and on-land deformation in the Ryukyu arc: implications for the kinematics of opening of the Okinawa Trough. *Journal of Geophysical Research: Solid Earth*, *106*, 13751–13768. doi: 10.1029/2001jb900010

Fowler, C. M. R., Fowler, C. M. R., & Fowler, M. (1990). *The solid earth: an introduction to global geophysics*. Cambridge University Press.

Franke, D. (2013). Rifting, lithosphere breakup and volcanism: comparison of magma-poor and volcanic rifted margins. *Marine and Petroleum Geology*, *43*, 63–87.

Geoffroy, L. (2005). Volcanic passive margins. *Comptes Rendus Geoscience*, *337*, 1395–1408. doi: 10.1016/j.crte.2005.10.006

Gerya, T. V., Stern, R. J., Baes, M., Sobolev, S. V., & Whattam, S. A. (2015). Plate tectonics on the Earth triggered by plume-induced subduction initiation. *Nature*, *527*, 221–225. doi: 10.1038/nature15752

Ghomsi, F. E. K., Tenzer, R., Njinju, E., & Steffen, R. (2022). The crustal configuration of the West and Central African Rift System from gravity and seismic data analysis. *Geophysical Journal International*, *230*, 995–1012. doi: 10.1093/gji/ggac089

Ghosh, A., Holt, W. E., & Wen, L. M. (2013). Predicting the lithospheric stress field and plate motions by joint modeling of lithosphere and mantle dynamics. *Journal of Geophysical Research: Solid Earth*, *118*, 346–368. doi: 10.1029/2012jb009516

Goes, S., Agrusta, R., van Hunen, J., & Garel, F. (2017). Subduction-transition zone interaction: a review. *Geosphere*, *13*, 644–664. doi: 10.1130/Ges01476.1

Gordon, R. G. (1998). The plate tectonic approximation: plate nonrigidity, diffuse plate boundaries, and global plate reconstructions. *Annual Review of Earth and Planetary Sciences*, *26*, 615–642. doi: 10.1146/annurev.earth.26.1.615

Gordon, R. G. (2023). Tectonic strain rates, diffuse oceanic plate boundaries, and the plate tectonic approximation. In J. C. Duarte (Ed.), *Dynamics of plate tectonics and mantle convection* (pp. 83–103). Cambridge, MA: Elsevier.

Gordon, R. G., & Stein, S. (1992). Global tectonics and space geodesy. *Science*, *256*, 333–342. doi: 10.1126/science.256.5055.333

Gouiza, M., & Paton, D. (2017). Magma-poor vs. magma-rich continental rifting and breakup in the Labrador Sea, AGU Fall. *Meeting Abstracts*, *2017*. T44A-05

Grevemeyer, I., Ranero, C. R., Papenberg, C., Sallares, V., Bartolome, R., Prada, M., et al. (2022). The continent-to-ocean transition in the Iberia Abyssal Plain. *Geology*, *50*, 615–619. doi: 10.1130/G49753.1

Ha, G., Montesi, L. G. J., & Zhu, W. L. (2020). Melt focusing along permeability barriers at subduction zones and the location of volcanic arcs. *Geochemistry, Geophysics, Geosystems*, *21*, e2020GC009253. doi: 10.1029/2020GC009253

Harris, R. N., Conder, J. A., & Heuret, A. (2014). The thermal structure of the subduction thrust within accretionary and erosive margins. *Tectonophysics*, *633*, 221–231. doi: 10.1016/j.tecto.2014.07.009

Hasterok, D., Halpin, J. A., Collins, A. S., Hand, M., Kreemer, C., Gard, M. G., & Glorie, S. (2022). New maps of global geological provinces and tectonic plates. *Earth-Science Reviews*, *231*, 104069. doi: 10.1016/j.earscirev.2022.104069

Hayford, J. F. (1911). The relations of isostasy to geodesy, geophysics and geology. *Science*, *33*, 199–208.

Heine, C., Zoethout, J., & Müller, R. D. (2013). Kinematics of the South Atlantic rift. *Solid Earth*, *4*, 215–253.

Heuret, A., & Lallemand, S. (2005). Plate motions, slab dynamics and back-arc deformation. *Physics of the Earth and Planetary Interiors*, *149*, 31–51. doi: 10.1016/j.pepi.2004.08.022

Holmes, A. (1931). Radioactivity and earth movements. *Transactions. Geological Society of Glasgow*, *18*, 559–606.

Hudson, M. R., & Grauch, V. J. S. (2013). Introduction. In M. R. Hudson & V. J. S. Grauch (Eds.), *New perspectives on Rio Grande rift basins: from tectonics to groundwater* (pp. v–xii). Geological Society of America. doi: 10.1130/2013.2494(00)

Jackson, M. P. A., Cramez, C., & Fonck, J. M. (2000). Role of subaerial volcanic rocks and mantle plumes in creation of South Atlantic margins: implications for salt tectonics and source rocks. *Marine and Petroleum Geology*, *17*, 477–498. doi: 10.1016/S0264-8172(00)00006-4

Karner, G. D., Johnson, C. A., & Figueredo, P. (2020). *Tectonic significance of passive margin breakup magmatic packages*. Paper presented at GFZ Potsdam, Rift and Rifted Margins Online Seminar.

Karson, J. A. (2017). The Iceland plate boundary zone: Propagating rifts, migrating transforms, and rift-parallel strike-slip faults. *Geochemistry, Geophysics, Geosystems*, *18*, 4043–4054. doi: 10.1002/2017gc007045

King, M. T., Welford, J. K., & Peace, A. L. (2020). Investigating the role of the Galicia Bank on the formation of the North West Iberian margin using deformable plate tectonic models. *Tectonophysics*, *789*, 228537. doi: 10.1016/j.tecto.2020.228537

Kneller, E. A., Johnson, C. A., Karner, G. D., Einhorn, J., & Queffelec, T. A. (2012). Inverse methods for modeling non-rigid plate kinematics: application to Mesozoic plate reconstructions of the Central Atlantic. *Computers & Geosciences*, *49*, 217–230. doi: 10.1016/j.cageo.2012.06.019

Kolzenburg, S., Kubanek, J., Dirscherl, M., Hamilton, C. W., Hauber, E., Scheidt, S. P., & Munzer, U. (2022). Solid as a rock: tectonic control of graben extension and dike propagation. *Geology*, *50*, 260–265. doi: 10.1130/G49406.1

Kumar, P., Mishra, A., Pitchika, V. K., Kumar, S., Dubey, K. M., Singh, D., & Chaubey, A. K. (2019). Integrated geophysical appraisal of crustal structure and tectonic evolution of the Angria Bank, western continental India. *Marine Geophysical Researches*, *40*, 433–449. doi: 10.1007/s11001-019-09383-9

Lavier, L. L., & Manatschal, G. (2006). A mechanism to thin the continental lithosphere at magma-poor margins. *Nature*, *440*, 324–328. doi: 10.1038/nature04608

Leat, P. T., Fretwell, P. T., Tate, A. J., Larter, R. D., Martin, T. J., Smellie, J. L., et al. (2016). Bathymetry and geological setting of the South Sandwich Islands volcanic arc. *Antarctic Science*, *28*, 293–303. doi: 10.1017/S0954102016000043

Lizarralde, D., Axen, G. J., Brown, H. E., Fletcher, J. M., González-Fernández, A., Harding, A. J., et al. (2007). Variation in styles of rifting in the Gulf of California. *Nature*, *448*, 466–469.

Macdonald, K. C. (1982). Mid-ocean ridges—fine scale tectonic, volcanic and hydrothermal processes within the plate boundary zone. *Annual Review of Earth and Planetary Sciences*, *10*, 155–190. doi: 10.1146/annurev.ea.10.050182.001103

Martin, P., van Hunen, J., Parman, S., & Davidson, J. (2008). Why does plate tectonics occur only on Earth? *Physics Education*, *43*, 144–150. doi: 10.1088/0031-9120/43/2/002

Martin-Barajas, A., Gonzalez-Escobar, M., Fletcher, J. M., Pacheco, M., Oskin, M., & Dorsey, R. (2013). Thick deltaic sedimentation and detachment faulting delay the onset of continental rupture in the Northern Gulf of California: analysis of seismic reflection profiles. *Tectonics*, *32*, 1294–1311. doi: 10.1002/tect.20063

Martinez, F., Okino, K., Ohara, Y., Reysenbach, A. L., & Goffredi, S. K. (2007). Back-arc basins. *Oceanography*, *20*, 116–127.

Martinez, F., & Taylor, B. (2003). Controls on back-arc crustal accretion: insights from the Lau, Manus and Mariana basins. *Geological Society, London, Special Publications*, *219*, 19–54.

Martinez, F., & Taylor, B. (2006). Modes of crustal accretion in back-arc basins: inferences from the Lau Basin. *Geophysical Monograph Series*, *166*, 5–30.

Michaud, F., Royer, J. Y., Bourgois, J., Dyment, J., Calmus, T., Bandy, W., et al. (2006). Oceanic-ridge subduction vs. slab break off: plate tectonic evolution along the Baja California Sur continental margin since 15 Ma. *Geology*, *34*, 13–16. doi: 10.1130/G22050.1

Molnar, P. (1988). Continental tectonics in the aftermath of plate-tectonics. *Nature*, *335*, 131–137. doi: 10.1038/335131a0

Moulin, M., Aslanian, D., & Unternehr, P. (2010). A new starting point for the South and Equatorial Atlantic Ocean. *Earth-Science Reviews*, *98*, 1–37.

Mutter, J. C., Talwani, M., & Stoffa, P. L. (1982). Origin of seaward-dipping reflectors in oceanic-crust off the Norwegian margin by subaerial sea-floor spreading. *Geology*, *10*, 353–357. doi: 10.1130/0091-7613(1982)10<353:Oosrio>2.0.Co;2

Naliboff, J. B., Conrad, C. P., & Lithgow-Bertelloni, C. (2009). Modification of the lithospheric stress field by lateral variations in plate-mantle coupling. *Geophysical Research Letters*, *36*. doi: 10.1029/2009gl040484

Nance, R. D., & Murphy, J. B. (2013). Origins of the supercontinent cycle. *Geoscience Frontiers*, *4*, 439–448. doi: 10.1016/j.gsf.2012.12.007

Nauret, F., Famin, V., Vlastelic, I., & Gannoun, A. (2019). A trace of recycled continental crust in the Reunion hotspot. *Chemical Geology*, *524*, 67–76. doi: 10.1016/j.chemgeo.2019.06.009

Nishimura, S., Hashimoto, M., & Ando, M. (2004). A rigid block rotation model for the GPS derived velocity field along the Ryukyu arc. *Physics of the Earth and Planetary Interiors*, *142*, 185–203. doi: 10.1016/j.pepi.2003.12.014

Nishizawa, A., Kaneda, K., Oikawa, M., Horiuchi, D., Fujioka, Y., & Okada, C. (2019). Seismic structure of rifting in the Okinawa Trough, an active backarc basin of the Ryukyu (Nansei-Shoto) island arc-trench system. *Earth, Planets and Space*, *71*. doi: 10.1186/s40623-019-0998-6

O'Neill, C., Marchi, S., Zhang, S., & Bottke, A. (2017). Impact-driven subduction on the Hadean Earth. *Nature Geoscience*, *10*, 793–797. doi: 10.1038/Ngeo3029

Ojakangas, R. W., Morey, G. B., & Green, J. C. (2001). The Mesoproterozoic Midcontinent Rift System, Lake Superior Region, USA. *Sedimentary Geology*, *141*, 421–442. doi: 10.1016/S0037-0738(01)00085-9

Parsons, B., & Sclater, J. G. (1977). Analysis of variation of ocean-floor bathymetry and heat-flow with age. *Journal of Geophysical Research*, *82*, 803–827. doi: 10.1029/JB082i005p00803

Perez-Gussinye, M., Collier, J. S., Armitage, J. J., Hopper, J. R., Sun, Z., & Ranero, C. R. (2023). Towards a process-based understanding of rifted continental margins. *Nature Reviews Earth and Environment*, *4*, 166–184. doi: 10.1038/s43017-022-00380-y

Peron-Pinvidic, G., Manatschal, G., Alves, T., Andersen, T., Andres-Martinez, M., Autin, J., et al. (2019). Rifted margins: state of the art and future challenges. *Frontiers in Earth Science*, *7*, 218.

Phipps Morgan, J., Parmentier, E. M., & Lin, J. (1987). Mechanisms for the origin of midocean ridge axial topography: implications for the thermal and mechanical structure of accreting plate boundaries. *Journal of Geophysical Research: Solid Earth*, *92*, 12823–12836. doi: 10.1029/JB092iB12p12823

Polteau, S., Mazzini, A., Hansen, G., Planke, S., Jerram, D. A., Millett, J., et al. (2019). The pre-breakup stratigraphy and petroleum system of the Southern Jan Mayen Ridge revealed by seafloor sampling. *Tectonophysics*, *760*, 152–164. doi: 10.1016/j.tecto.2018.04.016

Regenauer-Lieb, K., Yuen, D. A., & Branlund, J. (2001). The initiation of subduction: criticality by addition of water? *Science*, *294*, 578–580. doi: 10.1126/science.1063891

Ross, J. E., Kidwell, S. M., Dettman, D. L., Bright, J., Dorsey, R. J., & Jefferson, G. T. (2020). Evidence of pleistocene marine incursions into the salton basin. In D. M. Miller (Ed.), *Changing Facies* (Vol. *2020*). Desert Research Symposium.

Royer, J. Y., & Gordon, R. G. (1997). The motion and boundary between the Capricorn and Australian plates. *Science*, *277*, 1268–1274. doi: 10.1126/science.277.5330.1268

Saemundsson, K., Sigurgeirsson, M. A., & Fridleifsson, G. O. (2020). Geology and structure of the Reykjanes volcanic system, Iceland. *Journal of Volcanology and Geothermal Research*, *391*. doi: 10.1016/j.jvolgeores.2018.11.022

Sandwell, D. T. (2001). Plate tectonics: A martian view. In N. Oreskes (Ed.), *Plate tectonics: an insider's history of the modern theory of the Earth*. Cambridge, MA: Westview Press.

Sandwell, D. T., Muller, R. D., Smith, W. H. F., Garcia, E., & Francis, R. (2014). New global marine gravity model from CryoSat-2 and Jason-1 reveals buried tectonic structure. *Science*, *346*, 65–67. doi: 10.1126/science.1258213

Santos, R. V., Ganade, C. E., Lacasse, C. M., Costa, I. S. L., Pessanha, I., Frazao, E. P., et al. (2019). Dating Gondwanan continental crust at the Rio Grande Rise, South Atlantic. *Terra Nova*, *31*, 424–429. doi: 10.1111/ter.12405

Sapin, F., Ringenbach, J. C., & Clerc, C. (2021). Rifted margins classification and forcing parameters. *Scientific Reports-UK*, *11*.

Scotchman, I. C., Gilchrist, G., Kusznir, N. J., Roberts, A. M., & Fletcher, R. (2010). The breakup of the South Atlantic Ocean: formation of failed spreading axes and blocks of thinned continental crust in the Santos Basin, Brazil and its consequences for petroleum system development. *Petroleum Geology Conference Series*, *7*, 855–866. doi: 10.1144/0070855

Sdrolias, M., & Muller, R. D. (2006). Controls on back-arc basin formation. *Geochemistry, Geophysics, Geosystems*, *7*. doi: 10.1029/2005gc001090

Shillington, D. J., Scott, C. L., Minshull, T. A., Edwards, R. A., Brown, P. J., & White, N. (2009). Abrupt transition from magma-starved to magma-rich rifting in the eastern Black Sea. *Geology*, *37*, 7–10.

Sigmundsson, F., Einarsson, P., Hjartardóttir, Á. R., Drouin, V., Jónsdóttir, K., Árnadóttir, T., et al. (2020). Geodynamics of Iceland and the signatures of plate spreading. *Journal of Volcanology and Geothermal Research*, *391*, 106436.

Small, C., Buck, W. R., Delaney, P. T., Karson, J. A., & Lagabrielle, Y. (1998). *Global systematics of mid-ocean ridge morphology* (pp. 1–25). Washington, DC: American Geophysical Union.

Stamps, D. S., Kreemer, C., Fernandes, R., Rajaonarison, T. A., & Rambolamanana, G. (2021). Redefining East African Rift System kinematics. *Geology*, *49*, 150–155. doi: 10.1130/G47985.1

Stein, S., & Sella, G. F. (2002). Plate boundary zones: concept and approaches. In S. Stein & J. T. Freymueller (Eds.), *Plate boundary zones*. Washington, DC: American Geophysical Union.

Stein, S., Stein, C. A., Elling, R., Kley, J., Keller, G. R., Wysession, M., et al. (2018). Insights from North America's failed Midcontinent Rift into the evolution of continental rifts and passive continental margins. *Tectonophysics*, *744*, 403–421.

Taylor, B., Goodliffe, A., Martinez, F., & Hey, R. (1995). Continental rifting and initial sea-floor spreading in the Woodlark Basin. *Nature*, *374*, 534–537.

Taylor, B., Zellmer, K., Martinez, F., & Goodliffe, A. M. (1996). Sea-floor spreading in the Lau back-arc basin. *Earth and Planetary Science Letters*, *144*, 35–40.

Thatcher, W. (1995). Microplate versus continuum descriptions of active tectonic deformation. *Journal of Geophysical Research: Solid Earth*, *100*, 3885–3894. doi: 10.1029/94jb03064

Tugend, J., Gillard, M., Manatschal, G., Nirrengarten, M., Harkin, C., Epin, M.-E., et al. (2020). Reappraisal of the magma-rich versus magma-poor rifted margin archetypes. *Geological Society, London, Special Publications*, *476*, 23–47.

Turcotte, D. L., & Schubert, G. (2002). *Geodynamics* (2nd ed.). Cambridge: New York: Cambridge University Press.

Umhoefer, P. J. (2011). Why did the southern Gulf of California rupture so rapidly?—Oblique divergence across hot, weak lithosphere along a tectonically active margin. *GSA Today*, *21*, 4–10.

Umhoefer, P. J., Darin, M. H., Bennett, S. E., Skinner, L. A., Dorsey, R. J., & Oskin, M. E. (2018). Breaching of strike-slip faults and successive flooding of pull-apart basins to form the Gulf of California seaway from ca. 8–6 Ma. *Geology*, *46*, 695–698.

Umhoefer, P. J., Plattner, C., & Malservisi, R. (2020). Quantifying rates of "rifting while drifting" in the southern Gulf of California: the role of the southern Baja California microplate and its eastern boundary zone. *Lithosphere*, *12*, 122–132. doi: 10.1130/L1132.1

Unternehr, P., Péron-Pinvidic, G., Manatschal, G., & Sutra, E. (2010). Hyper-extended crust in the South Atlantic: in search of a model. *Petroleum Geoscience*, *16*, 207–215.

Uyeda, S. (1987). Chilean Vs. mariana type subduction zones with remarks on arc volcanism and collision tectonics. In J. W. H. Monger & J. Francheteau (Eds.), *Circum-pacific orogenic belts and evolution of the pacific ocean basin* (pp. 1–7). American Geophysical Union.

van der Lee, S., Regenauer-Lieb, K., & Yuen, D. A. (2008). The role of water in connecting past and future episodes of subduction. *Earth and Planetary Science Letters*, *273*, 15–27. doi: 10.1016/j.epsl.2008.04.041

van Wijk, J., Axen, G., & Abera, R. (2017). Initiation, evolution and extinction of pull-apart basins: implications for opening of the Gulf of California. *Tectonophysics*, *719*, 37–50. doi: 10.1016/j.tecto.2017.04.019

van Wijk, J. W., Heyman, S. P., Axen, G. J., & Persaud, P. (2019). Nature of the crust in the northern Gulf of California and Salton Trough. *Geosphere*, *15*, 1598–1616. doi: 10.1130/Ges02082.1

Von Herzen, R., Ruppel, C., Molnar, P., Nettles, M., Nagihara, S., & Ekstrom, G. (2001). A constraint on the shear stress at the Pacific-Australian plate boundary from heat flow and seismicity at the Kermadec forearc. *Journal of Geophysical Research. Solid Earth*, *106*, 6817–6833. doi: 10.1029/2000jb900469

Wallace, L. M., McCaffrey, R., Beavan, J., & Ellis, S. (2005). Rapid microplate rotations and backarc rifting at the transition between collision and subduction. *Geology*, *33*, 857–860. doi: 10.1130/G21834.1

Wang, M., & Shen, Z. K. (2020). Present-day crustal deformation of continental China derived from gps and its tectonic implications. *Journal of Geophysical Research. Solid Earth*, *125*. doi: 10.1029/2019JB018774

Wang, P., Huang, C.-Y., Lin, J., Jian, Z., Sun, Z., & Zhao, M. (2019). The South China Sea is not a mini-Atlantic: plate-edge rifting vs intra-plate rifting. *National Science Review*, *6*, 902–913.

Weeraratne, D., & Manga, M. (1998). Transitions in the style of mantle convection at high Rayleigh numbers. *Earth and Planetary Science Letters*, *160*, 563–568. doi: 10.1016/S0012-821x(98)00111-3

Wiens, D. A., Demets, C., Gordon, R. G., Stein, S., Argus, D., Engeln, J. F., et al. (1985). A diffuse plate boundary model for Indian-Ocean tectonics. *Geophysical Research Letters*, *12*, 429–432. doi: 10.1029/GL012i007p00429

Worthington, L. L., Shuck, B. D., Becel, A., Eilon, Z. C., & Lynner, C. (2021). Breaking up is hard to do, especially for continents. *Eos*, *102*. doi: 10.1029/2021EO155889

Zatman, S., & Richards, M. A. (2002). On the evolution of motion across diffuse plate boundaries. In S. Stein & J. T. Freymueller (Eds.), *Plate boundary zones*. Washington, DC: American Geophysical Union.

7

Rigidity-Seismicity Relation in the East African Rift System

Bernhard Weise[1] **and Rezene Mahatsente**[2]

ABSTRACT

The relationship between long-term mechanical behavior of the lithosphere and short-term deformation in the East African Rift System (EARS) is assessed based on thermal and rigidity models. The thermal and rigidity models were compared with earthquake distributions along the eastern and western branches of the EARS. The results of the comparison have shown that earthquakes along the Red Sea, the Afar region, and the rest of the eastern branch occur in regions of high temperature and low rigidity, whereas the earthquakes in the western branch are associated with the cold and rigid parts of the rift. Thus, earthquakes in the EARS can occur in regions of high rigidity and cold geothermic condition as well as in areas of low rigidity and hot geothermic condition. The density distribution in the EARS is consistent with predicted geothermic conditions. The broad residual mantle gravity anomalies in the eastern branch are distinctly negative, indicating a low-density mass in the uppermost mantle, whereas the less volcanic western rift is characterized by positive residual anomalies. This is attributed to a cold and thick lithosphere. The broad negative anomaly in the eastern branch may have a thermal origin and its southern limit is within the Kenyan Rift.

7.1. INTRODUCTION

The East African Rift System (EARS) is a series of several-thousand-kilometers-long aligned successions of rift zones. The rift system can be divided into two main branches. The eastern branch of the EARS stretches from the Afar region in the north through the Main Ethiopian Rift (MER) and Kenyan Rift toward the North Tanzanian Divergence in the south (Figure 7.1). The eastern branch of the EARS is evolving to form a divergent lithospheric boundary (Ghebreab, 1998). The western branch begins at Lake Albert in the north and runs through Lake Malawi in the south; this branch of the rift stretches along the basins of Lake Edward, Lake Kivu, Lake Tanganyika, and Lake Rukwa. The western branch turns from a NNE-SSW direction in the northern part to NW-SE in the south. The rift valleys are separated from each other by uplifted plateaus.

North of the Afar Triangle, rifting takes place throughout the Red Sea (Chorowicz, 2005). The EARS and the rift along the Red Sea form a rift system extending from Mozambique in the south to Israel, Jordan, and Syria in the north.

The EARS is highly segmented and exhibits along-strike variations in tectonics, volcanism, seismicity, and crustal thinning (cf. e.g., Benoit et al., 2006b; Ayele et al., 2006; Yang & Chen 2010, 2010; Ebinger et al., 2013; Halldórsson et al., 2014; Philippon et al., 2014; Civiero et al., 2016; Lavayssière et al., 2018; Glerum et al., 2020). The crust beneath the eastern branch is thinner than normal between the Ethiopian and Somalian plateaus. The thinning of the lithosphere in the northern MER toward the Afar depression may be caused by thermal erosion from mantle upwelling (Maguire et al., 2006; Domingues et al., 2013). The thermal anomaly shifts to the west with depth and may be part of the African Super plume. The origin of hot spot tectonism in Ethiopia is associated with thermal

[1] *School of Geography, Geology and the Environment, University of Leicester, Leicester, United Kingdom*

[2] *School of Geosciences, University of Louisiana at Lafayette, Lafayette, Louisiana, USA*

Extensional Tectonics: Rifting and Continental Extension, Geophysical Monograph 278, First Edition.
Edited by İbrahim Çemen and Elizabeth J. Catlos.

DOI: 10.1002/9781119773757.ch07

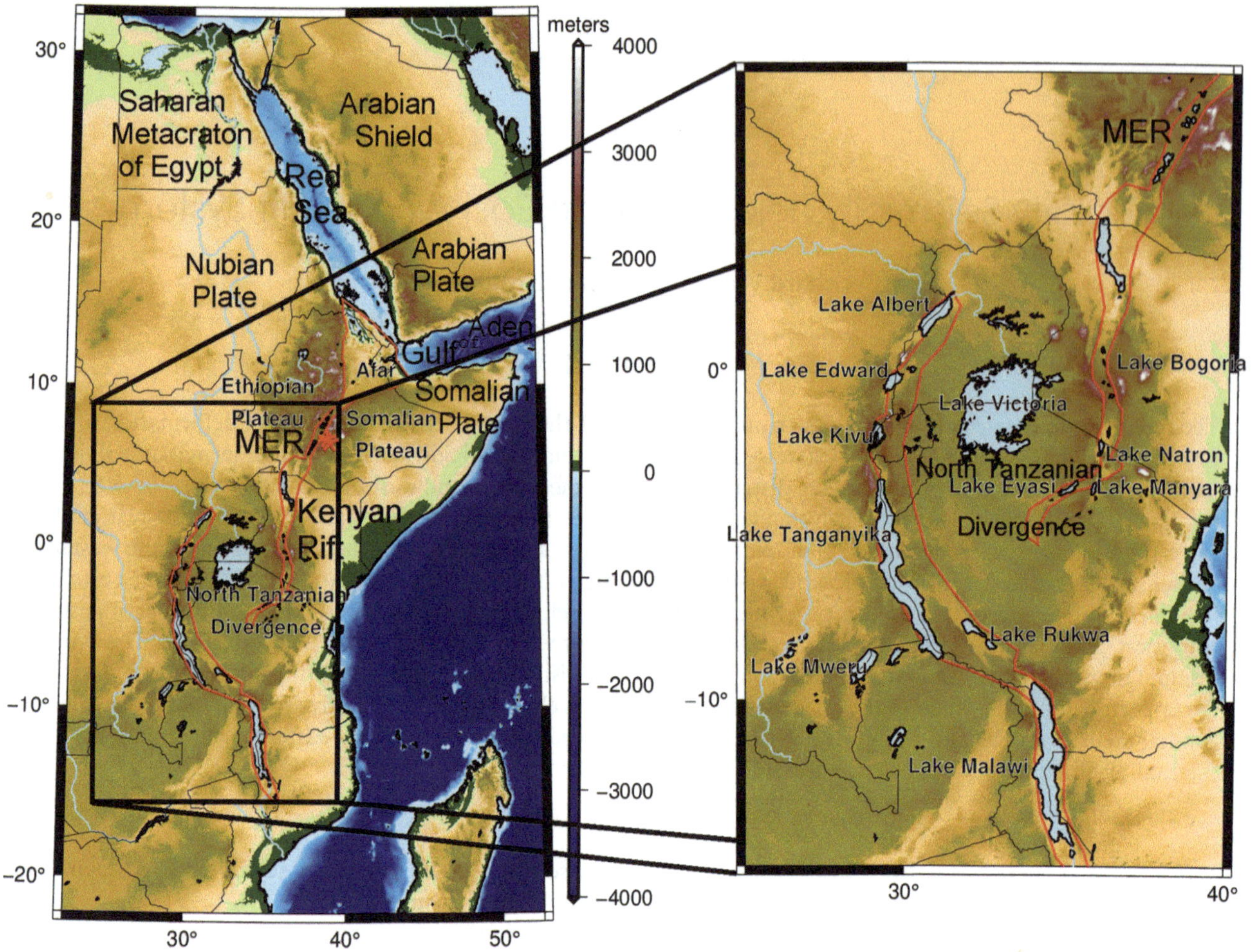

Figure 7.1 Topographic map of eastern Africa and Arabia. Red lines show the western and eastern branches of the East African Rift System (EARS). The elevation data are from the National Oceanic and Atmospheric Administration (Amante & Eakins, 2009). Red stars show locations of boreholes. MER, Main Ethiopian Rift.

upwelling (Benoit et al., 2006a; Civiero et al., 2016; Lavayssière et al., 2018). The existence of deep and large magma intrusions in the upper crust beneath the NW margin of the MER may be suggestive of a large-scale asthenospheric upwelling beneath the uplifted Africa (Mahatsente et al., 1999; Mahatsente et al., 2000; Cornwell et al., 2006; Whaler & Hautot, 2006; Mickus et al., 2007; Mammo, 2013). In contrast to the eastern branch, the western branch is much less volcanic and is underlain by a thick lithosphere (ca. 160 km) with localized low-velocity zones in the uppermost mantle (Fishwick & Bastow, 2011; Accardo et al., 2020). The localized low-velocity anomalies may be attributed to thermal alteration of the lithosphere by rifting (Accardo et al., 2020).

Seismic events and focal depths in the EARS vary along the long segment of the rift. The number of seismic events in the eastern branch is less compared to the western branch. This is due to the increased volcanic activities in the eastern branch (Calais et al., 2008). Although most of the earthquakes in the EARS nucleated in the crust, some deep earthquakes occurred in the mantle (Lindenfeld & Rüempker, 2011). The depths of earthquakes in the eastern branch are confined to the upper crust. The focal depths are <11 km in Afar, Red Sea, and Gulf of Aden and ~23 km beneath the NW Plateau of the MER and the southern part of the eastern branch (Yang & Chen, 2010). The earthquakes in the MER correlate with a strong crust (Muluneh et al., 2020). The strong upper crust beneath the Main Ethiopian and Kenia rifts contributes to the total lithospheric strength of the rifts (Sippel et al., 2017; Muluneh et al., 2018). The earthquakes in the less volcanic western branch have focal depths in the lower crust; the focal depths are >30 km beneath the Malawi and Tanganyika rifts (Yang & Chen, 2010). The deep lower crustal earthquakes in the EARS most likely nucleated in mafic rocks and propagated into surrounding felsic crust (Keir et al., 2009; Hellebrekers et al., 2019). The focal mechanism of earthquakes in the eastern and western branches is mainly extensional faulting (Craig, et al., 2011), and the state of stress is extensional (Mahatsente & Coblenz, 2015).

The seismicity of the EARS may be related to the long-term mechanical properties of the lithosphere. One widely suggested hypothesis is that the short-term deformation of the lithosphere in the EARS, such as earthquakes, may be linked to the total strength of the lithosphere (cf., e.g., Watts & Burov, 2003; Burov, 2011). But this relation is not well understood. In this chapter, we assess the relationship between rigidity and earthquake distribution in the EARS based on thermal and rigidity models and discuss the residual gravity anomalies of the mantle. There are two main goals: (a) to construct thermal and rigidity models of the EARS and (b) to use the models to assess the effects of thermal anomalies, and hence rigidity, on seismicity in the EARS. We present representative thermal and rigidity models of the EARS and the residual gravity anomalies of the mantle and discuss their primary features and seismic implications.

7.2. TECTONIC SETTING

The opening of the Gulf of Aden and the Red Sea started before 20–30 Ma (Ghebreab, 1998; Bosworth, 2015), and this period coincided with a significant increase in magmatism during the Oligocene (Ronney, 2017). Seafloor spreading in the Red Sea started 4–5 Ma ago (Ghebreab, 1998). The exact cause of the initial break-up, the process of extension in the rift, and the nature of the underlying rift floor are unclear. Today, seafloor spreading takes place in the southern region, whereas the northern Red Sea is still in the late continental rifting phase (Ghebreab, 1998; Ayele et al., 2006). The Gulf of Aden is a young ocean basin with the oldest oceanic crust dated back to ~ 20 Ma (Fantozzi & Sgavetti, 1998; D'Acremont et al., 2005). Oceanic rifting in the eastern Gulf of Aden started 18–19 Ma ago (Bosworth & Burke, 2005; Bosworth, 2015).

The Afar Triangle forms a triple junction on the African continent between the Nubian, Somalian, and Arabian plates (Figure 7.1). The elevations are low and at some places under sea level. The rift floor beneath the Afar region is a transition from typical continental to oceanic lithosphere (Makris & Ginzburg, 1987; Mahatsente et al., 1999; Mahatsente et al., 2000; Chorowicz, 2005). The basement is composed of both high-density oceanic and thin continental lithosphere.

The MER, which is part of the eastern branch of the EARS, connects to the southern corner of the Afar depression. Extensional strain affecting the plateau started approximately 11 Ma ago (Mickus et al., 2007). Since then, the MER deepened, and a series of graben segments developed. The rift marks the border between the Nubian and Somalian plates. The deformation migrated toward the center of the rift between 6.5 and 3.2 Ma ago, causing the break-up of the rift into a series of magmatic segments 1.8 Ma ago (Mickus et al., 2007). The rift is asymmetric, and the basin is largely covered with lava flows from volcanism over the past 1.6 Ma. The major fault lies along the eastern side, and the graben fill dips eastward (Casey et al., 2006; Mickus et al., 2007).

South of the MER are the Kenyan Rift and North Tanzanian Divergence where the eastern branch of the EARS splits. The first volcanism and formation of the low areas in the middle of Kenya occurred 23 Ma ago (Chorowicz, 2005). The Kenyan Rift splits into northern and central Kenyan Rift. The northern rift consists of two west-dipping half grabens, both ending at Lake Bogoria. The central rift starts at the southern end of the northern rift valley. In the North Tanzanian Divergence, south of Lake Natron, the Kenyan Rift splits into three branches: Lake Eyasi, Lake Manayara, and the southern plains (Figure 7.1). Southeast of the Kenyan Rift, the EARS continues offshore Tanzania as an undersea basin (Chorowicz, 2005; Saria et al., 2014).

The northernmost segment of the western branch of the EARS follows a NE-SW direction. It forms a succession of half-graben basins, many of them hosting a lake. The central segment of the western branch accommodates Lake Tanganyika and Lake Rukwa. The sediment thickness in Lake Tanganyika is about 4,000–5,000 m. Lake Rukwa is filled with up to 1,200 m of sediments. Southwest of Lake Tanganyika, the rift continues through the segments of Lake Mweru (Ebinger et al., 2017; Daly et al., 2020).

The Malawi Rift in the southern segment of the western branch of the EARS is mostly hidden by Lake Malawi. The rift itself is composed of four basins separated by NW-striking faults. The basins faults were reactivated during Jurassic times as extensional faults. South of Lake Malawi, the EARS is less developed (Albaric et al., 2009).

7.3. DATA

7.3.1. Surface Heat Flow

Surface heat flow is determined indirectly from measurements of temperature gradients in boreholes and thermal conductivities. The surface heat flow data for this study have been downloaded from the Global Heat Flow Database of the International Heat Flow Commission (Figure 7.2a). Along the EARS, there are over 340 data points in the Global Heat Flow Database. The heat flow

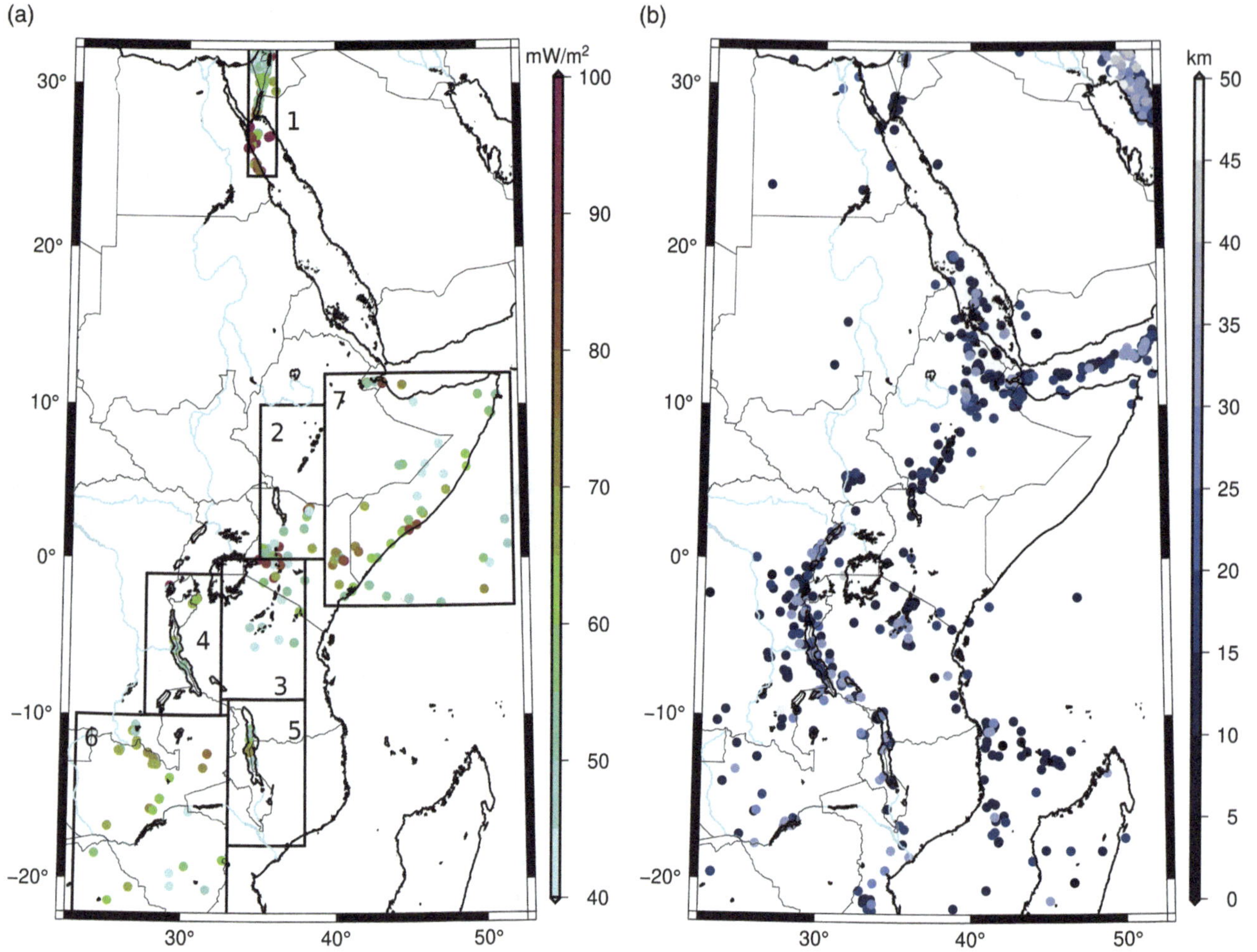

Figure 7.2 (a) Surface heat flow data in the EARS. Black rectangles indicate the compartments used for temperature and rigidity modeling. *Source:* Created using the data from International Heat Flow Commission. (b) Earthquakes registered in the EARS with magnitude 5.0 or higher. *Source:* USGS / Public Domain.

and temperature data in the EARS are from shallow and deep wells and lake-bottom probes. The database itself contains no information about the uncertainties of the heat flow determinations, but standard deviations of 10.7–19.4 mW/m^2 for the heat flow in Proterozoic crust and 37.5 mW/m^2 for Paleozoic regions have been estimated (Davies & Davies, 2010). Few borehole temperature measurements are available in the MER. The temperature data from the bottom of the boreholes drilled in the MER are included in this study (Di Paula, 1983; Hochstein, 1983; Woube, 1986).

7.3.2. Earthquake

Earthquake data were taken from the earthquake database of the U.S. Geological Survey (Figure 7.2b). The database consists of data from published papers and from various catalogs, including data from the National Geophysical Data Center of the National Oceanic and Atmospheric Administration. The data were mostly recorded far from the locations of earthquakes. There are ~4,400 earthquakes in the database for eastern Africa and Arabia. For this study, we considered 600 earthquakes with a magnitude of 5.0 or higher.

7.3.3. Gravity

The gravity data are from the EIGEN-6C4 model (European Improved Gravity model of the Earth; Förste et al., 2014). The EIGEN-6C4 model is based on the WGS84 reference ellipsoid and includes satellite data from GOCE, GRACE, and LAGEOS, as well as surface gravity data. It is calculated up to the 2,190th spherical harmonic degree and order (spatial resolution ~10 km). The data were downloaded from the ICGEM portal (International Center for Global Gravity Field Models).

As a basis for the calculation of the residual gravity anomaly of the mantle, gravity disturbance and the crust and sediment thicknesses in the EARS were used. The gravity disturbance (Figure 7.3) is calculated as the

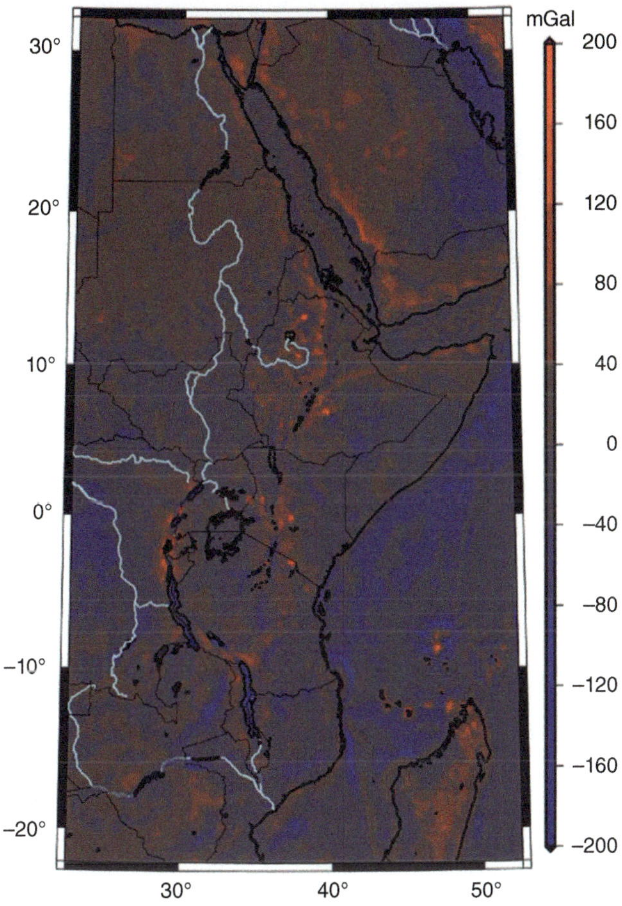

Figure 7.3 Gravity disturbance from the EIGEN 6C4 geopotential model (Förste et al., 2014 / Helmholtz-Zentrum Potsdam-Deutsches GeoForschungsZentrum GFZ / CC BY 4.0).

magnitude of the gradient of the potential at a given point minus the magnitude of the gradient of the normal potential (Barthelmes, 2009).

7.3.4. Crustal and Sediment Thicknesses

The crustal and sediment thicknesses used in this study are based on the LITHO 1.0 model (Pasyanos et al., 2014). This model of the crust and the uppermost mantle was obtained by fitting surface wave dispersion curves. It is tessellated on a 1° grid and contains layer thicknesses and densities.

7.4. METHODS

7.4.1. Lithospheric Geotherm

The model of the lithospheric strength of the EARS was developed using information about the thermal state of the rift and relevant rheological parameters. Therefore, geotherms for a three-layer model (upper and lower crust and lithospheric mantle) were computed from a solution to a one-dimensional steady-state heat conduction equation. (equation (7.1), Chapman, 1986; Turcotte & Schubert, 2002). The assumption of steady-state condition applies to areas where the age of the last tectonothermal event is in the order of tens of million years. To account for the transient response of the recent tectonothermal events in the EARS, the mantle heat flow has been increased. This is justified because the heat flow across the Moho is a function of the Moho temperature for a given crustal thickness and composition (Ranalli & Adams, 2013)

$$\frac{\partial}{\partial Z}\left(K\frac{\partial T}{\partial Z}\right)=-A \qquad (7.1)$$

where T is the temperature, A is the heat production rate, K is the thermal conductivity, and Z is depth.

We chose three distinct conductive layers (upper crust, lower crust, and lithospheric mantle) to develop the thermal models of the EARS because the continental geotherms are distinctively sensitive toward changes in the heat production in each layer. Moreover, detailed information on the composition and structure of the three layers in the EARS is available (Tables 7.1 and 7.2). The thicknesses of the layers are constrained by velocity and gravity models from previous seismic and gravity studies (Maguire et al., 2006; Mickus et al., 2007; Hosny & Nyblade, 2016; Saleh et al., 2006; Njinju et al., 2019; Borrego et al., 2018).

The study area was subdivided into seven different compartments dependent on the distribution of heat flow data. The sectors can have different properties, such as the composition and rheology of the three layers or compressional or extensional strain. Region 1 roughly covers the heat flow data at the Red Sea, compartments 2 and 3 span the eastern branch of the EARS, and areas 4 and 5 cover the western branch. Compartment 7 stretches over the Somalian Plateau, and compartment 6 is south of the main fault line of the western branch of the rift (Figure 7.2a, Table 7.2). In compartment 2, two borehole temperature measurements are available in the MER. The locations of the boreholes are shown in Figure 7.1. The recorded temperatures at the bottom of the first and second boreholes are 328 °C at 2,200.6 m and 315 °C at 2,144 m, respectively (Hochstein, 1983; Woube, 1986). The depth variation of temperature in the boreholes has been approximated using a linear temperature gradient equation:

$$T(z)=T_0+c\cdot z \qquad (7.2)$$

where T_0 is the temperature at the bottom of the borehole and c is the temperature gradient. The average temperature gradient in compartment 2 for a 100-km-thick continental lithosphere, assuming the temperature of the top of the thermal boundary from borehole measurements and a basal temperature of 1,600 K, is 13 K/km. This temperature gradient was used in equation (7.2).

Table 7.1 Material parameters used for the computation of geotherms and strength envelopes.

	Density (kg/m^3)	E (kJ/mol)	A[illegible] (MPa^n/s)	n	Thermal conductivity (W m^{-1} K^{-1})	Heat production, A (μW/m^3)	References
Quartzite (dry)	2,650	156	6.7×10^{-6}	2.4	2.5	2	Tesauro et al. (2007), Tejero and Ruiz (2002)
Quartzite (wet)	2,640	154	3.2×10^{-4}	2.3	2.5	1.4	Ranalli and Adams (2013)
Diabase (dry)	2,850	485	8.0	4.7	2.6	0.4	Ranalli and Adams (2013)
Diabase (wet)	2,850	260	2×10^{-4}	3.4	2.6	0.4	Ranalli and Adams (2013)
Peridotite (dry)	3,320	532	2.5×10^{4}	3.5	3.1	0.006	Ranalli et al. (2006), Ranalli and Adams (2013)
Peridotite (wet)	3,320	471	2.0×10^{3}	4	3.1	0.006	Ranalli and Adams (2013)

Table 7.2 Properties of the compartments shown in Figure 7.2a.

Region	Northern latitude	Western longitude	Southern latitude	Eastern longitude	Upper crust thickness (km)	Moho depth (km)	References
1	32°N	34.05°E	24.5°N	36°E	18	22	Saleh et al. (2006), Hosny and Nyblade (2016)
2	10°N	35°E	0°N	39.3°E	10	25	Maguire et al. (2006), Mickus et al. (2007)
3	0°N	32.5°E	9°S	38°E	20	35	Maguire et al. (2006), Mickus et al. (2007)
4	1°S	27.5°E	10°S	32.5°E	15	30	Njinju et al. (2019), Borrego et al. (2018)
5	9°S	33°E	18°S	38°E	15	30	Njinju et al. (2019), Borrego et al. (2018)
6	10°S	23°E	23°S	33°E	20	40	Njinju et al. (2019), Borrego et al. (2018)
7	12°N	39.3°E	3°S	52°E	25	35	Saleh et al. (2006), Hosny and Nyblade (2016)

7.4.2. Computation of Lithospheric Strength

The lithospheric strength in the EARS has been determined using thermal models, strain rate, and creep parameters of representative minerals in the continental upper crust, lower crust, and upper mantle. The thermal models are constrained by surface heat flow data. The strength of the lithosphere was computed at every depth, assuming brittle and ductile regimes:

$$\sigma_1 - \sigma_3 = \beta \rho g z (1 - \lambda) \tag{7.3}$$

$$\sigma_1 - \sigma_3 = \sqrt[n]{\left(\frac{\dot{\varepsilon}}{A_c}\right)} e^{\left(\frac{E}{nRT}\right)} \tag{7.4}$$

In equations (7.3) and (7.4), $\sigma_1 - \sigma_3$ is the critical stress difference for shear failure in the brittle and power-law creep in the ductile regime, respectively. β is a factor depending on material parameters (cohesion and friction coefficient) and the orientation of the stress field; ρ is the density of the rock, and λ is the ratio of pore fluid pressure to overburden pressure. The gravitational acceleration is g, and z is the depth. A_c, n, and E are creep parameters; R is the gas constant; $T(z)$ is the temperature; and $\dot{\varepsilon}$ represents the steady-state strain rate (Albaric et al., 2009; Ranalli & Adams, 2013). The strain rate is chosen at 10^{-15}/s, and a hydrostatic pore fluid pressure is assumed ($\lambda = 0.4$; Albaric et al., 2009). β varies from 0.75 to 3.0 for extensional and compressional faulting, respectively (Ranalli & Adams, 2013). Table 7.1 gives an overview of the values used for the various creep parameters. They are based on representative minerals in the continental upper crust, lower crust, and upper mantle. We used the creep parameters shown in Table 7.1 because the average composition of the upper crust is granodioritic to quartz-dioritic, the lower crust is composed of quartz-dioritic to basaltic, and the mantle is peridotitic (Fadaie & Ranalli, 1990). This composition is consistent with seismic velocities and densities in the region.

The strength at any depth is the minimum value provided by the dominant mechanism (brittle or power-law creep). The total depth-integrated lithospheric strength σ_T was determined as follows

$$\sigma_T(z_0) = \sum_{z=0}^{z_0} \sigma(z) \cdot \Delta z \tag{7.5}$$

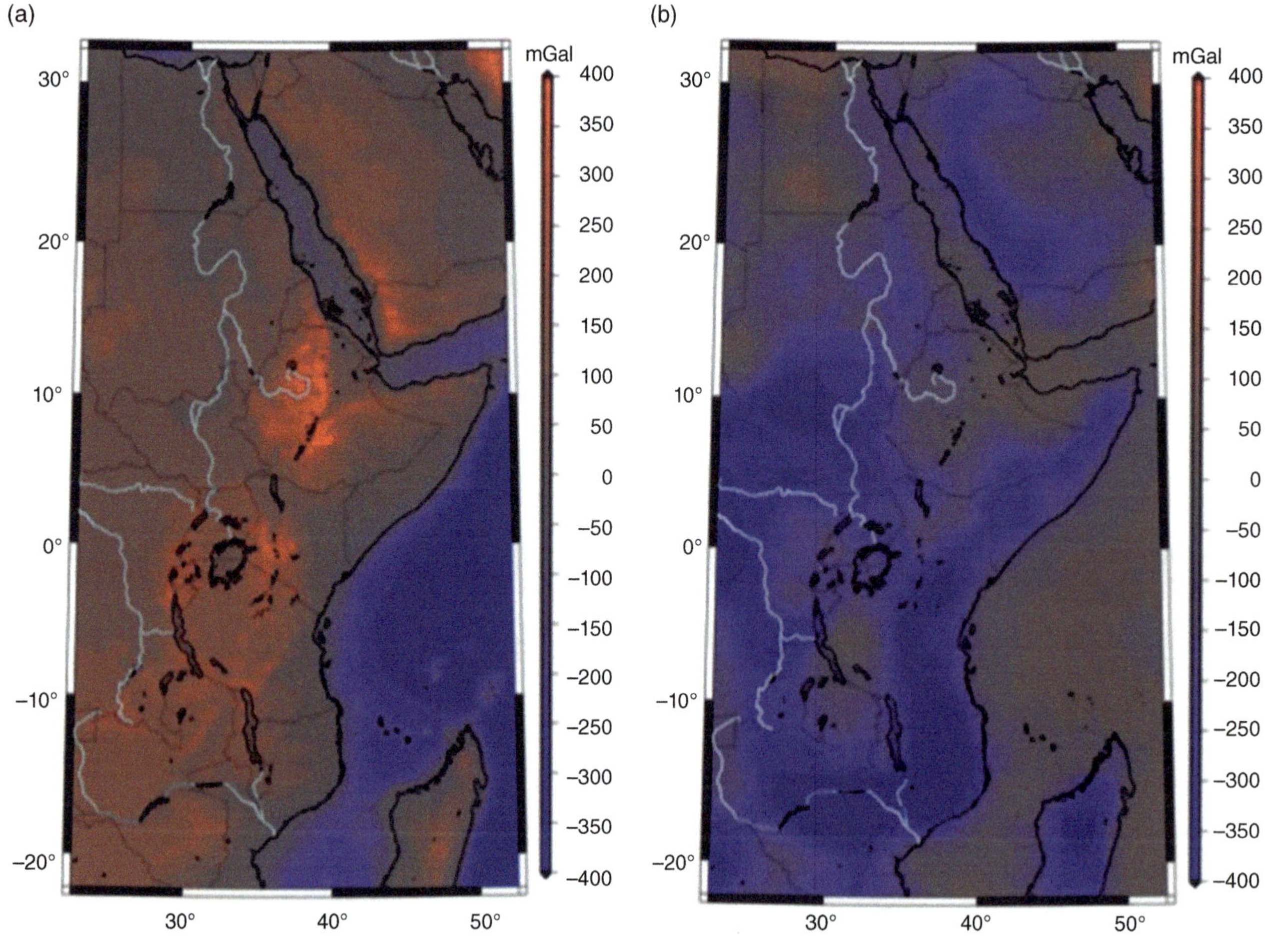

Figure 7.4 Gravity signals of the topography (a), crust (b), and sediment (c) in the EARS.

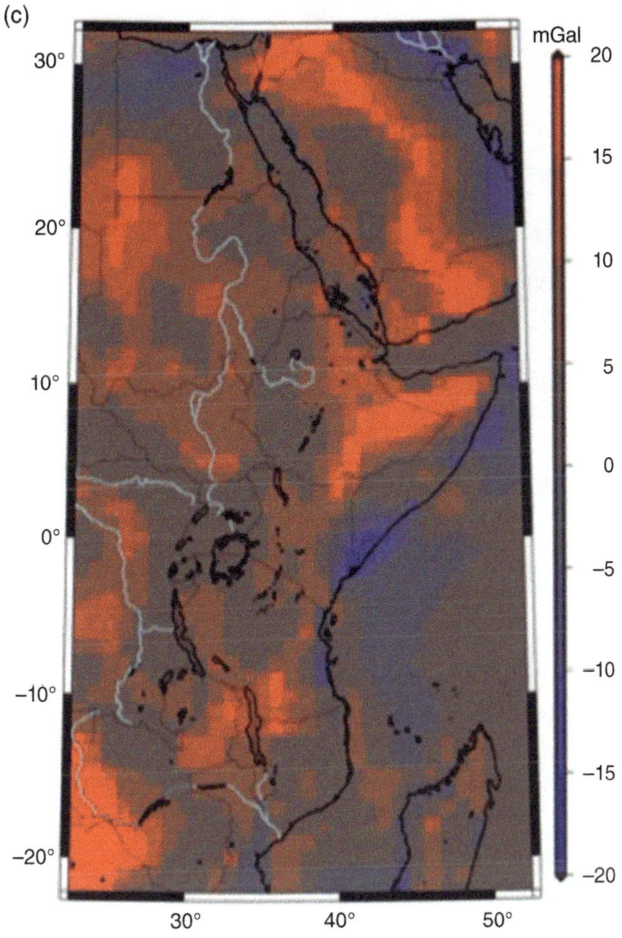

Figure 7.4 (Continued)

where z_0 is the depth at which the total strength was determined and $\sigma = \sigma_1 - \sigma_3$ (Fadaie and Ranalli, 1990).

7.4.3. Computation of Residual Mantle Gravity Anomaly

To retrieve the residual mantle gravity anomaly of the EARS, the gravity effects of the topography, crust, and sediment (Figure 7.4) were removed from the gravity disturbance (Figure 7.3). The topographic gravity effect was determined using the ETOPO1 topographic model (Amante & Eakins, 2009) and the Tesseriods software package (Uieda et al., 2016). A density of 2,670 kg/m^3 was assumed for the topography, and a mean crustal density of 1,630 kg/m^3 was assigned in the ocean. In the same way, using the sediment and crustal thicknesses and densities from the LITHO 1.0 model, the gravity signals of the sediment and the crust were calculated. The mean crustal and sediment densities from the LITHO 1.0 model were used as reference densities.

7.5. RESULTS AND DISCUSSION

7.5.1. Thermal Structure

The crust and lithospheric mantle beneath the eastern and western branches of the EARS have wet and dry rheology, and the two rheologies have different heat production values (Fadaie & Ranalli, 1990). For this reason, we determined the thermal structure of the rifts for dry and wet rheology. The thermal structure of the EARS at 20 and 50 km depth for dry rheology consisting of dry quartzite in the upper crust, dry diabase in the lower crust, and dry peridotite in the lithospheric mantle is shown in Figure 7.5. At 20 km depth, the temperature ranges from 400 to 800 K (Figure 7.5a). The highest temperatures are found along the Red Sea and the Afar Triangle. Slightly lower temperatures can be found in the Main Ethiopian and Kenyan Rifts. The high temperature shown in the region north of Lake Victoria is the effect of interpolation. There is no heat flow data for this region. The

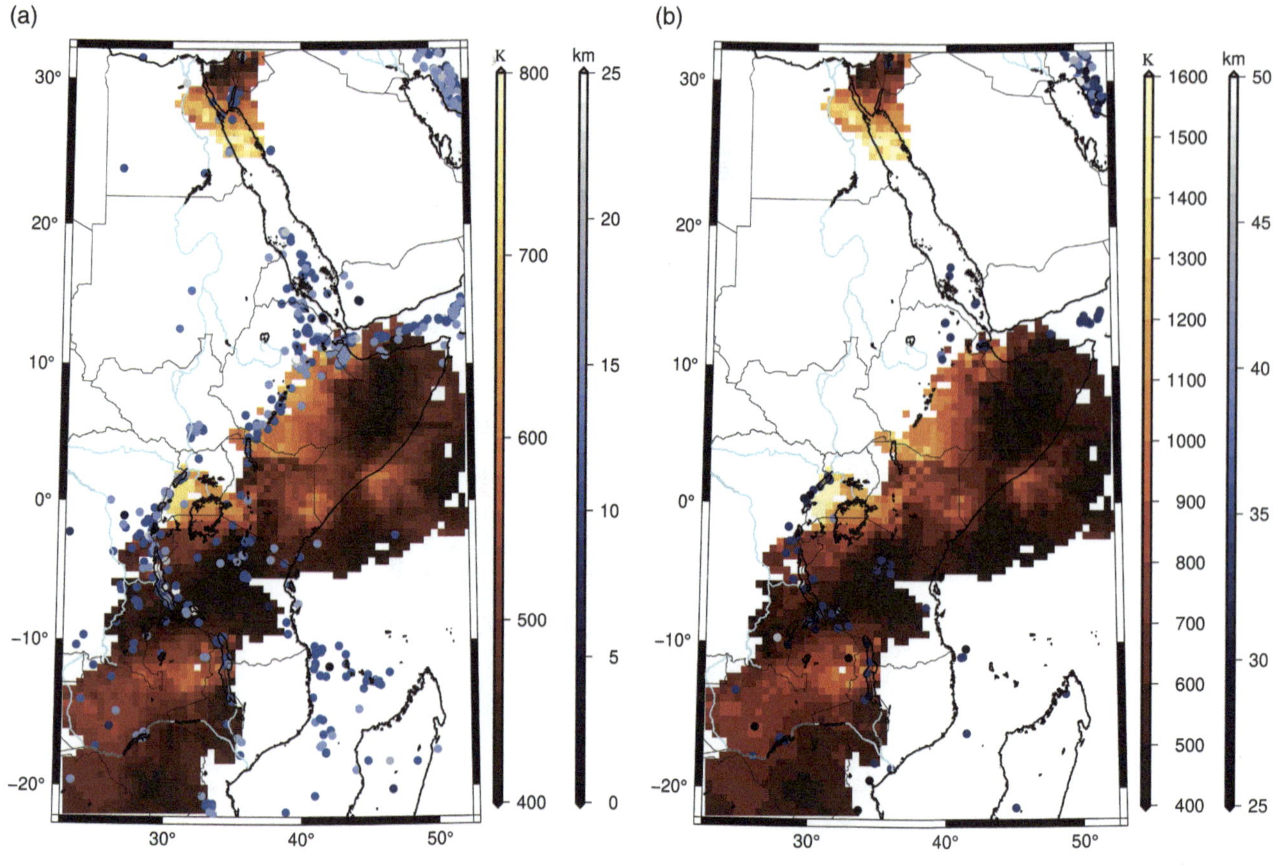

Figure 7.5 Temperature model at depths of 20 (a) and 50 km (b) for a dry rheology. Circles represent earthquakes registered in the EARS at various focal depths. *Source:* USGS / Public Domain.

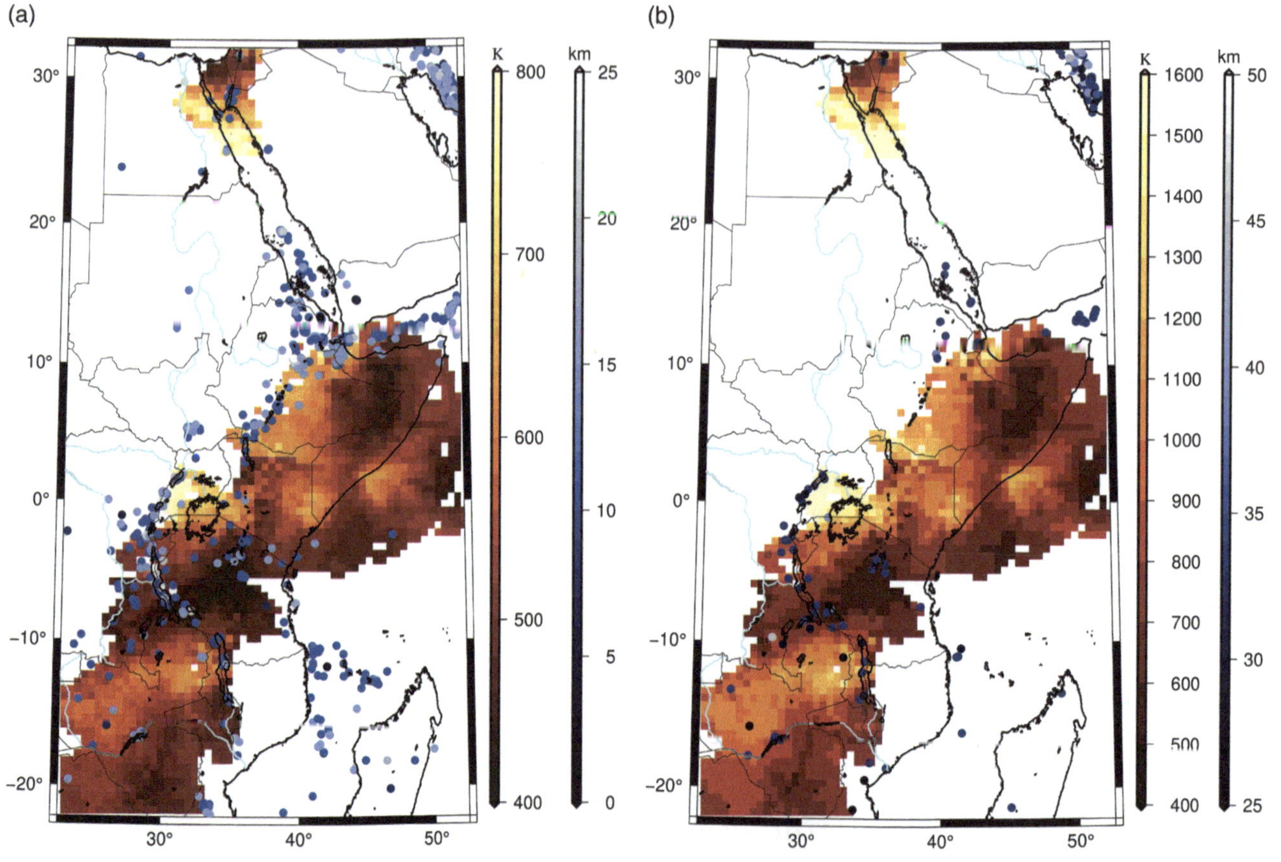

Figure 7.6 Temperature model at depths of 20 (a) and 50 km (b) for a wet rheology. Circles represent earthquakes registered in the EARS at various focal depths. *Source:* USGS / Public Domain.

Somalian Plateau has significantly lower temperatures, as does most of the western branch of the EARS. A similar pattern of low temperature would be suspected for the Ethiopian Plateau. It should be noted that there is no surface heat flow determination for the Ethiopian Plateau. This should be taken into consideration when interpreting the thermal structure and lithospheric strength of this region for all possible tectonic conditions and rheology. West of Lake Malawi, a slightly warmer spot can be seen. At 50 km depth, the pattern of the temperature stays much the same, but the magnitude ranges from 400 to 1,600 K (Figure 7.5b). In the colder regions, the temperature does not change much with depth, but there is a significant increase in the warm areas along the Red Sea and the eastern branch of the EARS. The highest temperatures (>1,600 K) lie well over the solidus temperature of rock and suggest that the lithosphere-asthenosphere boundary (LAB) may be shallower than 50 km in those areas.

For a wet rheology consisting of wet quartzite in the upper crust, wet diabase in the lower crust, and wet peridotite in the lithospheric mantle, the temperature pattern again is the same (Figure 7.6). However, the predicted temperature for wet rheology (Figure 7.6b) is higher by 200 K than the temperature for dry rheology (Figure 7.5b) along the Red Sea, the Afar Triangle, the MER, and the Kenyan Rift. This is not unexpected given some water content in wet quartzite and wet diabase in the upper and lower crust and the downward mobility of radio-elements. In the case of wet rheology, the heat-producing radio-elements can easily reach the base of the Moho and increase the heat production and temperature at the base of the lower crust. The high temperatures in the mantle can be explained by upwelling of hot asthenospheric mantle material under the Afar, northern rift, and Red Sea region. An elongated, wide, and deep-seated low-velocity zone has been imaged in various tomographic models beneath Afar and the northern part of the eastern branch of the EARS (cf. e.g., Benoit et al., 2006a, 2006b; Civiero et al., 2016). The low-velocity zone was interpreted as a thermal anomaly.

7.5.2. Sensitivity of Geotherms

To estimate the sensitivity of the thermal models toward changes in thermal parameters, geotherms were calculated for an average heat flow of 60 mW/m^2 and various heat-production values in the upper crust, lower crust, and lithospheric mantle. The interface between the upper and lower crust lies at 20 km depth, and the Moho depth is 35 km. During the computation of the geotherms, we varied the heat production values in each layer, but we kept the same heat flow value throughout the process.

As shown in Figure 7.7a,b, the geotherm is sensitive to changes in heat production in the upper and lower crust. A change in the heat production in the lithospheric mantle does not significantly change the geotherm (Figure 7.7c). A change in crustal heat production by 25% increases or decreases the estimate of temperature by less than 0.77 K/km (Figures 7.7a,b), which is much lower than the linear temperature gradient in the lithosphere (13.12 K/km for 100-km-thick continental lithosphere with surface and basal temperatures of 288 and 1,600 K, respectively). The same amount of change in heat production in the lithospheric mantle results in neither an increase nor a decrease in temperature (Figure 7.7c). Thus, the effects of the uncertainties in heat production values on thermal models are not significant.

7.5.3. Gravity Signatures of Thermal Anomalies

The residual mantle gravity anomaly map of the EARS, obtained by removing the gravity effects of the topography, crust, and sediment from the gravity disturbance, reflects compositional and thermal heterogeneities in the uppermost mantle (Figure 7.8). The East African Rift, the Arabian Shield, and the Saharan Metacraton in Egypt are characterized by negative residual mantle gravity anomaly. The rest of the regions surrounding the EARS have positive residual mantle gravity anomaly. Positive and negative residual mantle gravity anomalies indicate the presence of high-density mass and low-density mass in the uppermost mantle, respectively.

There are significant differences in the magnitude and shape of the residual mantle gravity anomaly between the eastern and western branches of the EARS. The residual mantle gravity anomaly over the eastern branch is broad, and the value ranges from intermediate to high negative values (ca. −50 to −350 mGal; anomaly A in Figure 7.8), whereas the western branch is characterized by intermediate positive residual mantle gravity value, and the anomaly is narrow and confined within the grabens (ca. 50 mGal; anomaly B in Figure 7.8).

The anomalies in the northern part of the eastern branch, the southern Red Sea, and the Afar Triangle are distinctively negative (Figure 7.8), indicating a low-density mass in the uppermost mantle. The anomalies are associated with asthenospheric upwelling beneath the rift, indicating that the anomalies may have a thermal origin. Tomographic images of the Afar Depression and the MER reveal the existence of an elongated low-velocity zone in the uppermost mantle (Benoit et al., 2006a; Civiero et al., 2016; Adams et al., 2018). The slow upper mantle was interpreted as a broad (>500 km wide) and deep-seated (extends to a depth of >660 km) thermal anomaly. This anomaly is associated with a low-velocity zone that has the core-mantle boundary as its origin (Benoit et al., 2006a; French & Romanowicz, 2015; Civiero et al., 2016; Adams et al., 2018). The thinning of the

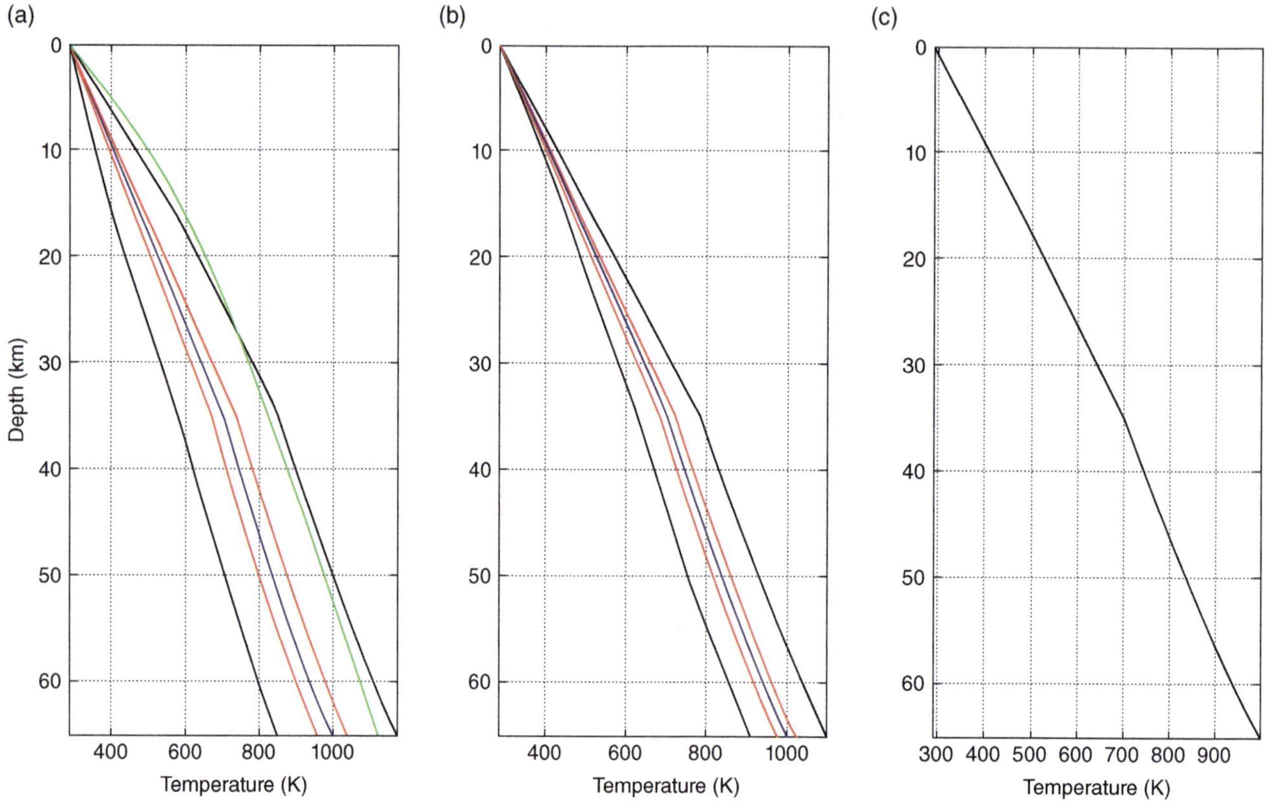

Figure 7.7 Sensitivity of geotherms with changes of the heat production value of $A \pm 25\%$ (red) and $A \pm 100\%$ (black) in the (a) upper crust, (b) lower crust, and (c) lithospheric mantle. The geotherm calculated with a constant thermal conductivity in the upper and lower crust is shown in green. The reference geotherm calculated with heat production values in Table 7.1 is shown in blue.

lithosphere in the eastern branch might be attributed to thermal erosion caused by an upwelling asthenosphere. The lower elevation of the MER and Afar Depression may be an isostatic response to lithospheric thinning observed in the regions (Benoit et al., 2006b; Cornwell et al., 2006; Whaler & Hautot, 2006; Mickus et al., 2007; Dugda et al., 2007). The crust beneath the MER (27–38 km) and Afar Depression (23–25 km) is thinner than the crust below the Ethiopian plateaus (33–44 km; Dugda et al., 2007; Lavayssière et al., 2018).

The western branch of the EARS is characterized by intermediate positive residual mantle gravity anomaly (ca. 50 mGal). The intermediate positive anomaly may be related to the increased lithospheric thickness beneath the western rifts (ca. 160 km; Fishwick and Bastow, 2011).

The residual mantle gravity anomaly along the EARS changes from strongly negative in the eastern branch of the rift to moderately positive in the western branch of the rift system. This may indicate that the broad low-velocity zone or thermal anomaly (>100 km) in the eastern branch may not be continuous beneath the western branch of the rift. The thermal anomaly may have its southern limit in the eastern branch of the EARS; this is probably south of the Kenyan Rift. This interpretation agrees with the results of seismic studies in the EARS. The relatively high Pn velocities (8.35–8.45 km/s) to depths of ~200 km beneath the western branch of the rift system and in the center of the Tanzanian Craton indicate that there are no broad thermal anomalies (>100 km) in the uppermost mantle. This suggests that the thermal modification of the uppermost mantle structures may be limited to the regions located north of the Kenyan Rift (Brazier et al., 2000; Adams et al., 2018).

7.5.4. Lithospheric Strength and the Occurrence of Earthquakes

The relation between rigidity and seismicity in the EARS depends on the thermal state of the lithosphere. We have determined the integrated lithospheric strength of the EARS based on thermal and representative rheological parameters and compared it with earthquake distributions along the eastern and western branches of the rift system.

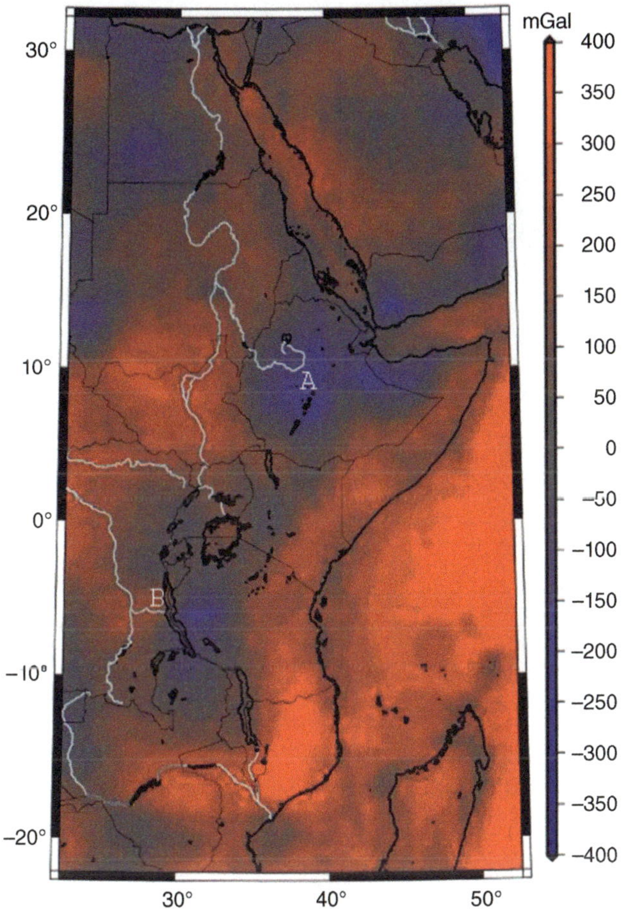

Figure 7.8 Residual mantle gravity anomalies in the EARS.

The rift zones are areas of volatile release. Therefore, hydration of crustal mineral assemblages can affect the geophysical properties of the crust, including the temperature. For this reason, we considered wet and dry rheology in the rigidity and temperature determination. The integrated lithospheric strength of the EARS for dry rheology and extensional tectonic setting is shown in Figure 7.9. The EARS exhibits considerable variations in lithospheric strength along the long segment of the rift. The differences in strength are mainly attributed to thermal and compositional variations. As expected, high rigidity is found in areas with low temperatures. At 20 km depth, low rigidity values can be found along the Red Sea, the Afar region, and the eastern branch of the EARS (Figure 7.9a), which agrees with the low effective elastic thickness determination for the Afar and MER (Pérez-Gussinyé et al., 2009). High strength can be found on the Somalian Plateau and the western branch of the EARS. At 50 km depth, most of the EARS has high rigidity; only the Red Sea and the Afar have lower rigidity (Figure 7.9b). The high rigidity shown in the region north of Lake Victoria is the effect of interpolation. There is no heat flow data for this region, and hence no temperature determination has been made.

The rigidity is weaker for a wet rheology. Especially in the Red Sea, the Afar Triangle, and the eastern branch of the EARS and west of Lake Malawi, the lithospheric strength at 20 km depth is low (Figure 7.10a). The Somalian Plateau has lower strength compared to a lithosphere with dry rheology. At 50 km depth, most parts of the eastern branch of the EARS are characterized by low rigidity compared to a lithosphere with dry rheology (Figure 7.10b).

The plateaus bordering the rifts may well be in compressional tectonics. For this reason, we considered compressional and extensional tectonics in the rigidity determination. We computed lithospheric strength models of the EARS for an extensional tectonic setting in all compartments but compartment 7 (containing the plateaus; see Figure 7.11). For this setting, the Somalian Plateau has a significantly higher rigidity than all other regions in the EARS. The lowest rigidity values can be found in the Red Sea and the Afar region, as well as in the eastern branch of the EARS (Figure 7.11a). The western branch has higher rigidity. At 50 km depth, the region of high rigidity stretches farther south from the Somalian Plateau, and the Afar region has higher rigidity than in other models (Figure 7.11b).

There are significant differences in rheological stratification, crustal structure, and thermal state between the eastern and western branches of the rift. These differences are reflected in the depth distributions of strengths in the upper and lower crustal layers. The strength of the upper crustal layer (at 20 km depth) in the eastern branch is lower than the strength of the same layer in the western branch (Figure 7.9a). This is mainly attributed to the higher temperature below the eastern branch compared to the relatively lower temperature in the western branch. The rheological stratification below the eastern branch may consist of a weak domain at the base of the upper crustal layer overlying the mantle, forming a jelly-sandwich model (Hirth & Kohlstedt, 2013). The strength distributions, however, change with increasing depth. At 50 km depth, the depth-integrated strength in the western branch is lower than the strength in the eastern branch (Figure 7.9b). This may be ascribed to a thick radioactive heat-generating crustal layer below the western branch. The western branch is underlain by a thick lithosphere (Accardo et al., 2020). We have compared our rigidity determination in the eastern rift with the Sippel et al. (2017) study in the same area. We find that the lithospheric strength for the Kenyan Rift ranges from 11 to 14 TN/m (1 Tera Newton = 10^{12} Newton; Figure 7.9b). This agrees with Sippel et al.'s rigidity determination for the Kenyan Rift (12–13.5 TN/m).

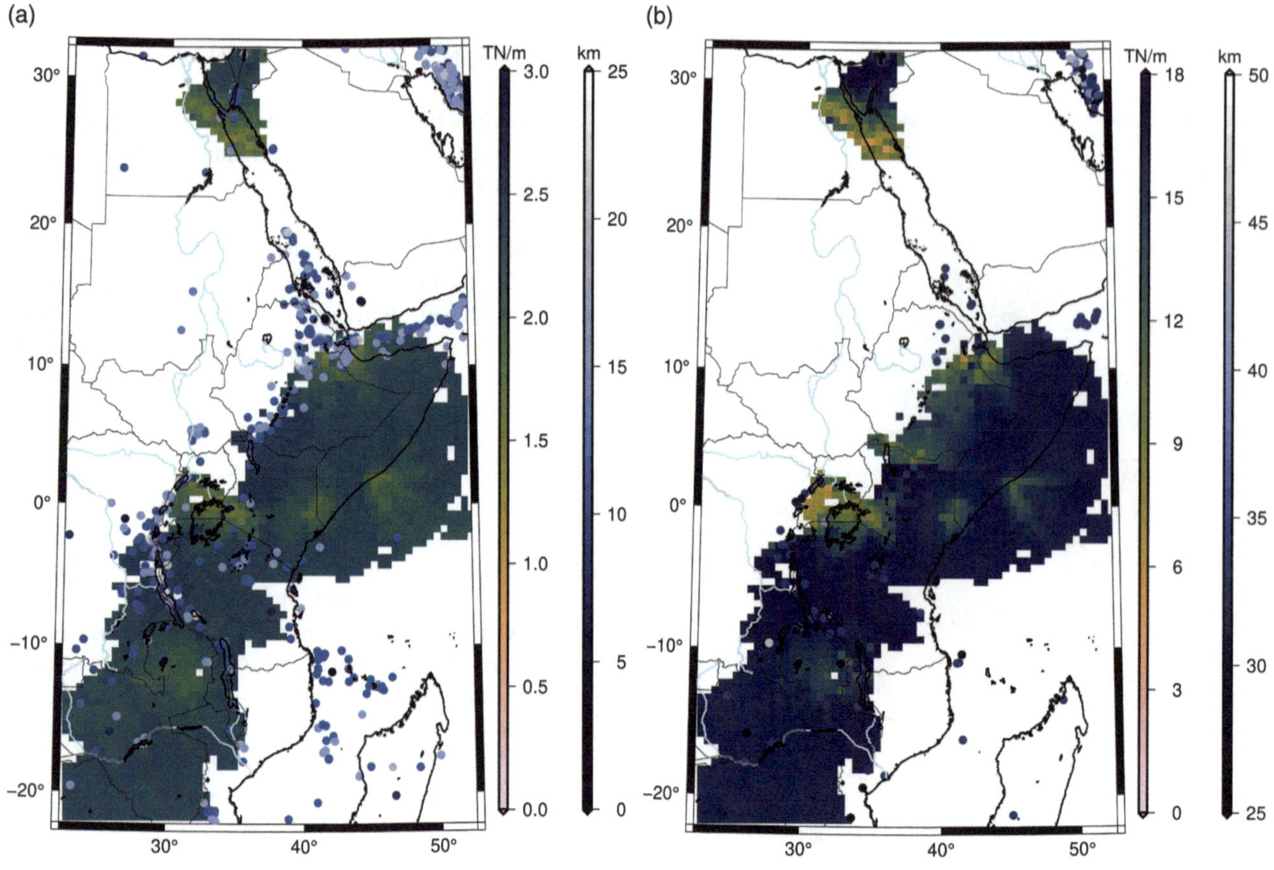

Figure 7.9 Integrated lithospheric strength at depths of 20 (a) and 50 km (b) for a dry rheology and extensional tectonic regime. Circles represent earthquakes registered in the EARS at various focal depths. *Source:* USGS.

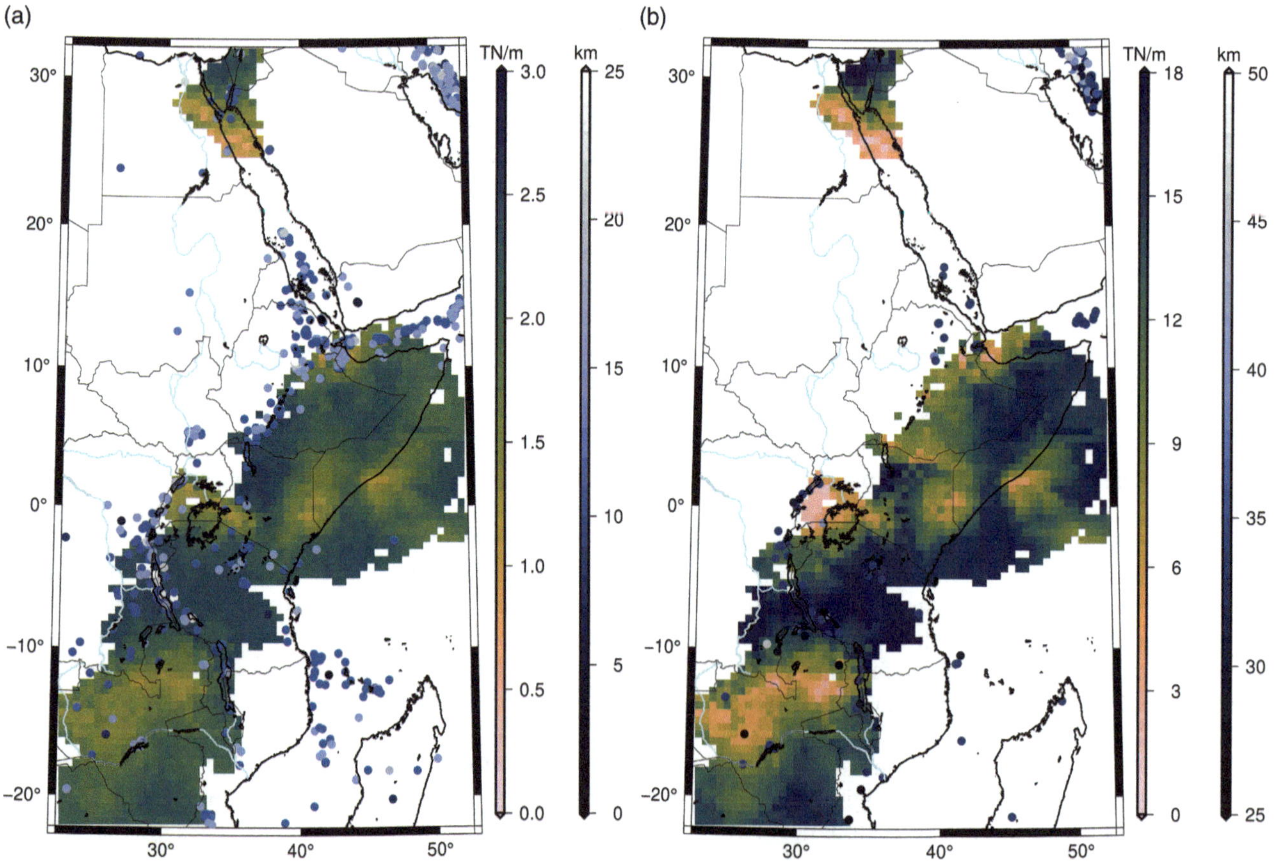

Figure 7.10 Integrated lithospheric strength at depths of 20 (a) and 50 km (b) for a wet rheology and extensional tectonic regime. Circles represent earthquakes registered in the EARS at various focal depths. *Source:* USGS.

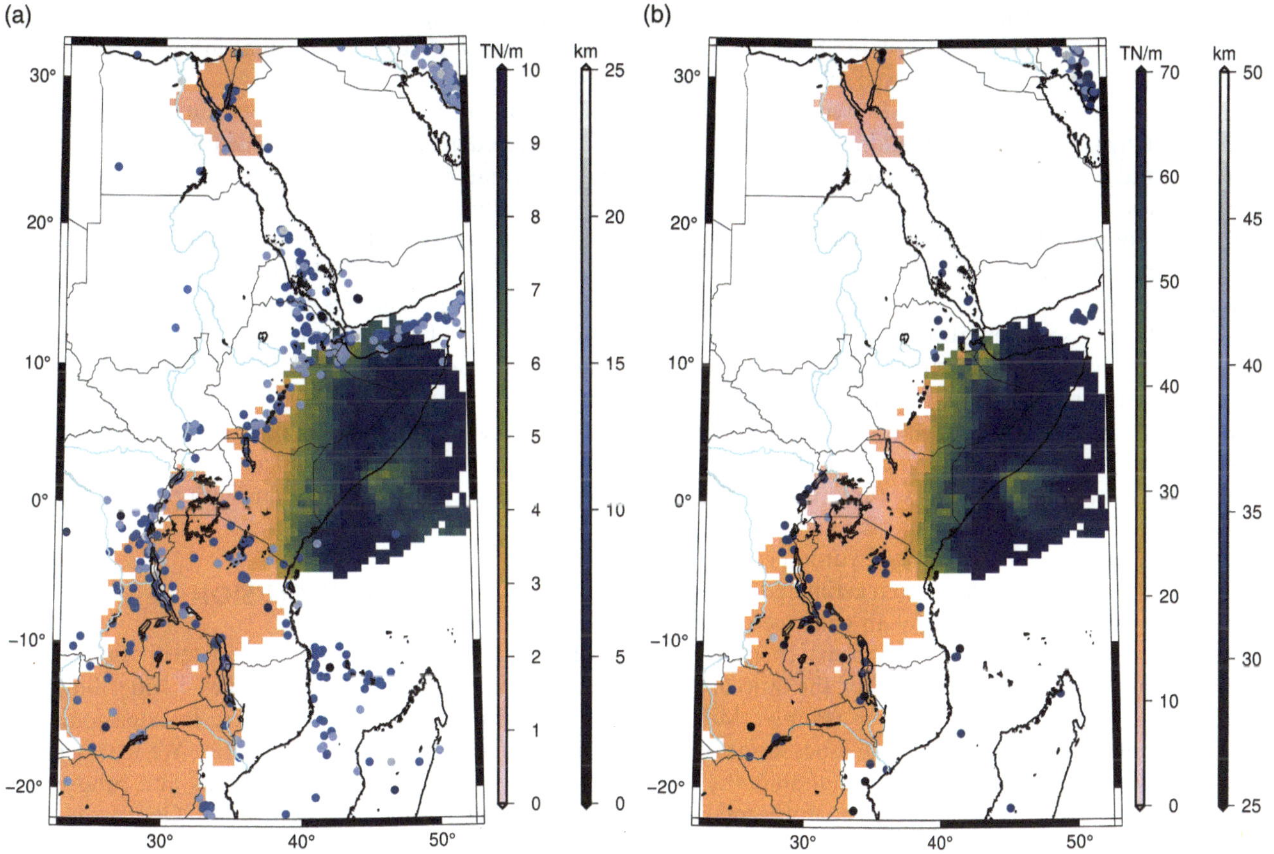

Figure 7.11 Integrated lithospheric strength at depths of 20 (a) and 50 km (b) for a dry rheology, compressional tectonics in compartment 7, and extensional in all other compartments. Circles represent earthquakes registered in the EARS at various focal depths. *Source:* USGS.

The rigidity of lithosphere in the EARS may be linked to seismicity. The Red Sea, the Afar region, and the rest of the eastern branch of the EARS exhibit high temperature, low rigidity, and modest seismic activities (cf. Figures 7.2, 7.5, 7.6, and 7.9–7.11). The less volcanic western branch of the EARS is characterized by low temperatures, high rigidity, and increased seismicity (cf. Figures 7.2, 7.5, 7.6, and 7.9–7.11). In all cases considered, earthquakes occur in regions of high rigidity and cold geothermic condition as well as in areas of low rigidity and hot geothermic condition. However, the number of seismic events and focal depths in the EARS vary along the long segment of the rift depending on the geothermic conditions. The eastern branch experiences a smaller number of seismic events than the less volcanic western branch (Calais et al., 2008). The focal depths of the earthquakes in the eastern branch are confined to the upper crust, whereas the seismic events in the western branch have focal depths in the lower crust (Yang and Chen, 2010).

In our analysis, we considered both wet and dry rheology and extensional and compressional tectonic settings. Consequently, the relation between lithospheric strength and earthquakes in the EARS simply points to the possibility that, if all conditions are met, the long-term mechanical properties of the lithosphere, such as rigidity, may be linked to seismicity.

7.6. MODEL UNCERTAINTY

There are significant variations in the observed heat flow values in the eastern and western branches. This may be related to volcanic activity, hydrothermal circulation, and sedimentary blanketing. Volcanic areas affected by hydrothermal circulation may show lower conductive heat flow values, which can affect predicted temperatures. Heat flux can also vary through time due to sediment compaction. Thick sedimentary deposit effectively works as a thermal blanket in sedimentary basins. The effects of thermal blanketing in sedimentary basins depend on the sedimentation rate. Regions with high sedimentation rates are covered with thick sediments, and this has effects on the observed surface heat flow values. Here we assess the effects of sedimentary blanketing and hydrothermal circulation on the predicted thermal and rigidity models and discuss model uncertainties.

To estimate the effects of sedimentary blanketing on temperature gradient, we assumed an average sedimentation rate of 0.2 cm/yr for the EARS (Degens et al., 1971). The sedimentation rate (0.2 cm/yr) used in this study has been estimated from ^{14}C measurements in the eastern and western branches (Degens et al., 1971). Given the thickness of sedimentary deposits from geophysical studies in the region (e.g., seismic) and sedimentation rate, corrected heat flow values can be determined. This approach is described in detail in Von Herzen et al. (1974). The effects of hydrothermal circulation on thermal models have been estimated by representing the role of convection by an enhanced thermal conductivity $k_e = k_o Nu$, where k_o is normal thermal conductivity and Nu is the Nusselt number, the ratio of convective to conductive heat transfer (Phipps Morgan et al., 1987; Roland et al., 2010). The exact range of the Nusselt number for hydrothermal circulation in the continental rift zones is unknown. For hydrothermal systems in the Mid Oceanic Ridges, the Nusselt number varies from 1 to 8 (Roland et al., 2010). In the present study, various values of Nusselt number Nu (1–5) have been tested. With the choice of $Nu = 2$, the maximum depth of brittle failure, as defined by the 620 K isotherm, is consistent with the observed depth of seismicity. The depth limit of hydrothermal circulation is controlled by the maximum depth of brittle failure (Roland et al., 2010).

We have selected four representative mean heat flow values (40, 60, 80, and 100 mW/m^2) from the eastern and western branches and corrected for effects of sedimentary blanketing and hydrothermal circulation. Figure 7.12 shows geotherms and lithospheric strengths with and without the effects of sedimentary blanketing for four mean heat flow values. The geotherms with corrected heat flow values for sedimentary blanketing have higher temperature gradients than the geotherms without correction (Figure 7.12a). This resulted in a reduction in the integrated lithospheric strength (Figure 7.12b).

The calculated geotherms with and without hydrothermal correction for four representative mean heat flow values are shown in Figure 7.13. The geotherms with corrected heat flow values for hydrothermal effects have higher temperature gradients than the uncorrected geotherms (Figure 7.13a). The corresponding strength envelops show a reduction in total lithospheric strength (Figure 7.13b).

The temperature difference between the uncorrected and corrected geotherms for sedimentary blanketing and hydrothermal circulation in the eastern and western branches ranges from 39 to 125 K. This indicates that the predicted thermal models presented in this study may be colder by an average of 83 K depending on the uncertainties of the parameters used to estimate the effects of sedimentary blanketing and hydrothermal circulation. The corresponding predicted lithospheric strength may be stronger by an average of 0.16 TN/m. The estimated uncertainties in temperature and lithospheric strength may only affect volcanic areas with well-developed hydrothermal systems and regions with rapid sedimentation rates.

7.7. CONCLUSION

The relationship between short-term lithospheric deformation, such as earthquakes, and long-term mechanical properties of lithosphere in extensional tectonic settings is not well understood. We assessed this relationship based on thermal models and lithospheric rigidity, using EARS as a case study. Geotherms were computed using surface heat flow data and relevant thermal parameters, thereby generating a temperature model for the EARS. From the temperature distribution, a model of rigidity was developed for dry and wet rheology and extensional and compressional tectonic settings. The temperature and rigidity models were compared to the earthquake distribution along the EARS. The results have shown that earthquakes occur in regions with high temperatures and low lithospheric strengths, such as the Afar region and other parts of the EARS where relatively low temperatures are observed. In the western branch of the EARS, earthquakes occur mostly along the colder and rigid parts of the rift. In all cases considered in this study, earthquakes occur in regions of high rigidity and low temperature as well as in areas of low rigidity and high temperature.

The residual mantle gravity anomalies of the eastern branch are distinctively negative, indicative of a low-density mass in the uppermost mantle. The broad (>100 km) negative residual mantle gravity anomaly in the eastern branch may have a thermal origin. The high temperature along the eastern branch and the Afar Triangle is attributed to an upwelling hot asthenosphere. The active volcanism, high heat flow, and thin lithosphere beneath this region can be considered as surface manifestations of asthenospheric upwelling.

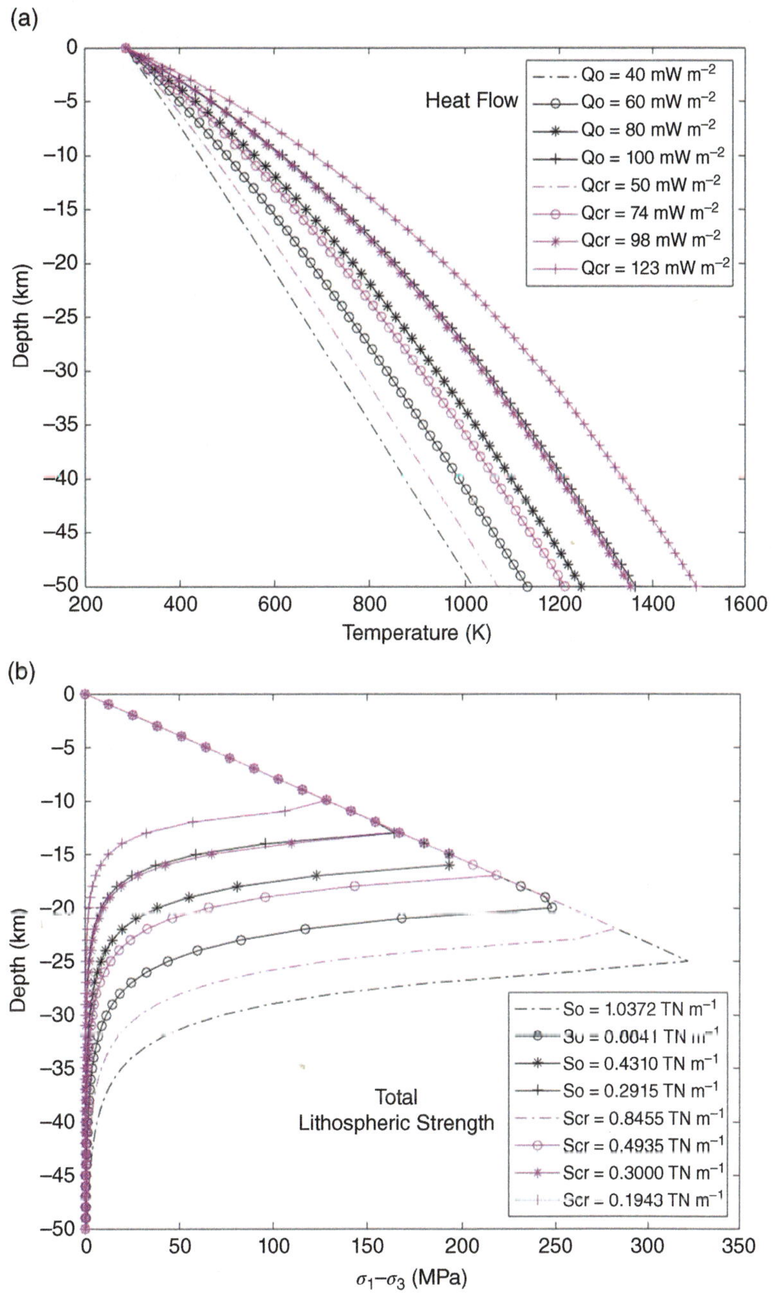

Figure 7.12 (a) Geotherms and (b) lithospheric strengths for four representative mean heat flow values with (magenta) and without (black) the effects of sedimentary blanketing. Q_O, uncorrected heat flow; Q_{Cr}, corrected heat flow; S_O, integrated lithospheric strength for uncorrected heat flow; S_{Cr}, integrated lithospheric strength for corrected heat flow.

The western branch of the EARS is characterized by low temperature and positive residual mantle gravity anomalies, indicative of a cold and thick lithosphere. The presence of positive mantel residual gravity anomalies in the western branch suggests that the broad thermal anomaly (>100 km) in the eastern branch may have a southern limit, probably south of the Kenyan Rift.

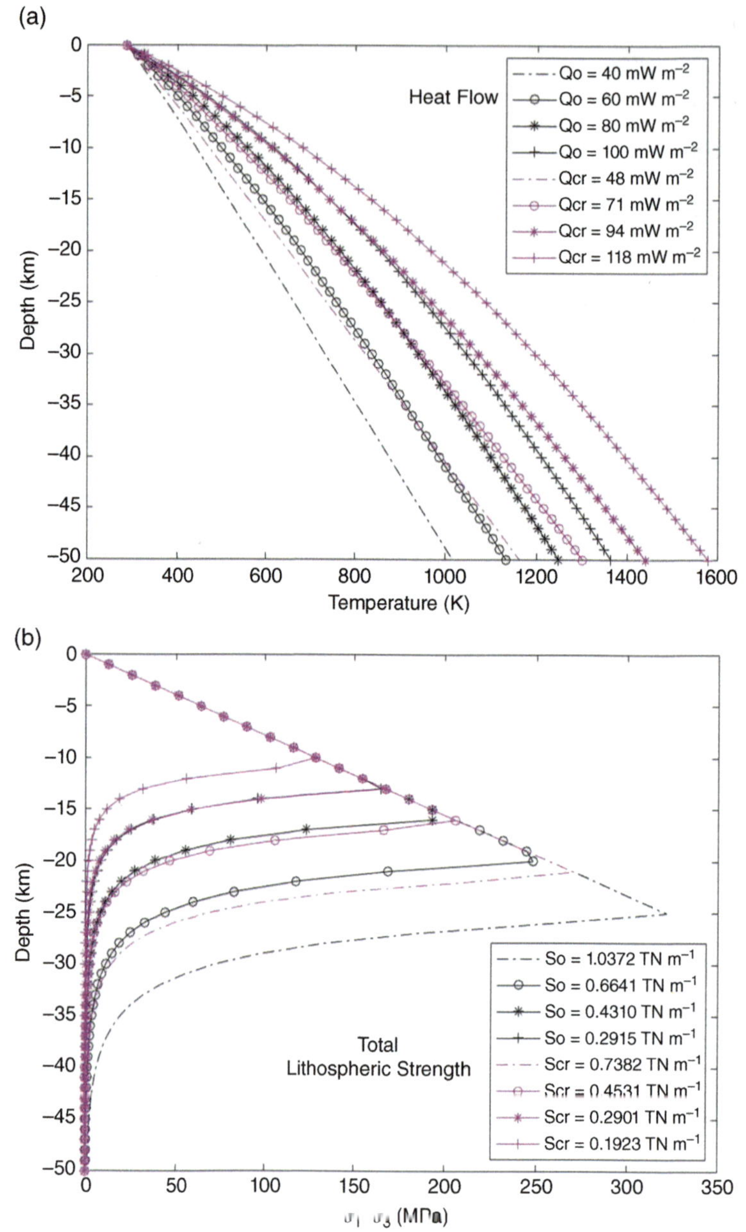

Figure 7.13 (a) Geotherms and (b) lithospheric strengths for four representative mean heat flow values with (magenta) and without (black) the effects of hydrothermal circulation. Abbreviations as in Figure 7.12.

ACKNOWLEDGMENTS

The authors thank Volume Editor Dr. Elizabeth Catlos, and two anonymous reviewers for their thoughtful and constructive comments.

REFERENCES

Accardo, N. J., Gaherty, J. B., Shillington, D. J., Hopper, E., Nyblade, A. A., Ebinger, C. J., et al. (2020). Thermochemical modification of the upper mantle beneath the northern Malawi Rift constrained from shear velocity imaging. *Geochemistry, Geophysics, Geosystems*, *21*, e2019GC008843. doi: 10.1029/2019GC008843

Adams, A., Miller, J., & Accardo, N. (2018). Relationships between lithospheric structures and rifting in the East African Rift System: a Rayleigh wave tomography study. *Geochemistry, Geophysics, Geosystems*, *19*, 3793–3810. doi: 10.1029/2018GC007750

Albaric, J., Déverchère, J., Petit, C., Perrot, J., & Le Gall, B. (2009). Crustal rheology and depth distribution of earthquakes: insights from the central and southern East African Rift System. *Tectonophysics*, *468*, 28–41.

Amante, C. & Eakins, B. W. W. (2009). *ETOPO1 – 1 arc-minute global relief model: procedures, data sources and analysis*

(NOAA Technical Memorandum NESDIS NGDC-24). National Geophysical Data Center, NOAA.

Ayele, A., Nyblade, A. A., Langston, C. A., Cara, M., & Leveque, J.-J. (2006). New evidence for Afro-Arabian plate separation in southern Afar. In G. Yirgu, C. Ebinger, & P. K. H. Maguire (Eds.), *The Afar Volcanic Province within the East African Rift System* (Vol. *259*, pp. 133–141). Geological Society, London, Special Publications.

Barthelmes, F. (2009). *Definition of functionals of the geopotential and their calculation from spherical harmonic models, theory and formulas used by the calculation service of the International Centre for Global Earth Models (ICGEM)* (Scientific Technical Report STR09/02). GFZ Potsdam.

Benoit, M. H., Nyblade, A. A., Owens, T. J., & Stuart, G. (2006a). Mantle transition zone structure and upper mantle S veloctity variations beneath Ethiopia: evidence for a broad, deep-seated thermal anomaly. *Geochemistry Geophysics Geosystems*, *7*, Q11013. doi: 10.1029/2006GC001398

Benoit, M. H., Nyblade, A. A., & Pasyanos, M. E. (2006b). Crustal thinning between the Ethiopian and East African plateaus from modeling Rayleigh wave dispersion. *Geophysical Research Letters*, *33*, L13301. doi: 10.1029/2006GL025687

Borrego, D., Nyblade, A. A., Accardo, N. J., Gaherty, J. B., Ebinger, C. J., Shillington, D. J., et al. (2018). Crustal structure surrounding the northern Malawi rift and beneath the Rungwe Volcanic Province. *East Africa. Geophysical Journal International*, *215*(2), 1410–1426. doi: 10.1093/gji/ggy331

Bosworth, W. (2015). Geological evolution of the Red Sea: historical background, review, and synthesis. In N. M. A. Rasul & I. C. F. Stewart (Eds.), *The Red Sea, Springer Earth system sciences*. Berlin Heidelberg: Springer-Verlag. doi: 10.1007/978-3-662-45201-1_3

Bosworth, W. & Burke, K. (2005). *Evolution of the Red Sea–Gulf of Aden rift system*. In P. J. Post et al. (Eds.), *Petroleum systems of divergent continental margin basins* (CD-ROM, pp. 342–372). SEPM Society for Sedimentary Geology.

Brazier, R. A., Nyblade, A. A., & Langston, C. A. (2000). Pn wave velocities beneath the Tanzania Craton and adjacent rifted mobile belts, East Africa. *Geophysical. Research Letters*, *27*(16), 2365–2368.

Burov, E. (2011). Rheology and strength of the lithosphere. *Marine and Petroleum Geology*, *28*, 1402–1443.

Calais, E., D'Oreye, N., Albaric, J., Deschamps, A., Delvaux, D., Déverchère, J., et al. (2008). Strain accommodation by slow slip and dyking in a youthful continental rift, East Africa. *Nature*, *456*(7223), 783787.

Casey, M., Ebinger, C., Keir, D., Gloaguen, R., & Mohamed, F. (2006). Strain accommodation in transitional rifts: extension by magma intrusion and faulting in Ethiopian rift magmatic segments. In G. Yirgu, C. Ebinger, & P. K. H. Maguire (Eds.), *The Afar Volcanic Province within the East African Rift System* (Vol. *259*, pp. 143–163). Geological Society, London, Special Publications.

Chapman, D. (1986). Thermal gradients in the continental crust. In J. Dawson, D. Carswell, J. Hall, & K. Wedepohl (Eds.), *The nature of the lower continental crust*, No (Vol. *24*, pp. 63–70). Geological Society Special Publication. Denver,.

Chorowicz, J. (2005). The East African Rift System. *Journal of African Earth Sciences*, *43*, 379–410.

Civiero, C., et al. (2016). Small-scale thermal upwellings under the northern East African Rift from S travel time tomography. *J. Geophys. Res. Solid Earth*, *121*, 7395–7408. doi: 10.1002/2016JB013070

Cornwell, D. G., Mackenzie, G. D., England, R. W., Maguire, P. K. H., Asfaw, L. M., & Oluma, B. (2006). Northern Main Ethiopian Rift crustal structure from new high-precision gravity data. In G. Yirgu, C. Ebinger, & P. K. H. Maguire (Eds.), *The Afar volcanic province within the East African Rift System* (Vol. *259*, pp. 307–321). Geological Society, London, Special Publications.

Craig, T. J., Jackson, J. A., Priestley, K., & McKenzie, D. (2011). Earthquake distribution patterns in Africa: their relationship to variations in lithospheric and geological structure, and their rheological implications. *Geophysical Journal International*, *185*(1), 403434. doi: 10.1111/j.1365-246X.2011.04950.x

D'Acremont, E., Leroy, S., Beslier, M. D., Bellahsen, N., et al. (2005). Structure and evolution of the eastern Gulf of Aden conjugate margins from seismic reflection data. *Geophysical Journal International*, *160*(3), 869–890. doi: 10.1111/j.1365-246X.2005.02524.x

Daly, M. C., Green, P., Watts, A. B., Davies, O., Chibesakunda, F., & Walker, R. (2020). Tectonics and landscape of the Central African Plateau and their implications for a propagating Southwestern Rift in Africa. *Geochemistry, Geophysics, Geosystems*, *21*, e2019GC008746. doi: 10.1029/2019GC008746

Davies, J. H., & Davies, D. R. (2010). Earth's surface heat flux. *Solid Earth*, *1*, 5–24.

Degens, E. T., Von Herzen, R. P., & Wong, H. K. (1971). Lake Tanganyika: water uncertainty, sediments, geological structure. *Naturwissnsch*, *58*, 229–241.

Di Paula, G. M. (1983). *Geology and related topics, government of Ethiopia project ETH/78/007 – development of geothermal resources*. UN, New York: Technical Review Meeting Mission Report.

Domingues, A., Chamussa, J., Silveira, G. M., Custodio, S., Lebedev, S., Chang, S., et al. (2013). *Ambient noise tomography of the East African Rift System in Mozambique*. American Geophysical Union, Fall Meeting 2013, abstract #T11F-03.

Dugda, M. T., Nyblade, A. A., & Julia, J. (2007). Thin lithosphere beneath the Ethiopian Plateau revealed by a joint inversion of Rayleigh wave group velocities and receiver functions. *Journal of Geophysical Research*, *112*, B08305. doi: 10.1029/2006JB004918

Ebinger, C. J., Keir, D., Bastow, I. D., Whaler, K., Hammond, J. O. S., Ayele, A., et al. (2017). Crustal structure of active deformation zones in Africa: implications for global crustal processes. *Tectonics*, *36*, 3298–3332. doi: 10.1002/2017TC004526

Ebinger, C. J., Van Wijk, J. W., Keir, D., Lindsey, N., Fischer, T. P., Kattenhorn, S. A., & Roecker, S. W. (2013). *Time and length scales of active deformation in weakly and strongly magmatic rift zones in East Africa (Invited)*. American Geophysical Union. Fall Meeting 2013, abstract #T11F-08.

Fadaie, K., & Ranalli, G. (1990). Rheology of the lithosphere in the East African Rift System. *Geophys. J. Int.*, *102*, 445–453.

Fantozzi, P. L., & Sgavetti, M. (1998). Tectonic and sedimentary evolution of the eastern Gulf of Aden continental margins: new structural and stratigraphic data from Somalia and Yemen. In B. H. Purser & D. W. J. Bosence (Eds.),

Sedimentation and Tectonics of Rift Basins: Red Sea–Gulf of Aden (pp. 56–76). London: Chapman & Hall.

Fishwick, S., & Bastow, I. D. (2011). Towards a better understanding of African topography: a review of passive-source seismic studies of the African crust and upper mantle. *Geological Society, London, Special Publications*, *357*(1), 343_371.

Förste, C., Bruinsma, S., Abrikosov, O., Lemoine, J. M., Marty, J. C., Flechtner, F., et al. (2014). *EIGEN-6C4 The latest combined global gravity field model including GOCE data up to degree and order 2190 of GFZ Potsdam and GRGS Toulouse. GFZ Data Services*. doi: 10.5880/icgem.2015.1

French, S. W., & Romanowicz, B. (2015). Broad plumes rooted at the base of Earth's mantle beneath major hotspots. *Nature*, *525*(7567), 95–99. doi: 10.1038/nature14876

Ghebreab, W. (1998). Tectonics of the Red Sea region reassessed. *Earth-Science Reviews*, *45*, 1–44.

Glerum, A., Brune, S., Stamps, D. S., & Strecker, M. R. (2020). Victoria continental microplate dynamics controlled by the lithospheric strength distribution of the East African Rift. *Nature Communications*, ***11***, 2881. doi: 10.1038/s41467-020-16176-x

Halldórsson, S. A., Hilton, D. R., Scarsi, P., Abebe, T., & Hopp, J. (2014). A common mantle plume source beneath the entire East African Rift System revealed by coupled helium-neon systematics. *Geophys. Res. Lett.*, *41*, 2304–2311. doi: 10.1002/2014GL059424

Hellebrekers, N., Niemeijer, A. R., Fagereng, Å., Manda, B., & Mvula, R. L. S. (2019). Lower crustal earthquakes in the East African Rift System: insights from frictional properties of rock samples from the Malawi Rift. *Tectonophyiscs*, ***767***, 228167. doi: 10.1016/j.tecto.2019.228167

Hirth, G., & Kohlstedt, D. (2013). Rheology of the upper mantle and the mantle wedge: a view from the experimentalist. In J. E. Tewelde (Ed.), *Inside the subduction factory* (Vol. *138*, p. 311). American Geophysical Union. doi: 10.1029/138GM06

Hochstein, M. P. (1983). *Aluto geothermal project* (Geothermal Institute report GIR 008). Geothermal Institute, University of Auckland.

Hosny, A., & Nyblade, A. (2016). The crustal structure of Egypt and the northern Red Sea region. *Tectonophysics*, *687*, 257–267.

Keir, D., Bastow, I. D., Whaler, K. A., Daly, E., Cornwell, D. G., & Hautot, S. (2009). Lower crustal earthquakes near the Ethiopian rift induced by magmatic processes. *Geochemistry, Geophysics, Geosystems*, *10*, Q0AB02. doi: 10.1029/2009GC002382

Lavayssière, A., Rychert, C., Harmon, N., Keir, D., Hammond, J. O. S., Kendall, J.-M., et al. (2018). Imaging lithospheric discontinuities beneath the northern East African Rift using S-to P receiver functions. *Geochemistry, Geophysics, Geosystems*, *19*, 4048–4062. doi: 10.1029/2018GC007463

Lindenfeld, M., & Rümpker, G. (2011). Detection of mantle earthquakes beneath the East African Rift. *Geophys. J. Int*, *186*(1), 1–5, 2011.

Maguire, P.K.H., Keller, G.R., Klemperer, S.L., Mackenzie, G.D., Keranen, K., Harder, S., O'Rerilly, B., Thybo, H., Asfaw, L., Khan, M.A., Amha, M., 2006, Crustal structure of the northern Main Ethiopian Rift from EAGLE controlled-source survey; a snapshot of incipient lithospheric break up. In: Yirgu, G., Ebinger, C. & Maguire, P.K.H. (eds) *The Afar volcanic province within the East African Rift System*. Geological Society, London, Special Publications, *259*, 269–291.

Mahatsente, R., & Coblenz, D. (2015). Ridge-push force and the state of stress in the Nubia-Somalia plate system. *Lithosphere*, *7*(5). doi: 10.1130/L441.1

Mahatsente, R., Jentzsch, G., & Jahr, T. (1999). Crustal structure of the Main Ethiopian Rift from gravity data: 3-dimensional modeling. *Tectonophysics*, *313*, 363–382.

Mahatsente, R., Jentzsch, G., & Jahr, T. (2000). Three-dimensional inversion of gravity data from the Main Ethiopian Rift. *Journal of African Earth Sciences*, *31*(2), 365–373.

Makris, J., & Ginzburg, A. (1987). The Afar depression: transition between continental rifting and sea floor spreading. *Tectonophysics*, *141*, 199–214.

Mammo, T. (2013). Crustal structure of the flood basalt province of Ethiopia from constrained 3-d gravity inversion. *Pure Appl. Geophys.*, *170*, 2185–2206.

Mickus, K., Tadesse, K., Keller, G. R., & Oluma, B. (2007). Gravity analysis of the main Ethiopian rift. *Journal of African Earth Sciences*, *48*, 59–69.

Muluneh, A. A., Brune, S., Illsley- Kemp, F., Corti, G., Keir, D., Glerum, A., et al. (2020). Mechanism for deep crustal seismicity: insight from modeling of deformation processes at the Main Ethiopian Rift. *Geochemistry, Geophysics, Geosystems*, *21*, e2020GC008935. doi: 10.1029/2020GC008935

Muluneh, A. A., Kidane, T., Corti, G., & Keir, D. (2018). Constraints on fault and crustal 760 strength of the Main Ethiopian Rift from formal inversion of earthquake focal 761 mechanism data. *Tectonophysics*, *731*, 172–180. doi: 10.1016/j.tecto.2018.03.010

Njinju, E. A., Atekwana, E. A., Stamps, D. S., Abdelsalam, M. G., Atekwana, E. A., Mickus, K. L., et al. (2019). Lithospheric structure of the Malawi Rift: implications for magma-poor rifting processes. *Tectonics*, *38*, 3835–3853. doi: 10.1029/2019TC005549

Pasyanos, M. E., Masters, T. G., Laske, G., & Ma, Z. (2014). LITHO1.0: an updated crust and lithospheric model of the Earth. *J. Geophys. Res. Solid Earth*, *119*, 2153–2173. doi: 10.1002/2013JB010626

Pérez-Gussinyé, M., Metois, M., Fernandez, M., Verges, J., Fullea, J., & Lowry, A. R. (2009). Effective elastic thickness of Africa and its relationship to other proxies for lithospheric structure and surface tectonics. *Earth and Planetary Science Letters*, *287*, 152–167.

Philippon, M., Corti, G., Sani, F., Bonini, M., Balestrieri, M.-L., Molin, P., et al. (2014). Evolution, distribution, and characteristics of rifting in southern Ethiopia. *Tectonics*, *33*. doi: 10.1002/2013TC003430

Phipps Morgan, J., Parmentier, E. M., & Lin, J. (1987). Mechanisms for the origin of mid-ocean ridge axial topography: implications for the thermal and mechanical structure of accreting plate boundaries. *J. Geophys. Res.*, *92*(12), 823–12836.

Ranalli, G., & Adams, M. (2013). Rheological contrast at the continental Moho: Effects of composition, temperature, deformation mechanism, and tectonic regime. *Tectonophysics*, *609*, 480–490. doi: 10.1016/j.tecto.2012.10.037

Ranalli, G., Piccardo, G. B., & Corona-Chávez, P. (2006). Softening of the subcontinental lithospheric mantle by asthenosphere melts and the continental extension/oceanic spreading transition. *Journal of Geodynamics*, *43*, 450–464.

Roland, E., Mark, D. B., & Greg, H. (2010). Thermal-mechanical behavior of oceanic transform faults: implications for the spatial distribution of seismicity. *Geochemistry. Geophysics, Geosytems*, *11*, 1–15.

Rooney, T. O. (2017). The Cenozoic magmatism of East-Africa: Part I — flood basalts and pulsed magmatism. *Lithos*, *286–287*, 264–301. doi: 10.1016/j.lithos.2017.05.014

Saleh, S., Jahr, T., Jentzsch, G., Saleh, A., & Abou Ashour, N. M. (2006). Crustal evaluation of the Northern Red Sea rift and Gulf of Suez, Egypt from geophysical data: 2-dimensional modeling. *J. Afr. Earth Sci.*, *45*, 257–278.

Saria, E., Calais, E., Stamps, D. S., Delvaux, D., & Hartnady, C. J. H. (2014). Present-day kinematics of the East African Rift. *J. Geophys. Res. Solid Earth*, *119*, 3584–3600. doi: 10.1002/2013JB010901

Sippel, J., Meeßen, C., Cacace, M., Mechie, J., Fishwick, S., Heine, C., et al. (2017). The Kenya rift revisited: insights into lithospheric strength through data-driven 3-D gravity and thermal modelling. *Solid Earth*, *8*(1), 45–81. doi: 10.5194/se-8-45-2017

Tejero, R., & Ruiz, J. (2002). Thermal and mechanical structure of the central Iberian Peninsula lithosphere. *Tectonophysics*, *350*, 49–62.

Tesauro, M., Kaban, M. K., Cloetingh, S. A. P. L., Hardebol, N. J., & Beekman, F. (2007). 3D strength and gravity anomalies of the European lithosphere. *Earth and Planetary Science Letters*, *263*, 56–73.

Turcotte, D. L., & Schubert, G. (2002). *Geodynamics* (2nd ed., p. 456). New York: Cambridge University Press.

Uieda, L., Barbosa, V., & Braitenberg, C. (2016). Tesseroids: forward-modeling gravitational fields in spherical coordinates. *GEOPHYSICS*, F41–F48. doi: 10.1190/geo2015-0204.1

Von Herzen, R. P., Finckh, P., & Hsu, K. J. (1974). Heat flow measurements in Swiss lakes. *J. Geophys.*, *40*, 141–172.

Watts, A. B., & Burov, E. (2003). Lithospheric strength and its relationship to the elastic and seismogenic layer thickness. *Earth Planet. Sci. Lett.*, *213*, 113e131.

Whaler, K. A., & Hautot, S. (2006). The electrical resistivity structure of the crust beneath the northern Main Ethiopian Rift. In G. Yirgu, C. Ebinger, & P. K. H. Maguire (Eds.), *The Afar Volcanic Province within the East African Rift System* (Vol. *259*, pp. 293–305). Geological Society, London, Special Publications.

Woube, Z. (1986). *Analysis of well test data from the Langano-Aluto-Geothermal field.* Reykjavik, Iceland, Part II: Ethiopia Geothermal training program.

Yang, Z., & Chen, W. P. (2010). Earthquakes along the East African Rift System: a multiscale, system-wide perspective. *Journal of Geophysical Research: Solid Earth*, *115*(12).

8

Edge-Driven Convection-Based Models for Evolution of the Mississippi Embayment and Associated Alkaline Magmatism

Samuel R. Walker[1], İbrahim Çemen[1], Matthew W. Wielicki[1], Berry H. Tew, Jr.[1,2], and Daniel Stockli[3]

ABSTRACT

The Mississippi Embayment (ME) is a south-plunging basin. The Arkansas Alkaline Province (AAP) lies along the ME's western border in central Arkansas and was emplaced in the mid-Cretaceous. The relationship between the ME and AAP is unclear. We investigated this relationship using a multidisciplinary approach that involves geochemistry, thermochronometry (U-Th/He), petrogenesis, and geometry of 11 igneous dikes from the AAP. These data reveal that dikes are potassic-ultrapotassic and are extremely enriched in incompatible elements relevant to OIBs and MORBs. The Nd isotopic ratio values demonstrate that lamprophyres are asthenosphere derived. Lamprophyres and remaining lithologies reveal enrichment in incompatible elements and high Sr and Nb contents. This decoupling between isotope ratios and trace element contents could represent low degrees of partial melting of a depleted asthenospheric source. We propose a non-plume evolution for the ME related to edge-driven convection (EDC) associated with regional plate tectonics. At the time of the AAP's emplacement (106-89 Ma), EDC would have been at its highest velocity when the region was uplifted, and melt was guided along pre-existing weaknesses. Results and interpretations from this study clarify a mechanism for alkaline magmatism during Cretaceous volcanism within the framework of global tectonics.

8.1. INTRODUCTION

The relationship between magmatism and tectonics in intraplate settings has proven enigmatic. It is well-documented that preexisting tectonic features such as rift zones can serve as pathways that guide melt into the overlying crust (Loper, 1998; Matton & Jebrak, 2009). However, there is still debate over the exact mechanism that allows magma to intrude so far away from plate boundaries, whether it be sudden changes in a regional stress field or a mantle plume that intrudes into the continental crust. The latter explanation—the plume hypothesis—involves an upwelling mass of material in the mantle (Condie, 2001). As a tectonic plate passes over a plume, magma is injected into the crust, leaving a long trail of igneous provinces terminating in a currently active "hotspot" region (Loper, 1998).

Igneous provinces along a hotspot track usually exhibit similar lithologies and geochemical signatures. Proposed hotspot tracks commonly feature igneous rocks of the alkaline variety, which can be found in anorogenic and within-plate settings and are related to rifting/extensional tectonics. Suites of alkaline igneous rocks are found in North and South America and Africa, with nearly all of them concentrated along pre-existing structures related to opening of the Iapetus

[1] *Department of Geological Sciences, University of Alabama, Tuscaloosa, Alabama, USA*

[2] *Geological Survey of Alabama, Tuscaloosa, Alabama, USA*

[3] *Jackson School of Geosciences, Department of Earth and Planetary Sciences, University of Texas at Austin, Austin, Texas, USA*

Extensional Tectonics: Rifting and Continental Extension, Geophysical Monograph 278, First Edition.
Edited by İbrahim Çemen and Elizabeth J. Catlos.

DOI: 10.1002/9781119773757.ch08

Ocean during the Late Proterozoic and Early Cambrian (Matton & Jebrak, 2009). These intrusive complexes mark a period of increased alkaline activity in the mid-Cretaceous that has been attributed to either deep mantle plume activity in the form of a "superplume" event (Larson, 1991), or periodic reactivation of preexisting lithospheric weaknesses due to edge-driven convection (EDC) processes (Matton & Jebrak, 2009).

A particular alkaline province in central Arkansas known as the Arkansas Alkaline Province (AAP) lies along proposed age-progressive tracks for the Bermuda hotspot (Figure 8.1) (Duncan, 1984; Muller, 1993). The proposed tracks comprise igneous intrusions in Kansas, central Arkansas, and Mississippi and are believed to terminate at the Bermuda swell in the central Atlantic Ocean. There are no known igneous rocks in Alabama or Georgia that contribute to or "complete" an age-progressive track of volcanism.

The AAP is also located at the junction between two structures: the Early Cambrian Reelfoot Rift (RfR) and the orthogonal Alabama-Oklahoma Transform (AOT), which were both formed during the opening of the Iapetus Ocean (Thomas, 2014). Unconformably overlying the RfR are Cretaceous and Cenozoic sedimentary rocks and alluvium of the Mississippi Embayment (ME), a south-plunging topographic depression that serves as a northward continuation of the Gulf of Mexico (GOM) coastal plain. Members of the AAP lie along the ME's western border in central Arkansas (Figure 8.2). Thus, the AAP serves as a natural laboratory for studying the emplacement mechanisms of alkaline provinces and examining the competing hypotheses that explain their evolution.

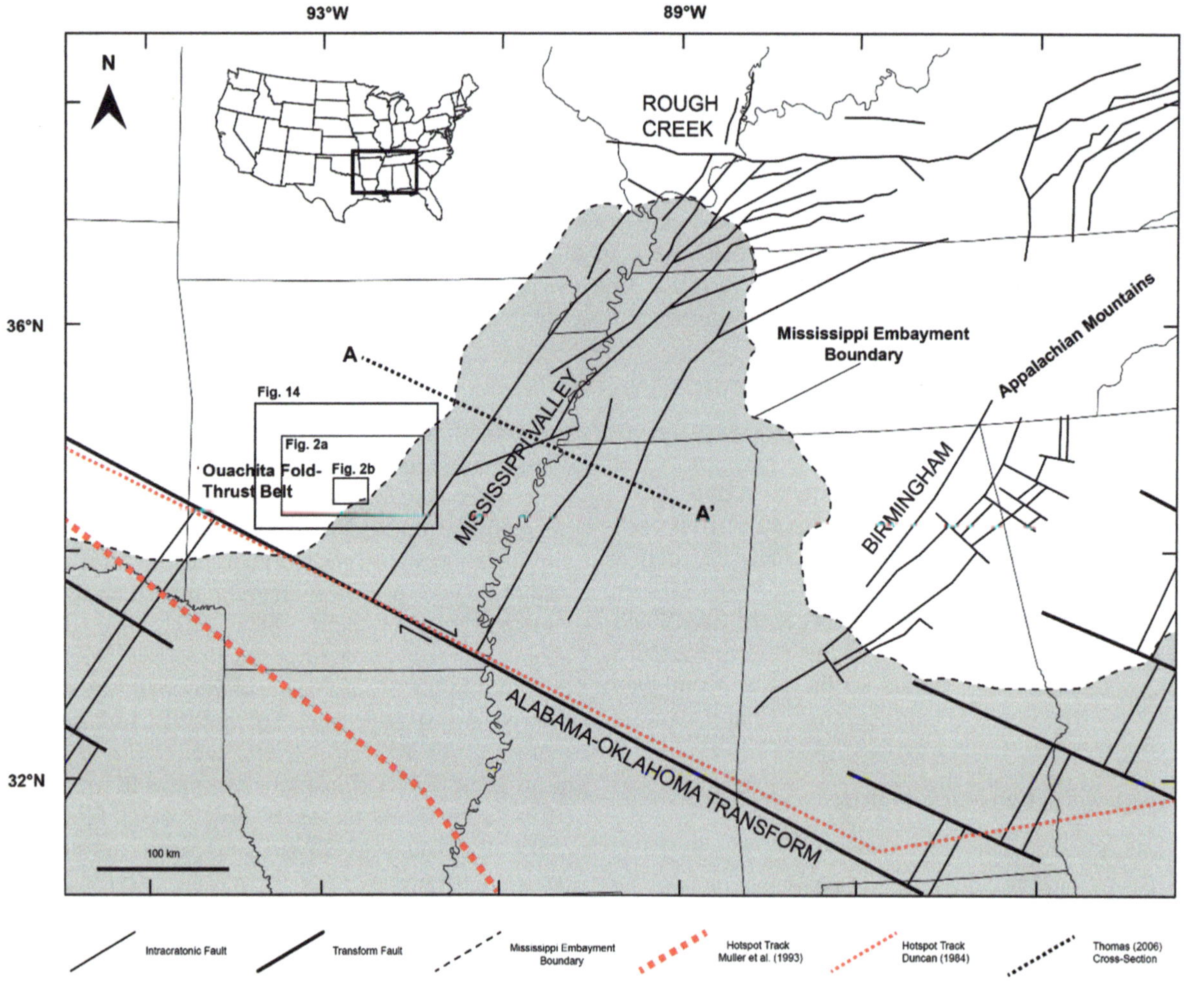

Figure 8.1 Map of the south-central United States, with regional tectonic features after Thomas (2006) and hotspot tracks after Muller (1993) and Duncan (1984). Locations of the Ouachita Fold-Thrust Belt and Appalachian Mountains are indicated as well as the location of extensional grabens (Birmingham, Mississippi Valley, and Rough Creek). The ME boundary is indicated, and the gulf coastal plain is shaded in gray. Locations of Figures 8.2 and 8.14 and a cross-section transect from Thomas (2014) are indicated. The map projection is NAD_1983_UTM_Zone_15N.

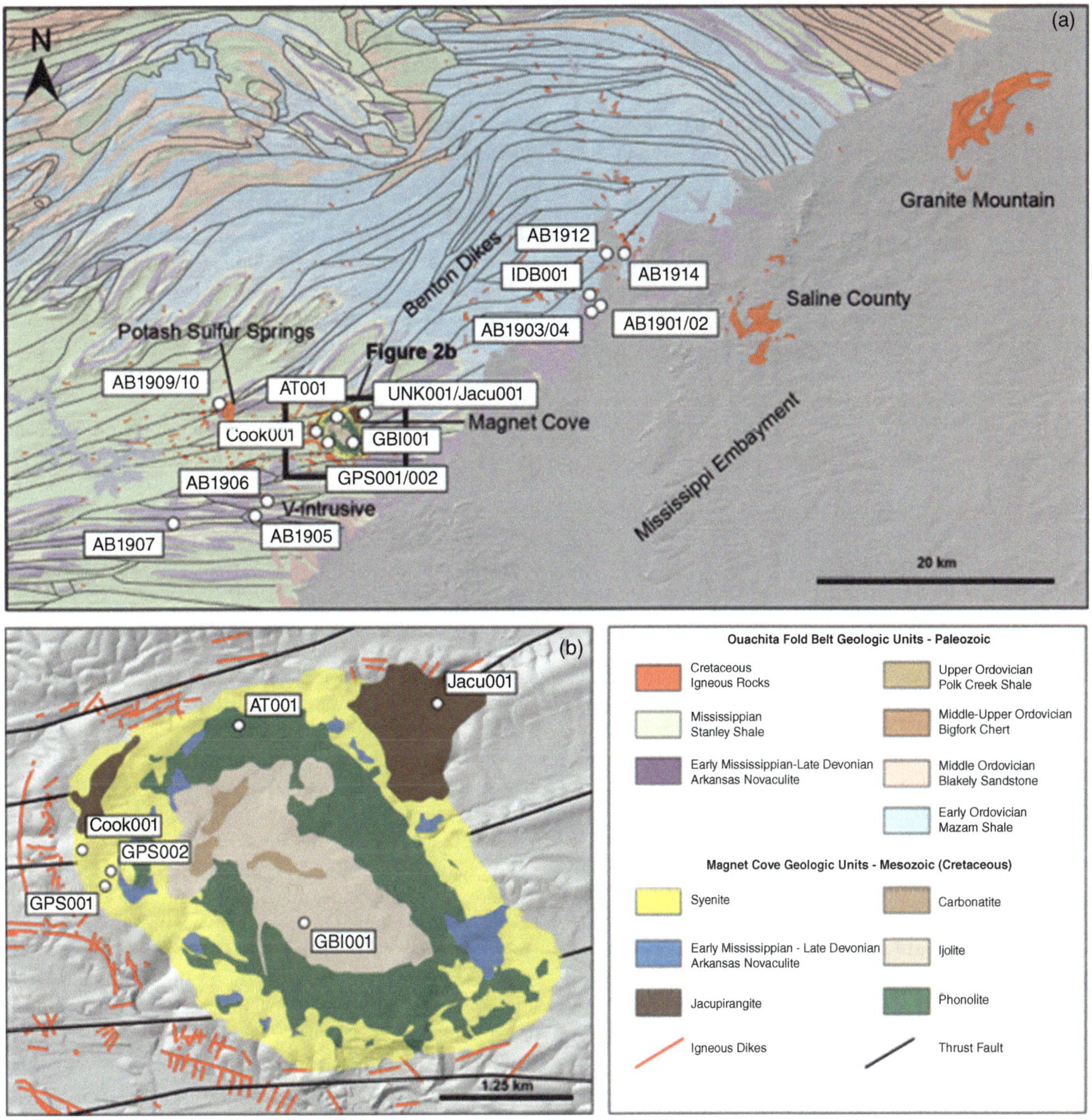

Figure 8.2 Digital elevation maps and geologic overlays for (a) the OFB with sample locations and each specified member of the AAP, and (b) a map of the Magnet Cove intrusive ring-dike complex with sample locations. Geologic overlays are from the adapted from Arkansas Geological Survey database and Erickson & Blade (1963). The digital elevation map is after the USGS National Map. Map projection is NAD_1983_UTM_Zone_15N.

The purpose of this study is to test two competing hypotheses regarding the ME's structural evolution. Hypotheses are tested by studying igneous intrusions of the AAP, with an emphasis on their petrogenesis and tectonic significance, especially their implications for the ME's formation. Geochemical data for 11 samples (collected in central Arkansas in April 2019; Figure 8.2) is presented in the form of whole-rock major oxides and trace elements. Whole-rock neodymium (Nd) isotopes from four representative samples are also presented and discussed to identify the origin of sampled rocks. Scanning electron microscopy (SEM) backscattered electron images, as well as energy dispersive spectroscopy (EDS), are used to investigate sample mineralogy and corroborate XRF results. In combination with the geochemical data, orientations of over 700 igneous dikes throughout the AAP are presented in rose diagrams. These diagrams were used to investigate stress regimes during AAP emplacement and demonstrate the tectonic significance of igneous intrusion orientations. Finally, thermochronology (zircon U-Th/He) data from five samples are also presented. Results from this study enhance the current understanding of the AAP's genesis and emplacement, and its associated effects on the structural evolution of the ME. Techniques and interpretations developed in this study can be applied to other rift zones in efforts to understand the interplay between extensional tectonics and alkaline magmatism related to basin evolution. Moreover, this study clarifies the mechanism by which numerous alkaline provinces across the world were emplaced in the crust during a "pulse" of increased alkaline magmatism in the Cretaceous Period.

8.2. GEOLOGIC OVERVIEW

8.2.1. The Mississippi Embayment

The ME is a southwest-plunging depression extending from southern Illinois to Louisiana (Figure 8.1). At its western border in Arkansas, the ME is bordered by the folded and faulted lower to middle Paleozoic rocks of the Ouachita fold-thrust belt (OFB) and Pennsylvanian rocks of the Arkoma Basin. To the east, the ME is bordered by the southern Appalachian Mountains in Alabama.

The southern half of the ME serves as the northward extension of the GOM coastal plain, containing ~1.5 km of Mesozoic and Cenozoic sedimentary rocks. The northern ME overlies Paleozoic cratonic shelf facies that range between Cambrian and Pennsylvanian in age. Units of the southern ME overlie rocks of the Ouachita orogeny, which were thrust over the Paleozoic craton. The basin fill comprises marine and deltaic sediments that are Late Cretaceous and Late Eocene in age. These sediments are thickest in the southern portion of the basin, where ~2 km of Upper Cretaceous to Upper Eocene sediments are overlain by Pliocene and Quaternary fluvial deposits (Stearns, 1957; Murray, 1961; Autin et al., 1991; Cox et al., 2002). These sediments are <1 km thick in the northern ME, where they overlie the RfR. Pre-Late Cretaceous strata of the ME's basin fill extend to central Arkansas and into western Mississippi, where they terminate into OFB rocks & Lower Cretaceous units of the Hosston Group in Mississippi (Murray, 1961; Schwalb, 1982;). Mooney and Prodehl, (1984) used seismic refraction to suggest that the crust of the ME comprises six primary layers, which include Mesozoic and Tertiary sediments, underlying Paleozoic clastic and carbonate rocks, Early Paleozoic sedimentary rocks, crystalline upper crustal rocks, lower crust, and a modified lower crust.

Structurally, the ME is a broad synclinal feature that plunges gently to the southwest. Its base forms an unconformable contact with the underlying RfR sedimentary rock units (usually referred to as the Mississippi Valley Graben [MVG]). The RfR is a NE-SW trending intracratonic rift that lies nearly perpendicular to the NW-SE trending AOT (Figure 8.1), which serves as the Late Precambrian-Cambrian Iapetan rifted margin in that area (Thomas, 2014). Normal faults of the RfR cut Precambrian metamorphic basement rocks, Lower and Middle Cambrian units, Upper Cambrian to Lower Ordovician units, Middle Ordovician to Middle Mississippian Units, and Pennsylvanian to Upper Mississippian sedimentary rock units of the OFB in central Arkansas (Figure 8.2a). Exact depths of normal faults are uncertain as there is no biostratigraphic documentation for deeper units (Thomas, 2014). The end of RfR fault movement is indicated by faulted strata of the Conasauga Formation that are blanketed by the Upper Cambrian–Lower Ordovician Knox Group (Thomas, 1991). The faulted units form a broad anticlinal structure that trends NE-SW and plunges southwest, similar to the ME's synclinal structure (Murray, 1961). The upper portion of the anticline's crest is nonexistent and forms an angular unconformity with overlying ME units (Figure 8.3).

8.2.2. The Arkansas Alkaline Province

The AAP is a northeast-trending suite of intrusive igneous bodies that lies adjacent to the ME's western border. Igneous members of the AAP intrude lower to middle Paleozoic rocks of the OFB. From the southwest in Murfreesboro, AR, to the northeast near Little Rock, the intrusive suites are referred to as Prairie Creek, Magnet Cove, V-intrusive, Potash Sulfur Springs, Benton Dikes, Saline County, and Granite Mountain, respectively (Figure 8.4). Thin carbonatite dikes are present ~80 km north of the Magnet Cove Complex in Morrilton and Perryville, AR (McCormick & Heathcote, 1979). AAP members occur as large plutonic bodies (Magnet Cove, Potash Sulfur Springs, Saline County, Granite Mountain) or smaller dikes (Benton, V-intrusive) that crosscut Paleozoic strata. Lithologies vary between carbonatite, ijolite, jacupirangite, lamproite, lamprophyre, kimberlite, and nepheline syenite (Morris & Fouad, 1987; Eby & Vasconcelos, 2009). A general geographic progression in lithologies is evident, with more ultramafic intrusions (lamprophyres, lamproites, ijolites, jacupirangites, kimberlites, carbonatites) to the southwest and more evolved, felsic rocks (nepheline syenites) found near the AAP's northeastern corner.

Multiple studies (Fig. 4) have used radiometric age-dating techniques (K-Ar, Ar-Ar, U-Pb, Rb-Sr, fission track) to determine the age of emplacement for the assorted AAP intrusions. Ages from these studies fall primarily within the middle Cretaceous Period (~106 to 88 Ma) (Arne, 1992; Baksi, 1997; Eby & Vasconcelos, 2009; Gogineni et al. 1978; Scharon & Hsu, 1969; Zartman & Howard, 1987). A rough SW-NE progression in age is evident for the AAP, with older intrusions (>100 Ma) in the

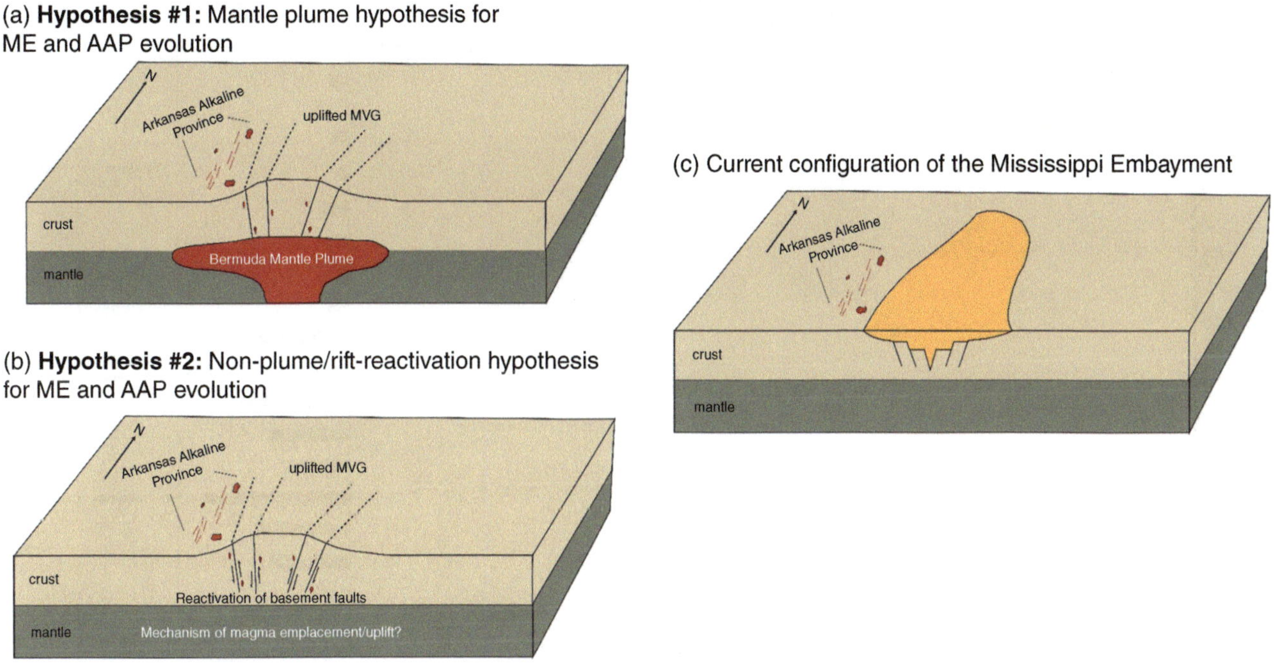

Figure 8.3 Schematic diagrams indicating the current structure of the Mississippi Embayment and alternate hypotheses concerning its evolution: (a) mantle plume hypotheses Cox and Van Arsdale (1996, 2002), which suggests that a mantle plume uplifted the proto-ME and is responsible for its evolution; (b) nonplume hypotheses for ME evolution, which relies on the reactivation of MVG faults and does not require a mantle plume for ME evolution (Ervin & McGinnis, 1975; Eby & Vasconcelos, 2009); (c) current configuration of the embayment. The transect for the cross-sections in the foreground of these schematics is indicated by the the A-A′ line in Figure 8.1.

southwest AAP and younger intrusive bodies (<90 Ma) to the northeast. Eby and Vasconcelos (2009) divided these ages into three primary time frames: >100 Ma, 100–90 Ma, and <90 Ma. A correlation between lithology and age is apparent, with emplacement ages of ~106 Ma for lamproites, 101–94 Ma for carbonatites and other ultramafic rocks, and ~88 Ma for nepheline syenite intrusions (Eby & Vasconcelos, 2009).

The genesis of AAP igneous bodies and their emplacement mechanisms remain unclear. Due to the diverse lithologies of AAP members, it is necessary to investigate the petrogenesis of each complex separately to comprehensively understand the greater magmatic system behind AAP evolution. Previous studies (Morris & Fouad, 1987; Tilton et al., 1987; Duke et al., 2014) have utilized isotopic data for this purpose. These studies indicate that multiple magma sources and processes were likely involved in AAP evolution. Sr, Nd, and Pb isotopic values of olivine lamproites from Prairie Creek ($\varepsilon Nd = -10$ to -13, $^{206}Pb/^{204}Pb = 16.61$ to 16.81, $^{87}Sr/^{86}Sr = 0.70627$ to 0.70829) and Dare Mine Knob ($\varepsilon Nd = -19$) reflect a subcontinental lithospheric mantle (SCLM) origin and involvement of a significant crustal component (Duke et al., 2014). Sr-Nd-Pb isotopic values from Magnet Cove carbonatites ($^{143}Nd/^{144}Nd = 0.51271$, $^{87}Sr/^{86}Sr = 0.70363$, $^{206}Pb/^{204}Pb = 15.549$) indicate little crustal involvement and an asthenospheric source that's depleted in large-ion lithophile elements (LILe) (Tilton et al., 1987).

8.2.3. Emplacement Mechanisms for Alkaline Provinces

Alkaline igneous provinces are ubiquitous and are found in anorogenic and intraplate settings, usually related to rifting/extensional tectonics. Similar to the AAP, a large portion of alkaline provinces were emplaced in the mid-Cretaceous Period, which Matton and Jebrak (2009) refer to as the Peri-Atlantic Alkaline Pulse (PAAP). Thus, the mid-Cretaceous Period represents an increased flare in alkaline igneous activity, which has been attributed to either the influence of mantle plumes or structural control related to the reactivation of preexisting lithospheric weaknesses.

Matton and Jebrak (2009) tested the validity of plume and structural models by examining PAAP provinces in eastern North America, eastern South America, and western Africa. Nearly all of the PAAP alkaline provinces are situated near preexisting zones of weakness, such as rift systems, including aulacogens (*senso-lotto*), grabens, large normal faults, and so on, which are pre-Cretaceous in age. This is also the case for the AAP, which is located at the junction between the NE-SW oriented RfR and the orthogonal AOT, two large-scale structures related to the opening of the Iapetus Ocean. Reactivation of these preexisting features has been documented for some PAAP members, including Pan-African related structures in Africa, Iapetus-related structures in eastern North America, Precambrian weaknesses in South America, and

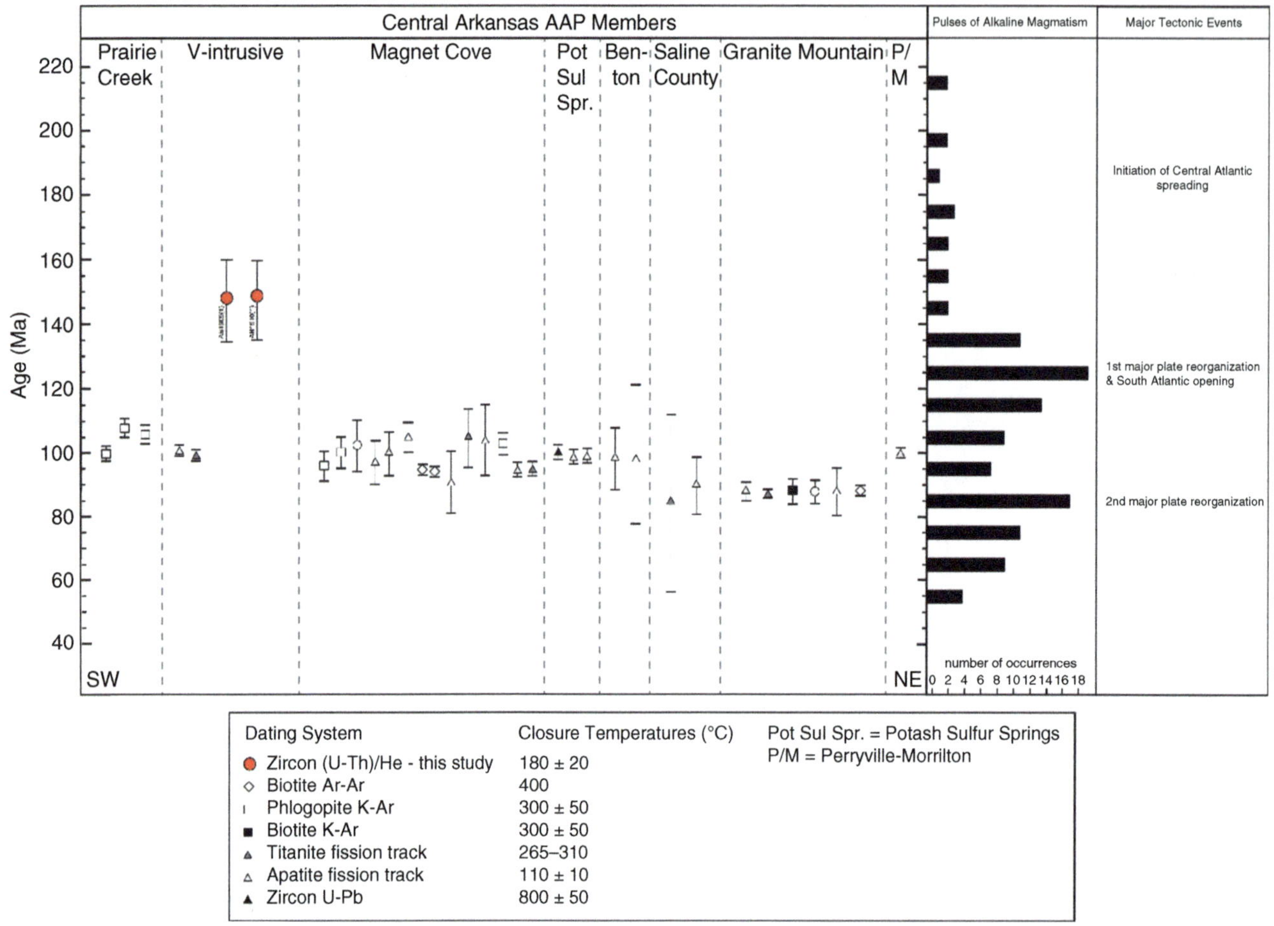

Figure 8.4 Summary plot of age data presented in previous studies (Biotite Ar-Ar: Baksi, 1997; Phlogopite K-Ar: Gogineni et al., 1978; Biotite K-Ar: Zartman & Howard, 1987; Titanite fission track: Scharon & Hsu 1969, Eby & Vasconcelos, 2009; Apatite fission track: Arne, 1992; Scharon & Hsu, 1969; Eby & Vasconcelos, 2009; Zircon U-Pb: Zartman & Howard, 1987) and (U-Th)/He data for this study. A histogram demonstrating timing and frequency of alkaline magmatism and associated Atlantic tectonic events is presented (from Eby & Vasconcelos, 2009).

Variscan-related lineaments in western Iberia (Matton & Jebrak, 2009). Comparing the emplacement ages of the AAP and the PAAP with the tectonic evolution of the Atlantic Ocean (see Figure 8.4), Matton and Jebrak (2009) suggested that the PAAP is best described by the reactivation of preexisting features due to EDC. EDC was initially described by King and Anderson (1998) as an instability that takes place between thick lithosphere and thinner lithosphere. This sharp change in the lithosphere-asthenosphere boundary provides a temperature gradient, which in turn creates EDC cells that carry hot mantle material toward the surface (Kim & So, 2020). The velocity of these cells increases with time as mantle viscosity decreases (King & Anderson, 1998). Partial melts and subsequent volcanism are promoted due to decompression melting caused by upwelling mantle, which lowers the solidus (Kim & So, 2020). In the EDC hypothesis, two major tectonic events at ~85 and ~125 Ma correlate strongly with spikes in PAAP occurrences. Thermodynamic requirements for melting and emplacement of PAAP provinces were provided by continental insulation—allowing accumulation of mantle heat beneath the lithosphere—in the Early Cretaceous Period during the breakup between South America and South Africa, followed by EDC between North America and West Africa (Matton & Jebrak, 2009).

An alternate explanation for the existence of alkaline provinces involves the influence of mantle plumes. Larson (1991) attributed the increase in mid-Cretaceous alkaline activity to a "superplume" event. A superplume is an abnormally large mantle plume head (~500 to 3,000 km in diameter) (Condie, 2001). Thus, a superplume event is a short-lived (<100 Ma) event characterized by increased convection and hotspot volcanism. During such an event, many superplumes and smaller plumes impact the base of the lithosphere and intrude into zones of weakness (Condie 2001). Larson (1991) used this model to explain a 50–75% increase in the rates of oceanic crust formation between 120 and 80 Ma. Other authors (e.g., Morgan, 1971; Burke & Dewey, 1973; Wilson, 1997; Leitch et al., 1998; Duncan, 1984; Muller, 1993) use the plume hypothesis to explain the opening of the Atlantic Ocean and emplacement of associated igneous bodies.

The origin of plumes has been speculated in the literature, with differing layer models of the mantle suggesting that plumes originate either from the core-mantle boundary or above the 660-km seismic discontinuity (Hofmann, 1997). With either model, rising columns of less-dense material rise to the surface and create "hotspots," or surface expressions of underlying mantle plumes. For purposes of this study, a plume is defined as a stationary upwelling of buoyant mantle material. As continents drift over plumes, upwelling material is periodically emplaced in the crust. Igneous material emplaced from mantle plumes should form an age-progressive track of volcanism that ends at an active hotspot, forming a "hotspot track" (Loper, 1998). Although there is no volcanic expression at its current location in the Atlantic Ocean (due to suggested volcanic inactivity over the last ~20 to 30 Ma), the Bermuda Rise (BR) has been explained by multiple mantle plume models (Davies, 1988; Sleep, 1990). Some researchers (Morgan, 1971; Duncan, 1984; Muller, 1993) link the BR with igneous provinces in North America, suggesting that these provinces form an age-progressive hotspot track terminating at the BR.

A more direct influence of a mantle plume on the AAP and ME was first proposed by Cox and Van Arsdale (1997, 2002). They suggested that the AAP lies along a southeastward age-progressive track of volcanism (Figure 8.1) ending at the BR. This track begins with ultramafic intrusions in eastern Kansas (115 Ma; Brookins & Naeser, 1971; Cullers et al. 1996) and continues within the ME with the AAP (106–88 Ma; Eby & Vasconcelos, 2009 and references therein) and the Jackson Dome (65 Ma; Dockery et al., 1997) intrusive in central Mississippi. Ultramafic rocks in Kansas are situated within the sedimentary cover above the Proterozoic Central North American Rift System (CNARS), similar to AAP intrusions, which lie just west of the Early Cambrian RfR. There are no known igneous rocks in Alabama or Georgia that contribute to an age-progressive track of volcanism. Although some plutonic igneous rocks have been encountered in wells in south Alabama, K-Ar ages of these rocks (267 ± 10 Ma for basalt and 334 ± 4.8 Ma for granite; Neathery & Thomas; 1975) are significantly older than the Jackson Dome province (65 Ma; Dockery et al., 1997) and therefore don't complete an age-progressive track. Based on this track, the model of Cox and Van Arsdale (1997, 2002) indicates westward passage of the North American continent over the Bermuda hotspot in the mid-Cretaceous Period. At this time, AAP intrusions were emplaced in the crust and the region of the proto-ME was uplifted on the order of 1–3 km. As North America moved away from the hotspot in the Late Cretaceous Period, the uplifted crust was eroded, and subsidence of the ME began.

8.3. METHODS

8.3.1. Sample Collection and Mineral Separation

The dataset produced for this study consists of geochemical data (whole-rock major oxides, trace elements, and Nd isotopes), petrographic data (thick sections), zircon (U-Th)/He thermochronology, and structural data (igneous dike geometries). Eleven samples were analyzed for their whole-rock major and trace element compositions, and backscattered electron images were produced with a JEOL FE-7000 SEM. Four of the 11 samples were analyzed for their Nd isotopic compositions. Due to the wide range in lithology of the AAP, samples were selected to adequately cover the mineralogical range and magmatic processes in a representative fashion. Five samples were also dated via zircon (U-Th)/He.

Sample locations form a roughly NE-SW transect from Little Rock just southwest of Magnet Cove, AR, along the ME's western border (Figure 8.3a). In conjunction with the Arkansas Geologic Survey, samples were collected from three AAP provinces: the Benton Dike Swarm, the V-intrusive dike swarm, and Magnet Cove (Figure 8.2).

8.3.2. (U-Th)/He Analyses

Zircons were separated from bulk rock samples by standard mechanical, heavy liquid, and magnetic separation techniques at The University of Alabama. For thermochronology analyses, euhedral, inclusion-free zircon grains were analyzed at the Geo- and Thermochronometry Laboratory at the University of Texas at Austin. Mineral grains were measured and photographed using a high-powered, cross-polarized stereographic microscope equipped with a digital camera. Grains were then placed into 1-mm Pt foil tubes that were crimped on both ends and heated for 10 min at 1,300 °C. ^{4}He was collected and measured from zircon grains using an ultra-high vacuum noble gas extraction and purification line with two Blazers prisma QMS-200 quadrupole mass spectrometers. Zircon grains were then dissolved with HF-HNO3 and HCL and analyzed for U, Th, and SM using a Thermo-Scientific Element-2 inductively coupled plasma mass spectrometer. Ages were calculated by applying standard α-ejection corrections with morphometric analyses (Reiners, 2005).

8.3.3. Major and Trace Elements

Bulk rock samples were prepared adhering to the double fusion technique with a 2:1 di-lithium tetraborate mixture as outlined in Johnson et al. (1999). Fresh, homogeneous portions of each sample were cut out using a diamond-tip saw before being pulverized into powder using a low-carbon steel ring and puck mill in a SPEX shatterbox. Powdered samples were then weighed and mixed with Spectromelt di-lithium tetraborate flux before fusion in graphite crucibles at 1,000 °C for 10 min. The glass beads were powdered again using the ring and puck mill and returned to the furnace for a second fusion. Major and trace element analyses were conducted at the Pomona X-Ray Fluorescence Laboratory at Pomona College with a Panalytical Axios XRF spectrometer, which allows for simultaneous measurements of both major and trace elements. Beads are directly irradiated and produce fluorescence, which is then measured with an energy-dispersive detector. Raw major element concentrations were normalized to volatile-free values, with iron expressed as Fe_2O_3.

8.3.4. Thick-Section Analyses

Representative portions of each sample were cut out using a diamond-tip saw and mounted in one-inch epoxy mounts that are herein referred to as "thick sections." Mounts were cured for 24 hr and then polished and gold-coated for analysis using SEM. Thick sections were analyzed at the Alabama Analytical Research Center at The University of Alabama with a JEOL 7000 FE SEM equipped with an Oxford X-Max 80 mm^2 drift detector. A working distance of 10 mm was used to maximize the efficiency of the EDS detector, which measures the energy of the samples' emitted photons in the X-ray electromagnetic spectrum. Accelerating voltages ranged between 15 and 20 kV with a spot size <1 μm. EDS spectra were primarily used to investigate the elemental composition of the samples, determine mineralogy, and corroborate XRF results. Backscattered electron (BSE) images were taken for two to three representative sites per thick section, and 10 to 20 EDS point spectra were analyzed per BSE image. EDS chemical maps of the representative sites were also used to investigate sample mineralogy and abundance.

8.3.5. Thermal Ion Mass Spectrometry

Representative slices were removed from four lamprophyre samples (AB1909, AB1910, AB1912, and AB1914), ground on diamond laps to remove weathered surfaces, and pulverized using a tungsten carbide ring and puck mill. Then 25–50 mg of the rock powders were weighed out and spiked (200 uL/50 mg of rock) with a mixed tracer enriched in ^{150}Nd and ^{147}Sm (University of North Carolina Basalt spike: UNC "B"). Rock powders were subsequently placed in Teflon screw-cap vials and dissolved in hot HF, HNO_3, and HCL. HCL was used to break down fluoride precipitates. Two-step ion chromatography was used to separate Sm and Nd. Bulk REE fractions were eluted in BioRad PolyPrep columns filled with resin and then a 2-methylactic acid (MLA) column filled with resin to collect Sm and Nd into separate Teflon vials. A final preparation step involved adding aqua regia to the vials containing eluted Sm and Nd, which were evaporated to remove MLA. The method used is explained in detail in Stowell and Tinkham (2003).

Sm and Nd isotope ratios were analyzed using a VG Sector 54 thermal ionization mass spectrometer at the UA Radiogenic Isotope Laboratory at The University of Alabama, Tuscaloosa. Nd samples were loaded onto Re filaments as NdO, and Sm samples were loaded onto Ta filaments as Sm^0. Isotopic concentrations of Nd were measured with seven Faraday collectors and a dynamic mode program. Filaments were heated to between 1,200 °C and 1,800 °C. At these temperatures, Nd is evaporated and ionized. Ionized Nd is then accelerated by an electric field into a mass spectrometer, where a large magnet separates the isotopes based on mass.

8.3.6. Paleostress Analysis

Igneous dikes and other structural features for the central Arkansas region were interpreted using ArcMap software (version 10.7.1). Shapefiles from the STATEMAP and COGEO mapping programs were provided by the AGS and downloaded for analyses. Each shapefile contains the location and extent of igneous dikes (line features) and larger igneous bodies (polygons), along with data for each feature (length and lithology). Shapefiles were combined and then divided into age groups according to the age boundaries outlined in Eby and Vasconcelos (2009), who used radiometric age data to separate AAP intrusions into three emplacement age ranges: >100 Ma, 100–90 Ma, and <90 Ma. Approximately 340 dikes were contained in the >100 age range and 340 dikes in the 100–90 Ma age range. The <90 Ma age range contains the Granite Mountain intrusion and lies primarily within the ME coastal plain.

Igneous dikes are tabular bodies and were therefore treated as planar features during the analyses in ArcMap. Dike orientation or "strike" was calculated for each dike using Python code. These values populated attribute tables for each age division. Attribute tables were then exported into Excel for analysis on rose diagrams. Dike orientations were imported into the GeoRose program (Yong Technology Inc., 2014) to analyze any observable trends. Rose diagrams, which indicate the cardinal directions of geologic features and the number of features that follow each direction, were generated in GeoRose.

8.4. RESULTS

8.4.1. Major and Trace Elements

The major and trace element geochemical data for each sample are given in Table 8.1. Studied samples represent a wide range of SiO_2 (34.6 to 56.7 wt.%), MgO (0.3 to 8.9 wt.%), TiO_2 (0.3 to 6.4 wt.%), and Al_2O_3 (6.9 to 21.8 wt.%) and CaO (0.57 to 20.0 wt.%) contents (Table 8.1). On a total alkalis-silica diagram (Figure 8.5), samples plot in the alkaline series field and vary between ijolites, phono-tephrite, tephriphonolite, and nepheline syenite. On a K_2O versus Na_2O diagram (Figure 8.6), samples are potassic to ultrapotassic, with three of those samples (AT001, GPS001, GPS002) having particularly high K_2O contents (>10 wt.%). Na_2O contents of the samples vary significantly (0.9 to 5.3 wt.%).

To fully understand the magmatic processes responsible for generation of the studied samples and identify major fractionating phases, MgO is used as an index of fractionation and plotted against specific major oxides and trace elements (Figure 8.7). A negative correlation is shown between MgO and Al_2O_3, whereas positive correlations are apparent with Fe_2O_3 and CaO. No correlations are apparent between MgO and TiO_2, Na_2O, or K_2O. Lamprophyre samples AB1909 and AB1910 (V-intrusive) systematically exhibit lower CaO and MgO compared to AB1914 (lamprophyre; Benton Dike Swarm) at a similar SiO_2 content (Figure 8.7b,d). A moderate negative correlation is evident between MgO and Al_2O_3. Ultramafic samples demonstrate high Al_2O_3 contents (11 to 16 wt.%) except for Jacu001, which contains 6.9 wt.%. Al_2O_3 contents are remarkably similar between ultramafic samples (AB1910, GBI001, IDB001, and Jacu001) and the higher-silica syenite sample Cook001. The remaining syenite samples (GPS 001 and GPS 002) represent the highest Al_2O_3 contents in the dataset (21.8 and 20.8 wt.%, respectively). MnO and P_2O_5 demonstrate no consistent trends with increasing MgO, although it is worth noting that GPS001 and GPS 001 contain very low P_2O_5 contents (0.06 and 0.08 wt.%).

On an AFM ternary diagram (Figure 8.8), the studied samples follow a general curvilinear trend, which begins at MgO and progresses to $Fe_2O_3^T$ and down to the alkali ($Na_2O + K_2O$) vertices, which is expected for evolving volcanic suites.

Trace element abundances vary between samples. Compatible trace elements for samples are shown in Figure 8.9 and range in composition: Sc = 0 to 39 ppm, V = 99 to 989 ppm, Cr = 0 to 290 ppm, and Ni = 10 to 93 ppm (Table 8.1). The compatible element Co is below detection limits (bdl) in all 11 samples. Ni and Sc demonstrate broad correlations (R^2 = 0.68 and 0.84, respectively) with MgO contents. There is no apparent correlation between Cr and MgO. Two samples (Cook001 and GBI001) contain no Cr, whereas sample Jacu001 represents the highest Cr content in the dataset with 290 ppm.

Trace element diagrams are shown in Figure 8.9 and are discussed separately according to sample lithology to provide clarity. Primitive mantle-normalized trace element spider diagrams for the samples are shown in Figure 8.9a–c. On a primitive mantle-normalized diagram, lamprophyre samples (AB1909, AB1910, AB1912, AB1914) are enriched in incompatible elements, including high field-strength elements (Zr, Hf, Nb, Ta), and especially so in LILe (Rb, Sr, Ba). A strong U anomaly is present in sample AB1914 but not in the remaining lamprophyre samples, which exhibit U contents below detection limits. Lamprophyres also exhibit a strongly negative Pb anomaly, as expected.

Similar to the lamprophyres, the phonolite sample (AT001) demonstrates a strongly positive U anomaly and a less-significant Pb anomaly. The phonolite exhibits extreme enrichment in incompatible elements, especially the LILe. High-field-strength element (HFSe) enrichment is present in Nb, Ta, and U but less so in Zr and Hf, with a negative Th anomaly. The syenite samples (Cook001, GPS001, GPS002) show strong enrichment in LILe, a positive U anomaly, and a negative Pb anomaly. Samples GPS001 and GPS002 lack Ta and Hf contents and demonstrate Y contents only 2× that of primitive mantle.

The jacupirangite sample (Jacu001) exhibits lesser enrichment in LILe compared to other samples, yet a negative Sr anomaly. Positive anomalies are present in light rare earth elements Pr, Nd, and Sm. The ijolite sample (GBI001) reveals a positive Sr anomaly, with multiple trace elements (Th, Ta, Pb, Hf) below detection limits.

Trace element data for each sample are also normalized to average values for mid-ocean ridge basalt (MORB; Sun & McDonough, 1989) and shown in Figure 8.9b. In this diagram, LILe (Rb, Ba) are plotted on the far left, and HFS elements are plotted along the rest of the axis toward the right of the diagram. The incompatibility of elements increases toward the left-center of the diagram (at Ba and Th). Average oceanic island basalt (OIB) values (Sun & McDonough, 1989) are also plotted for comparison lamprophyre samples (AB1909, AB1910, AB1912, and IDB001), demonstrating a gentle arched pattern at Ba-Th and a generally smooth negative slope, similar to the OIB. However, lamprophyres are enriched (compared to OIB) in most elements except where AB1912, Cook001, and IDB001 exhibit similar amounts of Zr, Hf, Sm, and Y. The remaining lamprophyre sample (AB1914) is distinguished by a strong Ba peak.

Syenite samples GPS 001 and GPS 002 demonstrate a similar pattern, with broadly negative slopes and much higher enrichment in incompatible elements compared to average OIB values and even higher enrichment compared to MORB. HFS elements are the most enriched in the syenites on the order of 1000x MORBS for Rb. Syenites

Table 8.1 Whole-rock analyses of igneous rocks from AAP numbers, central Arkansas.

Location sample	V-intrusive **AB1909**	V-intrusive **AB1910**	Benton **AB1912**	Benton **AB1914**	Benton **IDB001**	Magnet Cove **AT001**	Magnet Cove **COOK001**	Magnet Cove **GBI001**	Magnet Cove **GPS001**	Magnet Cove **GPS002**	Magnet Cove **JACU001**
wt.%											
Sio_2:	40.15	39.91	37.98	40.08	35.55	53.43	48.08	34.57	56.70	55.59	36.60
Tio_2	6.40	5.16	4.24	4.10	4.29	2.72	2.69	3.39	0.30	0.40	5.57
Al_2O3	13.10	15.85	10.90	12.51	15.74	15.12	16.09	15.63	21.85	20.81	6.93
FeO_3	18.85	17.68	15.08	15.85	17.44	10.21	10.92	16.71	5.04	6.29	18.01
MnO	0.55	0.28	0.26	0.15	0.28	0.26	0.26	0.34	0.45	0.43	0.29
MgO	4.50	5.04	8.14	7.89	7.94	2.54	2.70	5.58	0.30	0.33	8.98
CaO	9.04	9.95	17.52	13.14	12.39	1.86	8.51	16.44	0.57	1.07	19.97
Na_2O	3.01	1.60	1.75	2.07	0.48	0.99	5.30	3.59	2.66	2.80	1.31
K_2O	2.26	2.30	2.78	2.53	4.59	10.80	4.21	2.03	11.45	11.38	0.72
P_2O_5	1.56	1.66	0.87	0.83	0.81	1.38	0.57	1.06	0.06	0.08	1.18
Total	94.42	95.14	94.11	99.17	99.50	99.31	99.34	99.33	99.36	99.19	99.57
ppm											
Ta	390.2	390.2	317.1	198.5	159.4	149.8	196.0	0.0	0.0	99.8	366.0
Rb	69.5	99.0	93.0	76.3	148.2	222.1	122.6	82.5	535.5	506.3	20.0
Sr	1263.5	1307.0	1171.0	1112.9	488.0	715.4	1746.9	2314.6	828.8	1498.4	607.3
Ba	1233.0	1320.0	1541.0	1011.2	1379.1	3334.3	1838.3	1261.2	3067.9	3703.6	303.2
Zr	628.0	468.0	272.0	315.4	336.6	239.5	403.8	45.8	238.1	249.6	307.2
Y	41.5	44.0	21.0	30.5	33.8	32.7	45.2	39.7	9.2	13.3	37.0
Nb	297.0	229.0	150.0	109.9	143.8	188.3	206.9	266.9	169.6	209.7	173.1
Cs	248.5	139.0	41.0	105.8	17.4	0.0	14.1	0.0	0.0	0.0	0.0
Sc	13.0	12.0	39.0	32.6	34.9	4.1	9.0	5.1	0.0	0.0	35.0
V	639.0	628.0	442.0	469.0	503.3	360.2	286.3	989.2	99.1	133.0	510.3
Cr	9.5	18.0	70.0	36.6	43.6	55.3	0.0	0.0	9.2	35.8	290.1
Ni	26.0	26.0	53.0	87.5	92.6	32.7	18.1	14.3	10.2	15.3	90.0
Cu	116 5	124.0	170.0	15 0	175.4	222.1	54.2	69.3	24.5	28.6	117.1
Zn	197.5	178.0	94.0	122.1	111.1	137.1	147.7	97.8	156.4	199.4	113.1
Ga	30.5	27.0	18.0	20.3	28.3	20.5	24.1	21.4	25.5	22.5	20.0
La	136.5	151.0	94.0	97.7	111.1	52.2	135.6	58.1	68.5	82.8	153.1
Ce	285.0	322.0	189.0	179.0	214.6	90.1	264.2	99.8	107.3	107.4	321.2
Pr	30.0	33.0	22.0	20.3	25.1	12.3	29.1	13.2	14.3	17.4	33.0
Nd	124.0	139.0	87.0	83.4	90.4	35.8	103.5	29.5	20.4	24.5	137.1
Hf	15.5	12.0	10.0	8.1	12.0	4.1	9.0	0.0	0.0	0.0	8.0
Pb	10.0	5.0	2.0	8.1	6.5	6.1	8.0	0.0	0.0	4.1	15.0
Th	29.0	27.0	21.0	12.2	10.9	9.2	13.1	0.0	21.5	23.5	4.0
U	0.0	0.0	0.0	14.2	27.2	23.5	20.1	5.1	12.3	14.3	28.0
Sm	15.0	15.0	6.0	10.0	11.0	4.0	12.0	5.0	9.0	8.0	13.0
Ce/Pb	28.5	64.4	94.5	22	32.8	14.7	32.9	—	—	26.3	21.4
Nb/U	—	—	—	7.7	5.28	8.0	10.3	52.4	13.8	14.6	6.2
Ba/Rb	17.7	13.3	16.6	52.6	9.3	15.0	15.0	15.3	5.7	7.3	15.2
Rb/Sr	0.1	0.1	0.1	0.1	0.3	0.3	0.1	0.0	0.6	0.3	0.0
Zr/Y	15.1	10.6	13.0	10.3	10.0	7.3	8.9	1.2	25.9	18.8	8.3
Nb/Y	7.2	5.2	7.1	3.6	4.3	5.8	4.6	6.7	18.4	15.8	4.7
La/Sm	9.1	10.1	15.7	9.8	10.1	13.0	11.3	11.6	7.6	10.4	11.8

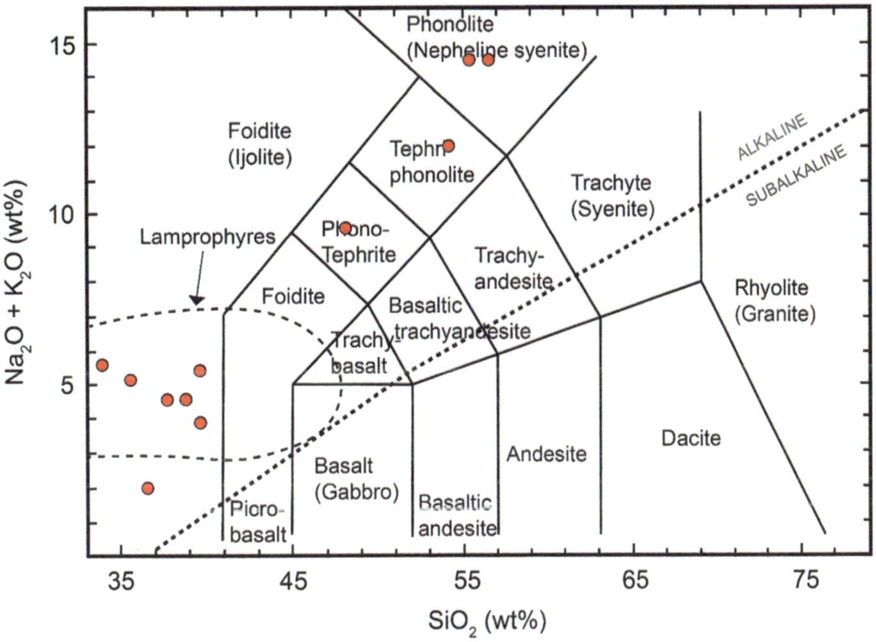

Figure 8.5 SiO_2 versus ($Na_2O + K_2O$) discrimination diagram adapted from Middlemost (1994).

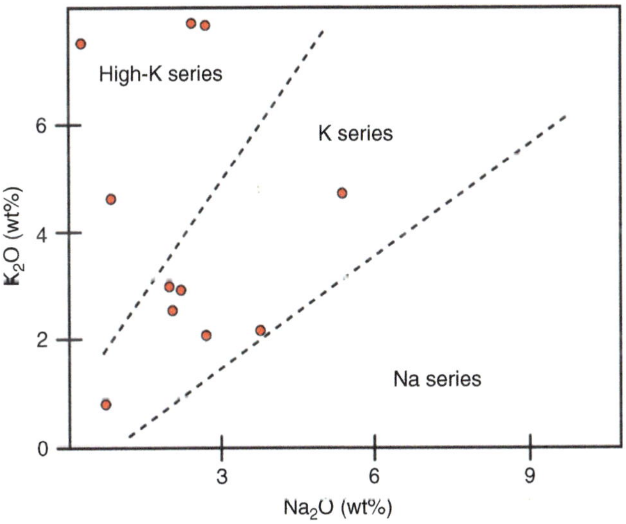

Figure 8.6 Na_2O versus K_2O discrimination diagram (fields adapted from Stoppa et al., 2014) that demonstrates the potassic nature of the studied samples.

demonstrate the most similarity with OIB trace element contents from the center of the diagram to the right (Ce-Y), except for Hf, which is bdl for both samples. The phonolite sample (AT001) demonstrates similar enrichment compared to the syenites but is characterized by positive anomalies at Ba and Nb and a negative anomaly at Th. Trace element contents for the phonolite intersect OIB values at Zr and are lower than OIB averages for HFS element Hf and REE Sm and Y.

Certain incompatible trace element ratios (TERs) (Ce/Pb and Nb/U) are known to remain constant throughout magmatic evolution and are thus referred to as "canonical" ratios in the literature (see Workman & Hart, 2005 and references therein). The constancy of these ratios is interpreted to be due to bulk partition coefficients that are nearly the same for each pair of elements during upper mantle melting, which prevents noticeable fractionation. Thus, these ratios are presumed to remain similar between the melt and the source.

Due to their presumed constancy during melting, canonical TERs are employed in this study to identify the sources from which they were derived. Ce/Pb, Nb/U, Ba/Rb, and Rb/Sr ratios are plotted against published values from the literature in Figures 8.10 and 8.11 and discussed shortly. Ce/Pb ratios for nine of the studied samples vary between 21.4 and 94.5, whereas the other two samples yield undefined Ce/Pb ratios due to bdl Pb contents. These ratios are quite scattered in comparison to published Ce/Pb values of 20 and 25 (see Workman & Hart, 2005 and references therein), although the average (30.6) is only slightly higher than the uniform value of 25±5 for both MORBs and OIBs published in Hofmann (1997). In Figure 8.10a, sample Ce/Pb ratios are plotted alongside the MORBs and OIBs average Ce/Pb values (Hofmann, 1997), average PM (Sun and McDonough, 1989), and average continental crust (Rudnick and Gao 2003) for comparison. Sample ratios plot to the right of the diagram with four of the samples within or bordering the MORB and OIB fields. One ratio

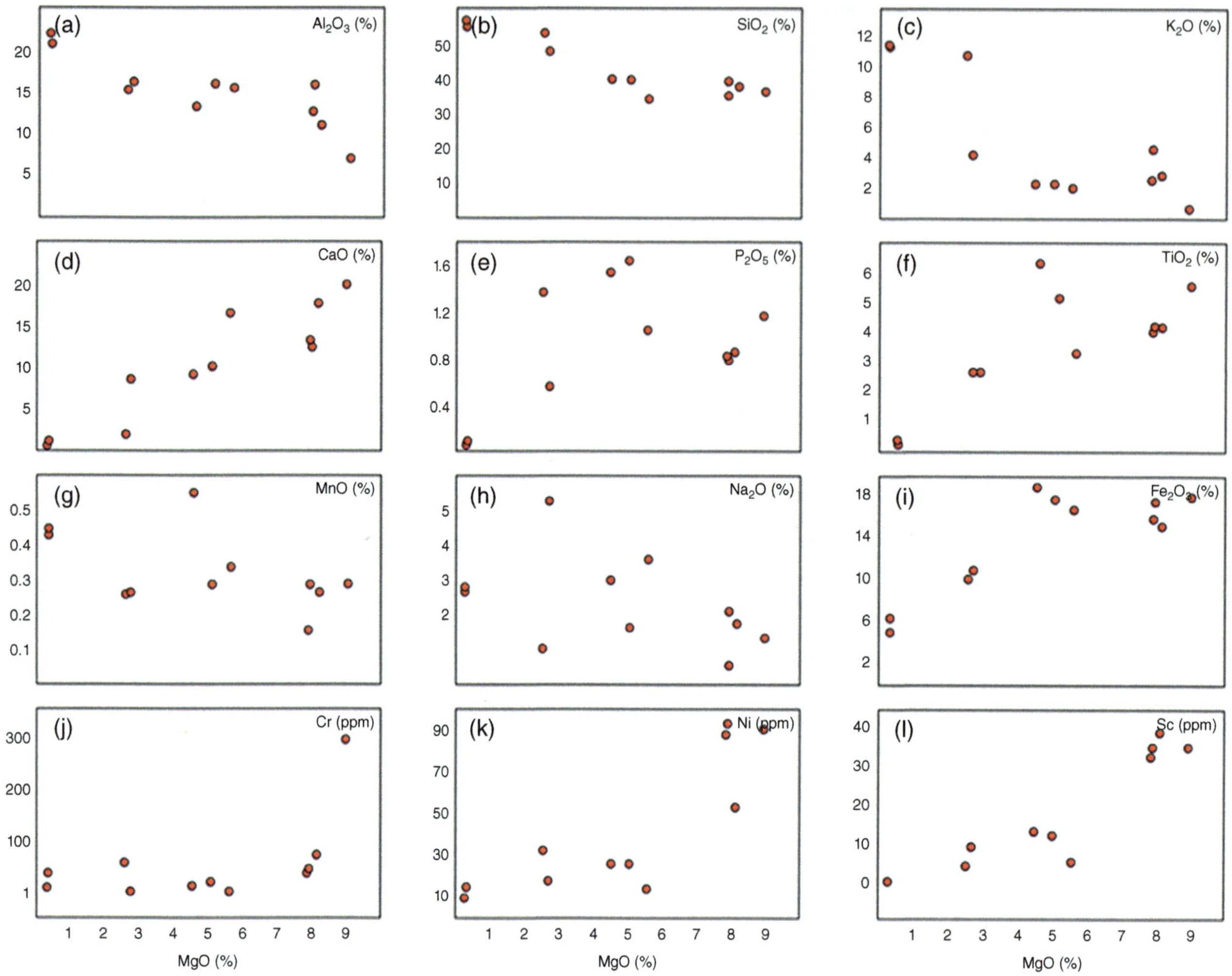

Figure 8.7 Harker plots of MgO (%) versus (a) Al_2O_3, (b) SiO_2, (c) K_2O, (d) CaO, (e) P_2O_5, (f) TiO_2, (g) MnO, (h) NaO, (i) Fe_2O_3T, (j) Cr, (k) Ni, and (l) Sc.

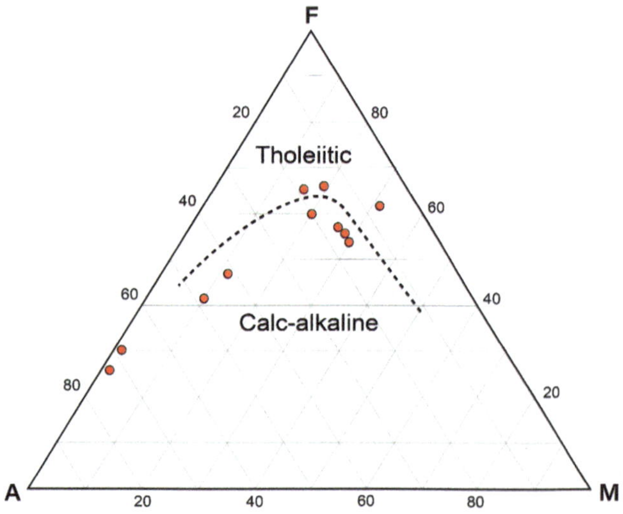

Figure 8.8 AFM diagram adapted from Irvine and Baragar (1971), with A = Na_2O+K_2O, F = $Fe_2O_3^T$, and M = MgO.

is slightly lower (~14.6) than the MORB-OIB minimum boundary line, and the remaining four are higher than the maximum bounding line. Lamprophyre samples AB1910 and AB1912 exhibit the highest Ce/Pb ratios, with 64.4 and 94.5, respectively.

Nb/U ratio values for the samples exhibit significant variation between 5.28 and 52.4, with three of the lamprophyre samples (AB1909, AB1910, and AB1912) yielding no ratios due to bdl U contents. The average ratio (14.8) for the samples is significantly lower than the average ratio of 47±10 taken from a suite of oceanic basalt data published in Hofmann (1997) and discussed in Sims and DePaolo (1997). Figure 8.10b indicates that Nb values for the samples do not vary significantly and thus scatter in Nb/U ratios is primarily controlled by differing amounts of U in the samples, which is known to be mobile during alteration and/or rock-fluid interaction (Dawood et al., 2004). The majority of ratios interestingly

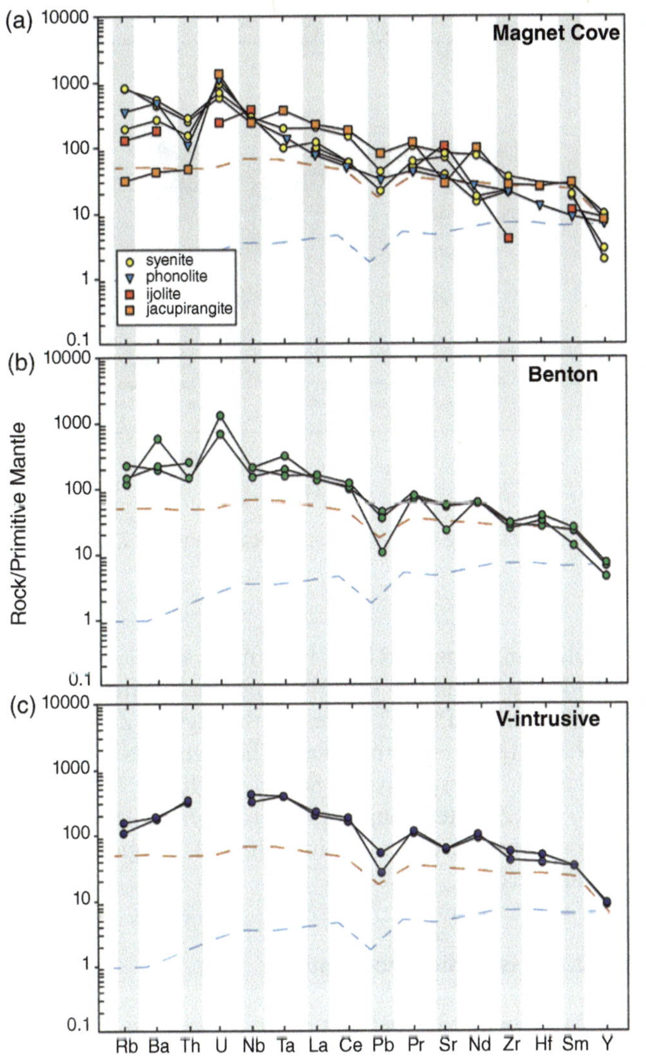

Figure 8.9 Trace element spider diagrams normalized to primitive mantle and MORB values adapted from McDonough (1989). Trace element abundances are normalized and separated by AAP province. OIB values are given by an orange dashed line, and EMORB values are given by a blue dashed line.

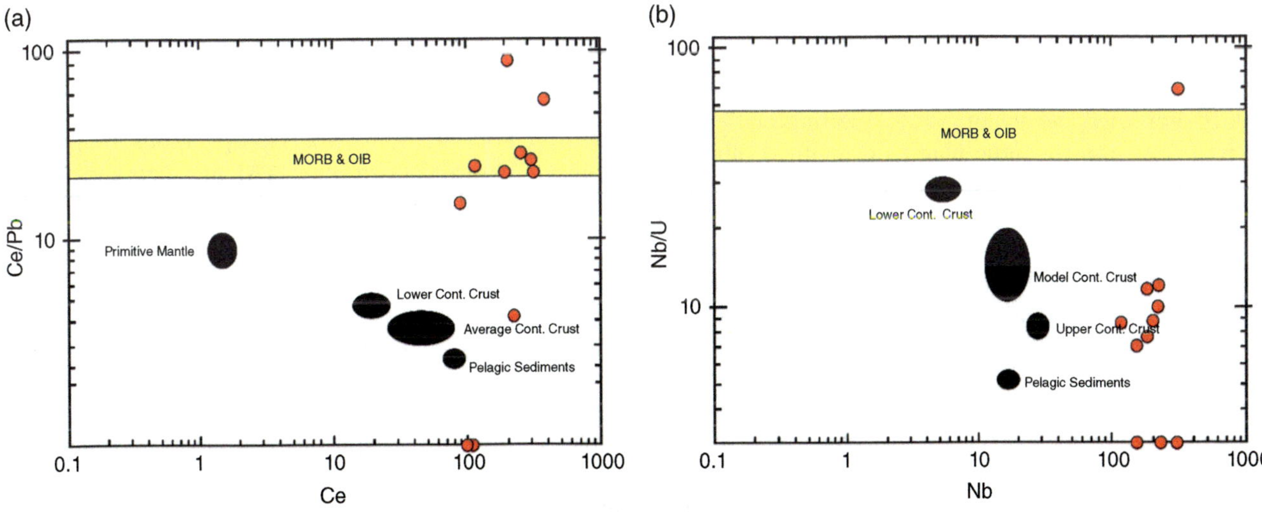

Figure 8.10 Trace element ratio diagrams for the studied samples including (a) Ce versus Ce/Pb and (b) Nb versus Nb/U. Plots and average values for MORB and OIB, continental crust, pelagic sediments, and primitive mantle are adapted from Sims and DePaolo (1997) and references therein.

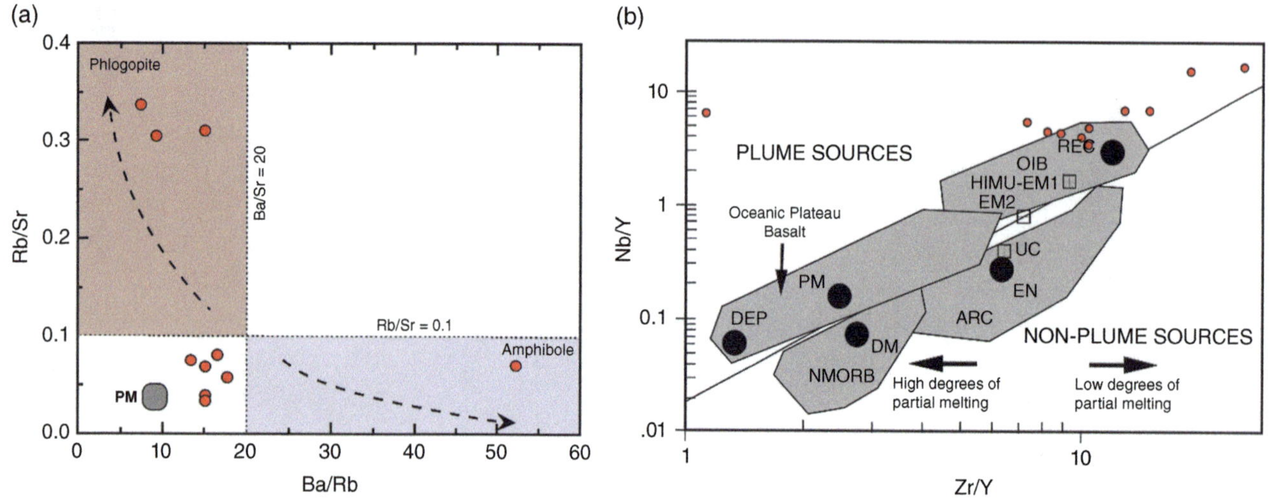

Figure 8.11 Trace element ratio diagrams for the studied samples including (a) Ba/Rb versus Rb/Sr and (b) Zr/Y versus Nb/Y adapted from Baksi (2001).

plot well below the minimum boundary line (37) for average MORB-OIB Nb/U ratios, and exhibit similar ratios to average continental crust after Rudnick and Gao (2003) but higher amounts of Nb. GBI001 is the only sample in this dataset with a Nb/U ratio that plots near the average MORB-OIB boundary lines, with a ratio of 52.4.

Ba/Rb and Rb/Sr ratios are variable for the samples (5.7–52.6 ppm and 0.03–0.6 ppm, respectively) (Figure 8.11a). Ten samples have lower Ba/Rb (< 20) whereas a single lamprophyre sample (AB1914) has a very high Ba/Rb of 52.3. On a Ba/Rb versus Rb/Sr diagram (e.g., Furman & Graham, 1999), six of the lower Ba/Rb samples plot near average primitive mantle Rb/Sr and Ba/Rb values (Sun & McDonough, 1989: Figure 8.11a). Three low Ba/Rb samples (AT001, GPS002, and IDB001) plot in the phlogopite source field, whereas the higher Ba/Rb sample (AB1914) plots in the amphibole source field.

Fitton et al. (1997) initially suggested that a deep depleted mantle source can be distinguished from a shallow MORB source using the TERs Zr/Y and Nb/Y. Baksi (2001) supported this hypothesis by plotting mafic lavas derived from the Icelandic plume on multiple Nb-Y-Zr plots. The same plot is used for this study and shown in Figure 8.11b. Zr/Y and Nb/Y sample ratios are plotted against average mantle compositional components (enriched mantle, depleted mantle, primitive mantle, and recycled component) and various tectonic settings (oceanic plateau, NMORB, subduction, volcanic arc, and OIBs) (Weaver, 1991; Condie, 2003). Each of the sample Nb/Y ratios plot above the ΔNb line that discriminates between plume and nonplume sources. The majority of samples plot to the far right of the diagram, with four samples plotting within the bounds of the OIB polygon and near the recycled component (REC). Four more samples plot just outside the bounds of the OIB polygon. Syenite samples GPS 001 and GPS 002 represent the highest Zr/Y ratios (25.9 and 18.8, respectively) and the highest Nb/Y ratios (18.4 and 15.8, respectively). Sample GBI001 (Nb/Y = 6.7, Zr/Y = 1.2) interestingly plots to the far left of the diagram, nowhere near compositional components or tectonic settings.

8.4.2. Nd Isotopic Geochemistry

Whole-rock Nd for four lamprophyre samples are listed in Table 8.2. Nd and Sm data are normalized to ^{146}Nd/^{144}Nd = 0.7219 and ^{149}Sm/^{152}Sm = 0.51685, respectively. Lamprophyre samples demonstrate a range of values for the Sm-Nd system, with ^{147}Sm/^{144}Nd values between 0.095128 and 0.101524 and more consistent ^{143}Nd/^{144}Nd values (0.512663 to 0.512798). Initial ^{143}Nd/^{144}Nd values define a range between 0.512648 and 0.512734. The lamprophyres yield positive εNd values between +2.66 and +4.37 and depleted mantle model ages (T_{DM}) of 346 to 467 Ma. Isotopic ratios are plotted against published data for other AAP members and mantle end members including enriched mantle (EMI, EMII), depleted mantle (DM), and primitive helium mantle (PHEM) (Figure 8.12).

8.4.3. (U-Th)/He Results

Zircon single grain (U-Th)/He data are reported in Table 8.3 for two lamprophyre samples AB1909 and AB1910. Two zircon crystals from sample AB1909 yielded ages ranging from 148.8 to 168.2 Ma with an average age

Table 8.2 Whole-rock Nd isotopic composition of AAP lamprophyres.

Location sample	V-intrusive AB1909	V-intrusive AB1910	Benton AB1912	Benton AB1914
Sm (ppm)	20.94	20.13	8.93	21.24
Nd (ppm)	129.32	127.96	55.98	126.50
Sm/Nd	0.161953	0.157345	0.159545	0.167925
$^{147}Sm/^{144}Nd$	0.097914	0.095128	0.096457	0.101524
$(^{143}Nd/^{144}Nd)_i$	0.512734	0.512732	0.512663	0.512648
$\pm 2\sigma$	0.000024	0.000024	0.000024	0.000024
$\varepsilon Nd(t)$	4.37	4.32	2.94	2.66
T_{DM}(Ma)	346.50	343.92	433.96	467.14

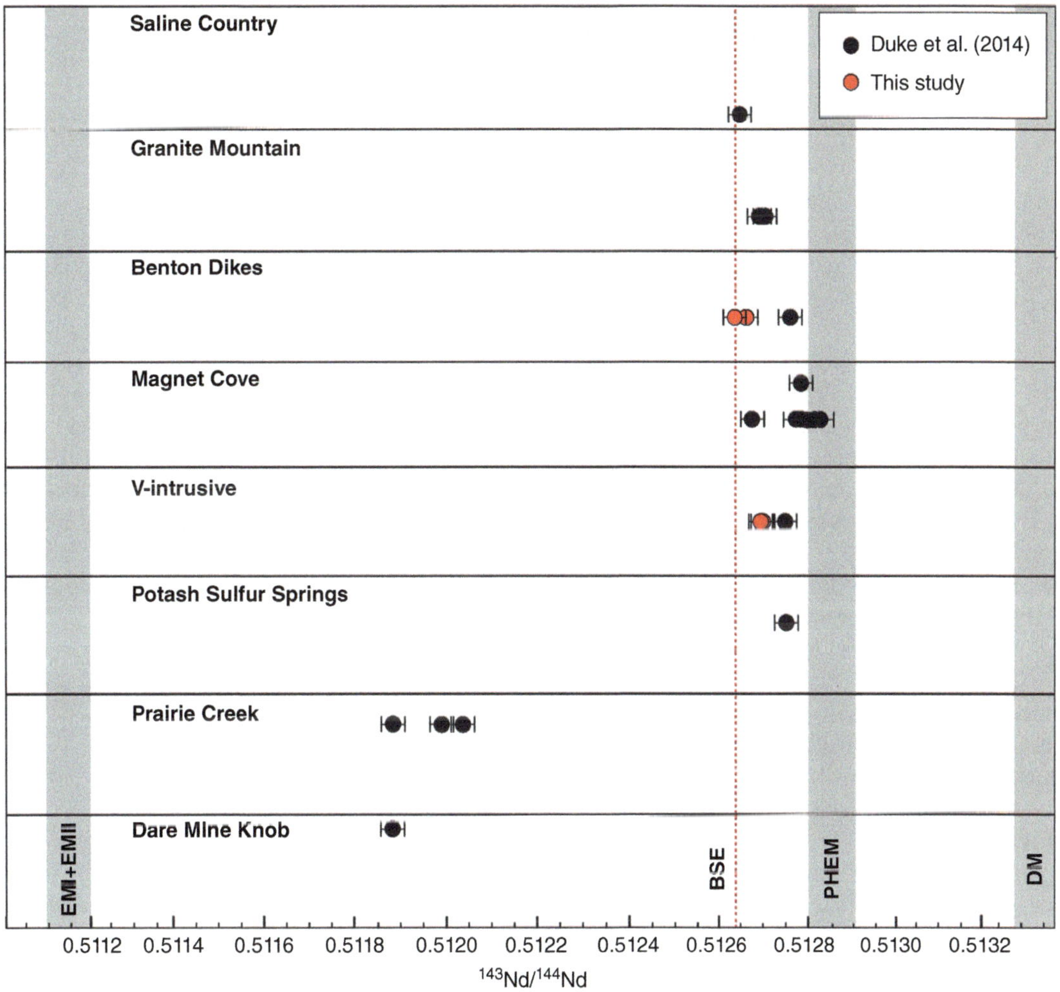

Figure 8.12 $^{143}Nd/^{144}Nd$ isotopic ratio plots for all eight provinces of the AAP. Various mantle reservoirs—EMI, EMII, PHEM, and DM—are plotted for comparison (EMI, EMII, DM, BSE: Rollinson, 1993). Values from Duke et al. (2014) are also plotted alongside data produced in this study.

of 158.5 Ma. Two zircon crystals from sample AB1910 yielded ages ranging from 147.0 to 175.1 Ma with an average age of 161.1 Ma.

Th/U ratios for the zircons are also reported in Table 8.3. The lamprophyre samples (AB1909 and AB1910) demonstrate Th/U values between 0.37 and 3.59, with an average Th/U = 1.25. Due to the possibility of inheritance and partial resetting of these grains from the host rock, we interpret these dates as maximum emplacement ages. Due to a low silica content and relatively low Zr content, it remains unclear if sampled dikes reached zircon saturation and crystallized zircon. If the youngest ages are in fact

Table 8.3 Summarized zircon (U-Th)/He data.

Lithology	Sample	Age (Ma)	error[a]	U (ppm)	Th (ppm)	Th/L	He (nmol/g)	Ft
Volcanic tuff	zAB-19-5-1	178.4	14.28	74.8	28.9	0.39	65.5	0.82
	zAB-19-5-2	212.6	17.01	98.4	58	0.59	111.8	0.86
	zAB-19-5-3	128.5	10.28	15.4	26.7	1.74	13.0	0.86
	zAB-19-6-1	192.0	15.36	85.5	96.0	1.12	93.6	0.82
	zAB-19-6-2	134.2	10.74	392.3	214.0	0.55	253.1	0.78
	zAB-19-6-3	157.1	12.57	83.7	32.3	0.39	65.0	0.83
	zAB-19-7-1	141.9	11.35	54.4	65.4	1.20	39.8	0.74
	zAB-19-7-2	186.1	14.89	129.9	73.7	0.57	107.6	0.72
	zAB-19-7-3	183.1	14.65	31.3	4.3.4	1.39	30.9	0.74
Lamprophyre	zAB-19-9-1	148.8	11.90	64.4	23.7	0.37	44.9	0.79
	zAB-19-9-3	168.2	13.45	69.5	40.1	0.58	52.8	0.73
	zAB-19-10-1	147.0	11.76	9.1	32.6	3.59	9.6	0.71
	zAB-19-10-2	175.1	14.01	53.5	23.7	0.44	42.4	0.75

[a] Calculated method error.

emplacement ages, then this would argue against any sort of age progression identified previously.

8.4.4. Petrography

Based on SEM thick-section imaging and EDS analyses, the discussed samples can be divided into four primary groups: lamprophyre, ijolite, jacupirangite, and syenite. Mineral assemblages are typical for each lithology, as discussed in the following sections. The lamprophyres are fine to coarse grained with a hypocrystalline porphyritic and panidiomorphic texture. Mineralogy of the lamprophyres varies, as one sample (AB1914) contains forsteritic olivine and the remaining four are olivine-free. Common minerals among every lamprophyre are amphibole and abundant Fe-Ti oxides.

Lamprophyre samples from the V-intrusive arc mineralogically similar with phases including apatite, Fe-Ti oxides (titanomagnetite and ulvospinel), amphibole (kaersutite and hornblende), plagioclase feldspar, and minor amounts of biotite (Figure 8.13a,b). Fe-Ti oxides are poikilitic with smaller amphibole inclusions. Biotite occurs primarily as microphenocrysts and inclusions in feldspar. The matrix of these samples comprises plagioclase, amphibole, and Ti oxides. Based on their mineral assemblage, these olivine-free lamprophyres are classified as camptonites (Rock, 1991).

Lamprophyres from the Benton Dike Swarm (AB1912 and IDB001) differ mineralogically from one another and from the discussed V-intrusive samples (Figure 8.13c,e). Sample AB1912 is characterized by amphibole (kaersutite), Fe-Ti oxides, apatite phenocrysts, plagioclase and alkali feldspar, rutile, and minor amounts of biotite, hauyne, and pyrite. Apatite is represented by large (~500 μm) bladed phenocrysts that exist as inclusions within feldspar and form radial nests (not shown). Rutile occurs as phenocrysts and as a vein (Figure 8.13c). Feldspar occurs in the groundmass. Sample IDB001 is characterized by subhedral Fe-Ti oxides (ulvospinel and titanomagnetite), large (>200 μm) biotite laths, ankerite, and minor zeolite (thomsonite), and pyrite. Interstitial feldspar is present within the biotite laths. The groundmass comprises primarily feldspar and minor amounts of zeolite. According to their mineral assemblages, these samples are classified as ouachitite (IDB001) and sannaite (AB1912) (Rock, 1991).

Sample AB1914 is characterized by forsteritic olivine phenocrysts (>300 μm), hornblende, pyroxene, Fe-Ti oxides (ulvospinel and titanomagnetite), and minor amounts of phlogopite, pyrite, and apatite. The groundmass comprises minerals similar to the phenocrysts. Fe-Ti oxides and biotite are found to be inclusions in pyroxene. Forsteritic olivine is abundant throughout the rock and is crosscut with iddingsite veins. Carbonate veins are also present in the rock and cut through amphibole but not olivine. This sample is classified as a monchiquite according to the classification scheme of Rocks (1991).

The remaining six samples were collected from the Magnet Cove intrusive body (see Figure 8.2b) and are discussed in terms of their mineralogy and petrographic characteristics in the following sections. These samples vary in lithology between jacupirangite, ijolite, phonolite, and syenite. The ijolite sample (GBI001) was collected from the interior of the Magnet Cove (Figure 8.3b). This sample is holocrystalline porphyritic with mostly anhedral grains. Minerals include anhedral masses of zeolite (thomsonite), biotite partially altered to phlogopite, clinopyroxene, and melanite garnet. Smaller amounts of perovskite, Fe-Ti oxides, and pyrite are also present in the rock. Zeolite is commonly formed through alteration of nepheline (Erickson & Blade, 1963). Pyrite and laths of mica occur as inclusions within zeolite. Subhedral calcite is also present as inclusions within zeolite.

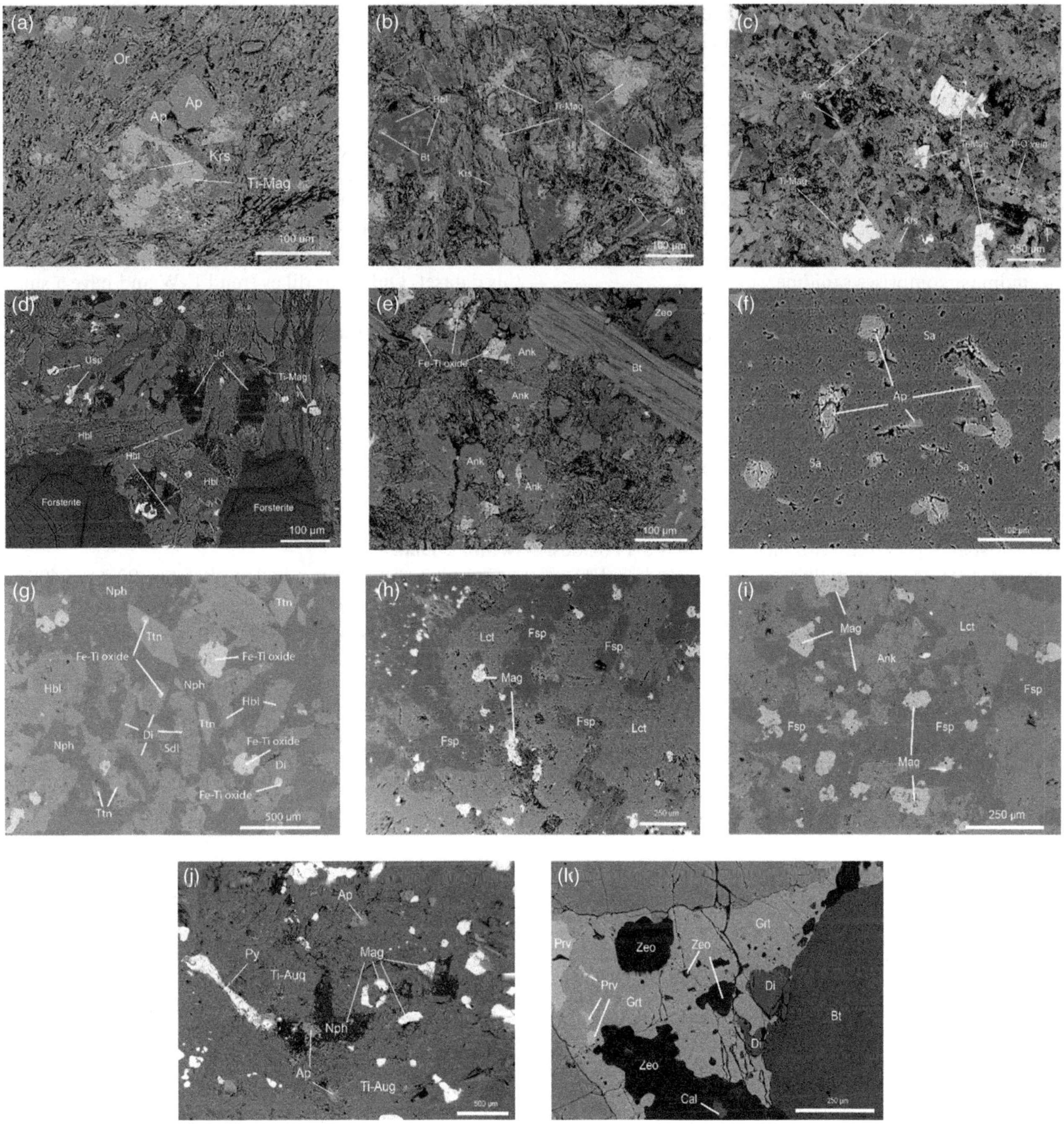

Figure 8.13 Backscattered electron images of 11 samples: (a) AB1909, (b) AB1910, (c) AB1912, (d) AB1914, (e) IDB001, (f) AT001, (g) Cook001, (h) GPS001, (i) GPS002, (j) Jacu001, and (k) GBI001. Ab, albite; Ank, ankerite; Ap, apatite; Aug, augite; Bt, biotite; Cal, calcite; Di, diopside; Fsp, Feldspar; Grt, garnet; Hbl, hornblende; Jd, jadeite; Krs, kaersutite; Lct, leucite; Mag, magnetite; Nph, nepheline; Or, orthoclase; Prv, perovskite; Sa, sanidine; Sdl, sodalite; Ttn- titanite; Usp, ulvospinel; Zeo, zeolite.

Due to its location within Magnet Cove and associated mineral assemblage, this sample is classified as a garnet biotite ijolite.

The jacupirangite sample (Jacu001) was collected from the northeast corner of the Magnet Cove province (Figure 8.3b) within the outer syenite ring, which is one of two locations within Magnet Cove where jacupirangite occurs. This sample is hypocrystalline porphyritic. Primary phenocrysts include abundant magnetite and fluorapatite, with lesser amounts of pyrite, nepheline, and perovskite. The anhedral nepheline masses are poikilitic with fine-grained apatite inclusions and finer magnetite inclusions. Nepheline is glomeroporphyritic, occurring in clusters and constrained to specific parts of the rock. The groundmass is primarily titanaugite.

Phonolite was sampled from the inner belt of Magnet Cove, which primarily comprises phonolite and trachyte. Sample AT001 is hypocrystalline porphyritic and exhibits significant weathering. Apatites are the only identified phenocrysts and are abundant throughout the rock. The groundmass comprises sanidine.

Three syenite samples were collected from the Cove's outer ring, which includes varieties of feldspathoidal syenite. Samples GPS 001 and GPS 002 were sampled in close proximity and thus contain some similar characteristics. Each of these syenites is holocrystalline porphyritic, with pseudoleucite megacrysts and smaller phenocrysts that vary between each sample. Besides the pseudoleucite crystals, GPS001 mostly comprises anhedral masses of k-feldspar and Fe-Ti oxides (magnetite and ilmenite). The feldspar is poikilitic with inclusions of Fe-Ti oxides. Minerals that occur in smaller amounts include anhedral masses of gahnite and ankerite, and spessartine-type euhedral garnet. GPS 002 contains a similar mineral assemblage but has larger amounts of ankerite and includes phases of biotite and sphalerite. Both GPS001 and GPS002 are classified as garnet-pseudoleucite syenites due to their characteristic pseudoleucite and garnet crystals.

The third syenite sample (Cook001) was collected from the outer ring at the western edge of the Magnet Cove complex. Like the other two syenites, this sample is holocrystalline porphyritic and characterized by euhedral sphene crystals, hornblende, clinopyroxene, subhedral Fe-Ti oxides, and anhedral masses of nepheline and sodalite. Both titanite and clinopyroxene are poikilitic with inclusions of Fe-Ti oxides. According to its mineral assemblage, this sample is identified as a sphene-nepheline syenite.

8.4.5. Dike Orientations

Igneous intrusions are separated into three early Late Cretaceous-age divisions (>100 Ma, between 100 Ma and 90 Ma, and <90 Ma). Dikes mapped west of the 100 Ma boundary line are referred to as "Group 1" dikes, and dikes between the 100 and 90 Ma boundary lines are "Group 2" dikes (Figure 8.14). The only mapped igneous body east of the 90 Ma line is the Granite Mountain province. Age divisions are taken from Eby and Vasconcelos (2009), which uses best ages from $^{40}Ar/^{39}Ar$ radiometric age-dating of phlogopite grains at Magnet Cove, fission-track dating of apatite and titanite from multiple AAP members, and previously published age data that include $^{40}Ar/^{39}Ar$ ages of Baksi (1997), K-Ar ages (Gogineni et al., 1978), and the U-Pb zircon age of Zartman and Howard (1987).

Group 1 Dikes

Group 1 dikes have a maximum length of 1,764 m and a minimum length of 107 m. In particular, the NE-SW-trending dikes reach lengths of ~900 m but are more typically between 100 and 400 m. NW-SE-trending dikes generally range between 100 and 500 m, with a single dike at 1,764 m, which represents the maximum length for this dataset. Only four dikes trend E-W and have a small range in lengths between 186 and 267 m. There are no dikes west of the 100 Ma boundary line that trend exactly N-S. The majority of dikes in Group 1 trend NE-SW, with 070° to 075° azimuths being the most prominent (Figure 8.13a). These dikes demonstrate no preferred location and randomly lie west of the 100 Ma boundary line. Additionally, there is no apparent correlation between length and azimuth for NE-SW trending dikes

Group 2 Dikes

Group 2 dikes range in length from 92 to 5,408 m. NE-SW-trending dikes reach up to ~5,400 m in length but are generally from 90 to 1,000 m in length. The NW-SE-trending dikes have a maximum length of 1,971 m with the majority of dikes ranging in length from 90 to 1,000 m, similar to NE-SW-trending dikes. Three dikes are trending exactly N-S and demonstrate a small range in length from 116 to 222 m. Four dikes trend exactly E-W and also demonstrate a small length range from 318 to 351 m. Dikes in Group 2 exhibit trends in nearly every direction, although there is a prominent trend from N085° to N095°(ENE-ESE) (Figure 8.15b). There is no apparent correlation between length and azimuth for ENE-ESE-trending dikes. Although the rose diagram indicates a variety of orientations, the dikes do not follow a specific pattern of orientations. Instead, dikes of varying orientations are found sporadically in between the 90 and 100 Ma bounding lines.

Thrust-Fault Orientations

Thrust faults are abundant throughout the OFB and represent a segment of a Pennsylvanian-Permian thrust system formed during the Ouachita orogeny and assembly of the Pangea Supercontinent (Whitaker & Engelder, 2006). Thrust orientations in the study area are divided into groups based on the age divisions of Eby and Vasconcelos (2009) to maintain consistency with rose diagrams of dikes. Because of their proximity to dikes, orientations of thrust faults are compared with orientations of dikes to investigate the control of local, preexisting planes of weakness on Cretaceous-age magma intrusion. Similar to the dike groupings, thrust faults west and east of the 100 Ma discriminatory line are classified as

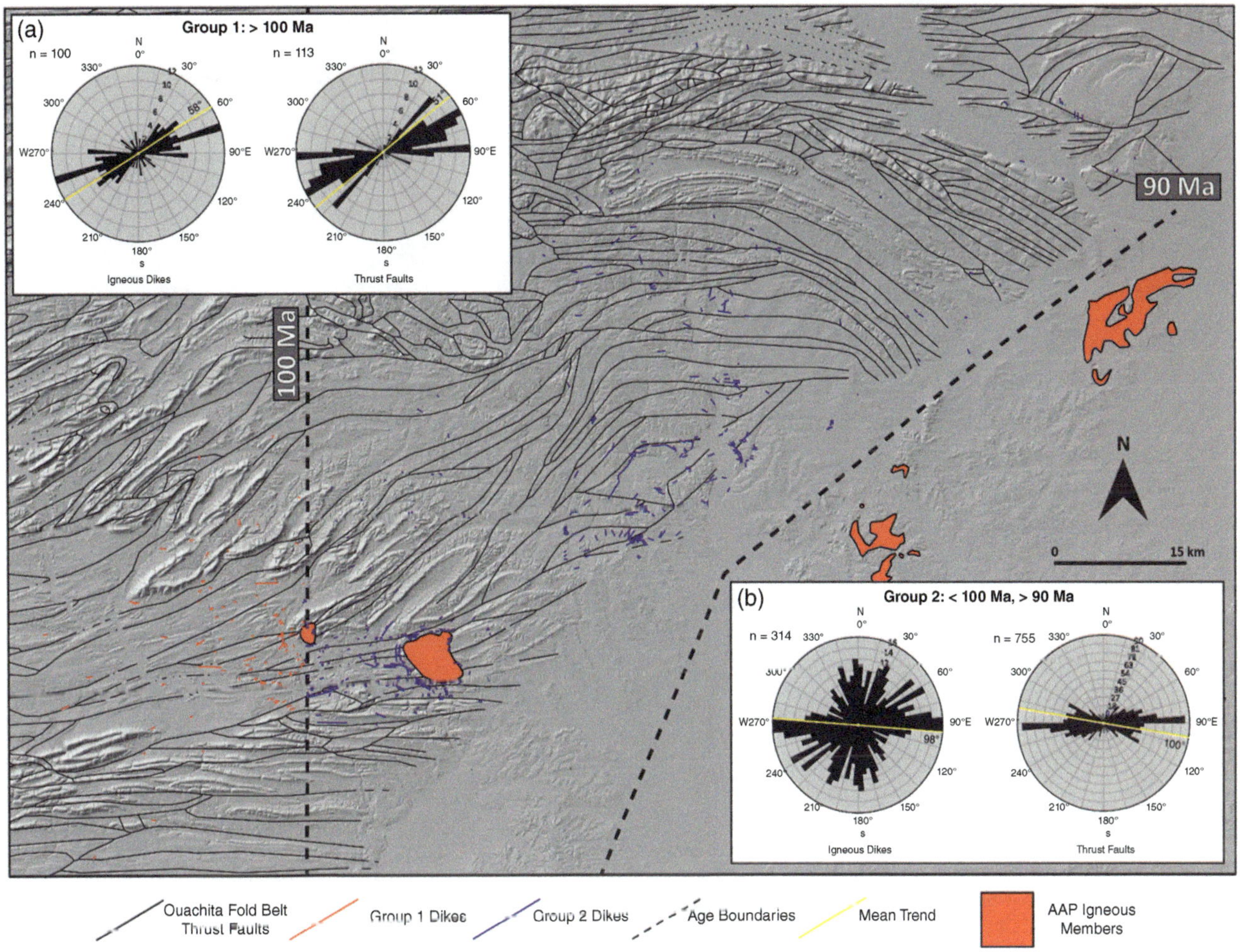

Figure 8.14 Digital elevation map of central Arkansas with overlain polygons of AAP igneous intrusions in red. Rose diagrams are shown for (a) igneous dikes greater than 100 Ma in age and associated thrust faults; (b) igneous dikes between 90 and 100 Ma in age and associated thrust faults. Mean orientation of each diagram is indicated in yellow. See Figure 8.1 for geographic location.

Group 1 and Group 2, respectively. Group 1 thrust faults strike primarily between 040° and 085° (NE-ENE). There are two prominent trends at 060° to 065° and 085° to 090°. This is generally consistent with Group 1 dike orientations (mean trend = 058°).

Basement Fault Orientations

The orientations of basement faults (locations taken from Thomas, 2014) were calculated and compared to dike trends. These faults (MVG normal faults, and the AOT transform fault) form a segment of the larger Laurentian continental margin and are related to Iapetan rifting in Late Precambrian to Cambrian time (~565 to 495 Ma) (Thomas, 2017). Thus, these lithospheric weaknesses existed prior to the emplacement of magmas that formed the Cretaceous-age AAP and are examined to investigate the control of preexisting structural fabrics on dike emplacement. Faults of the MVG generally strike NE-SW and lie between south-central Arkansas and southern Illinois. Two fault segments in Arkansas lie roughly 50 km to the southeast; their orientations are roughly 034° and 075°, respectively (Figure 8.15b). The AOT transform fault strikes at roughly 301° (Figure 8.15b).

8.5. DISCUSSION

8.5.1. Role of Partial Melting, Fractional Crystallization, and Crustal Contamination

Magmatic systems are affected by significant processes such as partial melting, fractional crystallization, and crustal contamination; these must be addressed before discussing the magma source. Trace element and major oxide geochemical data are used to investigate the influence of each process during magma evolution. Primary magmas (formed by partial melting of the upper mantle)

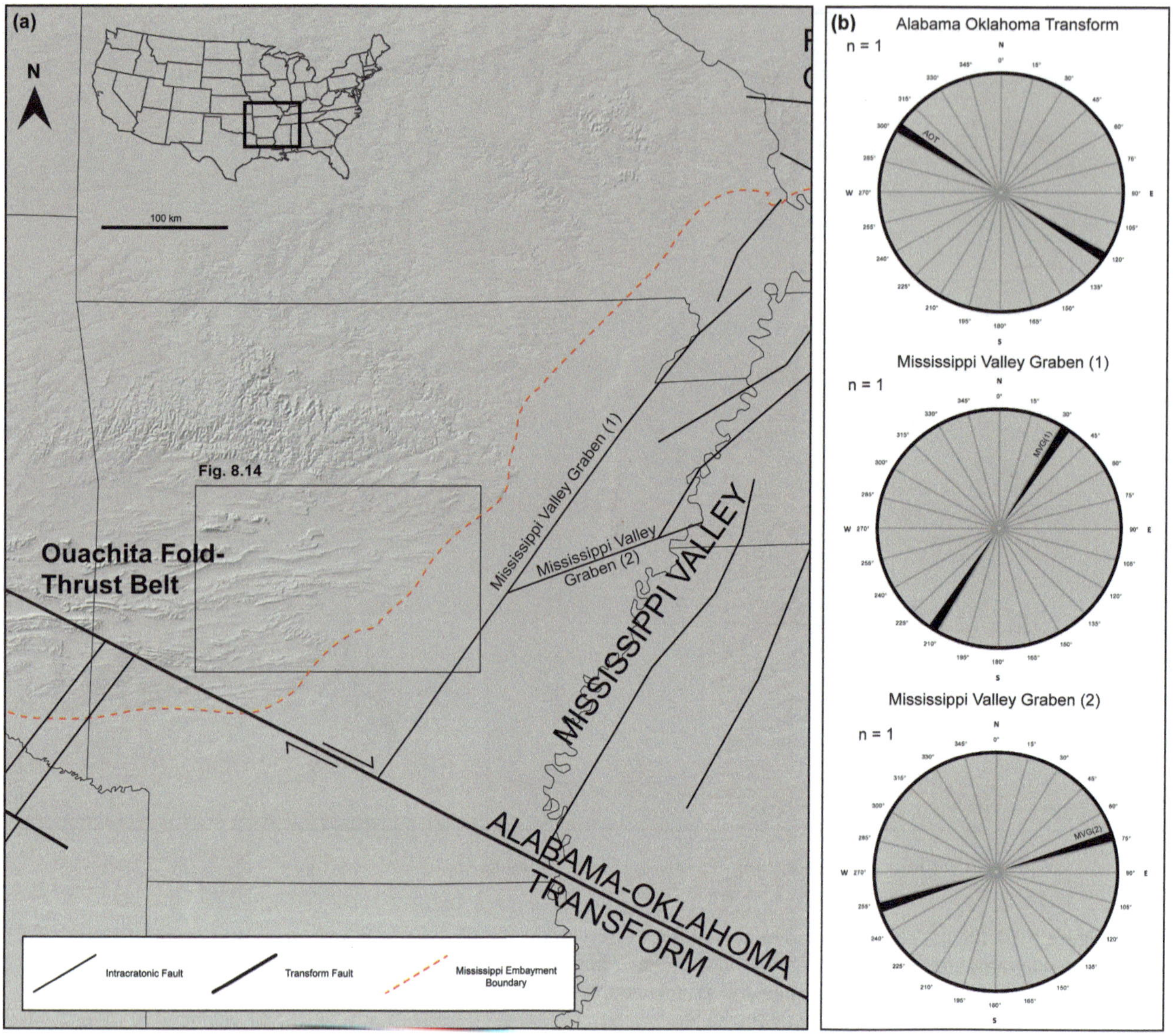

Figure 8.15 Digital elevation map of central Arkansas with basement faults (Alabama-Oklahoma Transform and Mississippi Valley Graben) shown in black and corresponding rose diagrams for individual faults shown on the right.

are expected to have high Ni (400 to 500 ppm) and Cr (>1,000 ppm) contents, with $SiO_2 \leq 50\%$ (Wilson, 1997). Although most of the samples analyzed during this study are silica-undersaturated, none demonstrate high Ni and Cr contents. Therefore, the samples must have developed from an evolved melt. Low degrees of partial melting are likely, suggested by extreme enrichment in incompatible elements (Figure 8.9a,b) and high Zr/Y (Figure 8.11b). This is further supported by enrichment in light-REE (Figure 8.9a,b). Plots of compatible elements and major oxides against MgO (Figure 8.7) indicate the effects of fractional crystallization. $Fe_2O_3^T$, CaO, and Sc all show positive correlations with MgO and thereby suggest a fractionated pyroxene phase. Ni shows a moderately positive correlation with MgO ($R^2 = 0.68$) and could indicate fractionation of olivine. A negative correlation with MgO and Al_2O_3 suggests that plagioclase was not an important fractionating phase. The lack of correlations between P_2O_5 and MgO indicates that apatite was not a significant fractionating phase. A weak correlation between TiO_2 and MgO could indicate that Fe-Ti oxides were fractionated.

The average Nb/U (14.8) is much lower than the canonical mantle ratio (Nb/U = 47 ± 10; Hofmann, 1997) and closer to the average continental crust (Figure 8.10b), reflecting some crustal contamination. However, the primary control on varying Nb/U ratios is varying U contents, which could be affected during fluid-rock interaction. Thus, Nb/U values must be viewed with caution. Ce/Pb ratios of the samples, on the other hand, are relatively high compared to crustal values, with one sample plotting near average continental crust (Figure 8.10a). Nd isotopes are also effective when determining the influence of crustal contamination. ^{143}Nd/^{144}Nd values (mean ^{143}Nd/^{144}Nd = 0.5127) are significantly higher than average continental crust (^{143}Nd/^{144}Nd = 0.5118; Hofmann, 1997), indicating little crustal contamination for the lamprophyres.

In summary, partial melting was an important process during magmatic evolution of the AAP, and some crustal contamination is evident. However, there is significant scatter for many elements, although some broad correlations are apparent between major oxides and trace elements. Given that the samples were collected from separate AAP outcrops, it is unlikely that the samples form a comagmatic suite.

8.5.2. Mantle Sources and Metasomatism

Nd isotope ratios reflect the history of mantle source reservoirs (Hofmann, 1997). The isotopic dataset presented in this study represents data for dikes from two separate provinces of the AAP (the V-intrusive and the Benton Dike Swarm); these plus six remaining provinces (Figure 8.3a) define the entire AAP suite. Thus, these data are discussed in unison with published geochemical data for other provinces in efforts to comprehensively understand magma sources that were potentially tapped during AAP evolution.

Nd isotopic data for the lamprophyre samples in the study area define a tight range of initial ^{143}Nd/^{144}Nd values (0.512648 to 0.512734; Table 8.2), indicating a similar source for dikes of the V-intrusive and Benton dikes. A mean ^{143}Nd/^{144}Nd ratio of 0.512694 is in agreement with ^{143}Nd/^{144}Nd ratios published for Magnet Cove carbonatites by Tilton (1987), but slightly lower than Granite Mountain syenites published in the same study. The ^{143}Nd/^{144}Nd data correspond to εNd(t) values of +2.66 to +4.37, which correspond well with carbonatite, ijolite, and jacupirangite samples Magnet Cove and Potash Sulfur Springs samples (εNd = +1.9 to +4.3) from Duke et al. (2014), which are interpreted to be derived from an asthenospheric source. These values are quite different from εNd(t) values for Prairie Creek and Dare Mine Knob (−19 to −10), which Duke et al. (2014) suggests is indicative of a lithospheric source. Nd isotopic data are roughly similar between the V-intrusive, Benton Dike Swarm, Potash Sulfur Springs, and Magnet Cove provinces and represents mantle-derived rocks. A mantle source for samples analyzed in this study and other published AAP isotopic data are corroborated by the presence of mantle minerals (olivine, pyroxene, garnet) in the samples (Figure 8.13).

Nd isotopic values demonstrate similarities to OIB-related mantle reservoirs. In comparison to other mantle reservoirs, ^{143}Nd/^{144}Nd isotopic values of the lamprophyres plot closely with ratio values for some low ^{4}He reservoirs (FOZO, PREMA, PHEM, and C) and are slightly higher than enriched mantle reservoirs (EMI and EMII), yet below average values for DM (Figure 8.12). Thus, a distinct mantle end-member source for the lamprophyres is not clear. However, trace element spider diagrams normalized to primitive mantle indicate extreme enrichment in incompatible elements for the lamprophyres (Figure 8.10a,b) indicating some enriched components in the source. Enrichment is especially evident in LILe and some HFSe (Figure 8.9a,b). On MORB-normalized spider diagrams, the lamprophyres are more enriched with respect to an average MORB and demonstrate a smooth slope similar to that of an average OIB (Sun & McDonough, 1989), with common OIB indicators such as strongly negative Pb anomalies and enrichment in both HFSe and LILe (Figure 8.9b).

Given that Nd isotopic data are only presented for the lamprophyre samples, the genesis of the remaining lithologies will be determined via TERs and spider diagram plots. Besides ratios of Nd isotopes, ratios of incompatible trace elements are also useful for fingerprinting the magmatic origin of the studied samples. Specific TERs are used due to their constancy during magma evolution. Figures 8.10 and 8.11 shows plots for a variety of TERs for the studied samples. In particular, Zr/Nb ratios for the samples are relatively low (<3), which is consistent with the Zr/Nb of OIBs (<10). The Nb/Y versus Zr/Y plot (Figure 8.11b) also suggests an OIB origin for the samples, as most samples plot near the OIB field and near the "recycled component" average, which could also explain the Nd isotopic ratios that lack a specific mantle end-member signature. Interestingly, the ijolite sample (which also shows a distinct trend on the MORB-normalized diagram; see Figure 8.9) plots to the far left of the diagram, which could indicate a garnet-bearing source.

A lack of mantle xenoliths in the studied samples limits direct information about the mantle source. Moreover, Morris & Fouad (1987) indicated the presence of granitoid, gneissic, and sedimentary xenoliths in some AAP members but not the presence of source xenoliths. High concentrations of incompatible trace elements in the samples (especially LILe; see Figure 8.9a,b) indicate that their mantle sources experienced some form of enrichment. In particular, Rb/Sr and Ba/Rb values indicate that phlogopite and amphibole were significant volatile-bearing phases for the samples (phlogopite for phonolite, syenite, and one lamprophyre sample, and amphibole for one lamprophyre; see Figure 8.11a). As discussed in Hawkesworth et al. (1984), these incompatible element ratios can be attributed to a mantle enrichment process,

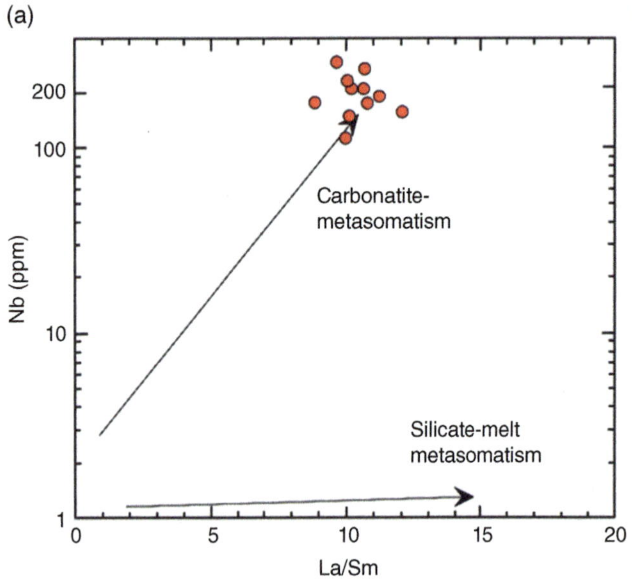

Figure 8.16 La/Sm versus Sr and Nb plots that demonstrate the effects of carbonatite versus silicate-melt metasomatism, adapted from Downes (2007).

such as metasomatism. Therefore, the presence of phlogopite or amphibole in the mantle sources (Figure 8.13a–c) indicates that metasomatism by fluids or volatile-rich melts occurred prior to melting. La/Sm versus Nb and Sr plots corroborate this, revealing that samples roughly plot along a line of carbonatite metasomatism (Figure 8.16a,b). Additional indicators of carbonatite metasomatism in whole-rock samples include the presence of apatite and a tendency toward wehrlite mineral assemblages (olivine, pyroxene, plagioclase, spinel, garnet, ilmenite, magnetite) which are seen in some BSE sample images (Figure 8.13) (Rudnick & Gao, 2003).

8.5.3. Paleostress Regimes Inferred from Dike Orientations

The orientations of mafic dike swarms can be used to investigate the paleostress fields that governed their emplacement in the crust. Dikes are generally emplaced parallel to the regional horizontal maximum compressive stress direction (σ_1) and perpendicular to the direction of extension (σ_3) (Pollard, 1987; Hoek & Seitz, 1995; Hou et al., 2010). Extension in the crust can result from uplift of the lithosphere above hot asthenosphere (potentially caused by a mantle plume) or through forces caused by plate tectonics (Hoek & Seitz, 1995). The trends of AAP dikes and associated crustal structures are discussed here in terms of their tectonic implications.

Dike orientations discussed in this study vary but demonstrate prominent trends that are used to infer Cretaceous-age directions of extension. Group 1 dikes (>100 Ma) show a prominent NE-SW trend with a mean strike of 058° (Figure 8.14a) and therefore reveal a NW-SE extensional stress regime at the time of their emplacement. These dikes follow the NE-SW-trending thrust faults of the Ouachita Mountains (mean strike 051°; Figure 8.14a); this suggests that dike emplacement in this area was controlled by local planes of weakness represented by potentially reactivated thrust planes during Cretaceous extension. Although Group 2 dikes (<100 Ma, >90 Ma) demonstrate a wider variety in trends (nearly every direction), a prominent E-W strike is apparent (Figure 8.14 b). A similar array of trends is observed for the thrust faults in the area, which also reveal a strong E-W strike, indicating that emplacement of Group 2 dikes was also controlled by local fabrics (Figure 8.14b).

It has been previously suggested that MVG basement faults served as pathways that guided melt into the overlying crust in the Cretaceous Period (Matton & Jebrak, 2009; Eby & Vasconcelos, 2009). A comparison of dike trends and the strikes of MVG rift faults does yield some similarities. The nearest MVG faults in Arkansas strike at N34°E and N75°E, respectively, which are roughly similar to the NE-SW-trending Group 1 dikes but less so with E-W-trending Group 2 dikes (Figure 8.15b). The NW-SE-trending AOT, which serves as the termination of MVG normal faults, does not demonstrate much similarity to dike trends (Figure 8.15). Rough similarities between dikes and MVG faults could indicate the influence of basement fabrics during their emplacement. However, due to stronger similarities between dikes and thrust faults, dikes appear to have been more influenced by local and regional stress directions compared to MVG faults.

Due to the similarities between Cretaceous dikes and Pennsylvanian-Permian-age thrust plane orientations, there is a case for the reactivation of Ouachita Mountain thrust faults during Cretaceous extension of the area. Extension must have reactivated thrust faults as normal faults, which allowed dikes to dilate along the strike of thrust faults. The reactivation of thrust faults caused by later crustal extension has been documented in several areas throughout the world (Bally et al., 1966; Çemen & Wright, 1990). Moreover, magma emplacement related to reactivated thrust faults has been discussed in detail in areas such as the Apennine Mountains, Italy (Musumeci et al., 2005; Dini et al., 2008), and the Mexican Fold and Thrust Belt (Ramirez-Pena et al., 2019). Dini et al. (2008) suggested that magmatic centers in the Apennine orogenic belt were focused by faults reactivated due to back-arc extension. Çemen and Wright (1990) determined that Cenozoic extension in the Death Valley area of the Basin and Range caused the reactivation of Mesozoic thrust planes.

8.5.4. AAP Petrogenesis

Competing hypotheses for mid-Cretaceous emplacement of the AAP and evolution of the adjacent ME each rely on differing emplacement mechanisms: (a) the existence of a mantle plume, presently situated in the Atlantic Ocean beneath the BR, or (b) a nonplume explanation which would involve tectonically induced melting and emplacement of magma forming the AAP. The acceptance of either of these hypotheses requires proving or disproving the existence of a mantle plume that would explain: isotopic mantle values for AAP igneous bodies, OIB-like trace element patterns, and significant uplift of the proto-ME region (Cox & Van Arsdale, 1997; 2002), and the apparent age-progressive hotspot track between Kansas, Arkansas, and Mississippi.

There is little evidence that a mantle plume (herein referred to as the Bermuda mantle plume [Bmp]) currently exists at its proposed location, 600 to 800 km east of the Eastern North American Margin. The location of the proposed plume, a topographic feature known as the BR, has been explained by multiple hotspot models (Davies, 1988; Sleep, 1990). Therefore, the Bmp's existence has not been exhaustively denied, and more investigation is necessary. Courtillot et al. (2003) proposed a set of criteria necessary for a mantle plume to exist: (a) an age-progressive chain of volcanoes, (b) a flood basalt at this track's origin, (c) a buoyancy flux, (d) high ratios of $^{4}He/^{3}He$, and (e) a low shear wave velocity in the underlying mantle. Courtillot et al. (2003) importantly distinguishes between "primary" (meeting all five criteria) plumes and "secondary" plumes, which do not meet all five criteria. An earlier paper by Courtillot et al. (2003) identified "tertiary" plumes, which are considered to be superficial in origin. Although an age-progressive track of volcanoes is argued, the Bmp is eligible for criteria (a) and can be termed a secondary plume.

OIB magmas are commonly believed to represent plume-derived melts (Hart & Zindler, 1986). Despite their noted heterogeneity, however, there are rough parameters for radiogenic isotope ratio values and trace element patterns for OIBs. As discussed in the previous section, trace element patterns exhibit smooth slopes similar to that of an average OIB and relatively less-enriched E-MORB on MORB-normalized diagrams (Figure 8.9b). However, trace element contents for the samples (excluding the ijolite) are significantly enriched over the OIB (Figure 8.9b). Nd isotope ratios for the studied lamprophyres and other AAP members are indicative of more than one mantle source yet roughly plot within the range of OIB values (Figure 8.12). This could be indicative of mantle mixing or a REC (likely an enriched component due to the trace element patterns).

It is worth noting that Nd isotopic ratios obtained during this study are relatively depleted for the lamprophyre samples, which interestingly demonstrate significant enrichment in incompatible elements. This decoupling between isotopes and trace elements is also noted in Hawaiian lavas and has been explained through multiple models that involve low degrees of partial melting of isotopically DM sources (which would result in chemically enriched melts) along with higher degrees of partial melting of isotopically enriched mantle sources (Winter, 2010, and references therein). This decoupling could be attained via shallow DM that is entrained in the margins of a mantle plume and enriched deep mantle with recycled components, which would produce tholeiites (see Figure 8.8). It is commonly understood that tholeiitic rocks can represent the hotter axis of a plume, whereas alkaline basalts should be more common toward the margins (Winter, 2010, and references therein). The close proximity of tholeiitic rocks and alkaline rocks could have been produced by upwelling of hot mantle material that was guided by previously existing lithospheric weaknesses.

Although potential OIB-like sources have been identified, one question remains: is it possible that the OIB signatures seen in the geochemical data were not derived from a mantle plume? It is noted in the literature that alkaline and transitional basalt in rift and nonrift settings are geochemically indistinguishable from true OIB (Fitton, 2007). Thus, OIB geochemical signatures for igneous rocks alone are not a diagnostic feature of mantle plumes, and the geochemical data must be viewed with caution when considering the plume hypothesis. It is necessary to consider other petrogenetic mechanisms that could produce similar geochemical signatures. In a specific example, McNutt (2005) suggested that oceanic island volcanoes described in Hirano et al. (2006) are proof that mantle plumes are unnecessary to explain linear, age-progressive chains of igneous rocks because the Pacific Plate's nearby subduction would likely block a rising mantle plume. Based on this line of reasoning, Fitton (2007) suggests that small degree melting of the upper mantle can produce OIB-like magma. Small degrees of melting are confirmed in the Nb/Y versus Zr/Y plot, where samples plot by a REC and by the extreme enrichment seen in trace element diagrams (Figure 8.11b). A relationship between recycled crustal components and a carbonatite metasomatizing agent has been suggested by Hart & Zindler, (1986) and Weaver (1991). Metasomatism would also explain the decoupling between less-enriched isotopic ratios and enriched trace element patterns, given that low degrees of upper mantle melting would yield enriched trace element patterns (see Figure 8.9a,b), yet less enriched isotopic values (see Figure 8.13). The effects of metasomatism were also significant during petrogenesis, due to the enrichment in incompatible trace elements and La/Sm versus Nb and Sr plots (Figure 8.16a,b).

The presence of AAP rocks that are isotopically distinct from the lamprophyre samples in the study area doesn't discredit this explanation. Although isotopic and trace element data for this study indicate an asthenospheric upper mantle origin for the AAP members (V-intrusive, Benton dikes, Magnet Cove, Potash Sulfur Springs), other studies (Tilton et al., 1987; Duke et al., 2014) provide isotopic data that indicate an additional lithospheric source for the remaining members (Prairie Creek, Dare Mine Knob). Thus, in summary, a comprehensive petrogenesis of the AAP involves low degrees of partial melting of two distinct sources (lithospheric and asthenospheric mantle) with enriched components provided by metasomatic processes. A mechanism that provides the necessary heat and metasomatizing agent to generate the observed geochemical signatures is still required to fully explain the AAP's origins.

8.5.5. A Nonplume Mechanism for AAP Emplacement and ME Evolution

Despite geochemical OIB signatures in the samples, the classification of the Bmp as a tertiary plume requires consideration of a nonplume mechanism. Proposed nonplume mechanisms for plume-like igneous provinces include extensional stresses which result in near-adiabatic rise, low-melting-point material transported upwards through convection, channelized fluids, leaky transform faults and fracture zones, and convective rise deflected by varying lithospheric thickness (Batiza, 1982; Anderson, 1998, 2005; Anderson & Schramm, 2005). In a particular example of nonplume mechanisms, Matton and Jebrak (2009) provide a comprehensive investigation of Cretaceous-age alkaline provinces across the world, known as the Peri-Atlantic Alkaline Pulse (PAAP). Due to a lack of convincing evidence for hotspots, Matton and Jebrak (2009) suggest that the PAAP (which includes the AAP) is controlled by lithospheric fracturing rather than mantle plume generation. In lieu of the mantle plume explanation, EDC and continental insulation are potential mechanisms used to explain the spikes in magmatic activity at 125 and 85 Ma (Matton & Jebrak, 2009) (Figure 8.17). Their primary lines of reasoning involve the proximity between alkaline provinces and lithospheric weaknesses and similar timing between magmatic spikes and major tectonic events. In this case, the AAP and associated ME lie near the AOT and MVG, both prominent lithospheric weaknesses. Moreover, rose diagrams for AAP dike orientations indicate a strong structural control on the dikes that are intruded into OFB thrust faults during Cretaceous extension (Figure 8.14). These dikes lack a typical "radial" stress pattern that would have been induced by an underlying mantle plume (Smith, 1987) over a thermally uplifted area. An additional point of evidence involves the proximity of the AAP and the margin of the North American craton, where significant lithospheric thinning is evident (Schaeffer & Lebedev, 2014) (Figure 8.17). As this contrast in lithospheric thickness developed between the North American Craton and the Atlantic Ocean, EDC cells were generated, and convective velocity increased (Kim & So, 2020). Partial melts generated by EDC could have then been guided along preexisting lithospheric weaknesses, as suggested in Matton and Jebrak (2009).

Published emplacement ages for the dikes (106–89 Ma; see Eby & Vasconcelos, 2009) correlate generally with major tectonic events related to opening of the Atlantic Ocean, especially when considering simulations of EDC by King and Anderson (1998), suggesting that convection reaches a peak velocity after about 80 to 100 Ma and subsequently decreases. This can potentially be explained by the gradual entrainment of colder, more viscous lithospheric material, which would decrease convective velocity (Kim & So, 2020), or by a change in the rate and direction of North American plate movement from 150 to 135 Ma (May et al., 1989). Based on this timeline, EDC related to central Atlantic opening would have been at its peak velocity between 120 and 80–90 Ma (Matton & Jebrak, 2009). This timing corresponds well with emplacement of Kansas (120 Ma) and Arkansas (106–89 Ma) igneous rocks, which both lie along the proposed Bmp track. Furthermore, EDC peak velocity would also explain

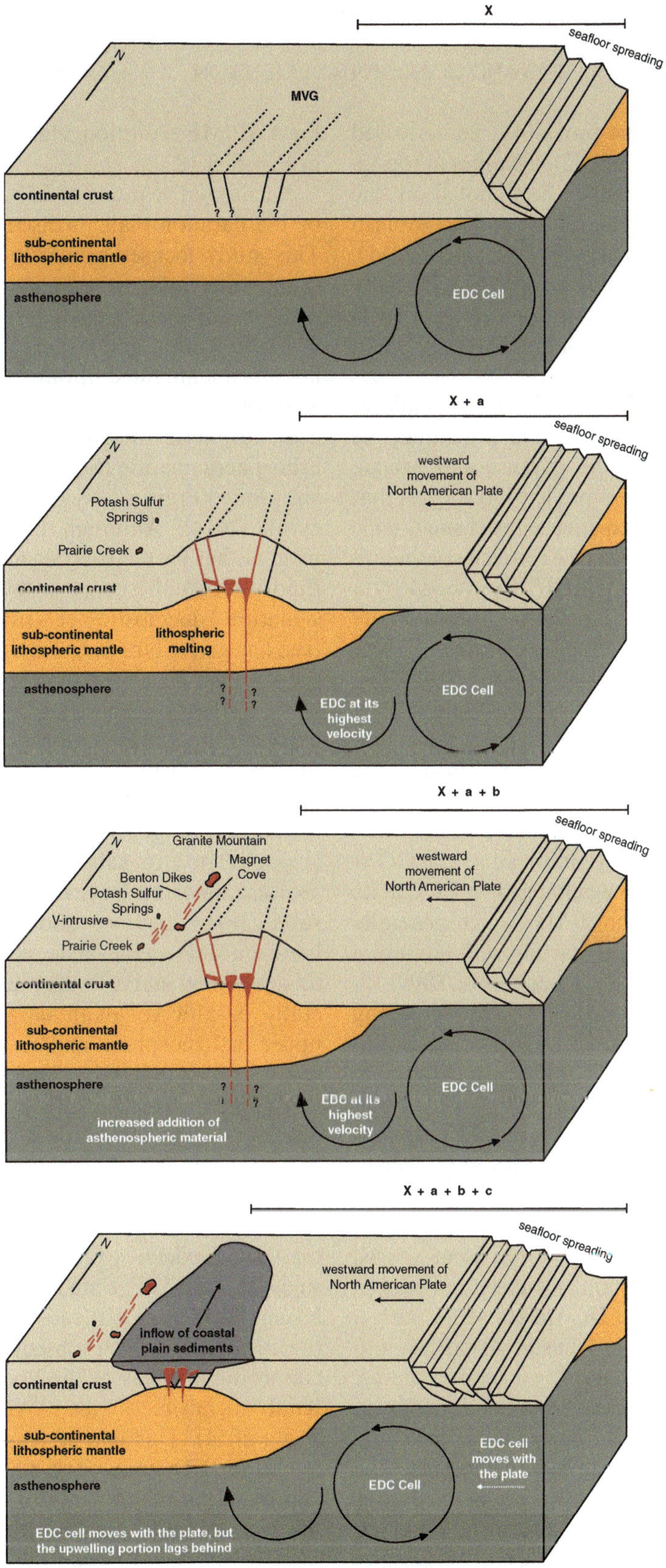

Figure 8.17 Schematic diagrams indicating a time-progressive evolution for the ME and AAP. X represents the distance between MVG and the Atlantic rift, with extension indicated for time 2 (a), time 3 (b), and time 4 (c). Modified after Matton and Jebrak (2009) and Duke et al. (2014). *Time 1 (~180–106 Ma):* spreading of the Atlantic Ocean begins and contrasting lithospheric thicknesses are sufficient to initiate EDC. *Time 2 (~106–105 Ma):* EDC reaches its highest velocity, causing mantle upwelling and provides metasomatizing carbonatite melt before melting; εNd SCLM (Dare Mine Knob and Prairie Creek). *Time 3 (~104–89 Ma):* metasomatism, melting and upwelling of + εNd magmas and emplacement of remaining AAP members. More asthenospheric material is incorporated in the melts at Time 3. *Time 4 (~89 Ma–present):* post-magma emplacement subsidence of the ME. Volcanic inactivity is attributed to a lag in the upwelling portion of the EDC cell as North America moves westward. Late Cretaceous sediments invade the ME.

the uplifted topography of the proto-ME, as discussed in Cox and Van Arsdale (1997, 2002). Changes in topography related to EDC have also been noted in the Colorado Plateau (van Wijk et al., 2010), and the western Atlantic (Shahnas, 2004; Ramsay & Pysklywec, 2011). The Jackson Dome intrusive (emplaced at 65 Ma; Dockery et al., 1997) could indicate a lag in the upwelling portion of the EDC cell (modeled in Shahnas 2004) as North America continued to move westward.

In contrast to Atlantic rifting, EDC-related emplacement and uplift of the AAP/ME could potentially be related to opening of the GOM. Krauss and Menke (2020) used seismic velocity measurements to suggest that the northern GOM may have experienced past small-scale convective upwelling near the North American continent's southeastern edge. Evidence for EDC is provided by a seismic anomaly known as the Northern Gulf Anomaly (NGA), which is delineated as a region of extremely low asthenospheric velocities (Krauss & Menke, 2020). The NGA is most intense on the Louisiana coast and in East Texas and forms a triangular wedge shape as far inland as 300 km (Krauss & Menke, 2020). In the context of the EDC model discussed in King and Anderson (1998), GOM evolution and AAP emplacement correlate well in geologic time. EDC could have initiated as the GOM began opening ~180 Ma, which would result in EDC at its highest velocity around 100 or 80 Ma (which generally overlaps with AAP emplacement starting with lamproite magmatism at ~106 Ma; Eby & Vasconcelos, 2009). A model for ME evolution associated with GOM spreading would involve more N-S oriented EDC, as compared to Atlantic EDC, which provides E-W EDC cells (Matton & Jebrak, 2009).

A potential issue with this model involves the proposed age of this anomaly and its relationship with emplacement ages of nearby alkaline provinces. The age of the anomaly itself has not been confirmed. Krauss and Menke (2020) suggest that, although the NGA could represent a fossil age associated with GOM opening, a younger age cannot be precluded. E-W oriented EDC makes sense with a rough age progression between Kansas, Arkansas, and Mississippi (suggested as evidence for a mantle plume; Eby & Vasconcelos, 2009), where continuous rifting in the Atlantic combined with a lag in magmatic upwelling creates an age-progressive track similar to a hotspot. The NW-SE orientation of the NGA (with a defined edge striking at 142°) could potentially explain the observed emplacement ages, as Kansas, Arkansas, and Mississippi provinces follow a general NW-SE trend. This model would involve emplacement of Kansas intrusives due to EDC, followed by Arkansas and then Mississippi volcanism as the GOM continues to open. A proposed model for AAP/ME evolution related to GOM opening is shown in Figure 8.18.

An alternative mode of EDC, facilitated by subduction of the Farallon Plate, is discussed in Duke et al. (2014). This study focuses primarily on mid-Cretaceous rocks (kimberlite, carbonatite, lamproite, etc.) between Montana and Arkansas and suggests that magmatism from 110 to 94 Ma in Kansas and Arkansas was generated as kimberlite magma migrated upward through tears in the eastern edge of the subducting Farallon Plate. The upwelling asthenospheric magma then melted subcontinental lithospheric mantle (SCLM) and crust, which generated lamproite (Prairie Creek) magmas in Arkansas (Duke et al., 2014). Although the generation of kimberlite magmas has been attributed to the release of slab-derived fluids (McCandless & Gurney, 1999), the trace element signature diagnostic of subduction-related magmas (i.e., a negative Nb-Ta anomaly) is not observed in any of the samples analyzed during this study (Figure 8.10). Moreover, the data indicate that the metasomatic agent is likely carbonatite in composition (Figure 6a,b) instead of fluid-related (which would be provided by a subducting slab).

This study argues that the generation of AAP magmas is not related to subduction of the Farallon Plate in geologic time and space due to the lack of a geochemical subduction signature and geochronologic similarities between AAP emplacement and proposed timing of EDC related to regional tectonic events. EDC magmas are generally basaltic to alkali basaltic in composition, with upper asthenosphere and/or lower lithospheric mantle sources (e.g., Kjarsgaard et al., 2017; Kaislaniemi & van Hunen, 2014; Rudzitis et al., 2016). This is accommodated by AAP isotopic data, which demonstrates both lithospheric ($\varepsilon Nd = -19$ to -10; Duke et al., 2014) and asthenospheric [$\varepsilon Nd = +2.7$ to $+4.4$, this study; $\varepsilon Nd = +1.9$ to $+4.3$; Duke et al. (2014)] sources. Based on the EDC model of Matton and Jebrak (2009), this study suggests that EDC related to rifting of the Atlantic Ocean brought upwelling asthenospheric material to the base of the lithosphere, which subsequently melted and produced lamproitic magmas of the Prairie Creek and Dare Mine Knob members of the AAP (as suggested in Duke et al., 2014). This line of reasoning is also applied to a model for EDC related to GOM opening (Figure 8.18) and based on the presence of a potentially EDC-related seismic anomaly in the northern Gulf (Krauss & Menke, 2020). Following the generation of lamproites, deeper asthenospheric sources were tapped to produce lamprophyres, carbonatites, ijolites, etc., of the V-intrusive, Benton Dike Swarm, Magnet Cove, and Potash Sulfur Springs provinces.

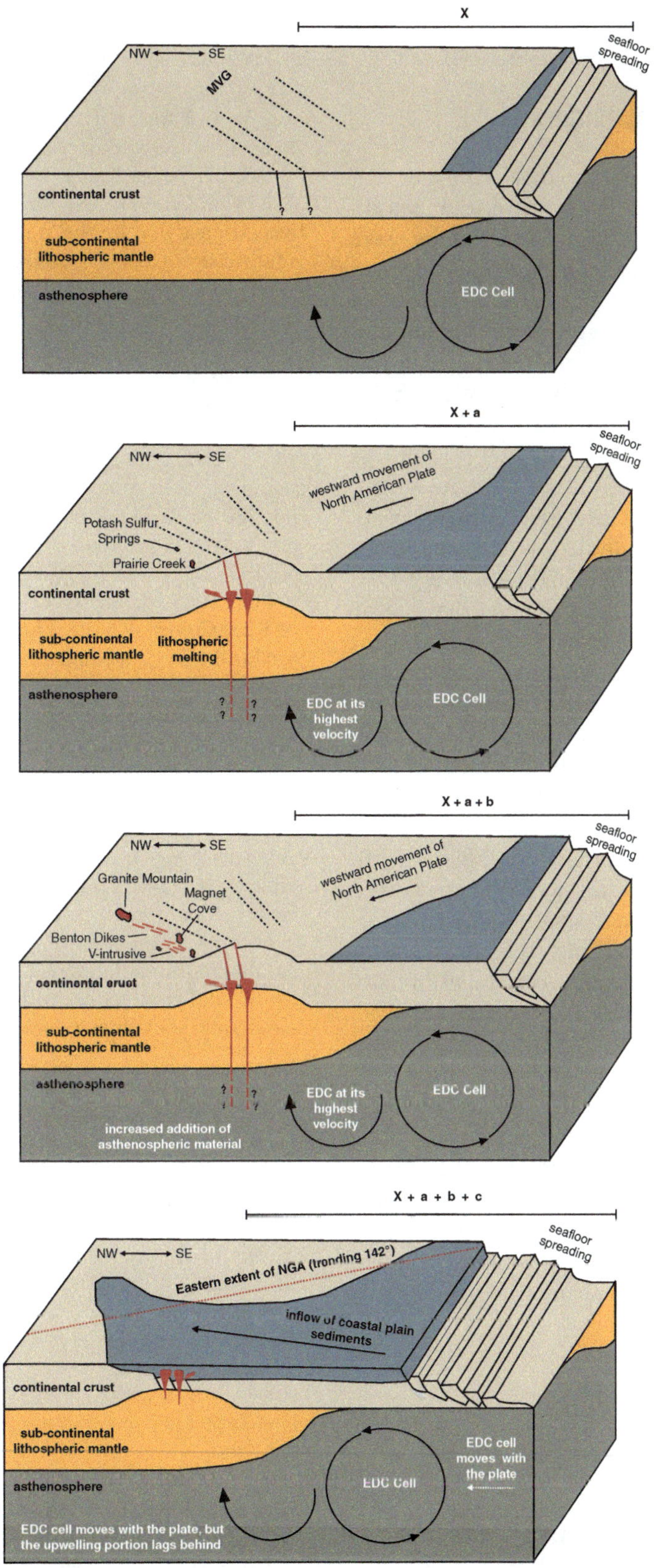

Figure 8.18 Schematic diagrams indicating a time-progressive evolution for the ME and AAP. X represents the distance between MVG and the GOM spreading center, with extension indicated for time 2 (a), time 3 (b), and time 4 (c). Matton and Jebrak (2009); Duke et al. (2014); Krauss & Menke (2020). *Time 1 (~180–106 Ma):* opening of the Gulf of Mexico begins, and contrasting lithospheric thicknesses are sufficient to initiate EDC. *Time 2 (~106–105 Ma):* EDC reaches its highest velocity, causing mantle upwelling, and provides metasomatizing carbonatite melt before melting; **ε**Nd SCLM (Dare Mine Knob and Prairie Creek). *Time 3 (~104–89 Ma):* metasomatism, melting and upwelling of + **ε**Nd magmas and emplacement of remaining AAP members. More asthenospheric material is incorporated in the melts at Time 3. *Time 4 (~89 Ma–present):* post-magma emplacement subsidence of the ME. Volcanic inactivity is attributed to a lag in the upwelling portion of the EDC cell as North America moves westward. Late Cretaceous sediments invade the ME.

8.6. CONCLUSIONS

Trace element and Nd isotopic data indicate that mid-Cretaceous lamprophyres (V-intrusive and Benton Dike Swarm) of the AAP in central Arkansas were generated by low degrees of partial melting of a depleted asthenospheric mantle source. The remaining sample lithologies (ijolite, jacupirangite, syenite, and phonolite) also demonstrate significant enrichment in incompatible trace elements and were likely derived from a similar asthenospheric source when compared with isotopic data from other studies (Duke et al., 2014). Both sources experienced metasomatism by a carbonatite agent (generated by a REC), which provided enrichment in incompatible elements, especially LILe. Although trace element and isotopic data plot similarly with rough OIB values, this study argues for a nonplume mechanism of melting resulting in the generation and emplacement of the magmas. Based on similar timing between igneous ages and regional plate tectonic events (opening of the Atlantic Ocean, and the GOM), we propose that EDC—at its peak velocity 80 to 100 Ma after initiation of Atlantic rifting and GOM opening—uplifted the region of the proto-ME and guided melt along attenuated lithosphere at the edge of the North American Craton (Figure 8.18). Lithospheric weaknesses such as the MVG and AOT guided melt into the crust, where smaller dike emplacement was governed primarily by preexisting thrust planes of the OFB. Upwelling asthenospheric material initially melted the base of the lithosphere resulting in lamproite magmas of the Prairie Creek and Dare Mine Knob members. Following lithospheric input, more asthenospheric material was incorporated in the melts, which resulted in the generation and emplacement of lamprophyres, ijolite, jacupirangite, phonolite, syenite, and so on of the Magnet Cove, V-intrusive, Benton Dike Swarm, and Potash Sulfur Springs provinces. Therefore, the tectonomagmatic evolution of the AAP and associated ME are best explained from a nonplume perspective involving EDC initiated by plate tectonics (Figure 8.18).

REFERENCES

Anderson, D. L. (1998). The edges of the mantle. *Journal of Geodynamics*, *28*, 255–271.

Anderson, D. L. (2005). *Scoring hotspots: the plume and plate paradigms* (Special Paper 388, pp. 31–54). Geological Society of America.

Anderson, D. L., & Schramm, K. A. (2005). *Global hotspot maps*. (Special Paper 388, pp. 19–29). Geological Society of America.

Arne, D. C. (1992). Evidence from apatite fission-track analysis for regional Cretaceous cooling in the Ouachita Mountain fold belt and Arkoma Basin of Arkansas. *AAPG Bulletin*, *76*, 392–402.

Autin, W. J., Burns, S. F., Miller, B. J., Saucier, R. T., & Snead, J. I. (1991). Quaternary geology of the lower Mississippi Valley. *Geological Society of America*, 547–582.

Baksi, A. K. (1997). The timing of Late Cretaceous alkalic igneous activity in the northern Gulf of Mexico basin, southeastern USA. *The Journal of Geology*, *105*, 629–644.

Baksi, A. K. (2001). Search for a deep-mantle component in mafic lavas using a Nb–Y–Zr plot. *Canadian Journal of Earth Sciences*, *38*, 813–824.

Bally, A. W., Gordy, P. L., & Stewart, G. A. (1966). Structure, seismic data, and orogenic evolution of southern Canadian Rocky Mountains. *Bulletin of Canadian Petroleum Geology*, *14*, 337–381.

Batiza, R. (1982). Abundances, distribution and sizes of volcanoes in the Pacific Ocean and implications for the origin of non-hotspot volcanoes. *Earth and Planetary Science Letters*, *60*, 195–206.

Brookins, D. G., & Naeser, C. W. (1971). Age of emplacement of Riley County, Kansas, kimberlites and a possible minimum age for the Dakota Sandstone. *Geological Society of America Bulletin*, *82*, 1723.

Burke, K., & Dewey, J. F. (1973). Plume-generated triple junctions: key indicators in applying plate tectonics to old rocks. *The Journal of Geology*, *81*, 406–433.

Çemen, I., & Wright, L. A. (1990). Effect of Cenozoic extension on Mesozoic thrust surfaces. *Basin and Range Extensional Tectonics Near the Latitude of Las Vegas Nevada*, *176*, 305.

Condie, K. C. (2001). *Mantle plumes and their record in earth history*. Cambridge University Press.

Condie, K. C. (2003). Incompatible element ratios in oceanic basalts and komatiites: Tracking deep mantle sources and continental growth rates with time: element ratios in oceanic basalts and komatiites. *Geochemistry, Geophysics, Geosystems*, *4*, 1–28.

Courtillot, V., Davaille, A., Besse, J., & Stock, J. (2003). Three distinct types of hotspots in the Earth's mantle. *Earth and Planetary Science Letters*, *205*, 295–308.

Cox, R. T., & Van Arsdale, R. B. (1997). Hotspot origin of the Mississippi embayment and its possible impact on contemporary seismicity. *Engineering Geology*, *46*, 201–216.

Cox, R. T., & Van Arsdale, R. B. (2002). The Mississippi Embayment, North America: a first order continental structure generated by the Cretaceous superplume mantle event. *Journal of Geodynamics*, *34*, 163–176.

Cullers, R. L., Dorais, M. J., Berendsen, P., & Chaudhuri, S. (1996). Mineralogy and petrology of cretaceous subsurface lamproite sills, southeastern Kansas, USA. *Lithos*, *38*, 185–206.

Davies, G. F. (1988). Ocean bathymetry and mantle convection: large-scale flow and hotspots. *Journal of Geophysical Research: Solid Earth*, *93*, 10467–10480.

Dawood, Y. H., Abd El-Naby, H. H., & Sharafeldin, A. A. (2004). Influence of the alteration processes on the origin of uranium and europium anomalies in trachyte, central Eastern Desert, Egypt. *Journal of Geochemical Exploration*, *81*, 15–27.

Dini, A., Westerman, D. S., Innocenti, F., & Rocchi, S. (2008). Magma emplacement in a transfer zone: the Miocene mafic Orano dyke swarm of Elba Island, Tuscany, Italy. *Geological Society, London, Special Publications*, *302*, 131–148.

Dockery, D. T., Marble, J. C., & Henderson, J. (1997). The Jackson Volcano. *Mississippi Geology*, *18*, 33–55.

Downes, H. (2007). Origin and significance of spinel and garnet pyroxenites in the shallow lithospheric mantle: ultramafic massifs in orogenic belts in Western Europe and NW Africa. *Lithos*, *99*, 1–24.

Duke, G. I., Carlson, R. W., Frost, C. D., Hearn, B. C., & Eby, G. N. (2014). Continent-scale linearity of kimberlite–carbonatite magmatism, mid-continent North America. *Earth and Planetary Science Letters*, *403*, 1–14.

Duncan, R. A. (1984). Age progressive volcanism in the New England Seamounts and the opening of the central Atlantic Ocean. *Journal of Geophysical Research: Solid Earth*, *89*, 9980–9990.

Eby, G. N., & Vasconcelos, P. (2009). Geochronology of the Arkansas Alkaline Province, southeastern United States. *The Journal of Geology*, *117*, 615–626.

Erickson, R. L. & Blade, L. V. (1963). *Map of bedrock geology of Magnet Cove igneous area, Hot Spring County, Arkansas* (Prof. Paper 426). U. S. Geological Survey.

Ervin, C. P., & McGinnis, L. D. (1975). Reelfoot rift: reactivated precursor to the Mississippi embayment. *Geological Society of America*, *86*, 1287–1295.

Fitton, J.G. (2007). The OIB paradox. *Special Paper 430: Plates, Plumes and Planetary Processes, Geological Society of America*, 430, 387–412.

Fitton, J. G., Saunders, A. D., Norry, M. J., Hardarson, B. S., & Taylor, R. N. (1997). Thermal and chemical structure of the Iceland plume. *Earth and Planetary Science Letters*, *153*, 197–208.

Furman, T., & Graham, D. (1999). Erosion of lithospheric mantle beneath the East African Rift system: geochemical evidence from the Kivu volcanic province. *Developments in Geotectonics*, *24*, 237–262.

Gogineni, S. V., Melton, C. E., & Giardini, A. A. (1978). Some petrological aspects of the Prairie Creek diamond-bearing kimberlite diatreme, Arkansas. *Contributions to Mineralogy and Petrology*, *66*, 251–261.

Hart, R. S., & Zindler, A. (1986) In search of a bulk-earth composition. *Chemical geology*, *57*, 247–267.

Hawkesworth, C. J., Rogers, N. W., van Calsteren, P. W. C., & Menzies, M. A. (1984). Mantle enrichment processes. *Nature*, *311*, 331–335.

Hirano, N., Takahashi, E., Yamamoto, J., Abe, N., Ingle, P. S., Kaneoka, I., et al. (2006). Volcanism in response to plate flexure. *Science*, *313*, 1426–1428.

Hock, J. D., & Seitz, A. (1995). Continental mafic dyke swarms as tectonic indicators: an example from the Vestfold Hills, Antarctica. *Precambrian Research*, *75*, 121–139.

Hofmann, A. W. (1997). Mantle geochemistry: The message from oceanic volcanism. *Nature*, *385*, 219–229.

Hou, G., Kusky, T. M., Wang, C., & Wang, Y. (2010). Mechanics of the giant radiating Mackenzie dyke swarm: a paleostress field modeling. *Journal of Geophysical Research*, *115*.

Irvine, T. N., & Baragar, W. R. A. (1971). A guide to the chemical classification of the common volcanic rocks. *Canadian Journal of Earth Sciences*, *8*, 523–548.

Johnson, E. S., Tate, C. M., & Fanning, C. M. (1999). New geologic mapping and SHRIMP U-Pb zircon data in the Peninsular Ranges batholith, Baja California, Mexico: evidence for a suture? *Geology*, *27*, 743–746.

Kaislaniemi, L., & Van Hunen, J. (2014). Dynamics of lithospheric thinning and mantle melting by edge-driven convection: application to Moroccan Atlas Mountains. *Geochemistry, Geophysics, Geosystems*, *15*, 3175–3189.

Kim, D. H., & So, B. D. (2020). Effects of rheology and mantle temperature structure on edge-driven convection: implications for partial melting and dynamic topography. *Physics of the Earth and Planetary Interiors*, *303*.

King, S. D., & Anderson, D. L. (1998). Edge-driven convection. *Earth and Planetary Science Letters*, *160*, 289–296.

Kjarsgaard, B. A., Heaman, L. M., Sarkar, C., & Pearson, D. G. (2017). The North America mid-Cretaceous kimberlite corridor: wet, edge-driven decompression melting of an OIB-type deep mantle source: Mid-Cretaceous North America Kimberlites. *Geochemistry, Geophysics, Geosystems*, *18*, 2727–2747.

Krauss, Z., & Menke, W. (2020). The Northern Gulf Anomaly: P- and S-wave travel time delays illuminate a strong thermal feature beneath the Northern Gulf of Mexico. *Earth and Planetary Science Letters*, *534*.

Larson, L. R. (1991). Geological consequences of superplumes. *Geology*, *19*, 963–966.

Leitch, A. M., Davies, G. F., & Wells, M. (1998). A plume head melting under a rifting margin. *Earth and Planetary Science Letters*, *161*, 161–177.

Loper, D. E. (1998). Mantle plumes and their effect on the Earth's surface: a review and synthesis. *Dynamics of Atmospheres and Oceans*, *27*, 35–54.

Matton, G., & Jébrak, M. (2009). The Cretaceous Peri-Atlantic Alkaline Pulse (PAAP): deep mantle plume origin or shallow lithospheric break-up? *Tectonophysics*, *469*, 1–12.

May, S. R., Beck, M. E., Jr., & Butler, R. F. (1989). North American apparent polar wander, plate motion, and left-oblique convergence: Late Jurassic-Early Cretaceous orogenic consequences. *Tectonics*, *8*, 443–451.

McCandless, T. E. & Gurney, J. J. (1999). *Kimberlites: mantle expressions of deep-seated subduction*. Paper presented at the 7th International Kimberlite Conference.

McCormick, G. & Heathcote, R. (1979). *Mineralogy of the Morrilton alvikite dike, Conway County, Arkansas*. Paper presented at the Geological Society of America.

McNutt, S. R. (2005). A review of volcanic seismology. *Annual Review of Earth and Planetary Science*, *33*, 461–491. doi: 10.1146/annurev.earth.33.092209.122459

Middlemost, A. K. E. (1994). Naming materials in the magma/igneous rock system. *Earth-Science Reviews*, *37*, 215–224.

Mooney, W. D., & Prodehl, C. (Eds.) (1984). *Proceedings of the 1984 Workshop of the IASPEI on the seismic modeling of laterally varying structures: contributions based on data from the 1978 Saudi Arabian refraction profile*. U.S. Geological Survey.

Morgan, W. J. (1971). Convection plumes in the lower mantle. *Nature*, *230*, 42–43.

Morris, J., & Fouad, T. (1987). ^{10}Be and ^{9}Be in mineral separates and whole rocks from volcanic arcs: Implications for sediment subduction. *Geochimica et Cosmochimica Acta*, *53*, 3197–3206.

Muller, R. D. (1993). Revised plate motions relative to the hotspots from combined Atlantic and Indian Ocean hotspot tracks. *Journal of Geology*, *21*, 275–278.

Murray, G. E. (1961). Geology of the Atlantic and Gulf Coastal Plain of United States and Mexico (symposium). *Geological Society of America Bulletin*, *63*, 1221–1228.

Musumeci, G., Mazzarini, F., Corti, G., Barsella, M., & Montanari, D. (2005). Magma emplacement in a thrust ramp anticline: the Gavorrano Granite (northern Apennines, Italy): magma emplacement in ramp anticline. *Tectonics*, *24*.

Pollard, D. D. (1987). Elementary fractures mechanics applied to the structural interpretation of dykes. In H. C. Halls & W. H. Fahrig (Eds.), *Mafic dyke swarms* (Vol. *34*, pp. 112–128). Geological Association of Canada.

Ramirez-Pena, C. F., Chavez-Cabello, G., Fitz-Diaz, E., Aranda-Gomez, J. J., & Sosa Valdes, R. (2019). Uplift and syn-orogenic magmatism in the Concecpcion del Oro Block: a thick-skinned (Laramide style?) contractional structure in the Mexican Fold-Thrust Belt. *Journal of South American Earth Sciences*, *93*, 242–252.

Ramsay, T., & Pysklywec, R. (2011). Anomalous bathymetry, 3D edge driven convection, and dynamic topography at the western Atlantic passive margin. *Journal of Geodynamics*, *52*, 45–56. doi: 10.1016/j.jog.2010.11.008

Reiners, P. W. (2005). Zircon (U-Th)/He thermochronometry. *Reviews in Mineralogy and Geochemistry*, *58*, 151–179.

Rock, N. M. S. (1991). *Lamprophyres. Springer Science & Business Media.*

Rollinson, H. R. (1993). *Using geochemical data: evaluation, presentation, interpretation.* Longman Scientific and Technical, Wiley.

Rudnick, R. L., & Gao, S. (2003). The composition of continental crust. In R. L. Rudnick (Ed.), *The crust* (pp. 1–64). Oxford: Elsevier.

Rudzitis, S., Reid, M. R., & Blichert-Toft, J. (2016). On edge melting under the Colorado Plateau margin. *Geochemistry, Geophysics, Geosystems*, *17*, 2835–2854.

Schaeffer, A. J., & Lebedev, S. (2014). Imaging the North American continent using waveform inversion of global and USArray data. *Earth and Planetary Science Letters*, *402*, 26–41.

Scharon, L., & Hsu, I. C. (1969). Paleomagnetic investigation of some Arkansas alkalic igneous rocks. *Journal of Geophysical Research*, *74*(10), 2774–2779.

Schwalb, H. R. (1982). *Paleozoic geology of the New Madrid Area.* Illinois State Geological Survey.

Shahnas, M. H. (2004). Anomalous topography in the western Atlantic caused by edge-driven convection. *Geophysical Research Letters*, *31*.

Sims, K. W. W., & DePaolo, D. J. (1997). Inferences about mantle magma sources from incompatible element concentration ratios in oceanic basalts. *Geochimica et Cosmochimica Acta*, *61*, 765–784.

Sleep, N. H. (1990). Hotspots and mantle plumes: some phenomenology. *Journal of Geophysical Research*, *95*.

Smith, C. G. (1987). *The geology of parts of the west coast of Palmer Land.* Cambridge: British Antarctic Survey.

Stearns, R. G. (1957). Cretaceous, Paleocene, and lower Eocene, geologic history of the northern Mississippi embayment. *Geological Society of America Bulletin*, *68*(9), 1077–1100.

Stoppa, F., Rukhlov, A. S., Bell, K., Schiazza, M., & Vichi, G. (2014). Lamprophyres of Italy: early cretaceous alkaline lamprophyres of southern Tuscany, Italy. *Lithos*, *188*, 97–112.

Stowell, H. H., & Tinkham, D. K. (2003). Integration of phase equilibria modelling and garnet Sm-Nd chronology for construction of P-T-t paths: examples from the Cordilleran Coast Plutonic Complex, USA. *Geological Society, London, Special Publications*, *220*, 119–145.

Sun, S. S., & McDonough, W. F. (1989). Chemical and isotopic systematics of oceanic basalts: implications for mantle composition and processes. *Geological Society, London, Special Publications*, *42*, 313–345.

Thomas, W. A. (1985). The Appalachian-Ouachita connection: Paleozoic orogenic belt at the southern margin of North America: Annual Review of Earth and Planetary Science Letters, pp. 175–199.

Thomas, W. A. (1991). The Appalachian-Ouachita rifted margin of southeastern North America. *Geological Society of America Bulletin*, *103*, 415–431.

Thomas, W. A. (2006). Tectonic inheritance at a continental margin. *GSA Today*, *16*(2), 4–11.

Thomas, W. A. (2014). The Iapetan rifted margin of southern Laurentia. *Geosphere*, *7*, 97–120.

Thomas, W. A., Gehrels, G. E., Greb, S. F., Nadon, G. C., Satkoski, A. M., & Romero, M. C. (2017). Detrital zircons and sediment dispersal in the Appalachian foreland. *Geosphere*, *13*, 2206–2230.

Tilton, G. R., Kwon, S. T., & Frost, D. M. (1987). Isotopic relationships in Arkansas Cretaceous alkalic complexes. *Geological Society of America Special Papers, Geological Society of America*, *215*, 241–248.

van Wijk, J. W., Baldridge, W. S., van Hunen, J., Goes, S., Aster, R., Coblentz, D. D., Grand, S. P. & Ni, J. (2010). Small-scale convection at the edge of the Colorado Plateau: Implications for topography, magmatism, and evolution of Proterozoic lithosphere. *Geology*, *38*, 611–614.

Weaver, B. L. (1991). Trace element evidence for the origin of ocean-island basalts. *Journal of Geology*, *19*, 123–126.

Whitaker, A. E., & Engelder, T. (2006). Plate-scale stress fields driving the tectonic evolution of the central Ouachita salient, Oklahoma and Arkansas. *Geological Society of America Bulletin*, *118*, 710–723.

Wilson, M. (1997). Thermal evolution of the Central Atlantic passive margins: continental break-up above a Mesozoic super-plume. *Journal of the Geological Society*, *154*, 491–495.

Winter, J. D. (2010). *Principles of Igneous and Metamorphic Petrology. Prentice Hall.*

Workman, R. K., & Hart, S. R. (2005). Major and trace element composition of the depleted MORB mantle (DMM). *Earth and Planetary Science Letters*, *231*, 53–72.

Yong Technology Inc. (2014). GeoRose (version 0.5.1).

Zartman, R. E., & Howard, J. M. (1987). Uranium-lead age of large zircon crystals from the Potash Sulfur Springs igneous complex, Garland County, Arkansas, *Geological Society of America Special Papers*, Geological Society of America, 215, 235.

Part III
Case Studies of Continental Extensional Tectonics

9

Rapid Versus Delayed Linkage and Coalescence of Interacting Rift Tips

Folarin Kolawole[1], Liang Xue[2], and Zuze Dulanya[3]

ABSTRACT

The linkage and coalescence of propagating continental rifts create through-going rift floors necessary to localize the breakup axis. Prior to linkage, interacting rift tips are separated by a topographic basement-high (rift interaction zone, RIZ), which is progressively dismembered and down-thrown by the propagation of rift-tip faulting and basin subsidence. Here, we explore the evolution of Middle-Shire and Nsanje RIZs located along three active, contiguous, amagmatic rifts in East Africa. The Middle-Shire RIZ is an overlapping-oblique divergent RIZ across which the southern Malawi Rift is propagating into the Lower Shire Graben (LSG) shoulder, and Nsanje RIZ is a tip-to-tip oblique RIZ across which the LSG has propagated into the Nsanje Graben tip. Utilizing field observations and landscape evolution models of contiguous RIZs with contrasting geometries, our results predict that with progressive extension and tip propagation, the Middle-Shire RIZ maintains minor basement down-throw and unequilibrated axial stream profile, which contrasts pronounced basement burial and equilibrated axial stream profiles across the Nsanje RIZ. Modeled stress distribution predicts compounding stress concentrations at tip-to-tip RIZs via synthetic border fault interactions, favoring rift coalescence, and stress relaxation at overlapping divergent RIZs favoring stalled coalescence, altogether demonstrating that RIZ geometries and kinematics strongly influence the pace of rift-tip coalescence.

9.1. INTRODUCTION

In continental lithosphere, the initiation of a divergent plate boundary is manifested by the development of a system of isolated graben and half-graben rift basins, each representing a rift segment (Figure 9.1a; e.g., Rosendahl et al., 1986; Rosendahl, 1987; Ebinger et al., 1987; Corti et al., 2007; Nelson et al., 1992; Kolawole et al., 2021a). Some rift segments may nucleate as a system of sub-basins with early establishment of the length of the entire border fault system (Rotevatn et al., 2018), whereas others nucleate by progressive lateral growth of the border fault (Agar & Klitgord, 1995). In narrow continental rift systems, although nucleating rift border faults may initially terminate at misoriented prerift basement shear zones or strong crustal blocks, stress buildup and magmatism at rift tips may drive renewed propagation of the rift basin (Ebinger et al., 1989; van Wijk & Blackman, 2005; Kolawole et al., 2022, 2024). With continued tectonic extension of the lithosphere, which may occur over multiple rift phases, rift basins deepen and lengthen by means of the three-dimensional (3D) propagation of their basin-bounding and intrabasinal fault systems, during which tectonic strain may migrate between the faults and sub-basins (e.g., Ebinger et al., 1989, 2000; Heilman et al.;

[1] *Department of Earth and Environmental Sciences, Columbia University, New York, USA*

[2] *Department of Earth and Environmental Sciences, Syracuse University, Syracuse, New York, USA*

[3] *Geography and Earth Sciences Department, University of Malawi, Zomba, Malawi*

Extensional Tectonics: Rifting and Continental Extension, Geophysical Monograph 278, First Edition.
Edited by İbrahim Çemen and Elizabeth J. Catlos.

DOI: 10.1002/9781119773757.ch09

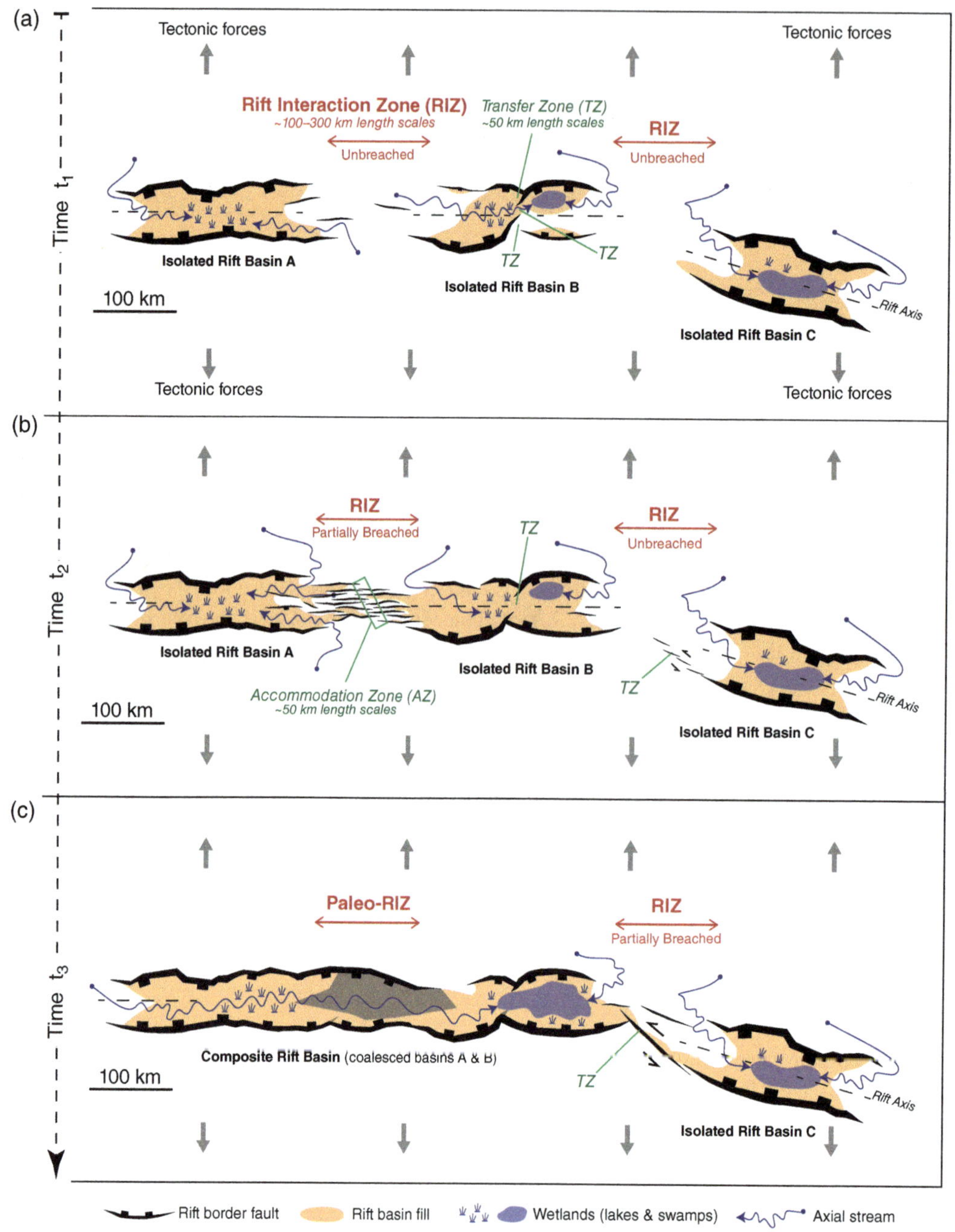

Figure 9.1 Cartoon illustrating the evolution of zones of rift segment interaction along narrow continental rift systems during (a) the initial nucleation of isolated rift basins, to (b) subsequent lateral propagation and interaction, and (c) linkage and coalescence of the basins. Prior to linkage and coalescence, the broad regions of interaction between the rifts (~100–300-km-length scales) are referred to as rift interaction zones (RIZs) (Nelson et al., 1992; Kolawole et al., 2021a). RIZs commonly host transfer and accommodation zones and their associated structures that localize and transfer strain across the RIZs (Bosworth, 1985; Ebinger et al., 1987; Morley et al., 1990; Nelson et al., 1992; Morley, 1995; Faulds & Varga, 1998). RIZ geometries can be defined by the relative trend and border fault dip polarity of the interacting rift basins (after Morley et al., 1990), such as tip-to-tip collinear (e.g., between rift basins A and B in panels (a) and (b)) and underlapping oblique convergent (e.g., between rift basins B and C in panels (a) and (b)) (Kolawole et al., 2021a).

2019; Fazlikhani et al., 2021; Scholz et al., 2020; Kolawole et al., 2022, 2024; Shaban et al., 2023). However, the inherited crustal rheology may temporarily stagnate the lateral propagation of a rift tip (e.g., van Wijk & Blackman, 2005; Kolawole et al., 2022; Shaban et al., 2023), during which mantle flow patterns beneath the rift tip promote stress buildup and shear zone formation (e.g., van Wijk & Blackman, 2005). As propagating rift tips encroach into unrifted basement areas in the intervening region between the rift segments, their faults may interact and transfer tectonic strain, leading to the hard and/or soft linkage of the faults (Figures 9.1b,c; McClay & Khalil, 1998; McClay et al., 2002; Fossen & Rotevatn, 2016; Kolawole et al., 2021a). These intervening regions of unrifted basement are usually of large spatial scales (~100–200-km length scales) and are referred to as rift interaction zones (RIZ in Figures 9.1a–c; Nelson et al., 1992; Koehn et al., 2008; Aanyu & Koehn, 2011; Sachau et al., 2016; Kolawole et al., 2021a). It is within this broad RIZ region that relatively smaller-scale (~50 km length) deformation and strain-transfer features such as "transfer zones" and "accommodation zones" develop to facilitate rift linkage (Figures 9.1a,b; Bosworth, 1985; Ebinger et al., 1987; Coffield & Schamel, 1989; Morley et al., 1990; Morley, 1995; Faulds & Varga, 1998; Withjack et al., 2002).

Eventually, the linkage of the propagating rift segments is succeeded by the coalescence of the rift basins, forming a composite rift basin with a through-going rift floor of pronounced crustal subsidence and significant basement burial along a localized narrow rift system (Figure 9.1c). RIZs constitute "sticky crustal blocks," and the significance of rift segment coalescence is founded on the grounds that a successful through-going breakup axis cannot develop along an evolving continental rift system without the structural breaching of its RIZs. During the process of rift linkage, a deforming RIZ transitions from an unbreached RIZ to partially breached and breached RIZ, where an additional stage of "recently breached" RIZ is observable in modern active continental rift systems (Kolawole et al., 2021a). Also, the maximum length of a rift border fault is determined by the effective elastic thickness of the lithosphere (Ebinger et al., 1999), such that after the coalescence of isolated rift basins, the resulting composite rift basin is made up of multiple border fault systems that flip polarity along-strike (e.g., Bosworth, 1985; Rosendahl et al., 1986; Ebinger et al., 1987; Rosendahl, 1987; Sander & Rosendahl, 1989; Patton et al., 1994; Laó-Dávila et al., 2015; Heilman et al., 2019). Segmentation zones of extensional systems also play important roles during the mature stages of divergent plate boundary development. For example, during continental breakup, delayed propagation of the breakup axis (oceanic ridge tip) promotes excess magmatism at the segmentation zones where the ridge propagation is stalled (Koopmann et al., 2014).

Analog and numerical modeling experiments of rift linkage, mostly focused on underlapping and overlapping parallel rift segments, show that various kinematics of faulting may develop in the zone of rift interaction, with geometrical and mechanical controls of inherited basement structure (e.g., Wilson, 1990; Acocella et al., 1999; Corti, 2004; Aanyu & Koehn, 2011; Corti, 2012; Neuharth et al., 2021; Wolf et al., 2022; Xue et al., 2024). A few recently published models incorporated obliquely oriented rifts (Molnar et al., 2019; Schmid et al., 2023). However, natural rift systems exhibit a broader variety of 3D RIZ geometries, which include tip-to-tip, oblique, and orthogonal plan-view geometries for rift pairs in which the rift border faults, although separated by up to 300 km, may have synthetic dip polarity defining a "convergent" geometry or are antithetic to one another, representing a "divergent" geometry (Kolawole et al., 2021a). Although RIZs are larger-scale structures than transfer zones, the convention of "convergent" and "divergent" terminology to describe the 3D structure of RIZs is adopted from the transfer zone classification of Morley et al. (1990). The complexity of RIZ geometries also encompasses triple and quadruple junctions that feature three or more juvenile rift segments (Kolawole et al., 2021a). Some of the enigmatic RIZ geometries are overlapping divergent RIZs in which the rifts are oriented obliquely or orthogonally to one another, such that the border faults of a rift segment are propagating into the uplifting footwall of the border faults (or rift flank) of the other rift basin. The diverse RIZ geometries demonstrate greater 3D complexity than the simple parallel and overlapping geometries commonly adopted in many published numerical and analog modeling studies of rift segment interaction.

In this chapter, we investigate the potential tectonic factors that may promote, inhibit, or delay the pace of rift linkage and coalescence and test the hypothesis that RIZ geometry and the associated kinematics may constitute one of these factors. We explore the structure, fault kinematics, displacement and stress transfer fields, and landscape evolution of two RIZs separating three contiguous actively propagating magma-poor rift segments in the southern East African Rift System: the southern Malawi Rift, LSG (LSG), and Nsanje Graben (Figure 9.2). The Middle Shire RIZ is an overlapping oblique divergent RIZ in which the NNE/N-trending southern Malawi Rift is propagating southward into the NE shoulder of the NW-trending LSG, whereas the Nsanje RIZ is a tip-to-tip oblique RIZ (convergent) in which the LSG has propagated southeast into the northern tip of the N-trending Nsanje Graben. The two RIZs provide natural

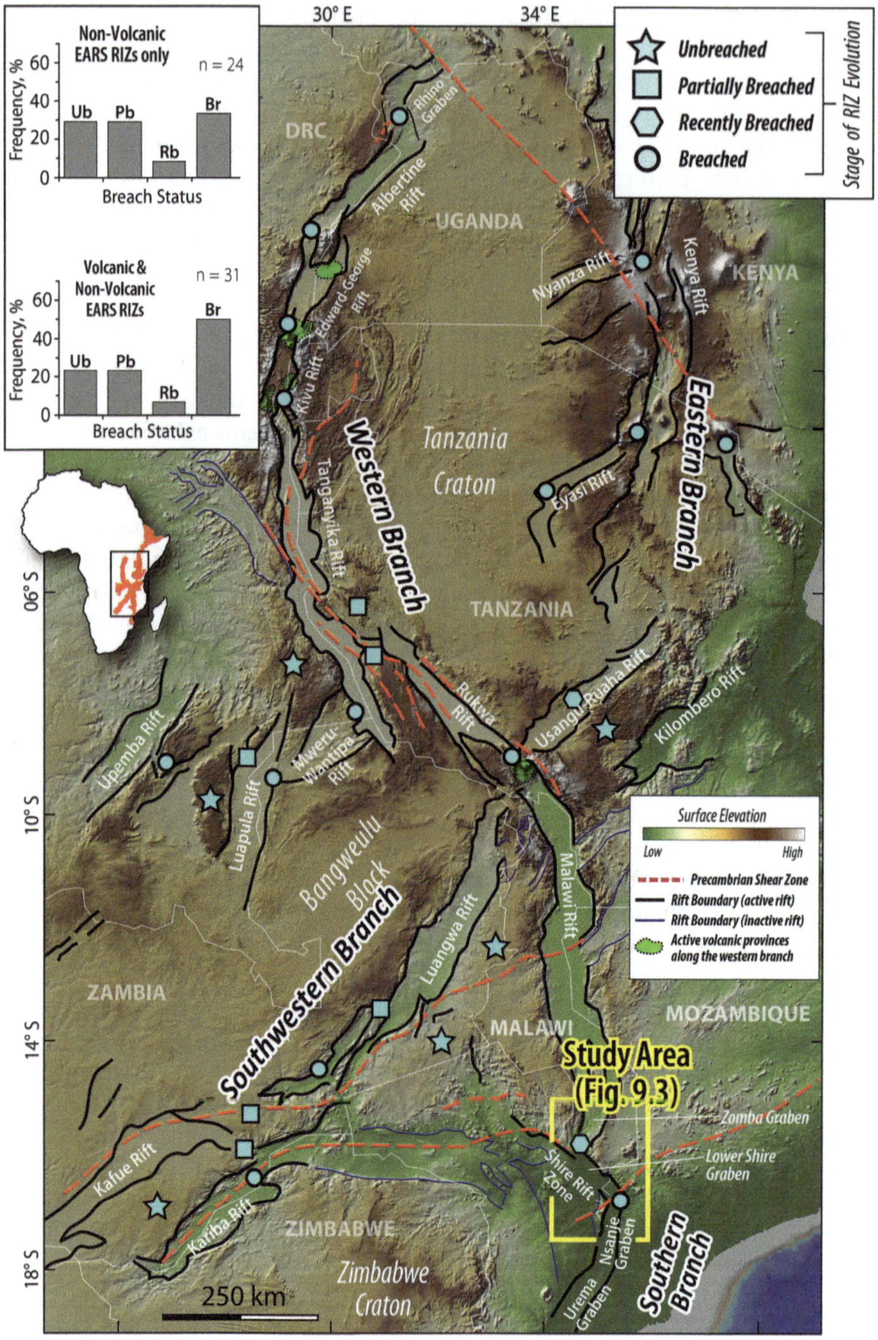

Figure 9.2 Topographic relief map of the East African Rift System (EARS; Source: Ryan et al., 2009), showing the magma-rich Eastern branch and magma-poor Western, Southwestern, and Southern branches. The symbols represent the stage of evolution of its rift interaction zones (RIsZ), inferred from spatial extents of fault connectivity and topographic and drainage morphology of the RIZs (Kolawole et al., 2021a). Histograms show the distribution of RIZ evolution stages along the rift system (adapted from Kolawole et al., 2021a). UB, unbreached; PB, partially breached; RP, recently breached; Br, breached. Note that the EARS is defined by the active rift segments (black lines). Bottom-right inset: map of Africa showing the EARS (red polygons) and region covered by the relief map (black rectangle).

examples to explore strain localization on rift tips and RIZ syn-rift evolution in the absence of magmatism. We utilize field observations, landscape evolution, and static stress distribution models to examine the potential influence of the contrasting RIZ geometries on the long-term evolution of the RIZs.

9.1.1. The Zomba–Lower Shire–Nsanje Rift Zone

Prerift Basement Geology

The East African basement is characterized by Archean and Paleoproterozoic cratonic blocks surrounded by Meso- to Neoproterozoic orogenic belts (Fritz et al., 2013). The orogenic belts are composed of exhumed igneous and metamorphic rocks of mostly amphibolite to granulite facies grades, with sporadic occurrences of eclogite facies rocks that are localized along ductile terrane boundary shear zones (e.g., Daly et al., 1991; Fritz et al., 2013). These orogenic belts developed during multiple Meso- to Neoproterozoic orogenic events that witnessed the amalgamation of Gondwana (Fritz et al., 2013). In the Phanerozoic, East Africa has experienced repeated phases of tectonic extension, in which the rift systems follow the mobile belts and avoid the cratonic blocks (Daly et al., 1989; Corti et al., 2007).

The southern Malawi Rift and Shire Rift Zone developed along tectonic boundaries that separate distinct Precambrian orogenic belts or terranes. These include the Southern Irumide Belt (1,060–950 Ma), Zambezi Belt (1,830–795 Ma), the Tete-Chipata Complex, and the Unango Complex (1060–950 Ma) (Barr & Brown, 1987; Hargrove et al. 2003; Fritz et al., 2013), whereas the Nsanje Graben developed within the Nampula Complex (1,025–1,075 Ma). The basement terranes along these mobile belts are composed of Mesoproterozoic crust, which has been reworked and overprinted by contractional structures and igneous intrusions of the Neoproterozoic Pan African Orogeny. Common rock types found in the mobile belts include schists, amphibolite and granulitic gneisses, deformed granites, granodiorites, syenites, gabbro, and anorthosites (e.g., Barr & Brown, 1987; Fritz et al., 2013).

Rift Structure and Tectonic History

The Phanerozoic Eon of eastern Africa has witnessed three distinct phases of tectonic extension that localized rift basins along the Precambrian mobile belts (Delvaux, 1989; Castaing, 1991; Chorowicz, 2005). The first rifting phase occurred in the Permian–Early Jurassic, known as the "Karoo" event; the second event occurred in the Late Jurassic–Cretaceous; and the third event is the current phase of tectonic extension, known as the East African Rift System (EARS; Delvaux, 1989; Castaing, 1991). The rift system is characterized by a network of rift segments that define multiple branches, mainly the Eastern, Western, Southern, and Southwestern branches (Figure 9.2). The Eastern Branch is magma-rich, manifested by widespread volcanism along the rift basins, whereas the Western, Southern, and Southwestern branches are largely magma-poor, with surface volcanism restricted to four rift interaction zones (RIZs; Figure 9.2). The Karoo and Cretaceous rift segments and associated faults were commonly reactivated in the Cenozoic rift phase (Castaing, 1991; Morley et al., 1999; Daly et al., 2020; Kolawole et al., 2018, 2021b, 2022). However, there is increasing evidence suggesting that the transition from Mesozoic to Cenozoic rifting phases was accompanied by the abandonment of some of the early rift fault segments, subbasins, and, in some cases, entire rift segments (e.g., blue lines of "inactive rifts" in Figure 9.2; Delvaux, 1989; Castaing, 1991; Ragon et al., 2019; Kolawole et al., 2022). Considering the history of pronounced magmatic Karoo and Cretaceous rifting in the southern Malawi-central Mozambique region, the "prerift basement" of the Cenozoic rift phase is characterized by a basal Precambrian metamorphic unit and its overlying volcano-sedimentary sequences of Mesozoic rift deposits (Castaing, 1991; Kolawole et al., 2021a, 2022).

The Zomba, Lower Shire, and Nsanje grabens define a system of three contiguous active rift basins that extend between the southern Malawi and central Mozambique regions of the EARS' magma-poor southern branch (Figure 9.2). Thermochronological modeling suggests that the border faults have been active since the Late Oligocene–Early Miocene, with a possible onset of Cenozoic rifting as early as the Eocene (Bicca et al., 2019; Ojo et al., 2024). Current geodetic extension rates in the region are ca. 2.2 mm/yr (Stamps et al., 2018), with a regional normal faulting stress regime of ENE-WSW-oriented minimum principal compressive stress axis, σ_3 (Williams et al., 2019) (Figure 9.3a). Geodetic extension rates vary along-trend of the Malawi Rift (Saria et al., 2014). Seismic- and gravity-derived estimates of crustal thickness beneath the grabens range ca. 40–43 km (Njinju et al., 2019a; Sun et al., 2021); lithospheric thickness is ca. 130 km but thins to ca. 110 km beneath the Shire Rift Zone (Njinju et al., 2019a). The distribution of aeromagnetic Curie Point depths suggests an elevated heat flow of ca. 76 m/Wm^2 within the grabens relative to the ca. 66 mW/m^2 in the rift flanks (Njinju et al., 2019b).

The NNE-trending Zomba Graben is the southernmost section of the Malawi Rift and is bounded by the Zomba Fault to the east and Lisungwe and Wamkurumadzi Fault systems to the west (Figure 9.3a). The border faults generally follow the large-scale strike of the Precambrian metamorphic fabrics, indicating basement control on border fault localization (Williams et al., 2019; Kolawole et al., 2021a). Rift sedimentation in the Zomba Graben is primarily localized in the northern, northeastern,

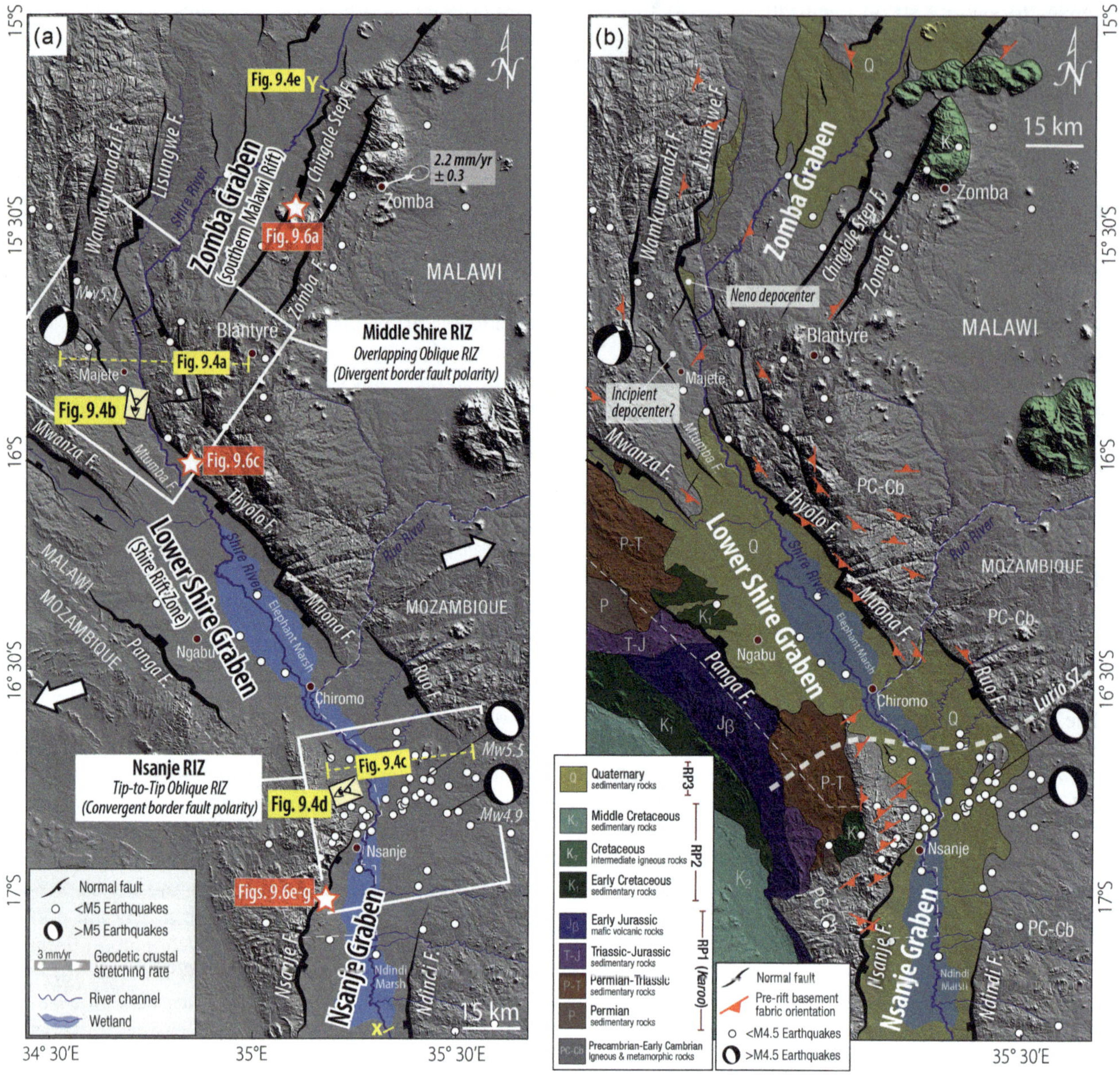

Figure 9.3 (a) Topographic relief map of southern Malawi and central Mozambique, showing the contiguous Zomba Graben (Malawi Rift), the Lower Shire Graben (Shire Rift Zone), and Nsanje Graben and their intervening Middle Shire and Nsanje rift interaction zones (RIZs). The main axial stream in the region, the Shire River and its tributaries, drains all three rift basins and flows further south into the Indian Ocean. Fault segments are surface-breaking active faults from Williams et al. (2022) and Kolawole et al. (2022). Earthquakes and focal mechanisms are from the U.S. Geological Survey earthquake catalog, Global CMT catalog, and Stevens et al. (2021) covering 1966–2020. Geodetic crustal stretching azimuth is from Stamps et al. (2008, 2018). Basement metamorphic fabric orientations are from Williams et al. (2019), Kolawole et al. (2022), Thomas et al. (2022), and field observations in this study. (b) Geologic map of the southern Malawi region and western Mozambique (Choubert et al., 1988; Kolawole et al., 2022).

and western parts of the graben (Figure 9.3b) and is dominated by unconsolidated Quaternary fluvial deposits and outcrops of clay-rich Matope paleo-lake sediments with characteristic shoreline gravels (Dulanya, 2017). The southern section of the graben is dominated by Precambrian basement exposures along the rift floor (Figure 9.3b; Kolawole et al., 2021a). The magnitudes of cumulative border fault offset along the Malawi Rift generally decrease southward, and the deepest syn-rift sedimentary sequences in the northern sub-basins of the rift appear to be absent in the south, suggesting an overall long-term southward propagation of the rift basin (Specht & Rosendahl, 1989; Scholz et al., 2020). South of Lake Malawi, shallow boreholes, electrical resistivity tomography, and depth-to-magnetic basement show rift sediment thickness of not more than 600 m (Bloomfield, 1965; Ojo et al., 2022b).

The NW-trending LSG is located south of the Zomba Graben and represents the easternmost sub-basin of the Shire Rift Zone, a multiphase rift basin in East Africa (Castaing, 1991; Kolawole et al., 2022). The LSG is bounded to the northeast by the Thyolo-Muona-Ruo/Camacho Fault System in the east and to the southwest by the Panga Fault. In the Early Pleistocene, the border fault accommodated initial accelerated slip on the Thyolo and Muona Segments, later followed by the initiation of the Ruo/Camacho segment, indicating an overall <2 Myrs timescale of southeastward growth and establishment of the fault system (Dulanya et al., 2024). The basin shows widespread coverage of Quaternary deposits (Figure 9.3b), with the thickest sections potentially occurring in the hanging wall of the Thyolo and Muona faults, dominated by the fluviolacustrine sequences of the Elephant Marsh and Shire River (Kolawole, 2020; Kolawole et al., 2022). The localization of the Elephant Marsh along the hanging wall of the Thyolo Fault and the spatial correlation of the lateral changes in the geometry of the marsh with prominent changes in the Thyolo Fault displacement profile suggest structural control on the development of the wetland (Kolawole, 2020). The rift axis hosts a buried segment of a large-offset fault, the southern segment of the Mwanza Fault, which is interpreted to be the eastern border fault of the Shire Rift Zone during the Mesozoic phases of extension (Kolawole et al., 2022). Similar to the Zomba Graben, the Mwanza and the Thyolo-Muona-Ruo border fault systems of the LSG generally follow the large-scale trends of the Precambrian metamorphic fabrics, indicating basement control of border fault localization (Wedmore et al., 2020a; Kolawole et al., 2022). However, the southeastern tip of the Ruo Fault crosscuts the NE-trending basement fabrics of the Precambrian Lurio Shear Zone (Kolawole et al., 2022). Outcrops of Mesozoic volcano-sedimentary sequences near the SW margin of the graben and confinement of basaltic aeromagnetic fabrics to the hanging wall of the buried Mwanza Fault segment suggest the presence of Mesozoic rift deposits beneath the Cenozoic sediments in the southwestern half of the graben (Kolawole et al., 2022). Depth-to-magnetic basement estimates suggest that the present-day NE border fault, the Thyolo-Muona-Ruo Fault System, may have accrued a larger post-Mesozoic vertical displacement (ca. 1.5 km) than the SW border fault, the Panga Fault (<200 m) (Kolawole et al., 2022). Despite up to 3 km of erosion of the early rift sediments during the Mesozoic rifting phases (Bicca et al., 2019), the present-day cumulative sediment thickness along the graben axis is estimated to be at least 2.4 km (Kolawole et al., 2022). This suggests a possible Cenozoic strain accommodation along the graben axis before the apparent "abandonment" and burial of the southern Mwanza Fault.

The N-to-NNE-trending Nsanje Graben is located south of the LSG, and it defines the northern continuation of the Urema Graben system of Mozambique (Figure 9.2; Pfaffling et al., 2009). The graben is bounded to the west by the Nsanje Fault and to the east by the Ndidi Fault (Figure 9.3a). Both the Nsanje and Ndindi border faults crosscut the large-scale NE-trending metamorphic fabrics of the host Precambrian basement (Bloomfield, 1958; Kolawole et al., 2022; Thomas et al., 2022), indicating a lack of large-scale basement inheritance on border fault development. Geologic maps show widespread coverage of Quaternary sediments across the entire graben, deposited in the Ndindi Marsh and floodplains of the Shire River (Figures 9.3b; Bloomfield, 1958; Choubert et al., 1988).

The Middle Shire and Nsanje Rift Interaction Zones

The Middle Shire RIZ, which represents the zone of transition from the Zomba Graben to the LSG, is characterized by a rotation of the Precambrian metamorphic basement fabrics from the NNE-SSW trend in the Zomba Graben area to a NW-SE trend in the LSG (Kolawole et al., 2021a). The Nsanje RIZ, which represents the zone of transition from the LSG to the Nsanje Graben, occurs at the NE-trending Lurio Shear Zone, an exhumed ductile shear zone separating the Unango and Nampula terrancs of the Precambrian Southern Irumide Belt (Fritz et al., 2013; Kolawole et al., 2022; Thomas et al., 2022). The Middle Shire RIZ defines an overlapping oblique divergent RIZ geometry in which the NNE/N-trending Malawi Rift is propagating southward into the NE shoulder of the NW-trending LSG, whereas the Nsanje RIZ is a tip-to-tip oblique RIZ ("convergent" border fault polarity pattern) in which the LSG has propagated southeast into the northern tip of the N-trending Nsanje Graben (Figure 9.3a).

In contrast to the Nsanje RIZ, which shows very sparse surface exposures of the Precambrian basement along the modern rift floor, the Middle Shire RIZ is dominated by widespread exposures of the basement (Figures 9.3b and 9.4a–d). In the Middle Shire RIZ, the rift floor tilts westward toward the escarpments of the Lisungwe and Wamkurumadzi Faults (e.g., Figure 9.4a). In the Nsaje RIZ, the rift floor also tilts to the southwest with a minor surface expression of faulting (Figures 9.3b and 9.4c,d). The longitudinal topographic relief profile of the rift axis shows that the Zomba Graben and Middle Shire RIZ are at a higher structural elevation than the Nsanje Graben (Figure 9.4e). Thus, based on the criteria of border fault connectivity, cross-over topographic and axial stream morphology, and directionality of axial stream flow, the Middle Shire RIZ is interpreted to be a recently breached RIZ and the Nsanje Graben a breached RIZ that is undergoing rift coalescence (Kolawole et al., 2021a).

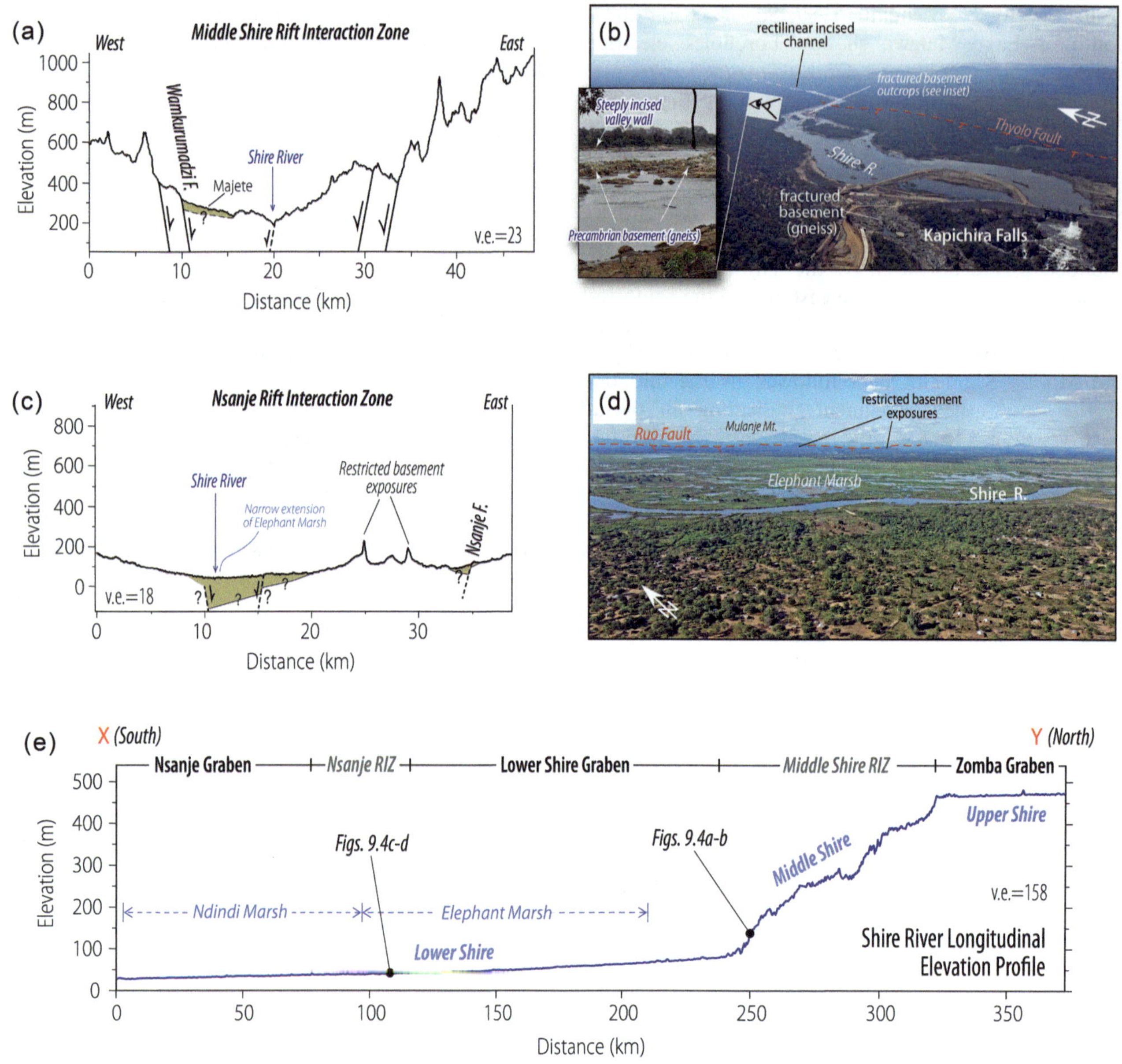

Figure 9.4 Topographic relief profiles. (a, c) 30 m resolution Shuttle Radar Topography Mission data and (b, d) field photographs showing rift morphology across the Middle Shire RIZ (a, b) and Nsanje RIZ (c, d). Sediment depocenters color-coded in yellow. (e) Longitudinal topographic relief profile of the Shire River (see Figure 9.3a for the start and end of the profile labeled x and y).

Modern Rift Basin Drainage Geomorphology

The southern Malawi–central Mozambique region is characterized by a tropical climate in which dry seasons generally extend from April to September (average annual rainfall of ~0.65 m at Tete) and wet seasons from October to March (Garzanti et al., 2022). The modern landscape of the Zomba–Lower Shire–Nsanje graben system is drained by the Shire River and its 18,000 km^2 catchment area (Figure 9.3a). The south-flowing, ca. 490-km-long axial stream represents the largest river in Malawi, the only outlet for Lake Malawi, and one of the major tributaries of the Zambezi River (Price, 1966; Palamuleni et al., 2011; Dulanya, 2017). Zambezi River Delta represents one of the largest sediment supply points into the Mozambique Channel arm of the Indian Ocean, highlighting the significance of the Shire River for sediment contributions into the ocean basin depositional sink (Garzanti et al., 2022). Although the Shire River represents the only modern fluvial link between Lake Malawi and the Indian Ocean, the trunk stream is punctuated by major sediment sinks, which includes Elephant Marsh in the LSG and Ndindi Marsh in the Nsanje Graben (Figure 9.3a). The Elephant Marsh and Ndindi Marsh represent wetlands that have developed in the axial depocenters of the Lower Shire and Nsanje grabens but are connected across the Nsanje RIZ, and the transition between the two wetlands is defined by a zone of decrease in width of the wetland within the Nsanje RIZ (Figure 9.3b).

The river course is divided into the "Upper Shire" section at ca. 500 m elevation in the Zomba Graben, the "Middle Shire" section in the Middle Shire RIZ where the river drops across a ca. 80 m escarpment over 72 km distance, and the "Lower Shire" section at ca. 45–50 m elevation in the LSG (Figure 9.4e). The rectilinear planform geometry, the large-scale relief morphology of the longitudinal elevation profile of the Middle Shire section, as well as the clustering of rapids and waterfalls (i.e., knickpoints) within the section (see Figure 9.4b) suggests that the Middle Shire section is not equilibrated with the downstream sections of the stream profile (Dulanya, 2017; Kolawole et al., 2021a; Dulanya et al., 2022). More importantly, in contrast to the axial stream profile within the Middle Shire RIZ, the "flat" longitudinal relief morphology and sinuous planform geometry of the axial stream across the Nsanje RIZ (Figure 9.4d,e) suggests an equilibrated profile. These observations are also consistent with the contrasts in the axial stream planform morphology between the two RIZs, whereby in the Middle Shire RIZ, the axial stream exhibits significantly lower channel sinuosity and channel width in contrast to the Nsanje RIZ (Kolawole et al., 2021a). Furthermore, the highly sinuous axial stream planform morphology of the Lower Shire section generally continues southward through the Nsanje RIZ and the Nsanje Graben.

Investigation of the knickpoint initiation times in the Middle Shire River catchment suggested that the Upper and Lower Shire were not integrated into a single south-flowing stream until around the Mid-Pleistocene (Dulanya et al. 2022). The Middle Shire region is interpreted to have been an elevated basement block that previously isolated the LSG from the Zomba Graben, when the Lower Shire River section flowed south and the Upper Shire hosted a lake (paleo-lake Matope) that flowed north into Lake Malawi with alternating periods flooding of Lake Malawi into the Zomba Graben (Dulanya, 2017). The subsequent integration of the axial stream across the Middle Shire is likely to have been facilitated by the continued brittle deformation and down-faulting of the Middle Shire basement by the tectonic interaction between the northwestward propagation of the Thyolo Fault and possibly a simultaneous propagation of the southern Zomba Graben faults (Kolawole et al., 2021a). Thus, the structural breaching of the Middle Shire RIZ and the linkage of the Zomba and LSGs is inferred to have taken place as recently as the Mid-Pleistocene, relative to the Oligocene-Miocene initiation of Cenozoic rifting in the region (Dulanya et al., 2022).

9.2. METHODS

9.2.1. Field Observations as Constraints on Rift Kinematics

To constrain the kinematics of rifting in southern Malawi, we perform field investigations on the outcrops of border faults in the Lower Shire and Nsanje grabens (locations are red stars in Figure 9.3a). The natural observations of fault kinematics from the study area constrain our numerical model (see "Modeling Landscape Evolution"). Our fieldwork involved the collection of measurements of slip indicators on the slip surfaces of the border faults, which include strike and dip of fault surfaces and the trend and plunge of the highest-quality slip stria on the surfaces. As for the Zomba Graben, we utilize published field measurements (from Wedmore et al., 2020b) taken along the largest intrabasinal fault in the basin, the Chingale Step Fault. These measurements (see Table 9.1) allow us to estimate a single kinematic strain tensor compatible with the kinematics of three rift basins in the current regional tectonic stress field. For this analysis, we utilize the FaultKin program (Marrett & Allmendinger, 1990; Cladouhos & Allmendinger, 1993), adopting a uniform weight moment tensor summation. The program determines the principal strain directions using slip surfaces with known geometries and striae with known slip senses as input while assuming that the deformation is coaxial. The analysis is a standard technique that produces a fault plane solution with two orthogonal nodal planes, the bisectors of which produce P- and T-axes of the incremental strain tensor. The P- and T-axes, respectively, represent the principal shortening (ε_3) and principal extension (ε_1) strain axes of the array of input faults.

Table 9.1 Field data on slip vectors on border faults used for the kinematic tensor calculation.

Rift basin	Border fault	Strike/dip	Plunge/trend	Source
Zomba Graben	Chingale Step F.	189°/54°	52°/301°	Wedmore et al. (2020b)
Lower Shire Graben	Thyolo F.	160°/69°	80°/231°	This study
Nsanje Graben	Nsanje F.	059°/79°	44°/193°	This study

9.2.2. Modeling Landscape Evolution

We use the Fastscape algorithm (Braun & Willett, 2013) to solve the stream power law to estimate the response of surface processes to rift-tip propagation in different RIZs. The Fastscape code simulates bedrock river incision and deposition by (Yuan et al., 2019a,b):

$$\frac{\partial h}{\partial t}=U-KA_mS_n+\frac{G}{A}\int_A\left(U-\frac{\partial h}{\partial t}\right)dA \text{ where } h\geq h_{\text{base}} \quad (9.1)$$

$$\frac{\partial h}{\partial t}=K_b\nabla^2 h+\int_A\left(\frac{U-\partial\frac{h}{\partial}t}{\partial x\partial y}\right)dA \quad \text{where } h<h_{\text{base}} \quad (9.2)$$

where h and h_{base} are the elevation and elevation of the rift lake, respectively, so equations (9.1) and (9.2) represent the landscape evolution on the land; in the rift lake, t is time; U, A, and S are uplift rate, drainage area, and local slope, respectively. K and G reflect the lithology erodibility and deposition coefficient; K_b is the diffusion coefficient in the rift lake, presumably linearly corresponding to the local slope, with a value of 30 m, n are poorly constrained constants that mostly depend on lithology and climate. We use $m = 0.5$, $n = 0.1$, $K = 4\text{e-}6$, $G = 1\text{e-}3$ based on previous work (e.g., Croissant & Braun, 2014, Guerit et al., 2019; Yuan et al., 2019a,b).

The domain is 500 km by 500 km with a cell size of 1 km² (Figure 9.5a–c). Focusing on the effects of rift segment tip propagation, we simplify the impacts of climate and lithology in the model and assume a constant rainfall

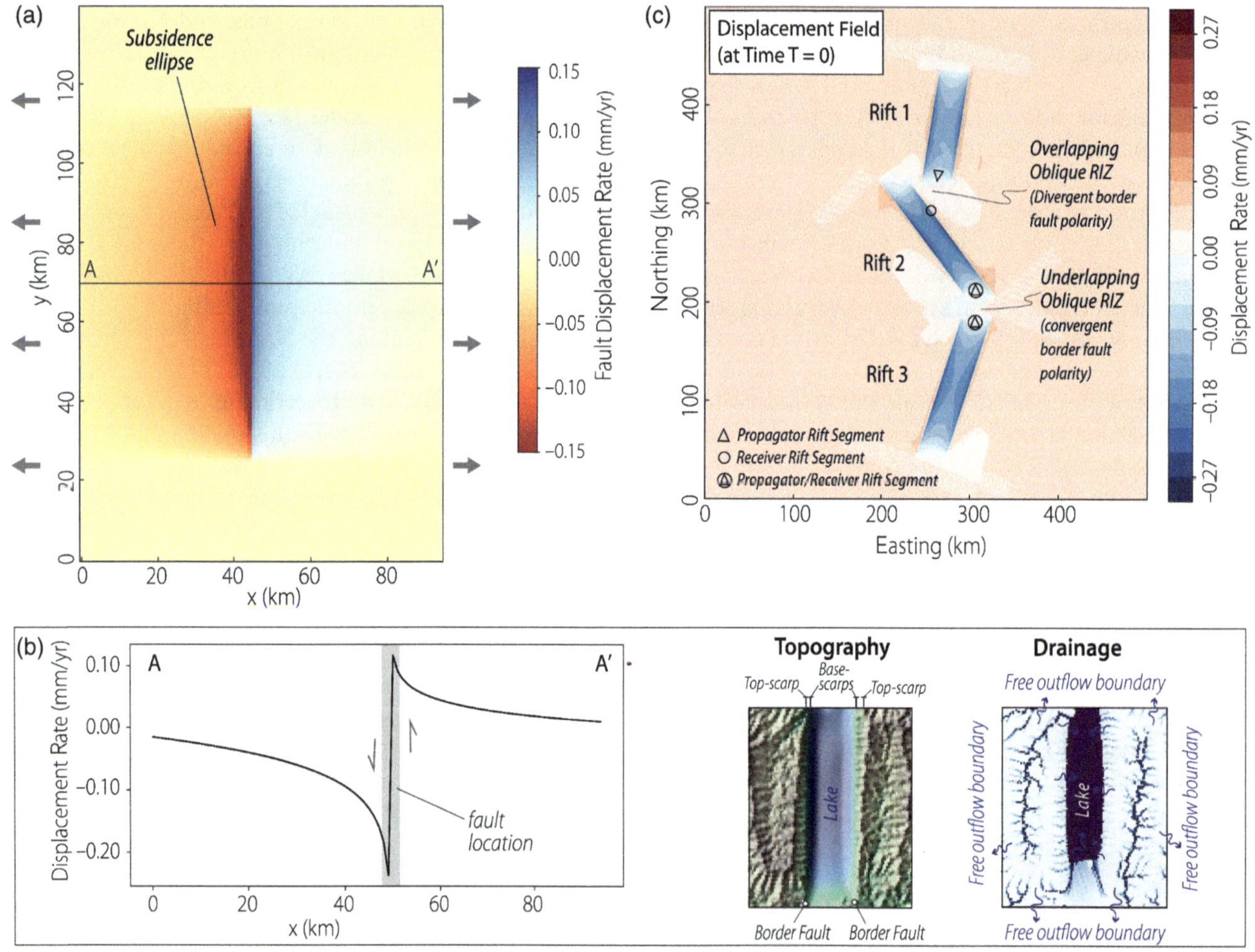

Figure 9.5 Model setup to simulate landscape evolution across two contiguous rift interaction zones with geometries similar to those of the Middle Shire and Nsanje rift interaction zones. (a, b) Fault segment uplift field with hanging wall subsidence and footwall uplift. Inset maps in panel (b) show the resulting topography and drainage for a graben configuration. (c) Initial displacement field applied in the model to estimate the fault development and rift evolution.

of 1 m/yr in the study area and a homogenous lithology. We apply the fixed value boundary condition for the four borders. The model time is 10 m.y. with a timestep of 1 k.y., and we present results for the 5 and 10 m.y. timesteps. The model domain contains three contiguous grabens: NNE-trending Rift 1 in the north, NW-trending Rift 2 in the center, and NNE-trending Rift 3 in the south (Figure 9.5c). The orientation of each graben-bounding fault reflects the location of border faults in the Zomba Graben (southern Malawi Rift), Lower Shire Graben, and Nsanje Graben, thus defining an oblique overlapping divergent RIZ between Rifts 1 and 2, and an underlapping oblique convergent RIZ between Rifts 2 and 3 (Figure 9.5c). Based on field observations along the major faults of the Zomba–Lower Shire–Nsanje graben system (see "Fault Kinematics along the Zomba–Lower Shire–Nsanje Graben System"; Figure 9.6a–h), we impose a normal faulting tectonic stress regime on each border fault in the model, such that each model rift basin undergoes orthogonal extension. The displacement of graben-bounding (border) normal faults is added as time-dependent uplift fields on topography. The maximum displacement, posted at the center of each fault, is 0.8 mm/yr and it decreases to zero at the tips of the faults, following a sinusoidal curve. The half-extension rate and tip growth rate are 1 and 5 mm/yr, respectively, for all border faults in the study area (Figure 9.4), based on the observed low extension rate of 2.2 mm/yr (Stamps et al., 2018). Positive displacement indicates uplift, and negative displacement indicates subsidence; the background displacement rate in the model is 0.03 mm/yr (Figure 9.5a,b). In the model, each rift propagates bidirectionally. Thus, following the convention of Kolawole et al. (2021a), because Rift 1 is located on the flank of Rift 2 in the orthogonal rift interaction model setup, Rift 1 serves as the propagator and Rift 2 the receiver segment across their intervening RIZ, whereas, at the underlapping oblique RIZ, both Rifts 1 and 2 serve as propagator and receiver segments. Note that during the model run, the underlapping oblique convergent RIZ evolved into a tip-to-tip oblique RIZ geometry.

9.2.3. Modeling Static Tectonic Stress Distribution

Although the Fastscape algorithm can project a topography at the end of a model run, it is limited in that it cannot resolve the static stress field associated with tectonic deformation. This is relevant for understanding the influence of compounding stress distribution on the displacement field observed in the landscape evolution model. Thus, we simulate the static Coulomb stress distribution across the modeled RIZ geometries, using the same input parameters of model domain size, fault geometry, and kinematics as in the landscape evolution model, and we implement the simulation in Coulomb 3.3 available from the USGS website (Toda et al., 2005; Lin & Stein, 2004). The software utilizes an elastic half-space model to calculate static stress changes in space and time within a model domain due to slip on a given source fault. The general Coulomb stress change formulation is as follows:

$$\Delta\sigma_c = \Delta\tau_s + \mu'\Delta\sigma_n \tag{9.3}$$

where $\Delta\sigma_c$ is the Coulomb stress change, $\Delta\tau_s$ is the shear stress change, $\Delta\sigma_n$ is the normal (clamping) stress change on receiver normal faults due to slip on a source normal fault, and μ' is the effective coefficient of friction on the faults. We use μ' of 0.55, appropriate for the southern Malawi region (Williams et al., 2019), and fault rupture depth of 5 km. We use the total fault slip in 5 m.y. as prescribed displacement on the fault plane and model cases involving the activation of each border fault as the source fault while other faults act as the receiver faults; then, we model a single case in which all the border faults are the source faults (i.e., all rupture in the same event) and calculate stress change on N-S strike faults with a rake of −90. For simplicity, each fault is represented as a single rectilinear planar surface. By implementing the static stress model, we can assess the predominant type of Coulomb stress change (i.e., positive or negative) transferred into an oblique overlapping divergent RIZ area versus tip-to-tip RIZ.

9.2.4. Model Limitations

The active northeastern border fault system of the LSG (Thyolo-Muona-Ruo Fault System) is characterized by a staggered, side-stepping geometry that developed over multiple rift phases (Kolawole et al., 2022; Figure 9.3a,b). Similarly, a major strand of the western border fault system of the Zomba Graben (Lisungwe Fault) shows zig-zag geometry (Figure 9.3a,b). However, for simplicity, we implemented this fault system as a single planar through-going fault in our model (Figure 9.5c). We assume the modeled faults have uniform tip propagation and extension rates due to a lack of constants on the timing and rate of fault-slip rates. We note a limitation of the poor constraints of fault-slip rates in southern Malawi and the shorter duration of model tectonic extension versus nature. We note that although our model uses a constant rainfall rate throughout the model run, paleoclimate studies in the region (e.g., Scholz et al., 2011; Beuning et al., 2011; Konecky et al., 2011; Dulanya et al., 2014) show records of alternating wet and dry periods in as low as 100-year intervals. The limited constraints on fault-slip rates, rainfall rates, and the paleotopographic (prerift) variations might have resulted in the

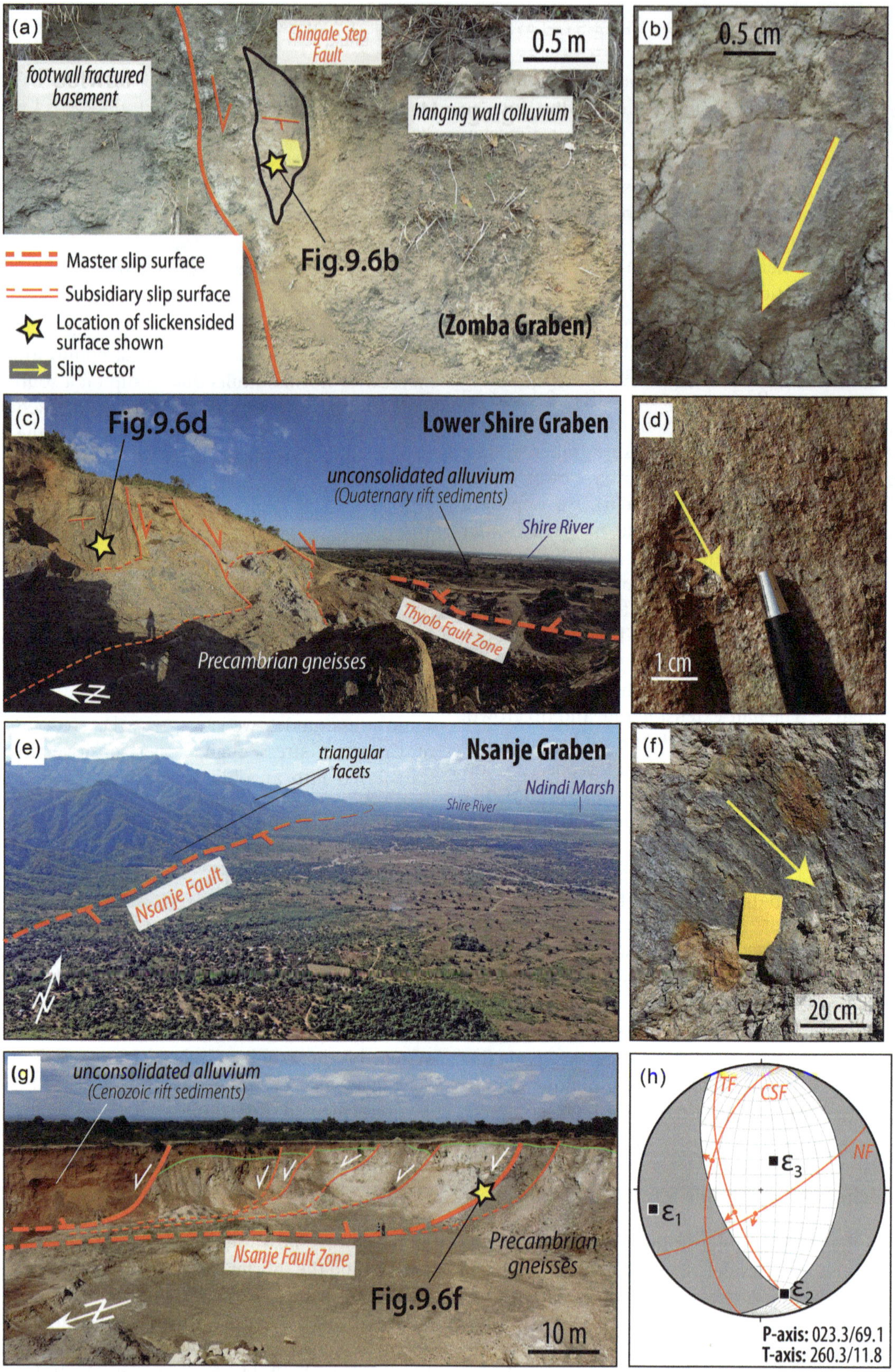

Figure 9.6 Field photographs showing slip surfaces along the Chingale Step Fault, Zomba Graben (a, b; photographs from Wedmore et al. (2020b) / John Wiley & Sons / CC BY 4.0), Thyolo Fault, Lower Shire Graben (c, d), Nsanje Fault, and Nsanje Graben (e–g). See Figure 9.3a for the location of the photos and look-direction of the landscape shots. (h) Kinematic tensor solution produced from the combined geological slip vectors measured along the three faults (i.e., Figure 9.7b,d,f; see the "Methods" section). ε_1, ε_2, and ε_3 represent the principal strain axes 1 (extension), 2 (intermediate), and 3 (shortening), respectively. The solution shows a predominantly normal faulting regime.

mismatch between the absolute values of surface elevation in southern Malawi and those in our model results.

We do not implement the NNW rotation of the northwestern tip of the Thyolo Fault as it is found to have been controlled by the Precambrian basement metamorphic fabrics (Kolawole et al., 2021a). We also acknowledge that the static Coulomb stress change model only assumes a single instance of slip involving the rupture of an entire border fault and does not capture the very complex interactions of the border and intrabasinal fault segments over single and multiple cycles of earthquake activity in active rift basins (e.g., Wedmore et al., 2017; Njinju et al., 2024).

9.3. RESULTS

9.3.1. Fault Kinematics Along the Zomba–Lower Shire–Nsanje Graben System

Field observations of fault-slip striae on exposed fault surfaces in the Zomba, Lower Shire, and Nsanje grabens (Table 9.1; Figure 9.6a–g) show a strong component of normal slip on the faults but with the Nsanje data showing a minor obliquity. This is given by the 301°/52° striae (average trend/plunge) on the Zomba Fault slip surface (189°/54° strike/dip; Wedmore et al., 2020b; Figure 9.8a,b), 231°/80° striae (trend/plunge) on the Thyolo Fault slip surface (160°/69°SW strike/dip; Figure 9.6c,d), and 193°/44° striae (trend/plunge) on the Nsanje Fault slip surface (059°/79°SE strike/dip; Figure 9.6e–g). However, these slip vectors altogether produce a kinematic strain tensor of 023.3°/69.1° P-axis and 260.3°/11.8° T-axis, suggesting a predominantly normal faulting regime, although with a very minor strike-slip component. The 260.3° (80.3°) (ENE-WSW) T-axis trend is parallel to the known regional geodetically resolved tectonic extension azimuth of 086° ± 5° (Stamps et al., 2008, 2018; Figure 9.3a) and the regional σ_3 stress direction of 072° ± 14° from earthquake focal mechanism inversion (Williams et al., 2019). Similarly, the strain tensor solution is consistent with the largely normal faulting style of focal mechanism solutions of the recent earthquakes in the Nsanje RIZ (i.e., the northern tip of the Nsanje Graben and the southeastern tip of the LSG) and those in the Middle Shire RIZ (i.e., the southern tip of the Zomba Graben) (Figure 9.3a,b). These results constrain the normal slip kinematics imposed on the faults in our landscape evolution model.

9.3.2. Overlapping Oblique RIZ: Model Displacement Field, Rift Topography, and Drainage Morphology

After 5 m.y. of model run, Rift 1 has propagated into the footwall of Rift 2, and the western border fault of Rift 1 is hard-linked with the northeastern border fault of Rift 2 across their intervening overlapping oblique RIZ (Figure 9.7a). The integrated fault displacement rate fields represent the distribution of rift basin subsidence and flank uplift in RIZs. Particularly, Rift 1 is characterized by uniform displacement ellipses on its two border faults with up to 0.12 mm/yr footwall uplift rate, ca. 0.2 mm/yr maximum hanging wall subsidence rate. At the overlapping RIZ, the southern tip of Rift 1 extends into a narrow region dominated by a background displacement rate of ca. 0.03 mm/yr or less. To the south of the RIZ, the northeastern border fault of Rift 2 shows a more pronounced hanging wall subsidence rate of >0.32 mm/yr. At the region of hard linkage with the western border fault of Rift 1, the fault shows a localized zone of significant displacement rate at the western margin of the RIZ, defined by the region of fault intersection. More specifically, the localized subsidence occurs at the intersection of the southern extension of Rift 1's western border fault with the northwestern extension of Rift 2's northeastern border fault. Also, we note that, in general, Rift 2 shows a greater axial subsidence rate (0.12–0.16 mm/yr) than Rift 1 (0.04–0.08 mm/yr).

The topography of RIZs is largely modified by the rift fault propagation, resulting in distinct topography and drainage patterns due to different types of RIZs. The surface topography of the overlapping oblique RIZ shows a narrow, uplifted block of up to 12–25 m, and the rift lake in Rift 2 is deepest (ca. 40 m) to the south of the uplift (Figure 9.7b), consistent with the patterns in the displacement field map (Figure 9.7a). Although the drainage system is characterized by well-developed rift lakes in Rifts 1 and 2, the lakes remain disconnected across the overlapping RIZ (Figure 9.7c). These results indicate that Rifts 1 and 2 remain isolated basins, and their fluviolacustrine depositional environments are disconnected. The cross-sectional view of rift structure across the RIZ features a paired graben–half graben morphology in which there is a significantly exposed basement at the rift floor with a graben structure to the west and half-graben to the east (cross-section B-B′ in Figure 9.7d). Also, the section shows a thinner syn-rift cover (ca. 5 m) on the hanging wall of Rift 1's eastern border fault and a thicker cover (up to ca. 40 m) on Rift 2's eastern border fault. In the region of transition from the overlapping RIZ into Rift 2, the model shows the presence of a prominent, isolated depression at the top-basement surface, which is also represented at the model top-syn-rift sediment surface (Figure 9.9a); this represents the fault intersection-related depocenter that developed early and continued to be active with progressive tectonic extension.

However, at the end of the 10 m.y. model run, the displacement field results show that the southward propagating tip of Rift 1 has finally breached the overlapping RIZ, and the rifts are now structurally hard-linked and

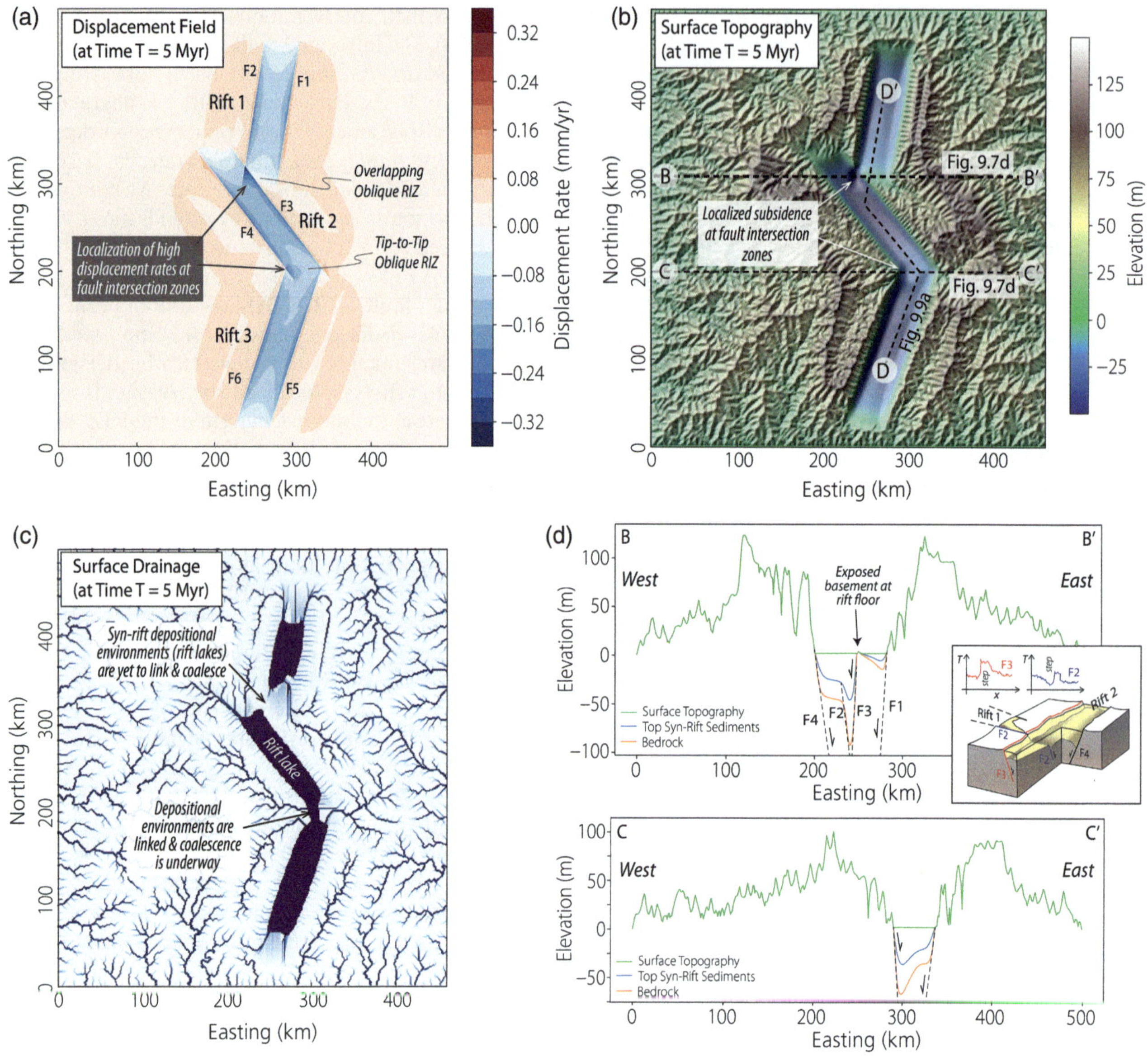

Figure 9.7 (a) Displacement field, (b) modeled topography, and (c) drainage map at 5 m.y., showing the large-scale fault structure and surface morphology of the laterally propagating and interacting rift tips. (d) Cross-sections A-A′ across the overlapping oblique rift interaction zone (RIZ) and B-B′ across the tip-to-tip oblique RIZ. Note that the tip-to-tip oblique RIZ was initially an underlapping oblique RIZ at T = 0 m.y. (Figure 9.5c). (d inset) 3D schematic showing the large-scale structure of antithetically linked faults as seen in the model results (a, b) and idealized throw-distance (T-x) plots inspired by the model results and natural examples (e.g., Adapted from Duffy et al., 2015).

coalesced across the RIZ (Figure 9.8a). In addition, both the surface topography and drainage system (Figure 9.8b,c) show that the depositional environments of rift basins 1 and 2 are fully connected and that a through-going axial rift floor has been established across the RIZ. Moreover, the model results also show that the displacement rates on the border faults have now decreased due to the overlapped uplift and subsidence and that the zone of rift intersection is now the focus of accelerated displacement (Figure 9.8a) and surface subsidence (Figure 9.8b). This zone of focused displacement is the fault intersection depocenter bounded by faults that trend parallel to the border faults of the interacting rift pair.

9.3.3. Tip-to-Tip Oblique RIZ: Model Displacement Field, Rift Topography, and Drainage Morphology

In contrast to the overlapping RIZ, at the end of the 5 m.y. model run, the rift tips of both Rift 2 and Rift 3 have propagated toward each other, and their axial depocenters have merged, forming a tip-to-tip oblique RIZ (Figure 9.7a). The displacement field at the end of 5 m.y.

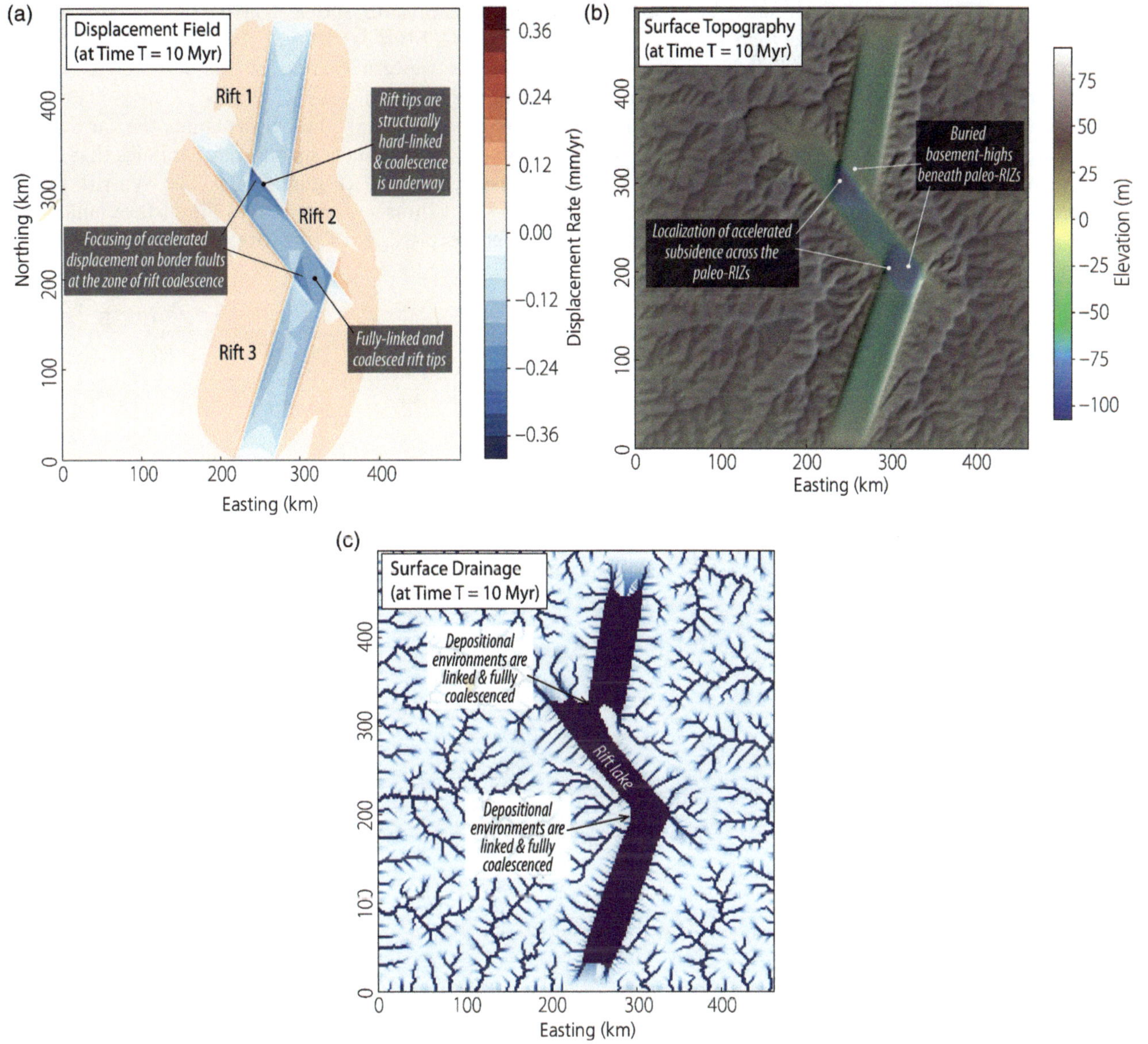

Figure 9.8 (a) Displacement field, (b) modeled topography, and (c) drainage map at 10 m.y. showing complete rift coalescence across the tip-to-tip RIZ and established hard linkage with ongoing coalescence across the overlapping RIZ. Interestingly, the results also highlight that once rift linkage is established, rift coalescence is accompanied by the focusing of accelerated fault displacement and basin subsidence across the deforming RIZs.

presents a different pattern than the one of overlapping RIZ, where the border faults of Rift 3 commonly show a footwall uplift rate of ca. 0.04–0.08 mm/yr with a maximum of 0.12 mm/yr, and a hanging wall subsidence rate of up to 0.28 mm/yr. The basin shows an axial subsidence rate of 0.12–0.16 mm/yr, similar to Rift 2. More importantly, a depocenter of 0.08–0.12 mm/yr subsidence rate extends from the axes of both rifts and merges across the intervening tip-to-tip oblique RIZ. The model result shows the development of localized subsidence at the zone of intersection of the southeastern extension of Rift 2's western border fault with the northern extension of Rift 3's western border fault. Similarly, the southeastern extension of Rift 2's northeastern border fault intersects with the northern extension of Rift 3's eastern border fault (Figure 9.7a,b). These fault intersections represent a hard linkage of the border faults, forming a broad obtuse-angled hanging wall sub-basin along the eastern margin of the RIZ and a reflex-angled sub-basin along the western margin. The reflex-angled linkage zone localized more subsidence rate (up to 0.24 mm/yr), which decreases eastward to 0.08 in the obtuse-angled linkage zone.

In the tip-to-tip RIZ, the basinal surface topography features a wide low elevation area (ca. 12 m), which deepens westward to ca. 30 m in the reflex-angled linkage zone (Figure 9.7b). The model drainage map (Figure 9.7c) shows that the rift lakes of Rift 2 and Rift 3 are connected across the RIZ and deepen from 12 m westward to ca. 25 m in the western reflex-angled RIZ margin. In essence, the fluviolacustrine depositional environments of the two rifts have linked and become open to one another. The cross-section of the model tip-to-tip RIZ shows a west-dipping asymmetric graben morphology, an absence of basement exposure, and a syn-rift sedimentary wedge that thickens from ca. 10 m at the eastern margin to ca. 30 m at the western margin of the RIZ (cross-section C-C′ in Figure 9.7d).

Along the axes of the model rift zones, the variation in top-syn-rift sedimentary surface relief mimics that of the top-basement relief (Figure 9.9a). The syn-rift fill of model Rift 1 thins southward, and the surface onlaps the basement exposures of the overlapping oblique RIZ.

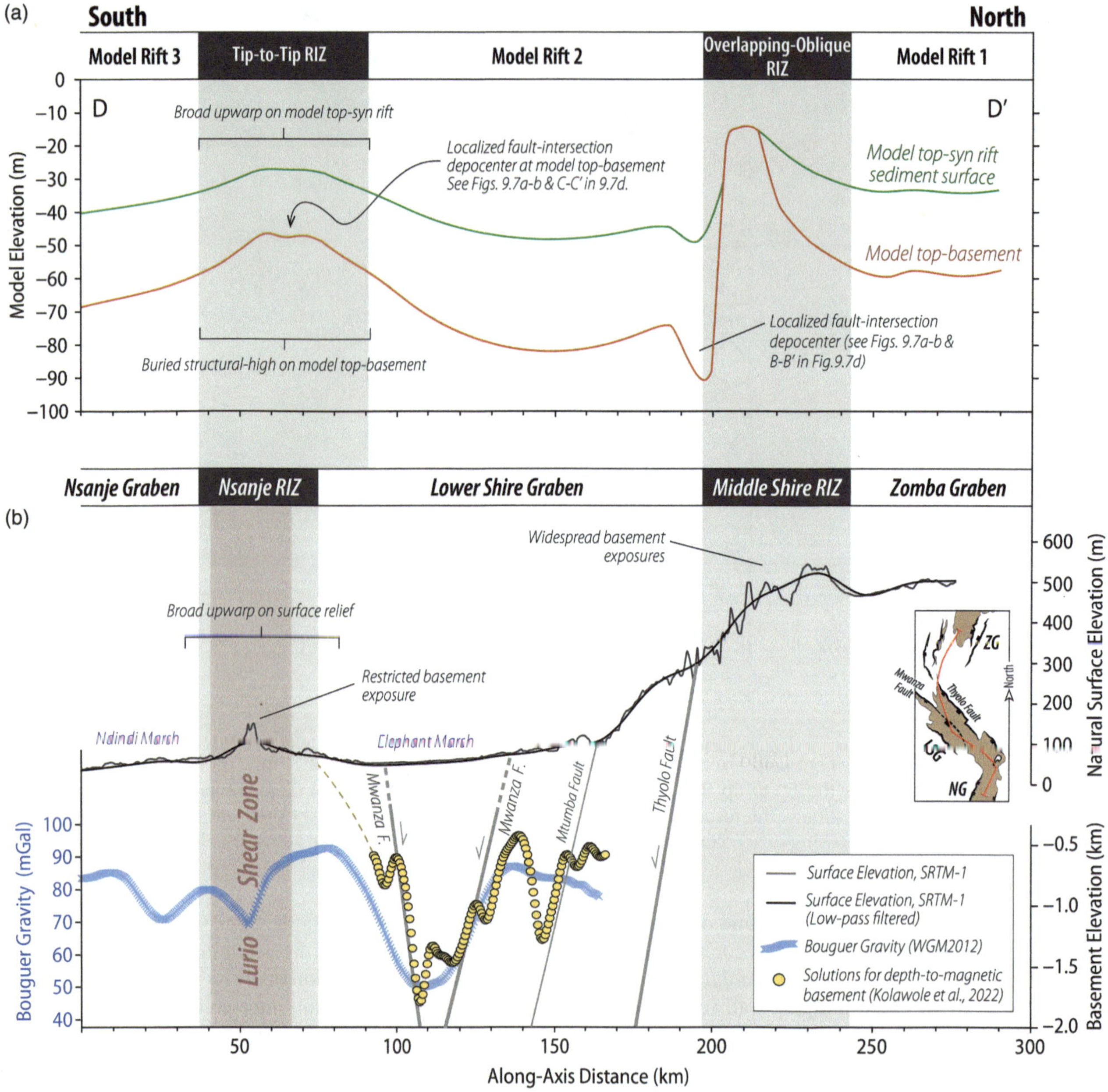

Figure 9.9 (a) Plot comparing the along-axis model surface relief, model top-syn-rift surface relief, and model top-basement relief (at 5 m.y. time step) with the natural along-axis variation in surface topographic relief in the studied rifts. See transect for profile D-D′ in panel (b) (modeled topography). (b) Along-axis plot of natural surface relief along the rifts, Bouguer gravity anomaly, and variation of basement elevation (transect in inset map). NG, Nsanje Graben; LSG, Lower Shire Graben; ZG, Zomba Graben. The natural topography data are from 30 m-resolution Shuttle Radar Topography Mission (SRTM), Bouguer gravity data from World Gravity global model, WGM2012 (Bonvalot et al., 2012), and basement depths calculated from aeromagnetic grid (Kolawole et al., 2022).

South of the RIZ, the model surface drops steeply into Rift 2 at a location where the syn-rift cover sequence is thickest among all three rifts. Interestingly, although the basement is buried across the tip-to-tip oblique RIZ, a broad upwarp exists at the top basement above which the syn-rift sequence also defines a structural high. Also, the model tip-to-tip RIZ shows an isolated depression at the top-basement surface, but which is nonexistent at the model top-syn-rift sediment surface (Figure 9.9a); this represents the fault intersection-related depocenter that developed early and later became buried as the rift tips coalesced across the RIZ.

Overall, before the coalescence of the rifts (10 m.y. time step, Figure 9.8), the comparison of the along-axis relief morphology styles across the rift interaction zones in the initial model time step and the study area (Figure 9.9a,b) shows a correlation of "steep" topographic down-step across the overlapping oblique RIZ and a gentle surface arching across the tip-to-tip RIZ. However, we note that the absolute values of model surface displacement or elevation are not the same as those of the natural RIZ (Figure 9.9a,b), likely due to the poor constraints on the slip rate on border faults from southern Malawi, paleotopography, and possibly the imposed constant rainfall rates and duration of tectonic extension in the models. The model predicts that there is no exposed basement across the tip-to-tip RIZ, whereas natural observation shows the presence of a minor exposure of the basement within the RIZ (Figure 9.9a). In addition, the model predicts that the rift depocenters are yet to link across the overlapping RIZ, which contrasts with the Middle Shire RIZ across which the Shire River connects the Zomba and Lower Shire grabens (Figure 9.3a,b).

The 10 m.y. timestep shows a paleo-RIZ across which rift coalescence is even more advanced than it was at the 5 m.y. timestep (Figure 9.8a,c). However, similar to the overlapping RIZ, the tip-to-tip RIZ also shows a focusing of greater displacement rates and subsidence at the intersection zone of the border faults and relatively lower displacement rates on the border faults themselves (Figure 9.8a,b).

9.3.4. Predicted Static Tectonic Stress Distribution Across the RIZs

The modeled stress field shows different patterns between tip-to-tip oblique RIZ and overlapping oblique RIZ. In each case of single border fault rupture (Figure 9.10a–f), the regions of positive Coulomb stress change (stress concentration) are at the tips of the faults, and the regions of negative Coulomb stress change (stress relaxation) are in the hanging wall and footwall of the faults. Across the overlapping RIZ, the lobes of stress concentration at the southern tips of Rift 1 border faults (faults 1 and 2) generally extend into the footwall of Rift 2's northeastern border fault (Figure 9.10a,b). The proximity of the southern tip of Rift 1's western fault (fault 2) to the northern tip of Rift 2's northeastern fault (fault 3) allows for the repeat stress concentration events at the intersection zone of the two border faults (Figure 9.10b,c).In contrast, the lobes of stress concentration at the southern tip of Rift's eastern fault (fault 1) always only intersect with the large lobe of stress relaxation in the footwall of Rift 2's northeastern fault (fault 3) (Figure 9.10a,c). Coulomb stress change transfer to fault 3 due to slip on fault 1 shows a minor increase, mainly focused onto the central section of fault 3's slip surface (3D fault plane view in Figure 9.10a). In contrast, stress change transfer to fault 3 due to slip on fault 2 shows a significant increase, and it is focused onto a large portion of the northwestern slip surface of fault 3 (3D fault plane view in Figure 9.10b).

Across the tip-to-tip RIZ, all events of slip on any of Rift 2 and Rift 3's border faults (faults 3–6) concentrate positive Coulomb stress changes in the RIZ (Figure 9.10c–f). The lobes of positive stress change transfer due to slip on Rift 2's border faults (faults 3 and 4) extend onto the northern tips of Rift 3's border faults (faults 5 and 6). Similarly, the lobes of positive stress change transfer due to slip on faults 5 and 6 extend far onto the southwestern sections of Rift 2's faults 3 and 4. Overall, the results can be summarized in a single model case where all the faults are activated as source faults (Figure 9.10g), showing that the overlapping RIZ is largely a zone of stress relaxation and the tip-to-tip RIZ a zone of compounded stress concentration.

9.4. DISCUSSION

9.4.1. Comparison of Model Results with Natural Observations

Rift Interaction Zone Morphology

The general patterns of the surface and top-basement displacement and drainage morphology of the model RIZs (Figure 9.9a,b) provide important insights relevant to understanding the structure and pace of evolution of the Middle Shire RIZ and Nsanje RIZ. In the model, at some time during the propagation of the rifts (e.g., the 5 m.y. time step), unlike the tip-to-tip RIZ, the rift floor of overlapping RIZ retained widespread basement exposure, which is consistent with the current widespread exposure of the Precambrian basement in the Middle Shire RIZ. Similarly, the broad upwarp of the model syn-rift surface across the model tip-to-tip RIZ is collocated with the region of upwarp in the long wavelength surface elevation across the Nsanje RIZ (Figure 9.9a,b). The model shows an early coalescence of the rift axial lakes across the tip-to-tip RIZ, consistent with the extension of a wetland drained by the Shire River across the Nsanje RIZ (Elephant and Ndindi Marshes, Figures 9.3a,b, 9.4d, and 9.7c). Overall, in contrast to the tip-to-tip RIZ, the

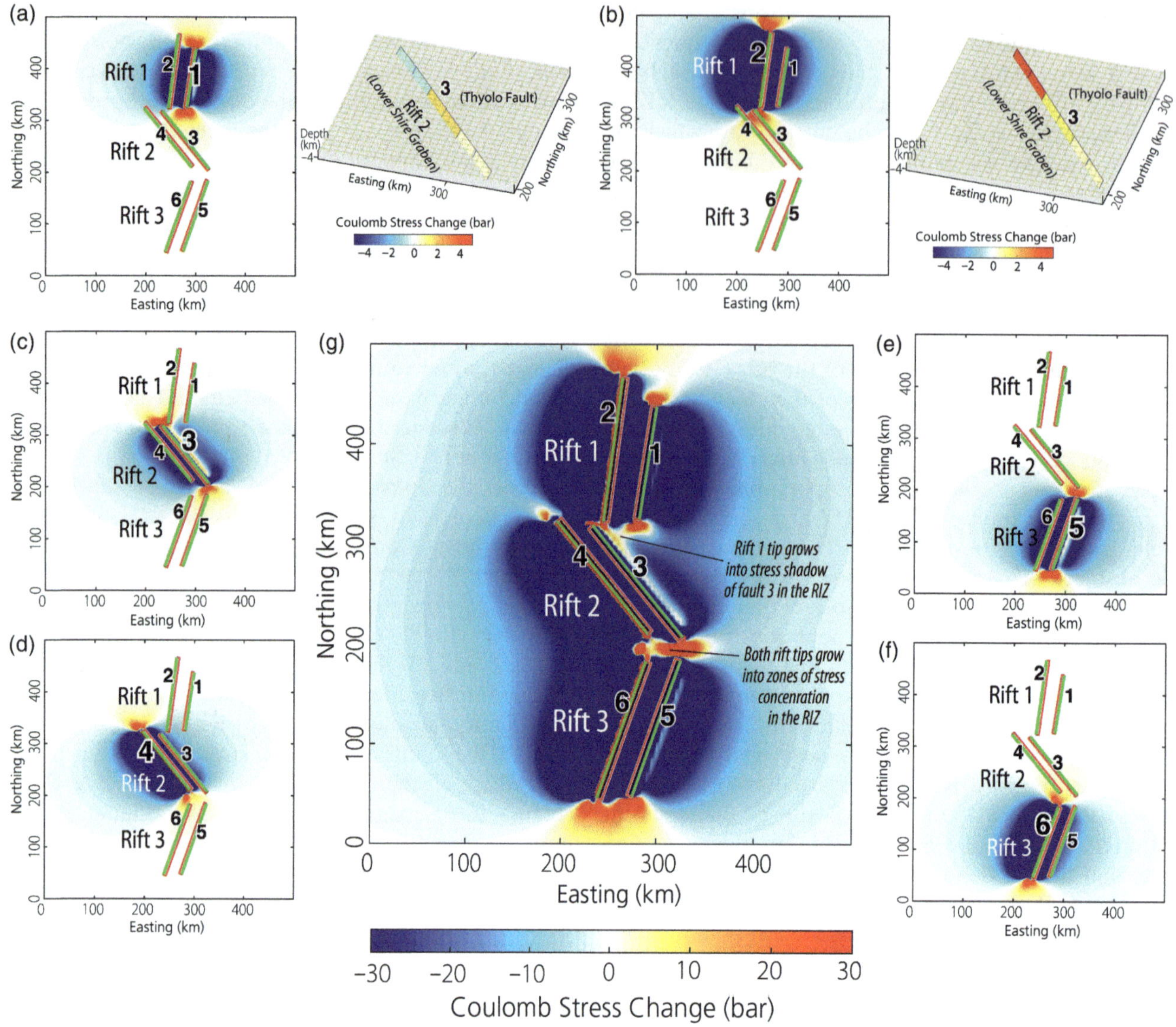

Figure 9.10 (a–f) Model results of predicted static Coulomb stress change distribution for instances of slip on each of the border faults (as source fault, enlarged fault number) of the model rift basins, maintaining the same input rift geometries and model parameters used for the Landscape evolution model. We assume a coefficient of friction of 0.55, appropriate for the southern Malawi region (Adapted from Williams et al., 2019), and fault rupture depth of 5 km. In each model, the entire border fault length ruptures in a single event. Note that numbers 1 to 6 represent fault identifiers. 3D views in panels (a) and (b) show Coulomb stress changes on fault 3 with slip on source fault 1 (a) and fault 2 (b), illustrating the difference effects of these two fault-slip events on the stress field of fault 3 in the RIZ. (g) Static Coulomb stress change distribution for the same model domain as shown in (f), but in which all the border faults act as source faults. Coulomb stress change is calculated for N-S striking faults with a potential rake of −90 (normal dip-slip). Altogether, the models show that in overlapping RIZs, the propagating rift tip grows into a stress relaxation zone, which contrasts with tip-to-tip RIZs where propagating rift tips grow into a compounding stress concentration zone.

model predicts delayed coalescence of the depositional environments across the overlapping oblique divergent RIZ (see the disconnected rift lakes in Figure 9.7c). In nature, although the depositional environments of the Zomba and LSGs are already connected (Shire River flows across the Middle Shire RIZ; Figure 9.3a), published geomorphological analysis and knickpoint age estimates suggest a recent (Mid. Pleistocene) breaching of the Middle Shire RIZ (Kolawole et al., 2021a; Dulanya et al., 2022) relative to the Late Oligocene initiation of Cenozoic rifting in the region (Ojo et al., 2024). Furthermore, both the model and natural surface relief show the presence of the broad topographic "step" from the Zomba Graben (Rift 1) into the LSG (Rift 2) (Figure 9.9a,b).

The Bouguer gravity model of the LSG (WGM2012; Bonvalot et al., 2012; Figure 9.9b) shows a prominent rift-parallel gravity-low (ca. 25 mGal) anomaly near the center of the basin. This anomaly has been observed in published airborne gravity map of the graben, interpreted to represent a deep depocenter in the basin (Ngabu depocenter; Chisenga et al., 2018). This gravity-low anomaly is colocated with a zone of localized >2.5 km subsidence on the hanging wall of the buried southern Mwanza Fault along the axis of the LSG, which records multiple phases of tectonic extension (Kolawole et al., 2022). To the southeast, this gravity-low anomaly transitions into a broad rift-orthogonal gravity-high (ca. 92 mGal) anomaly extending far into the Nsanje RIZ, which then decreases again to <85 mGal in the Nsanje Graben (Figure 9.9b). Similarly, the along-rift variation in the trend of the basement depth (Figure 9.9b) shows decreasing basement depths toward the southeast. Both the aeromagnetic basement depth and gravity anomaly trends, together, suggest that the basement shallows into the Nsanje RIZ and deepens southward into the Nsanje Graben (Figure 9.9b). The results indicate that there exists a shallowly-buried basement-high beneath the Nsanje RIZ, which is also predicted by the model in the tip-to-tip RIZ (Figure 9.9a).

Our interpretation of shallow but widespread basement burial in the Nsanje RIZ is reflected by the relatively small vertical surface scarp height on the Ruo Fault (<40 m; Figure 9.S2b; Kolawole, 2020), the broad upwarp of the surface topographic relief across the RIZ (low-pass filtered SRTM plot in Figure 9.9b), and the restricted basement exposure on the rift floor (Figures 9.3b, 9.4c,d, and 9.9b). Interestingly, the Nsanje RIZ surface tilts and deepens to the southwest without any visible surface fault scarp at the western margin of the RIZ (Figure 9.4c,d). However, we note the presence of a subtle SW-dipping intra-rift fault, which may be assisting the uplift of the restricted basement exposure. We speculate that the westward tilting and deepening of the rift floor across the Nsanje RIZ may be due to the presence of a blind active NW-trending NE-dipping fault near the western margin of the RIZ (see dashed fault in Figures 9.4c). Potentially, higher slip rates on this blind fault relative to the Ruo Fault could have created the apparent half-graben geometry of the RIZ, in which the Ruo Fault hanging wall is on the uplifted margin. The uplift of the northeastern margin of the RIZ relative to the southwestern margin is highlighted by the presence of incising west-flowing transverse streams that cut into the hanging wall of the Ruo Fault.

The inferred blind fault at the SW margin of the Nsanje RIZ may be responsible for some of the earthquake clusters in the Nsanje RIZ (see Figure 9.3a) but likely remains blind due to sedimentation rates outpacing the slip rate on the fault. Modern fault-slip rates are considered to be generally low in southern Malawi, but quantitative constraints are lacking (Williams et al., 2022). Also, there is no quantitative constraint on sedimentation rates in the LSG due to the absence of high-resolution subsurface imaging and well controls. However, the Shire River is a major river that drains the region and channels large amounts of sediments into the Lower Shire and Nsanje grabens and represents a major sediment contributor to the Zambezi River, with the Mozambique Channel being the sink of the river system (e.g., Dulanya et al., 2022; Garzanti et al., 2022). Furthermore, the high sedimentation inference is supported by the broad swampy terrain of the southwestern margin of the Nsanje RIZ (Figure 9.4d) being dominated by the Shire River flood plains and the Elephant Marsh beneath which the fault is located (Figure 9.4c,d). In Malawi region, areas of high sedimentation along the margins of the rifts may host active faults that are partially or completely blind due to high sedimentation rates (e.g., Williams et al., 2022; Kolawole et al., 2021a, 2022). For example, the 2009 Mw6 Karonga earthquake in Northern Malawi ruptured the St. Mary Fault system which was hitherto an unknown blind basement-rooted fault beneath the flood plains and deltas of the Mbalizi, North Rukuru, Kasisi, Lufira, and Kibwe rivers in the highly-sedimented Karonga area (Kolawole et al., 2018). In summary, at some time in the past, the currently buried basement high beneath the Nsanje RIZ sediments would have been a broad exposed basement arch that may have isolated the Lower Shire Graben from the Nsanje Graben. With progressive tectonic extension and rift propagation into the RIZ, this elevated basement block subsequently experienced down-faulting, erosion, and burial, a process that facilitated the structural breaching of the RIZ and linkage of the two rift basins (Kolawole et al., 2021a).

Significance of Fault Intersection for the Deformation of Rift Interaction Zones

The landscape evolution model results predict a localized zone of elevated displacement rates and associated basement subsidence at the western margin of the overlapping divergent RIZ and another one at the western margin of the tip-to-tip oblique RIZ, both colocated with regions of fault intersection (Figures 9.7a,b,d, and 9.9a). We interpret this modeled subsidence to be consistent with the development of isolated fault-bounded depocenters in the Middle Shire RIZ and the potential presence of a blind high-strain fault in the western margin of the Nsanje Graben. Near the western margin of the Middle Shire RIZ, the surface geology shows the presence of a ca. 17 km^2 isolated depocenter of Quaternary sediments, herein referred to as the "Neno depocenter" (Figure 9.3b). This localized depocenter is elongated, NW-trending, and fault-bounded. It sits at the intersection of the southern extension of the NNE-trending Lisungwe Fault and a NW-trending fault that trends parallel to the Thyolo border fault of the LSG (Figure 9.3b). The basement subsidence is caused by overlapped and enhanced displacement of rift faults, implying the possible control on the formation of its natural equivalent, the Neno depocenter. Moreover, the topographic surface of the depocenter is relatively "smoother" than the surrounding basement-dominated topography (Figure 9.3a), consistent with a dominance of unconsolidated sediments.

Just south of the Neno depocenter, in Majete, we observe an even broader (ca. 42 km^2) NW-trending fault-bounded area with similar surface smoothness as the Neno depocenter, possibly representing an incipient localized depocenter (Figure 9.3b,c). This potential depocenter in Majete sits at the intersection of the southern extension of the NNE-trending Wamkurumadzi Fault and a NW-trending fault parallel to the Thyolo Fault. The cross-sectional topographic relief profile of the RIZ (Figure 9.4a) shows a ca. 35-km-wide graben structure, with a surface that deepens toward the Lisungwe and Wamkurumadzi faults to the west and is bounded to the east by a system of NW-trending faults. Generally, both antithetic and synthetic fault linkage create localized amalgamated depocenters on the hanging wall of both faults (synthetic interaction; Figures 9.7a and 9.8a) or in the hanging wall of the dominant fault segment (antithetic interactions; Figure 9.7a,d inset; Duffy et al., 2015). The depocenters are often well represented as prominent anomalies in the throw-distance (T-x) profiles of the interacting faults with their characteristic abrupt "throw steps" (3D sketch and T-x plots in Figure 9.7d inset; Duffy et al., 2015). However, based on insights of fault intersection depocenters from our modeling results (Figures 9.7 and 9.8), we interpret these isolated fault-bounded depocenters in the Middle Shire RIZ as distributed local subsidence zones controlled by the intersection and interaction of fault segments extending from the Zomba and Lower Shire rifts into the RIZ. Notably, recent analog models of linkage between obliquely oriented rifts show the linkage of antithetic faults at the propagating fault tips (Schmid et al., 2023).

On the dominant dip of faults that host the fault intersection depocenters, we note that the overlapping divergent RIZ model results present the fault intersection-related depocenter to be localized on a NW-dipping fault, whereas natural observations show the isolated depocenters in the Middle Shire RIZ to be localized on NE-dipping faults. We suggest that the inherited heterogeneity of the prerift basement of the Middle Shire RIZ may play an important role in determining the dominant fault dip direction, which our models are not designed to assess. Nevertheless, the model results suggest that localities of fault intersection within deforming RIZs are zones that localize subsidence where significant tectonic strain associated with progressive rift linkage may accumulate. Further, these fault intersection zones are zones where strain is transferred between the interacting rift tips as they propagate across a deforming RIZ. The early localization of subsidence in the model tip-to-tip RIZ is facilitated by the early intersection of propagating border fault tips into the RIZ, and the model predicts that even during the advanced stages of rift coalescence, the continued deepening of the RIZ sub-basin is most intense in the location of the early fault intersection depocenter (Figure 9.7d). Thus, we suspect that the westward deepening of the Nsanje RIZ could indicate either a buried fault intersection-related localized depocenter or a NW-trending, NE-dipping normal fault at the southwestern margin of the RIZ, potentially facilitating the incursion of wetlands (Elephant Marsh) from the LSG into the western margin of the Nsanje RIZ (Figures 9.3a,b and 9.4c,d). Altogether, the landscape evolution model results and natural observations suggest that after the structural breaching of a RIZ (i.e., post-rift linkage), rift coalescence is initiated by the localization of distributed incipient depocenters as overlapped hanging wall subsidence. Such depocenters may preferentially develop at localities of fault intersections within the RIZ.

Tectonic Stress Transfer and Active Faulting in the Rift Interaction Zones

The seismicity patterns in the region reflect a relatively greater clustering of earthquakes in the Nsanje RIZ compared to the Middle Shire RIZ (Figure 9.3a,b). The clustered events in the Nsanje RIZ include a Mw 5.5 event and its aftershocks that likely ruptured the Ruo Fault and the northern extension of the Nsanje and Ndindi faults. The largest magnitude events in the Nsanje RIZ (Mw 5.5 and 4.9; source: USGS earthquake catalog) occurred near the eastern margin of the RIZ. The focal mechanism nodal planes generally trend N-to-NNW and NW, the former being parallel to the trend of the Nsanje Graben border faults (Nsanje and Ndindi faults) and the latter being parallel to the trend of the Ruo border fault segment (Figure 9.3a,b). These seismicity patterns likely reflect a more pronounced active crustal deformation with significant moment release and localization of brittle strain within the Nsanje RIZ. Although instrument records show the occurrence of at least one Mw > 5 event in the Middle Shire RIZ (Figure 9.3a), the events seem relatively more sporadic than those in the Nsanje RIZ. Morphometric and stream power inversion analysis of drainage networks along the Thyolo-Muona-Ruo/Camacho Fault System shows a southeastward growth of the fault system within <2 Myrs timescale (Dulanya et al., 2024).

The Coulomb static stress transfer models predict a concentration of positive Coulomb stress changes within the tip-to-tip RIZ produced by normal slip on the rift border faults, which contrasts with the case of overlapping oblique divergent RIZ where a colocation of positive and negative stress changes is predicted (Figure 9.8). In essence, in the tip-to-tip RIZ, the propagating rift tips grow into a zone of stress concentration within the RIZ, whereas in the overlapping divergent RIZ, the propagating rift tip grows into a zone of stress relaxation. These results highlight the significant role of border fault dip polarity in determining how static stress changes are transferred across deforming RIZs. Indeed, antithetic dips of interacting border faults inhibit stress concentrations compared to synthetically interacting border faults. The long-term compounding effect of stress relaxation in overlapping RIZs suggests that such RIZs host a persistent stress shadow where strain relaxation dominates. Altogether, these model results suggest that over the repeated cycles of fault-slip events at the propagating rift tips, the time-averaged brittle strain accumulation will likely be greater in tip-to-tip RIZs than in overlapping divergent RIZs. Although the constraints on fault-slip history are sparse, we can still interpret that depending on RIZ geometry, an RIZ may localize compounding stress concentration, favoring rapid rift linkage and coalescence as in the case of tip-to-tip RIZs, or may experience compounding stress relaxation favoring delayed rift coalescence, as in the case of overlapping divergent RIZs.

The southern tip of the Zomba Fault appears to be "stagnated" as it does not extend across the entire Middle Shire RIZ (Figure 9.3a) relative to the western border fault system of the Zomba Graben (Lisungwe-Wamkurumadzi fault system), which has propagated close to and is soft-linked with the northeastern border fault of the LSG (Thyolo Fault and its subsidiary footwall faults) (Figure 9.3a). We suggest that the attainment of hard linkage between the Lisungwe-Wamkurumadzi and Thyolo faults can be explained by positive Coulomb stress change transfer onto the Thyolo Fault surface due to slip on the Lisungwe-Wamkurumadzi fault system. The Coulomb stress change transfer calculations show that slip on the Zomba Fault is not able to transfer significant positive stress change onto the Thyolo Fault plane across the overlapping RIZ (3D plot in Figure 9.10a). In contrast, the proximity of the tips of Fault 2 (representing the Lisungwe-Wamkurumadzi fault system) and Fault 3 (representing the Thyolo Fault) predicts compounding stress concentration in the area between the northwestern end of the Thyolo Fault and southern ends of the Lisungwe and Wamkurumadzi faults (Figure 9.10a,b). Thus, the interaction of the fault tips allows the localization of brittle deformation, facilitating the northwestward propagation of the Thyolo Fault and breaching of the RIZ. The two fault systems are hard-linked by a cluster of NW-to-NNW-trending fault clusters, guided by the NW-SE to NNW-SSE rotation of the prerift basement metamorphic fabrics (Kolawole et al., 2021a).

9.4.2. Controls of Rift Interaction Zone Geometry on Early versus Delayed Rift Coalescence

The striking contrast in the stages of evolution of the two model RIZs at the end of the 5 m.y. (relative to the 10 m.y.) model runs and the observations of present-day natural rift morphology at the Middle Shire and Nsanje RIZs (e.g., Figure 9.9a,b) provide compelling evidence for a difference in the pace of rift linkage with variation in RIZ geometry. This interpretation agrees with the predicted patterns of crustal stress distribution across overlapping divergent (antithetic border fault interactions) and tip-to-tip RIZs (synthetic border fault interactions) (Figure 9.10). The results predict delayed rift coalescence across an overlapping divergent RIZ and demonstrate that it can be explained by the propagation of rift tips into a persistent stress relaxation zone in the RIZ. Although the persistent footwall uplift of the flank of the "receiver" rift segment (e.g., Rift 2) inhibits subsidence across the overlapping divergent RIZ; the kinematic opening direction of such a rift induces compression that acts normal to the direction of growth of the "propagator" rift segment tip (i.e., the southern tip of Rift 1). Thus, compression normal to rift-tip propagation direction can also contribute to a delayed advancement of the propagator rift tip into the overlapping divergent RIZ. This is consistent with observations of stagnation of propagating mid-oceanic ridges by compressive tectonic loading in the direction of ridge propagation (Le Pourhiet et al., 2018).

Strong crustal blocks beneath a RIZ, either crustal-scale or only in the lower crust, can inhibit rapid propagation of a rift tip (van Wijk & Blackman, 2005; Le Pourhiet et al., 2018). However, this is not likely to be the case in our study area as basement geologic maps (e.g., Bingen et al., 2009; Fritz et al., 2013; Thomas et al., 2022) and large-scale basement metamorphic fabric trends (Kolawole et al., 2021a, 2022) in the region do not show the presence of a discrete basement block separating the southern Malawi Rift from the LSG. However, numerical models show that significant activation of rift stalling by lower crustal rheological blocks is in fact, most significantly activated when there is compressive tectonic loading ahead of the propagating rift tip (Le Pourhiet et al., 2018). Also, a regional weak lower crust beneath interacting rifts and their intervening RIZ may also delay rift linkage, and low surface erosional efficiency may delay the coalescence of depositional environments of interacting rifts (Wolf et al., 2022). Nonetheless, although the Nsanje RIZ developed over the NE-trending Precambrian Lurio Shear Zone (Kolawole et al., 2022), there is no evidence on the lower crustal property of the shear zone that might suggest that it is significantly stronger than the lower crust of the Middle Shire RIZ. Therefore, we cannot strongly claim that some unknown lower crustal mechanical property influenced the delayed rift linkage and coalescence across the Middle Shire RIZ.

Thus, we argue that in nonvolcanic active rift settings, the geometry of rift interaction zones strongly influences the pace of rift linkage and coalescence by modulating the overall static stress distribution and tectonic loading patterns across the RIZ. As a consequence, RIZ geometry may modulate the evolution of the syn-rift depositional environment, as well as patterns of landscape evolution in actively deforming zones of rift interaction. This was initially speculated by Kolawole et al. (2021a) based on a conceptual understanding of idealized evolution of surface topography from a 'basement high' RIZ into a relatively 'flat' sediment-covered area defining a paleo-RIZ. It was also speculated that paleo-RIZs can be identified by the presence of buried large basement blocks at zones of lateral changes in the along-trend geometry of a continental rift or rifted margin (Kolawole et al., 2021a). However, this current study provides natural and model examples that validate these speculations, as well as a detailed evaluation of how RIZ geometry may control the pace of transformation of a RIZ with progressive tectonic deformation.

Models of rift propagation commonly test for the controls of extension direction (i.e., oblique versus orthogonal) on the deformation patterns across RIZs (e.g., Brune, 2014; Zwaan & Schreurs, 2017; Zwaan et al., 2022). However, little is known about the relationships between extension obliquity and the pace of lateral rift propagation and associated coalescence of interacting rift segments. Although our study is primarily focused on a rift zone with predominantly normal faulting kinematics (Figure 9.6h; Williams et al., 2019), we note that it is also possible that the minor strike-slip component of rift kinematics in the region may have promoted the rapid lateral propagation of the rifts into the Nsanje RIZ. In addition, we note that elsewhere along the magma-poor segments of the EARS, ascending mantle fluids may be promoting crustal weakening and lateral propagation of the rift tips (Ajala et al., 2024; Kolawole & Ajala, 2024).

In general, we propose that our observations and model results for the evolution of the Middle Shire and Nsanje RIZs can be applied to RIZs elsewhere. For example, the ca. 750-km-long Malawi Rift and 730-km-long Tanganyika Rift exhibit greater rift lengths than many other rift segments in the EARS. The two rifts were proposed to have evolved over multiple pulses of lateral propagation (e.g., Specht & Rosendahl, 1989; Scholz et al., 2020; Kolawole et al., 2021a; Shaban et al., 2023). The Kavala Island Ridge, buried beneath the central Tanganyika Rift, is an example of a basement block at which rift segments define overlapping RIZ geometry and in which their border faults define a divergent overlapping transfer zone (Specht & Rosendahl 1989; Morley et al., 1990; Muirhead et al., 2019; Shaban et al., 2023). This ridge is argued to be a long-lived structural high in the Tanganyika Basin, which served as a paleo-drainage divide during periods of lowstands (Scholz & Rosendahl, 1988; Shaban et al., 2023). Such large structurally controlled paleo-drainage divides commonly modulate biodiversity patterns in active continental rift environments (e.g., Russell et al., 2012; Dommain et al., 2022), demonstrating the roles of RIZ evolution on sedimentary depositional environment, landscape evolution, and floral and faunal speciation gradients.

Similarly, along the Malawi Rift, published maps of the basement topography (Specht & Rosendahl, 1989; Scholz et al., 2020) reveals the presence of a large 4,700 km^2 basement block, the "Likoma-Lipichilli Block," buried beneath younger syn-rift sediments in the central section of the rift. This basement block separates the northern Malawi Rift from the southern Malawi Rift and could have represented a major structural high on the paleo-topography of the rift, possibly serving as a paleo-drainage divide during periods of low stands. For example, the block is collocated with an inferred earlier termination zone of the northern Malawi Rift during the long-term southward growth of the rift (Scholz et al., 2020).

9.5. CONCLUSIONS

We applied geologically constrained landscape evolution and static stress models to evaluate the surface processes and stress state in two contiguous nonvolcanic RIZs of contrasting end-member RIZ geometries, the Zomba–Lower Shire–Nsanje graben system at the southern branch of the EARS. The Middle Shire RIZ is an overlapping oblique divergent RIZ in which the NNE/N-trending southern Malawi Rift is propagating into the NE shoulder of the NW-trending LSG, where widespread basement exposure dominates the rift floor. The Nsanje RIZ is a tip-to-tip oblique RIZ in which the LSG has propagated into the northern tip of the N-trending Nsanje Graben, where the rift floor is dominated by widespread sediment deposition with minor basement exposure.

In general, the landscape evolution model results provide insights into the evolution of surface morphology and rift structure across the RIZs in the study area. The model results show that with progressive extension and tip growth, the overlapping oblique divergent RIZ maintains minor basement down-throw and an unequilibrated axial stream profile, which contrasts with the widespread basement burial and equilibrated axial stream profile across tip-to-tip RIZ. Further, static tectonic stress distribution models suggest compounding stress concentrations at tip-to-tip RIZs, thus implying that brittle strain localization and rift coalescence are favored in such RIZs. In contrast, the model predicts compounding stress relaxation at overlapping oblique divergent RIZs, favoring stalled rift coalescence and providing a kinematic explanation of the distribution of deformation in these two RIZs. These findings indicate that antithetically

interacting border faults inhibit stress concentration within RIZs and that synthetically interacting border faults are stress concentrators within their intervening RIZs. The results show that the kinematics of the rift border faults across the overlapping divergent RIZ induces compression normal to the rift-tip propagation direction, promoting delayed advancement of the "propagator" rift tip into the overlapping RIZ. Thus, we argue that in the absence of magmatism, RIZ geometry strongly influences the pace of rift-tip coalescence by modulating the spatial distribution of tectonic stresses necessary to promote rift-linking deformation. The field observations and model results in Middle Shire and Nsanje RIZs presented in this study provide a better understanding of how the geometry of rift interaction zones elsewhere may influence the pace of rift linkage and rift-tip coalescence.

ACKNOWLEDGMENTS

We thank Elizabeth Catlos for the editorial handling of our paper and reviewers Bailey Lathrop and two anonymous reviewers for their comments and suggestions that have helped to improve the quality of the manuscript. Thanks to Chikondi Chisenga for assisting during the fieldwork. We also thank the Columbia Climate School for providing the research funds that supported the field component of this project.

AUTHOR CONTRIBUTIONS

F.K. and L.X. conceptualized the project. F.K. performed the fieldwork. L.X. conducted the numerical modeling. F.K. and L.X. interpreted the results. F.K. wrote the manuscript. L.X. and Z.D. revised the manuscript.

AVAILABILITY STATEMENT

All the datasets supporting the analysis performed in this work are either already in the public domain or provided in the manuscript, and none are proprietary. Fastscape is available via Zenodo (doi: 10.5281/zenodo.4435110). Coulomb 3.3 is accessible from USGS (https://pubs.usgs.gov/of/2011/1060/).

REFERENCES

Aanyu, K., & Koehn, D. (2011). Influence of pre-existing fabrics on fault kinematics and rift geometry of interacting segments: analogue models based on the Albertine Rift (Uganda), Western Branch-East African Rift System. *Journal of African Earth Sciences*, *59*(2–3), 168–184.

Acocella, V., Faccenna, C., Funiciello, R., & Rossetti, F. (1999). Sand-box modelling of basement-controlled transfer zones in extensional domains. *Terra Nova-Oxford*, *11*(4), 149–156. doi: 10.1046/j.1365-3121.1999.00238.x

Agar, S. M., & Klitgord, K. D. (1995). Rift flank segmentation, basin initiation and propagation: a neotectonic example from Lake Baikal. *Journal of the Geological Society*, *152*(5), 849–860.

Ajala, R., Kolawole, F., & Menke, W. (2024). Blind magmatism abets nonvolcanic continental rifting. *Communications Earth & Environment*, *5*(1), 80. doi: 10.1038/s43247-024-01244-7

Barr, M. W. C., & Brown, M. A. (1987). Precambrian gabbro–anorthosite complexes, Tete Province, Mozambique. *Geological Journal*, *22*(S2), 139–159.

Beuning, K. R. M., Zimmerman, K. A., Ivory, S. J., & Cohen, A. S. (2011). Vegetation response to glacial–interglacial climate variability near Lake Malawi in the southern African tropics. *Palaeogeography, Palaeoclimatology, Palaeoecology*, *303*, 81–92.

Bicca, M. M., Jelinek, A. R., Philipp, R. P., & Jamal, D. L. (2019). Mesozoic-Cenozoic landscape evolution of NW Mozambique recorded by apatite thermochronology. *Journal of Geodynamics*, *125*, 48–65.

Bingen, B., Jacobs, J., Viola, G., Henderson, I. H. C., Skår, Ø., Boyd, R., et al. (2009). Geochronology of the Precambrian crust in the Mozambique belt in NE Mozambique, and implications for Gondwana assembly. *Precambrian Research*, *170*(3–4), 231–255.

Bloomfield, K. (1958). The geology of the Port Herald area. *Bulletin of the Geological Survey Malawi*, *9*, 75.

Bloomfield, K. (1965). The geology of the Zomba area. *Bulletin of the Geological Survey Malawi*, *16*, 193.

Bonvalot, S., Balmino, G., Briais, A., Kuhn, M., Peyrefitte, A., Vales, N., et al. (2012). *World gravity map* (1:50000000 map). CCGM, BGI, CNES, IRD.

Bosworth, W. (1985). Geometry of propagating continental rifts. *Nature*, *316*(6029), 625–627.

Braun, J., & Willett, S. D. (2013). A very efficient O (n), implicit and parallel method to solve the stream power equation governing fluvial incision and landscape evolution. *Geomorphology*, *180*, 170–179.

Brune, S. (2014). Evolution of stress and fault patterns in oblique rift systems: 3D numerical lithospheric-scale experiments from rift to breakup. *Geochemistry, Geophysics, Geosystems*, *15*(8), 3392–3415.

Castaing, C. (1991). Post-Pan-African tectonic evolution of South Malawi in relation to the Karroo and recent East African rift systems. *Tectonophysics*, *191*(1–2), 55–73.

Chisenga, C., Dulanya, Z., & Yan, J. (2018). The structural re-interpretation of the Lower Shire Basin using edge detection and gravity inversion methods of basement topography. *Journal of African Earth Science*, *149*, 280–290.

Chorowicz, J. (2005). The East African Rift System. *Journal of African Earth Sciences*, *43*(1–3), 379–410.

Choubert, G., Faure-Muret, A., Chanteux, P., Roche, G., Simpson, E. S. W., Shackleton, L., et al. (1988). *International geological map of Africa* (scale 1: 5,000,000; commission for the Geological Map of the World). UNESCO.

Cladouhos, T. T., & Allmendinger, R. W. (1993). Finite strain and rotation from fault-slip data. *Journal of Structural Geology*, *15*(6), 771–784.

Coffield, D. Q., & Schamel, S. (1989). Surface expression of an accommodation zone within the Gulf of Suez rift. *Egypt. Geology*, *17*(1), 76–79.

Corti, G. (2004). Centrifuge modelling of the influence of crustal fabrics on the development of transfer zones: Insights into the mechanics of continental rifting architecture. *Tectonophysics*, *384*(1–4), 191–208. doi: 10.1016/j.tecto.2004.03.014

Corti, G. (2012). Evolution and characteristics of continental rifting: analog modeling-inspired view and comparison with examples from the East African Rift System. *Tectonophysics*, *522*, 1–33. doi: 10.1016/j.tecto.2011.06.010

Corti, G., van Wijk, J., Cloetingh, S., & Morley, C. K. (2007). Tectonic inheritance and continental rift architecture: numerical and analogue models of the East African Rift system. *Tectonics*, *26*(6).

Croissant, T., & Braun, J. (2014). Constraining the stream power law: a novel approach combining a landscape evolution model and an inversion method. *Earth Surface Dynamics*, *2*(1), 155–166.

Daly, M. C., Chorowicz, J., & Fairhead, J. D. (1989). Rift basin evolution in Africa: the influence of reactivated steep basement shear zones. *Geological Society, London, Special Publications*, *44*(1), 309–334.

Daly, M. C., Green, P., Watts, A. B., Davies, O., Chibesakunda, F., & Walker, R. (2020). Tectonics and landscape of the Central African Plateau, and their implications for a propagating Southwestern Rift in Africa. *Geochemistry, Geophysics, Geosystems*, *21*, e2019GC008746.

Daly, M. C., Lawrence, S. R., Kimun'a, D., & Binga, M. (1991). Late Palaeozoic deformation in central Africa: a result of distant collision? *Nature*, *350*(6319), 605–607.

Delvaux, D. (1989). The Karoo to recent rifting in the western branch of the East-African Rift System: a bibliographical synthesis. In Royal Museum African Center, Tervuren (Belgium), Department of Geology and Mining, *Annual Report* (pp. 63–83).

Dommain, R., Riedl, S., Olaka, L. A., deMenocal, P., Deino, A. L., Owen, R. B., et al. (2022). Holocene bidirectional river system along the Kenya Rift and its influence on East African faunal exchange and diversity gradients. *Proceedings of the National Academy of Sciences*, *119*(28), e2121388119.

Duffy, O. B., Bell, R. E., Jackson, C. A. L., Gawthorpe, R. L., & Whipp, P. S. (2015). Fault growth and interactions in a multiphase rift fault network: Horda Platform, Norwegian North sea. *Journal of Structural Geology*, *80*, 99–119.

Dulanya, Z. (2017). A review of the geomorphotectonic evolution of the south Malawi rift. *Journal of African Earth Sciences*, *129*, 728–738.

Dulanya, Z., Croudace, I., Reed, J. M., & Trauth, M. H. (2014). Palaeolimnological reconstruction of recent environmental change in Lake Malombe (S. Malawi) using multiple proxies. *Water SA*, *40*(4), 717–728.

Dulanya, Z., Gallen, S. F., Kolawole, F., Williams, J. N., Wedmore, L. N., Biggs, J., & Fagereng, Å. (2022). Knickpoint morphotectonics of the Middle Shire River basin: implications for the evolution of rift interaction zones. *Basin Research*, *34*(6), 1839–1858.

Dulanya, Z., Kolawole, F., Gallen, S. F., & Williams, J. N. (2024). On the timescales of border fault growth: Pleistocene slip acceleration and lateral border fault propagation in the Lower Shire Graben, East Africa. *Tectonics*, *43*(9), e2023TC008232.

Ebinger, C. J., Deino, A. L., Drake, R. E., & Tesha, A. L. (1989). Chronology of volcanism and rift basin propagation: Rungwe volcanic province, East Africa. *Journal of Geophysical Research: Solid Earth*, *94*(B11), 15785–15803.

Ebinger, C. J., Jackson, J. A., Foster, A. N., & Hayward, N. J. (1999). Extensional basin geometry and the elastic lithosphere. *Philosophical Transactions of the Royal Society of London, Series A: Mathematical, Physical and Engineering Sciences*, *357*(1753), 741–765.

Ebinger, C. J., Rosendahl, B. R., & Reynolds, D. J. (1987). Tectonic model of the Malawi Rift, Africa. *Tectonophysics*, *141*(1–3), 215–235.

Ebinger, C. J., Yemane, T., Harding, D. J., Tesfaye, S., Kelley, S., & Rex, D. C. (2000). Rift deflection, migration, and propagation: linkage of the Ethiopian and Eastern rifts, Africa. *Geological Society of America Bulletin*, *112*(2), 163–176.

Faulds, J. E., & Varga, R. J. (1998). The role of accommodation zones and transfer zones in the regional segmentation of extended terranes. *Geological Society of America Special Papers*, *323*, 1–45.

Fazlikhani, H., Aagotnes, S. S., Refvem, M. A., Hamilton-Wright, J., Bell, R. E., Fossen, H., et al. (2021). Strain migration during multiphase extension, Stord Basin, northern North Sea rift. *Basin Research*, *33*(2), 1474–1496.

Fossen, H., & Rotevatn, A. (2016). Fault linkage and relay structures in extensional settings—a review. *Earth-Science Reviews*, *154*, 14–28.

Fritz, H., Abdelsalam, M., Ali, K. A., Bingen, B., Collins, A. S., Fowler, A. R., et al. (2013). Orogen styles in the East African Orogen: a review of the Neoproterozoic to Cambrian tectonic evolution. *Journal of African Earth Sciences*, *86*, 65–106.

Garzanti, E., Bayon, G., Dinis, P., Vermeesch, P., Pastore, G., Resentini, A., et al. (2022). The segmented Zambezi sedimentary system from source to sink: 2. Geochemistry, clay minerals, and detrital geochronology. *The Journal of Geology*, *130*(3), 171–208.

Guerit, L., Yuan, X. P., Carretier, S., Bonnet, S., Rohais, S., Braun, J., & Rouby, D. (2019). Fluvial landscape evolution controlled by the sediment deposition coefficient: Estimation from experimental and natural landscapes. *Geology*, *47*, 853–856. doi: 10.1130/G46356.1

Hargrove, U. S., Hanson, R. E., Martin, M. W., Blenkinsop, T. G., Bowring, S. A., Walker, N., & Munyanyiwa, H. (2003). Tectonic evolution of the Zambezi orogenic belt: geochronological, structural, and petrological constraints from northern Zimbabwe. *Precambrian Research*, *123*(2–4), 159–186.

Heilman, E., Kolawole, F., Atekwana, E. A., & Mayle, M. (2019). Controls of basement fabric on the linkage of rift segments. *Tectonics*, *38*(4), 1337–1366. doi: 10.1029/2018tc005362

Koehn, D., Aanyu, K., Haines, S., & Sachau, T. (2008). Rift nucleation, rift propagation and the creation of basement micro-plates within active rifts. *Tectonophysics*, *458*(1–4), 105–116.

Kolawole, F. (2020). *The roles of structural inheritance in regions of induced seismicity and active tectonics* (PhD Dissertation). University of Oklahoma. Retrieved from https://hdl.handle.net/11244/324859

Kolawole, F., & Ajala, R. (2024). Propagating rifts: the roles of crustal damage and ascending mantle fluids. *EGU-Solid Earth*, *15*, 747–762. doi: 10.5194/se-15-747-2024

Kolawole, F., Atekwana, E. A., Láo-Dávila, D. A., Abdelsalam, M. G., Chindandali, P. R., Salima, J., & Kalindekafe, L. (2018). Active deformation of Malawi rift's north basin hinge zone modulated by reactivation of preexisting Precambrian shear zone fabric. *Tectonics*, *37*(3), 683–704.

Kolawole, F., Firkins, M. C., Al Wahaibi, T. S., Atekwana, E. A., & Soreghan, M. J. (2021a). Rift interaction zones and the stages of rift linkage in active segmented continental rift systems. *Basin Research*, *33*(6), 2984–3020.

Kolawole, F., Phillips, T. B., Atekwana, E. A., & Jackson, C. A. L. (2021b). Structural inheritance controls strain distribution during early continental rifting, Rukwa Rift (p. 670). *Frontiers in Earth Science*.

Kolawole, F., Vick, T., Atekwana, E. A., Laó-Dávila, D. A., Costa, A. G., & Carpenter, B. M. (2022). Strain localization and migration during the pulsed lateral propagation of the Shire Rift Zone, East Africa. *Tectonophysics*, *839*, 229499.

Konecky, B. L., Russell, J. M., Johnson, T. C., Brown, E. T., Berke, M. A., Werne, J. P., & Huang, Y. (2011). Atmospheric circulation patterns during late Pleistocene climate changes at Lake Malawi, Africa. *Earth and Planetary Science Letters*, *312*, 318–326.

Koopmann, H., Brune, S., Franke, D., & Breuer, S. (2014). Linking rift propagation barriers to excess magmatism at volcanic rifted margins. *Geology*, *42*(12), 1071–1074.

Laó-Dávila, D. A., Al-Salmi, H. S., Abdelsalam, M. G., & Atekwana, E. A. (2015). Hierarchical segmentation of the Malawi Rift: the influence of inherited lithospheric heterogeneity and kinematics in the evolution of continental rifts. *Tectonics*, *34*(12), 2399–2417.

Le Pourhiet, L., Chamot-Rooke, N., Delescluse, M., May, D. A., Watremez, L., & Pubellier, M. (2018). Continental break-up of the South China Sea stalled by far-field compression. *Nature Geoscience*, *11*(8), 605–609.

Lin, J., & Stein, R. S. (2004). Stress triggering in thrust and subduction earthquakes, and stress interaction between the southern San Andreas and nearby thrust and strike-slip faults. *Journal of Geophysical Research*, *109*, B02303. doi: 10.1029/2003JB002607

Marrett, R., & Allmendinger, R. W. (1990). Kinematic analysis of fault-slip data. *Journal of Structural Geology*, *12*(8), 973–986.

McClay, K., & Khalil, S. (1998). Extensional hard linkages, eastern Gulf of Suez, Egypt. *Geology*, *26*(6), 563–566.

McClay, K. R., Dooley, T., Whitehouse, P., & Mills, M. (2002). 4-D evolution of rift systems: insights from scaled physical models. *AAPG Bulletin*, *86*(6), 935–959.

Molnar, N. E., Cruden, A. R., & Betts, P. G. (2019). Interactions between propagating rifts and linear weaknesses in the lower crust. *Geosphere*, *15*(5), 1617–1640.

Morley, C. K. (1995). Developments in the structural geology of rifts over the last decade and their impact on hydrocarbon exploration. *Geological Society, London, Special Publications*, *80*(1), 1–32.

Morley, C. K., Nelson, R. A., Patton, T. L., & Munn, S. G. (1990). Transfer zones in the East African rift system and their relevance to hydrocarbon exploration in rifts. *AAPG Bulletin*, *74*(8), 1234–1253.

Morley, C. K., Wescott, W. A., Harper, R. M., & Cunningham, S. M. (1999). Geology and geophysics of the Rukwa Rift. geoscience of rift systems-evolution of East Africa. *AAPG Studies in Geology*, *44*, 91–110.

Muirhead, J. D., Wright, L. J., & Scholz, C. A. (2019). Rift evolution in regions of low magma input in East Africa. *Earth and Planetary Science Letters*, *506*, 332–346.

Nelson, R. A., Patton, T. L., & Morley, C. K. (1992). Rift-segment interaction and its relation to hydrocarbon exploration in continental rift systems (1). *AAPG Bulletin*, *76*(8), 1153–1169.

Neuharth, D., Brune, S., Glerum, A., Heine, C., & Welford, J. K. (2021). Formation of continental microplates through rift linkage: numerical modelling and its application to the Flemish Cap and Sao Paulo Plateau. *Geochemistry, Geophysics, Geosystems*, *22*, 1–22.

Njinju, E.A., Atekwana, E.A., Stamps, D.S., Abdelsalam, M.G., Atekwana, E.A., Mickus, K.L., Fishwick, S., Kolawole, F., Rajaonarison, T.A. and Nyalugwe, V.N. (2019a). Lithospheric structure of the Malawi rift: implications for magma-poor rifting processes. *Tectonics*, *38*(11), 3835 -3853. doi:10.1029/2019TC005549

Njinju, E. A., Kolawole, F., Atekwana, E. A., Stamps, D. S., Atekwana, E. A., Abdelsalam, M. G., & Mickus, K. L. (2019b). Terrestrial heat flow in the Malawi Rifted Zone, East Africa: implications for tectono-thermal inheritance in continental rift basins. *Journal of Volcanology and Geothermal Research*, *387*, 106656.

Njinju, E. A., Kolawole, F., Stamps, D. S., Atekwana, E. A., Ghomsi, F. E. K., & Atekwana, E. A. (2024). Intrarift fault interactions: Insights from coseismic stress redistribution from large seismogenic segment ruptures, Northern Malawi Rift. *Journal of Structural Geology*, 105326.

Ojo, O., Thomson, S. N., & Laó-Dávila, D. A. (2024). Neogene–Quaternary initiation of the southern Malawi Rift linked to the reactivated the Carboniferous–Jurassic Shire Rift. *Tectonophysics*, *877*, 230270.

Ojo, O. O., Ohenhen, L. O., Kolawole, F., Johnson, S. G., Chindandali, P. R., Atekwana, E. A., & Laó-Dávila, D. A. (2022b). Under-displaced normal faults: Strain accommodation along an early-stage rift-bounding fault in the Southern Malawi Rift. *Frontiers in Earth Science*, *10*, 846389.

Palamuleni, L. G., Ndomba, P. M., & Annegarn, H. J. (2011). Evaluating land cover change and its impact on hydrological regime in Upper Shire River catchment, Malawi. *Regional Environmental Change*, *11*(4), 845–855.

Patton, T. L., Moustafa, A. R., Nelson, R. A., & Abdine, S. A. (1994). Tectonic evolution and structural setting of the Suez rift. In S. M. Landon & A. B. Coury (Eds.), *Interior rift basins* (Memoir 59, pp. 7–55). American Association of Petroleum Geologists.

Pfaffling, A., Monstad, S., Groom, R. W., & Rudd, J. (2009). Airborne-EM hydrocarbon mapping in Mozambique. *ASEG Extended Abstracts*, *2009*(1), 1–6.

Price, T. (1966). Shire, Shirwa, and Nyasa. *The Society of Malawi Journal*, 15–19.

Ragon, T., Nutz, A., Schuster, M., Ghienne, J. F., Ruffet, G., & Rubino, J. L. (2019). Evolution of the northern Turkana depression (East African Rift System, Kenya) during the Cenozoic rifting: new insights from the Ekitale Basin

(28-25.5 Ma). *Geological Journal*, *54*(6), 3468–3488. doi: 10.1002/gj.3339

Rosendahl, B. R. (1987). Architecture of continental rifts with special reference to East Africa. *Annual Review of Earth and Planetary Sciences*, *15*(1), 445–503.

Rosendahl, B. R., Reynolds, D. J., Lorber, P. M., Burgess, C. F., McGill, J., Scott, D., et al. (1986). Structural expressions of rifting: lessons from Lake Tanganyika, Africa. *Geological Society, London, Special Publications*, *25*(1), 29–43.

Rotevatn, A., Kristensen, T., Ksienzyk, A., Wemmer, K., Henstra, G., Midtkandal, I., et al. (2018). Structural inheritance and rapid rift-length establishment in a multiphase rift: the East Greenland rift system and its Caledonian orogenic ancestry. *Tectonics*, *37*, 1858–1875.

Russell, J. M., Cohen, A. S., Johnson, T. C., & Scholz, C. A. (2012). Scientific drilling in the East African Rift Lakes: a strategic planning workshop. *Scientific Drilling*, *14*, 49–54.

Ryan, W. B. F., Carbotte, S. M., Coplan, J., O'Hara, S., Melkonian, A., Arko, R., et al. (2009). Global multi-resolution topography (GMRT) synthesis data set. *Geochemistry, Geophysics, Geosystems*, *10*, Q03014.

Sachau, T., Koehn, D., Stamps, D. S., & Lindenfeld, M. (2016). Fault kinematics and stress fields in the Rwenzori Mountains, Uganda. *International Journal of Earth Sciences*, *105*, 1729–1740.

Sander, S., & Rosendahl, B. R. (1989). The geometry of rifting in Lake Tanganyika, East Africa. *Journal of African Earth Sciences (and the Middle East)*, *8*(2–4), 323–354.

Saria, E., Calais, E., Stamps, D. S., Delvaux, D., & Hartnady, C. J. H. (2014). Present-day kinematics of the East African Rift. *Journal of Geophysical Research: Solid Earth*, *119*(4), 3584–3600.

Schmid, T. C., Brune, S., Glerum, A., & Schreurs, G. (2023). Tectonic interactions during rift linkage: insights from analog and numerical experiments. *EGU Solid Earth*, *14*, 389–407.

Scholz, C. A., Cohen, A. S., Johnson, T. C., King, J., Talbot, M. R., & Brown, E. T. (2011). Scientific drilling in the Great Rift Valley: the 2005 Lake Malawi scientific drilling project—an overview of the past 145,000 years of climate variability in Southern Hemisphere East Africa. *Palaeogeography, Palaeoclimatology, Palaeoecology*, *303*(1–4), 3–19.

Scholz, C. A., & Rosendahl, B. R. (1988). Low lake stands in Lakes Malawi and Tanganyika, East Africa, delineated with multifold seismic data. *Science*, *240*(4859), 1645–1648.

Scholz, C. A., Shillington, D. J., Wright, L. J., Accardo, N., Gaherty, J. B., & Chindandali, P. (2020). Intrarift fault fabric, segmentation, and basin evolution of the Lake Malawi (Nyasa) Rift, East Africa. *Geosphere*, *16*(5), 1293–1311.

Shaban, S. N., Kolawole, F., & Scholz, C. A. (2023). The deep basin and underlying basement structure of the Tanganyika Rift. *Tectonics*, *42*(7), e2022TC007726.

Specht, T. D., & Rosendahl, B. R. (1989). Architecture of the Lake Malawi Rift, East Africa. *Journal of African Earth Sciences*, *8*(2–4), 355–382. doi: 10.1016/s0899-5362(89).80032-6

Stamps, D. S., Calais, E., Saria, E., Hartnady, C., Nocquet, J. M., Ebinger, C. J., & Fernandes, R. M. (2008). A kinematic model for the East African Rift. *Geophysical Research Letters*, *35*, L05304. doi: 10.1029/2007GL032781

Stamps, D. S., Saria, E., & Kreemer, C. (2018). A geodetic strain rate model for the East African Rift system. *Scientific Reports*, *8*(1), 1–8.

Stevens, V. L., Sloan, R. A., Chindandali, P. R., Wedmore, L. N., Salomon, G. W., & Muir, R. A. (2021). The entire crust can be seismogenic: evidence from Southern Malawi. *Tectonics*, *40*(6), e2020TC006654. doi: 10.1029/2020TC006654

Sun, M., Gao, S. S., Liu, K. H., Mickus, K., Fu, X., & Yu, Y. (2021). Receiver function investigation of crustal structure in the Malawi and Luangwa rift zones and adjacent areas. *Gondwana Research*, *89*, 168–176.

Thomas, R. J., Fullgraf, T., Macey, P. H., Boger, S. D., Hölttä, P., Lach, P., et al. (2022). The Mesoproterozoic Nampula subdomain in southern Malawi: completing the story from Mozambique. *Journal of African Earth Sciences*, *196*, 104667.

Toda, S., Stein, R. S., Richards-Dinger, K., & Bozkurt, S. (2005). Forecasting the evolution of seismicity in southern California: animations built on earthquake stress transfer. *Journal of Geophysical Research*, *110*, B05S16. doi: 10.1029/2004JB003415

Wedmore, L. N., Faure Walker, J. P., Roberts, G. P., Sammonds, P. R., McCaffrey, K. J. W., & Cowie, P. A. (2017). A 667 year record of coseismic and interseismic Coulomb stress changes in central Italy reveals the role of fault interaction in controlling irregular earthquake recurrence intervals. *Journal of Geophysical Research: Solid Earth*, *122*(7), 5691–5711.

Wedmore, L. N., Williams, J. N., Biggs, J., Fagereng, Å., Mphepo, F., Dulanya, Z., et al. (2020a). Structural inheritance and border fault reactivation during active early-stage rifting along the Thyolo fault, Malawi. *Journal of Structural Geology*, *139*, 104097.

Wedmore, L. N. J., Biggs, J., Williams, J. N., Fagereng, Å., Dulanya, Z., Mphepo, F., & Mdala, H. (2020b). Active fault scarps in southern Malawi and their implications for the distribution of strain in incipient continental rifts. *Tectonics*, *39*(3), e2019TC005834.

Williams, J. N., Fagereng, Å., Wedmore, L. N., Biggs, J., Mphepo, F., Dulanya, Z., et al. (2019). How do variably striking faults reactivate during rifting? Insights from southern Malawi. *Geochemistry, Geophysics, Geosystems*, *20*(7), 3588–3607.

Williams, J. N., Wedmore, L. N., Scholz, C. A., Kolawole, F., Wright, L. J., Shillington, D. J., et al. (2022). The Malawi active fault database: an onshore-offshore database for regional assessment of seismic hazard and tectonic evolution. *Geochemistry, Geophysics, Geosystems*, *23*(5), e2022GC010425.

Wilson, D. S. (1990). Kinematics of overlapping rift propagation with cyclic rift failure. *Earth and Planetary Science Letters*, *96*(3–4), 384–392.

Withjack, M. O., Schlische, R. W., & Olsen, P. E. (2002). Rift-basin structure and its influence on sedimentary systems. In R. W. Renaut & G. M. Ashley (Eds.), *Sedimentation in continental rifts (Special Publication)* (Vol. *No. 73*, pp. 57–81). Society for Sedimentary Geology.

Wolf, L., Huismans, R. S., Wolf, S. G., Rouby, D., & May, D. A. (2022). Evolution of rift architecture and fault linkage during continental rifting: investigating the effects of tectonics and surface processes using lithosphere-scale 3D coupled numerical models. *Journal of Geophysical Research: Solid Earth*, *127*, e2022JB024687.

Van Wijk, J. W., & Blackman, D. K. (2005). Dynamics of continental rift propagation: the end-member modes. *Earth and Planetary Science Letters*, *229*(3–4), 247–258.

Xue, L., Moucha, R., Kolawole, F., Muirhead, J. D., & Scholz, C. A. (2024). The influence of the strength of pre-existing crustal-scale weak zones on rift geometry and strain localization. *Tectonophysics*, *890*, 230472. doi: 10.1016/j.tecto.2024.230472

Yuan, X. P., Braun, J., Guerit, L., Simon, B., Bovy, B., Rouby, D., et al. (2019a). Linking continental erosion to marine sediment transport and deposition: a new implicit and O(N) method for inverse analysis. *Earth and Planetary Science Letters*, *524*, 115728. doi: 10.1016/j.epsl.2019.115728

Yuan, X. P., Braun, J., Guerit, L., Rouby, D., & Cordonnier, G. (2019b). A new efficient method to solve the stream power law model taking into account sediment deposition. *Journal of Geophysical Research: Earth Surface*, *124*, 1346–1365. doi: 10.1029/2018JF004867

Zwaan, F., Chenin, P., Erratt, D., Manatschal, G., & Schreurs, G. (2022). Competition between 3D structural inheritance and kinematics during rifting: insights from analogue models. *Basin Research*, *34*(2), 824–854.

Zwaan, F., & Schreurs, G. (2017). How oblique extension and structural inheritance influence rift segment interaction: insights from 4D analog models. *Interpretation*, *5*(1), SD119–SD138.

10

Root Problem of Mid-Tertiary Cordilleran Detachment Faults: Deciphering the Evolution of the Whipple Mountains Detachment Shear Zone in Southeastern California

Valeria Jaramillo[1], Elizabeth J. Catlos[2], Elizabeth A. Bell[1], Axel K. Schmitt[3,4], and An Yin[1]

ABSTRACT

The North American Cordilleran metamorphic core complex belt provides insight into the tectonometamorphic evolution of North America. Garnet-bearing assemblages have been used to generate pressure-temperature (P-T) constraints on the metamorphic history in its northern and central segments. Such datasets are scarce in its southern segment. We review existing metamorphic conditions and timing for these core complexes. Conditions from the Whipple detachment shear zone in southeastern California were also ascertained using conventional and isopleth thermobarometry. A sample from the easternmost Whipple Mountains shows monazite in reaction with allanite trapped in garnet. The monazite was dated in situ using an ion microprobe and yield 62±9 Ma, 67±7 Ma, and 80±6 Ma (Th-Pb dates, ±1s). P-T conditions recorded by garnet and matrix minerals from an adjacent rock are high-grade (rim, 680-750°C and 7-9.4 kbar and core, 750-800°C and 6.2-7.8 kbar). The results suggest exhumation from depths deeper than previously thought. With this new information, a new pressure-temperature-time path may be established for the rocks in the Whipple detachment and may suggest a deeper initiation or multiple metamorphic events not previously analyzed.

10.1. INTRODUCTION

The North American Cordillera hosts a suite of metamorphic core complexes that extend from Canada to northern Mexico (Figures 10.1–10.3) (Armstrong, 1982). The orogenic nature of the North American Cordillera Anatectic Belt provides insights into the interaction between plutons, metamorphosed country rocks, and tectonics (Coney et al., 1980; Armstrong, 1982; Chapman et al., 2021). The various competing models for the initiation, style, and kinematic evolution of detachment-fault systems (Figure 10.4) make testable predictions on the pressure-temperature-time (P-T-t) paths, timing and duration of faulting, and structural associations requiring specific fault geometry and kinematics can be tested by using field-based investigations coupled with geochemical and geochronological analytical methods.

In this chapter, we review existing P-T-t data from the North American Cordilleran metamorphic core complexes (Table 10.1; Figure 10.1). Core-complex extension and timing of metamorphism are well-constrained in the northern and central segments of the North American Cordillera; however, the southern segment lacks this data (Table 10.1; Chapman et al., 2021). One core complex in particular, the Whipple Mountains, contains

[1] *Department of Earth and Space Sciences, University of California, Los Angeles, California, USA*

[2] *Jackson School of Geosciences, Department of Earth and Planetary Sciences, University of Texas at Austin, Austin, Texas, USA*

[3] *Institute of Earth Sciences, Heidelberg University, Heidelberg, Germany*

[4] *John de Laeter Centre, Curtin University, Bentley, Australia*

Extensional Tectonics: Rifting and Continental Extension, Geophysical Monograph 278, First Edition.
Edited by İbrahim Çemen and Elizabeth J. Catlos.

DOI: 10.1002/9781119773757.ch10

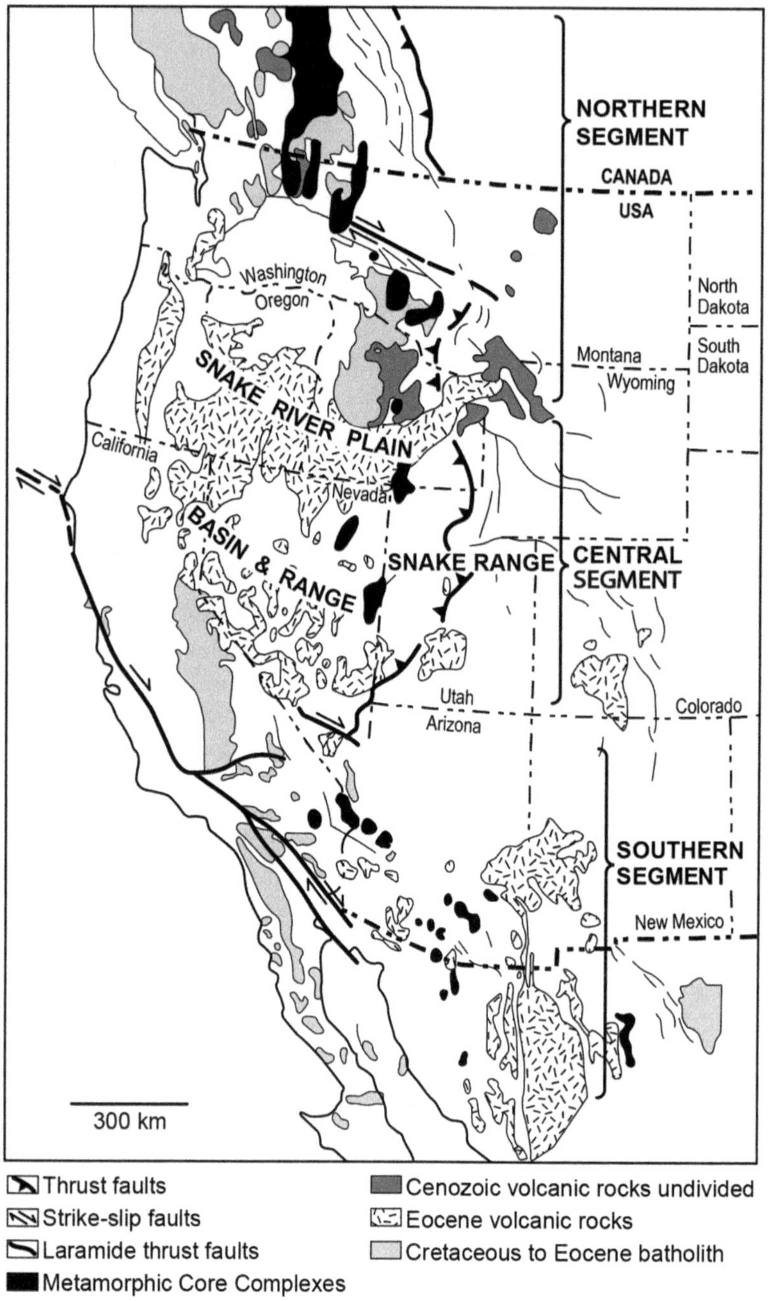

Figure 10.1 Map of the North American Cordilleran Anatectic Belt and metamorphic core complex locations, Sevier thrust belt front, and Laramide deformation front. Map projection: UTM, NAD 83 Zone 12N. Gébelin et al. (2011) / American Geophysical Union.

amenable assemblages for P-T-t determinations. The Whipple detachment fault system consists of a regionally extensive detachment fault (the Whipple detachment fault; Dickey et al., 1980), a metamorphic footwall, and a highly extended hanging wall (Davis et al., 1980, 1982; Davis, 1988; Figure 10.3). The Whipple mylonitic shear zone is capped by the Whipple detachment fault on top without exposing its base (Davis et al., 1980, 1982). The Whipple detachment fault system has been studied extensively, yet how the detachment roots into the crust is poorly understood (Behr & Platt, 2011; Gans & Gentry, 2016; Axen, 2020).

Here we present newly acquired P-T-t data from a rock from the mid-Tertiary Whipple detachment shear zone in the Whipple Mountains core complex. Thermodynamic modeling of garnet-bearing rocks allows for the acquisition of precise and reliable metamorphic records (de Capitani & Petrakakis, 2010; Moynihan & Pattison, 2013; Gaidies

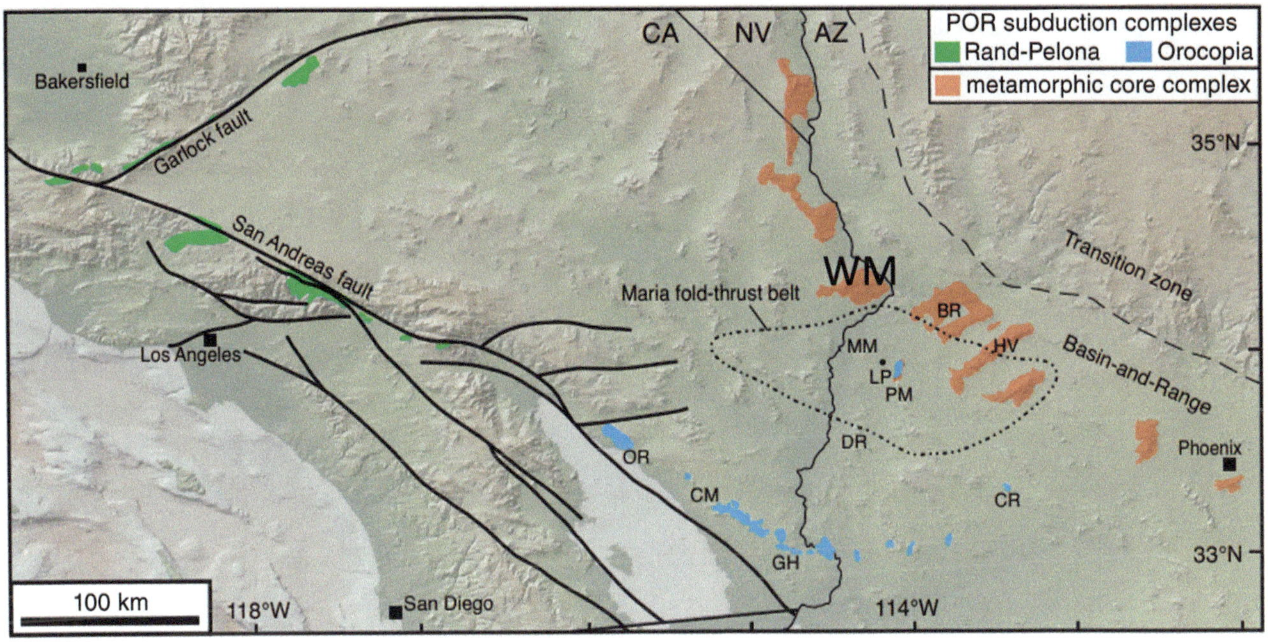

Figure 10.2 Map illustrating the distribution of Pelona-Orocopia-Rand Schists (POR) subduction complexes in southern California and southwestern Arizona (USA), and locations of the Whipple Mountains (WM) and other ranges mentioned in the text (Haxel et al., 2015). Base map is colored by elevation and derived from GeoMapApp (http://www.geomapapp.org). Maria fold-and-thrust belt dotted outline is modified from Spencer and Reynolds (1990). Orange polygons indicate Miocene metamorphic core complexes. Bold black lines are major Quaternary strike-slip faults. BR, Buckskin-Rawhide Mountains; CM, Chocolate Mountains; CR, Cemetery Ridge; DR, Dome Rock Mountains; GH, Gavilan Hills; HV, Harcuvar Mountains; MM, Mesquite Mountains; OR, Orocopia Mountains; PM, Plomosa Mountains. Modified from Strickland et al. (2018).

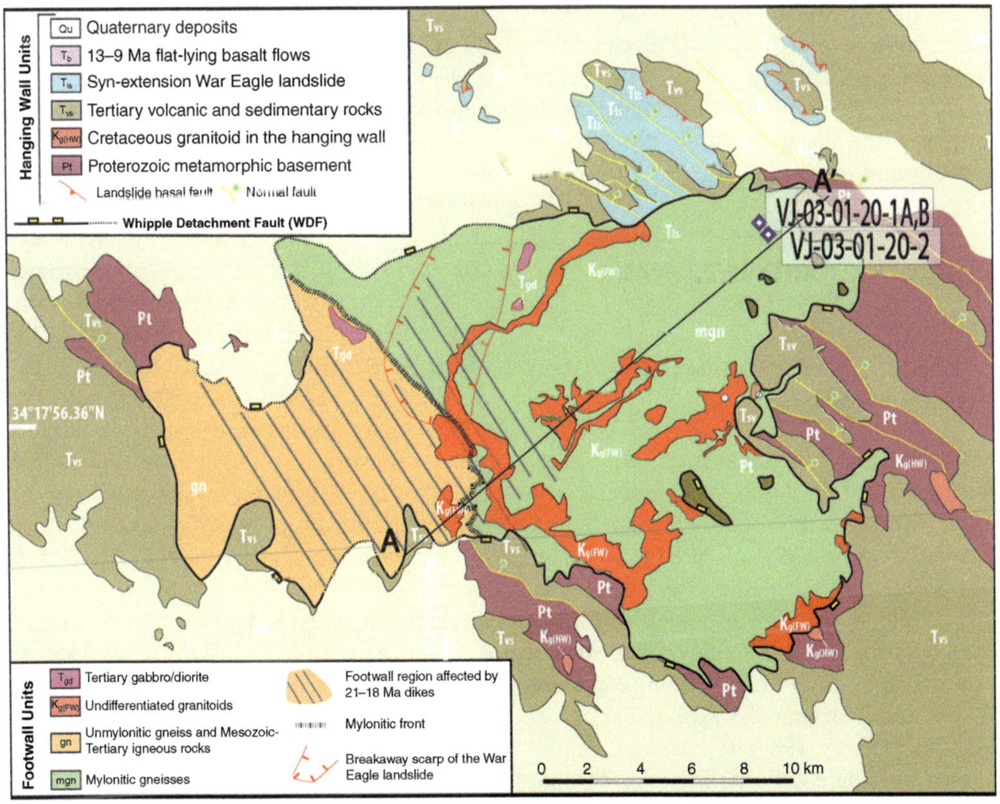

Figure 10.3 Simplified geologic map of the Whipple Mountains based on a compilation of Davis (1988), Yin and Dunn (1992), Forshee and Yin (1995), Behr and Platt (2011), and Gans and Gentry (2016). The diamond corresponds to sample locations VJ-03-01-20-1A, VJ-03-01-20-1B, and VJ-03-01-20-2.

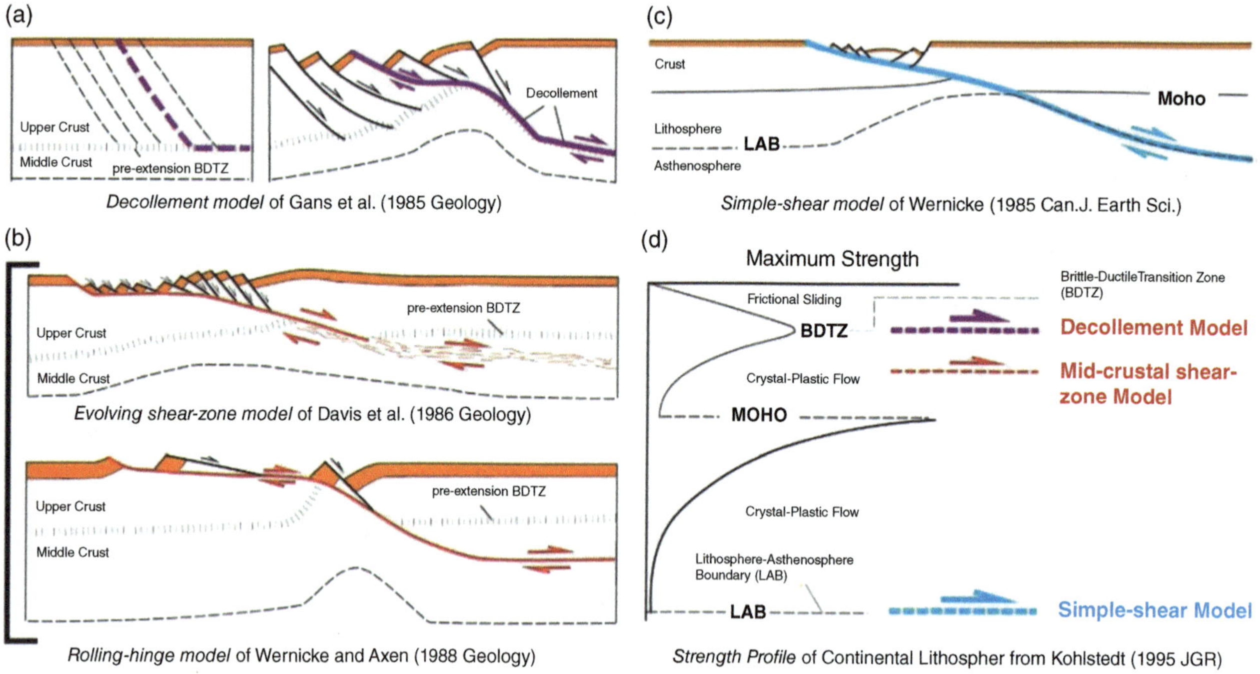

Figure 10.4 (a–c) Models of detachment-fault-system evolution: (a) Gans et al. (1985), (b) Wernicke and Axen (1988), (c) Wernicke (1985). (d) Inferred strength profile of the continental lithosphere from Kohlstedt (1995).

et al., 2008), including those exposed in extended terranes (Etzel et al., 2019; Etzel & Catlos, 2021). The objectives of this chapter are to highlight the lack of such data in the southern North American Cordillera metamorphic core complexes and to present a petrochronological approach to developing P-T-t data in the Whipple Mountains using thermodynamic modeling of garnet-bearing rocks and in situ (in thin section) monazite geochronology.

10.2. GEOLOGIC AND REGIONAL BACKGROUND

10.2.1. Problems and Models for Continental Extension in North America

Continental extension is an integral part of plate tectonics, in part because of its role in the spatiotemporal transition from continental rifting to mid-ocean-ridge spreading, leading to the creation of new oceanic lithosphere (e.g., Wilson, 1966; Sleep, 1971; Lister et al., 1986; Hempton, 1987; White & McKenzie, 1989; Bosworth et al., 2005; Brothers et al., 2009). In addition, continental extension is a crucial stage of mountain building expressed by topographic collapse (Molnar & Tapponnier, 1978; Coney & Harms, 1984; Dewey, 1988), typically accommodated by distributed intraplate deformation (e.g., Molnar & Tapponnier, 1975; Atwater & Stock, 1998). Despite its importance, how continental extension operates in the ductile middle and lower crust remains poorly understood. This issue is well illustrated by the debate on how the brittle detachment faults extend downward into the deeper ductile crust during the development of the ~2,000-km-long and 100–200-km-wide mid-Tertiary North American metamorphic core complexes (Figure 10.1).

Low-angle detachment faulting is a global phenomenon documented in the North American Cordillera and many other parts of the world (Davis & Coney, 1979; Çemen et al., 2006; Coney et al., 1980; Whitney et al., 2013; Gibbs, 1984; Lister et al., 1984; Morley, 1989; Lister et al., 1991; Harrison et al., 1995; Reston et al., 1996; Davis, 1996; Gessner et al., 2001; Murphy et al., 2002; McNulty & Farber, 2002; Robinson et al., 2004; Kapp et al., 2000, 2008; Chen et al., 2013; Worthington et al., 2020). However, despite its wide occurrence, the mechanical origin of low-angle detachment faults with their average dip in the upper crust at ~30° or lower is poorly understood. Specifically, it is unclear whether subhorizontal midcrustal flow (Yin, 1989; Melosh, 1990; Westaway, 1999), thermally triggered gravitational spreading (Coney & Harms, 1984; Coney, 1987), fault-zone mechanics (Axen, 1992, 2004; Axen & Selverstone, 1994; Selverstone et al., 2012), or crustal-scale anisotropies (Axen, 2020) has played a controlling role in initiating regionally extensive, crustal-scale, low-angle extensional detachment faults.

Figure 10.4 illustrates competing detachment-fault models. Differentiating which applies to the Cordilleran

Table 10.1 List of metamorphic core complexes associated with the southwestern portion of the North American Cordillera[a].

Core complex	References[b]	Core-complex extension timing		Metamorphism timing and peak conditions		
Name		Rapid extension (Ma) (±1σ)	Extension onset (Ma)	Peak timing (Ma)	Peak P (kbar)	Peak T (°C)
		California				
Whipple	Anderson and Cullers (1990) and Gans and Gentry (2016)	19(1)	20	—	~4.5	~450
Sacramento	Carter et al. (2006), Pease et al. (1999), Foster et al. (1990), Campbell-Stone et al. (2000)	18(2)	23	—	—	500
Chemehuevi	John and Foster (1993), Carter et al. (2006)	18(2)	23	—	3.5–4.5	450–500
Iron Mountain	Boettcher and Walker (1993), Wells et al. (2002)	24	21	—	4.8	< ~670
Old Woman-Piute	Miller et al. (1990), Miller and Wooden (1994)	—	—	—	4-5	~650
		Arizona				
Mesquite		—	—	—	—	—
Buckskin-Rawhide	Singleton et al. (2014)	19(2)	21	—	—	450–500
Plomosa	Spencer et al. (2018), Singleton et al. (2022)	18(2)	19	—	4.6–6	459–590
Harcuvar	Bryant and Wooden (2008), Wong et al. (2011), DeWitt and Reynolds (1990)	17(3)	24	—	—	—
Harquahala	Isachsen et al. (1999), Prior et al. (2016), Richard et al. (1990)	19(2)	21	—	—	>300
White tank	Shafiqullah et al. (1980), Prior et al. (2016)	20(5)	—	—	—	—
South Mountain	Fitzgerald et al. (1993)	20(5)	22	—	—	—
Picacho	Gottardi et al. (2018), Spencer et al. (2003), Schaper (2016)	20(2)	—	—	—	—
Tortolita	Creasey et al. (1976)	22(3)	—	—	1.4–2.2	~661–682
Gunnery Range	Arnold (1986)	—	—	—	—	>225
Pinaleno	Long et al. (1995)	25(5)	—	—	—	300–415; 673–802
Santa Catalina	Fornash et al. (2013), Fayon et al. (2000), Terrien (2012), Peterman et al. (2014), Davis et al. (2019), Ducea et al. (2020)	25(2)	—	—	1.4–2.2	~661–682
Rincon		—	—	—	—	—

(Continued)

Table 10.1 (Continued)

Core complex	References[b]	Core-complex extension timing		Metamorphism timing and peak conditions		
Name		Rapid extension (Ma) (±1σ)	Extension onset (Ma)	Peak timing (Ma)	Peak P (kbar)	Peak T (°C)
Texas Canyon	Chapman et al. (2018), Shafiqullah et al. (1980)					>225
Coyote	Wright and Haxel (1982), Gottardi et al. (2020)	29(2)	30	—	<5	~400–500
Northern Mexico						
Pozo Verde	Goodwin and Haxel (1990)	—	25	—	UG–LA	UG–LA
Guacomea	Anderson et al. (1980), Nourse et al (1994)	—	—	—	LA	LA
Tubutama-Mezquital	Mallery et al. 2018), Nourse et al. (1994)	—	—	—	A	A
Carnero	Hayama et al. (1984), Jacobson et al. (2019)	17(1)	—	—	LA	350–400
Jarillas Portrero-Tortuga		—	—	—	UA	UA
Magdalena Madera	Salas (1968)	23.5(1.5)	25	—	LA	>500
El Babizo	González-León et al. (2011)	—	—	—	A	A
Aconchi	Wong et al. (2010), Roldán-Quintana (1991), González-León et al. (2011)	23(3)	26	—	—	250–400
Puerto del Sol	González-Becuar et al. (2017), Wong and Gans (2008)	—	—	—	>5	900
Sierra Mazatan	Wong and Gans (2008), Anderson et al., (1980)	18(2)	25	—	>5	—

A, amphibolite facies; GS, greenschist facies; U, upper; L, lower.

[a] See Figure 10.5 for locations. Adapted from Chapman et al. (2021). Note that "—" indicates that no data are available from this location.

[b] Most recent references for further information about the core complex.

metamorphic core complexes is difficult because these conceptual kinematic models are qualitative, and their space-time predictions of fault-system evolution are ambiguous. Figure 10.4a predicts that the mid-Tertiary Cordilleran detachment faults become horizontal decollement zones along the base of the upper crust (Miller et al., 1983; Gans et al., 1985; Lee et al., 2017). Figure 10.4b suggests that these faults flatten into mid-crustal ductile-shear zones (Davis et al., 1986; Wernicke & Axen, 1988; Lister & Davis, 1989; Cooper et al., 2017; Lutz et al., 2021). Alternatively, Figure 10.4c shows that they cut obliquely across the entire lithosphere (Wernicke, 1985). Differentiating these models assesses whether extensional detachment faulting occurs under normal or abnormal continental-strength-profile conditions (Behr & Platt, 2011; Brun et al., 2018). The abnormal strength profile assumes an inviscid-fluid lower crust (Wernicke, 1995; Kohlstedt et al., 1995; Brun et al., 2018).

These proposed models facilitate our understanding of the fundamental mechanisms controlling continental extension. Several studies used numerical simulations assuming realistic continental thermal conditions and using laboratory-determined rheological parameters to understand their feasibility and applicability to the North American Cordillera. This class of research has been focused on whether the style of detachment faulting depends on the boundary and thermal conditions of a continental extensional system (e.g., Buck, 1991; Rey et al., 2009) or lithospheric rheology (e.g., Lavier & Manatschal, 2006; Gueydan et al., 2008; Korchinski et al., 2018). Some suggest the presence of partial crustal melting (Rey et al., 2009; Schenker et al., 2012) or lateral crustal heterogeneities in mechanical properties (e.g., Huet et al., 2011). The variability of Moho morphologies has been explored (Wu et al., 2015), as has crustal-density distributions (e.g., Tirel et al., 2008; Korchinski et al., 2018). The potential for lower-crustal flow (Block & Royden, 1990; Rey et al., 2011) and the impact of rift-basin sedimentation are also concerns (Korchinski et al., 2021). Overall, the formation of low-angle detachment faults could have been induced by diverse geodynamic processes involving various thermal, rheological, and mechanical parameters.

10.2.2. North American Cordillera Metamorphic Core Complexes

The North American Cordilleran orogen occupies a ~5,000-km-long section of the larger circum-Pacific orogenic belt (Figure 10.1; e.g., Dickinson et al., 2009). The mid-Tertiary (ca. 45–10 Ma) Cordilleran metamorphic core-complex belt extending from southern Canada to northern Mexico is part of the North American Cordilleran orogen (Davis & Coney, 1979; Coney et al., 1980). The core-complex belt consists of about 25 individual metamorphic complexes that can be divided into the northern, central, and southern segments (Table 10.1; Coney et al., 1980; Wernicke et al., 1987; Dickinson, 2004; Hildebrand, 2013; Cassel et al., 2018; Chapman et al., 2021).

The fundamental structural association of a metamorphic core complex consists of a low-angle decollement associated with mylonitic fabrics that separate a metamorphic basement in the footwall and unmetamorphosed cover rocks in the hanging wall (Figure 10.4; e.g., Coney et al., 1980; Armstrong, 1982; Brun et al., 2018). These complexes are characterized by their uplifted, dome-like detachments and metamorphic and plutonic basement rocks that are overprinted by low-dipping, lineated, and foliated mylonitic gneisses. The crystalline basement rocks are separated from the overlying, unmetamorphosed, faulted, and tilted younger strata by a subhorizontal detachment fault (Coney et al., 1980; Çemen et al., 2006; Gans & Gentry, 2016; Brun et al., 2018).

The core-complex belt coincides with the root zone of the Mesozoic Cordilleran thrust belt that had a crustal thickness of 40–65 km (Coney & Harms, 1984; Chapman et al., 2015, 2020; Zuza et al., 2021), a Late Cretaceous anatectic plutonic belt that is characterized by the lack of an extrusive counterpart (Miller & Bradfish, 1980; Chapman et al., 2021), and local sites of Late Cretaceous to Paleogene metamorphism (Miller & Gans, 1989; Cooper et al., 2010; Levy et al., 2021), all in the hinterland of the Cordilleran orogen between the Cordilleran Mesozoic magmatic arc in the west and the Cordilleran Mesozoic foreland fold-thrust belt in the east (Chapman et al., 2021; Figure 10.1). Mesozoic metamorphic rocks and anatectic plutons are exposed in the footwalls of major detachment faults associated with the Cordilleran metamorphic core complexes (e.g., Anderson et al., 1988; Miller & Gans, 1989; John & Wooden, 1990; Anderson and Cullers, 1990; Zuza et al., 2021). The Cordilleran Anatectic Belt is dominated by peraluminous granitoids from partial melting of the North American crust at 675–775 °C induced by decompression, crustal thickening, fluid fluxing, or increased heat flux from the mantle (Miller & Bradfish, 1980; Anderson & Cullers, 1990; John & Wooden, 1990; Chapman et al., 2020; Chapman et al., 2021).

10.2.3. Metamorphic Core Complexes in the Southern Segment

The southern segment of the North American Cordilleran core-complex belt hosts metamorphic complexes in California, Arizona, and northern Mexico (Figure 10.2; Table 10.1). This southern segment also hosts the lower Colorado River extensional corridor, a 400-km-long section of east- and northeast-dipping extensional faults between Las Vegas, NV, and Phoenix, AZ (Spencer et al., 2018). Detachment surfaces characteristic of metamorphic core complexes here may record extension direction in the brittle hanging wall and mylonitic footwall rocks (Axen, 2020). Typical hanging-wall rocks include sedimentary or metasedimentary rocks (Buckskin-Rawhide, Whipple Mountains, Plomosa, Pinaleno) or volcanic flows (Harquahala, Whipple Mountains, Harcuvar, Picacho) (Prior et al., 2016; Gottardi et al., 2018; Strickland et al., 2018; Scoggin, 2022; Singleton et al., 2022). Most of the metamorphic core complexes in eastern California and western Arizona record NE-directed hanging-wall displacement during early Oligocene-Miocene exhumation, whereas core complexes further east record SW-directed hanging-wall displacement (Reynolds & Spencer, 1985; Dickinson et al., 1987; Spencer, 2006). In southern Arizona and northern Mexico, the transport direction of the hanging-wall rocks along the detachment faults is consistently southwestward except for the NNE-directed detachment faults in the Coyote, Pinaleno, Alvarez, Mezquital, and Tubutama core complexes (Figure 10.2; Dickinson et al., 1982; Reynolds & Spencer, 1985; Spencer, 2006).

A 70- to 100-km-wide region of extended crust from southern Nevada, eastern California, and western Arizona (Faulds et al., 2001) provides insights into the evolution of Miocene extension in the area (Howard & John, 1987; Davis, 1988; Davis & Lister, 1988; Lister & Davis, 1989). The region forms part of the southern Cordillera core-complex belt, and the onset of extension along the corridor is constrained to occur between 26–22 Ma in the south to 20–18 Ma in the north (Foster & John, 1999; Spencer et al., 2018; Strickland et al., 2018). The extensional corridor partially overlaps the Big Maria fold-and-thrust belt (Spencer & Reynolds, 1990) and locally exposes the Orocopia schist in a few detachment-fault footwalls (Haxel et al., 2015; Jacobson et al., 2017; Seymour et al., 2018; Strickland et al., 2018). This schist consists of quartzofeldspathic, metachert (Fe- and Mn-bearing quartz), and metabasite (metabasalt), with pods of actinolite, talc, and serpentinite (Strickland et al., 2018).

The exposure of Orocopia schist in the extensional corridor has been associated with Miocene detachment faulting (e.g., Jacobson et al., 2017; Strickland et al., 2018). However, the mechanism for predetachment emplacement of the schist in the middle crust during the flat-subduction Laramide orogeny (Dickinson et al., 1978) has been attributed to eastward displacement of the lower crust and/or the mantle lithosphere of the North American plate accompanied by basal accretion of the subducted mélange complex and forearc-basin strata below the North American middle crust or the tectonically eroded/thinned North American mantle lithosphere (see Bird, 1988; Livaccari & Perry, 1993; Yin et al., 2002; Chapman, 2017; Ducea & Chapman, 2018; Axen et al., 2018; Figure 10.5a). Alternatively, the corridor may be the result of ductile injection of subducted mélange complex and forearc-basin strata as a sheet into the North American middle crust (Strickland et al., 2018; Figure 10.5b). A model of diapiric intrusion of the subduction mélange complex and forearc-basin strata has also been proposed (Figure 10.5c; Chapman et al., 2021).

The role of oceanic plateau subduction has also been invoked as the cause of Laramide flat subduction (Livaccari et al., 1981; Saleeby, 2003; Chapman, 2017). East-west extension along the axis of the Mesozoic U.S. Cordilleran orogen may have occurred immediately before (late Cretaceous) or during the flat slab subduction (Jacobson et al., 1996; Saleeby, 2003; Wells & Hoisch, 2008; Chapman et al., 2020). The cause of this extensional event has been attributed to partial mantle-lithospheric delamination of the North American plate (Wells & Hoisch, 2008; Wells et al., 2012; Chapman et al., 2020).

A consistent finding of the exhumed Orocopia schist against the San Andreas fault in the west and those exposed in the lower Colorado River extensional corridor in the east is the presence of 65–40 Ma metamorphic zircons (Grove et al., 2003; Jacobson et al., 2017; Seymour et al., 2018). Paleogene zircon growth implies that the schist remained under amphibolite facies conditions long after underplating. The persistent heating of the schist during flat subduction is counterintuitive as the process would have exerted a refrigeration effect that cooled the overriding plate materials (Dumitru et al., 1991). The schist's unexpected and sustained Paleogene heating has been attributed to frictional heating (Jacobson et al., 2017) or extensional tectonics (Seymour et al., 2018) during the Laramide flat subduction.

10.2.4. P-T Conditions of Core-Complex Development

Although P-T constraints exist for the peak metamorphic conditions for rocks exposed in the northern segment of the North America Cordillera, conditions in the southern segment remain poorly constrained (see also Chapman et al., 2021). Table 10.1 lists the approaches to generate these conditions, which have primarily focused on applying classic thermobarometric techniques to garnet-bearing rocks, relying on exchange thermometry and net-transfer barometry.

Metamorphic core complexes in British Columbia and Washington (Figure 10.1), record peak conditions of 5–12 kbar and 700–850 °C (see Table 10.1 for references). The Frenchman's Cap metamorphic core complex records the highest conditions in the North American Cordillera of 10–12 kbar at 800 °C, whereas the Okanogan Complex records 9–10 kbar and 700–850 °C (Hansen & Goodge, 1988; Armstrong & Ward, 1991; Carlson et al., 1991; Crowley et al., 2001; Crowley et al., 2008; Kruckenberg et al., 2008; Chapman et al., 2021).

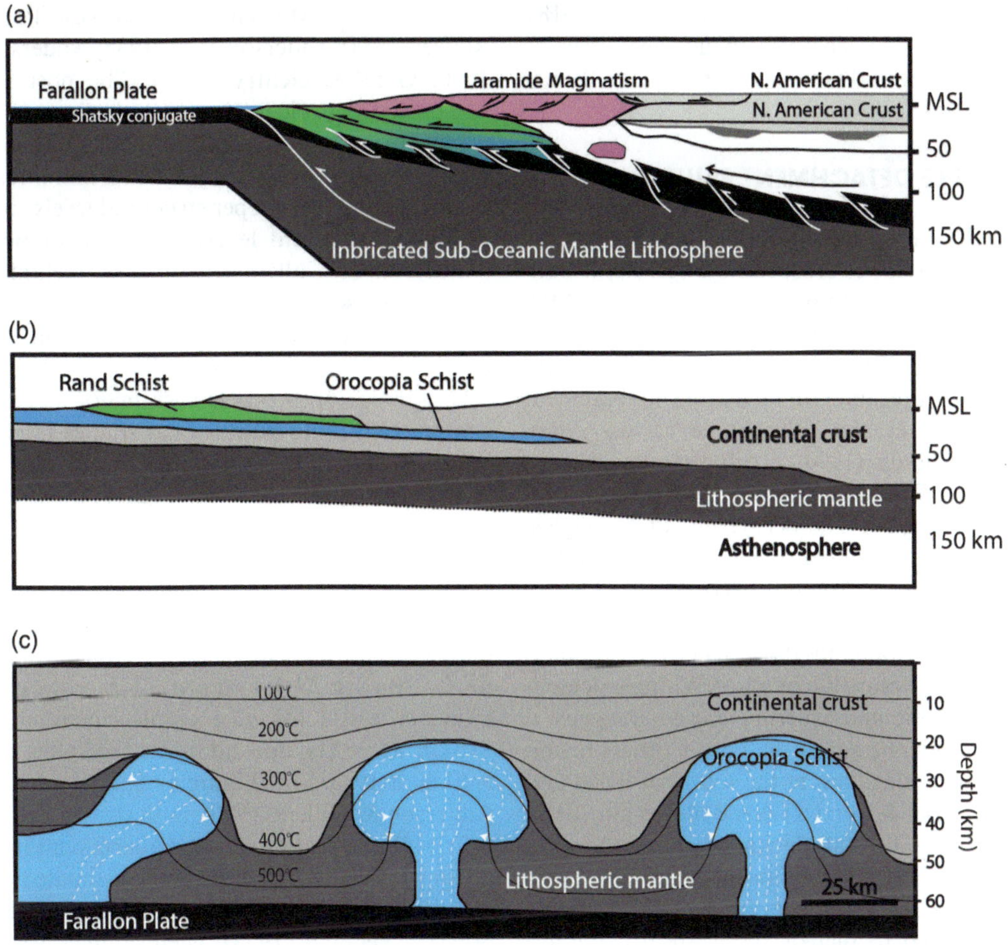

Figure 10.5 Three end-member models for the emplacement of the Orocopia schist below Mojave Desert. (a) Shallow-angle subduction and schist underplating (adapted from Chapman et al., 2020). (b) Proposed mechanism of underplating of the Orocopia Schist without subduction (Strickland et al., 2018 / The Geological Society of America). (c) Diapiric-style emplacement of Orocopia Schist (Adapted from Chapman et al., 2021).

Those in the central segment from Idaho to southern Nevada record between 4–11 kbar and 500–800 °C. The lowest P recorded in this central region is in the Pioneer Mountains in Idaho at 4 kbar from the Pioneer Intrusive Suite (Silverberg, 1990; Vogl et al., 2012; Chapman et al., 2021). Estimates from the Ruby-E Humboldt complex in Nevada are 8–10 kbar and 700–800 °C using garnet rim thermobarometry with matrix minerals on pelitic samples (Hodges et al., 1992; McGrew & Snee 1994; Lee et al., 2003; Howard et al., 2011).

In contrast to the northern segment of the North American Cordillera, peak P-T conditions of the southernmost metamorphic core complexes are not clearly constrained. Foster (1989) suggests that the emplacement of the Old Woman–Piute batholith in the Old Woman Mountains, CA, triggered high-grade metamorphism and reports estimates of 3.5–4.5 kbar and 530 °C to >650 °C. Analysis of rocks from the Coyote core complex in Arizona yield 450 ± 50 °C at <5 kbar (Wright & Haxel, 1982; Gottardi et al., 2020). Recent thermobarometric studies on the Plomosa core complex (Arizona) suggest that the rocks may have reached 6 kbar and between 549 and 590 °C (Spencer et al., 2018; Singleton et al., 2022).

Core complexes may record metamorphism due to pluton emplacement or mylonitization (Anderson, 1988). For example, the Catalina, Tortolita, and Pinaleno plutons are suggested to be related to thermal conditions that range between ~650 and 800 °C (Long et al., 1995; Fayon et al., 2000; Wong and Gans, 2008). The Picacho core complex records mylonitization T in the range of 300–630 °C and the Mazatan core complex between 300 and 450 °C (Terrien, 2012; Fornash et al., 2013; Peterman et al., 2014; Schaper, 2016; Gottardi et al., 2018; Davis et al., 2019; Ducea et al., 2020). Studies from core complexes in northern Mexico, closest to the border with Arizona, indicate metamorphism at greenschist- to lower-amphibolite facies, amphibolite facies, and/or upper-amphibolite facies (Anderson et al., 1988; Goodwin and

Haxel, 1990; Nourse et al., 1994; Mallery et al., 2018). The Puerta del Sol core complex in northern Mexico reached 900 °C (Wong and Gans, 2008; Gonzalez-Becuar et al., 2017).

10.3. WHIPPLE DETACHMENT FAULT SYSTEM

The crustal thickness in the Whipple Mountains area is 26–28 km, and the lower-crust seismic velocity is consistent with a felsic composition (McCarthy et al., 1991). The Whipple Mountains and neighboring regions lack the Paleozoic strata as Tertiary volcanic and sedimentary strata lie on top of Precambrian metamorphic-basement rocks (e.g., Dickey et al., 1980; Stone et al., 1983). Spencer and Reynolds (1990) attributed the absence of Neoproterozoic-Paleozoic strata to Mesozoic crustal shortening in the region. The occurrence of Paleogene gravel deposits in an eastward-flowing drainage system (Young, 1982; Elston & Young, 1991) in the western Colorado Plateau requires a higher elevation and thicker crust in the Whipple Mountains than that of the Colorado Plateau immediately before the onset of Whipple detachment faulting. Seismic and experimental data suggest a 40–48 km crustal thickness near the Four Corners region compared to ~35 km thickness at the southern Basin and Range region (Hauser & Lundy, 1989; Bailey et al., 2012). Because the current crustal thickness of the western Colorado Plateau is ~45 km (Hauser & Lundy, 1989; Bailey et al., 2012), the higher elevation of the Whipple Mountains compared to that of the western Colorado Plateau in the Paleogene implies that the preextension Whipple crust was >45 km. Comparing the current crustal thickness (26–28 km) requires ~50% of post-Paleogene crustal extension in the Whipple Mountains area.

10.3.1. Footwall Rocks

Like all core complexes in the North American Cordilleran orogen, the Whipple detachment fault system consists of a regionally extensive detachment fault (the Whipple detachment fault, Dickey et al., 1980), a metamorphic footwall, and a highly extended hanging wall (Davis et al., 1980, 1982; Davis, 1988; Figure 10.3). The Whipple mylonitic shear zone is capped by the Whipple detachment fault on top without exposing its base, and a cross-section projection requires the shear zone to be >3.9 km thick (Davis et al., 1980, 1982). Foliation surfaces in the mylonitic shear zone are domal-shaped, expressed by a radially dipping direction from the center of the mountains; this shape mimics the geometry of the overlying Whipple detachment fault (Davis, 1988). Both the mylonitic and non mylonitic metamorphic rocks in the footwall are composed of Proterozoic gneisses (1.4–1.2 Ga) intruded by mylonitic and nonmylonitic Cretaceous and Miocene plutons/dikes/sills (Anderson & Rowley, 1981; Anderson et al., 1988; Anderson & Cullers, 1990; Gans & Gentry, 2016). The host rocks of the Mesozoic-Cenozoic intrusions in the footwall are gneisses, which are composed of interlayered amphibolite, garnet-muscovite gneiss and schist, and aluminosilicate with kyanite exposed at deeper structural levels and sillimanite at higher structural levels in the mylonitic shear zone (Anderson & Cullers, 1990). The metamorphic units are locally associated with migmatites, whereas the plutons are associated with aplite sills and dikes (Anderson & Rowley, 1981; Anderson & Cullers, 1990).

Oxygen isotopes, whole-rock major and trace element compositions, and petrological studies indicate that the Whipple mylonitic rocks experienced two episodes of hydrothermal alteration: a high-T (>~600 °C) episode involving a metamorphic or magmatic fluid and a low-T (~350 °C) episode involving meteoric fluids (Morrison, 1994; Table 10.1). Interestingly, footwall intrusions show a progressive decrease in their emplacement depths with time. Granite was emplaced at a depth of 33 ± 4 km at 89 ± 3 Ma, at 29 ± 1 km at 73 ± 3 Ma (Figure 10.6). Evidence exists for dike emplacement at a depth of ~16 km at 26 Ma and additional granites and dikes at 5–6 km at 19-17 Ma (Figure 10.6; Anderson et al., 1988; Anderson & Cullers, 1990).

Crystal-plastically deformed quartz veins in the Whipple mylonitic shear zone were used to estimate the magnitude of differential stresses during the ductile-shear deformation (Hacker et al., 1992). Their samples were collected at the two sites, with one directly below the Whipple detachment fault in the east and another directly below the Whipple mylonitic front in the west (Figure 10.3). Using grain-size piezometers and kinetic laws of thermally activated grain growth, the grain-size cluster at 32–61 μm and the cooling history of the Whipple shear zone require the formation of the mylonitic veins under ~40–150 MPa of differential stresses. The cooling history used in Hacker et al. (1992) for grain-growth calculations was obtained from the work of Anderson et al. (1988), assuming the entire mylonitic shear zone was created in the Miocene during Whipple detachment faulting.

Using updated grain-size piezometers coupled with Ti-in-quartz thermometry, Behr and Platt (2011) estimated 0.10–1.40 kbar of differential stresses over a T range of 310–550 °C from multiphase mylonitic-pluton samples across the Whipple shear zone. Their estimated stress magnitudes did not consider the grain-growth effect during post-deformation annealing (cf., Hacker et al., 1992). In addition, the potential pinning effect on quartz grain deformation and growth (Evans et al., 2001; Hiraga et al., 2010; Cross et al., 2015; Tasaka et al., 2017) was not considered. Because the timing of the measured paleostress magnitudes is unknown, the question remains of

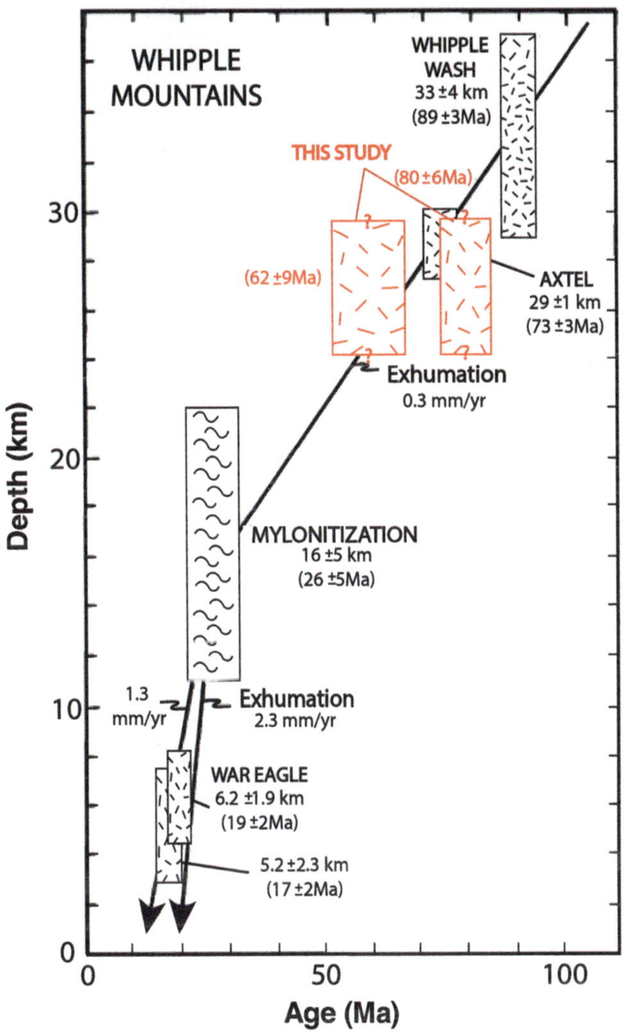

Figure 10.6 Plot of age (Ma) versus depth (km), aiming to illustrate the transport of crustal sections in the Whipple Mountains as a function of age. The 16 ± 5 Ma mylonitization date is refined to 24.1 ± 0.5 Ma by Foster and John (1999). Red boxes illustrate the monazite dates and their approximate depth. Modified from Anderson et al. (1988).

whether the apparent T-dependent stress magnitudes across the Whipple shear zone indicate a vertical stress profile at a particular geologic time or a composite section that different stress magnitudes related to different geologic events. Furthermore, independent P estimates under which their samples were mylonitized have not been reported.

The west-dipping, transitional boundary between mylonitized and nonmylonitized metamorphic rocks in the footwall is referred to as the "mylonitic front" by Davis et al. (1980, 1982) and Davis (1988). This dipping contact was extrapolated westward to a depth of ~15 km based on analyzing and interpreting a seismic-reflection profile across the western Whipple Mountains (Flueh & Okaya, 1989). The Whipple shear zone imaged by seismic-reflection surveys is ca. 15 km thick above a low-reflectivity and higher seismic-velocity region in the lower crust (Henyey et al., 1987; Flueh & Okaya, 1989).

The cooling history of the Whipple footwall rocks was constrained by a $^{40}Ar/^{39}Ar$ hornblende age at ~20 Ma (DeWitt et al., 1986) and an eastward younging trend in the minimum $^{40}Ar/^{39}Ar$ K-feldspar ages from ~21 to ~15 Ma across the footwall (Foster & John, 1999). Apatite (U–Th)/He ages from the Whipple footwall rocks are 26.6 ± 1.1 Ma to 13.7 ± 0.4 Ma, and zircon (U–Th)/He ages are 46.8 ± 1.2 Ma to 13.6 ± 0.3 Ma, with an inflection point at 21.5 ± 1.0 Ma (Stockli et al., 2006).

10.3.2. Whipple Detachment Fault and Mylonitic Shear Zone

Although motion along low-angle (dip <30°) normal faults requires fault-zone weakening (e.g., Axen, 1992, 2004, 2007), patterns of syn-detachment brittle structures directly below the Whipple detachment fault are interpreted to indicate a normal coefficient of static friction and a nearly hydrostatic pore-fluid P at the time of fault motion (Axen & Selverstone, 1994; Axen et al., 2015). The Whipple fault zone consists of an alteration-induced assemblage of secondary epidote + titanite + chlorite, + calcite + feldspar formed during the terminal phase of mylonitization and cataclasis at 380–420 °C (Selverstone et al., 2012). Fluid-inclusion studies of the Whipple fault-zone rocks show that its pore-fluid P may have evolved from having a lithostatic value below and a hydrostatic value above a depth of ~10 km during its development (Selverstone et al., 2012). Oxygen isotope compositions of epidote and quartz in the Whipple fault-zone breccias require an extreme thermal gradient of 82 °C over 38 m (2,160 °C/kilometer) across the fault from the footwall to the hanging wall, interpreted as a result of fluid-induced fault-zone refrigeration (Morrison & Anderson, 1998). The thermal gradient was determined by analyzing quartz-epidote samples and calculating the δ18O of quartz and epidote, which yield a linear relationship between the values and temperature; in this case, a steep slope represents equilibrium exchange among the two minerals, which suggests that hydrothermal fluid interaction occurred readily in the system (Matthews, 1994; Morrison & Anderson, 1998).

The Whipple mylonitic shear zone is >3.9 km thick (Davis et al., 1980) and lies directly below the Whipple detachment fault in the eastern Whipple Mountains (Figure 10.3). To the west, the mylonitic shear zone and nonmylonitic footwall metamorphic rocks are separated by a west-dipping, transitional boundary that was termed the Whipple mylonitic front (Davis et al., 1980, 1982; Davis, 1988; Anderson & Cullers, 1990). The mylonitic front is locally cut across by the ~8-km-wide, NNW-trending,

21–18 Ma Chambers Well dike swarm. Detailed mapping by Gans and Gentry 2016; (see their figure) shows that the dikes have variable northeasterly dips from 20° to 80°. The Whipple detachment truncates the Chambers Well dike swarm and its country metamorphic rocks (Davis et al., 1980, 1982; Davis, 1988; Anderson & Cullers, 1990). Their hanging-wall counterpart is interpreted to have been offset for ~ 40 km to the east (Howard et al., 1982, 1999).

10.3.3. Hanging-Wall Rocks of the Whipple Detachment Fault

The hanging wall of the Whipple detachment fault consists of Precambrian basement rocks overlain by mid-Tertiary sedimentary and volcanic strata (Davis et al., 1980, 1982; Davis, 1988). The basement rocks are dominantly banded quartzofeldspathic gneisses interlayered with amphibolite (Anderson & Cullers, 1990). The mid-Tertiary strata in the Whipple Mountains and the neighboring core complexes are 22–12 Ma syn-extension basin fills (Miller & John, 1988, 1999; Nielson & Beratan, 1990, 1995; Beratan, 1991; Fedo & Miller, 1992; Dorsey & Becker, 1995; Faulds et al., 2001; Prior et al., 2018; Spencer et al., 2018). Tertiary strata in the Whipple Mountains consist of four units (Nelson & Beratan, 1990). Unit 1 is the oldest on top of the basement rocks, and its ages range from 23 ± 2.4 Ma (zircon fission track ages) to 18.5 ± 0.5 Ma ($^{40}Ar/^{39}Ar$ sanidine ages of the Peach Springs Tuff) (Beratan, 1991; Kuniyoshi & Freeman, 1974). The unit is dominated by mafic to intermediate volcanic rocks. Unit 2 consists of the regionally extensive 18.5 Ma Peach Springs Tuff. Unit 3 is equivalent to the Copper Basin Formation (Davis et al., 1980), which is composed of sedimentary strata, limestone, and minor basalt. Unit 4 is the youngest unit characterized by a bimodal suite of mostly flat-lying basaltic and rhyolitic flows with ages from 12.2 Ma (K-Ar, sanidine) to 8.6 Ma (K-Ar, whole rock) (Nielson & Beratan, 1990; Suneson & Lucchitta, 1983). The age was determined by K-Ar, whole-rock dating of one of the basalt flows (Kuniyoshi & Freeman, 1974). The sedimentology and growth-strata relationship of the mid-Tertiary strata in the northeastern Whipple Mountains were examined by Dorsey and Becker (1995). These authors show southward thickening and downward steepening of a syn-extensional sequence that was deposited before 18.5 Ma and lasted until at least ~14.5 Ma.

10.4. METHODS

We conducted geological mapping and sampling along two traverses across the Whipple mylonitic shear zone (Figures 10.3, 10.7, and 10.8). Three samples (nomenclature of VJ-*date-number*) were collected at different structural levels along the Whipple ductile-shear zone (Table 10.2). Thin sections of each sample were analyzed for mineral identification and textural relationships using a petrographic microscope (Figure 10.9). Bulk rock compositions of each sample (Table 10.2) were obtained from rock chips using inductively coupled plasma (ICP) mass spectrometry at Activation Laboratories Ltd. (Table 10.2). Elemental distributions in Mn, Mg, Ca, and Fe of each thin section were mapped using the JEOL 8200 microprobe (EMP), which is part of the Geomaterials Characterization and Imaging Facility at the Department of Earth and Planetary Sciences at the University of Texas at Austin. The X-ray maps were used to determine where garnet and matrix mineral compositions (plagioclase, biotite, chlorite, and muscovite) were obtained using the electron microprobe (Figure 10.10). All compositional data and X-ray element maps are available in the data repository.

Conventional and isochemical phase diagram approaches were used to determine the P-T conditions experienced by sample VJ-03-01-20-2. For the conventional approach, we applied garnet-biotite (GB) thermometry (Ferry & Spear, 1978; Berman, 1990) and garnet-plagioclase-biotite-muscovite (GPBM) barometry (Hoisch, 1990). We used the lowest Mn and lowest Fe/(Fe+Mg) garnet paired with matrix minerals (biotite, plagioclase, and muscovite) for these conventional rim P-T conditions. To determine the garnet core conditions, we used the program Theriak-Domino for the isochemical phase diagrams (de Capitani & Brown, 1987; de Capitani & Petrakakis, 2010) and the Holland and Powell (1998) thermodynamic dataset with updates to solution models through 2010. Relevant mixing models in the system MnO-Na_2O-CaO-K_2O-FeO-MgO-Al_2O_3-SiO_2-TiO_2-H_2O were used. The solid-solution models pertinent in this study are biotite (Powell & Holland, 1999; White et al., 2000), chlorite (Holland et al., 1998), chloritoid (Mahar et al., 1997; White et al., 2000), feldspar (Baldwin et al., 2005; Holland & Powell, 2003), garnet (Mahar et al., 1997; Zeh & Holness, 2003), ilmenite (Ideal Mn-Mg-Fe solution), staurolite (Holland & Powell, 1998; Mahar et al., 1997), and white mica (Coggon & Holland, 2002). The system was considered to form in the presence of water (activity of H_2O = 1.0), and the iron oxidation state was 2+. We assume H_2O saturation (i.e., H_2O (100) in Theriak-Domino). Whole-rock molar wt% data were used as direct input for developing the isochemical phase diagrams. Primary garnet compositions are modified due to diffusion, as shown by the flat Mn zoning, but the grossular and pyrope appeared to follow trends across the garnet that may indicate some degree of primary preservation (Figure 10.10b). As Ca diffuses more slowly in garnet compared to Fe, Mn, and Mg (Spear, 1993), isopleths of grossular and pyrope (± 0.2-mole fraction) were overlain on the P-T diagrams to evaluate potential conditions of the garnet core.

Samples with monazite inclusions in garnets (VJ-03-01-20-2 and VJ-03-01-20-1A) were cut and mounted with

Figure 10.7 (a) Landscape photograph of the Whipple Mountains highlighting a 400-m-thick section of exposed shear zone rocks. Outcrop photographs illustrate the styles of deformation within the Whipple Mountains: (b) boudinage, (c) garnets along foliation planes, (d) garnets in a mica schist, (e) gabbro blocks, (f) amphibolite, and (g) mylonitic gneiss with blocky feldspar grains as kinematic indicators.

a monazite Th-Pb age standard (554) in a 1-inch epoxy ring for secondary ion mass spectrometry (SIMS) U/Th-Pb analysis at the Institute of Earth Sciences at Heidelberg University. Monazite ages were obtained using the CAMECA ims1280-HR ion microprobe at Heidelberg University (Table 10.3). The reference used was monazite 554 (45.3 ± 1.4 Ma, ±2σ; Harrison et al., 1995). A common ^{204}Pb correction after stripping the contribution of $^{144}Nd^{232}Th^{16}O_2^{++}$ estimated from $^{143}Nd^{232}Th^{16}O_2^{++}$ and the isotopic ratio of $^{144}Nd/^{143}Nd = 1.95$ was applied for all grains. To test for potential matrix effects that are known for monazite with different Th abundances (Stern & Berman, 2001; Fletcher et al., 2010), a series of eight reference monazites was cross-calibrated using 44069 monazite as the primary reference. An analytical procedure using high-energy ions (−40 eV relative to

Figure 10.8 Outcrop photographs illustrate the styles of deformation within the Whipple Mountains. (a) Mylonitic, leucocratic, ~1.5-m-thick, sheet-like granitic intrusion. (b) Leucocratic sills are locally isoclinally folded into parallelism to the mylonitic foliation. (c) Folded and foliated garnet-bearing leucogranite bodies. (d) Curviplanar and shear-offset leucocratic veins are present in some boudinaged, but internally undeformed amphibolite bodies characterized by a salt-pepper texture. (e) Unfoliated amphibolite bodies also contain pockets of coarse-grained, undeformed mafic materials.

Table 10.2 Bulk rock compositions (molar wt %) from samples analyzed in this study.

Sample/analysis	VJ-03-01-20-1A	VJ-03-01-20-1B	VJ-03-01-20-2
Latitude	34°22′34.41″N	34°22′34.41″N	34°22′32.98″N
Longitude	114°19′24.30″W	114°19′24.30″W	114°19′16.00″W
SiO_2	48.127	78.333	46.559
Al_2O_3	22.788	7.463	28.608
Fe_2O_3	8.602	6.873	7.834
MnO	0.466	0.587	0.165
MgO	4.858	3.549	3.370
CaO	3.173	0.517	2.146
Na_2O	5.443	0.298	3.960
K_2O	5.764	2.199	6.438
TiO_2	0.755	0.154	0.887
P_2O_5	0.024	0.026	0.034
Total	100.0	100.0	100.0

the nominal accelerating voltage of +10 kV) minimized monazite matrix effects to within ±2% (relative) of reported Th-Pb ages for most reference monazites, except for very high-Th monazite. The approximately 10× lower abundance of high-energy secondary ions, however, precluded taking this approach for the comparatively young monazite samples. Instead, a conventional (Harrison et al., 1995) setup was used, for which an additional age uncertainty of ~5% (relative) is estimated because of the slightly higher Th abundances indicated by primary-beam

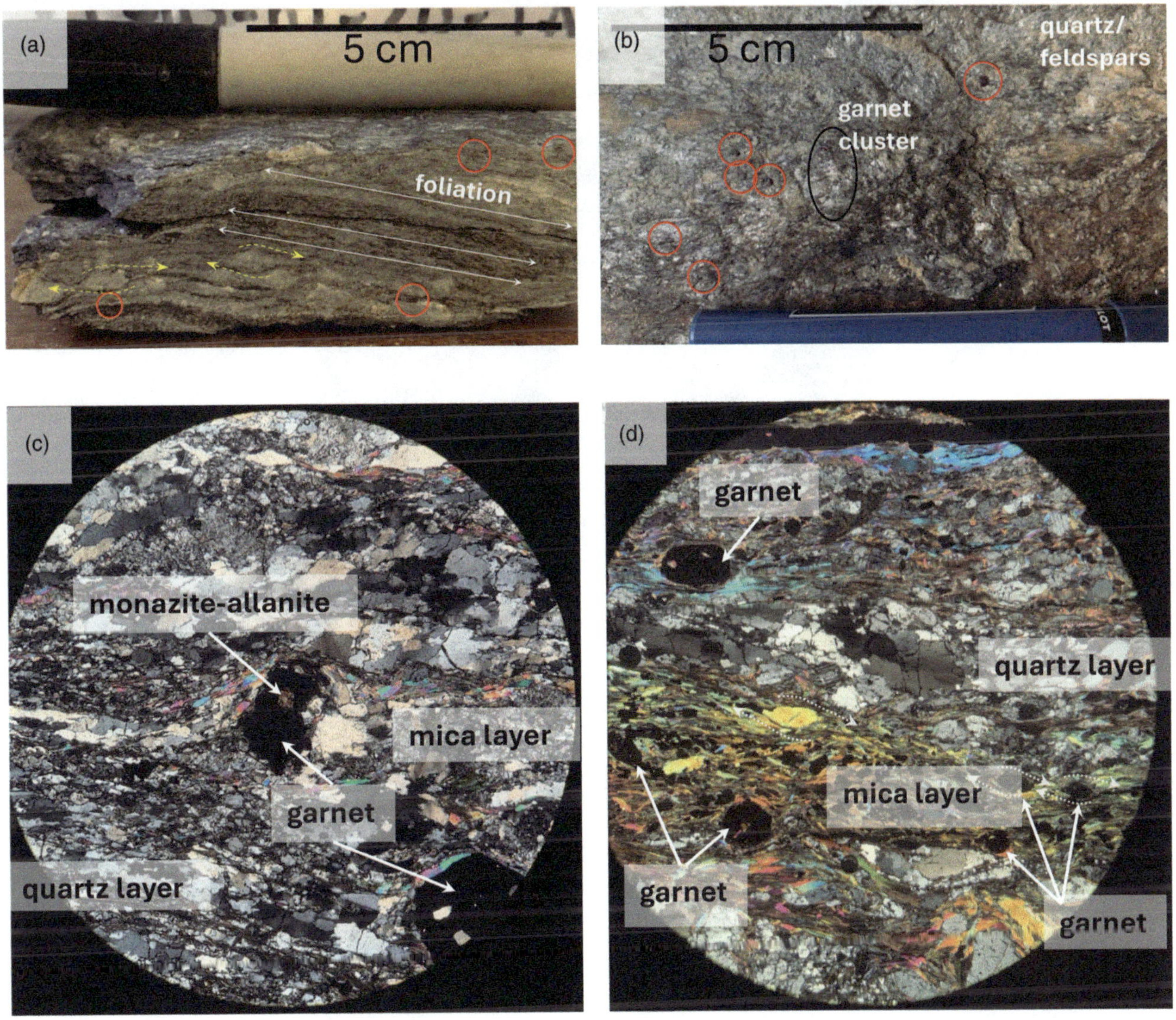

Figure 10.9 Garnet-bearing metapelite samples. (a) Sample VJ-03-01-20-1A, a two-mica, garnet-bearing metapelite. The white lines with arrows on both ends illustrate the foliation direction in the sample; crenulated cleavage is visible between the bottom two white lines. The dashed yellow lines highlight feldspar kinematic indicators with top-to-the-right sense of shear. Garnet grains are enclosed in red circles. See Table 10.3 for monazite dates from this rock. (b) Sample VJ-03-01-20 2, a garnet schist. Individual garnet grains are enclosed in red circles. A cluster of garnet grains is enclosed in the black oval. The matrix is composed of mica, quartz, and feldspar grains. (c) Petrographic image of sample VJ-03-01-20-1A. Monazite-allanite reaction texture enclosed in a garnet was used for Th-Pb age data. Bottom-right garnet grain highlights various inclusions. Oriented quartz grains are visible in the quartz layer and mica layer surrounding the top of the garnet grain. (d) Petrographic image of sample VJ-03-01-20-2. Garnet grain at top left contains monazite used for Th-Pb age data (Table 10.3). Mica (muscovite, biotite, chlorite) and quartz banding clearly visible. Garnet grains in the bottom-right are deformed and appear to have top-to-the-right sense of shear, as highlighted by the dotted white arrows.

normalized ThO_2^+ count rates for the unknowns relative to the reference monazites. This bias is typically within the stated uncertainties based on propagating counting and calibration curve uncertainties for the young monazites studied here. Although the monazite standards are sufficiently old to determine U-Pb and Th-Pb at reasonable age precision, there is also a caveat that for the comparatively young monazites studied here, only Th-Pb ages are relevant due to high uncertainties of U-Pb ages, and the potential of U-Pb (in particular $^{206}Pb/^{238}U$) ages being affected by initial disequilibrium (i.e., ^{230}Th excess). The U-Pb calibration using monazite 554 results in geologically meaningless ages, and thus we do not report U-Pb ages or Concordia plots from these grains.

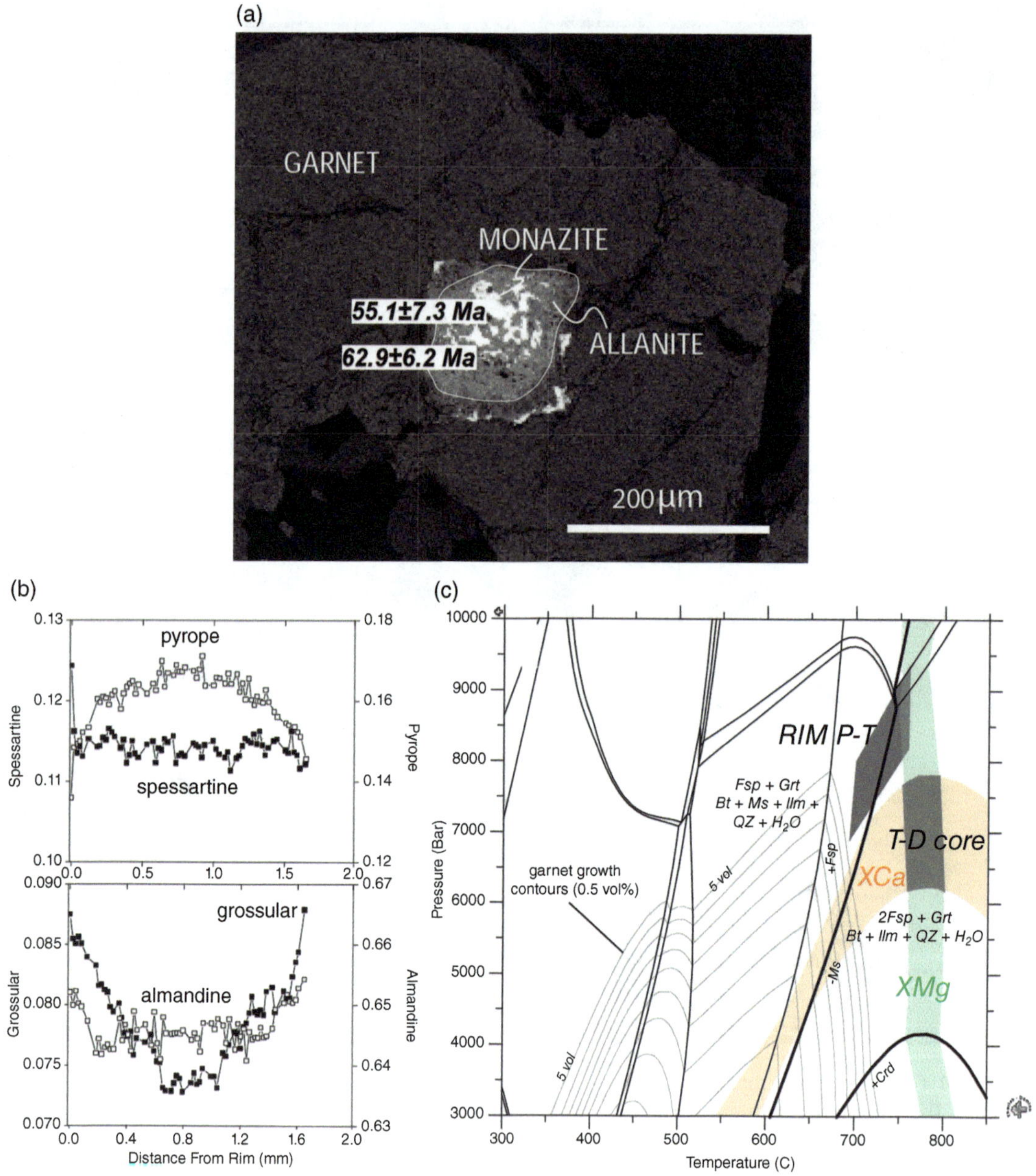

Figure 10.10 (a) Backscattered electron (BSE) image of a Whipple Mountains garnet with an allanite-monazite reaction texture occluded. Th-Pb SIMS ages are indicated ($\pm 1\sigma$). (b) Compositional transects of spessartine, pyrope, grossular, and almandine across a garnet from rim to rim. (c) The sample's isochemical phase diagram using the rock bulk composition and the program Theriak-Domino (see proposal text). Garnet growth contours (0.5 vol%) and the conventional rim P-T estimate are overlain on the diagram. Colored bars are the garnet mole fraction Mg (XMg) and Ca (XCa) from the garnet core (±0.2 mole fraction). These bars intersect our best estimate for the garnet core conditions. The rim P-T conditions are from conventional thermobarometry (see text).

10.5. RESULTS

10.5.1. Lithologic and Structure Data

The eastern limb of the shear zone exposes banded, quartzofeldspathic gneisses locally interlayered with boudinaged amphibolite (Figure 10.7b). The shear zone also exposes garnet-bearing schist (Figure 10.7c,d) and meter-scale boudinaged gabbro blocks (Figure 10.7e). At deeper structural levels away from the overlying Whipple detachment fault, both amphiboles and feldspars are aligned in the NE-SW direction, parallel to the regional mylonitic lineation in the Whipple shear zone. In contrast, feldspars at the highest structural level directly below the Whipple detachment fault are cataclastically fractured in crystal-plastically deformed quartz-mica matrixes (Figure 10.7g).

The western limb of the Whipple shear zone exposes mylonitic, leucocratic, sheet-like granitic intrusions (feature 1 marked by white arrows in Figure 10.8a), which either truncate an older set of mylonitic foliation (feature 2 marked by yellow arrows in Figure 10.8a) or are concordant with the hosting foliation (see features 1 for the sill and 2 for the mylonitic foliation in Figure 10.8b). Leucocratic sills are isoclinally folded into parallelism to the mylonitic foliation (feature 3 in Figure 10.7b). Leucocratic intrusions that crosscut the mylonitic foliation may also be mylonitically foliated (e.g., feature 4 in Figure 10.8b). Garnet-bearing leucogranite bodies are folded and foliated in the shear zone (Figure 10.8c). Curviplanar and shear-offset leucocratic veins are present in some boudinaged but internally undeformed amphibolite bodies characterized by a salt-pepper texture (Figure 10.8d). The nonfoliated amphibolite bodies also contain pockets of coarse-grained, undeformed mafic materials (Figure 10.8d).

The rock units in this study area within the Whipple mylonitic shear zone (Figure 10.7a) include a two-mica garnet-bearing metapelite (Figure 10.7c), garnet schist (Figure 10.7d), quartzofeldspathic gneiss (Figure 10.7g), amphibolite with interlayered quartz veins (Figure 10.7f), and gabbro blocks/boudins (Figure 10.7b,e). Key mineral assemblages for some garnet-bearing metapelites include sillimanite and/or kyanite. Garnets found in metapelites throughout the Whipple mylonitic shear zone may be as large as ~1 cm across (Figure 10.7c,d); however, a leucogranite vein was found with abundant garnets >1 cm (Figure 10.8c). Feldspar crystals in the mylonitic gneiss fabric record top-to-the-northeast sense of shear, as is shown in Figure 10.8e.

Generally, the foliation along transect A-A′ (Figure 10.3) follow a continuous dome-shaped trend in dip direction. Foliation dips ~25° north-east in the northeastern section of the map, foliation flattens toward the center of the mountain range and dips radially, and foliation dips southwest in the southwestern section of the map. Stretching lineation parallel to foliation is consistent with the dip direction of the foliation. Boudinaged gabbro blocks display stretching in an east-west direction (Figure 10.7b). Figure 10.8a,b highlights mylonitic, sheet-like, granitic intrusions found throughout the entire shear zone, some of which are several meters thick; others are folded with the fold axial planes parallel to the mylonitic foliation.

Sample VJ-03-01-20-1A (Figure 10.9a) was collected from a two-mica garnet metapelite that contains garnet + biotite + muscovite + chlorite + quartz and accessory minerals xenotime + monazite + allanite + zircon + ilmenite + rutile. Petrographic images highlight zones of banded mica-rich versus quartz-rich layers (Figure 10.9c). Porphyroblastic garnets are broken and altered, have reaction rims, and appear surrounded mainly by mica layers (Figure 10.9d). Monazite inclusions in a garnet range from ~15 to 50 mm in size but appear fully inside the garnet grains with no visible traces of fluid interaction along cracks. Sample VJ-03-01-20-2 (Figure 10.9d) was collected from a garnet schist that contains garnet + biotite + muscovite + chlorite + plagioclase + quartz and accessory minerals monazite + ilmenite + zircon. Most porphyroblastic garnet grains are surrounded by a mica matrix, and a smaller subset of garnet grains are surrounded by quartz. The garnet porphyroblasts are aligned parallel to the foliation, providing useful kinematic indicators of delta and sigma clasts (Figure 10.9a,d). Monazite inclusions in garnet are ~10–20 μm, but some were also found in the matrix within banded mica veins. Petrographic images in all samples highlighted the fine- to medium-grained lepidoblastic mica layers and banded to slightly crenulated mylonitic fabric.

10.5.2. P-T-t Data

Figure 10.7c shows sample VJ-03-01-20-1A collected from the Whipple Mountains. Garnet in the sample trapped an allanite-monazite reaction texture during crystallization (Figure 10.10a). The texture is similar to observations elsewhere that document monazite appeared after allanite decomposition (Broska et al., 2000, Catlos et al., 2002). SIMS dates from monazite grains within the center of the texture are Paleocene–Early Eocene (62 ± 9 Ma, 67 ± 7 Ma, and 80 ± 6 Ma (Th-Pb dates, ±1σ)) (Table 10.3). Th-Pb dates from another monazite inclusion in garnet from the same rock are older (1.01 ± 0.08 Ga; ±1σ).

Compositional transects across a garnet collected nearby (VJ-03-01-20-2) yield flat zoning in almandine and spessartine but high pyrope and lower grossular contents in the core than the rim (Figure 10.10b). Conventional P-T estimates (Ferry and Spear, 1978; Hoisch, 1989; Berman, 1990) using the garnet rim and matrix minerals from VJ-03-01-20-2 are rim, 680–750 °C and 7–9.4 kbar; and core, 750–800 °C and 6.2–7.8 kbar (Figure 10.10c).

Table 10.3 Results of Th-Pb monazite ages for samples analyzed in this study.

Sample	Age (Ma)	Age (Ma)	% Radiogenic	Age (Ma)	Age (Ma)
	$^{208}Pb/^{232}Th$	$^{208}Pb/^{232}Th$ 1 s.e.	^{208}Pb	$^{207}Pb/^{206}Pb$	$^{207}Pb/^{206}Pb$ 1 s.e.
VJ-03-01-20-2_VJ1_2.ais	79.8	6.0	81.5	—	—
VJ-03-01-20-1A_VJ2_1a_1.ais	1,011.0	82.3	99.1	1,684	57
VJ-03-01-20-1A_VJ2_1a_2.ais	62.1	9.1	81.2	—	—
VJ-03-01-20-1A_VJ2_1a_2b.ais	66.5	6.8	86.4	—	—

We overlie these conditions on the sample's isochemical phase diagram, and the predicted mineral assemblage overlaps the rock's observed major mineral assemblage (garnet + feldspar (plagioclase) + biotite + muscovite + ilmenite + quartz + H_2O) (Figure 10.10c). We also overlie the grossular and pyrope compositions from the garnet core onto the isochemical phase diagram because they may have preserved their prograde conditions best. In this case, the garnet core yields conditions of 750–800 °C and 6–7.5 kbar, consistent with the garnet crystallizing at higher-grade conditions. These conditions lack muscovite with the assemblage of garnet + feldspar (plagioclase) + biotite + ilmenite + quartz + H_2O (Figure 10.10c).

10.6. DISCUSSION

10.6.1. Magmatism and Exhumation

Each core-complex system along a narrow zone of Laramide-thickened orogenic axial zone from southern Canada to northern Mexico provides insight into the development of low-angle detachment faults (Coney & Harms, 1984). At a regional scale, there is evidence that crustal melting during the Late Cretaceous to Eocene time, extension, and core-complex formation may be linked (Chapman et al., 2021). In some cases, magmatism in the Cordilleran Anatectic Belt may be directly related to the timing of core-complex extension. For example, in the Valhalla complex (Figure 10.1), the duration of magmatism at 60–50 Ma directly overlaps with the onset of extension at 60 Ma and rapid extension at 50 ± 2 Ma, and the peak timing of metamorphism also spans that duration of time (60–57 Ma, Spear & Parrish, 1996; Holk & Taylor Jr, 2000; Spear, 2004; Gordon et al., 2008; Chapman et al., 2021; Table 10.1).

In contrast, the Cordilleran Anatectic Belt Late Cretaceous-age magmatism at 86–70 Ma recorded in plutons from the Snake Range core complex in Nevada is consistent with the timing of peak metamorphism at 90–70 Ma, different from estimates of Tertiary extension with an onset at 45 ± 3 Ma and rapid extension at 20±2 Ma (Table 10.1; Lee & Christiansen, 1983; King et al., 2004; Henry et al., 2011; Hallett & Spear, 2014, 2015; Evans et al., 2015; Gottlieb, 2017; Lee et al., 2017; Chapman et al., 2021). Similarly, in the Whipple Mountains, the Whipple Wash granite pluton emplacement occurred at 89 ± 3 Ma and mylonitization due to extension between 26.6 ± 1.1 Ma and 13.7 ± 0.4 Ma (Anderson et al., 1988; Stockli et al., 2006). These previous estimates for the latest Tertiary mylonitization were the expected monazite-in-garnet ages that would have represented a single-phase of extension during middle to late Cenozoic time (Wong et al., 2023). However, the monazite dates from the metapelite rock yielded much older ages, which suggests a larger time span over which the deformation occurred. Thus, the monazite ages from this study suggest a Late Cretaceous to Paleogene metamorphic event that has not been accounted for in past studies. Recent studies propose a multiphase deformation history for the Whipple Mountains shear zone (Seymour et al., 2016; Flansburg & Stockli, 2023; Kapp et al., 2023; Wong et al., 2023). Thus, we tentatively interpret the younger monazite ages of 80–62 Ma to date peak metamorphism of the garnet schist in the Whipple ductile shear zone. Garnet often appears during the allanite → monazite reaction (Figure 10.10a) during the metamorphism of pelitic bulk rock compositions (e.g., Smith and Barreiro, 1990; Catlos et al., 2002). Within error, pluton emplacement in the Whipple Mountains overlaps with the monazite dates (Table 10.3) and suggests a possible relationship between crustal melting and metamorphism at depths >25 km (Anderson et al., 1988). We note that when correlating whole-rock geochemistry to crustal thickness, crustal contamination of magma bodies and their influence on quantitative outputs should be considered (Chapman et al., 2015, 2020).

10.6.2. Past Models

Models describing how low-angle detachments operate in the middle crust suggest that mylonitic zones within the detachments accommodate strain and, thus, areas of extreme extension (Singleton et al., 2022). Most of the metamorphic core complexes record NE-directed hanging-wall extension in the southern segment of the North American Cordillera. In the northern segment, east-west extension is more prominent and is perpendicular to the N-S-trending margin of the Sevier Thrust Belt front (Figure 10.1) but parallel to shear strain and flattening direction (e.g., Frenchmans Cap) (Gervais et al., 2010). Reconstructions of detachment faulting in the North American Cordillera core complexes have been possible by examining dike swarms (Richard et al., 1990; Foster et al., 1993; Gans & Gentry, 2016). Strain alone may help determine the deformation mechanism in an area, especially if the area has undergone multiple deformation events, thus differentiating competing kinematic models have yielded ambiguous results.

Behr and Platt (2011) combined structural, thermobarometric, and geochronological data to infer differential stress versus T and differential stress versus depth profiles for samples from the Whipple detachment to constrain their profiles consistent with previously suggested P-T conditions. The thermal model assumptions include an initial geothermal gradient of 25 °C/km (Foster & John, 1999), density of upper crust of 2,700 kg/m^3, density of lower crust of 2,800 kg/m^3, conductivity of 2.5 W/mK, heat capacity of 1,000 J kg K^{-1}, and heat production of 2.3e-6 W/m^3. The slip rate of 5.9 mm/yr used for the Whipple fault is from average estimates reported by Foster and John (1999) and Stockli et al. (2006). The dip angle of the Whipple fault was estimated at 25° based on data from Davis (1988). Anderson et al. (1988) provided an estimate of peak metamorphism of 4.6 ± 0.9 kbar, which Behr and Platt (2011) used to constrain a stress profile through the middle crust. Their studies showed that the peak strength at the brittle-ductile transition zone in the Whipple Mountains reached stresses ranging from 1.04 to 1.36 kbar at a depth of 9 km (Anderson et al., 1988; Behr & Platt, 2011). The Behr and Platt (2011) model for determining a stress profile follows normal continental-strength profile conditions. However, a stress magnitude at its current location cannot be used as the stress value at its initial deformation location when the P-T-time paths are unknown because maximum shear stress varies in magnitude with time and space (Behr & Platt, 2011).

10.6.3. P-T Constraints and Future Work

Our garnet rim P-T estimate from sample VJ-03-01-20-2 of 7–9.4 kbar and 680–750 °C and core condition of 6.2–7.8 kbar and 750–800 °C (Figure 10.10c) suggests that rocks now exposed at the surface were deformed at conditions close to those in the middle crust. Divalent-cation compositional zoning in garnet grains <100 μm has long been known to alter prograde compositions at T>600 °C (e.g., Tracy et al., 1976; Florence and Spear, 1991; Ganguly, 2010), which is consistent with the observation of flat garnet zoning in Mn and Fe in sample VJ-03-01-20-2 (Figure 10.10b). Conventional thermobarometry, isochemical phase diagrams, and isopleth thermobarometry approaches that are used to determine P-T conditions prescribe that the samples experienced closed system behavior and no modification of garnet chemistry after growth (e.g., Lanari & Engi, 2017). An additional assumption is that the sample experienced equilibrium, which can never be proven for any rock (Spear & Peacock, 1989; Lanari & Duesterhoeft, 2019). Sources of uncertainty for the thermobarometric approach applied here include the accuracy of electron microprobe analyses, calibration errors, variations in activity models, compositional heterogeneity, and uncertainty associated with the thermodynamic properties inherent in the choice of an internally consistent database (Kohn & Spear, 1991; White et al., 2014; Lanari & Duesterhoeft, 2019). Because the error suggested by the grid created due to overlapping mineral compositional isopleths likely underestimates the actual uncertainty, we suggest applying uncertainty values (±25 °C and ±1 kbar) to the overlapping isopleth conditions as is commonly reported for conventional thermobarometry (e.g., Spear & Peacock, 1989; Kohn, 1993; Kohn et al., 2001). The approach to generating P-T conditions discussed here transforms the sample into a model representing the true rock and mineral assemblage. It also restricts its behavior as if it was in a closed system. Confidence in conventional P-T conditions increases when conditions agree with mineral assemblages and are consistent with garnet zoning. A value of the isopleth approach is that a user can detect when systems stray from the equilibrium and closed system assumptions (Catlos et al., 2020, 2022). The conditions obtained here are consistent with the rock's mineral assemblage and garnet zoning. If these conditions record garnet growth during metamorphism and not a previous metamorphic cycle, the results suggest that the rolling-hinge (Figure 10.1b) and simple-shear (Figure 10.1c) models are more applicable to the Whipple Mountains (Wernicke, 1985; Wernicke and Axen, 1988).

Overall, the P-T data suggest that a detachment along the base of the upper crust is too shallow. Models that place the shear zone at a depth of ~15 km, 5 kbar, and 350–400 °C, barely in the ductile deformation regime of the crust, appear inapplicable to the Whipple Mountains (Figure 10.4). The results permit the detachment fault to have extended at least into the middle crust. The data cannot address whether the fault could have extended farther into the lower crust and even the mantle lithosphere. The most important insight is that (a) the maximum shear stress varies in magnitude with time and space, and (b) a measured stress magnitude at its current location cannot be used as the stress value at its initial launching point when the P-T-t path of the hosting rock mass is unknown (cf., Behr & Platt, 2011, 2014). With this new information, a new P-T-t path may be established for the rocks in the Whipple detachment and may suggest a deeper initiation or multiple metamorphic events not previously analyzed.

It is important to note that the P-T estimates were determined by analyzing sample VJ-03-01-20-2. Meanwhile, the monazite inclusion ages were determined by analyzing sample VJ-03-01-20-1A. Both samples are collected near each other (Table 10.2), although one is structurally higher than the other by a couple hundred meters. Abundant garnets found throughout the study area provide a starting point for determining the tectono-metamorphic evolution of the Whipple ductile-shear zone. Locally, systematic sampling across the Whipple detachment shear zone has not been conducted. However, the P-T-t data from a single garnet suggests that the approach has the potential to tremendously improve existing ideas regarding the conditions experienced. New thermobarometric approaches to determine P-T estimates for the Whipple detachment shear zone garnets must be carried out. Further studies may be able to differentiate between existing models that have attempted to explain how the rocks reached said conditions (Strickland et al., 2018; Chapman et al., 2020; Chapman et al., 2021).

10.7. CONCLUSIONS

Extensive studies have been conducted on metamorphic core complexes of the northern and central segments of the North American Cordilleran metamorphic core-complex belt. However, much more work is needed in the southern segment of the belt to better constrain the mechanisms of core-complex development in this region. We present an approach to constrain the deformation history along the Whipple detachment shear zone. P-T data obtained from the Whipple detachment shear zone indicate that the mylonitic shear-zone deformation occurred under higher-grade conditions than previously thought. These P-T conditions indicate deformation at midcrustal levels, which may eliminate kinematic models that suggest deformation occurred solely in the upper crust. The monazite-in-garnet inclusion ages at 62 ± 9 Ma, 67 ± 7 Ma, and 80 ± 6 Ma (Th-Pb dates, $\pm 1\sigma$) do not represent Oligocene shear zone development; instead, they highlight a possible metamorphic event during Paleogene time that has not been accounted for in the models until recently. Therefore, the suggested systematic and detailed approach with these new P-T-t conditions may be used to refine previous models and continue to constrain the spatiotemporal evolution of the Whipple mylonitic shear zone. Future work and coupled studies between P-T estimates and the age determination of the P-T history may further highlight how continental extension plays a role in low-angle detachment faulting. This approach would determine whether there have been multiple deformation events in the Whipple ductile-shear zone, plus it would apply to other shear zones across the North American Cordilleran metamorphic core complexes.

ACKNOWLEDGMENTS

I am grateful for my former adviser, late Professor An Yin, whose life was unexpectedly cut short. His passion for geology has impacted many lives, and the lessons will live with us for years to come. Constructive comments from Jay Chapman, Brandon Lutz, and an anonymous reviewer greatly improved the original manuscript.

AVAILABILITY STATEMENT

The authors confirm that the data supporting the analysis and conclusions are available at the UCLA DRYAD data repository (doi: https://doi.org/10.5061/dryad.ht76hdrs4).

REFERENCES

Anderson, J. L. (1988). Core complexes of the Mojave-Sonoran Desert: Conditions of plutonism, mylonitization, and decompression. *Metamorphism and Crustal Evolution of the Western United States, Rubey*, *7*, 503–525.

Anderson, J. L., Barth, A. P., & Young, E. D. (1988). Mid-crustal Cretaceous roots of Cordilleran metamorphic core complexes. *Geology*, *16*(4), 366–369.

Anderson, J. L., & Cullers, R. L. (1990). Middle to upper crustal plutonic construction of a magmatic arc; an example from the Whipple Mountains metamorphic core complex. In J. L. Anderson (Ed.), *The nature and origin of Cordilleran magmatism* (Vol. *174*, pp. 47–69). Geological Society of America.

Anderson, J. L., & Rowley, M. C. (1981). Synkinematic intrusion of peraluminous and associated metaluminous granitic magmas, Whipple Mountains, California. *The Canadian Mineralogist*, *19*(1), 83–101.

Anderson, T. H., Sílver, L. T., & Salas, G. A. (1980). Distribution and U-Pb isotope ages of some lineated plutons, northwestern Mexico. In J. M. Crittenden, et al. (Eds.), *Cordilleran metamorphic core complexes* (Vol. *153*, pp. 269–283). Geological Society of America.

Armstrong, R. L. (1982). Cordilleran metamorphic core complexes-from Arizona to southern Canada. *Annual Review of Earth and Planetary Sciences*, *10*, 129.

Armstrong, R. L., & Ward, P. (1991). Evolving geographic patterns of Cenozoic magmatism in the North American Cordillera: the temporal and spatial association of magmatism and metamorphic core complexes. *Journal of Geophysical Research: Solid Earth*, *96*(B8), 13201–13224.

Arnold, A. H. (1986). *Geologic implications of a geo-chemical study of three two-mica granites in Southern Arizona* (MS Thesis). The University of Arizona, Arizona. Retrieved from https://repository.arizona.edu/handle/10150/231236

Atwater, T., & Stock, J. (1998). Pacific-North America plate tectonics of the Neogene southwestern United States: an update. *International Geology Review*, *40*(5), 375–402.

Axen, G. J. (1992). Pore pressure, stress increase, and fault weakening in low-angle normal faulting. *Journal of Geophysical Research: Solid Earth*, *97*(B6), 8979–8991.

Axen, G. J. (2004). Mechanics of low-angle normal faults. In G. D. Karner, et al. (Eds.), *Rheology and deformation of the lithosphere at continental margins* (pp. 46–91). Columbia University Press.

Axen, G. J. (2007). Research focus: significance of large-displacement, low-angle normal faults. *Geology*, *35*(3), 287–288.

Axen, G. J. (2020). How a strong low-angle normal fault formed: the Whipple detachment, southeastern California. *GSA Bulletin*, *132*(9–10), 1817–1828.

Axen, G. J., Luther, A., & Selverstone, J. (2015). Paleostress directions near two low-angle normal faults: testing mechanical models of weak faults and off-fault damage. *Geosphere*, *11*(6), 1996–2014.

Axen, G. J., & Selverstone, J. (1994). Stress state and fluid-pressure level along the Whipple detachment fault, California. *Geology*, *22*(9), 835–838.

Axen, G. J., van Wijk, J. W., & Currie, C. A. (2018). Basal continental mantle lithosphere displaced by flat-slab subduction. *Nature Geoscience*, *11*(12), 961–964.

Bailey, I. W., Miller, M. S., Liu, K., & Levander, A. (2012). VS and density structure beneath the Colorado Plateau constrained by gravity anomalies and joint inversions of receiver function and phase velocity data. *Journal of Geophysical Research: Solid Earth*, *117*, B2.

Baldwin, J. A., Powell, R., Brown, M., Moraes, R., & Fuck, R. A. (2005). Modelling of mineral equilibria in ultrahigh-temperature metamorphic rocks from the Anápolis–Itauçu Complex, central Brazil. *Journal of Metamorphic Geology*, *23*(7), 511–531.

Behr, W. M., & Platt, J. P. (2011). A naturally constrained stress profile through the middle crust in an extensional terrane. *Earth and Planetary Science Letters*, *303*(3–4), 181–192. Bulletin, 105(1), 56-76

Behr, W. M., & Platt, J. P. (2014). Brittle faults are weak, yet the ductile middle crust is strong: implications for lithospheric mechanics. *Geophysical Research Letters*, *41*(22), 8067–8075.

Beratan, K. K. (1991). Miocene synextension sedimentation patterns, Whipple Mountains, southeastern California: implications for the geometry of the Whipple detachment system. *Journal of Geophysical Research: Solid Earth*, *96*(B7), 12425–12442.

Berman, R. G. (1990). Mixing properties of Ca-Mg-Fe-Mn garnets. *American Mineralogist*, *75*(3–4), 328–344.

Bird, P. (1988). Formation of the Rocky Mountains, western United States: a continuum computer model. *Science*, *239*(4847), 1501–1507.

Block, L., & Royden, L. H. (1990). Core complex geometries and regional scale flow in the lower crust. *Tectonics*, *9*(4), 557–567.

Boettcher, S. S., & Walker, J. D. (1993). Geologic evolution of Iron Mountain, central Mojave Desert. *California. Tectonics*, *12*(2), 372–386.

Bosworth, W., Huchon, P., & McClay, K. (2005). The Red Sea and Gulf of Aden basins. *Journal of African Earth Sciences*, *43*(1–3), 334–378.

Broska, I., Petrík, I., & Williams, C. T. (2000). Coexisting monazite and allanite in peraluminous granitoids of the Tribec Mountains, Western Carpathians. *American Mineralogist*, *85*(1), 22–32.

Brothers, D. S., Driscoll, N. W., Kent, G. M., Harding, A. J., Babcock, J. M., & Baskin, R. L. (2009). Tectonic evolution of the Salton Sea inferred from seismic reflection data. *Nature Geoscience*, *2*(8), 581–584.

Brun, J. P., Sokoutis, D., Tirel, C., Gueydan, F., Van Den Driessche, J., & Beslier, M. O. (2018). Crustal versus mantle core complexes. *Tectonophysics*, *746*, 22–45.

Bryant, B., & Wooden, J. L. (2008). *Geology of the Northern Part of the Harcuvar Complex, West-Central Arizona* (U.S. Geological Survey Professional Paper No. 1752) (p. 52). Retrieved from https://pubs.usgs.gov/pp/1752/

Buck, W. R. (1991). Modes of continental lithospheric extension. *Journal of Geophysical Research: Solid Earth*, *96*(B12), 20161–20178.

Campbell-Stone, E., John, B. E., Foster, D. A., Geissman, J. W., & Livaccari, R. F. (2000). Mechanisms for accommodation of Miocene extension: low-angle normal faulting, magmatism, and secondary breakaway faulting in the southern Sacramento Mountains, southeastern California. *Tectonics*, *19*(3), 566–587.

Carlson, D. H., Fleck, R., Moye, F. J., & Fox, K. F. (1991). Geology, geochemistry, and isotopic character of the Colville Igneous Complex, northeastern Washington. *Journal of Geophysical Research: Solid Earth*, *96*(B8), 13313–13333.

Carter, T. J., Kohn, B. P., Foster, D. A., Gleadow, A. J., & Woodhead, J. D. (2006). Late-stage evolution of the Chemehuevi and Sacramento detachment faults from apatite (U-Th)/He thermochronometry—evidence for mid-Miocene accelerated slip. *Geological Society of America Bulletin*, *118*(5–6), 689–709.

Cassel, E. J., Smith, M. E., & Jicha, B. R. (2018). The impact of slab rollback on Earth's surface: Uplift and extension in the hinterland of the North American Cordillera. *Geophysical Research Letters*, *45*(20), 10–996.

Catlos, E. J., Dubey, C. S., & Etzel, T. M. (2022). Imbrication and erosional tectonics recorded by garnets in the Sikkim Himalayas. *Geosciences*, *12*(4), 146.

Catlos, E. J., Gilley, L. D., & Harrison, T. M. (2002). Interpretation of monazite ages obtained via in situ analysis. *Chemical Geology*, *188*(3–4), 193–215.

Catlos, E. J., Perez, T. J., Lovera, O. M., Dubey, C. S., Schmitt, A. K., & Etzel, T. M. (2020). High-resolution P-T-time paths across Himalayan faults exposed along the Bhagirathi transect NW India: implications for the construction of the Himalayan orogen and ongoing deformation. *Geochemistry, Geophysics, Geosystems*, *21*(12), e2020GC009353.

Çemen, I., Catlos, E. J., Göğüs, O., & Özerdem, C. (2006). Postcollisional extensional tectonics and exhumation of the Menderes massif in the Western Anatolia extended terrane, Turkey. In Y. Dilek & S. Pavlides (Eds.), *Postcollisional tectonics and magmatism in the Mediterranean region and Asia* (Vol. *409*, pp. 353–379). Geological Society of America. 10.1130/2006.2409(18)

Chapman, A. D. (2017). The Pelona–Orocopia–Rand and related schists of southern California: a review of the best-known archive of shallow subduction on the planet. *International Geology Review*, *59*(5–6), 664–701.

Chapman, J. B., Dafov, M. N., Gehrels, G., Ducea, M. N., Valley, J. W., & Ishida, A. (2018). Lithospheric architecture and tectonic evolution of the southwestern U.S. Cordillera: Constraints from zircon Hf and O isotopic data. *GSA Bulletin*, *130*(11–12), 2031–2046. https://doi.org/10.1130/B31937.1

Chapman, A. D., Rautela, O., Shields, J., Ducea, M. N., & Saleeby, J. (2020). Fate of the lower lithosphere during shallow-angle subduction: the Laramide example. *GSA Today*, *30*(1), 4–10. Cordillera. Geology, 43(10), 919-922.

Chapman, J. B., Ducea, M. N., DeCelles, P. G., & Profeta, L. (2015). Tracking changes in crustal thickness during orogenic evolution with Sr/Y: an example from the North American Cordillera. *Geology*, *43*(10), 919–922.

Chapman, J. B., Greig, R., & Haxel, G. B. (2020). Geochemical evidence for an orogenic plateau in the southern US and northern Mexican Cordillera during the Laramide orogeny. *Geology*, *48*(2), 164–168.

Chapman, J. B., Runyon, S. E., Shields, J. E., Lawler, B. L., Pridmore, C. J., Scoggin, S. H., et al. (2021). The north American cordilleran anatectic belt. *Earth-Science Reviews, 215*, 103576.

Chen, L., Gerya, T., Zhang, Z., Zhu, G., Duretz, T., & Jacoby, W. R. (2013). Numerical modeling of eastern Tibetan-type margin: influences of surface processes, lithospheric structure and crustal rheology. *Gondwana Research, 24*(3–4), 1091–1107.

Coggon, R., & Holland, T. J. B. (2002). Mixing properties of phengitic micas and revised garnet-phengite thermobarometers. *Journal of Metamorphic Geology, 20*(7), 683–696.

Coney, P. J. (1987). The regional tectonic setting and possible causes of Cenozoic extension in the North American Cordillera. *Geological Society, London, Special Publications, 28*(1), 177–186.

Coney, P. J., Crittenden, M. D., & Davis, G. H. (1980). Cordilleran metamorphic core complexes: an overview. In J. M. Crittenden, et al. (Eds.), *Cordilleran metamorphic core complexes* (Vol. *153*, pp. 7–31). Geological Society of America.

Coney, P. J., & Harms, T. A. (1984). Cordilleran metamorphic core complexes: Cenozoic extensional relics of Mesozoic compression. *Geology, 12*(9), 550–554.

Cooper, F. J., Platt, J. P., & Behr, W. M. (2017). Rheological transitions in the middle crust: insights from Cordilleran metamorphic core complexes. *Solid Earth, 8*(1), 199–215.

Cooper, F. J., Platt, J. P., Platzman, E. S., Grove, M. J., & Seward, G. (2010). Opposing shear senses in a subdetachment mylonite zone: implications for core complex mechanics. *Tectonics, 29*(4).

Creasey, S. C., Banks, N. G., Ashley, R. P., & Theodore, T. G. (1976). *Middle Tertiary plutonism in the Santa Catalina and Tortolita Mountains, Arizona* (US Geological Society Open-File Report No. 76–262) (p. 20). Arizona. Retrieved from https://pubs.usgs.gov/publication/ofr76262

Cross, A. J., Kidder, S., & Prior, D. J. (2015). Using microstructures and TitaniQ thermobarometry of quartz sheared around garnet porphyroclasts to evaluate microstructural evolution and constrain an Alpine Fault Zone geotherm. *Journal of Structural Geology, 75*, 17–31.

Crowley, J. L., Brown, R. L., Gervais, F., & Gibson, H. D. (2008). Assessing inheritance of zircon and monazite in granitic rocks from the Monashee Complex, Canadian Cordillera. *Journal of Petrology, 49*(11), 1915–1929.

Crowley, J. L., Brown, R. L., & Parrish, R. R. (2001). Diachronous deformation and a strain gradient beneath the Selkirk allochthon, northern Monashee complex, southeastern Canadian Cordillera. *Journal of Structural Geology, 23*(6–7), 1103–1121.

Davis, G. A. (1988). Rapid upward transport of mid-crustal mylonitic gneisses in the footwall of a Miocene detachment fault, Whipple Mountains, southeastern California. *Geologische Rundschau, 77*(1), 191–209.

Davis, G. A. (1996). Mesozoic deformation and plutonism in the Yunmeng Shan: a metamorphic core complex north of Beijing, China. *Tectonic Evolution of Asia*, 253–280.

Davis, G. A., Anderson, J. L., Frost, E. G., & Shackelford, T. J. (1980). Mylonitization and detachment faulting in the Whipple-Buckskin-Rawhide Mountains terrane, southeastern California and western Arizona. In J. M. Crittenden, et al. (Eds.), *Cordilleran metamorphic core complexes* (Vol. *153*, pp. 79–130). Geological Society of America.

Davis, G. A., Anderson, J. L., Martin, D. L., Krummenacher, D., Frost, E. G., & Armstrong, R. L. (1982). Geologic and geochronologic relations in the lower plate of the Whipple detachment fault, Whipple Mountains, southeastern California: a progress report. In E. G. Frost & D. L. Martin (Eds.), *Mesozoic-Cenozoic tectonic evolution of the Colorado River region, California, Arizona, and Nevada* (pp. 408–432). San Diego, CA: Cordilleran Publishers.

Davis, G. A., & Lister, G. S. (1988). Detachment faulting in continental extension; perspectives from the southwestern US Cordillera. In S. P. Clark, et al. (Eds.), *Processes in continental lithospheric deformation* (Vol. *218*, pp. 133–159). Geological Society of America.

Davis, G. A., Lister, G. S., & Reynolds, S. J. (1986). Structural evolution of the Whipple and South Mountains shear zones, southwestern United States. *Geology, 14*(1), 7–10.

Davis, G. H., & Coney, P. J. (1979). Geologic development of the Cordilleran metamorphic core complexes. *Geology, 7*(3), 120–124.

Davis, G. H., Spencer, J. E., & Gehrels, G. E. (2019). Field-trip guide to the Catalina-Rincon metamorphic core complex, Tucson, Arizona. *Geologic Excursions in Southwestern North America: Geological Society of America Field Guide, 55*, 1–38.

de Capitani, C., & Brown, T. H. (1987). The computation of chemical equilibrium in complex systems containing non-ideal solutions. *Geochimica et Cosmochimica Acta, 51*(10), 2639–2652.

de Capitani, C., & Petrakakis, K. (2010). The computation of equilibrium assemblage diagrams with Theriak/Domino software. *American Mineralogist, 95*(7), 1006–1016.

Dewey, J. F. (1988). Extensional collapse of orogens. *Tectonics, 7*(6), 1123–1139.

DeWitt, E., & Reynolds, S. J. (1990). Late Cretaceous plutonism and cooling in the Maria fold and thrust belt, west-central Arizona. *Geological Society of America Abstracts with Programs, 22*(3), 18.

DeWitt, E., Sutter, J. F., Davis, G. A., & Anderson, J. L. (1986). 40Ar/39Ar age-spectrum dating of Miocene mylonitic rocks, Whipple Mountains, southeastern California. *Geological Society of America Abstracts with Programs, 18*(6), 584.

Dickey, D. D., Carr, W. J., & Bull, W. B. (1980). Geologic map of the Parker NW, Parker, and parts of the Whipple Mountains SW and Whipple Wash quadrangles, California and Arizona (No. 1124). U.S. Geological Survey.

Dickinson, W. R. (2004). Evolution of the North American Cordillera. *Annual Review of Earth and Planetary Sciences, 32*, 13–45.

Dickinson, W. R., Goodlin, T. C., Grover, J. A., Mark, R. A., & Shafiqullah, M. (1987). Low-angle normal-fault system along the range front of the southwestern Galiuro Mountains in southeastern Arizona. *Geology, 15*(8), 727–730.

Dickinson, W. R., Ingersoll, R. V., Cowan, D. S., Helmold, K. P., & Suczek, C. A. (1982). Provenance of Franciscan graywackes in coastal California. *Geological Society of America Bulletin, 93*(2), 95–107.

Dickinson, W. R., Kay, S. M., & Ramos, V. A. (2009). Anatomy and global context of the North American Cordillera. In S. M. Kay, et al. (Eds.), *Backbone of the Americas: shallow subduction, plateau uplift, and ridge and terrane collision* (Vol. *204*, pp. 1–29). Geological Society of America.

Dickinson, W. R., Snyder, W. S., & Matthews, V. (1978). Plate tectonics of the Laramide orogeny. In V. Matthews, III (Ed.), *Laramide folding associated with basement block faulting in the western United States* (Vol. *151*, pp. 355–366). Geological Society of America.

Dorsey, R. J., & Becker, U. L. F. (1995). Evolution of a large Miocene growth structure in the upper plate of the Whipple detachment fault, northeastern Whipple Mountains, California. *Basin Research*, *7*(2), 151–163.

Ducea, M. N., & Chapman, A. D. (2018). Sub-magmatic arc underplating by trench and forearc materials in shallow subduction systems; a geologic perspective and implications. *Earth-Science Reviews*, *185*, 763–779.

Ducea, M. N., Triantafyllou, A., & Krcmaric, J. (2020). New timing and depth constraints for the Catalina metamorphic core complex, southeast Arizona. *Tectonics*, *39*(8), e2020TC006383.

Dumitru, T. A., Gans, P. B., Foster, D. A., & Miller, E. L. (1991). Refrigeration of the western Cordilleran lithosphere during Laramide shallow-angle subduction. *Geology*, *19*(11), 1145–1148.

Elston, D. P., & Young, R. A. (1991). Cretaceous-Eocene (Laramide) landscape development and Oligocene-Pliocene drainage reorganization of transition zone and Colorado Plateau, Arizona. *Journal of Geophysical Research: Solid Earth*, *96*(B7), 12389–12406.

Etzel, T. M., & Catlos, E. J. (2021). Garnet chemical zoning based thermobarometry: method evaluation and applications in the Menderes Massif, Western Turkey. *Geosciences*, *11*(12), 505.

Etzel, T. M., Catlos, E. J., Ataktürk, K., Lovera, O. M., Kelly, E. D., Çemen, I., & Diniz, E. (2019). Implications for thrust-related shortening punctuated by extension from P-T paths and geochronology of garnet-bearing schists, Southern (Çine) Menderes Massif, SW Turkey. *Tectonics*, *38*(6), 1974–1998.

Evans, B., Renner, J., & Hirth, G. (2001). A few remarks on the kinetics of static grain growth in rocks. *International Journal of Earth Sciences*, *90*(1), 88–103.

Evans, S. L., Styron, R. H., van Soest, M. C., Hodges, K. V., & Hanson, A. D. (2015). Zircon and apatite (U-Th)/He evidence for Paleogene and Neogene extension in the southern Snake Range, Nevada, USA. *Tectonics*, *34*(10), 2142–2164.

Faulds, J. E., Feuerbach, D. L., Miller, C. F., & Smith, E. I. (2001). *Cenozoic evolution of the northern Colorado River extensional corridor, southern Nevada and northwest Arizona* (Publication GB 78, pp. 239–272). Pacific Section of the American Association of Petroleum Geologists.

Fayon, A. K., Peacock, S. M., Stump, E., & Reynolds, S. J. (2000). Fission track analysis of the footwall of the Catalina detachment fault, Arizona: tectonic denudation, magmatism, and erosion. *Journal of Geophysical Research: Solid Earth*, *105*(B5), 11047–11062.

Fedo, C. M., & Miller, J. M. (1992). Evolution of a Miocene half-graben basin, Colorado River extensional corridor, southeastern California. *Geological Society of America Bulletin*, *104*(4), 481–493.

Ferry, J. T., & Spear, F. S. (1978). Experimental calibration of the partitioning of Fe and Mg between biotite and garnet. *Contributions to Mineralogy and Petrology*, *66*(2), 113–117.

Fitzgerald, P. G., Reynolds, S. J., Stump, E., Foster, D. A., & Gleadow, A. J. W. (1993). Thermochronologic evidence for timing of denudation and rate of crustal extension of the south mountains metamorphic core complex and Sierra Estrella, Arizona. *Nuclear Tracks and Radiation Measurements*, *21*(4), 555–563. https://doi.org/10.1016/1359-0189(93)90196-G

Flansburg, M. E., & Stockli, D. F. (2023). Progressive Miocene unroofing of the Big Maria and Riverside Mountains (southeastern California, USA) along the southwestern margin of the Colorado River extensional corridor. *Geosphere*, *19*(3), 676–694.

Fletcher, I. R., McNaughton, N. J., Davis, W. J., & Rasmussen, B. (2010). Matrix effects and calibration limitations in ion probe U–Pb and Th–Pb dating of monazite. *Chemical Geology*, *270*(1–4), 31–44.

Florence, F. P., & Spear, F. S. (1991). Effects of diffusional modification of garnet growth zoning on PT path calculations. *Contributions to Mineralogy and Petrology*, *107*(4), 487–500.

Flueh, E. R., & Okaya, D. A. (1989). Near-vertical and intermediate offset seismic reflection data from west of the Whipple Mountains, SE California. *Journal of Geophysical Research: Solid Earth*, *94*(B1), 625–636.

Fornash, K. F., Patchett, P. J., Gehrels, G. E., & Spencer, J. E. (2013). Evolution of granitoids in the Catalina metamorphic core complex, southeastern Arizona: U–Pb, Nd, and Hf isotopic constraints. *Contributions to Mineralogy and Petrology*, *165*(6), 1295–1310.

Forshee, E. J., & Yin, A. (1995). Evolution of monolithological breccia deposits in supradetachment basins, Whipple Mountains, California. *Basin Research*, *7*(2), 181–197. https://doi.org/10.1111/j.1365-2117.1995.tb00103.x

Foster, D. A. (1989). *Mesozoic and Cenozoic thermal history of the eastern Mojave Desert, California and western Arizona, with emphasis on the Old Woman Mountains area and the Chemehuevi metamorphic core complex*. Albany, NY (United States): State University of New York.

Foster, D. A., Gleadow, A. J., Reynolds, S. J., & Fitzgerald, P. G. (1993). Denudation of metamorphic core complexes and the reconstruction of the transition zone, west central Arizona: constraints from apatite fission track thermochronology. *Journal of Geophysical Research: Solid Earth*, *98*(B2), 2167–2185.

Foster, D. A., Harrison, T. M., Miller, C. F., & Howard, K. A. (1990). The $^{40}Ar/^{39}Ar$ thermochronology of the eastern Mojave Desert, California, and adjacent western Arizona with implications for the evolution of metamorphic core complexes. *Journal of Geophysical Research: Solid Earth*, *95*(B12), 20005–20024. https://doi.org/10.1029/JB095iB12p20005

Foster, D. A., & John, B. E. (1999). Quantifying tectonic exhumation in an extensional orogen with thermochronology: examples from the southern Basin and Range Province. *Geological Society, London, Special Publications*, *154*(1), 343–364.

Gaidies, F., De Capitani, C., & Abart, R. (2008). THERIA_G: a software program to numerically model prograde garnet growth. *Contributions to Mineralogy and Petrology*, *155*, 657–671.

Ganguly, J. (2010). Cation diffusion kinetics in aluminosilicate garnets and geological applications. *Reviews in Mineralogy and Geochemistry*, *72*(1), 559–601.

Gans, P., Miller, E. L., McCarthy, J., & Ouldcott, M. L. (1985). Tertiary extensional faulting and evolving ductile-brittle transition zones in the northern Snake Range and vicinity: new insights from seismic data. *Geology*, *13*(3), 189–193.

Gans, P. B., & Gentry, B. J. (2016). Dike emplacement, footwall rotation, and the transition from magmatic to tectonic extension in the Whipple Mountains metamorphic core complex, southeastern California. *Tectonics*, *35*(11), 2564–2608.

Gébelin, A., Mulch, A., Teyssier, C., Heizler, M., Vennemann, T., & Seaton, N. C. (2011). Oligo-Miocene extensional tectonics and fluid flow across the Northern Snake Range detachment system, Nevada. *Tectonics*, *30*(5).

Gervais, F., Brown, R. L., & Crowley, J. L. (2010). Tectonic implications for a Cordilleran orogenic base in the Frenchman Cap dome, southeastern Canadian Cordillera. *Journal of Structural Geology*, *32*(7), 941–959.

Gessner, K., Ring, U., Johnson, C., Hetzel, R., Passchier, C. W., & Güngör, T. (2001). An active bivergent rolling-hinge detachment system: central Menderes metamorphic core complex in western Turkey. *Geology*, *29*(7), 611–614.

Gibbs, A. D. (1984). Structural evolution of extensional basin margins. *Journal of the Geological Society*, *141*(4), 609–620.

González-Becuar, E., Pérez-Segura, E., Vega-Granillo, R., Solari, L., González-León, C. M., Solé, J., & López Martínez, M. (2017). Laramide to Miocene syn-extensional plutonism in the Puerta del Sol area, central Sonora, Mexico. *Revista Mexicana de Ciencias Geológicas*, *34*(1), 45–61.

González-León, C. M., Solari, L., Solé, J., Ducea, M. N., Lawton, T. F., Bernal, J. P., et al. (2011). Stratigraphy, geochronology, and geochemistry of the Laramide magmatic arc in north-central Sonora. *Mexico. Geosphere*, *7*(6), 1392–1418.

Goodwin, L. B., & Haxel, G. B. (1990). Structural evolution of the Southern Baboquivari Mountains, south-central Arizona and north-central Sonora. *Tectonics*, *9*(5), 1077–1095.

Gordon, S. M., Whitney, D. L., Teyssier, C., Grove, M., & Dunlap, W. J. (2008). Timescales of migmatization, melt crystallization, and cooling in a Cordilleran gneiss dome: Valhalla complex, southeastern British Columbia. *Tectonics*, *27*(4).

Gottardi, R., McAleer, R., Casale, G., Borel, M., Iriondo, A., & Jepson, G. (2020). Exhumation of the Coyote Mountains metamorphic core complex (Arizona): implications for orogenic collapse of the southern North American Cordillera. *Tectonics*, *39*(8), e2019TC006050.

Gottardi, R., Schaper, M. C., Barnes, J. D., & Heizler, M. T. (2018). Fluid–rock interaction and strain localization in the Picacho Mountains detachment shear zone, Arizona, USA. *Tectonics*, *37*(9), 3244–3260.

Gottlieb, E. S. (2017). *Geologic insights from zircon inheritance* (Doctoral dissertation). Stanford University.

Grove, M., Jacobson, C. E., Barth, A. P., Vucic, A., Johnson, S. E., Paterson, S. R., et al. (2003). Temporal and spatial trends of Late Cretaceous-early Tertiary underplating of Pelona and related schist beneath southern California and southwestern Arizona. *Geological Society of America Special Papers*, *374*, 381–406.

Gueydan, F., Morency, C., & Brun, J. P. (2008). Continental rifting as a function of lithosphere mantle strength. *Tectonophysics*, *460*(1–4), 83–93.

Hacker, B. R., Yin, A., Christie, J. M., & Davis, G. A. (1992). Stress magnitude, strain rate, and rheology of extended middle continental crust inferred from quartz grain sizes in the Whipple Mountains, California. *Tectonics*, *11*(1), 36–46.

Hallett, B. W., & Spear, F. S. (2014). The P–T history of anatectic pelites of the Northern East Humboldt Range, Nevada: evidence for tectonic loading, decompression, and anatexis. *Journal of Petrology*, *55*(1), 3–36.

Hallett, B. W., & Spear, F. S. (2015). Monazite, zircon, and garnet growth in migmatitic pelites as a record of metamorphism and partial melting in the East Humboldt Range, Nevada. *American Mineralogist*, *100*(4), 951–972.

Hansen, V. L. & Goodge, J. W. (1988). *Metamorphism, structural petrology, and regional evolution of the Okanogan complex, northeastern Washington*. Paper presented at the Rubey Colloquium on Metamorphism and Crustal Evolution of the Western United States.

Harrison, T. M., Copeland, P., Kidd, W. S. F., & Lovera, O. M. (1995). Activation of the Nyainqentanghla shear zone: implications for uplift of the southern Tibetan Plateau. *Tectonics*, *14*(3), 658–676.

Hauser, E. C., & Lundy, J. (1989). COCORP deep reflections: Moho at 50 km (16 s) beneath the Colorado Plateau. *Journal of Geophysical Research: Solid Earth*, *94*(B6), 7071–7081.

Haxel, G. B., Jacobson, C. E., & Wittke, J. H. (2015). Mantle peridotite in newly discovered far-inland subduction complex, southwest Arizona: initial report. *International Geology Review*, *57*(5–8), 871–892.

Hayama, Y., Shibata, K., & Takeda, H. (1984). K-Ar ages of the low-grade metamorphic rocks in the Altar massif, northwest Sonora. *Mexico. Chishitsugaku Zasshi (Japan)*, *90*(8), 589–596.

Hempton, M. R. (1987). Constraints on Arabian plate motion and extensional history of the Red Sea. *Tectonics*, *6*(6), 687–705.

Henry, C. D., McGrew, A. J., Colgan, J. P., Snoke, A. W., & Brueseke, M. E. (2011). Timing, distribution, amount, and style of Cenozoic extension in the northern Great Basin. In J. Lee & J. P. Evans (Eds.), *Geologic field trips to the basin and range, Rocky Mountains, Snake River Plain, and Terranes of the U.S. Cordillera* (Vol. *21*, pp. 27–66). Geological Society of America.

Henyey, T. E., Okaya, D. A., Frost, E. G., & McEvilly, T. V. (1987). CALCRUST (1985) seismic reflection survey, Whipple Mountains detachment terrane, California: an overview. *Geophysical Journal International*, *89*(1), 111–118.

Hildebrand, R. S. (2013). *Mesozoic assembly of the North American cordillera* (GSA Special Papers, Vol. 495). Geological Society of America.

Hiraga, T., Tachibana, C., Ohashi, N., & Sano, S. (2010). Grain growth systematics for forsterite±enstatite aggregates: effect of lithology on grain size in the upper mantle. *Earth and Planetary Science Letters*, *291*(1–4), 10–20.

Hodges, K. V., Snoke, A. W., & Hurlow, H. A. (1992). Thermal evolution of a portion of the Sevier hinterland: the northern Ruby Mountains-East Humboldt range and Wood Hills, northeastern Nevada. *Tectonics*, *11*(1), 154–164.

Hoisch, T. D. (1989). A muscovite-biotite geothermometer. *American Mineralogist*, *74*(5–6), 565–572.

Hoisch, T. D. (1990). Empirical calibration of six geobarometers for the mineral assemblage quartz+ muscovite+ biotite+ plagioclase+ garnet. *Contributions to Mineralogy and Petrology*, *104*(2), 225–234.

Holk, G. J., & Taylor, H. P., Jr. (2000). Water as a petrologic catalyst driving 18O/16O homogenization and anatexis of the middle crust in the metamorphic core complexes of British Columbia. *International Geology Review*, *42*(2), 97–130.

Holland, E. C., Hively, W. P., DePinho, R. A., & Varmus, H. E. (1998). A constitutively active epidermal growth factor receptor cooperates with disruption of G1 cell-cycle arrest pathways to induce glioma-like lesions in mice. *Genes & Development*, *12*(23), 3675–3685.

Holland, T., & Powell, R. (2003). Activity–composition relations for phases in petrological calculations: an asymmetric multicomponent formulation. *Contributions to Mineralogy and Petrology*, *145*(4), 492–501.

Holland, T. J. B., & Powell, R. T. J. B. (1998). An internally consistent thermodynamic data set for phases of petrological interest. *Journal of Metamorphic Geology*, *16*(3), 309–343.

Howard, K. A., Goodge, J. W., & John, B. E. (1982). Detached crystalline rocks of the Mohave, Buck, and Bill Williams mountains, western Arizona. In E. G. Frost & D. L. Martin (Eds.), *Mesozoic-Cenozoic tectonic evolution of the Colorado River Region, California, Arizona and Nevada* (pp. 377–390). San Diego, Cordilleran.

Howard, K. A., & John, B. E. (1987). Crustal extension along a rooted system of imbricate low-angle faults: Colorado River extensional corridor, California and Arizona. *Geological Society, London, Special Publications*, *28*(1), 299–311.

Howard, K. A., Nielson, J. E., Wilshire, W. G., Nakata, J. K., Goodge, J. W., Reneau, S. L., et al. (1999). Geologic map of the Mohave Mountains area, Mohave County, western Arizona (Map I-2308, scale 1: 48,000). U.S. Geological Survey.

Howard, K. A., Wooden, J. L., Barnes, C. G., Premo, W. R., Snoke, A. W., & Lee, S. Y. (2011). Episodic growth of a Late Cretaceous and Paleogene intrusive complex of pegmatitic leucogranite, Ruby Mountains core complex, Nevada, USA. *Geosphere*, *7*(5), 1220–1248.

Huet, B., Le Pourhiet, L., Labrousse, L., Burov, E., & Jolivet, L. (2011). Post-orogenic extension and metamorphic core complexes in a heterogeneous crust: the role of crustal layering inherited from collision. Application to the Cyclades (Aegean domain). *Geophysical Journal International*, *184*(2), 611–625.

Isachsen, C. E., Gehrels, G. E., Riggs, N. R., Spencer, J. E., Ferguson, C. A., Skotnicki, S. J., & Richard, S. M. (1999). *U-Pb geochronologic data from zircons from eleven granitic rocks in central and western Arizona* (Arizona Geological Survey Open File Report No. OFR-99-05) (p. 27). Retrieved from http://hdl.handle.net/10150/630032

Jacobson, C. E., Hourigan, J. K., Haxel, G. B., & Grove, M. (2017). Extreme latest Cretaceous–Paleogene low-angle subduction: Zircon ages from Orocopia Schist at Cemetery Ridge, southwestern Arizona, USA. *Geology*, *45*(10), 951–954.

Jacobson, C. E., Jacques-Ayala, C., Barth, A. P., y Barragán, J. C. G., Pedrick, J. N., & Wooden, J. L. (2019). Protolith age of the Altar and Carnero complexes and Late Cretaceous–Miocene deformation in the Caborca–Altar region of northwestern Sonora, Mexico. *Revista Mexicana de Ciencias Geológicas*, *36*(1), 95–109. 10.22201/cgeo.20072902e.2019.1.784

Jacobson, C. E., Oyarzabal, F. R., & Haxel, G. B. (1996). Subduction and exhumation of the Pelona-Orocopia-Rand schists, southern California. *Geology*, *24*(6), 547–550.

John, B. E., & Wooden, J. (1990). Petrology and geochemistry of the metaluminous to peraluminous Chemehuevi Mountains Plutonic Suite, southeastern California. In J. L. Anderson (Ed.), *The nature and origin of Cordilleran magmatism* (Vol. *174*, pp. 71–98). Geological Society of America.

Kapp, P., Jepson, G., Carrapa, B., Schaen, A. J., He, J. J., & Wang, J. W. (2023). Laramide bulldozing of lithosphere beneath the Arizona transition zone, southwestern United States. *Geology*, *51*(10), 952–956.

Kapp, P., Taylor, M., Stockli, D., & Ding, L. (2008). Development of active low-angle normal fault systems during orogenic collapse: insight from Tibet. *Geology*, *36*(1), 7–10.

Kapp, P., Yin, A., Manning, C. E., Murphy, M., Harrison, T. M., Spurlin, M., et al. (2000). Blueschist-bearing metamorphic core complexes in the Qiangtang block reveal deep crustal structure of northern Tibet. *Geology*, *28*(1), 19–22.

King, E. M., Valley, J. W., Stockli, D. F., & Wright, J. E. (2004). Oxygen isotope trends of granitic magmatism in the Great Basin: location of the Precambrian craton boundary as reflected in zircons. *Geological Society of America Bulletin*, *116*(3–4), 451–462.

Kohlstedt, D. L., Evans, B., & Mackwell, S. J. (1995). Strength of the lithosphere: constraints imposed by laboratory experiments. *Journal of Geophysical Research: Solid Earth*, *100*(B9), 17587–17602.

Kohn, M. J. (1993). Uncertainties in differential thermodynamic (Gibbs' method) PT paths. *Contributions to Mineralogy and Petrology*, *113*(1), 24–39.

Kohn, M. J., Catlos, E. J., Ryerson, F. J., & Harrison, T. M. (2001). Pressure-temperature-time path discontinuity in the Main Central thrust zone, central Nepal. *Geology*, *29*(7), 571–574.

Kohn, M. J., & Spear, F. S. (1991). Error propagation for barometers; 1, accuracy and precision of experimentally located end-member reactions. *American Mineralogist*, *76*(1–2), 128–137.

Korchinski, M., Rey, P. F., Mondy, L., Teyssier, C., & Whitney, D. L. (2018). Numerical investigation of deep-crust behavior under lithospheric extension. *Tectonophysics*, *726*, 137–146.

Korchinski, M., Teyssier, C., Rey, P. F., Whitney, D. L., & Mondy, L. (2021). Single-phase vs two-phase rifting: numerical perspectives on the accommodation of extension during continental break-up. *Marine and Petroleum Geology*, *123*, 104715.

Kruckenberg, S. C., Whitney, D. L., Teyssier, C., Fanning, C. M., & Dunlap, W. J. (2008). Paleocene-Eocene migmatite crystallization, extension, and exhumation in the hinterland of the northern Cordillera: Okanogan dome, Washington, USA. *Geological Society of America Bulletin*, *120*(7–8), 912–929.

Kuniyoshi, S., & Freeman, T. (1974). Potassium-argon ages of Tertiary volcanic rocks from the eastern Mojave Desert. *Geological Society of America Abstracts with Programs*, *6*(3), 204.

Lanari, P., & Duesterhoeft, E. (2019). Modeling metamorphic rocks using equilibrium thermodynamics and internally consistent databases: past achievements, problems and perspectives. *Journal of Petrology*, *60*(1), 19–56.

Lanari, P., & Engi, M. (2017). Local bulk composition effects on metamorphic mineral assemblages. *Reviews in Mineralogy and Geochemistry*, *83*(1), 55–102.

Lavier, L. L., & Manatschal, G. (2006). A mechanism to thin the continental lithosphere at magma-poor margins. *Nature*, *440*(7082), 324–328.

Lee, D. E., & Christiansen, E. H. (1983). The granite problem as exposed in the southern Snake Range, Nevada. *Contributions to Mineralogy and Petrology*, *83*(1–2), 99–116.

Lee, J., Blackburn, T., & Johnston, S. (2017). Timing of mid-crustal ductile extension in the northern Snake Range metamorphic core complex, Nevada: evidence from U/Pb zircon ages. *Geosphere*, *13*(2), 439–459.

Lee, S. Y., Barnes, C. G., Snoke, A. W., Howard, K. A., & Frost, C. D. (2003). Petrogenesis of Mesozoic, peraluminous granites in the Lamoille Canyon area, Ruby Mountains, Nevada, USA. *Journal of Petrology*, *44*(4), 713–732.

Levy, D. A., Zuza, A. V., Haproff, P. J., & Odlum, M. L. (2021). Early Permian tectonic evolution of the Last Chance thrust system: an example of induced subduction initiation along a plate boundary transform. *GSA Bulletin*, *133*(5–6), 1105–1127.

Lister, G. S., Banga, G., & Feenstra, A. (1984). Metamorphic core complexes of Cordilleran type in the Cyclades, Aegean Sea, Greece. *Geology*, *12*(4), 221–225.

Lister, G. S., & Davis, G. A. (1989). The origin of metamorphic core complexes and detachment faults formed during Tertiary continental extension in the northern Colorado River region, USA. *Journal of Structural Geology*, *11*(1–2), 65–94.

Lister, G. S., Etheridge, M. A., & Symonds, P. A. (1986). Detachment faulting and the evolution of passive continental margins. *Geology*, *14*(3), 246–250.

Lister, G. S., Etheridge, M. A., & Symonds, P. A. (1991). Detachment models for the formation of passive continental margins. *Tectonics*, *10*(5), 1038–1064.

Livaccari, R. F., Burke, K., & Şengör, A. M. C. (1981). Was the Laramide orogeny related to subduction of an oceanic plateau? *Nature*, *289*(5795), 276–278.

Livaccari, R. F., & Perry, F. V. (1993). Isotopic evidence for preservation of Cordilleran lithospheric mantle during the Sevier-Laramide orogeny, western United States. *Geology*, *21*(8), 719–722.

Long, K. B., Gehrels, G. E., & Baldwin, S. L. (1995). Tectonothermal evolution of the Pinaleno-Jackson Mountain core complex, southeast Arizona. *Geological Society of America Bulletin*, *107*(10), 1231–1240.

Lutz, B. M., Ketcham, R. A., Axen, G. J., Beyene, M. A., Wells, M. L., van Wijk, J. W., et al. (2021). Thermo-kinematic modeling of detachment-dominated extension, northeastern Death Valley area, USA: implications for mid-crustal thermal-rheological evolution. *Tectonophysics*, *808*, 228755.

Mahar, E. M., Baker, J. M., Powell, R., Holland, T. J. B., & Howell, N. (1997). The effect of Mn on mineral stability in metapelites. *Journal of Metamorphic Geology*, *15*(2), 223–238.

Mallery, C., Barth, A., Roldan-Quintana, J., Haxel, G., Wooden, J., & Jacobson, C. (2018). *A geochemical model for the origin of the Pan Tak granite, the granite of Presumido Peak & the granite of Sierra San Juan during the laramide orogeny, Southern Arizona & Northern Sonora*. Paper presented at the GSA Annual Meeting in Indianapolis, Indiana.

Matthews, A. (1994). Oxygen isotope geothermometers for metamorphic rocks. *Journal of Metamorphic Geology*, *12*(3), 211–219.

McCarthy, J., Larkin, S. P., Fuis, G. S., Simpson, R. W., & Howard, K. A. (1991). Anatomy of a metamorphic core complex: seismic refraction/wide-angle reflection profiling in southeastern California and western Arizona. *Journal of Geophysical Research: Solid Earth*, *96*(B7), 12259–12291.

McGrew, A. J., & Snee, L. W. (1994). 40Ar/39Ar thermochronologic constraints on the tectonothermal evolution of the northern East Humboldt Range metamorphic core complex, Nevada. *Tectonophysics*, *238*(1–4), 425–450.

McNulty, B., & Farber, D. (2002). Active detachment faulting above the Peruvian flat slab. *Geology*, *30*(6), 567–570.

Melosh, H. J. (1990). Mechanical basis for low-angle normal faulting in the Basin and Range province. *Nature*, *343*(6256), 331–335.

Miller, C. F., & Bradfish, L. J. (1980). An inner Cordilleran belt of muscovite-bearing plutons. *Geology*, *8*(9), 412–416.

Miller, C. F., Wooden, J. L., Bennett, V. C., Wright, J. E., Solomon, G. C., & Hurst, R. W. (1990). Chapter 6: Petrogenesis of the composite peraluminous-metaluminous Old Woman-Piute Range batholith, southeastern California; Isotopic constraints. *Geological Society of America Memoirs* (Vol. *174*, pp. 99–110). Geological Society of America. https://doi.org/10.1130/MEM174-p99

Miller, C. F., & Wooden, J. L. (1994). Anatexis, hybridization and the modification of ancient crust: mesozoic plutonism in the Old Woman Mountains area, California, *Lithos*, *32*(1–2), 111–133. https://doi.org/10.1016/0024-4937(94)90025-6

Miller, E. L., & Gans, P. B. (1989). Cretaceous crustal structure and metamorphism in the hinterland of the Sevier thrust belt, western US Cordillera. *Geology*, *17*(1), 59–62.

Miller, E. L., Gans, P. B., & Garing, J. (1983). The Snake Range decollement: an exhumed mid-Tertiary ductile-brittle transition. *Tectonics*, *2*(3), 239–263.

Miller, J. M., & John, B. E. (1988). Detached strata in a Tertiary low-angle normal fault terrane, southeastern California: a sedimentary record of unroofing, breaching, and continued slip. *Geology*, *16*(7), 645–648.

Miller, J. M., & John, B. E. (1999). Sedimentation patterns support seismogenic low-angle normal faulting, southeastern California and western Arizona. *Geological Society of America Bulletin*, *111*(9), 1350–1370.

Molnar, P., & Tapponnier, P. (1975). Cenozoic Tectonics of Asia: effects of a Continental Collision: Features of recent continental tectonics in Asia can be interpreted as results of the India-Eurasia collision. *Science*, *189*(4201), 419–426.

Molnar, P., & Tapponnier, P. (1978). Active tectonics of Tibet. *Journal of Geophysical Research: Solid Earth*, *83*(B11), 5361–5375.

Morley, C. K. (1989). Extension, detachments, and sedimentation in continental rifts (with particular reference to East Africa). *Tectonics*, *8*(6), 1175–1192.

Morrison, J. (1994). Downward circulation of meteoric water into the lower plate of the Whipple Mountains metamorphic core complex, California. *Journal of Metamorphic Geology*, *12*, 827–840.

Morrison, J., & Anderson, J. L. (1998). Footwall refrigeration along a detachment fault: implications for the thermal evolution of core complexes. *Science*, *279*(5347), 63–66.

Moynihan, D. P., & Pattison, D. R. M. (2013). An automated method for the calculation of P–T paths from garnet zoning, with application to metapelitic schist from the Kootenay Arc, British Columbia, Canada. *Journal of Metamorphic Geology*, *31*(5), 525–548.

Murphy, M. A., Yin, A., Kapp, P., Harrison, T. M., Manning, C. E., Ryerson, F. J., et al. (2002). Structural evolution of the Gurla Mandhata detachment system, southwest Tibet: implications for the eastward extent of the Karakoram fault system. *Geological Society of America Bulletin*, *114*(4), 428–447.

Nielson, J. E., & Beratan, K. K. (1990). Tertiary basin development and tectonic implications, Whipple detachment system, Colorado River extensional corridor, California and Arizona. *Journal of Geophysical Research: Solid Earth*, *95*(B1), 599–614.

Nielson, J. E., & Beratan, K. K. (1995). Stratigraphic and structural synthesis of a Miocene extensional terrane, southeast California and west-central Arizona. *Geological Society of America Bulletin*, *107*(2), 241–252.

Nourse, J. A., Anderson, T. H., & Silver, L. T. (1994). Tertiary metamorphic core complexes in Sonora, northwestern Mexico. *Tectonics*, *13*(5), 1161–1182.

Pease, V., Foster, D., Wooden, J., O'Sullivan, P., Argent, J., & Fanning, C. (1999). The Northern Sacramento Mountains, southwest United States. Part II: exhumation history and detachment faulting. *Geological Society, London, Special Publications*, *164*(1), 199–238.

Peterman, E. M., Hourigan, J. K., & Grove, M. (2014). Experimental and geologic evaluation of monazite (U Th)/He thermochronometry: Catnip Sill, Catalina Core Complex, Tucson, AZ. *Earth and Planetary Science Letters*, *403*, 48–55.

Powell, R., & Holland, T. (1999). Relating formulations of the thermodynamics of mineral solid solutions; activity modeling of pyroxenes, amphiboles, and micas. *American Mineralogist*, *84*(1–2), 1–14.

Prior, M. G., Singleton, J. S., & Stockli, D. F. (2018). Late-stage slip history of the Buckskin-Rawhide detachment fault and temporal evolution of the Lincoln Ranch supradetachment basin: new constraints from the middle Miocene Sandtrap Conglomerate. *GSA Bulletin*, *130*(9–10), 1747–1760.

Prior, M. G., Stockli, D. F., & Singleton, J. S. (2016). Miocene slip history of the Eagle Eye detachment fault, Harquahala Mountains metamorphic core complex, west-central Arizona. *Tectonics*, *35*(8), 1913–1934.

Reston, T. J., Krawczyk, C. M., & Klaeschen, D. (1996). The S reflector west of Galicia (Spain): evidence from prestack depth migration for detachment faulting during continental breakup. *Journal of Geophysical Research: Solid Earth*, *101*(B4), 8075–8091.

Rey, P. F., Teyssier, C., Kruckenberg, S. C., & Whitney, D. L. (2011). Viscous collision in channel explains double domes in metamorphic core complexes. *Geology*, *39*(4), 387–390.

Rey, P. F., Teyssier, C., & Whitney, D. L. (2009). Extension rates, crustal melting, and core complex dynamics. *Geology*, *37*(5), 391–394.

Reynolds, S. J., & Spencer, J. E. (1985). Evidence for large-scale transport on the Bullard detachment fault, west-central Arizona. *Geology*, *13*(5), 353–356.

Richard, S. M., Fryxell, J. E., & Sutter, J. F. (1990). Tertiary structure and thermal history of the Harquahala and Buckskin Mountains, west central Arizona: implications for denudation by a major detachment fault system. *Journal of Geophysical Research: Solid Earth*, *95*(B12), 19973–19987.

Robinson, D., Bevins, R. E., Aguirre, L., & Vergara, M. (2004). A reappraisal of episodic burial metamorphism in the Andes of central Chile. *Contributions to Mineralogy and Petrology*, *146*(4), 513–528.

Roldán-Quintana, J. (1991). Geology and chemical composition of the Jaralito and Aconchi batholiths in east-central Sonora, México. In E. Perez-Segura & C. Jacques-Ayala (Eds.), *Studies of Sonoran geology* (Vol. *254*, pp. 69–80). Geological Society of America.

Salas, G. A. (1968). Areal geology and petrology of the igneous rocks of the Santa Ana region, Northwest Sonora. *Boletín de La Sociedad Geológica Mexicana*, *31*(1), 11–63.

Saleeby, J. (2003). Segmentation of the Laramide slab—evidence from the southern Sierra Nevada region. *Geological Society of America Bulletin*, *115*(6), 655–668.

Schaper, M. C. (2016). *Fluid-rock Interaction within the Picacho Mountains Detachment Shear Zone* (Doctoral dissertation). University of Louisiana at Lafayette.

Schenker, F. L., Gerya, T., & Burg, J. P. (2012). Bimodal behavior of extended continental lithosphere: modeling insight and application to thermal history of migmatitic core complexes. *Tectonophysics*, *579*, 88–103.

Scoggin, S. H. (2022). *Cenozoic tectonic evolution of the Pinaleño Mountains region, Southeastern Arizona* (Doctoral dissertation). University of Wyoming.

Selverstone, J., Axen, G. J., & Luther, A. (2012). Fault localization controlled by fluid infiltration into mylonites: formation and strength of low-angle normal faults in the midcrustal brittle-plastic transition. *Journal of Geophysical Research: Solid Earth*, *117*(B6).

Seymour, N. M., Strickland, E. D., Singleton, J. S., Stockli, D. F., & Wong, M. S. (2018). Laramide subduction and metamorphism of the Orocopia Schist, northern Plomosa Mountains, west-central Arizona: insights from zircon U-Pb geochronology. *Geology*, *46*(10), 847–850.

Seymour, N. M., Wong, M., & Singleton, J. S. (2016). *TitaniQ analyses of mylonites from metamorphic core complexes in west-central Arizona: implications for two phases of extension in the lower Colorado River Extensional Corridor*. Paper presented at the AGU Fall Meeting.

Shafiqullah, M., Damon, P. E., Lynch, D. J., Reynolds, S. J., Rehrig, W. A., & Raymond, R. H. (1980). K-Ar geochronology and geologic history of southwestern Arizona and adjacent areas. *Studies in Western Arizona: Arizona Geological Society Digest*, *12*, 201–260.

Silverberg, D. S. (1990). *The tectonic evolution of the Pioneer metamorphic core complex, south-central Idaho* (Doctoral dissertation). Massachusetts Institute of Technology.

Singleton, J. S., Seymour, N. M., & Strickland, E. D. (2022). The Buckskin-Rawhide and northern Plomosa Mountains metamorphic core complexes, west-central Arizona, USA. In G. Jiang & C. Dehler (Eds.), *Field excursions from Las Vegas, Nevada: guides to the 2022 GSA Cordilleran and Rocky Mountain Joint Section Meeting* (Vol. *63*, pp. 85–108). Geological Society of America.

Singleton, J. S., Stockli, D. F., Gans, P. B., & Prior, M. G. (2014). Timing, rate, and magnitude of slip on the Buckskin-Rawhide detachment fault, west central Arizona. *Tectonics, 33*(8), 1596–1615. https://doi.org/10.1002/2013TC003517

Sleep, N. H. (1971). Thermal effects of the formation of Atlantic continental margins by continental break up. *Geophysical Journal International, 24*(4), 325–350.

Smith, H. A., & Barreiro, B. (1990). Monazite U-Pb dating of staurolite grade metamorphism in pelitic schists. *Contributions to Mineralogy and Petrology, 105*(5), 602–615.

Spear, F. S. (1993). Metamorphic phase equilibria and pressure-temperature-time paths. *Mineralogical Society of America Monograph*, 352–356.

Spear, F. S. (2004). Fast cooling and exhumation of the Valhalla metamorphic core complex, southeastern British Columbia. *International Geology Review, 46*(3), 193–209.

Spear, F. S., & Parrish, R. R. (1996). Petrology and cooling rates of the Valhalla complex, British Columbia, Canada. *Journal of Petrology, 37*(4), 733–765.

Spear, F. S., & Peacock, S. M. (1989). *Metamorphic pressure-temperature-time paths* (Vol. *7*). American Geophysical Union.

Spencer, J. E. (2006). *Application of critical-taper theory to Oligo-Miocene extension and core-complex exhumation in Western Arizona, USA*. Paper presented at the AGU Fall Meeting.

Spencer, J. E. Isachsen, C. E., Ferguson, C. A., Richard, S. M., Skotnicki, S. J., Wooden, J., & Riggs, N. R. (2003). *U-Pb isotope geochronologic data from 23 igneous rock units in central and southeastern Arizona* (Open File Report, OFR-03-08). Arizona Geological Survey.

Spencer, J. E., & Reynolds, S. J. (1990). Relationship between Mesozoic and Cenozoic tectonic features in west central Arizona and adjacent southeastern California. *Journal of Geophysical Research: Solid Earth, 95*(B1), 539–555.

Spencer, J. E., Singleton, J. S., Strickland, E., Reynolds, S. J., Love, D., Foster, D. A., & Johnson, R. (2018). Geodynamics of Cenozoic extension along a transect across the Colorado River extensional corridor, southwestern USA. *Lithosphere, 10*(6), 743–759.

Stern, R. A., & Berman, R. G. (2001). Monazite U–Pb and Th–Pb geochronology by ion microprobe, with an application to in situ dating of an Archean metasedimentary rock. *Chemical Geology, 172*(1–2), 113–130.

Stockli, D. F., Brichau, S., Dewane, T. J., Hager, C., & Schroeder, J. (2006). Dynamics of large-magnitude extension in the Whipple Mountains metamorphic core complex. *Geochimica et Cosmochimica Acta. Supplement., 70*(18), A616–A616.

Stone, P., Howard, K. A., & Hamilton, W. (1983). Correlation of metamorphosed Paleozoic strata of the southeastern Mojave Desert region, California and Arizona. *Geological Society of America Bulletin, 94*, 1135–1147.

Strickland, E. D., Singleton, J. S., & Haxel, G. B. (2018). Orocopia Schist in the northern Plomosa Mountains, west-central Arizona: a Laramide subduction complex exhumed in a Miocene metamorphic core complex. *Lithosphere, 10*(6), 723–742.

Suneson, N. H., & Lucchitta, I. (1983). Origin of bimodal volcanism, southern Basin and Range province, west-central Arizona. *Geological Society of America Bulletin, 94*(8), 1005–1019.

Tasaka, M., Zimmerman, M. E., & Kohlstedt, D. L. (2017). Rheological weakening of olivine+ orthopyroxene aggregates due to phase mixing: 1. Mechanical behavior. *Journal of Geophysical Research: Solid Earth, 122*(10), 7584–7596.

Terrien, J. J. (2012). *The role of magmatism in the Catalina metamorphic core complex, Arizona: insights from integrated thermochronology, gravity and aeromagnetic data* (Doctoral dissertation). Syracuse University.

Tirel, C., Brun, J. P., & Burov, E. (2008). Dynamics and structural development of metamorphic core complexes. *Journal of Geophysical Research: Solid Earth, 113*(B4), B04403.

Tracy, R. J., Robinson, P., & Thompson, A. B. (1976). Garnet composition and zoning in the determination of temperature and pressure of metamorphism, central Massachusetts. *American Mineralogist, 61*(7–8), 762–775.

Vogl, J. J., Foster, D. A., Fanning, C. M., Kent, K. A., Rodgers, D. W., & Diedesch, T. (2012). Timing of extension in the Pioneer metamorphic core complex with implications for the spatial-temporal pattern of Cenozoic extension and exhumation in the northern US Cordillera. *Tectonics, 31*(1), 31.

Wells, M. L., Spell, T. L., & Grove, M. (2002). *Late Cretaceous intrusion and extensional exhumation of the Cadiz Valley batholith, Iron Mountains, southeastern California* (Vol. 34, p. 178). Presented at the Geological Society of America, Denver, Colorado. Retrieved from https://gsa.confex.com/gsa/2002AM/webprogram/Paper40694.html

Wells, M. L., & Hoisch, T. D. (2008). The role of mantle delamination in widespread Late Cretaceous extension and magmatism in the Cordilleran orogen, western United States. *Geological Society of America Bulletin, 120*(5–6), 515–530.

Wells, M. L., Hoisch, T. D., Cruz-Uribe, A. M., & Vervoort, J. D. (2012). Geodynamics of synconvergent extension and tectonic mode switching: Constraints from the Sevier-Laramide orogen. *Tectonics, 31*(1).

Wernicke, B. (1985). Uniform-sense normal simple shear of the continental lithosphere. *Canadian Journal of Earth Sciences, 22*(1), 108–125.

Wernicke, B. (1995). Low-angle normal faults and seismicity: a review. *Journal of Geophysical Research: Solid Earth, 100*(B10), 20159–20174.

Wernicke, B., & Axen, G. J. (1988). On the role of isostasy in the evolution of normal fault systems. *Geology, 16*(9), 848–851.

Wernicke, B. P., England, P. C., Sonder, L. J., & Christiansen, R. L. (1987). Tectonomagmatic evolution of Cenozoic extension in the North American Cordillera. *Geological Society, London, Special Publications, 28*(1), 203–221.

Westaway, R. (1999). The mechanical feasibility of low-angle normal faulting. *Tectonophysics, 308*(4), 407–443.

White, R., & McKenzie, D. (1989). Magmatism at rift zones: the generation of volcanic continental margins and flood basalts. *Journal of Geophysical Research: Solid Earth*, *94*(B6), 7685–7729.

White, R. W., Powell, R., Holland, T. J. B., & Worley, B. A. (2000). The effect of TiO2 and Fe2O3 on metapelitic assemblages at greenschist and amphibolite facies conditions: mineral equilibria calculations in the system K2O-FeO-MgO-Al2O3-SiO2-H2O-TiO2-Fe2O3. *Journal of Metamorphic Geology*, *18*(5), 497–511.

White, R. W., Powell, R., & Johnson, T. E. (2014). The effect of Mn on mineral stability in metapelites revisited: new a–x relations for manganese-bearing minerals. *Journal of Metamorphic Geology*, *32*(8), 809–828.

Whitney, D. L., Teyssier, C., Rey, P., & Buck, W. R. (2013). Continental and oceanic core complexes. *Bulletin*, *125*(3–4), 273–298.

Wilson, J. T. (1966). Did the Atlantic close and then re-open? *Nature*, *211*, 676–681.

Wong, M. S., & Gans, P. B. (2008). Geologic, structural, and thermochronologic constraints on the tectonic evolution of the Sierra Mazatán core complex, Sonora, Mexico: new insights into metamorphic core complex formation. *Tectonics*, *27*(4), TC4013.

Wong, M. S., Gans, P. B., & Scheier, J. (2010). The 40Ar/39Ar thermochronology of core complexes and other basement rocks in Sonora, Mexico: implications for Cenozoic tectonic evolution of northwestern Mexico. *Journal of Geophysical Research: Solid Earth*, *115*(B7), B07414.

Wong, M. S., O'Brien, H. P., Bunting, K. C., & Gans, P. B (2011). 40Ar/39Ar K-feldspar thermochronology of the Harcuvar core complex, western Arizona: new insight into the timing of extension and degree of footwall tilt. *Geological Society of America Abstracts with Programs*, *43*, 53.

Wong, M. S., Singleton, J. S., Seymour, N. M., Gans, P. B., & Wrobel, A. J. (2023). Late Cretaceous-Early Paleogene extensional ancestry of the Harcuvar and Buckskin-Rawhide metamorphic core complexes, western Arizona. *Tectonics*, *42*(2), e2022TC007656.

Worthington, J. R., Ratschbacher, L., Stübner, K., Khan, J., Malz, N., Schneider, S., et al. (2020). The Alichur dome, South Pamir, western India–Asia collisional zone: detailing the Neogene Shakhdara–Alichur syn-collisional gneiss-dome complex and connection to lithospheric processes. *Tectonics*, *39*(1), e2019TC005735.

Wright, J. E., & Haxel, G. (1982). A garnet-two-mica granite, Coyote Mountains, southern Arizona: geologic setting, uranium-lead isotopic systematics of zircon, and nature of the granite source region. *Geological Society of America Bulletin*, *93*(11), 1176–1188.

Wu, G., Lavier, L. L., & Choi, E. (2015). Modes of continental extension in a crustal wedge. *Earth and Planetary Science Letters*, *421*, 89–97.

Yin, A. (1989). Origin of regional, rooted low-angle normal faults: a mechanical model and its tectonic implications. *Tectonics*, *8*(3), 469–482.

Yin, A., & Dunn, J. F. (1992). Structural and stratigraphic development of the Whipple-Chemehuevi detachment fault system, southeastern California: implications for the geometrical evolution of domal and basinal low-angle normal faults. *Geological Society of America Bulletin*, *104*(6), 659–674.

Yin, A., Rumelhart, P. E., Butler, R., Cowgill, E., Harrison, T. M., Foster, D. A., et al. (2002). Tectonic history of the Altyn Tagh fault system in northern Tibet inferred from Cenozoic sedimentation. *Geological Society of America Bulletin*, *114*(10), 1257–1295.

Young, R. A. (1982). Paleogeomorphic evidence for the structural history of the Colorado Plateau margin in western Arizona. In E. G. Frost & D. L. Martin (Eds.), *Mesozoic–Cenozoic tectonic evolution of the Colorado River region, California, Arizona, and Nevada* (pp. 29–39). San Diego: Cordilleran.

Zeh, A., & Holness, M. B. (2003). The effect of reaction overstep on garnet microtextures in metapelitic rocks of the Ilesha Schist Belt, SW Nigeria. *Journal of Petrology*, *44*(6), 967–994.

Zuza, A. V., Henry, C. D., Dee, S., Thorman, C. H., & Heizler, M. T. (2021). Jurassic–Cenozoic tectonics of the Pequop Mountains, NE Nevada, in the North American Cordillera hinterland. *Geosphere*, *17*(6), 2078–2122.

11

Büyük Menderes Graben in Western Turkey: An Example of a Rift Basin Developed in Back-Arc Extension

Osman Merey[1], İbrahim Çemen[1], and James Granath[2]

ABSTRACT

During this study, eight N-S and three E-W trending seismic reflection profiles in the Büyük Menderes Graben in Western Turkey were interpreted to determine the subsurface structural geometry. Eight N-S structural cross-sections were restored to determine the original geometry of the normal faults during their initiation in the early Miocene and their structural evolution throughout the Neogene sedimentation in the Graben. The cross-section restoration based on our structural interpretation of the seismic reflection profiles reveals that a rollover geometry did not develop on the hanging wall of the south-dipping normal fault along the northern margin of the Graben. Furthermore, the kinematic modeling created by backstripping the cross sections and tectonic subsidence rates suggest that the evolution of the Büyük Menderes Graben was controlled by two oppositely dipping normal faults on its northern and southern margins. Therefore, we suggest that the Büyük Menderes Graben formed as a rift basin during the early Miocene back-arc extension, which developed due to the subduction rollback along the Hellenic subduction zone. The Graben continued to receive sediments as a rift basin from early Miocene to Present.

11.1. INTRODUCTION

Extensional tectonics in the western Turkey region is responsible for the exhumation of the Menderes Metamorphic Core Complex (MMCC) and the development of multiple grabens. The MMCC is divided into northern, central, and southern sections by two major east-west trending grabens: the Büyük Menderes Graben (BMG) to the south and the Alaşehir Graben (AG) to the north (Figure 11.1). These E-W trending grabens are morphologically prominent features and have been considered symmetrical supradetachment basins that are filled with Neogene sediments (Seyitoğlu et al., 2000, 2002, 2004; Gessner et al., 2001, 2013; Çemen et al., 2006; Çiftçi & Bozkurt, 2009). The grabens were considered to develop under the control of two opposite dipping detachment surfaces, which were initially formed as high-angle normal faults and rotated to low-angle detachment faults (Gessner et al., 2001, 2013; Seyitoğlu et al., 2000, 2002, 2004; Çemen et al., 2006).

Field-oriented geological research, together with subsurface studies based on structural interpretation of 2D seismic reflection profiles and well-log data, suggest that the AG (Figures 11.2 and 11.3) contains a well-developed fault-bend rollover structure related to the north-dipping Alaşehir detachment fault (Seyitoğlu et al., 2000, 2002; Çemen et al., 2006; Çiftçi & Bozkurt, 2009; Gessner et al., 2013). A similar geometry has been proposed for the BMG based on surface geological studies (Seyitoğlu et al., 2000, 2002; Göğüş, 2004; Çemen et al., 2006; Gessner et al., 2013), which suggests that the graben

[1] *Department of Geological Sciences, University of Alabama, Tuscaloosa, Alabama, USA*

[2] *Granath and Associates Consulting Geology, Highland Ranch, Colorado, USA*

Extensional Tectonics: Rifting and Continental Extension, Geophysical Monograph 278, First Edition.
Edited by İbrahim Çemen and Elizabeth J. Catlos.

DOI: 10.1002/9781119773757.ch11

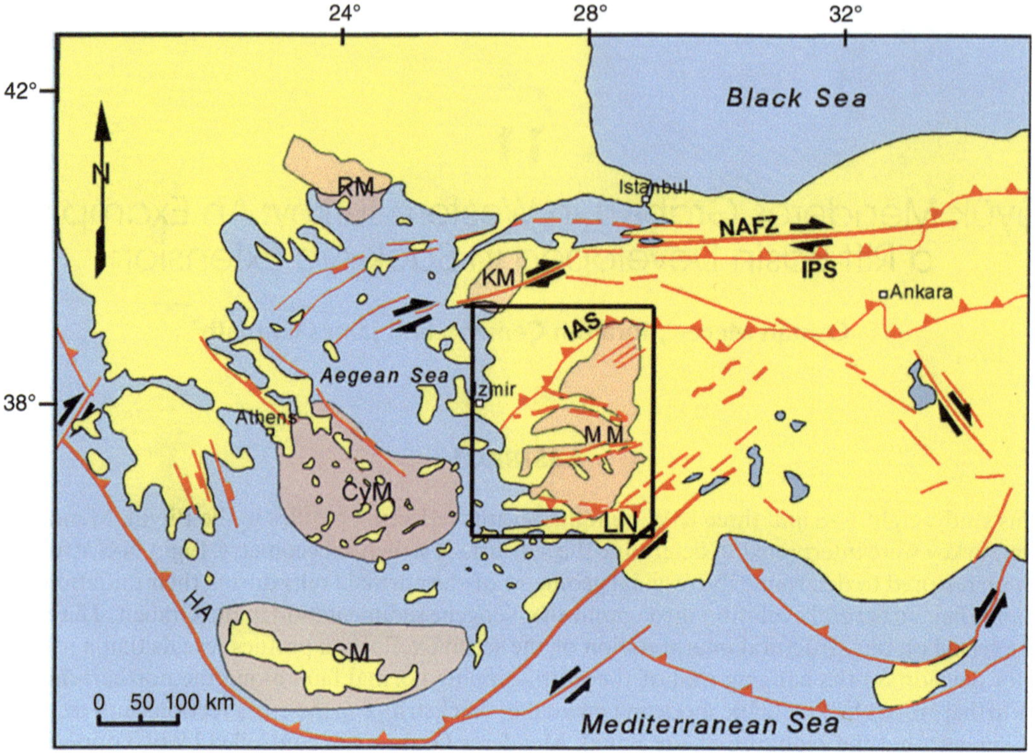

Figure 11.1 Generalized tectonic map of the Aegean extensional province. The rectangle shows the area covered in Figure 11.2. AG, Alaşehir Graben; BMG, Büyük Menderes Graben; CM, Crete Massif; CyM, Cylades Massif; HA, Hellenic Arc; IAS, Izmir-Ankara Suture; IPS, Intra Pontid Suture; KMM, Kazdag Massif; NAFZ, North Anatolian Fault Zone; RMM, Rhodop Massif; SWASZ, Southwest Anatolian Shear Zone (from Çemen et al., 2006).

evolved along the hanging wall of the south-dipping low-angle Büyük Menderes detachment fault located on the northern margin of the graben (Figures 11.2, 11.3, and 11.4). However, the subsurface geology of the BMG, including its stratigraphy and structural geology, remained in question because previous studies were based on surface geological mapping.

During this study, we examined eight N-S and three E-W seismic reflection profiles (Figure 11.4) provided by the Turkish Petroleum Corporation (TPAO) for Osman Merey's MS thesis project. We also investigated the kinematic evolution of the BMG based on structural interpretation of 2D seismic profiles, well-log data, and a structural restoration constructed using 2D Move software available for educational use in the UA Department of Geological Sciences.

Our interpretation of the available subsurface data and kinematic modeling of Cenozoic extensional faults in the BMG provided a much-needed subsurface geological understanding of the BMG (Figure 11.4). We determined the geometry of the normal faults that control the structural evolution of the BMG. We were mainly interested in whether the graben was developed on the hanging wall of the south-dipping detachment surface and developed rollover structures on its hanging wall similar to the AG to the north (Figures 11.3 and 11.4). Our subsurface structural interpretations provided (a) a new structural model for the kinematic evolution of the BMG and (b) a revised tectonics model for the Cenozoic extensional tectonics in western Turkey. They provided a better understanding of extensional tectonics of the central part of the MMCC within the Aegean back-arc extensional tectonics setting.

11.2. GEOLOGIC OVERVIEW

The western Turkey extended terrane is located in the eastern part of the Aegean extensional province (Figure 11.1). The tectonics of the region are dominantly constrained by the continental collision of the African and Eurasian plates, the Hellenic-Cyprian subduction zone, and the right-lateral strike-slip movement along the North Anatolian Fault zone (NAFZ) (Le Pichon & Angelier, 1979, 1981; Şengör et al., 1985; Meulenkamp et al., 1988; Seyítoğlu & Scott, 1996; Çemen et al., 2006) (Figure 11.1). The extension is responsible for the exhumation of Menderes Massif and associated major extensional structural features, which from north to South are the Simav Graben, Alaşehir Graben, Küçük, Büyük Menderes Graben and Gulf of Gökova Graben (Figure 11.2).

Since the 1970s, many researchers have proposed explanations for the cause of the initiation of Cenozoic

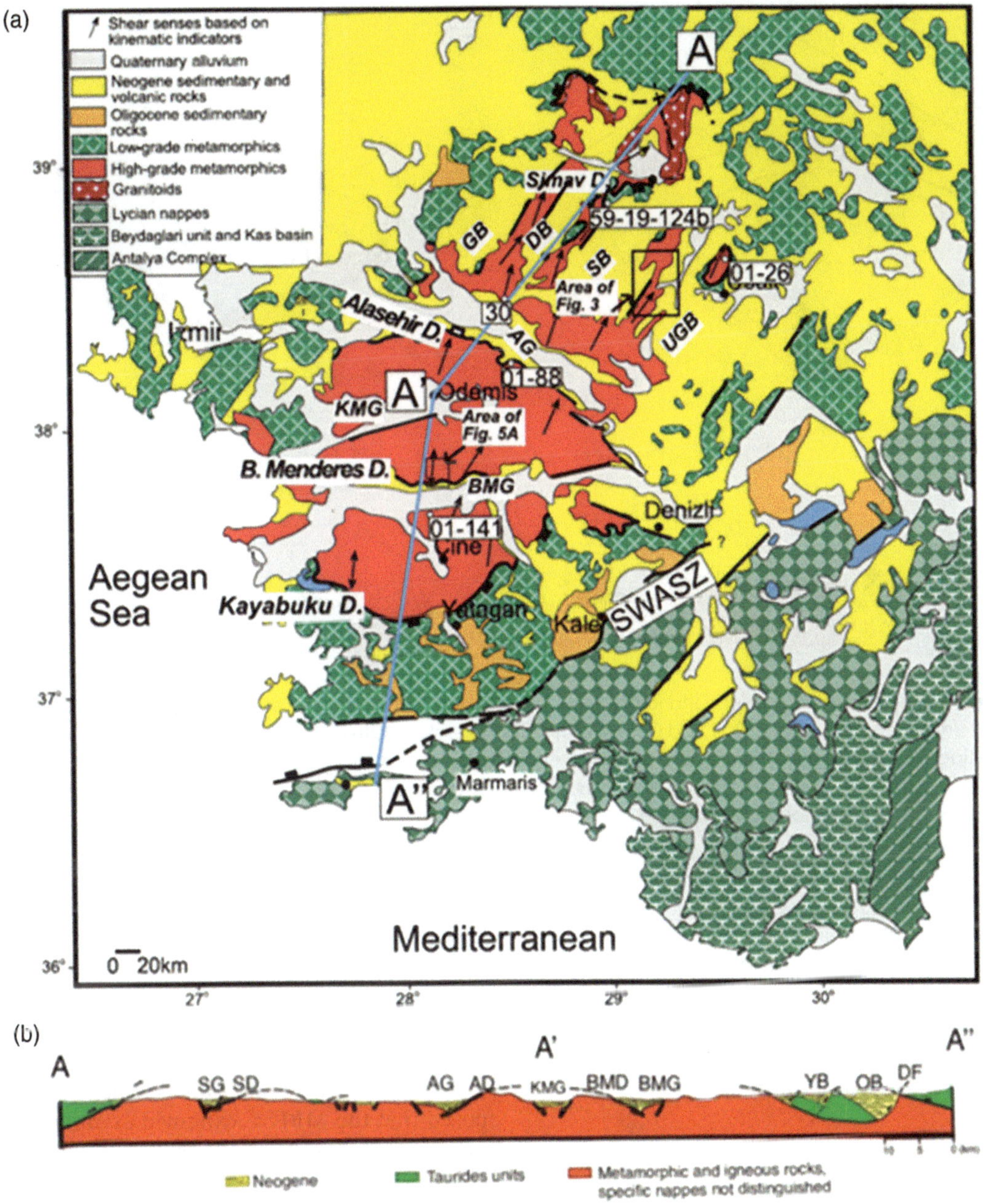

Figure 11.2 (a) Generalized geological map of western Turkey. (b) Geological cross-section along the line A-A′ (Çemen et al., 2006 / with permission of Geological Society of America); Alasehir D, Alaşehir Detachment; B. Menderes D, Büyük Menderes Detachment; Kayabuku D, Kayabuku Detachment; Simav D, Simav Detachment; SWASZ, Southwest Anatolian Shear Zone.

extension in the region. These include (a) tectonic escape of the Anatolian microplate along the NAFZ and associated lateral extrusion (Dewey & Şengör, 1979; Şengör, 1979; Şengör & Yılmaz, 1981; Şengör et al., 1985; Çemen et al., 1993, 1999, Sumer et al., 2020); (b) back-arc extension in response to subduction roll-back of the African Plate and migration of the Hellenic-Cyprian Trench system (McKenzie 1978; Le Pichon & Angelier, 1979, 1981; Meulenkamp et al., 1988; Spakman et al., 1988; Jolivet et al., 2013; Göğüş, 2014); (c) orogenic collapse of spreading and thinning of thickened crust as a result of Paleocene–Eocene Alpine collision (Dewey, 1988; Seyítoğlu & Scott, 1996); and (d) a combination of the three mechanisms proposing that extension in western Turkey was initiated in the late Oligocene as a result of orogenic collapse and progressively developed in three stages following the initiation of a major extensional shear zone named the Southwest Anatolian Shear Zone (SWASZ) (Çemen et al., 2006; Gessner et al., 2013; Ersoy et al., 2014). Since the

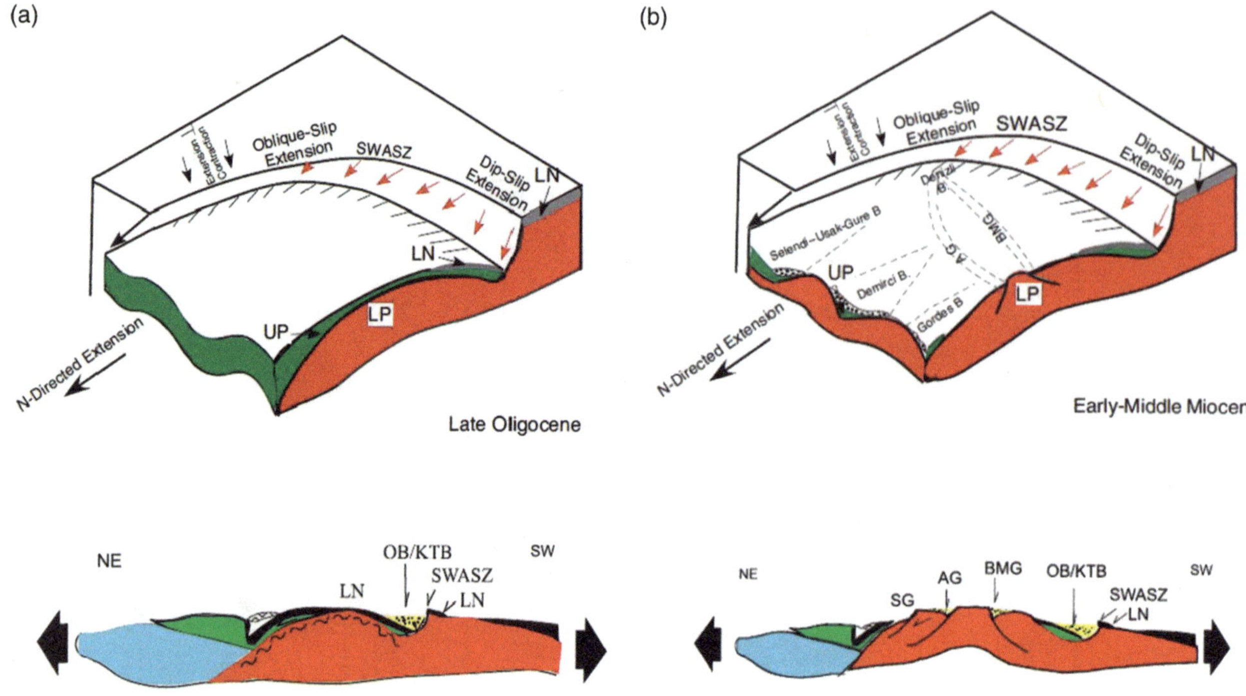

Figure 11.3 3D tectonic models and cross-sections showing the Cenozoic extensional evolution of western Turkey (from Seyitoğlu et al., 2004 and Çemen et al., 2006). (a) Late Oligocene extension on SWASZ. (b) Early- to Mid-Miocene extension that causes the development of the E-W and N-S trending grabens. AG, Alaşehir graben; BMG, Büyük Menderes graben; KTB, Kale-Tavas Basin; LN, Lycian Nappes; ÖB, Ören Basin; SG, Simav Graben; SWASZ, Southwest Anatolian Shear Zone.

recognition of metamorphic core complexes in the Basin and Range extensional province of western North America, the Menderes Massif has been interpreted as a metamorphic core complex (Bozkurt & Park, 1994; Seyítoğlu & Scott, 1996; Işik & Tekeli, 2001; Catlos & Çemen, 2005; Çemen et al., 2006).

11.2.1. Exhumation of the Menderes Metamorphic Core Complex

The Cenozoic extension in western Turkey is characterized by the exhumation of the MMCC and its extensional deformation. The Izmir-Ankara suture zone borders the MMCC to the north and the NE-SW trending SWASZ to the south and southeast (Çemen et al., 2006). The MMCC is divided into northern, central, and southern parts by E-W trending horst and graben system (Figures 11.1, 11.2, and 11.3). The exhumation mechanism of MMCC has been discussed in several papers in terms of symmetrical/asymmetrical metamorphic core complex development.

Gessner et al. (2001, 2013) and Ring et al. (2003) suggested that the MMCC is a symmetrical core complex whose exhumation initiated in the Oligocene between the north-dipping Simav detachment and south-dipping Lycian detachment surfaces, followed by extension of the central part of the MMCC along two symmetrical breakaway zones that were initially high angle: the Büyük Menderes detachment to the south and the Alaşehir detachment to the north.

Seyitoğlu et al. (2004) and Çemen et al. (2006) proposed that the MMCC exhumed asymmetrically along a main simple-shear breakaway zone that extends from the Datça fault located along the southern margin of the Gulf of Gökova and continuing northeastward to the Kale/Tavas and Acigöl basins (Figures 11.2 and 11.3). Çemen et al. (2006) observed that the breakaway zone extends northeastward and constitutes the southern boundary between the western Turkey extended terrane and the western Taurids. They named this structure the SWASZ, which contains the Datça, Kale-Tavas, and Acıgöl fault segments (Figures 11.2 and 11.3). Çemen et al. (2006) proposed that the SWASZ is a large listric detachment surface and was initiated as a primary breakaway in Southwestern Turkey. They suggested that ongoing extension along the SWASZ and associated tectonic uplift caused the exhumation of the MMCC and development of the symmetrical Alaşehir and Büyük Menderes grabens on the central part of the MMCC in the early Miocene (Figure 11.3).

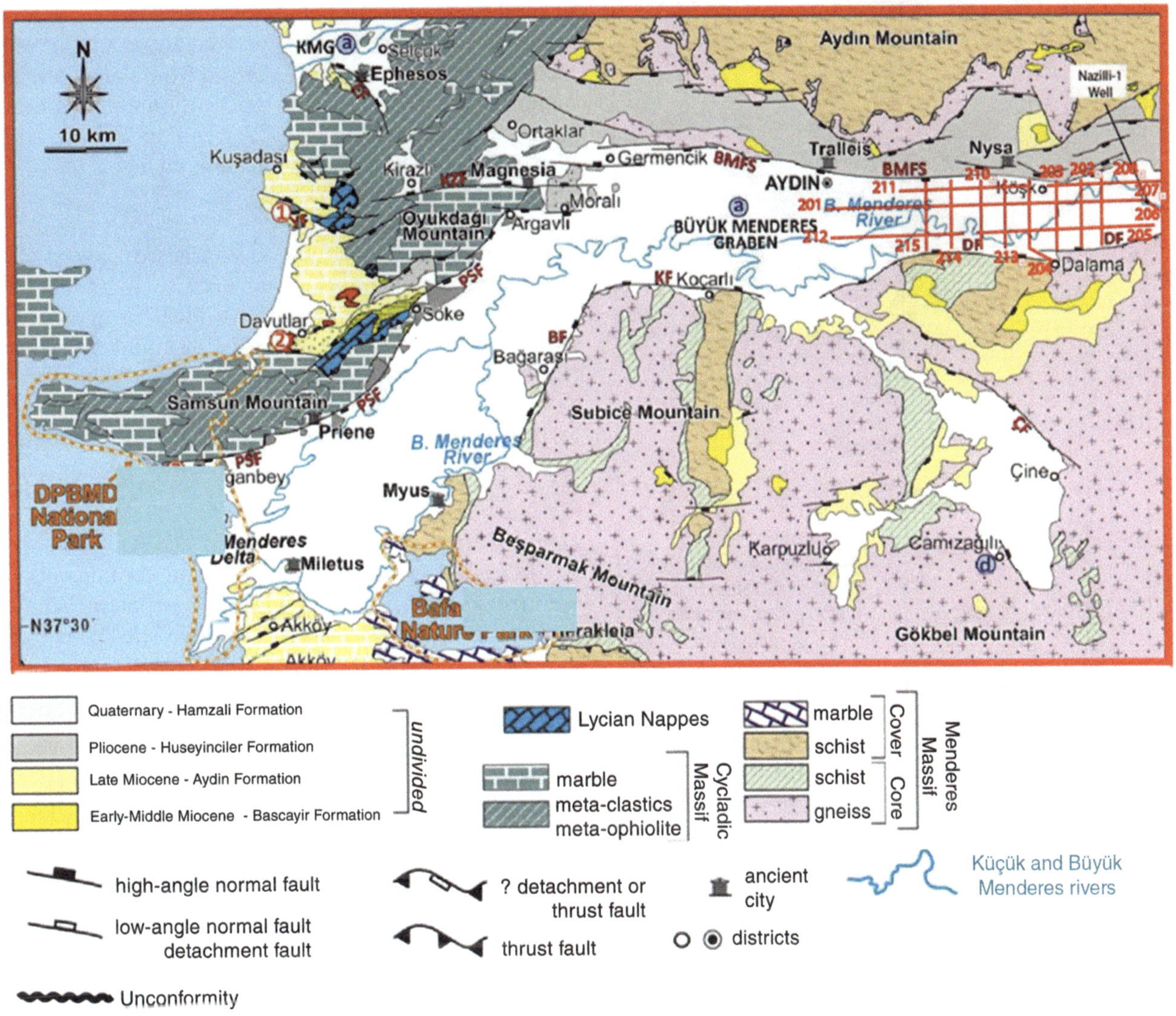

Figure 11.4 Generalized geological map of the BMG and surrounding areas in western Turkey (Inaner et al., 2009 / with permission of Springer Nature) showing the approximate location of the eight N-S and three E-W seismic reflection profiles and the Nazilli-1 well used for subsurface structural interpretations during this study.

11.3. STRATIGRAPHY AND STRUCTURAL OVERVIEW OF THE BMG

Sedimentary rock units of the BMG are either adjacent to the well-developed normal faults along the graben or are disconformably overlying the Pre-Neogene metamorphic basement rocks (Cohen et al., 1995; Emre & Sözbilir, 1997; Bozkurt 2000; Göğüş, 2004; Çemen et al., 2006; Şen & Seyitoğlu, 2009; Kazancı et al., 2009). Figure 11.5 is a simplified stratigraphical columnar section of the BMG area.

11.3.1. Pre-Neogene Metamorphic Basement Rocks

The Pre-Neogene metamorphic basement rocks of the graben consist of metamorphic rocks of MMCC, which was exhumed as a result of extensional tectonics in the region (Gessner et al., 2001, 2013; Seyitoğlu et al., 2002; Çemen et al., 2006; Ersoy et al., 2014). This investigation is concerned with the Neogene sedimentary units of the BMG. Therefore, the reader is referred to other publications for detailed descriptions of the metamorphic basement rocks in the graben (e.g., Şengör et al., 1984; Konak et al., 1987; Candan et al., 2011).

Neogene Sedimentary Rocks

Sedimentary rocks of the BMG are mainly composed of lacustrine, fluvial Neogene sediments, and Quaternary alluvium (Sözbilir & Emre, 1991; Göğüş, 2004; Çemen et al., 2006; Şen & Seyitoğlu, 2009). Neogene sedimentary successions that overlie the metamorphic basement consist of the early to middle Miocene Bascayir, late Miocene

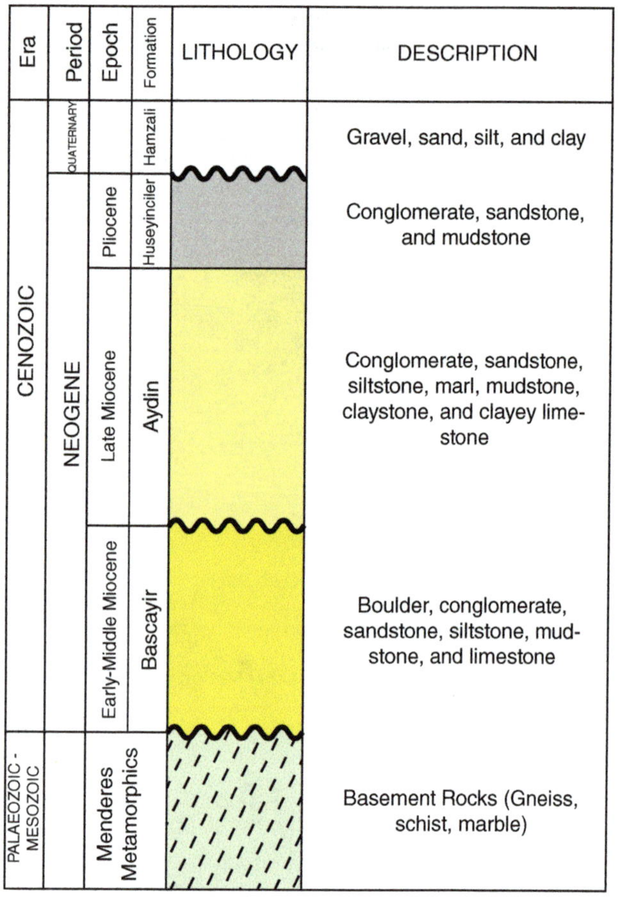

Figure 11.5 Generalized stratigraphic column of the Büyük Menderes Graben area. Modified form Yılmaz et al., (2000), Göğüş (2003), Kazancı et al. (2009), and Sert (2015).

Aydin, and Pliocene Huseyinciler formations. These units are overlain by the Quaternary Hamzali Formation (Kazancı et al., 2009; see Figure 11.5).

Bascayir Formation

The Bascayir Formation represents the oldest sedimentary rocks (early-middle Miocene) related to extensional tectonics and overlies the metamorphic basement in the graben with a major disconformity (Göğüş, 2004; Çemen et al., 2006). The basal sedimentary unit of the formation is composed of coarse-grained conglomerate, which is overlain by a shale-dominated sequence filled with alluvial fan delta and lacustrine deposits (Sözbilir & Emre, 1991; Cohen et al., 1995). The presence of cross-bedding and the growth faulting in this formation suggest that the deposition of the Bascayir Formation is syn-extensional (Göğüş, 2004; Çifçi et al., 2010).

Aydin Formation

The late Miocene Aydin Formation is composed of coarse-grained conglomerate with interbedded layers of sandstone, mudstone, and claystone (Sözbilir & Emre 1991; Bozkurt 2000; Şen & Seyitoğlu, 2009). Clastic sediments in the formation were derived from the underlying metamorphic basement and the early-middle Miocene Bascayir Formation. The E-W trending growth faults also control the formation's thickness change.

Huseyinciler Formation

The Huseyinciler Formation consists of lateral alluvial clastic sediments that are massive to moderately bedded, poorly compacted boulder conglomerates with interbedded layers of sandstones and mudstones (Emre & Sözbilir 1997; Çemen et al., 2006; Şen & Seyitoğlu, 2009; Çifçi et al., 2010). This late Pliocene age formation unconformably overlies the Aydin Formation (Ünay et al., 1995).

Hamzali Formation

The Quaternary Hamzali Formation contains clastic sediments and alluvium deposited from lateral alluvial fans that are sourced from N-S directed streams perpendicular to the axis of the Büyük Menderes River (Kazancı et al., 2009; Çifçi et al., 2010).

11.4. STRUCTURAL GEOLOGY

The BMG is the largest E-W trending graben in western Turkey (Figures 11.1 and 11.2) and is about 200 km long and 8–12 km wide. It has been characterized by an extensional fault system that controls the graben architecture and sedimentary deposition (Yılmaz et al., 2000; Göğüş, 2004; Çemen et al., 2006; Şen & Seyitoğlu, 2009). The south-dipping low-angle fault on the northern flank of the graben (Figures 11.2, 11.3, and 11.4) was interpreted as a detachment fault that separates the Neogene sedimentary rocks from the exhumed metamorphic basement rocks (Seyitoğlu et al., 2000; Gessner et al., 2001, 2013; Göğüş, 2004; Çemen et al., 2006).

The BMG contains a north-dipping normal fault along its southern boundary where Neogene sedimentary rocks are found at fault contact with the metamorphic rocks of the southern Menderes Massif (also called Cine Massif). However, the structural relationship with the northern and southern boundary faults remained in question for a long time because no subsurface geological study has been conducted along the BMG.

11.5. DATA AND METHODOLOGY

11.5.1. Dataset

TPAO provided eight N-S and three E-W 2D seismic reflection profiles along the BMG (Figure 11.4). The

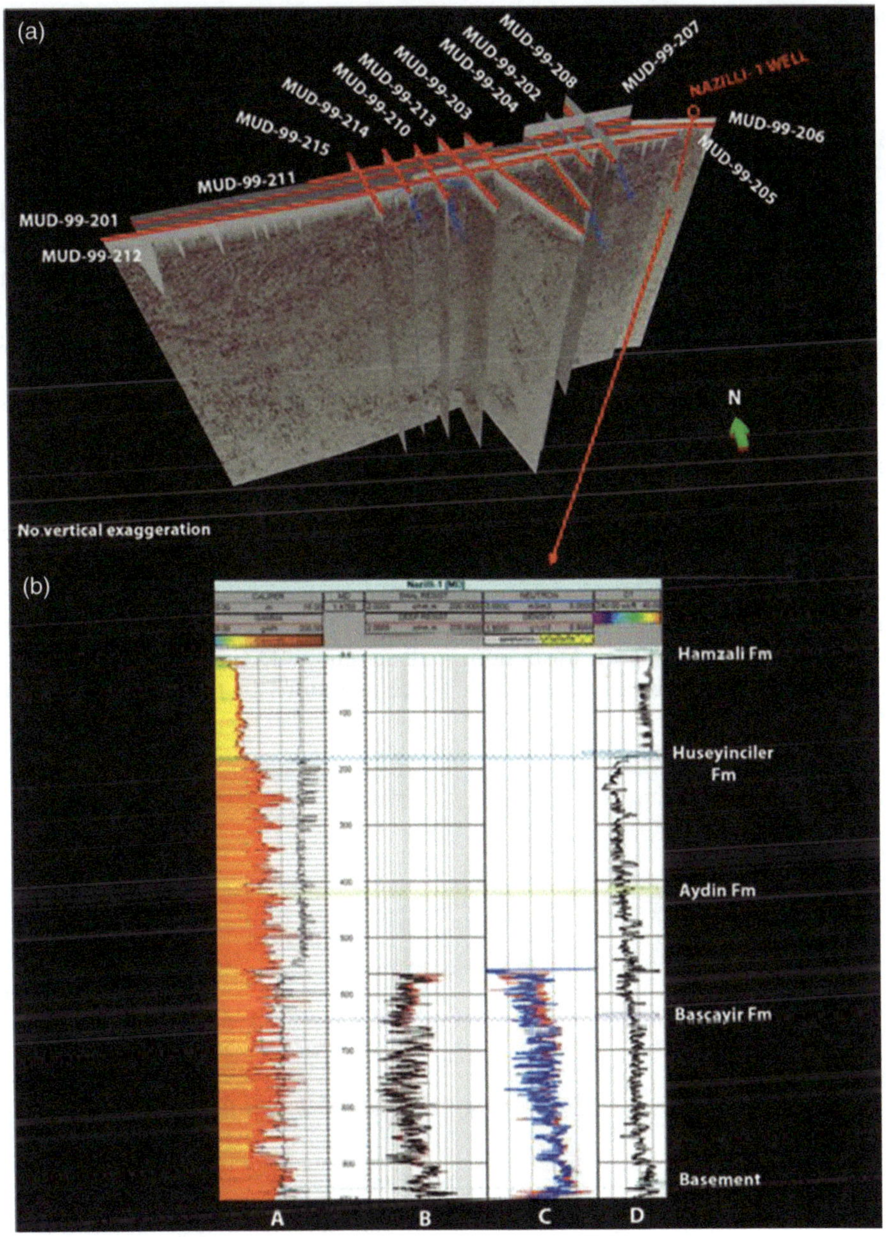

Figure 11.6 (a) 3D view of the 11 seismic reflection profiles from a PETREL; Schlumberger seismic interpretation project in the time domain. (b) Geophysical logs from the Nazilli-1 well showing gamma-ray and caliper logs (A), shallow and deep resistivity logs (B), neutron porosity and bulk density logs (C), and sonic log (D). The database of the well does not include resistivity, density, and neutron porosity logs above 560 m.

TPAO also provided a database for the Nazilli-1 well, which was drilled in the study area by TPAO and completed as a dry well in 2003. The well is located along an approximately N-S trending seismic profile (Figure 11.4). It is spotted at the Quaternary Hamzali Formation at the surface and penetrated the Pliocene Huseyinciler Formation (240 m), the late Miocene Aydin Formation (225 m), and the early Miocene Bascayir Formation (320 m) before reaching the metamorphic basement at 965 m below the surface (Figure 11.6). The Nazilli-1 database provided for this study consists of gamma-ray, caliper, resistivity, density, neutron porosity, and sonic logs. The resistivity, density, and neutron porosity logs are available only for the lower part of the Aydin and Bascayir formations within the 560–965 m interval (Figure 11.6).

11.5.2 Data Availability

There are no supplementary data available for this study.

11.5.3. Methodology

During this study, three main methods were used to provide a better insight into the structural/kinematic evolution of the BMG. The methods are (a) stratigraphic and structural interpretation of seismic reflection profiles (Figures 11.2 and 11.6), (b) structural cross-section restoration of depth-converted 2D seismic profiles, and (c) 3D kinematic modeling.

11.5.4. Seismic Interpretations

During seismic interpretations, we first generated a synthetic seismogram (Figure 11.7) for the well to seismic tie, resulting in a preliminary horizon and fault interpretation in the time domain. Time-depth correlations for the well to seismic tie are generated using a synthetic seismogram (Figure 11.7).

Second, we created a velocity model and structural interpretation in the depth domain. During this study, we determined the velocity interval for each formation. We used average velocity values during the depth convergence, which is accomplished using a velocity model generated by defining a mathematical function for each geological formation after a complete horizon interpretation in the time domain (Yılmaz, 2001; Robein, 2003).

Surfaces created for the tops of all formations in the study area were input into the velocity model to describe zone intervals. The formations are, from youngest to oldest, Hamzali, Huseyinciler, Aydin, and Bascayir sedimentary units and the metamorphic basement. Each formation in the velocity model was defined with specific interval velocities. The first and highest interval is the 180 m of the Hamzali Formation, defined with 1,700 m/s interval velocity. The second interval is the 244 m of the Huseyinciler Formation, defined with 1,880 m/s velocity. The third and fourth intervals, 224 m of the Aydin and 324 m of the Bascayir formations, were defined as 2,010 and 2,177 m/s velocity intervals, respectively. A constant velocity of 3,200 m/s was assigned for depth conversion in the metamorphic basement below the sedimentary section. This velocity model (Figure 11.8) was applied to all seismic reflection lines, including horizons and fault interpretations in the time domain. Consequently, the

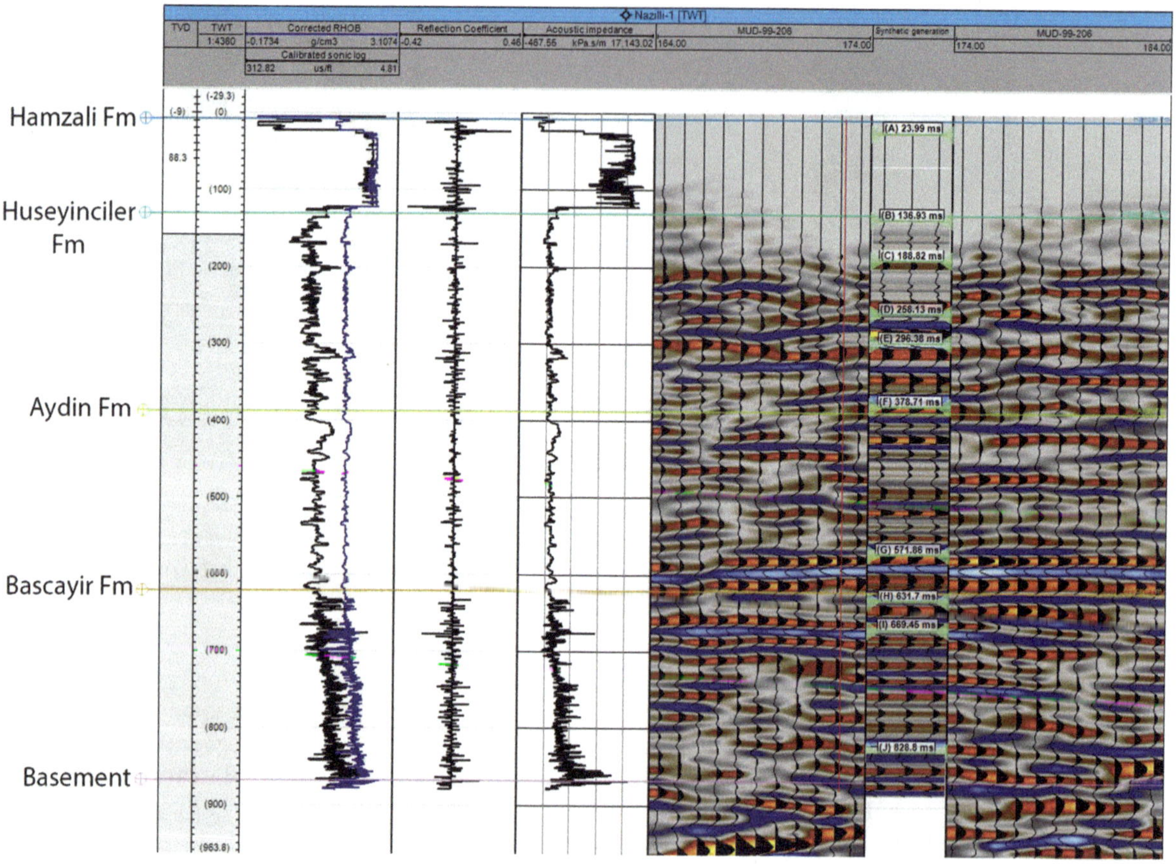

Figure 11.7 Synthetic seismogram generation. The synthetic is shown with the calibrated sonic log, corrected density log (blue), acoustic impedance, reflection coefficient, and seismic image of the seismic reflection profile where the Nazilli-1 well is located.

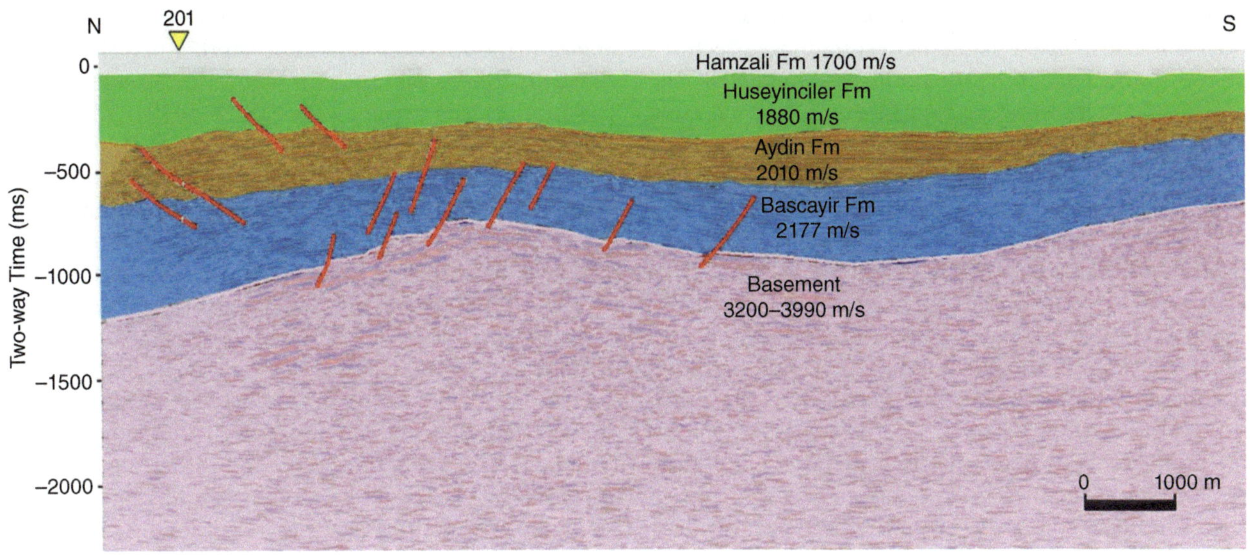

Figure 11.8 Velocity values used for the velocity model for each formation on the seismic reflection profile MUD-99-206.

position of the horizons and fault orientations were checked in the depth domain to correct the velocity model defined in the time domain.

Four Neogene formations (Figure 11.5) are identified in the well logs of the Nazilli-1 well and interpreted on the seismic reflection profiles. Envelope attribute analysis, which is also useful in highlighting main sequence boundaries, was used to contribute to seismic stratigraphic interpretation. The analysis shows that the metamorphic basement can be observed around 860 m/s with strong, high-amplitude reflections. The early Miocene Bascayir and the late Miocene Aydin formations above the basement rocks are mainly composed of coarse-grained sediments that are characterized by strong lens-shaped reflection geometry on the seismic line, whereas the Pliocene Huseyinciler Formation that overlies the Miocene units documents relatively weak seismic reflection due to its poorly sorted clastic rock composition as well as the youngest Quaternary Hamzali Formation that represents alluvial deposits of the Büyük Menderes River.

11.5.5. Structural Interpretation and Structural Restoration

During this study, eight N-S and three E-W trending 2D seismic reflection profiles were interpreted to explain sedimentary rock accumulations and their thickness changes along the graben from east to west in addition to the structural geometry related to the Cenozoic extensional tectonics in western Turkey. In the following section, the structural geology of the BMG will be discussed based on our interpretation of a representative N-S seismic reflection profile, MUD 99-202.

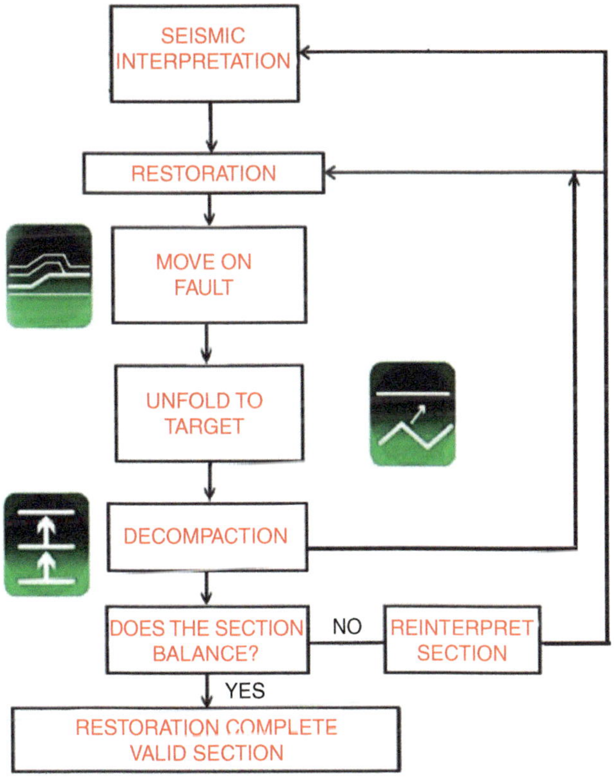

Figure 11.9 The structural cross-section restoration workflow in 2D Move software (modified from 2D Move Manual, 2012).

After structural seismic interpretation of all of the N-S seismic lines, we conducted structural restoration on MUD 99-202 and other seismic lines using the workflow shown in Figure 11.9. A restoration process is used to construct admissible structural cross-sections from depth-converted seismic reflection profiles. A 3D kinematic model was constructed using the restored cross-sections to better understand the structural development of the BMG in the early Miocene, late Miocene, and Pliocene geologic epochs. Structural restoration and kinematic evolution will be discussed in detail. We will also discuss the isochore maps we prepared based on our stratigraphic and structural interpretation of the N-S and E-W seismic reflection profiles.

The restoration process starts with digitizing interpreted seismic horizons to form polygons for each point. Then, each point in the hanging wall is moved the same horizontal distance, following a path parallel to the fault. During this process, overlays and mismatches on the faults' hanging wall due to misinterpretation are fixed. The process validates the interpretation by correcting the relationship between fault orientation and depositional geometry during restoration. Horizons moved along the faults are unfolded to remove deformation in the surface during deposition, and fixed horizons are decompacted by assuming that the basin was uplifted due to isostatic rebound. Decompaction of the restored unit allowed us to delineate the geometry of the basin at the time when each formation was deposited (Allen & Allen 2013). Each step in this process is repeated for each geological time period until a valid section is obtained (2D Move, 2012).

11.5.6. Backstripping

Backstripping processes rely on the decompaction of the sediment column, reconstruction of its original thickness at the time of deposition, and correction of paleobathymetric and eustatic variations (Allen & Allen, 2013). Decompaction allows for calculation of the effects of each sediment volume change due to porosity loss related to depth during deposition (Sclater & Christie, 1980). A porosity-depth relationship can be constructed from values measured in well logs or from a theoretical porosity-depth curve calculated from the deformation tests (Sclater & Christie, 1980). During this study, the porosity-depth relationship was created using calculated porosity values from the density log of the Nazilli-1 well.

11.6. SEISMIC INTERPRETATIONS

11.6.1. Structural Seismic Interpretation and Restoration

Along the eight N-S trending seismic profiles, the BMG has a symmetrical geometry bounded by two major south and north-dipping normal faults. Figure 11.10 shows an uninterpreted and interpreted seismic reflection profile, MUD-99-202. Along the northern flank of the graben, the south-dipping boundary fault propagates an almost-planar normal fault geometry with ~38–45° dip angles without flattening out at depth along the graben. Figure 11.10c shows the depth-converted structural interpretation of the seismic line MUD-99-202. The south-dipping boundary fault becomes steeper at depth toward the western part of the study area. The north-dipping boundary fault dips ~35° along the southern flank of the graben and joins the south-dipping boundary fault around 3,500 m below sea level (BSL) in the eastern part of the study area along seismic profile MUD-99-202 (Figure 11.10), at about 2,300 m.

The fault orientation of the synthetic and antithetic normal faults shows that in the center of the graben, the normal faults on the hanging wall of the two boundary faults dip with high angles (65–78°), whereas those that are located on the edges have shallower dip angles (50–60° along the graben from east to west within the study area). This geometry suggests that the faults located on both flanks of the graben developed first and experienced fault rotation due to continuous extension and footwall uplift. However, there was no major rotation along either the northern or southern boundary fault to cause development of rollover structures that are well observed along the Alaşehir Graben to the north (Çemen et al., 2006; Çifçi et al., 2010).

All of the N-S 2D seismic reflection profiles show that the graben fill is offset by synthetic and antithetic normal faults traced with a displacement on the hanging wall units. The faults merge into the boundary faults at ~3,400 m base sea level (BSL) at the eastern part and about 2,000 m BSL at the western part of the study area (Figures 11.11–11.13). Seismic profiles show more sedimentary rock deposition on the northern margin than the southern margin of the graben in the southern part of the study area, whereas this thickness difference becomes more balanced toward the west. The Neogene succession thickens toward the center of the graben and gets thinner and progressively higher in elevation toward the northern and southern margins of the graben (Figure 11.10). This suggests that the margins of the graben have experienced basement uplift preferentially to the center, which caused sediment accumulation from the northern and southern margins to the center of the basin due to ongoing Cenozoic extension. The thickness change also suggests that the extension might have initiated along the northern margin of the graben and initiation of the south-dipping boundary fault and created more accommodation space for sedimentary rock deposition on the northern margin of the graben. Then the graben experienced a polarity change toward the west due to the north-dipping boundary fault activity on the southern margin of the graben. Therefore, based on the geometry observed in the cross-sections (e.g., Figure 11.10), it can be suggested that the pattern of sediment deposition shifted to the north-dipping boundary fault along the southern margin of the graben after considerable sedimentation along the northern margin. This geometry suggests that both boundary faults were active at different times and disagrees with the idea that BMG is a supradetachment basin mainly controlled by the movement and rotation of the hinge along the south-dipping boundary fault.

The previous statement is evidenced by the fault rotation along the seismic profiles from the eastern to the western part of the study area. For instance, in seismic

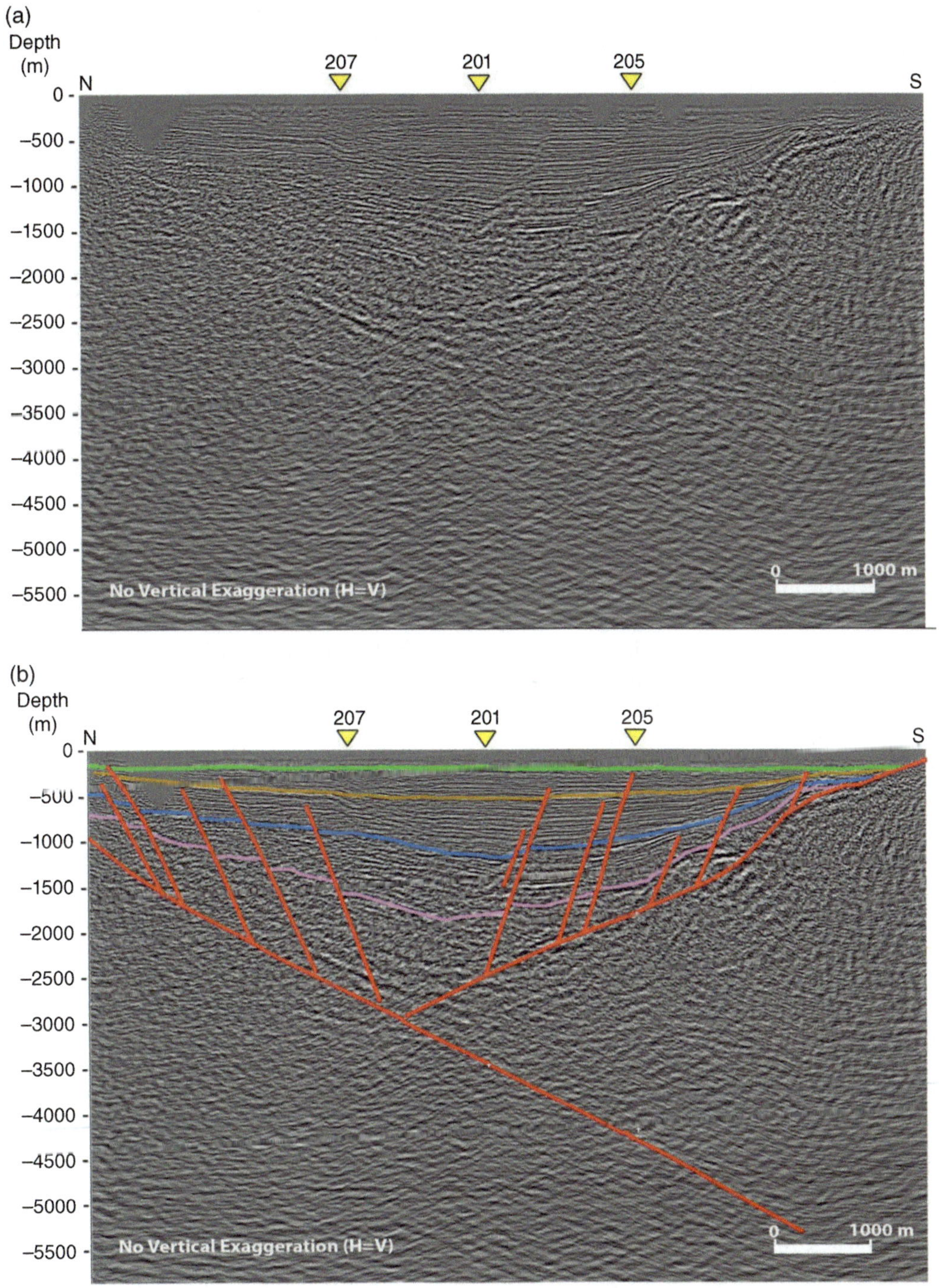

Figure 11.10 Depth converted uninterpreted (a) and interpreted (b) and balanced structural cross-section (c) along the N-S oriented MUD-99-202 seismic reflection profile showing main boundary faults and their synthetic splays. The yellow triangles show the locations of cross lines MUD-99-207, 201, and 205. Color codes show formation tops (surface, Hamzali Formation; green, Huseyinciler Formation; orange, Aydin Formation; blue, Bascayir Formation; pink, basement; red, normal faults).

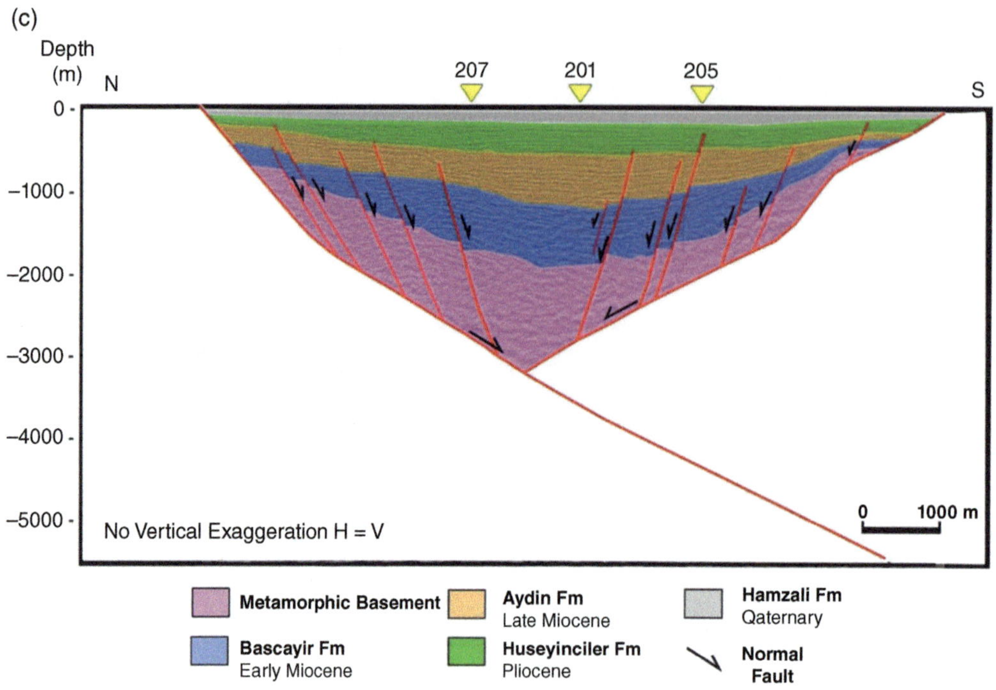

Figure 11.10 (Continued)

profile MUD-99-202, the synthetic and antithetic faults located on the northern margin of the graben dip are shallower than those on the southern flank (Figure 11.10). This in turn suggests that the graben was initiated in the study area under the control of the south-dipping boundary fault, most likely during the deposition of the early Miocene Bascayir Formation. Then the north-dipping fault became more dominant and controlled the sedimentation during the late Miocene Aydin Formation deposition.

Based on field-oriented studies, it was proposed that the BMG was developed along a low-angle south-dipping detachment fault (Seyitoğlu et al., 2004; Göğüş, 2004; Çemen et al., 2006; Çifçi et al., 2010). However, the structural interpretation of the available seismic reflection profiles indicates that sedimentary rocks on the hanging wall of the south dipping normal fault do not contain a rollover structure related to detachment faulting (Figures 11.9 and 11.10). This geometry suggests that the margins of the graben have experienced basement uplift preferentially to the center, which caused sediment accumulation from the northern and southern margins to the center of the basin due to ongoing Cenozoic extension. The thickness change also suggests that the extension might have initiated along the northern margin of the graben with the initiation of the south-dipping boundary fault and created more accommodation space for sedimentary rock deposition on the northern margin of the graben. Then, the graben experienced a polarity change toward the west due to the north-dipping boundary fault activity on the southern margin of the graben. Therefore, based on the geometry toward the northern and southern margins of the BMG, it can be suggested that the south-dipping normal fault along the northern margin of the graben does not contain a rollover structure related to detachment faulting (Figures 11.11 and 11.13).

Structural restoration of the depth-converted seismic profiles played an important role in determining the study area's subsurface structural geometry. Three restoration steps were applied to depth-converted seismic reflection profiles: (a) move on fault, (b) unfold to horizon, and (c) decompaction. Each step in the restoration workflow was repeated for each geological cross-section until a structurally valid model was obtained.

Figure 11.12 shows the restoration steps of the balanced section along the seismic section MUD-99-202 with the best-fit dip angle of the south-dipping boundary fault at depth. The figure demonstrates no gaps or overlaps in the section when horizons on the synthetic and antithetic faults are restored, and then the Aydin Formation is unfolded to the surface, rotating hanging wall, and footwall blocks. During the restoration process, first the Quaternary and Pliocene formations were decompacted. Then the late Miocene Aydin Formation was restored, assuming the hanging wall of each fault block was deformed by simple shear (Figure 11.11). Calculating the shear angle of each fault along both synthetic and antithetic faults, each point in the hanging wall was moved the same horizontal distance following a path parallel to the fault using a fault algorithm in 2D Move software.

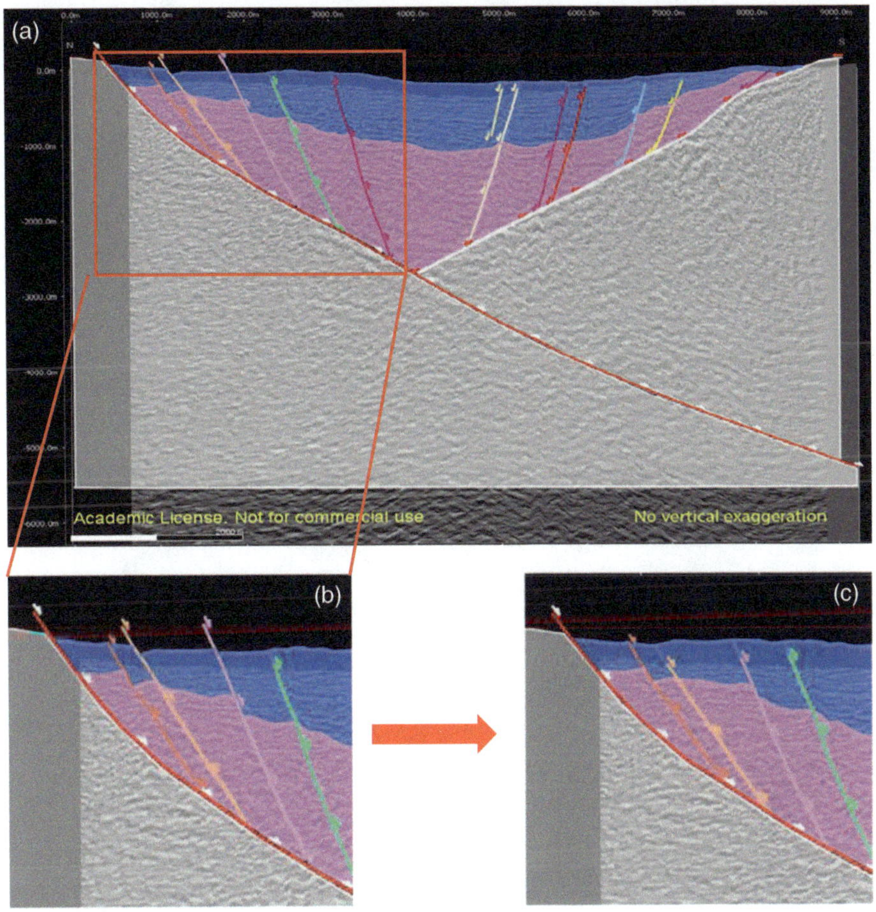

Figure 11.11 Restoration steps showing (a) polygon creation for seismic profile MUD-99-202, (b) moving horizons on the hanging wall of the faults, and (c) balanced horizon (blue) without any overlays and mismatches.

To verify the boundary fault interpretation, the main fault geometry at depth, including listric shape, was modeled by changing the fault dip angle until a balanced section was obtained. First, the profile was restored, assuming a listric geometry to test whether the sediment orientation on the hanging wall of the south-dipping fault matched the interpreted subsurface geometry. When the late Miocene Aydin Formation was unfolded and the boundary faults along both sides were rotated, the listric geometry of the graben was not balanced because the rotated hanging wall block overlays the basement, and the south-dipping boundary fault did not fill the gap due to its geometry (Figure 11.12). This suggests that the lower part of the south-dipping fault should be steeper at depth, and the graben floor should have more space for the basement unit at the intersection point of the boundary faults.

Because the listric fault geometry did not produce a balanced structural cross-section, the fault dip was rotated at 5° increments until the graben geometry was balanced to properly restore the section without gaps and overlaps (Figure 11.12). As a result of this process, the south-dipping fault with a 42 to ~35° dip angle was found to be the best fit to balance the cross-section.

After the decompaction of the restored late Miocene Aydin Formation, the same steps were applied to the middle Miocene Bascayir Formation to finalize the restoration. The movement of the horizons on the hanging wall of the faults and rotation of the boundary faults at both margins show the same geometric consistency during the restoration of the Bascayir Formation. The balanced structural cross-sections (Figures 11.14 and 11.15) demonstrate that the south-dipping boundary fault is not becoming horizontal at depth. On the contrary, the fault maintains its dip angle until the reliable depth of the seismic section (~5 km). This process allowed for validation of the geometry of the south-dipping boundary fault and helped to understand the role of the north-dipping boundary fault and its antithetic and synthetic splays during the Miocene extension, which will be discussed later, in the kinematic evolution section. Validating the graben geometry resulting from the restoration process applied to seismic line MUD-99-202 enabled the same steps for the other N-S-directed seismic reflection profiles

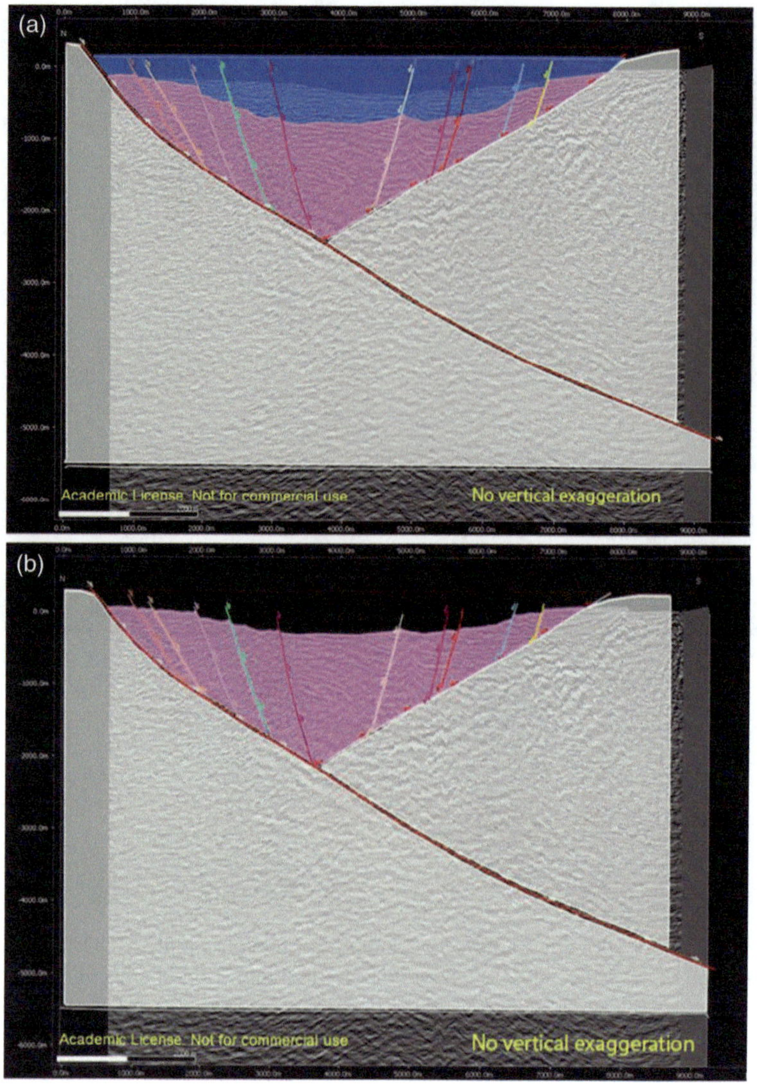

Figure 11.12 (a) Unfolding of balanced horizons to remove the deformation of a surface during deposition. (b) Decompaction of the unfolded horizon (blue).

to delineate the graben geometry from the early Miocene to the present in 3D.

11.6.2. Isochore Maps

The isochore maps constructed from depth-converted seismic lines support the previous interpretation. They illustrate that the thickness of the Neogene sedimentary units varies along the graben. The isochore map (Figure 11.15a) showing the distribution of the Miocene sediments (the Aydin and Bascayir formations) along the graben indicates that the thickness of the Miocene sedimentary units reaches a maximum of about 1,500 m in the eastern part of the graben and becomes thinner along the northern and southern margins of the graben.

The Miocene thickness decreases gradually to the west along the graben and reaches about 1,000 m, which may indicate that the graben has become shallower toward the west. The westward stratigraphic thinning might also suggest that the deposition period of the Miocene sediments may be shorter toward the west because the opening of the BMG was initiated from the eastern part.

The Pliocene-Quaternary isochore map shown in Figure 11.15b indicates that the sedimentary thickness is decreasing slightly toward the western part of the study area, except between seismic profiles MUD-99-203 and MUD-99-204, where the N-S trending Bozdogan Graben is located (Figure 11.4). This slight thickness change suggests that the graben has had more uniform thickness

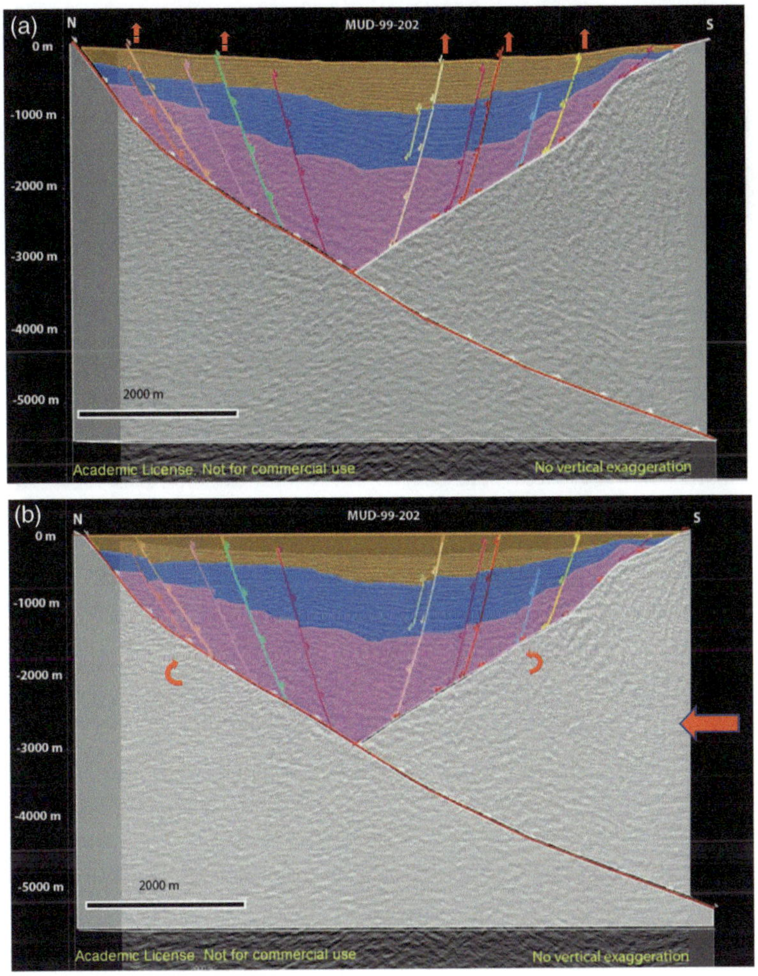

Figure 11.13 (a) Restoration of the late Miocene Aydin Formation (orange unit), moving the horizons on the faults. (b) Balanced section after moving horizons on the fault rotating hanging wall and footwall blocks and unfolding the formation.

distribution in the Plio-Pleistocene time compared to the Miocene time. The Plio-Pleistocene thickness distribution also suggests that the graben has experienced more Pliocene sedimentation and Quaternary alluvial deposition toward the western part of the study area. This might be caused by the polarity change to the west due to movement along the north-dipping boundary fault at the southern margin of the graben.

11.7. KINEMATIC EVOLUTION AND SUBSIDENCE HISTORY

A 3D kinematic model was created using balanced structural cross-sections and tectonic subsidence rates calculated from the backstripping technique to understand the BMG's evolution. The model allowed us to delineate the geometry of the boundary faults on both margins associated with their synthetic splays and the syn-extensional deposition rate of sedimentary rocks through geological time (Figures 11.16–11.20).

The model reveals that in the earliest Miocene, the M.G. BMG initially started to open along its northern margin with a normal fault dipping ~70° to the south during the early Miocene (Figure 11.16). Continued N-S extension caused downward warping along the south-dipping normal fault and formed the north-dipping boundary fault (~62°) as an antithetic fault of the south-dipping fault in the early Miocene. As the extension continued and the graben enlarged, the south-dipping fault rotated to ~54° as a result of footwall uplift. This process caused the metamorphic basement to deform and form steeper, high-angle synthetic growth faults (Figure 11.17a).

The lowest part of the Bascayir Formation may have started to deposit as a clastic unit soon after the south-dipping normal fault and its antithetics were formed. Continued rotation of the northern margin triggered

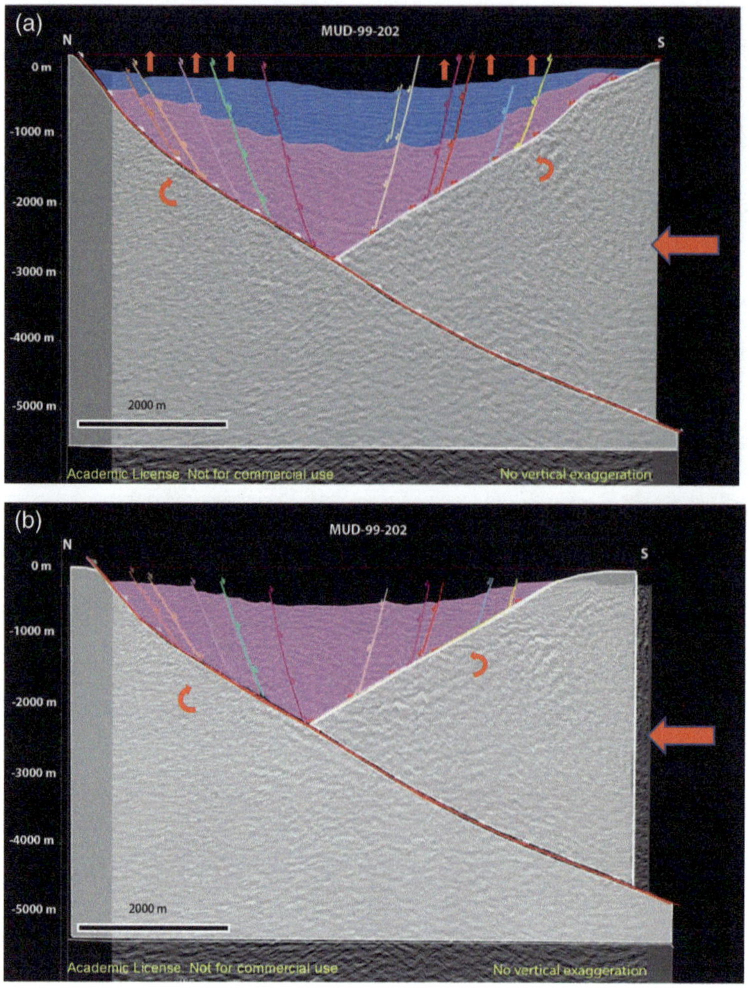

Figure 11.14 (a) Restoration of the middle Miocene Bascayir Formation (blue unit) showing rotation of the synthetic and antithetic faults and moving horizons. (b) Position of the metamorphic basement (pink unit) after decompaction of the Bascayir Formation.

rotation of the southern margin to ~50°, which resulted in developing synthetic faults of the north-dipping boundary fault. This alternating deformation on both sides created more accommodation space for the Bascayir Formation deposition in the early Miocene (Figure 11.17b). Footwall uplift and associated rotation of the south and north-dipping faults on both margins of the graben and differential loading due to the Bascayir Formation sediments enabled growth faulting along the synthetic faults and their subsequent rotation. This caused sedimentary units of the graben to thicken basinward on the hanging wall of the boundary faults and get thinner toward the basin margins (Figure 11.18).

By the end of the early Miocene, sedimentary rocks of the Bascayir Formation continued to be deposited almost symmetrically as a result of continued N-S extension and associated rotation of the boundary faults on the northern and southern margins (Figure 11.17). At the eastern part of the study area, the south-dipping fault rotated to a dip angle of ~46° that gradually decreased at depth to ~38°. The rotation lessened toward the west, and the width of the graben floor became narrower. During this time, the south-dipping boundary fault and its synthetic splays were formed, allowing more sediment accumulations on the northern margin of the graben due to the creation of more accommodation space (Figure 11.17). Syn-extensional growth faulting also provided the formation of new accommodation space on the downthrown side of the growth faults, which caused the Bascayir Formation to thicken toward the center of the graben and gradually thin toward the edges of the graben (Figure 11.17). This extensional geometry caused the graben to start to display its wedge shape geometry at the end of the middle Miocene (Figure 11.18a).

During the middle Miocene to late Miocene, the north-dipping boundary fault rotated more than (~5°) the

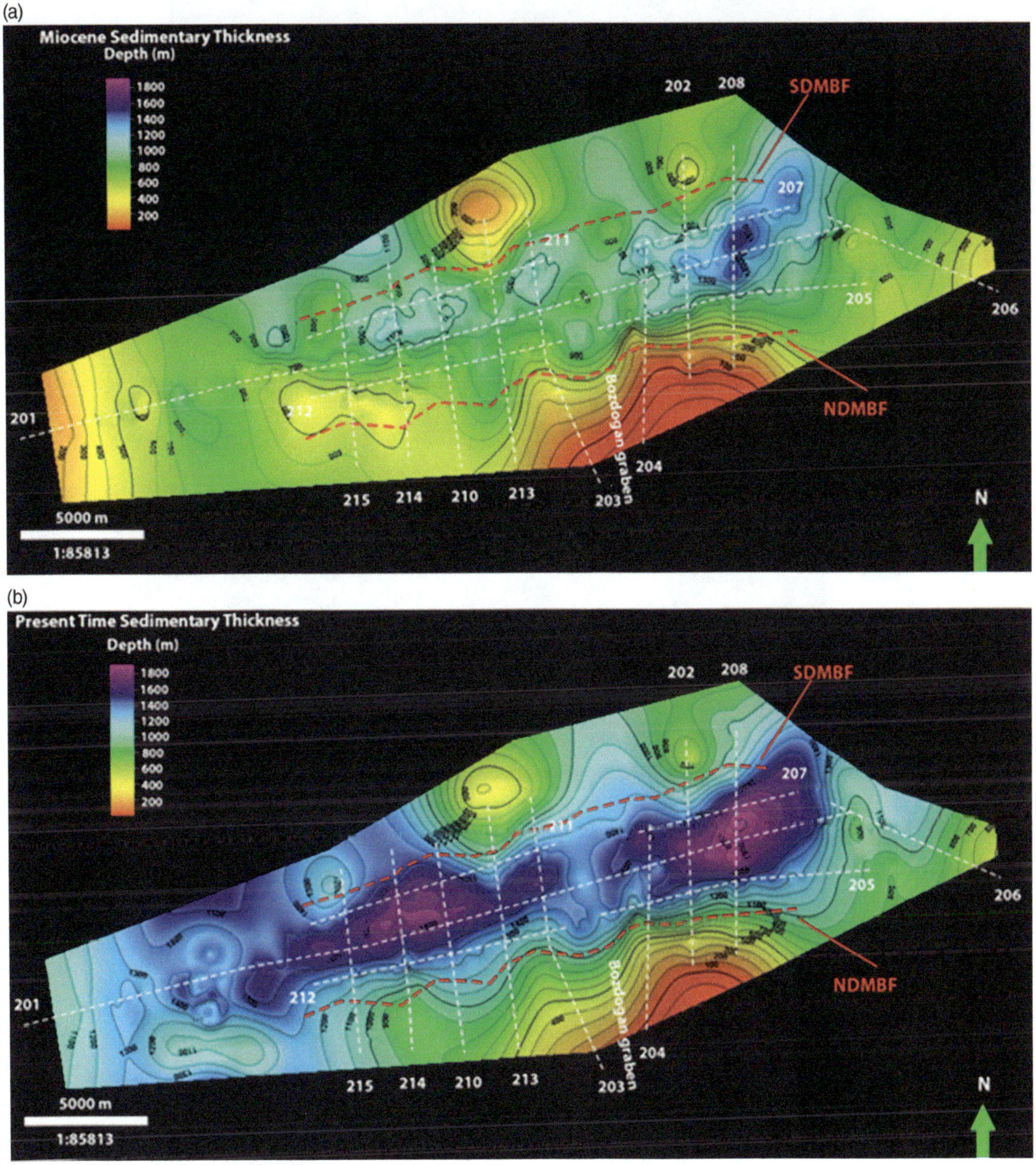

Figure 11.15 Isochore maps showing Miocene sedimentary (Aydin and Bascayir formations) thickness (a) and the total sedimentary rock thickness (b) variation along the BMG within the study area. The map shows that (a) sediment thickness decreases toward the western part of the study area during the Miocene time, and (b) the sediment thickness distribution is almost symmetrical along the graben within the study area after the Miocene time. SDMBF, south-dipping main boundary fault; NDMBF, north-dipping main boundary fault.

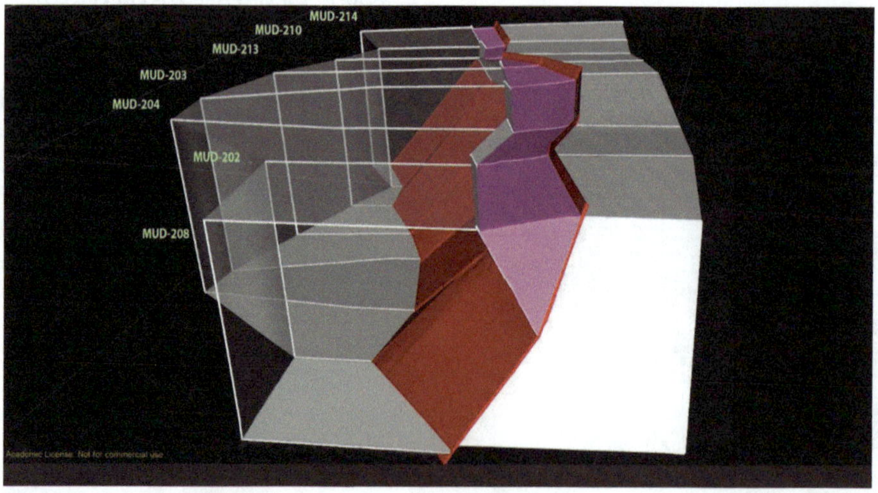

Figure 11.16 Initiation of the south-dipping boundary fault and its antithetic, the north-dipping fault, during the extension initiation in earliest Miocene. The pink unit represents the metamorphic basement rocks.

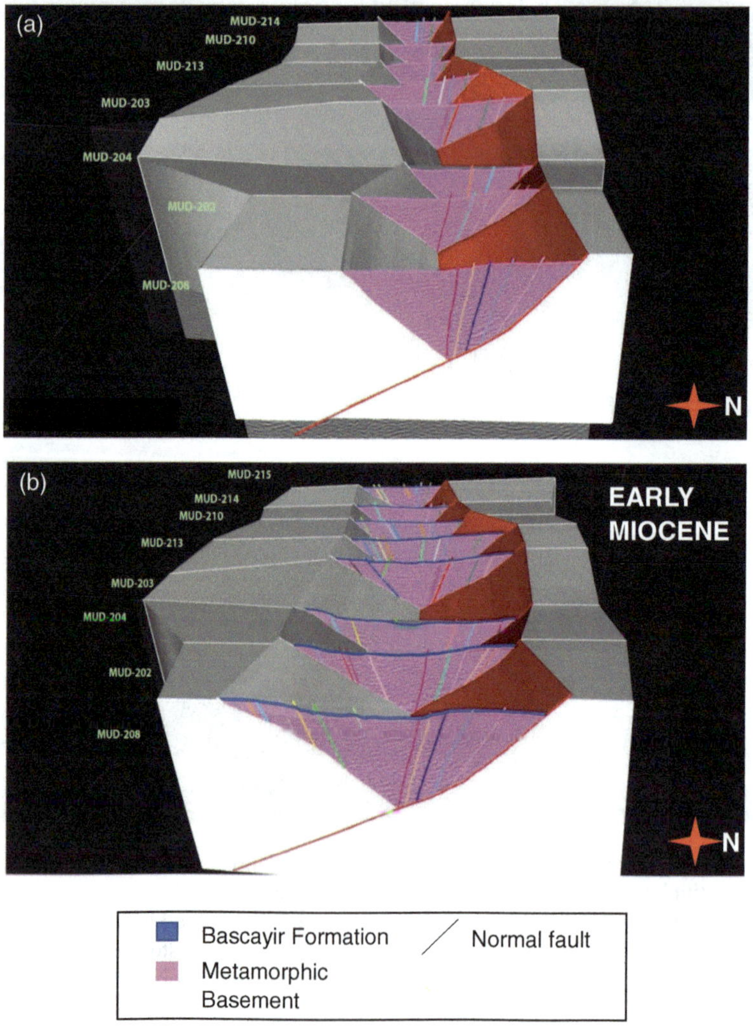

Figure 11.17 (a) Formation of the high-angle synthetic faults that superimposed onto the south-dipping boundary fault as a result of footwall uplift during the early stages of extension in early Miocene. (b) Rotation of the north-dipping boundary fault caused the formation of the synthetic fault along the southern margin, creating accommodation space for the Bascayir Formation deposition in early Miocene.

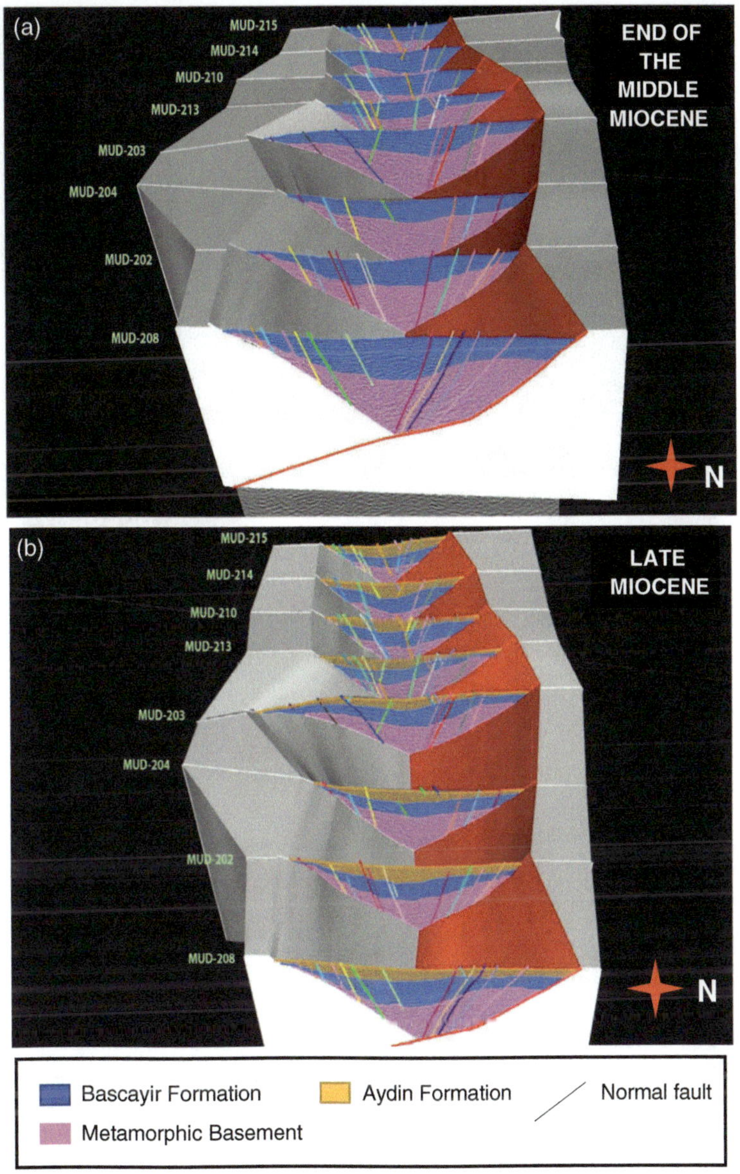

Figure 11.18 (a) Deposition of the Bascayir Formation, thickening toward the south-dipping boundary fault at the end of the middle Miocene. (b) Deposition of the Aydin Formation, showing the depocenter shift toward the south due to the rotation of the north-dipping boundary fault.

south-dipping boundary fault. Therefore, the synthetic growth faults located on the southern margin of the graben during this time created more accommodation space for the late Miocene Aydin Formation deposition. Hence, the graben has experienced a depocenter shift to the north-dipping boundary fault along its southern margin (Figure 11.18b).

Finally, from the Pliocene to Quaternary, the Huseyinciler and Hamzali formations were deposited slightly symmetrical along the graben, preserving the wedge-shaped geometry of the BMG. During the Pliocene to Quaternary, subsidence rates decreased due to the graben margins having experienced less rotation than during the Miocene (Figure 11.19).

11.7.1. Tectonic Subsidence

Tectonic subsidence analyses calculated from back-stripped cross-sections along the BMG are used to delineate sedimentation rates in the early Miocene, middle Miocene, and Plio-Pleistocene The diagram in Figure 11.20 shows the tectonic subsidence rates calculated within the study area from the east to the west and from the northern margin to the southern margin. It

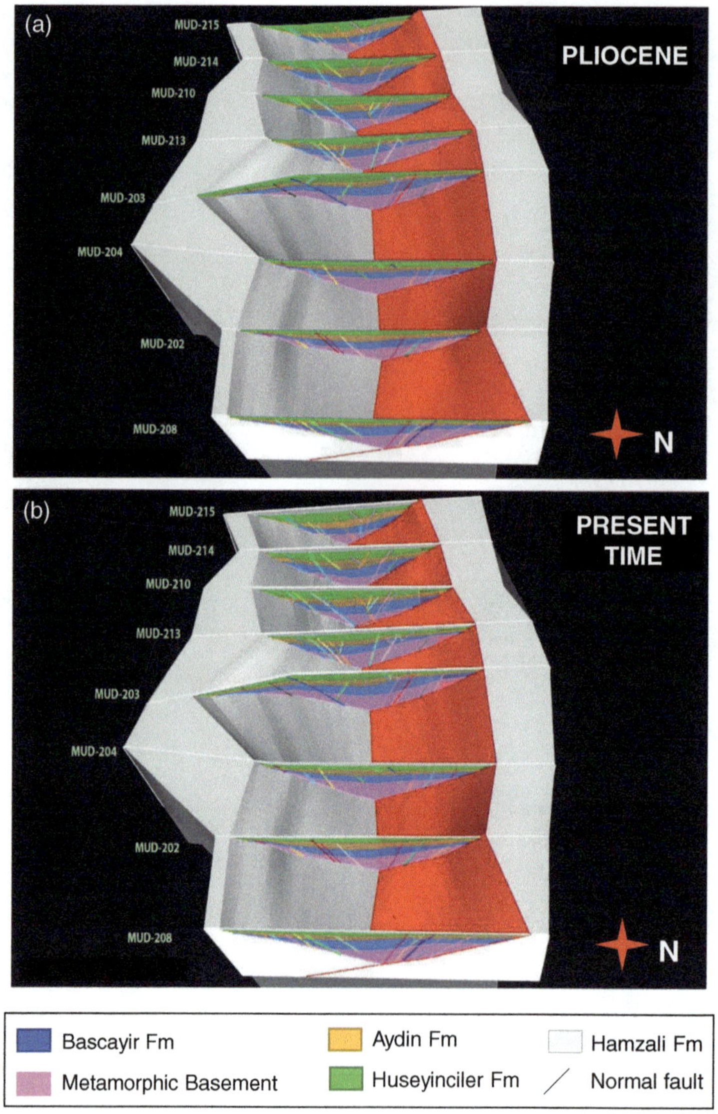

Figure 11.19 (a) Deposition of the Huseyinciler Formation in the Pliocene. (b) Quaternary (present-day) geometry of the BMG.

indicates that the graben has more subsidence in the central part, and the subsidence decreases toward the edges, contributing to wedge shape geometry and growth faulting associated with the Cenozoic extension. Toward the western part of the study area, the general trend of subsidence curves for the central part of the graben shows a decrease toward an intrabasinal high, which is located between seismic profiles MUD-99-203–213, and an increase toward the profile MUD-99-215. This indicates that graben kinematics have varied within the study area from the east to the west. The subsidence at the northern and southern flanks increases due to the graben floor's width toward the seismic reflection profile MUD-99-215.

The subsidence curves, representing sediment accumulation at the center of the graben at the end of the early and late Miocene (~23 and ~12 Ma), show a general decrease toward the western part of the study area except between profiles MUD-99-203 and MUD-99-213, where an intrabasinal high is located. This indicates that the extension might have initiated from the eastern part of the study area and continued to the west. The diagram also shows that the northern flank experienced more sediment deposition during the early Miocene. This shows a relationship with the balanced structural cross-sections, indicating that the south-dipping boundary fault along the northern margin rotated more than the north-dipping boundary fault along the southern margin of the graben, supporting the interpretation that the development of the graben was controlled by the south-dipping boundary fault on the northern flank during the deposition of the Bascayir Formation in the early Miocene.

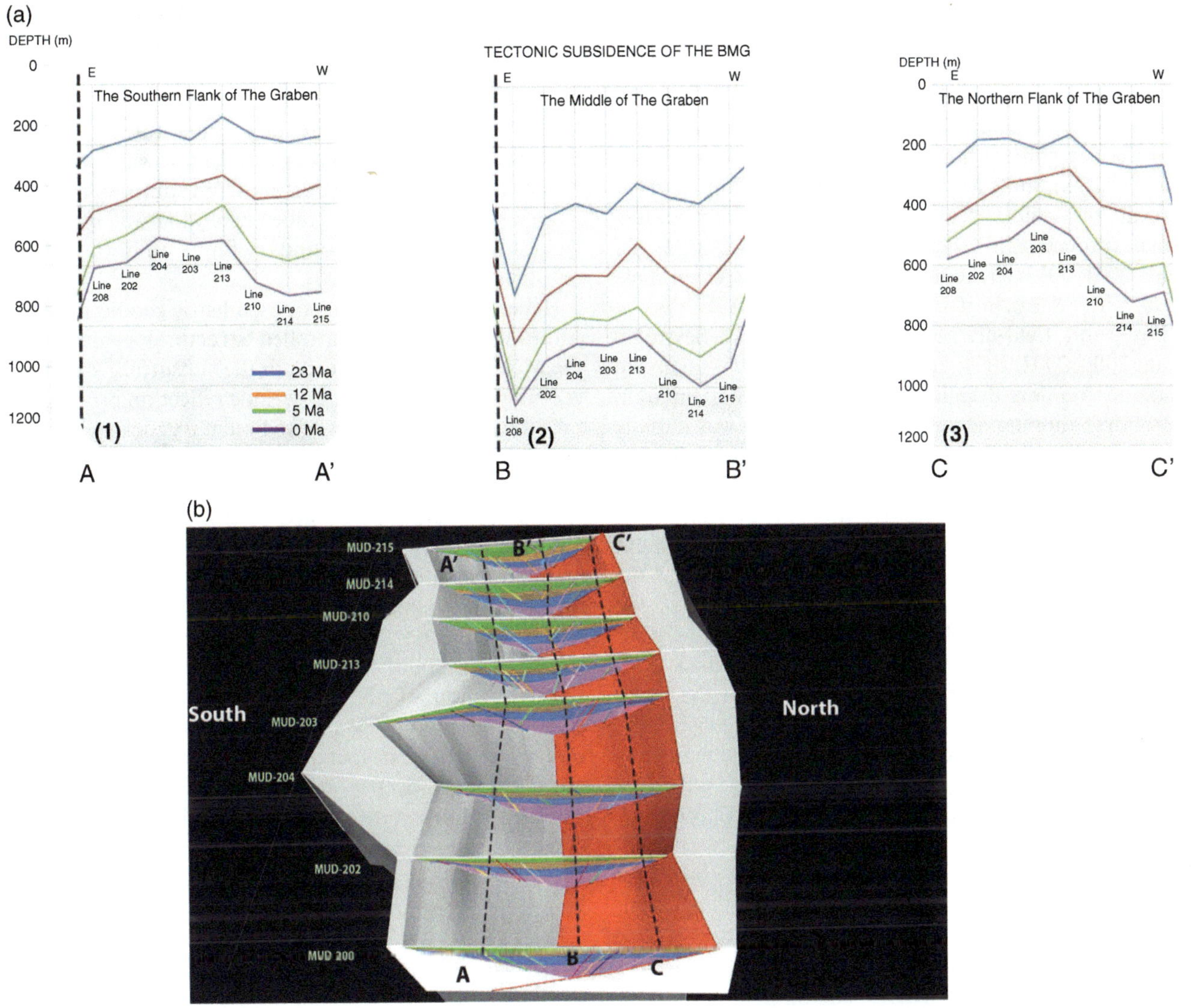

Figure 11.20 (a) Tectonic subsidence rates during the deposition of each formation along E-W line A-A′ (1) line, B-B′ (2), and C-C′ (3). The colored lines show the end of the deposition. Blue, end of the middle Miocene, the Bascayir Formation; orange, end of the late Miocene, the Aydin Formation; yellow, end of the Plionce, Huseyinciler Formation; purple, Quaternary, Hamzali Formation. (b) N-S cross-sections constructed in the study area based on our structural seismic interpretations.

The transition to the Aydin Formation deposition in the late Miocene (~12 Ma) is characterized by almost the same increase in the subsidence rates of the Bascayir Formation deposition. However, during this time, more sediment accumulation occurred in the southern flank compared to the northern flank, indicating that the north-dipping boundary fault became more dominant than the south-dipping fault (Figure 11.4).

After the late Miocene, the graben experienced a slow subsidence rate at the end of the Pliocene, specifically at both margins compared to the center of the graben, and the subsidence rates have gradually decreased to the present time. This suggests that the extension rate of the graben associated with boundary fault rotation decreased after the Miocene (Figure 11.3).

11.8. DISCUSSION

Previous studies proposed several different models to explain the extensional deformation of the MMCC in western Turkey. These models came to the conclusion that the two major E-W trending grabens, the Alaşehir and Büyük Menderes Grabens, are structurally symmetrical and developed as supradetachment basins under the control of opposite dipping detachment faults (Gessner et al., 2001, 2013; Seyitoğlu et al., 2000, 2002; Çemen et al., 2006; Çiftçi & Bozkurt 2009). However, the initial geometry of the detachment faults remained controversial. Some researchers suggested that the detachment faults of the E-W trending grabens were formed initially as low-angle listric faults (Hetzel et al., 1995b; Kocyiğit et al., 1999; Sözbilir, 2001). Others, however, suggested that the detachment faults were initially formed as high-angle normal faults and rotated to their low dip angle as a result of isostatic rebound and an associated rolling-hinge process in their footwall blocks (Gessner et al., 2001, 2013; Seyitoğlu et al., 2000, 2002; Çemen et al., 2006; Çiftçi & Bozkurt 2009; Ersoy et al., 2014).

Based on field-oriented geological work, Seyitoğlu et al. (2000, 2002, 2004) and Çemen et al. (2006) suggested that the AG was developed along its southern margin by a north-dipping detachment fault, which was initially formed as a high-angle normal fault and rotated to a low angle due to the rolling-hinge process causing a large rollover structure in the Early Miocene to Pleistocene sedimentary rocks on the hanging wall of the detachment surface. Interpreted 2D seismic reflection profiles in the AG and determined that the flat-ramp geometry of the detachment fault caused the formation of the large rollover structure on the southern margin of the graben. Both surface and subsurface studies conclude that the interpreted structural geometry of the AG agrees with the supradetachment basin development in the region regardless of the initial dip angle of the detachment faults.

A similar structural geometry was proposed by Seyitoğlu et al. (2004), Göğüş (2004), and Çemen et al. (2006) for the development of the M.G. BMG based on field mapping along the northern margin of the graben. These studies proposed that Neogene sedimentary rock accumulation in the BMG is controlled by a south-dipping, low-angle detachment fault along the northern margin of the graben. In addition, Çifçi et al. (2010) proposed that the south-dipping detachment fault of the BMG has a listric geometry at ~2.5 km and becomes horizontal about 10 km at depth, although they did not have seismic lines or other geophysical data.

Our structural interpretation of available seismic reflection profiles and balanced structural cross-sections do not support the previous interpretation for the following reasons:

1. The balanced structural cross-sections reveal that the south-dipping fault along the northern margin of the M.G. BMG shows nearly planar geometry (Figures 11.10 and 11.11).
2. The cross-sections clearly demonstrate that the M.G. BMG does not have a rollover structure on the hanging wall of the south-dipping boundary fault surface along the northern margin of the graben (Figures 11.10, 11.11, and 11.13–11.15).
3. Kinematic modeling suggests that after the initial accumulation of sedimentary rocks during the early to middle Miocene (Bascayir Formation), there was a depocenter shift toward the southern side of the graben, which is seen in available seismic reflection profiles in the study area (Figures 11.17–11.20).
4. The Neogene sedimentary units thicken toward the center of the graben and thin toward the edges of the graben on its northern and southern margins, suggesting that the graben evolution was separately controlled by two boundary faults on both margins (Figures 11.18–11.20).

Hence, the previously proposed listric geometry of the south-dipping fault controlled structural evolution of the M.G. BMG cannot explain the subsurface structural geometry shown on the 2D seismic reflection profiles and the pattern of deposition of sedimentary rocks within the graben in the study area.

These lines of evidence suggest that the BMG has experienced rifting controlled by two oppositely dipping planar boundary faults (Figures 11.18–11.20). The kinematic model constructed during this study suggests that the depocenter shift to the southern side of the graben is related to the north-dipping boundary fault along the southern margin of the graben being more active during the deposition of the late Miocene Aydin Formation (Figure 11.18). The south-dipping boundary fault may flatten out at the deeper part of the graben in the mid crust and join a deeper structure as a result of the domal rise related to isostatic rebound along the SWASZ.

11.8.1. Proposed Extensional Model for the Central Menderes Metamorphic Core Complex

The difference in subsurface geometry of the E-W trending BMG and AG bordering the Central Menderes metamorphic core complex (CMMCC) discussed in this study requires that previously proposed extensional mechanisms for the CMMCC be updated. Therefore, a new model was developed for the extensional evolution of CMMCC and E-W trending graben formation in the region.

Our model (Figure 11.21) suggests that the Cenozoic extension in the MMCC may have been initiated in the late Oligocene following the Eocene Alpine collision and formation of the Izmir-Ankara suture zone (Figure 11.21a,b, Şengör et al., 1984; Çemen et al., 2006; Seyitoğlu et al., 2004; Seyitoğlu & Işık, 2015). This extension was controlled by the SWASZ as a main breakaway and resulted in a domal uplift beneath the region (Figure 11.21b; Seyitoğlu et al., 2004; Çemen et al., 2006).

The domal uplift caused the reactivation of Lycian nappes as an extension structure in opposite directional movement and the first exhumation of the high-grade metamorphic rocks of the Menderes Massif in the late Oligocene (Seyitoğlu et al., 2004; Çemen et al., 2006;

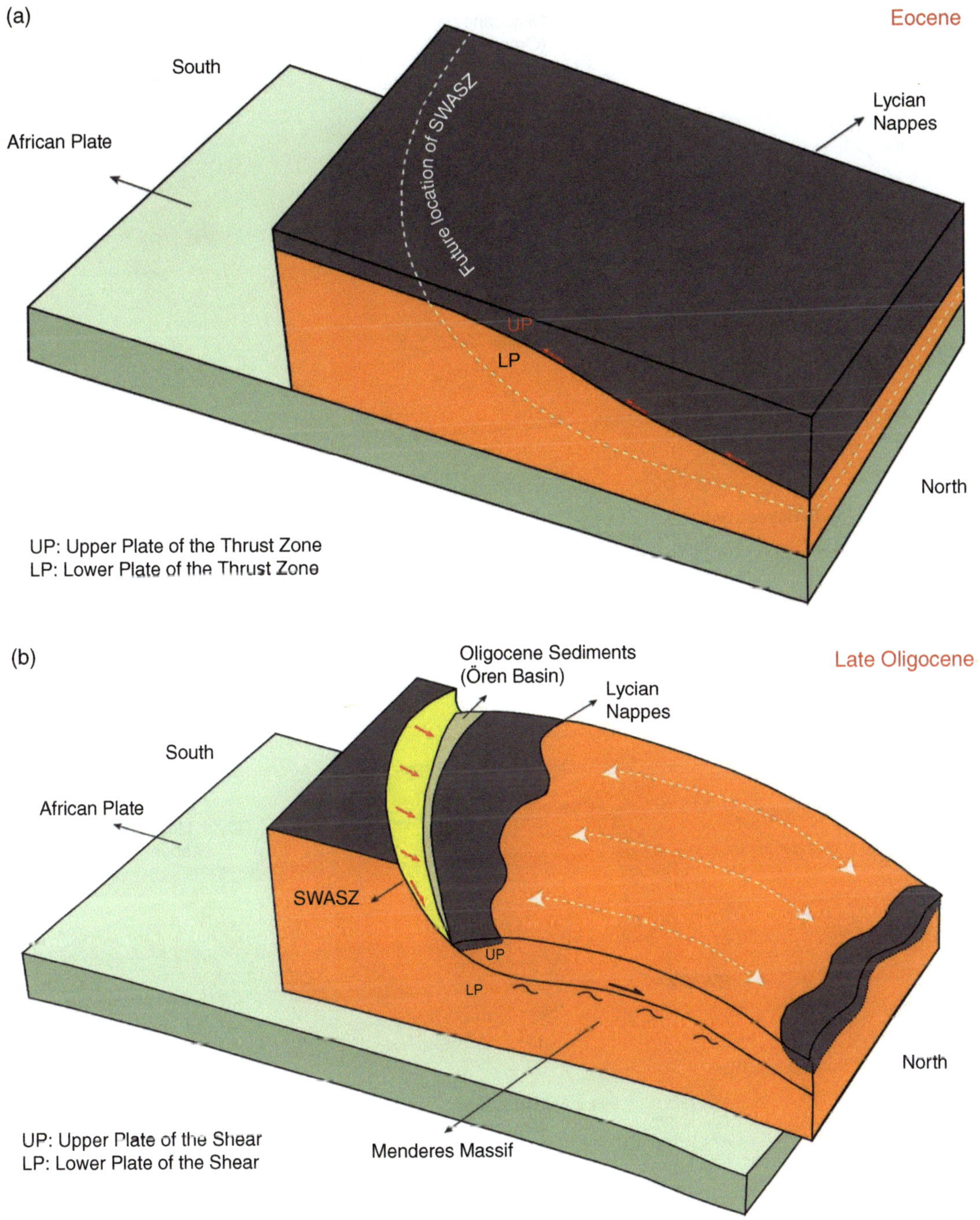

Figure 11.21 3D diagrams showing extensional evolution of the Menderes metamorphic core complex (MMCC) and the formation of the E-W trending grabens. (a) Eocene: formation of the Izmir-Ankara suture zone and the Lycian Nappes. (b) Late Oligocene: initiation of the Cenozoic extension in western Turkey and asymmetric exhumation of the Menderes core complex along the SWASZ (Çemen et al., 2006; Seyitoğlu et al., 2004). (c) Early Miocene: development of the Alaşehir Graben along a secondary breakaway as a supradetachment basin and the BMG as a small rift basin. (d) Present: continuing extension along the AG and BMG grabens and development of the Küçük Menderes Graben due to the westward escape of the Anatolian plate.

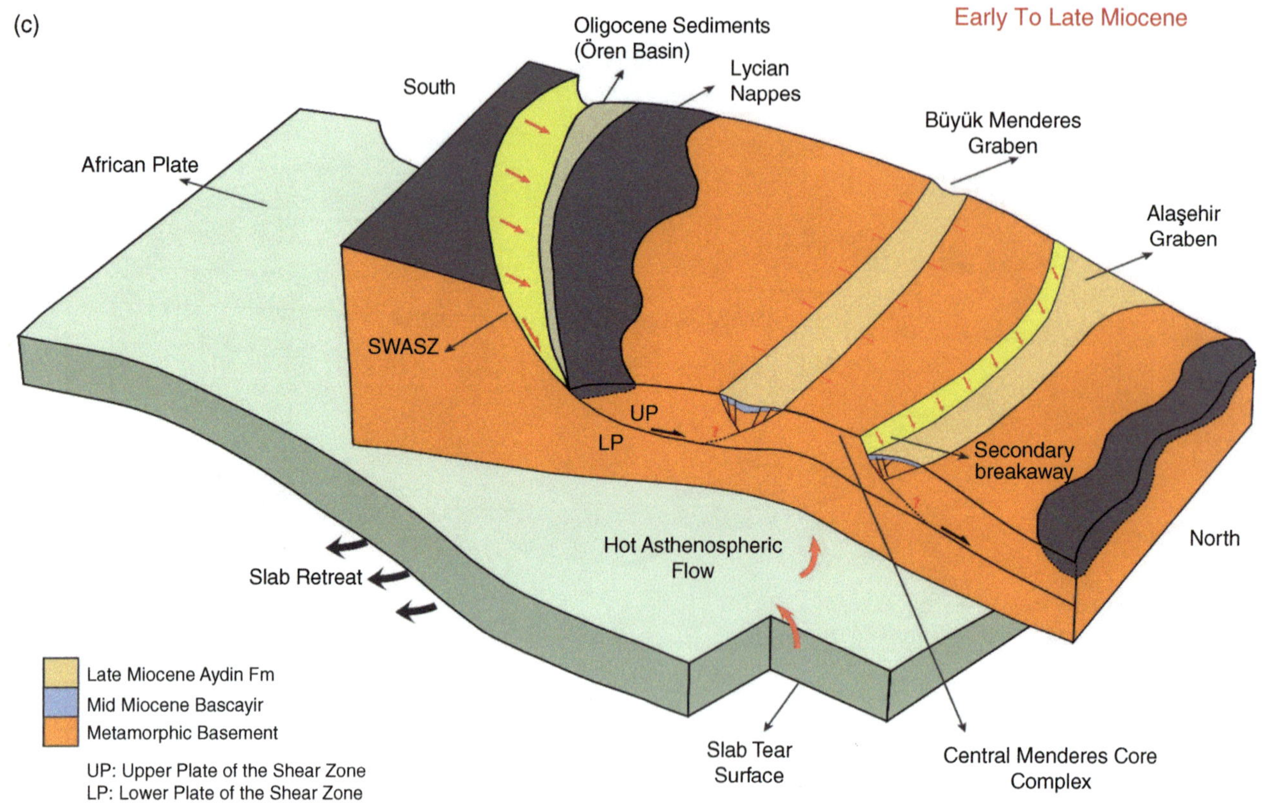

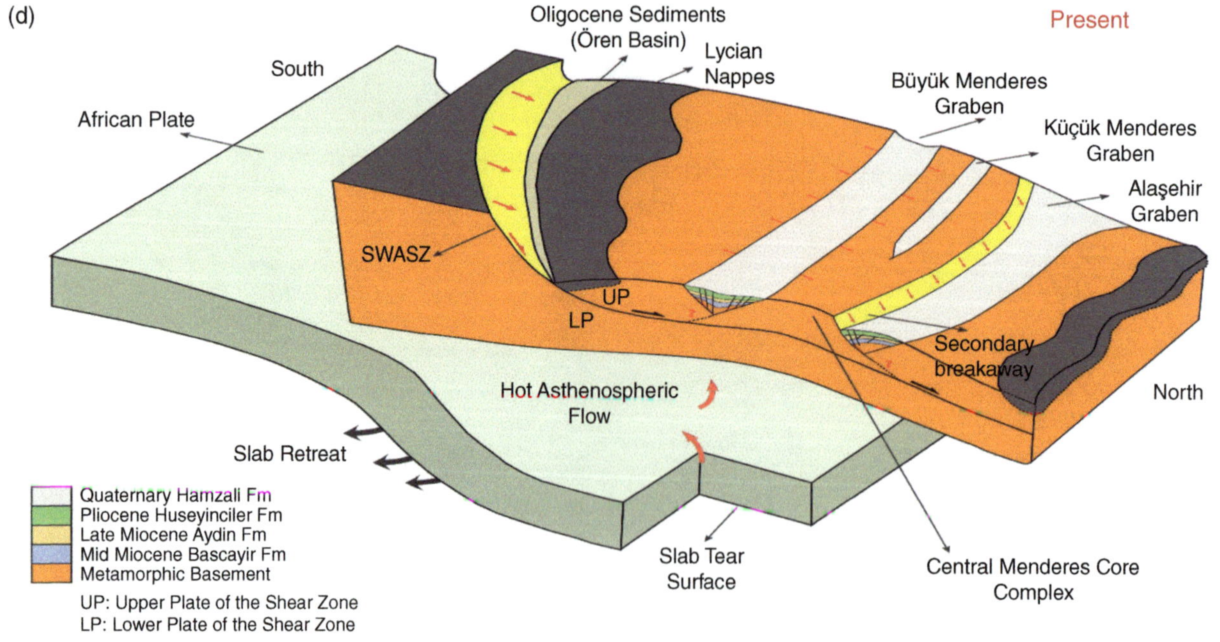

Figure 11.21 (Continued)

Figure 11.21b). Three lines of evidence support this interpretation: (a) the presence of Lycian nappes on the footwall of the SWASZ, containing the Taurids Marbles and Ophiolites (Seyitoğlu et al., 2004; Çemen et al., 2006); and (b) the presence of conglomerate fragments of the Taurids Marbles and Ophiolites in the Oligocene sedimentary rocks of the Ören and Kale-Tavas basins in the southern Menderes Massif formed along the hanging wall of the SWASZ (Seyitoğlu et al., 2004; Çemen et al., 2006; Figure 11.21b). These fragments were eroded from the upper plate (cover rocks) of the Menderes Massif during the Oligocene extension (Seyitoğlu et al., 2004; Çemen et al., 2006; Ersoy et al., 2014). (c) The extension on the SWASZ can also be seen from the seismic reflection profile interpreted by Kurt et al., 1999. The seismic profile shows the listric geometry of the southern segment of the SWASZ related to the formation of a rollover structure and the late Oligocene basin fill over the basement (Kurt et al., 1999; Çemen et al., 2006).

The domal uplift of the Menderes Massif on the hanging wall of the SWASZ experienced a second stage of extension in the early Miocene, which was distributed nonuniformly along the central part of the Menderes Massif (Figure 11.21c). This extension may be related to the slab tear of the African Plate between the Hellenic and Cyprus arcs that caused hot asthenospheric flow beneath the crust (Çemen et al., 2006; Biryol et al., 2011; Gessner et al., 2013; Ersoy et al., 2014; Mahatsente et al., 2017).

During the Miocene extension, the high-angle Alaşehir Detachment surface formed in the hanging wall of the SWASZ as a secondary breakaway zone (Figure 11.21c). This may have caused the AG to evolve on the limb of the dome as a supradetachment basin. Meanwhile, the M.G. BMG has experienced a rift-type deformation due to the presence of the hot asthenospheric flow during the formation of the secondary breakaway (Figure 11.21c). The rifting caused the M.G. BMG to be initiated as a bowl-shaped depression along the south-dipping, high-angle normal fault with its synthetic splays as a result of footwall uplift of the northern margin. This geometry, together with the continued extension, formed the north-dipping normal fault as the antithetic of the primary fault, most probably as a result of rapid thinning of the lithosphere due to hot asthenospheric material beneath western Turkey during the Miocene Cenozoic extension (Lavier & Buck 2002; Göğüş, 2014; Mahatsente et al., 2017). As the footwall was uplifted by isostatic rebound during extension, the north- and south-dipping boundary faults formed the current wedge shape geometry in the BMG (Figure 11.21c,d). This rift-type extension is evidenced by the Miocene sediments filling the basin center almost symmetrically and gradually thinning to the edges and rotation of both boundary faults as interpreted in available seismic lines (Figures 11.10, 11.12, and 11.13).

During the early Pliocene, extension in western Turkey was triggered by westward lateral extrusion of the Anatolian Plate related to the formation of the NAFZ about 5 Ma ago (Şengör, 1979; Şengör & Yılmaz, 1981; Şengör et al., 1984; Çemen et al., 1993,1999, 2006). This, together with the ongoing retreat of the African slab, produced an extension along the CMMCC and most probably caused the formation of the Küçük Menderes Graben (Figure 11.21d).

11.9. CONCLUSIONS

The major conclusions of this study are summarized here:

- The BMG has experienced a rift-type extension under the control of two main boundary faults that were active during the Miocene extension along the northern and southern margins of the graben.
- The Neogene sedimentary units thicken toward the central part of the graben and become thinner toward the northern and southern margins. This indicates that the BMG has experienced syn-extensional deformation related to growth faulting.
- Tectonic subsidence rates and the kinematic model constructed during this study show that the depocenter shift to the southern margin is related to the normal movement along the north-dipping boundary fault, which was also active along the southern margin of the graben during the deposition of the Aydin Formation in the late Miocene.
- The kinematic model constructed during this study proposes that the graben initiation started along the high-angle south-dipping boundary fault on the northern margin of the BMG. Due to rapid uplift during ongoing N-S extension, the north-dipping boundary fault was formed as an antithetic of the south-dipping boundary fault. Afterward, the two boundary faults were active and rotated to create accommodation space for the sedimentation along the BMG.
- The rotation of the boundary faults and their splays vary along the graben from east to west within the study area. The southern flank shows more rotation than the northern flank toward the west after the intrabasinal high, which indicates that the tectonic activity of the southern flank after the early Miocene might have changed the polarity of extension and depocenter of sedimentary rocks deposition from east to the west.
- The isochore maps constructed during this study show that the Miocene sedimentary thickness of the graben (the Bascayir and Aydin formations) decreases toward the west, which might indicate that the graben opening initiated from the eastern side and propagated westward. However, the Paleocene-Quaternary isochore map shows that the graben has had almost a uniform thickness of sedimentary rocks from east to west. This indicated that the graben had experienced more post-Miocene sedimentary rock deposition toward the west.

- The fault orientation of the synthetic and antithetic normal faults of the boundary faults indicates that the normal faults on the edges of the margin have shallower dip angles than those at the central part of the graben. This suggests that the faults on both flanks developed first and experienced footwall uplift and fault rotation due to continuous extension.

ACKNOWLEDGMENTS

We sincerely thank TPAO for providing 2D seismic data and well data for this study. TPAO also provided financial support during Osman Merey's Master's degree program at the University of Alabama, Department of Geological Sciences. This chapter is mostly based on Osman Merey's MS thesis project.

REFERENCES

2D Move. (2012). *Manual*. Midland Valley Company.

Allen, A. P., & Allen, R. J. (2013). *Basin analysis, principles and application to petroleum play assessment* (3th ed.). Hoboken: Wiley-Blackwell.

Biryol, C. B., Beck, S. L., Zandt, G., & Ozacar, A. A. (2011). Segmented African lithosphere beneath the Anatolian region inferred from teleseismic P- wave tomography. *Geophysical Journal International*, *184*, 3, 1037–1057.

Bozkurt, E. (2000). Timing of extension on the Büyük Menderes Graben, western Turkey, and its tectonic implications. *Geological Society, London, Special Publications*, *173*(1), 385–403.

Bozkurt, E., & Park, R. G. (1994). Southern Menderes Massif: an incipient metamorphic core complex in western Anatolia, Turkey. *Journal of the Geological Society*, *151*(2), 213–216.

Candan, O., Koralay, E., Akal, C., Kaya, O., Oberhansli, R., Dora, O. O., et al. (2011). Supra-Pan- African unconformity between core and cover series of the Menderes Massif/Turkey and its geological implications. *Precambrian Research*, *184*, 1–23.

Catlos, E. J., & Çemen, İ. (2005). Monazite ages and rapid exhumation of the Menderes Massif, western Turkey. *International Journal of Earth Sciences*, *94*, 204–217.

Çemen, I. (2010). Extensional tectonics in the basin and range, the Aegean, and western Anatolia: Introduction. *Tectonophysics*, *488*(1), 1–6.

Çemen, I., Catlos, E. J., Gogus, O., & Özerdem, C. (2006). *Postcollisional extensional tectonics an exhumation of the Menderes Massif in the western Anatolia extended terrane* (p. 409). GSA, Special Paper: Turkey.

Çemen, I., Göncüoğlu, M. C., Erler, A., Kozlu, H., & Perinçek, D. (1993). Indentation tectonics and associated lateral extrusion in east, southeast and central Anatolia. *Geological Society of America Abstracts with Programs*, *25*(7), A116.

Çemen, I., Göncüoğlu, M. C., & Dirik, K. (1999). Structural evolution of the Tuzgölu Basin in Central Anatolia, Turkey. *Journal of Geology*, *107*, 693–706. doi: 10.1086/314379

Çifçi, G., Pamukçu, O., Çoruh, C., Çopur, S., & Sözbilir, H. (2010). Shallow and deep structure of a supradetachment basin based on geological, conventional deep seismic reflection sections and gravity data in the Büyük Menderes Graben, Western Anatolia. *Geophysics*, *32*, 271–290.

Çiftçi, N. B., & Bozkurt, E. (2009). Pattern of normal faulting in the Gediz Graben, SW Turkey. *Tectonophysics*, *473*, 234–260.

Cohen, H. A., Dart, C. J., Akyüz, H. S., & Barka, A. A. A. (1995). Syn-rift sedimentation and structural development of Gediz and Büyük Menderes Graben, western Turkey. *Journal of the Geological Society of London*, *152*, 629–638.

Dewey, J. F. (1988). Extensional collapse of orogens. *Tectonics*, *7*, 1123–1139.

Dewey, J. F., & Şengor, A. M. C. (1979). Aegean and surrounding regions: complex multiplate and continuum tectonics in a convergent zone. *Geological Society of America Bulletin*, *90*, 84–92.

Emre, T. & Sözbilir, H. (1997). *Field evidence for metamorphic core complex, detachment faulting and accommodation faults in the Alaşehir and Büyük Menderes grabens (western Anatolia)*. Paper presented at the International Earth Science Colloquium: Aegean Region, Izmir.

Ersoy, E. Y., Çemen, I., Helvacı, C., & Billor, Z. (2014). Tectono-stratigraphy of the Neogene basins in western Turkey: implications for tectonic evolution of the Aegean Extended Region. *Tectonophysics*, *635*, 33–58.

Gessner, K., Gallardo, L. A., Markowitz, V., Ring, U., & Thomson, S. N. (2013). What caused the denudation of the Menderes Massif: review of crustal evolution, lithosphere structure, and dynamic topography in southwest Turkey. *Gondwana Research*, *24*, 243–274.

Gessner, K., Ring, U., Johnson, C., Hetzel, R., Passchier, C. W., & Güngör, T. (2001). An active bi-vergent rolling-hinge detachment system: Central Menderes metamorphic core complex in western Turkey. *Geology*, *29*, 611–614.

Göğüş, O. H. (2004). *Geometry and tectonic significance of the Büyük Menderes Detachment in the Bascayir area, Büyük Menderes Graben, western Turkey* (Master's thesis). Oklahoma State University.

Göğüş, O. H. (2014). Rifting and subsidence following lithospheric removal in continental back arcs. *Geology*, *43*(1), 3–6.

Hetzel, R., Ring, U., Akal, C., & Troesch, M. (1995b). Miocene NNE-directed extensional unroofing in the Menderes massif, southwestern Turkey. *Journal of the Geological Society of London*, *152*, 639–654.

İnaner, H., Sümer, Ö., & Akbulut, M. (2019). New geosite candidates at the western termination of the Büyük Menderes Graben and their importance on science education. *Geoheritage*, ***11***, 1291–1305. doi: 10.1007/s12371-019-00406-w

Işık, V., & Tekeli, O. (2001). Late orogenic crustal extension in the northern Menderes Massif (western Turkey); evidence for metamorphic core complex formation. *International Journal of Earth Sciences*, *89*, 757–765.

Jolivet, L., Faccenna, C., Huet, B., Labrousse, L., Le Pourhiet, L., Lacombe, O., et al. (2013). Aegean tectonics: Strain localisation, slab tearing and trench retreat. *Tectonophysics*, *597–598*, 1–33.

Kazancı, N., Dündar, S., Mehmet, C. A., & Gürbüz, A. (2009). Quaternary deposits of the Büyük Menderes Graben in western Anatolia, Turkey: implications for river capture and the longest Holocene estuary in the Aegean Sea. *Marine Geology*, *264*, 165–176.

Kocyiğit, A., Yusufoglu, H., & Bozkurt, E. (1999). Evidence from the Gediz graben for episodic two-stage extension in western Turkey. *Journal of the Geological Society of London*, *156*, 605–616.

Konak, N., Akdeniz, N., & Öztürk, E. M. (1987). Geology of the south of Menderes massif, correlation of Variscan and pre-Variscan events of the Alpine Mediterranean Mountain Belt (guide book for the field excursion along western Anatolia, Turkey). *IFCP Project, 5*: 42–53.

Kurt, H., Demirbağ, E., & Kuşçu, I. (1999). Investigation of submarine active tectonism in the Gulf of Gökova, southwest Anatolia-southeast Aegean Sea, by multi-channel seismic reflection data. *Tectonophysics*, *305*, 477–496. doi: 10.1016/S0040-1951999000037-2

Lavier, L. L., & Buck, W. R. (2002). Half graben versus large-offset low angle normal fault: importance of keeping cool during normal faulting. *Journal of Geophysical Research*, *107*(B6), 2122. doi: 10.1029/2001JB000513

Le Pichon, X., & Angelier, J. (1979). The Hellenic arc and trench system: a key to the neotectonic evolution of the Eastern Mediterranean area. *Tectonophysics*, *60*, 1–42. doi: 10.1016/0040-1951(79)90131-8

Le Pichon, X., & Angelier, J. (1981). The Aegean Sea. *Philosophical Transactions of the Royal Society of London*, *300*, 357–372.

Mahatsente, R., Alemdar, S., & Çemen, İ. (2017). Effect of slab-tear on crustal structure in Southwestern Anatolia. In İ. Çemen & Y. Yılmaz (Eds.), *Active global seismology* (pp. 103–119). Geophysical Monograph Series.). American Geophysical Union. doi: 10.1002/9781118944998.ch4

McKenzie, D. (1978). Active tectonics of the Alpine-Himalayan belt: the Aegean Sea and surrounding regions. *Geophysical Journal of the Royal Astronomical Society*, *55*, 217–254.

Meulenkamp, J. E., Wortel, W. J. R., Van Wamel, W. A., Spakman, W., & Hoogerduyn Strating, E. (1988). On the Hellenic subduction zone and geodynamic evolution of Crete since the late Middle Miocene. *Tectonophysics*, *146*, 203–215.

Ring, U., Johnson, C., Hetzel, R., & Gessner, K. (2003). Tectonic denudation of a Late Cretaceous-Tertiary collisional belt-regionally symmetric cooling patterns and their relation to extensional faults in the Anatolide belt of western Turkey. *Geological Magazine*, *140*, 1–21.

Robein, E. (2003). *Velocities, time-imaging and depth-imaging in refection seismics*. EAGE.

Sclater, J. G., & Christie, P. A. F. (1980). Continental stretching; an explanation of the post-Mid- Cretaceous subsidence of the central North Sea basin. *Journal of Geophysical Research*, *85*, 3711–3739.

Sert, S. (2015). *Subsurface structural geology of the eastern part of the Büyük Menderes Graben, western Turkey: implications for structural evolution of the Alaşehir and Büyük Menderes Grabens* (Master's thesis). University of Alabama.

Seyítoğlu, G., & Scott, B. C. (1996). The cause of north-south extensional tectonics in western Turkey: tectonic escape vs. back-arc spreading vs. orogenic collapse. *Journal of Geodynamics*, *22*, 145–153. 10.1016/ 0264-3707(96)00004-X

Seyitoğlu, G., Çemen, I., & Tekeli, O. (2000). Extensional folding in the Alaşehir (Gediz) Graben, western Turkey. *Journal of the Geological Society*, *157*(6), 1097–1100.

Seyitoğlu, G., Işık, V., & Çemen, I. (2004). Complete Tertiary exhumation history of the Menderes massif, western Turkey: an alternative working hypothesis. *Terra Nova*, *16*, 358–364. doi: 10.1111/j,1365-3121.2004.00574.x

Seyitoğlu, G. & Işık, V. (2015). Late Cenozoic extensional tectonics in western Anatolia: exhumation of the Menderes core complex and formation of related basins. *Bulletin of Mineral Research and Exploration, 151*, 2. doi: https://doi.org/10.19111/bmre.49951

Seyitoğlu, G., Tekeli, O., & Çemen, I. (2002). The role of the flexural rotation/rolling hinge model in the tectonic evolution of the Alaşehir Graben, western Turkey. *Geological Magazine*, *139*, 15–26.

Şen, Ş. & Seyitoğlu, G. (2009). Magnetostratigraphy of early-middle Miocene deposits from E-W trending Alaşehir and Büyük Menderes grabens in western Turkey, and its tectonic implications. In D. J. J. van Hinsbergen et al. (Eds.), *Geodynamics of collision and collapse at the Africa-Arabia-Eurasia subduction zone* (Special Publications 311, pp. 321–342). Geological Society London.

Şengör, A. M. C. (1979). The North Anatolian Transform Fault: its age, offset and tectonic significance. *Journal of the Geological Society of London*, *13*, 269–282.

Şengör, A. M. C., Görür, N., & Saroğlu, F. (1985). Strike-slip faulting and related basin formation in zones of tectonic escape: Turkey as a case study. In K. T. Biddle & N. Christie-Blick (Eds.), *Strike-slip deformation basin formation and sedimentation* (Special Publication (Vol. 37, pp. 227–264). Society of Economic Paleontologists and Mineralogists.

Sengör, A. M. C., Satir, M., & Akkök, R. (1984). Timing of tectonic events in the Menderes Massif, western Turkey: implications for tectonic evolution and evidence for Pan-African basement in Turkey. *Tectonics*, *3*(7), 693–707.

Şengör, A. M. C., & Yilmaz, Y. (1981). Tethyan evolution of Turkey: a plate tectonic approach. *Tectonophysics*, *75*, 181–241.

Sozbilir, H., & Emre, T. (1991). Extensional tectonics and the geometry of related macroscopic structures: field evidence from the Gediz detachment, western Turkey. *Turkish Journal of Earth Sciences*, *10*(2), 51–67.

Spakman, W., Wortel, M. J. R., & Vlaar, N. J. (1988). The Hellenic subduction zone: a tomographic image and its geodynamic implications. *Geophysical Research Letters*, *15*, 60–63.

Sumer, Ö., Akbulut, M., & İnaner, H. (2020). New geosite candidates from Urla (İzmir, Western Anatolia, Turkey): a list of geological assets nested with antique and modern cultural heritage. *Turkish Journal of Earth Sciences*, *29*(6), 8. doi: 10.3906/yer-2002-6

Ünay, E., Göktaş, F., Hakyemez, H. Y., Avşar, M., & San, O. (1995), Dating of the sediments exposed at the northern part of the Büyük Menderes graben (Turkey) on the basis of Arvicolidae (Rodentia, Mammalia). *Geological Bulletin of Turkey*, 38, 75–80.

Withjack, O. M., Schlische, W. R., & Olsen, P. (2002). Rift-basin structure and its influence on sedimentary systems. (Special Publication 73, pp. 57–81). Society for Sedimentary Geology.

Yılmaz, O. (2001). *Seismic data analysis*. Society of Exploration Geophysics.

Yılmaz, Y., Genç, S. C., Gürer, F., Bozcu, M., Yilmaz, K., Karacik, Z., et al. (2000). When did the western Anatolian grabens begin to develop?: Geological Society. *London, Special Publications*, *173*(1), 353–384.

Zhu, L., Mitchell, J. B., Akyol, N., Çemen, I., & Kekovali, K. (2006). Crustal thickness variations in the Aegean region and its implications for the extension of continental crust. *Journal of Geophysical Research*, *111*(B1). doi: 10.1029/2005JB003770

12

Transfer Zones in Extended Terranes: Examples from Western Anatolia

Elizabeth J. Catlos[1] and İbrahim Çemen[2]

ABSTRACT

Transfer zones, locations where strain is transferred from one structural element to another, are common features in the Aegean and Anatolian microplates. These features link fault systems and accommodate broader-scale tectonic processes in the area, including the extrusion of the microplates and extension driven by the rollback of the African (Nubian) subduction slab. Transfer zones outcrop at local scales, where transfer faults activate due to regional tectonic forces (e.g., in the Sea of Marmara, Thrace Basin, and Menderes Massif). Broader transfer zones are more complex and may have different origins, connections, and overlaps. Two regional-scale Western Anatolian transfer zones, the İzmir–Balıkesir Transfer Zone and Uşak-Muğla Transfer Zone, are depicted as ~150 km long strike-slip faults. However, their specific dimensions and relationship to fault systems exposed within the zones they encompass are debated. Here, we identify and describe transfer zones mapped in specific locations in Western Anatolia. These zones are significant across diverse geological processes, from identifying hazards to regions of hydrocarbon deposition and mineralizations. In addition, they shed light on lithosphere dynamics as the microplates experience the fundamental tectonic processes of rollback, slab tear, extension, and extrusion.

12.1. INTRODUCTION

Transfer zones occur in settings where faults die out along tip lines and deformation is accommodated by the displacement between adjacent faults (Figure 12.1) (Dahlstrom, 1970; Gibbs, 1984; Bosworth, 1985; Rosendahl, 1987; Morley et al., 1990; Twiss & Moores, 1992; Morley, 1995; Faulds & Varga, 1998; Gawthorpe & Hurst, 1993; Gürboğa, 2014). Transfer zones play a significant role in accommodating deformation in the Aegean and Western Anatolian regions (Figure 12.2) (Barbot & Weiss, 2021). In these zones, strain transfers from one structural element to another, and displacement changes between individual fault and basin segments (Gawthorpe & Hurst, 1993; Barbot & Weiss, 2021). Despite their importance in accommodating subduction and extrusion dynamics, where specific transfer zones occur across Western Anatolia is debated (Sangu et al., 2020; Seyitoğlu et al., 2022). How and when these zones initiate and why they occur are critical for understanding diverse geological processes, including the tectonic impacts of the subducting African (Nubian) slab, the localization of geohazards, processes facilitating hydrocarbon deposits, and mineralizations.

This chapter aims to review the present-day knowledge of some transfer zones documented in surface exposures

[1] *Jackson School of Geosciences, Department of Earth and Planetary Sciences, University of Texas at Austin, Austin, Texas, USA*

[2] *Department of Geological Sciences, University of Alabama, Tuscaloosa, Alabama, USA*

Extensional Tectonics: Rifting and Continental Extension, Geophysical Monograph 278, First Edition.
Edited by İbrahim Çemen and Elizabeth J. Catlos.

DOI: 10.1002/9781119773757.ch12

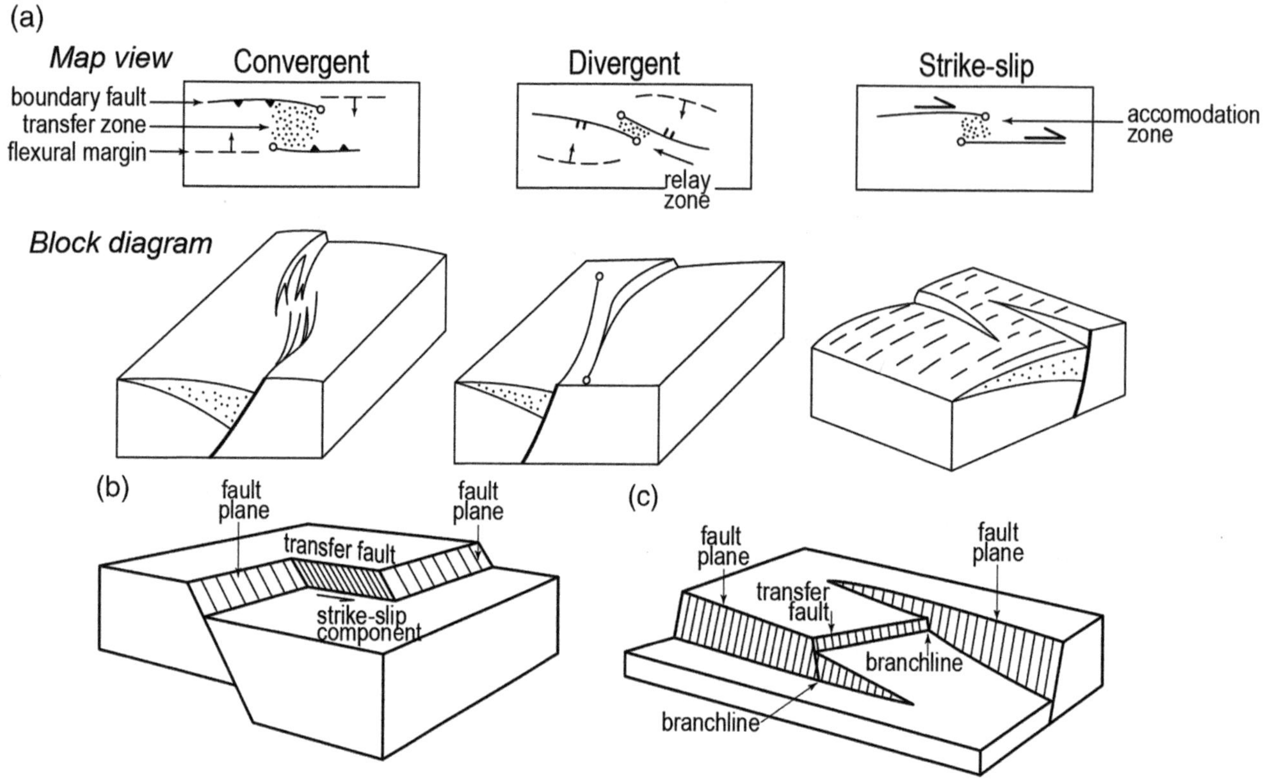

Figure 12.1 Examples of some transfer zone geometries. (a) Morley et al. (1990) and Morley (1995) describe the three types of transfer zones between half grabens. (b) Extension transfer fault that transitions displacement between two faults during extension. (c) Connecting geometry with additional complexity after Peacock & Sanderson (1994).

in Western Anatolia at the local and regional scales. We also summarize outstanding questions regarding the origin of some specific larger-scale transfer zones and their role in accommodating extrusion tectonics.

12.2. NOMENCLATURE AND BACKGROUND OF TRANSFER ZONES

The term *transfer zone* implies that adjacent faults were active simultaneously and motion transferred from one fault system to another as the system propagated (Morley, 1995). Transfer zones allow stress to be accommodated between individual faults or basin segments along the length of a deformation zone. These zones form critical structural elements in sedimentary basins that developed due to extension in the Aegean region (Morley et al., 1990; Coskun, 1997; Konstantinovskaya et al., 2007; Çiftçi et al., 2010).

Transfer-zone geometries are classified in specific ways (Figure 12.1) (Faulds & Varga, 1998; Gamal et al., 2021). However, in nature, transfer zones have various permutations depending on the faults' orientation, sense of motion, linking elements, scale, lithology, and nature and timing of deformation. Transfer zones can be interbasin, linking individual half grabens, also termed half-graben to half-graben (H-H) transfer zones. Intrabasin transfer zones link individual fault segments and are termed fault-to-fault (F-T-F) transfer zones (Moustafa, 2002). The transfer zones in Figure 12.1 are stepover zones (fault jogs) where fault motion transfers from one fault pane to another. Step-up zones, where high-angle faults physically transect moderately angled ones, and step-down zones, where concave depressions develop in a footwall block of a normal fault zone, are also common in the Aegean (Stewart & Hancock, 1991).

Transfer faults with a strike-slip sense of motion, ramps, or relay zones often accommodate deformation in transfer zones (e.g., Peacock & Sanderson, 1994; Whipp et al., 2017). Transfer faults are considered hard linkage (Bally, 1981; Gibbs, 1984), whereas ramps and relay zones are soft linkages between *en échelon* fault tip lines (Rosendahl et al., 1986; Leeder & Gawthorpe, 1987; Larsen, 1988). Transfer-zone geometries may contain hard and soft linkages (Morley et al., 1990).

Transfer zones are divided into those that affect individual fault strands or basins (local scale) and those that link major structural elements (>120 km) over broad areas (20–50 km, regional scale; Gibbs, 1984; Morley et al., 1990; Faulds & Varga, 1998; Moustafa, 2002). Local-scale transfer zones within a basin system affect sediment transport near the fault but would have little impact on basin stratigraphy (Gawthorpe & Hurst, 1993). Broad

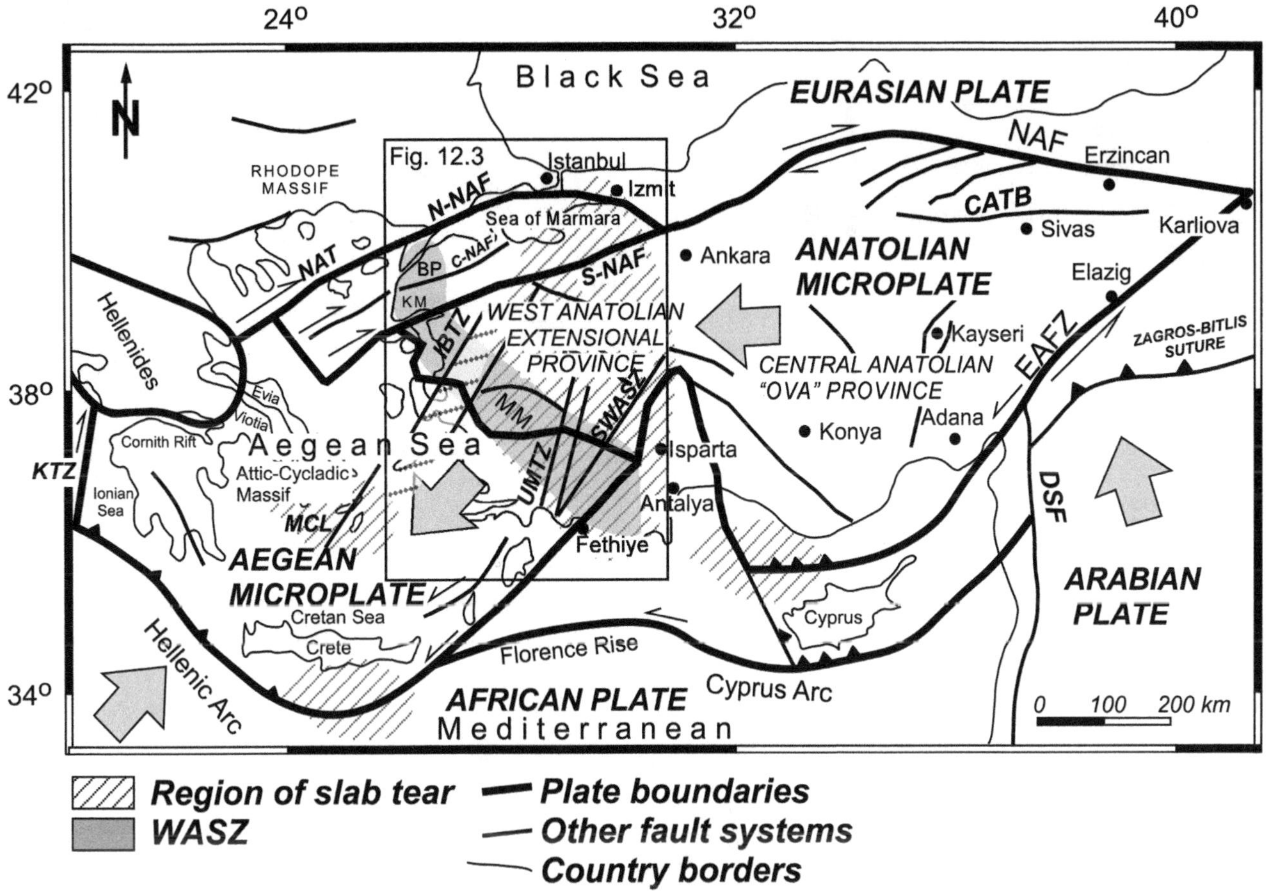

Figure 12.2 Tectonic map of the Aegean and Anatolian microplates. Plate boundaries after McClusky et al. (2000), Nyst & Thatcher (2004), Piper et al. (2010), Harrison et al. (2013), and Tan (2013). Only some major fault systems are labeled. EAFZ, East Anatolian fault zone; CATB, Central Anatolian Thrust Belt; DSF, Dead Sea Fault; KM, Kazdağ Massif; KTZ, Kephalonia Transform Zone; MCL, Mid-Cycladic lineament; MM, Menderes Massif; İBTZ, İzmir–Balıkesir transfer zone; NAT, North Aegean Trough; NAF, North Anatolian Fault (N- northern, C- central, and S- southern segments); UMTZ, Uşak-Muğla Transfer Zone; SWASZ, South West Anatolian Shear Zone. Region of slab tear in western Turkey and the Aegean after Jolivet et al. (2015), near Crete (Özbakır et al., 2013), Cyprus (Woodside et al., 2002), and between the Aegean domain and the Menderes Massif (Roche et al., 2019). Boundaries between Central and Western Anatolia after Şengör et al. (1984). The inset shows the location of Figure 12.3.

transfer zones develop when fault orientations exhibit low variability, as indicated by a b-factor of 1.5 (Morley, 1995). Broad transfer zones have fault systems often manifesting as an *en échelon* pattern. They frequently divide regions into domains with different amounts of extension, fault orientation, or tilt (e.g., Faulds & Varga, 1998).

Specific categories for transfer zones exist. Relay structures, zones, or ramps exist between extensional faults (Figure 12.1a; Peacock & Sanderson, 1991; Whipp et al., 2017). These are a type of synthetic segment boundary in that their sense of displacement is like that of the main fault (Whipp et al., 2017). In extensional systems, transfer zones often contain strike-slip or oblique-slip faults that trend parallel to the extension direction and facilitate a strain transfer between extended terrains (Faulds & Varga, 1998). Accommodation zones or spaces exist between stepovers in strike-slip faults (Dooley & McClay, 1997; Sadiki et al., 2021). Boundary faults in transform zone systems can be scissor faults, *en échelon* arrays of Riedel shears that change their sense of motion along strike as the hanging wall changes to the footwall (Naylor et al., 1986).

12.3. GEOLOGICAL BACKGROUND

12.3.1. Assembly

The Aegean and Anatolian microplates (Figure 12.2) developed as a series of terranes separated by ophiolite remnants as ancient oceans collided along subduction zones. Continental fragments, separated by the Paleo- and Neo-Tethyan Oceans, converged during the Late Cretaceous–Eocene period. For example, the Vardar–İzmir–Ankara Ocean closed during the Pontide Orogeny.

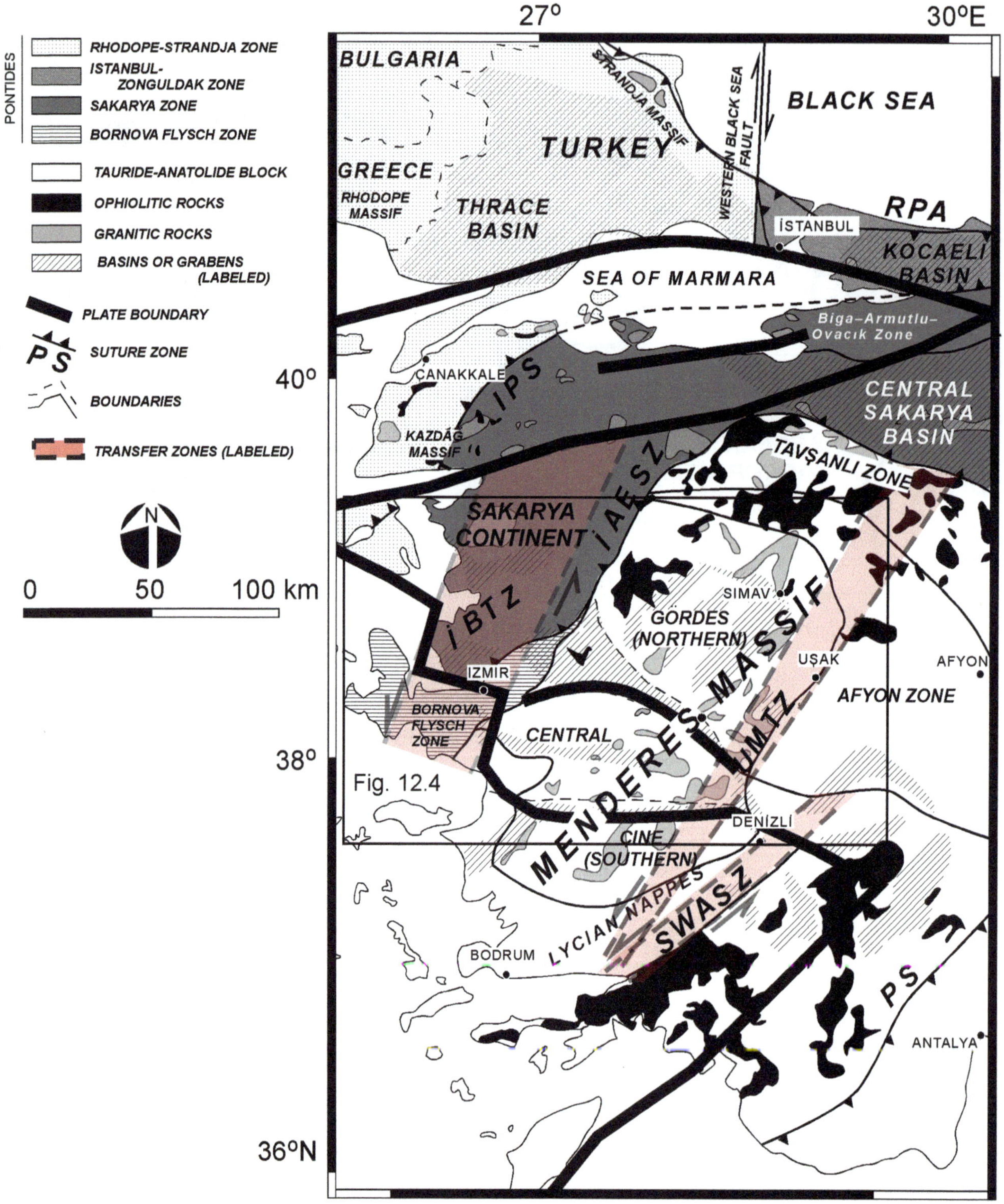

Figure 12.3 Geological map of Western Anatolia focusing on the ophiolite and granite assemblages along the boundary between the Aegean and Anatolia microplates. Plate boundary after Nyst & Thatcher (2004). Terrane boundaries, major fault systems, and transfer zones after Okay (2008), Akbayram et al. (2017), Oner et al. (2010), and Karaoğlu and Helvacı (2014). İAESZ, İzmir-Ankara-Erzincan suture zone; İBTZ, İzmir-Balıkesir Transfer Zone; IPS, Intra-Pontide suture zone; RPA, Rhodope-Pontide arc; PS, Pamphylian suture zone; SWASZ, South West Anatolian Shear Zone; UMTZ, Uşak-Muğla Transfer Zone. See Figure 12.2 for the location of this map in its regional context. The inset shows the location of Figure 12.4.

Ophiolite and high-pressure/low-temperature (HP/LT) rock formations identify suture zones (Figure 12.3; Şengör & Yılmaz, 1981; Okay, 2008; Moix et al., 2008; Okay & Tüysüz, 1999; Pourteau et al., 2016; Okay et al., 2020).

The Intra-Pontide suture zone (IPS) within the Pontide belt exists between the Istanbul-Zonguldak Unit (also known as the Istanbul-Zonguldak Zone, Istanbul Nappe, or Istanbul Zone) and the Sakarya continental zone (Figure 12.3; Yiğitbaş et al., 2004). The nature of the IPS remains mysterious, described alternatively as an accretionary complex, a suprasubduction zone, or a relic of a former ocean basin that once extended into eastern Europe (Okay et al., 1996; Robertson & Ustaömer, 2004; Göncüoğlu et al., 2012, 2014; Marroni et al., 2014; Akbayram et al., 2017; Sayit et al., 2016; Frassi et al., 2018). Mafic IPS formations are linked to a Middle Jurassic–Early Cretaceous Ocean (Göncüoğlu et al., 2014; Sayit et al., 2016).

The Sakarya Composite Terrane (or zone) of the Pontide region is separated from the Tauride Anatolide block by the İzmir Ankara Erzincan suture zone (İAESZ) (Şengör & Yılmaz, 1981; Okay & Tüysüz, 1999; Tekin et al., 2002; Göncüoğlu, 2010). This feature is concealed beneath the Aegean Sea, obscuring its connection to ophiolite formations in Greece. The suture has been associated with the Axios-Vardar suture zone, marking the closure of the Vardar Ocean (Channell & Kozur, 1997; Okay & Tüysüz, 1999; Tekin et al., 2002; Moix et al., 2008). However, the Axios-Vardar suture zone may link to the IPS (Şengör & Yılmaz, 1981; Okay & Satir, 2000; Okay et al., 2001; Beccaletto & Jenny, 2004; Okay et al., 2010; d'Atri et al., 2012; Di Rosa et al., 2019). Numerous ophiolite fragments associated with the Vardar Ocean in continental Greece suggest a multistage Late Jurassic–Early Cretaceous closure (Sharp & Robertson, 2006; Saccani et al., 2008; Papanikolaou, 2021).

The İAESZ exposes numerous isolated blocks of ophiolite-related suites and is identified as the North Ophiolite Belt in Western Anatolia. A Late Cretaceous estimate (104 ± 1 Ma) for the initiation of the Hellenic arc is derived from Lu-Hf ages obtained from prograde garnet in the Tavşanlı Zone, a segment of metamorphic soles from the Anatolia ophiolite, which is in the North Ophiolite Belt near the İAESZ (Figure 12.3; Pourteau et al., 2019; Crameri et al., 2020). The Tavşanlı Zone is one of the largest continuous expanses of exhumed subducted crust (Davis & Whitney, 2006; Okay, 2002; Shin et al., 2013). Pressures peaking at ~26 kbar are documented in blueschists and lawsonite-eclogites of the Tavşanlı Zone (Davis & Whitney, 2006). These rocks offer insight into the exhumed northern continental margin of the Anatolide-Tauride Block influenced by the closure of a branch of the Neo-Tethyan Ocean (Shin et al., 2013).

The initiation and duration of subduction along the Hellenic arc are debated (LePichon et al., 2019; Catlos & Çemen, 2023). A Late Cretaceous–Jurassic subduction zone initiation age for the Hellenic arc implies that the continental lithospheric fragments and the ancient oceans between them are accreted without creating a new subduction zone. This model aligns more with a "Cordilleran" emplacement model, where ophiolites are progressively placed over subduction complexes through accretionary processes (Wakabayashi & Dilek, 2003). Note that a Late Cenozoic (Eocene-Pliocene) subduction initiation age is also proposed for the Hellenic arc based on the analysis of topography, estimates of slab age and depth, paleomagnetism, and the timing of metamorphism, volcanic activity, and sedimentation (Meulenkamp et al., 1988; Spakman et al., 1988; Papadopoulos, 1997; Brun & Sokoutis, 2010; Le Pichon et al., 2019). This age follows an induced-transference subduction model, where a new subduction zone initiates following the jamming of an older subduction zone by buoyant crust due to regional compression, uplift, and underthrusting (Stern, 2004; Stern & Gerya, 2018). This "Tethyan" subduction model (Wakabayashi & Dilek, 2003) places ophiolites over passive continental margins or microcontinents due to collisional events.

The southward paleogeographic extension of the Tavşanlı zone is known as the Afyon Zone (referred to as the Afyon-Bolkardağ Zone by Okay [1986] and Özdamar et al. [2013] or Ören-Afyon Zone by Pourteau et al. [2013]) (Figure 12.3; Candan et al., 2005; Pourteau et al., 2010; Akal et al., 2013; Özdamar et al., 2013). This zone contains Paleozoic-Mesozoic Tauride metasedimentary and metavolcanic rocks, which exhibit regional greenschist to blueschist facies metamorphism (Fe-Mg carpholite and glaucophane) (Okay, 1984; Candan et al., 2005; Pourteau et al., 2010; Özdamar et al., 2013). The precise southern boundary of the Afyon Zone is unclear, and a section may also be exposed between the southern Menderes (Çine) Massif and Lycian Nappes (Okay, 1986; Candan et al., 2005; Pourteau et al., 2013; Ustaömer et al., 2020). HP/LT metamorphism in the Afyon Zone is believed to have occurred at 70–65 Ma concurrent with the closure of the Neo-Tethyan Ocean (Pourteau et al., 2010, 2013; Özdamar et al., 2013; Plunder et al., 2013). Sections of the Afyon Zone may have subducted beneath the Tavşanlı Zone during the Late Cretaceous, as indicated by zircon ages of granites intruding Tavşanlı Zone blueschist and altered ophiolitic assemblages (Speciale et al., 2012; Shin et al., 2013).

The Lycian and Antalya (Tekirova) ophiolites are also part of southern Turkey's Tauride Belt (Figure 12.3). The Antalya ophiolite includes harzburgitic tectonites, ultramafic to mafic cumulates, isotropic gabbros, and sheeted dikes. Based on geochemical analyses, the ophiolite is comprised of basic magmas, island arc tholeiite (IAT) isolated dikes and isotropic gabbros, low-Ti boninitic isolated dikes, and Ocean Island Basalt (OIB)-type gabbros (Bağcı & Parlak, 2009). The OIB-related assemblages are linked to a late-stage asthenospheric activity due to slab break-off or subduction of a ridge system before emplacement (Bağcı & Parlak, 2009). The Lycian ophiolites form the Lycian nappes' upper thrust sheets, and the metamorphic sole rocks are intensely deformed (Çelik & Chiaradia, 2008). These ophiolites are coarse-grained ultramafic tectonites, dunite, chromitite, and harzburgite (Engin & Hirst, 1970; Collins & Robertson, 1998; Faruk Çelik & Delaloye, 2003; Çelik & Chiaradia, 2008) with cumulates of olivine-gabbro, gabbro-norite, anorthositic gabbro, anorthosite, and pyroxenite (Çelik & Chiaradia, 2008). Geochemical data suggest that they are associated with a suprasubduction zone setting (Collins & Robertson, 1998). Early dating studies using K-Ar methods indicated that these ophiolites experienced Late Cretaceous metamorphism at 94 ± 4 to 102 ± 7 Ma (Antalya) and 83 ± 3 to 102 ± 4 Ma (Lycian) (Thuizat et al., 1981). Older amphibole K-Ar (114.4 ± 11.0 to 130.4 ± 11.5 Ma) and younger (48.7 ± 9.7 and 49.5 ± 9.9 Ma) ages were also reported from the Antalya amphibolite and gabbro, respectively (Yilmaz & Maxwell, 1984). More recent $^{40}Ar/^{39}Ar$ dates from amphiboles and white micas from the metamorphic sole of these ophiolites are Late Cretaceous (90.7 ± 0.5 to 93.8 ± 1.7 Ma, Lycian; and 91.2 ± 2.3 to 93.6 ± 0.8 Ma, Antalya; Çelik et al., 2006). Similar hornblende plateau ages from the amphibolites of the Lycian ophiolites (near Köyceğiz) and Antalya and are interpreted as recording a similar metamorphic event (Çelik et al., 2006).

12.3.2. Core Complex Development

The Aegean and Anatolian microplates underwent post-collisional extension, leading to the emergence of several Alpine metamorphic core complexes, including the Menderes, Crete, Attic-Cycladic, Rhodope, and Kazdağ Massifs (Lister et al., 1984; van Hinsbergen & Meulenkamp, 2006; Figures 12.2–12.4). The commencement of this extension is associated with the rollback of the African (Nubian) slab beneath the Hellenic arc. It is estimated that the penetration of the subducting slab to depths of 1,200–1,400 km occurred around 50 Ma, potentially triggering the onset of extension in the region by 30–35 Ma (Jolivet & Faccenna, 2000; Brun & Faccenna, 2008; Capitanio et al., 2010). Moreover, the exhumation of core complexes might have begun in the Rhodope Massif (Bulgaria), which shares lithological correlations and timing of metamorphism with the Kazdağ Massif in the Biga Peninsula of western Turkey

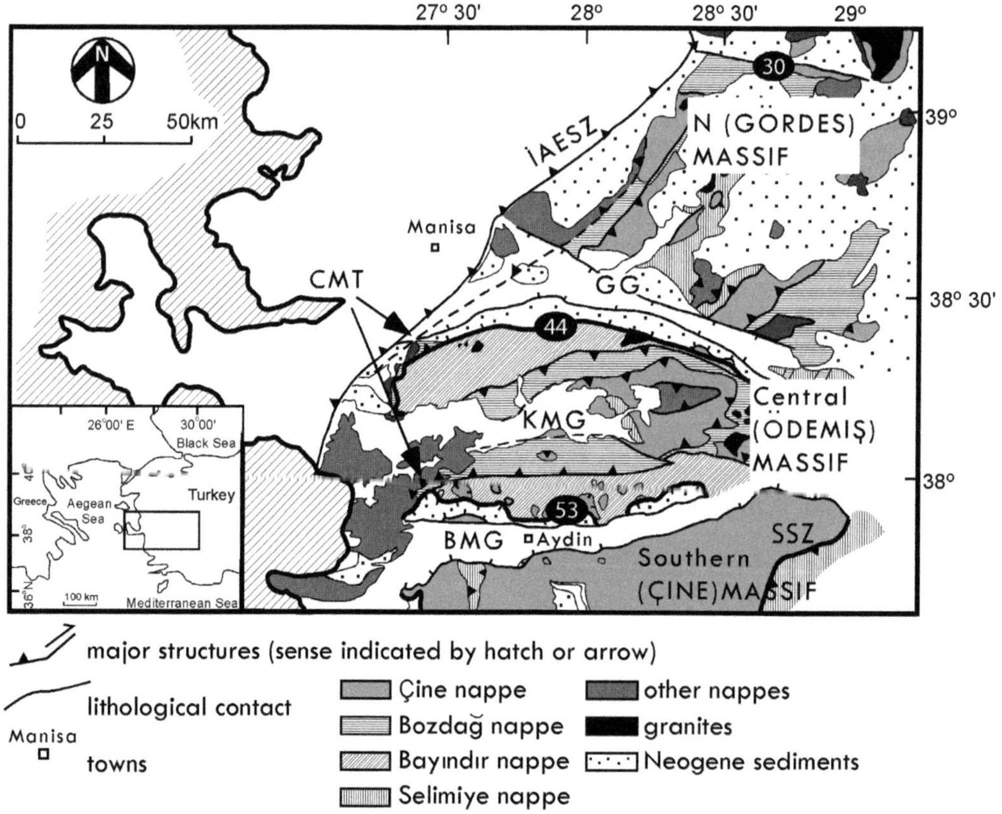

Figure 12.4 Map of the Menderes Massif. İAESZ, İzmir-Ankara-Erzincan suture zone; CMT, Cyclades Menderes Thrust; GG, Gediz Graben; KMG, Küçük Menderes Graben; BMG, Büyük Menderes Graben. See Table 12.1 for the names of the numbered faults (Adapted from Emre et al., 2018).

(Beccaletto & Jenny, 2004; Okay & Satır, 2000; Okay et al., 1996). These complexes underwent regional metamorphism and rapid exhumation during the Late Oligocene–Early Miocene period (Okay & Satır, 2000; Bonev et al., 2009; Cavazza et al., 2009).

The Menderes Massif is the largest metamorphic core complex in the region and exposes ~40,000 km^2 of metamorphic and igneous rocks (Figure 12.4; Şengör et al., 1984; Bozkurt & Park, 1994; Hetzel et al., 1995; Yılmaz et al., 2001; Işik & Tekeli, 2001; Catlos & Çemen, 2005; Etzel et al., 2019, 2020). Low-angle detachment faults and high-angle normal faults bound sedimentary basins and separated the massif into northern (Gördes), central (Ödemiş), and southern (Çine) submassifs. The massif has also been mapped as a large-scale recumbent fold (Okay, 2001), a series of nappes stacked during south-directed thrusting (Ring et al., 1999; Ring, 2001; Gessner et al., 2001), or north-directed thrusting (Hetzel et al., 1995a, 1995b; Gessner et al., 2013). Peak Barrovian metamorphism of pelitic and crystalline rocks, termed the Main Menderes Metamorphism (MMM), occurred during the late Eocene–early Oligocene (Satır & Friedrichsen, 1986; Bozkurt & Satır, 2000; Lips et al., 2001; Catlos & Çemen, 2005; Etzel et al., 2019, 2020). The region then transitioned from shortening to extension by the Mid-Oligocene (<30 Ma), driven by the rollback of the African (Nubian) slab (Etzel et al., 2020; Catlos & Çemen, 2023). Granites exposed in the northern portion of the Menderes Massif rapidly cooled below ~250 °C by 20 Ma during a pulse of regional cooling due to rapid denudation as N-NNE movement along the regional Simav fault (or detachment, #30 in Emre et al., 2018) (Ring et al., 2003; Gessner et al., 2013; Etzel et al., 2020). This chapter uses the numbered nomenclature established for active faults in Turkey by Emre et al. (2018) (Table 12.1). In the central (Ödemiş) Menderes Massif,

Table 12.1 Faults, fault zones, segments, depressions, and grabens discussed in this chapter.

Number[a]	Structure name	Reference[b]	Figure[c]	Sense of motion[d]
North Anatolian fault zone				
1-25	Düzce segment	Yousefi-Bavil et al. (2022)	12.5	RL
1-37	Ganos segment	Janssen et al. (2009)	12.5	RL
4	Evciler fault	Sözbilir et al. (2016)	12.5	RL
5	Edremit fault zone	Sözbilir et al. (2016)	12.5	N
Havran-Balya fault				
23-1	Havran segment	Sözbilir et al. (2016)	12.8	RL
23-2	Osmanlar segment	Sümer et al. (2018)	12.8	RL
23-3	Turplu segment	Ece and Schroeder (2007)	12.8	RL
23-4	Ovacık segment	Sözbilir et al. (2016)	12.8	RL
Balıkesir fault zone				
24-1	Gökçeyazı segment	Koçyiğit and Gürboğa (2021)	12.8	RL
24-2	Kepsut segment	Seyitoğlu et al. (2022)	12.8	RL + N
Zeytindağ fault zone				
25a	Zeytindağ Graben	Cengiz et al. (2021)	12.8	RL + N
25b	Kıranlı fault zone	Sangu et al. (2020)	12.8	RL
25c	Kaşıkçı fault zone	Cengiz et al. (2021)	12.8	N
25d	Yenişakran fault zone	Sangu et al. (2020)	12.8	LL
31	Yenifoça fault	Ozener et al. (2013)	12.8	N
Bergama fault zone				
26a	Bergama Graben	Polat et al. (2008)	12.8	N
26b	Örenli-Eğiller depression (graben)	Yang et al. (2021)	12.8	N
26c	Dikili fault zone	Sangu et al. (2020)	12.8	LL
26d	Kınık-Turgutalp fault zone	Cengiz et al. (2021)	12.8	RL
Soma-Kırkağaç (Kırkağaç) fault zone				
27a	Soma basin	Westerweel et al. (2020)	12.8	N
27b	Bigadiç basin	Koç-Taşgın and Diniz-Akarca (2018)	12.8	N
Gelenbe fault zone				
28-1	Eastern segment	Seyitoğlu et al. (2022)	12.8	RL
28-2	Western segment	Seyitoğlu et al. (2022)	12.8	RL
29	Düvertepe fault zone	Ece et al. (2013)	12.6	RL + N

(Continued)

Table 12.1 (Continued)

Number[a]	Structure name	Reference[b]	Figure[c]	Sense of motion[d]
Simav fault zone				
30-1	Sındıgrı segment	Caglayan et al. (2021)	12.6, 12.8	RL + N
30-2	Çaysimav segment	Günay et al. (2021)	12.6	RL + N
30-3	Şaphane segment	Gürboğa (2014)	12.6, 12.10	RL + N
30-4	Abide segment	Gürboğa et al. (2013)	12.10	RL + N
30-5	Banaz segment	Inel et al. (2013)	12.10	RL + N
30-6	Elvanpaşa segment	Akyar (2019)	12.10	RL + N
30-7	Sinanpaşa segment	Akiska and Varol (2020)	12.10	RL + N
İzmir area faults				
38	İzmir fault zone (E-W trending)	Drahor and Berge (2017)	12.8	N
36	Güzelhisar fault zone	Mozafari et al. (2021)	12.8	RL
37	Menemen fault zone	Ozener et al. (2013)	12.8	RL
39	Tuzla fault	Ozener et al. (2013)	12.8	RL
41	Dagkızılca fault	Sözbilir et al. (2011)	12.8	RL
Gediz Graben (Detachment)				
44-1	Alaşehir fault, segment, detachment	Etzel et al. (2020)	12.4, 12.6, 12.7, 12.8, 12.10	N
44-2	Salihli segment	Catlos et al. (2010)	12.6, 12.7, 12.8, 12.10	N
44-3	Akçapinar segment	Kent et al. (2021)	12.8, 12.10	N
44-4	Armutlu segment	Tepe and Sözbilir (2017)	12.8	N
44-5	Nifdağı segment	Tepe and Sözbilir (2017)	12.8	N
44-6	Kemalpaşa fault	Tepe and Sözbilir (2017)	12.8	N
44-7	Manisa fault, fault zone	Duran et al. (2021)	12.8	N
44-8	Killik fault	Van Noten et al. (2019)	12.6, 12.10	N
44-9	Çapaklı segment	Van Noten et al. (2019)	12.6, 12.8, 12.10	N
44-10	Kemerdamları	Doğan et al. (2022)	12.6, 12.8	N
44-12	Halitpasa fault	Gürer (2022)	12.8	N
44-13	Ozanca fault	Eyidoğan (2020)	12.8	N
44-14	Gölmarmara fault	Eski et al. (2020)	12.8, 12.10	N
44-15	Akselendi fault	Eyidoğan (2020)	12.8	N
44-16	Akhisar fault	Eski et al. (2020)	12.8	N
44-17	Muradiye fault	Özkaymak and Sözbilir (2008)	12.8	N
48	Köprübaşı fault zone	Akıska et al. (2019)	12.6	RL
49	Kiraz fault	Gokten et al. (2001)	12.10	RL
Büyük Menderes Graben system				
53-3	Umurlu segment	Altunel (1999)	12.4, 12.10	N
53-4	Atça segment	Sümer et al. (2020)	12.4, 12.10	N
53-5	Pamukören segment	Haklıdır and Şengün (2020)	12.4, 12.10	N
53-6	Buharkent segment	Sümer et al. (2020)	12.4, 12.10	N
54	Çine fault	Özsayın et al. (2023)	12.10	N
55	Bozdoğan fault	Özsayın et al. (2023)	12.10	N
Milas Fault				
58-1	Karakuyu segment	Kırkan et al. (2023)	12.10	RL
58-2	Beçin segment	Akyüz et al. (2019)	12.10	RL
59	Yatağan fault	Basmenji et al. (2021)	12.10	N
60	Muğla fault	Aksoy et al. (2022)	12.10	N
62	Gökova fault zone	Dikbaş et al. (2022)	12.10	N
76	Beyağaç fault	Aydin and Tuncer (2021)	12.10	N
80	Tekkeköy (Babadağ) fault	Kumsar et al. (2016)	12.10	RL + N
81	Cankurtaran fault	Oudeika et al. (2020)	12.10	N
Pamukkale fault zone				
82-1	Akköy segment	Hançer (2013)	12.7, 12.10	N + LL
82-2	Gölemezli segment	Çakir (1999)	12.10	N + LL

Table 12.1 (Continued)

Number[a]	Structure name	Reference[b]	Figure[c]	Sense of motion[d]
82-3	Tripolis segment	Uysal et al. (2019)	12.10	N + LL
82-4	Hierapolis segment	Hancock and Altunel (1997)	12.10	N + LL
82-5	Serinyayla segment	Hancock et al. (1999)	12.10	N
82-6	Aydoğdu segment	Çakir (1999)	12.10	N + LL
82-7	İsmailbey segment	Hancock et al. (1999)	12.10	N
106	Çivril (-Dinar) fault	Özalp et al. (2018)	12.10	N
116	Güre fault zone	Kadir and Erkoyun (2013)	12.6, 12.10	N
117	Selendi fault zone	Ersoy et al. (2010)	12.6, 12.10	RL + N
118	Rahmanlar fault zone	Ersoy et al. (2010)	12.6, 12.10	RL + N
119	Kızılüzüm fault	Emre et al. (2018)	12.6, 12.10	N
120-2	Emet-Gediz fault zone	Mutlu (2020)	12.10	N
121	Çukurören fault	Doğan et al. (2017)	12.10	RL
127	Naşa fault zone	Mutlu (2020)	12.6	N

[a] Numbers after Emre et al. (2018).
[b] Reference that discusses the particular structure or feature.
[c] Figure that shows this structure or feature.
[d] N, normal; RL, right lateral strike slip; LL, left-lateral strike slip. More than two senses of motion indicates that this type of motion has been attributed to the feature.

extensional-related granites along the Gediz (Alaşehir, Kuzey, or Karadut) detachment (#44, Figure 12.4) experienced a more prolonged cooling/exhumation history, cooling below 250 °C during the Late Miocene–Pliocene (Etzel et al., 2020).

The Attic-Cycladic Massif forms a curved belt of metamorphic rocks parallel to the Hellenic trench (Figure 12.2; Robertson & Dixon, 1984; Keay et al., 2001). These rocks share a similar geological history with those in Western Anatolia, where metamorphic formations document Miocene continental extension following collision (Vandenberg & Lister, 1996). However, Cyclades assemblages typically exhibit HP (blueschist and eclogite facies) metamorphism alongside Barrovian prograde pressure-temperature (P-T) metamorphism (Lister et al., 1984; Beaudoin et al., 2015). The HP history is absent in portions of Naxos (Lamont et al., 2020) and Paros Island (Malandri et al., 2017). Certain areas also show evidence of intrusion by Late Miocene–age granites (such as the Mykonos-Delos-Rhenia metamorphic core complex, Denèle et al. [2011]; Serifos, Rabillard et al. [2015]; and Ikaria Island, Beaudoin et al. [2015]). The concluding phases of the exhumation of south Aegean core complexes are documented in rocks exposed on Crete. Metamorphic assemblages indicate extensional exhumation from the Late Miocene until the Pliocene (van Hinsbergen & Meulenkamp, 2006).

12.3.3. Extrusion

The Aegean microplate and the western part of the Anatolian microplate are currently transitioning from a stress regime primarily characterized by extension to transform tectonics (Papanikolaou & Royden, 2007; Cavazza et al., 2009). The active extensional strain has been reported to be largely absent except in specific localized areas (Brooks & Ferentinos, 1984; Papanikolaou & Royden, 2007; Chousianitis et al., 2013, 2015). The movement of the Anatolian and Aegean microplates, prompted by the collision of the Arabian plate, exemplifies tectonic escape, with associated lateral extrusion representing a fundamental process in the evolution of continents throughout Earth's history (Figure 12.2; e.g., Şengör & Yilmaz 1981; Burke & Şengör, 1986; Çemen et al., 1999). In the eastern part of the Anatolian microplate, collision occurs along the Zagros-Bitlis suture. The subducting Hellenic arc acts as a free lateral boundary, accommodating the motion of the Anatolian microplate toward the Aegean microplate. The right-lateral strike-slip systems of the North Anatolian Shear Zone (NASZ) and the North Anatolian Fault (NAF) play a significant role in accommodating the movement of the extruding plate and dominate the present-day geodynamics of the region (Figure 12.5; fault #1 in Emre et al., 2018).

GPS vectors show that the extrusion of the Anatolian microplate is widespread throughout the plate (Reilinger et al., 2006; Simão et al., 2016), and displacement is often attributed to the NASZ, NAF, and East Anatolian fault zone (EAFZ) (Figure 12.2). The NASZ and NAF extend for ~1,200 km from the Karlıova triple junction through the Sea of Marmara and Biga Peninsula (Armijo et al., 1999; Şengör & Zabcı, 2019). In western Turkey, the NASZ contains the NAF, a right-lateral structure that has accommodated from 25 to 110 km of displacement, depending on location, since the Late Miocene (Figure 12.5; Westaway, 1994; Yoshioka, 1996; Armijo et al., 1999; Hubert-Ferrari et al., 2002; Şengör & Zabcı, 2019).

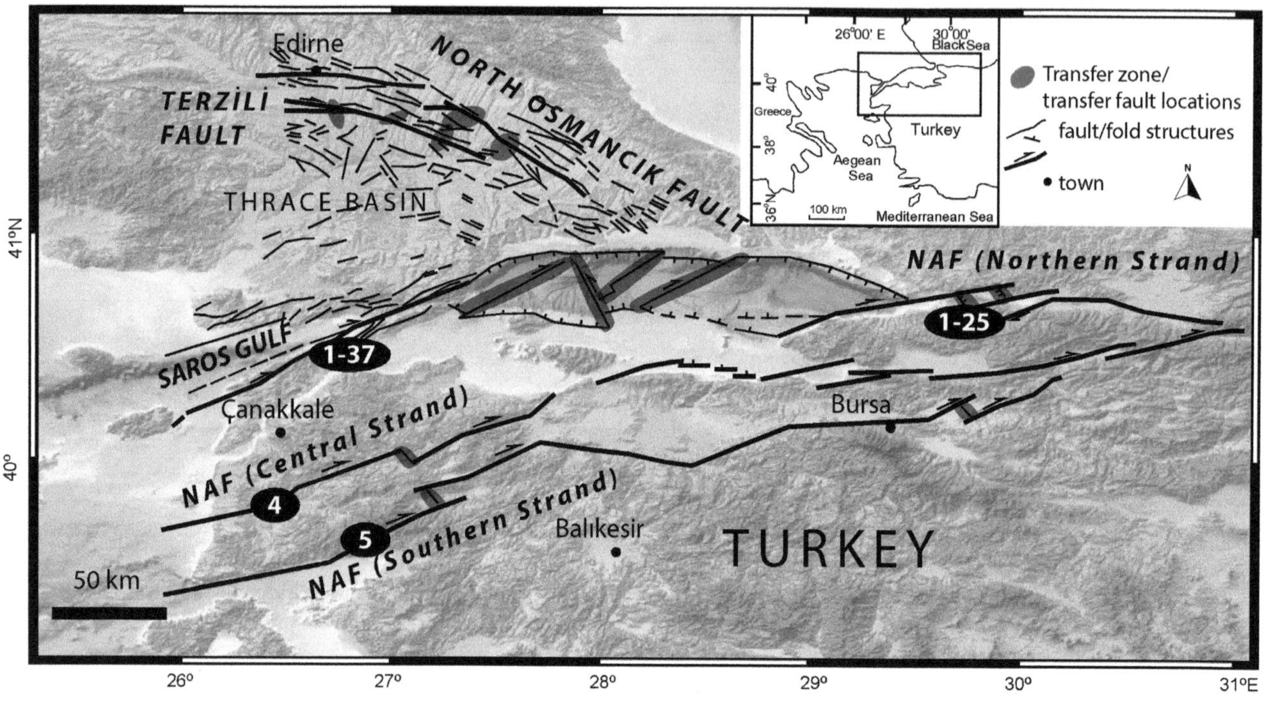

Figure 12.5 A map of northern Turkey and the NASZ shows major strands' locations and transfer zones (ArcGIS Online map hosted by Esri). Numbered faults after Emre et al. (2018) are listed in Table 12.1.

The NAF is divided into western transtentional, central translational, and eastern transpressional sections (e.g., Fraser et al., 2010). The NAF may have initiated in the east at 10–12 Ma (Perinçek & Çemen, 1990) in central Anatolia at 7–6 Ma and appeared in the Aegean region by the Pliocene (Şengör et al., 2005; Faccenna et al., 2006). In northern Turkey, the NASZ accommodates ~24 mm/year of slip (McClusky et al., 2000; Bulut et al., 2018). The NAF is thought to extend into the lithospheric mantle and has an expression controlled by older and structurally weak suture zones at depth (Fichtner et al., 2013).

Besides the NASZ and NAF, the Simav fault (#30 in Emre et al., 2018), which bounds the northern (Gördes) Menderes Massif, has been mapped as an active right-lateral strike-slip fault that accommodates extrusion (Figure 12.6; Ambraseys & Tchalenko, 1972; Şaroğlu et al., 1992). This sense of motion is also supported by offsets of metamorphic zones east of Simav (Seyitoğlu, 1997) and its relationship to the formation of the NAFZ (Doğan & Emre, 2006). However, analysis of earthquakes near the fault consistently indicates a normal mechanism (Görgün, 2014; Yolsal-Çevikbilen et al., 2014; Demirci et al., 2015; Karasözen et al., 2016; Bello et al., 2017; Mutlu, 2020). Speculation also exists that the Simav fault was initiated as a strike-slip system but switched to extension after the Late Miocene (Oygür & Erler, 2000). Alternatively, the Simav fault may have been generated as normal, but its latest motion has a strike-slip component. The Simav fault may also connect to the Muratdağ Fault near the town of Gediz in an *en échelon* pattern, which supports a right-lateral system (Ambraseys & Tchalenko, 1972).

In Western Anatolia, transfer zones outcropped locally within its extensional and strike-slip systems and more regionally at the boundaries between the Aegean and Anatolian microplates and those separating specific domains. Others describe transfer-zone systems within this region at depth (see review by Jolivet et al., 2021). Here, we document and describe the expression of transfer zones that outcrop at local and regional scales.

12.4. SOME LOCAL-SCALE TRANSFER ZONES (FAULTS)

Local-scale transfer zones or faults are at the scale of a specific location with unique physical features or topography. For example, the NASZ has several named local-scale transfer faults of significant concern regarding their hazards. In Western Anatolia, the Karadere-Düzce (or Düzce, #1–25, Emre et al., 2018) fault helped to form the Düzce pull-apart basin (Figure 12.5; Akyüz, 2002; Hitchcock et al., 2003; Ben-Zion et al., 2003; Şengör et al., 2005). The structure marks the beginning of a transition between a linear portion of the northern segment of the NAF and the western portion, where it bifurcates into different fault strands (Ben-Zion et al., 2003; Seyitoğlu & Işik, 2015; Sözbilir et al., 2016; Rost et al., 2021; Yousefi-Bavil et al., 2022). The fault geometry

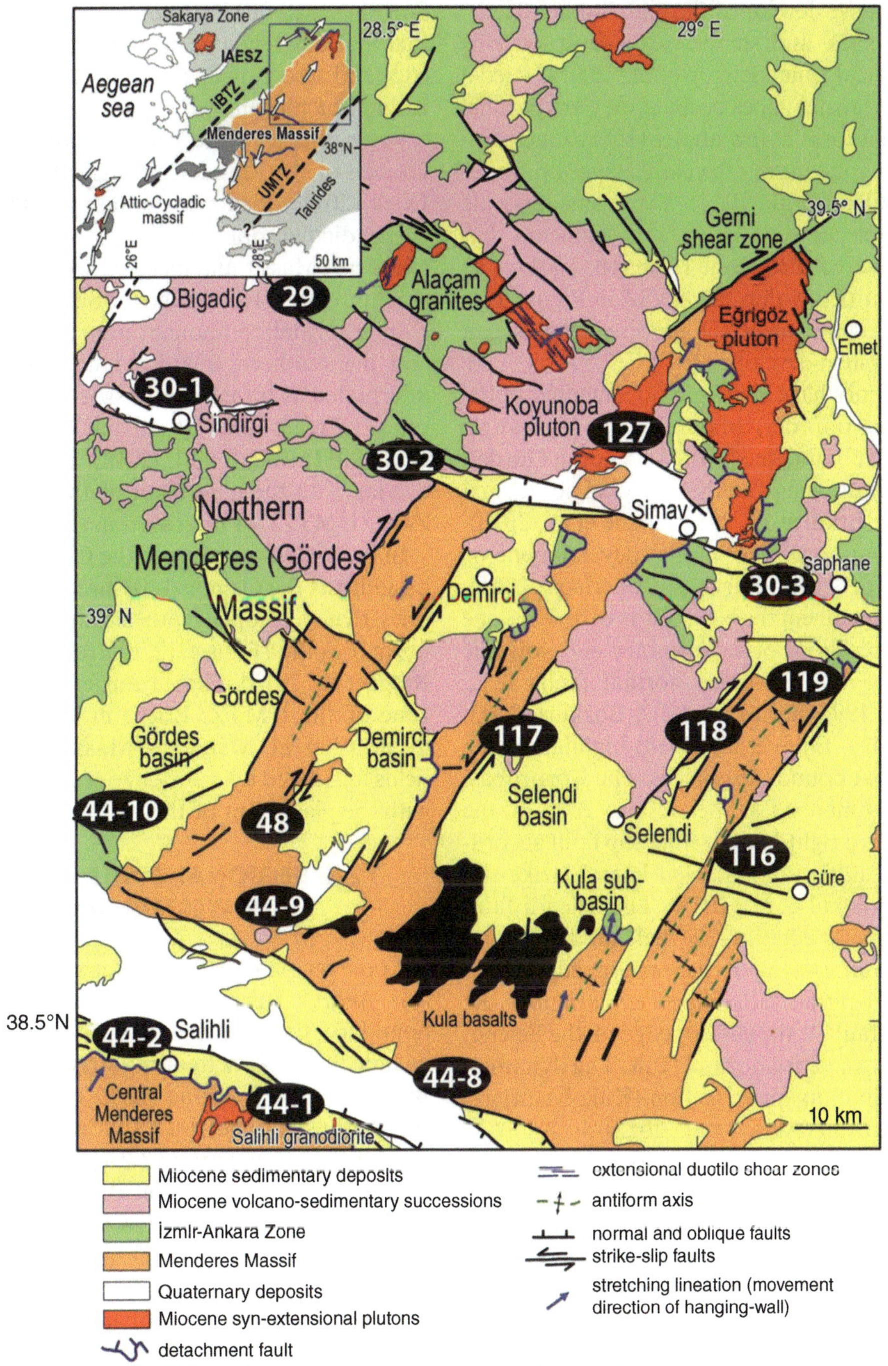

Figure 12.6 Map of the northern (Gördes) and portions of the central (Ödemiş) Menderes Massif after Erkül et al. (2017). Numbered faults after Emre et al. (2018) are listed in Table 12.1.

of the Sea of Marmara is also modeled as a pull-apart basin that transitioned to a system of strike-slip faults (Armijo et al., 1999, 2002; Rangin et al., 2004; Kuşçu, 2009). Dextral transfer faults identified within the Sea of Marmara accommodated motion along the NAF and inverted to form sediment depocenters (Wong et al., 1995; Rangin et al., 2004). These fault strands are active and provide conduits for gas emissions (Géli et al., 2008). The Thrace Basin also contains right-lateral wrench faults that became inactive by the Pliocene as strike-slip motion transferred to the NASZ and faults within the Marmara Sea became active (Figure 12.5; e.g., Perinçek, 1991). The basin has oil and gas reservoirs developed by active transfer zones during the Miocene (Coskun, 1997).

The NASZ may have developed through linking Riedel shears, tension gashes, and smaller-scale fault systems linked over geological time (Şengör et al., 2005). Riedel shears differ from transfer zones because they are parallel arrays of faults that form at low angles (15°) to the principal displacement zone during the early stages of strike-slip formation (Wilcox et al., 1973). These smaller-scale structures are visible as *en échelon* fault segments that link to form a principal shear zone (Ahlgren, 2001). The transfer zones identified within the NASZ in Figure 12.5 may instead be the outcome of the processes that began early in the fault system development.

Local transfer zones have also been documented in the northern Menderes (Gördes) Massif, where high-angle strike-slip faults that bound the NE-SW trending Gördes, Demirci, and Selendi basins connect low-angle normal faults (Figure 12.6; Erkül et al., 2017; Roche et al., 2018). However, this is only one view of the development of these basins (Bozkurt, 2003; Purvis & Robertson, 2005), and most authors document the basin fill as early Miocene (24.2–16.3 Ma) successions of sedimentary and volcanic assemblages bound by high-angle normal faults (e.g., Seyitoğlu & Scott, 1994; Seyitoğlu, 1997; Bozkurt, 2003; Bozkurt et al., 2011). Emre et al. (2018) highlight some active structures that bound the grabens. The Körprübaşı fault zone (#48) is mapped at the western edge of the Demirci basin and is a right-lateral strike-slip fault according to Emre et al. (2018) but is a left-lateral strike-slip fault according to Erkül et al. (2017). The Selendi fault (#117) at the western edge of the Selendi basin and the Rahmanlar fault (#118) at its eastern edge show both normal and right-lateral components, whereas the Kızılüzüm (#119) fault at the eastern edge of the Selendi basin is mapped as an active normal fault. The dynamics of the Selendi basin in its lower section (Kula basin) are dominated by extension (Aksay et al., 2022).

One of the more debated contacts is located along the northwestern edge of the Eğrigöz and Koyunoba plutons in the northern (Gördes) Menderes Massif (Figure 12.6). The Germi shear zone is a Miocene-age right-lateral shear zone that links two extensional domains with the Eğrigöz and Koyunoba magmatic domes (Erkül et al., 2017). However, this contact has long been considered part of the Simav fault system (#30 in Emre et al. [2018]). The earliest maps of the contact by Işik and Tekeli (2001) suggest that a normal fault is present on the northern portion of the Eğrigöz pluton. Erkül (2010), Ring and Collins (2005), and Işik et al. (2004) indicate that a normal fault is exposed along the western edges of both the Eğrigöz and Koyunoba plutons. Seyitoğlu et al. (2004) place a detachment within the central portion of the Eğrigöz pluton. In contrast, Ersoy et al. (2010) suggest that a detachment follows the outer boundaries of the Eğrigöz and Koyunoba bodies and is paralleled by a thrust fault along their entire perimeter (also mapped by Sözbilir et al. 2011). Thomson and Ring (2006) suggest that the detachment is prominent only along the northern edge and a portion of the central and eastern edges of the Koyunoba pluton.

Some smaller-scale transfer faults exist in the Gediz Graben of the Menderes Massif. For example, Oner and Dilek (2011) document numerous NNE-striking, basin-perpendicular oblique-slip hinge faults that divide the Menderes Massif and its basin sediments into fault blocks (Figure 12.7a). High-angle N-S trending strike-slip faults with a normal component are reported within both this and the southern Büyük Menderes Graben, which are interpreted as transfer or scissor faults between the structures (Çiftçi & Bozkurt, 2010; Oner & Dilek, 2011; Çifçi et al., 2011; Roche et al., 2018). Some of these structures overlap with the placement of the Uşak-Muğla Transfer Zone (UMTZ), as discussed in the next section.

In the eastern portion of the Gediz Graben, the Denizli basin hosts the Cambazlı fissure ridge comprising travertines that developed with a transfer zone between two normal faults (Figure 12.7b; Temiz & Eikenberg, 2011; Brogi et al., 2014). These transfer faults also lie within the zone of the UMTZ. Roche et al. (2018) document the localization of Menderes Massif Grabens' geothermal fields facilitated by transfer zones, some of which coincide with the placement of the regional-scale UMTZ.

12.5. REGIONAL-SCALE TRANSFER ZONES IN WESTERN ANATOLIA

At the regional scale, transfer zones within the Anatolia and Aegean microplates are primarily located at the boundary between the two microplates. These zones include the NE-SW trending strike-slip dominated İzmir–Balıkesir Transfer Zone (İBTZ), UMTZ (also transtensional transfer zone), and Southwestern Anatolian Shear Zone (SWASZ) (Figures 12.2 and 12.3; Çemen et al., 2006; Oner et al., 2010; Sözbilir et al., 2011; Gessner et al., 2013; Özkaymak et al., 2013; Uzel et al., 2013; Karaoğlu & Helvacı, 2014; Seghedi et al., 2015). Some of these transfer zones are considered significant portions of the larger Western Anatolian Shear Zone (WASZ) or Western Anatolian Extensional Province and may have developed due to mantle processes related to the subduction of the Aegean slab (e.g., Gessner et al., 2013; Uzel et al., 2020). In addition, Western Anatolian transfer zones are locations for ore deposits (Rabayrol et al., 2019; Rabayrol & Hart, 2021).

12.5.1. İzmir-Balıkesir Transfer Zone

The İBTZ is a wide zone located from the western edge of the Menderes Massif to near the Aegean coast and consists of several Miocene to recent transtensional basins with volcanic and magmatic intrusions (Figures 12.3, 12.8,

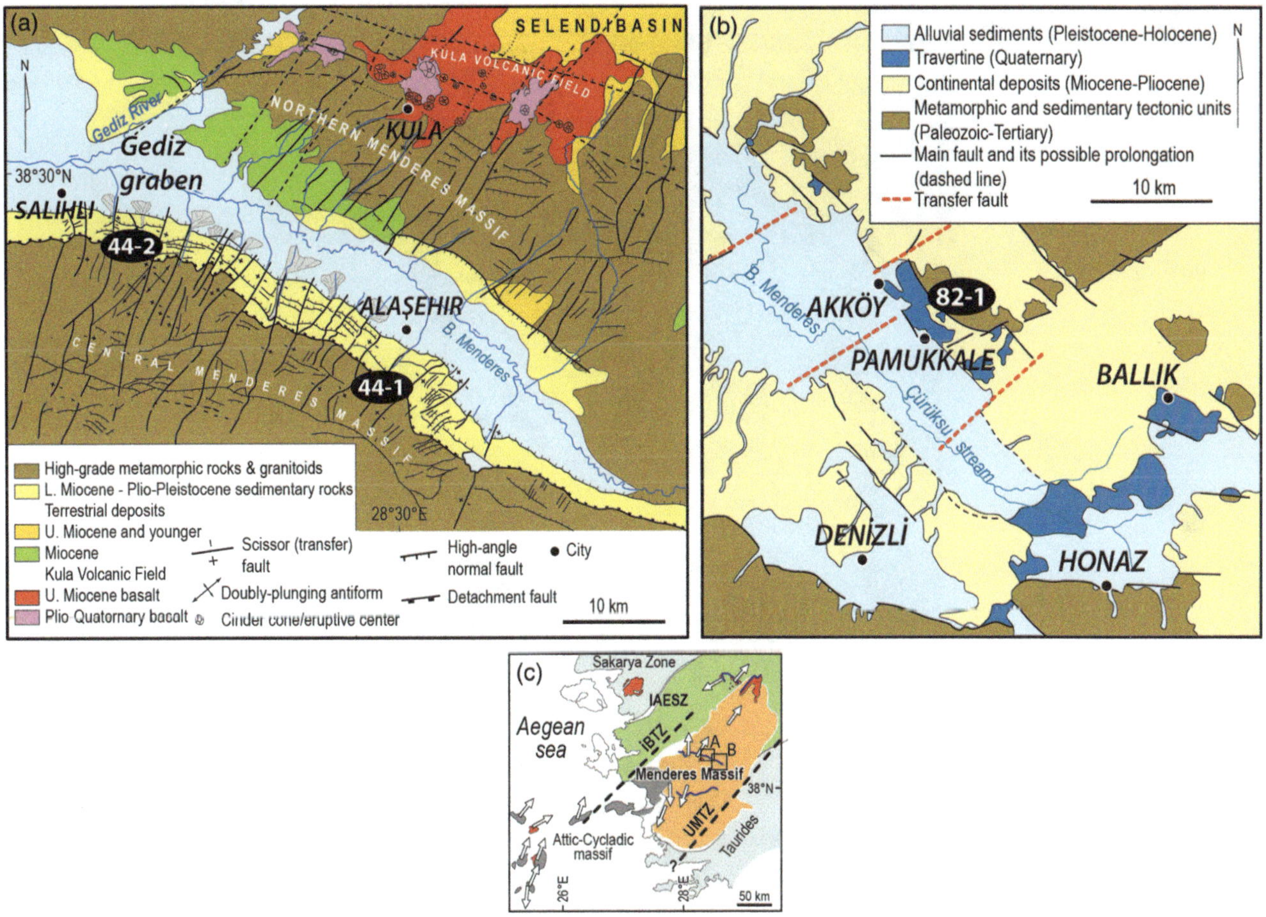

Figure 12.7 (a) Map across Gediz Graben documenting WNW-oriented normal and NNE-striking oblique, hinge (scissor) faults, which behave as boundary faults in a transfer zone system (Oner & Dilek, 2013 / Elsevier). (b) Map of the Denizli subbasin of the Gediz Graben. Interpretations of the transfer zones and travertine deposits are adapted from Brogi et al. (2014). (c) The location of (a) and (b) in the context of Western Anatolia.

and 12.9; Erkül et al., 2005; Ring et al., 1999; Kaya et al., 2007; Uzel & Sözbilir, 2008; Ersoy et al., 2011; Sözbilir et al., 2011; Uzel et al., 2013; Uzel et al., 2015; Uzel et al., 2020; Eyubagil et al., 2021). The zone is sometimes called the Western Anatolia Transfer Zone (Gessner et al., 2013, 2018). In its initial descriptions, the İBTZ was defined as a NE-trending dextral strike-slip zone (Sözbilir et al., 2003) and a transtensional brittle shear zone that deforms pre-Neogene basement rock units, early-middle Miocene volcano-sedimentary units, and Plio-Quaternary continental units (Uzel et al., 2013). When depicted as a single feature, it is shown as a ~150-km-long strike-slip fault (Eyubagil et al., 2021). Westerweel et al. (2020) bracket the İBTZ between the Çandarli-Karaburun and Gelembe-Kemapaşa (Kemalpaşa) lineaments (Figure 12.9b). Lineaments are defined as linear geomorphic features that are landscape expressions of underlying geologic structures such as faults, fractures, joints, continental margins, or submarine ridges (Kaiser, 1950; Bakliwal & Ramasamy, 1987; Muthumaniraja et al., 2019). Unless definitive evidence is provided, lineaments are not considered faults, or else they would be termed as such.

Figure 12.8 and Table 12.1 outline only some of the numerous fault systems and components within the İBTZ between the cities of İzmir and Balıkesir. We divide the main active structures into seven major fault zones (Havran-Balya, Balıkesir, Zeytindağ, Bergama, Soma-Kırkağaç, Gelenbe, and Simav) and those associated with the Gediz Graben system and around the İzmir region. We attempted to maintain the terminology of Emre et al. (2018), which summarizes active fault systems throughout Turkey. If we identified additional strands, we added a letter to their assigned number system.

Sangu et al. (2020) indicate that the İBTZ is not sufficiently understood, as the geological details of the zone are missing and only general descriptions are often provided for this area. Seyitoğlu et al. (2022) note that the specific terminal ends of the transfer zone are unclear. As an example, the northern edge of the

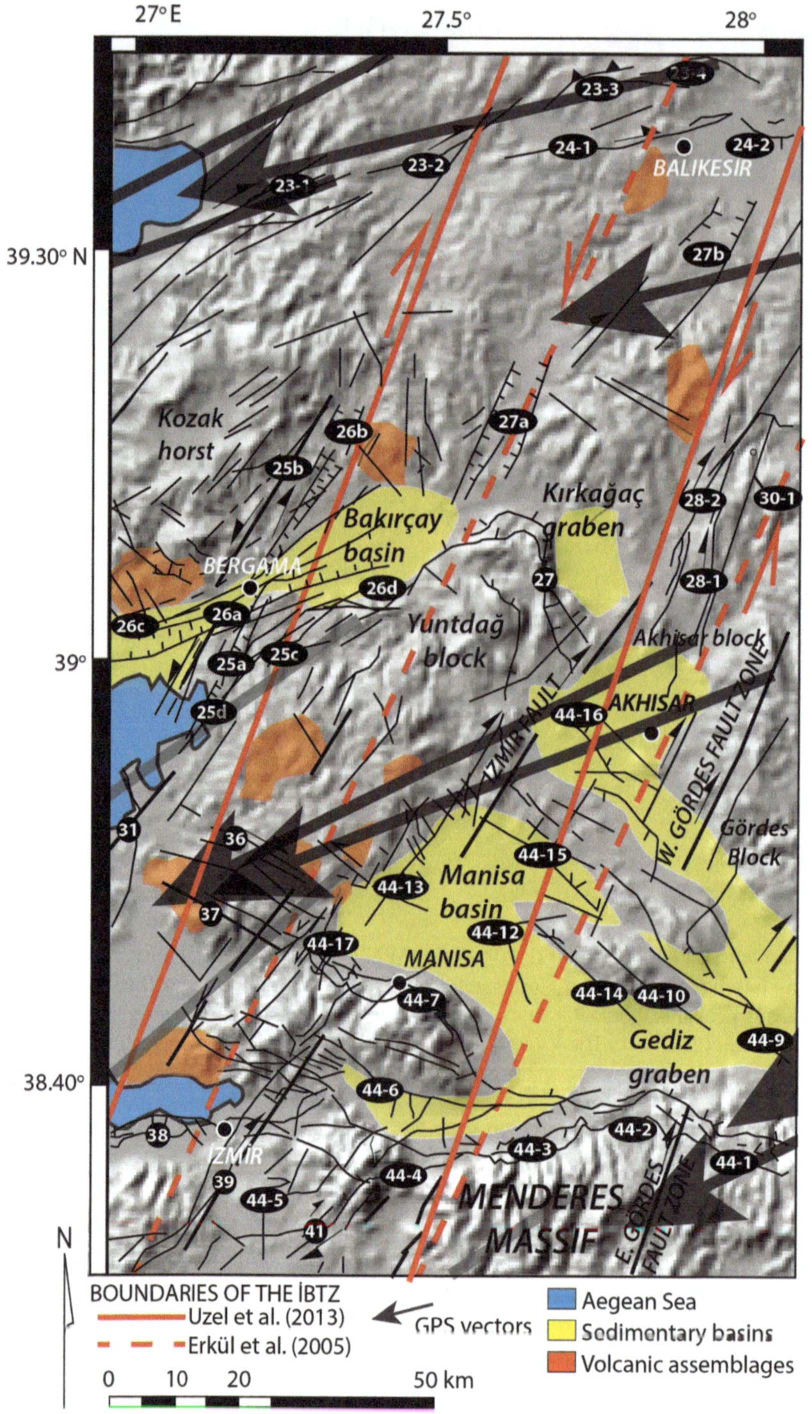

Figure 12.8 Map of the İBTZ from the cities of İzmir and Balıkesir only. Boundaries of the zone are shown in red and are after Uzel et al. (2013) and Erkül et al. (2005). Active faults are numbered and are listed in Table 12.1. The İzmir, western, and eastern Gördes fault zones are after Kaya et al. (2007). Volcanic assemblages are indicated after Karaoğlu (2014). GPS vectors have a length of 10 mm/yr and are relative to Eurasia after Simão et al. (2016).

İBTZ may be the NASZ (Figure 12.3; Uzel et al., 2013; Seghedi et al., 2015; Westerweel et al., 2020; Eyubagil et al., 2021) or the Havran-Balya and Balıkesir fault zone that parallels the NASZ to the south (Figure 12.8) (#23, #24, Emre et al., 2013; Sözbilir et al., 2016; Sümer et al., 2018). These fault zones are sometimes combined into the Havran-Balıkesir fault zone, a 120-km-long, 10–12-km-wide right-lateral strike-slip system that consists of two roughly parallel fault strands, the Havran-Balya (#23) to the north and the Balıkesir (#24) to the south (Figure 12.8). These faults are expressed as several Riedel shears with their own names (#23–1, 23–2, 23–3, 23–4, 24–1, 24–2; Table 12.1; Emre et al., 2018). Activity may have been initiated during the Miocene or older to accommodate NW-SE extension, followed by N-S trending contraction and E-W trending extension related to the initiation of NASZ in the Plio-Quaternary (Sümer et al., 2018). The fault zones are presently controlled by transpressional tectonics associated with NE-SW striking extension and NW-SE contraction (Sümer et al., 2018). Note that GPS vectors overlie this region and show largely southwesterly motion (Figure 12.8; e.g., Simão et al., 2016).

The western edge of the İBTZ is often depicted as an unnamed large right-lateral (Figure 12.9; Uzel et al., 2013; Tepe & Sözbilir, 2017; Sangu et al., 2020) or left-lateral strike-slip fault (Erkül et al., 2005; Seghedi et al., 2015; Helvaci, 2015). In some cases, only one structure is depicted (Özkaymak & Sözbilir, 2008; Eyubagil et al., 2021). The zone may have experienced and recorded multiple senses of motion at different time frames (Özkaymak & Sözbilir, 2008; Uzel et al., 2015; Westerweel et al., 2020). Uzel (2013) indicates that the İBTZ evolved from a more extensive shear zone into a relatively narrow discrete fault zone by the late Pliocene.

The northwestern edge of the İBTZ is sometimes delineated by the Kozak horst (Figure 12.8) (Bergama or Foça block), which developed following the emplacement of the Kozak granite (Altunkaynak & Yilmaz, 1999; Kaya et al., 2007; Yilmaz et al., 2007; Black et al., 2013; Cengiz et al., 2021). Oligocene to late-Miocene oblique faults transect the block and create several NE-SW trending horsts and grabens (Kaya et al., 2007; Yilmaz et al., 2007; Hou et al., 2015). The fault-bounded graben-horst structures here and throughout the İBTZ are promising geothermal field resources and are thus extensively studied for their geometry and sense of motion (Özürlan et al., 2006). Grabens surrounding the uplifted region near the Kozak pluton include the Bergama (#26a), Zeytindağ

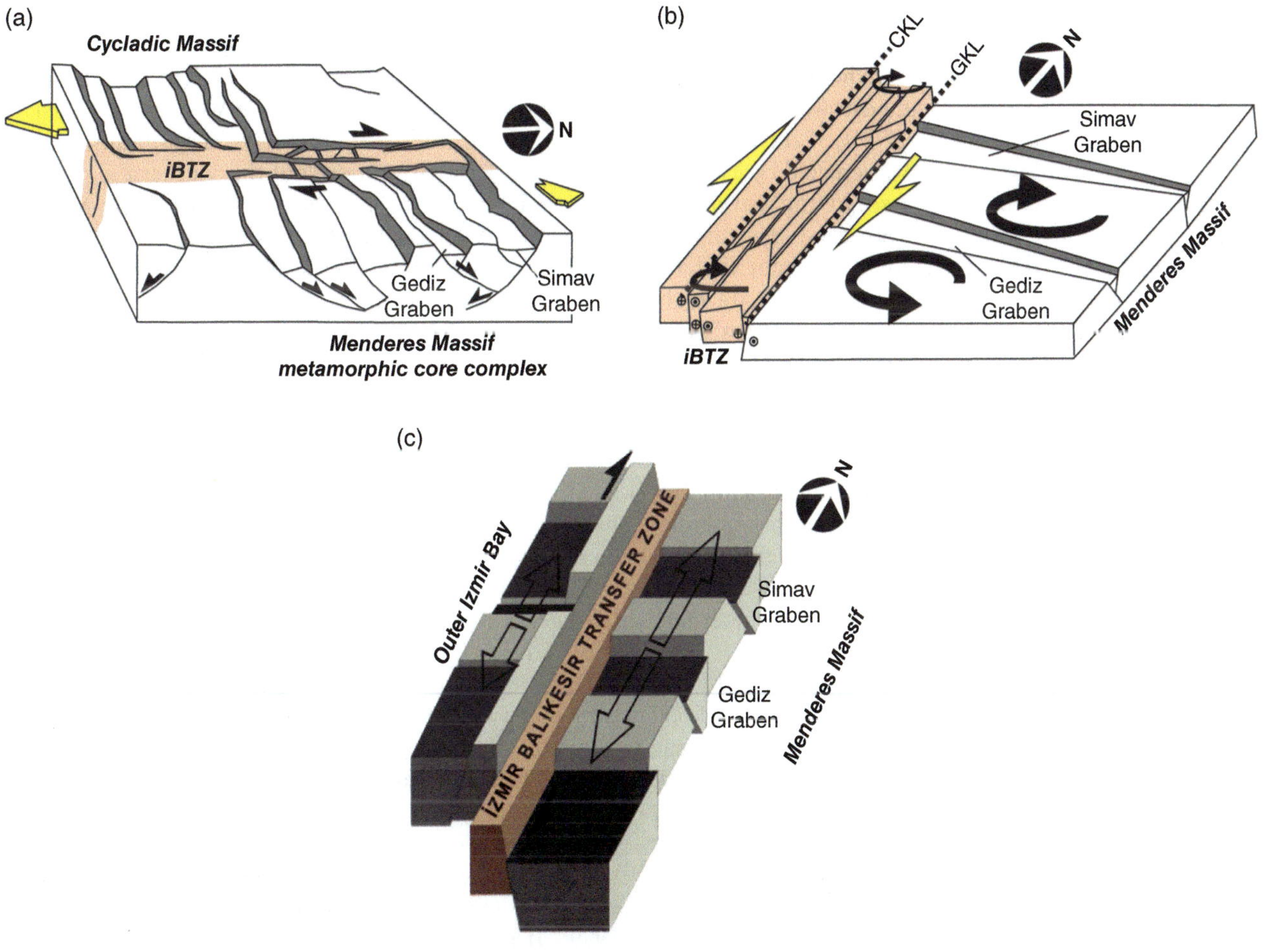

Figure 12.9 Block diagrams showing the nature of the İBTZ. (a) Uzel et al. (2015) / with permission of Elsevier, (b) Westerwheel et al. (2020) / with permission of Elsevier, and (c) Uzel et al. (2013) / with permission of Elsevier. CKL, Çandarli-Karaburun Lineament; GKL, Gelembe-Kemapaşa (Kemalpaşa) Lineament.

(#25a), and Örenli-Eğiller (#26b) (Figure 12.8). Cengiz et al. (2021) identify three specific crustal blocks in the area (Bergama, Dikili, and Zeytindağ) divided by grabens and associated fault systems. The Bergama block, of which the Kozak horst is a part, rotates counterclockwise due in part to right-lateral strike-slip motion along the Bergama (#26a) and Kıranlı (#25b) fault zones. The Dikili and Zeytindağ blocks rotate clockwise due to fault motion along the Kaşikçi (#25c), Yenişakran (#25d), Dikili (#26c), and Kinik-Turgutalp (#26d) fault systems. The Bergama and the Kıranlı fault zones form the larger left-lateral Zeytindağ fault zone (#25) that extends from this region to south of İzmir (Şaroğlu et al., 1992; Kaya et al., 2007; Ocakoğlu et al., 2005). This structure is of particular concern due to low b-values that suggest sizable earthquakes are likely to be generated in the future (Polat et al., 2008; Öztürk, 2015).

The Bergama Graben and fault (#26) coincide with the Bakırçay basin (Figure 12.8), which initiated in the Pliocene during extension but records transtensional segments during the Quaternary (Sangu et al., 2020). In some interpretations, the İBTZ transects basin, whereas in others, the İBTZ coincides with the Soma basin (#27a), which appears as an extension of the Bakırçay basin (Incı, 1998, 2002). The Soma basin hosts western Turkey's most significant coal deposits (738 Mt; Karayiğit et al., 2017). Although the Soma basin was created by Miocene extensional tectonics, faults related to the basin may have accommodated strike-slip Neogene deformation (Westerweel et al., 2020). Inci (2002) indicates that the significant tectonic structures in the coalfield are high-angle oblique-slip normal faults with minor lateral components. To the southeast of the Soma basin (#27a), the Kırkağaç Graben is bounded by the Soma-Kırkağaç fault zone (#27), which was initially described as a strike-slip fault but is recognized today as a normal fault system with N-E trending strike-slip (right and left lateral, depending on location), E-W trending oblique-slip, and N-S trending normal faults with several folds (Uzel, 2016). Sediments within the Soma (#27a) and Kırkağaç (#27) Grabens originated from Quaternary debris flows (Arpaliyigit, 2004).

North of the Soma and Kırkağaç Grabens, the Bigadiç Graben (#27b) is a NE-SW trending basin but is filled primarily by late Oligocene–early Miocene-Neogene volcanogenic sedimentary assemblages (Helvaci & Alaca, 1991; Erkül et al., 2006; Koç-Taşgın & Diniz-Akarca, 2018). The graben is part of the Soma-Kırkağaç fault zone (#27) (Emre et al., 2018). The northwestern edge of the İBTZ has been mapped near the western edge of the Bigadiç Graben (#27b) and assigned as a left-lateral strike-slip system (Erkül et al., 2005). In other interpretations, the northeastern edge of the İBTZ is parallel to the eastern edge of the Bigadiç Graben. Uzel et al. (2013) show the İBTZ extending from the cities of Balıkesir, Bigadiç, and İzmir to the east of the Greek Island of Samos. In other interpretations, the NNE-SSW trending right-lateral Gelenbe fault zone (#28-1, #28-2) coincides with the eastern edge of the İBTZ (Şaroğlu et al., 1992; Seyitoğlu et al., 2022). The structure, along with the Bergama, Zeytindağ, and Soma-Kırkağaç fault zones, have low-velocity variations to ~14 km at depth (Çubuk-Sabuncu et al., 2017). The sense of motion of the Gelenbe fault zone is more consistent with interpretations of the İBTZ having a right-lateral sense of motion on its eastern boundary (Emre et al., 2018).

Horst and graben structures form most of the İBTZ central sections, with several uplifted blocks and depressions identified as separated by NE-SW trending structures (Kaya et al., 2007). The Kozak horst (Bergama or Foça block) is separated from the Yuntdağ block (volcanic center) by the Zeytindağ fault system (#25a–d). The NE-SW trending İzmir fault separates the Akhisar block from the Yuntdağ block (Kaya et al., 2007). Note that a different İzmir fault (#38; Emre et al., 2018) is identified south of the İzmir Gulf and is a two-segment, E-W trending strike-slip fault with a minor dip-slip component (Emre & Barka, 2000; Aktuğ & Kılıçoğlu, 2006). The NE-SW trending İzmir fault of Kaya et al. (2007) has two parallel fault systems, the Western and Eastern Gördes fault zones (Figure 12.8). These faults share a left-lateral sense of motion, but the southern segment of the İzmir fault was reactivated after the Early Miocene as a thrust (Cumalı thrust). In some interpretations, the Western Gördes fault zone would mark the eastern limit of the İBTZ (e.g., Erkül et al., 2005). However, the horst and graben system continues, as evidenced by the parallel Eastern Gördes fault zone that cuts through the Menderes Massif (Figure 12.8).

The southwestern edge of the İBTZ often parallels the Yeniceova fault (#31) or intersects the Menemen (#37) and Güzelhisar (#36) fault systems, which are thought to be related to the extensional dynamics of the Gediz Graben system (Emre et al., 2018). The Yeniceova fault (#31) is mapped as a normal (Emre et al., 2018) or a left-lateral strike-slip system (Altun et al., 2012; Ozener et al., 2013).

The precise southern termination of the İBTZ as it extends past İzmir into the Aegean is unclear. However, on land, the Gediz Graben dominates the tectonics (Figure 12.8 and Table 12.1). The graben marks the division of the northern (Gördes) and central Menderes Massif and is one of several east-west trending sedimentary sections developed in the massif due to large-scale extension (Figure 12.4; Emre, 1996; Koçyiğit et al., 1999). It is one of the youngest continental extensional basins exposed in Western Anatolia and comprises Miocene to recent sediments (e.g., Poyraz et al., 2019). The graben evolved from half to full graben after the Miocene. The Gediz Graben merges with the Manisa basin to the east (Figure 12.8; Özkaymak & Sözbilir, 2008; Özkaymak et al., 2013). This basin is bound by the Manisa fault zone (#44-7), which has had a multistage history with at least three different motions since the Miocene, with its earliest history recording strike-slip motion and trending into extension since the Quaternary (Özkaymak & Sözbilir, 2008).

Several strike-slip fault systems south of İzmir exist, which suggest that the İBTZ may extend into the Mid-Cycladic Lineament (MCL) in the Cyclades (Figure 12.2; Uzel et al., 2013; Seghedi et al., 2015; Westerweel et al., 2020). The MCL is only exposed near or on the island of Paros, where it records orogen-parallel extension or transform fault motion (Morris & Anderson, 1996; Avigad et al., 1998; Walcott & White, 1998a, 1998b; Pe-Piper et al., 2002; Gessner et al., 2013; Philippon et al., 2014; Beniest et al., 2016; Malandri et al., 2017).

12.5.2. Uşak-Muğla Transfer Zone

The eastern portion of the Menderes Massif is characterized by the left-lateral UMTZ that parallels the İBTZ in an NE-SW orientation (Figures 12.3 and 12.10; Karaoğlu & Helvacı, 2012). The zone connects extensional deformation within the Menderes Massif to the İAESZ along the eastern margin of the massif (Karaoğlu & Helvacı, 2014). In this sense, the fault system extends far past the town of Uşak. Figure 12.10 shows the zone between the towns of Uşak and Muğla only. As with the İBTZ, the exact location of this transfer zone within the context of Anatolia is inconsistent. For example, some geologists place the northern edge of the transfer zone along the Simav fault (#30), particularly its Şaphane, Abide, Banaz, and Elvanpaşa segments (Figure 12.10 and Table 12.1; Ersoy & Palmer, 2013; Rabayrol et al., 2019). Here, the structure is also in close association with normal faults from the Emet-Gediz fault zone (#120) and Aslıhanlar fault (#122), as well as the right-lateral Çukurören fault (#121).

In the northern Menderes (Gördes) Massif, the western extent of the UMTZ coincides with the eastern edge of the Selendi basin (Karaoğlu & Helvacı, 2014) or the eastern edge of the Güre portion of the Uşak-Güre basin (Karaoğlu & Helvacı, 2012; Figure 12.10). In addition, several local-scale transfer faults in the central and northern Menderes Massif coincide with the UMTZ (Figure 12.6). In its northwest section, the fault coincides with the dip-slip (normal) right-lateral Selendi (#117) and Rahmanlar faults (#118). However, not all authors agree with the sense of motion on the Selendi fault (Figure 12.6; Erkül et al. 2017). The Güre fault also shows a normal sense of motion in this region. In the eastern interpretation of the zone, the UMTZ coincides with normal faults of the Çivril Graben system (#106) and Sivaslı fault (#107).

Further to the south, the feature transects extensional faults associated with the Büyük Menderes Graben system (#53). Although the UMTZ is speculated to end within the Menderes Massif, in some interpretations, it extends into the Milas fault zone (#58), a right-lateral strike-slip system. Its southeastern extent also coincides and overlaps with the NE-SW trending left-lateral SWASZ (Çemen et al., 2006). However, their influence on each other is unclear and needs further investigation.

The UMTZ may have operated since the Middle Miocene and accommodated differential stretching possibly caused by crustal-scale extension during the Middle-Late Miocene. However, activity had transferred to the Simav fault by the Pliocene-Quaternary (Karaoğlu & Helvaci, 2012). The location of the UMTZ suggests that it may also be affected or initiated by the tear of the African (Nubian) slab by the Late Miocene (11 Ma) or later (Karaoğlu & Helvacı, 2014; Rabayrol et al., 2019). Volcanism is suggested to have migrated southwest along the transfer zone from 17–12 Ma (Erkül et al., 2005; Karaoğlu et al., 2010; Ersoy et al., 2014).

12.6. DISCUSSION

This chapter reviews the present-day knowledge of some local and regional-scale transfer zones documented in outcrops throughout Western Anatolia to summarize outstanding questions regarding their origin, relationship to each other, and roles in accommodating extrusion tectonics. Western Anatolia transfer zones are important for ascertaining seismic hazards, understanding the development of mineral resources, and developing ideas regarding how the crust operates under the influence of extrusion and extensional tectonics. At the local scale, we emphasized the development of transfer zones associated with the NAF (Figure 12.5) and Menderes Massif (Figure 12.3) because they exemplify all these factors.

Western Anatolia regional-scale transfer zones have challenges ascertaining their connections and relationships with active and past fault systems (Sangu et al., 2020; Seyitoğlu et al., 2022). As outlined in this chapter, the İBTZ and UMSZ have unknown dimensions. The specific strike-slip faults that bound their segments are debated regarding their sense of motion and location. Transfer zones are not faults but rather regions of deformation associated with fault systems (Figure 12.1). However, the İBTZ and UMSZ are drawn as long, parallel twin strike-slip faults or lineaments (Figures 12.2 and 12.3). A problem arises when no specific active or past strike-slip fault or lineament exists to mark these locations. The depiction of these structures extending >100 km on the edge of western Turkey can mislead our present-day understanding of the region's dynamics.

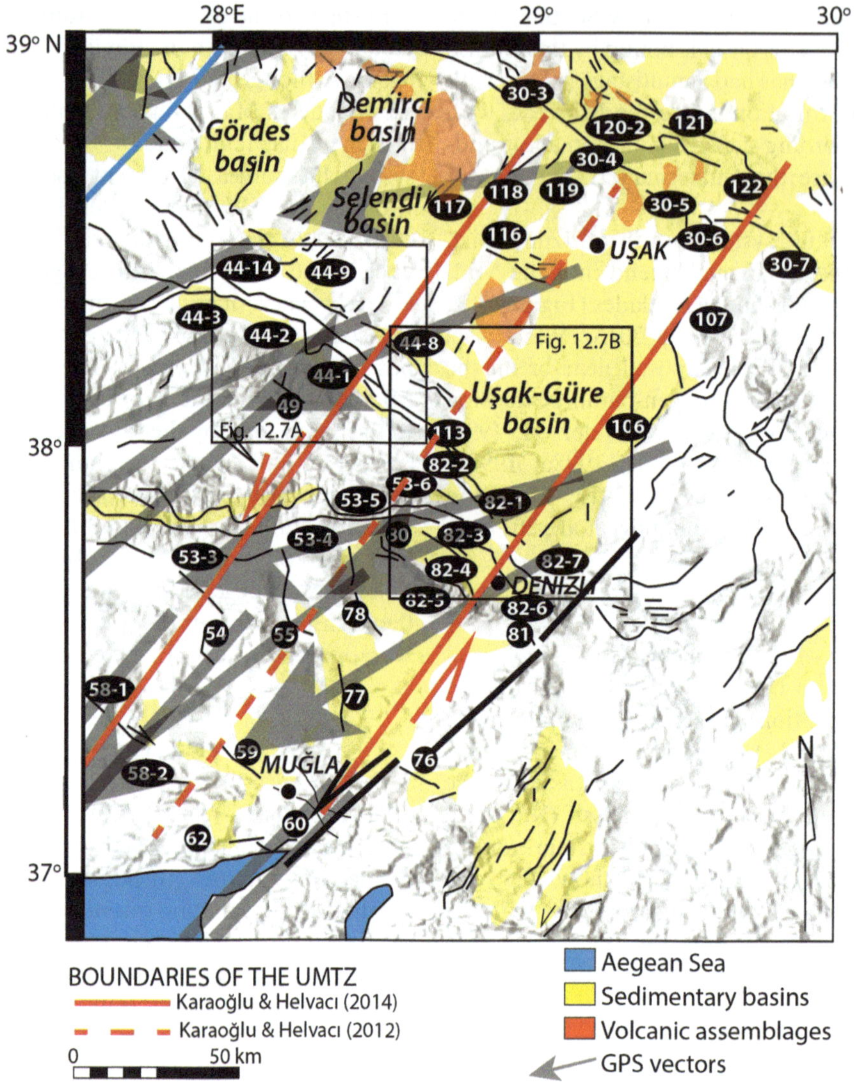

Figure 12.10 Map of the UMTZ from the cities of Uşak and Muğla only. The zone's boundaries are shown in red and are after Karaoğlu and Helvacı (2014) and Karaoğlu and Helvacı (2012). Active faults are numbered and listed in Table 12.1. Volcanic assemblages are indicated after Karaoğlu (2014). GPS vectors have a length of 10 mm/yr and are relative to Eurasia after Simão et al. (2016).

For example, fault length provides critical information regarding depth and hazard (e.g., Wells & Coppersmith, 1994; Kirkpatrick et al., 2008). Therefore, an assumption is made that these transfer zones extend to great depths and are regions of tear in the African (Nubian) slab (Gessner et al., 2013; Jolivet et al., 2013; Uzel et al., 2015; Sümer et al., 2018; Eyubagil et al., 2021). Slab tear would suggest that activity within the İBTZ and UMTZ played a factor in concentrating volcanism, creating critical economic deposits (Rabayrol et al., 2019). In addition, slab tear typically results in asthenosphere-derived (OIB-like) Na-alkaline basalts, which are only exposed in northern Menderes (Gördes) Massif (Kula basalts) (Figure 12.7; Ersoy et al., 2008). However, other volcanic assemblages are scattered throughout the İBTZ and UMTZ. Uzel et al. (2020) suggest that volcanic activity in İBTZ follows the trend of the transfer zone (Figure 12.8), and the lack of $^{40}Ar/^{39}Ar$ ages from igneous assemblages between 15.97 and 13.82 Ma is due to a pulse of core complex exhumation and a change in partitioning extension between the Cyclades and Menderes Massif. The volcanic assemblages outlined in Figures 12.8 and 12.10 are suggested to have been emplaced along N-E trending strike-slip and normal faults (Erkül et al., 2005; Ersoy & Palmer, 2013; Rabayrol et al., 2019). Geochemical compositions of the Miocene-age (17.48–14.94 Ma) volcanoes within the transfer zone indicate their origins are decompression melting of the upper mantle/lower crust, consistent with the outcome of regional transtensional movements in a post-collisional setting (Seghedi et al., 2015).

The İBTZ and UMSZ developed partly due to the recognition of extrusion and rotation of Western Anatolia, as this portion accommodates tectonic escape. Current GPS velocities (20 mm/yr) and orientations were used as evidence that the İBTZ operates as a transfer fault under an extensional tectonic regime (Aktuğ et al., 2009; Geçkin et al., 2021; Eyubagil et al., 2021). However, GPS vectors documenting the extrusion process appear at an angle to their orientations (Figures 12.8 and 12.10). The connection between GPS data and faults that accommodate that motion is unclear in this portion of Turkey. The block may be laterally translated, but motion does not require the presence of more extensive, regional-scale faults. How GPS velocities relate to fault dynamics can be complex, sometimes leading to fault systems assuming to behave with motions incompatible with surface observations (Flerit et al., 2003). Figures 12.8 and 12.10 show that some local-scale, presently active faults in the region covered by the İBTZ and UMSZ are inconsistent with the speculated sense of motion (strike-slip). Some smaller-scale transfer zones are present within these more extensive features, but how they relate to the broader dynamics is unclear.

Another impact of identifying regional-scale transfer zones is reassigning the sense of motion of faults within their designated boundaries. For example, this practice has been done in the northern Menderes (Gördes) Massif, where fault systems that bound the massif's grabens have changed from extensional to strike-slip (compare Purvis & Robertson [2005] and Ersoy et al. [2010]). The Gerni shear zone (Figure 12.6) is an example of a recently identified local-scale transfer zone with a debated sense of motion in Western Anatolia. This zone has also switched from an extensional to a strike-slip feature (compare Işik & Tekeli [2001] and Erkül et al. [2017]).

A potential explanation for the remarkable conflicting kinematics between the dominant strike-slip tectonics suggested by the İBTZ and UMSZ compared to the extensional nature of the fault systems in the Menderes Massif and its vicinity may lie in the timing of when these fault systems were active and how they developed. The timing of the initiation age of the NAF and NASZ is controversial (see the review in Karabacak et al., 2022). However, it would be reasonable to assume that these strike-slip systems are younger than the extensional dynamics of Western Anatolia. The transfer-zone kinematics may not have yet developed to an extent to noticeably disrupt the extensional sense of motion of fault systems in Western Anatolia. We note that transfer-zone fault systems changing kinematics to the east of the UMTZ have also been documented. For example, the Dinar fault zone is thought to have transformed from a Neogene transfer zone to a normal fault during the Quaternary (Alçiçek et al., 2013).

Overall, the sense of motion expressed by the İBTZ is controversial—either left or right lateral strike-slip. Disagreements about the sense of motion experienced by the İBTZ may be due to changes in its sense of motion from initiation to the present day. For example, the İBTZ may have initiated during the Late Cretaceous as a deep crustal transform fault and changed to a transtensional transfer fault zone during the Neogene (Okay & Siyako, 1993; Okay et al., 1996; Ring et al., 1999; Uzel & Sözbilir, 2008; Sözbilir et al., 2011). During the Miocene, the transfer zone may have operated as a sinistral wrench corridor (Ring et al., 1999; Karaoğlu & Helvacı, 2012).

The exhumation of the Menderes Massif is also ascribed to motion along İBTZ and UMTZ (Erkül et al., 2005; Uzel & Sözbilir, 2008; Ersoy et al., 2011; Karaoğlu & Helvacı, 2014; Rabayrol et al., 2019). The İBTZ, in particular, is located between large-scale metamorphic core complexes (Figure 12.9a) and is speculated to have played a role in accommodating differential deformation between two lithospheric extensional systems. In this scenario, the transfer zones are surface expressions of Menderes Massif rotation and Cyclades metamorphic core complex differential exhumation triggered by the rollback of the African (Nubian) slab and tearing in the Neogene (Figure 12.9b). The İBTZ is also interpreted to express a transition between extensional and strike-slip dynamics due to the southward rotation rollback of the African (Nubian) slab along the Hellenic arc subduction zone (Ersoy & Palmer, 2013; Özkaymak et al., 2013; Ersoy et al., 2014; Ersoy et al., 2017; Uzel et al., 2020).

If the northern edge of the İBTZ is a strike-slip system, in the traditional sense of a transfer zone, the southern termination of the İBTZ should perhaps share this sense of motion (Figure 12.10). Although its southern extent is unknown, speculation that the İBTZ extends toward the Greek island of Samos links the transfer zone to active seismicity in that region (Fountoulakis et al., 2023). The zone is linked to active fault systems southwest of İzmir (Pamukçu et al., 2014). In some interpretations, the İBTZ trends directly into the MCL (Figure 12.2). The island of Paros is the only location where the MCL is exposed, and it records E-W stretching lineation in the north and N-S to the south (Walcott & White, 1998; Pe-Piper et al., 2002; Malandri et al., 2017). Undeformed Miocene granites and rhyolite dikes intrude into the stretching lineations (Malandri et al., 2017). Instead of a strike-slip system, the MCL appears to be an extensional shear zone that is no longer active (Malandri et al., 2017; Pe-Piper et al., 2002).

A critical question is whether regional-scale transfer zones exist in this portion of Western Anatolia. These

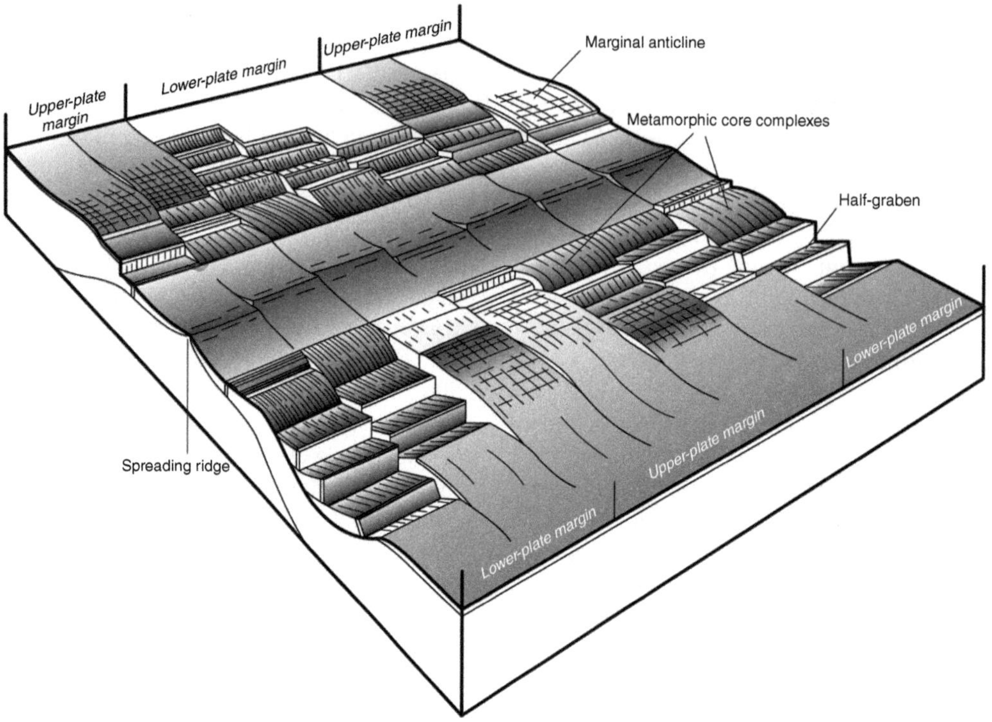

Figure 12.11 Cartoon of how extension can be accommodated as the process progresses from the upper to lower plate adapted from Lister et al. (1986). The cartoon assumes that an ocean basin has formed. Transfer faults mark changes from upper-plate margins to lower-plate margins. Metamorphic core complexes are identified. The İBTZ and UMTZ would mark boundaries between the core complexes in this scenario.

zones should be expected to occur as strike-slip systems could, in principle, separate the Menderes and Attic-Cycladic massifs as these core complexes extend at different rates and accommodate different amounts of extension (Figure 12.9). Normal faults in extensional terranes commonly terminate at orthogonal strike-slip or shear zones, like oceanic transform faults (e.g., Lister et al., 1986; Figure 12.11). However, the İBTZ and UMTZ shown as regional-scale strike-slip dynamics imply that the Menderes and Cycladic Massifs share the same stretching orientation relative to extension direction (b-type in the terminology of Jolivet et al., 2004). According to Jolivet et al. (2004), only the Menderes Massif is classified as b-type, but the Attic-Cycladic metamorphic core complex can be b- or a-type (parallel to stretching lineation; Le Pourhiet et al., 2012).

Overall, the sentiment of Sangu et al. (2020) that the İBTZ is not sufficiently understood should be considered when developing ideas about the dynamics of Western Anatolia. If they are present, which particular fault systems are associated with these features should be more clearly documented by those who advocate for these as tectonic features. In addition, the vocabulary of strike-slip tectonics for these transfer zones may be inapplicable. The areas designated to the zones transect multiple fault features and have an unclear or unknown connection between specific boundary faults. They do not appear to link boundary faults that share a sense of motion as is required for a transfer-zone system.

12.7. CONCLUSIONS

This chapter reviews the present-day knowledge of selected regional and local-scale transfer zones in Western Anatolia. These structures accommodate the extrusion of the Anatolian and Aegean microplates. They assist with transferring motion and stress from adjacent faults as deformation propagates. At the local scale, they are studied for their hazards and associations with mineralizations and are critical depocenters for oil and gas. We outline only a few smaller-scale transfer zones along the NAF and Menderes Massif. At the regional scale, we outline the İBTZ and UMTZ, depicted as long, parallel strike-slip transfer-zone systems that largely follow GPS vectors tracking the extrusion of the Anatolian plate. Many outstanding questions exist for these larger transfer zones. These questions include their dimensions, sense of motion, locations, and which specific ancient and active fault systems are associated with their strike-slip sense of motion. Despite these many unknowns, they are invoked as major features controlling Western Anatolia's tectonic history, seismicity, and mineralizations. Finally, this chapter echoes the sentiments of a few researchers who suggest that the İBTZ and UMTZ are not sufficiently understood.

ACKNOWLEDGMENTS

No real or perceived financial conflicts of interest exist for any author. We appreciate drafting assistance from Jeffrey S Horowitz. We appreciate reviews from Konstantinos Soukis (National and Kapodistrian University of Athens, Greece) and an anonymous reviewer.

AVAILABILITY STATEMENT

No new data were generated, analyzed, or presented in the manuscript.

REFERENCES

Ahlgren, S. G. (2001). The nucleation and evolution of Riedel shear zones as deformation bands in porous sandstone. *Journal of Structural Geology*, *23*(8), 1203–1214. doi: 10.1016/S0191-8141(00)00183-8

Akal, C., Helvacı, C., Prelević, D., & Van Den Bogaard, P. (2013). High-K volcanism in the Afyon region, western Turkey: From Si-oversaturated to Si-undersaturated volcanism. *International Journal of Earth Sciences*, *102*(2), 435–453. doi: 10.1007/s00531-012-0809-9

Akbayram, K., Şengör, A. M. C., & Özcan, E. (2017). The evolution of the Intra-Pontide suture: Implications of the discovery of Late Cretaceous–early Tertiary mélanges. In R. Sorkhabi (Ed.), *Tectonic Evolution, Collision, and Seismicity of Southwest Asia<subtitle>In Honor of Manuel Berberian's Forty-Five Years of Research Contributions</subtitle>*. Geological Society of America. 10.1130/2016.2525(18)

Akıska, E., Karakaş, Z., & Öztürk, C. (2019). Uranium, thorium and rare earth element deposits of Turkey. In F. Pirajno, T. Ünlü, C. Dönmez, & M. B. Şahin (Eds.), *Mineral resources of turkey* (Vol. *16*, pp. 655–679). Cham: Springer International Publishing. doi: 10.1007/978-3-030-02950-0_14

Akiska, E., & Varol, B. E. (2020). Alluvial-lacustrine sedimentation and volcaniclastic deposition in an intracontinental tectonic graben: Paleoenvironmental evolution of the Neogene Sinanpaşa Basin,west-central Turkey. *Turkish Journal of Earth Sciences*, *29*(2), 295–324. doi: 10.3906/yer-1907-28

Aksay, S., Schoorl, J. M., Veldkamp, A., Demir, T., Aytaç, A. S., & Maddy, D. (2022). Structurally controlled landscape evolution in kula badlands, Western Turkey. *Geosciences*, *12*(10), 390. doi: 10.3390/geosciences12100390

Aksoy, M. E., Türe, O., Yılmaz-Aksoy, Ö., Çetin-Kasa, E., & Arık, H. K. (2022). Tectonic Geomorphology and Paleoseismology of the Muğla-Yatağan Fault (SW Turkey). In A. Çiner, S. Grab, E. Jaillard, D. Doronzo, A. Michard, M. Rabineau, & H. I. Chaminé (Eds.), *Recent research on geomorphology, sedimentology, marine geosciences and geochemistry* (pp. 9–12). Cham: Springer International Publishing. doi: 10.1007/978-3-030-72547-1_2

Aktuğ, B., & Kılıçoğlu, A. (2006). Recent crustal deformation of İzmir, Western Anatolia and surrounding regions as deduced from repeated GPS measurements and strain field. *Journal of Geodynamics*, *41*(5), 471–484. doi: 10.1016/j.jog.2006.01.004

Aktug, B., Nocquet, J. M., Cingöz, A., Parsons, B., Erkan, Y., England, P., et al. (2009). Deformation of western Turkey from a combination of permanent and campaign GPS data: Limits to block-like behavior. *Journal of Geophysical Research: Solid Earth*, *114*(B10), 2008JB006000. doi: 10.1029/2008JB006000

Akyar, B. E. (2019). *Deprem Tekrarlama Periyotlarının Jeodezik Verilerle Belirlenmesi : Banaz ve Elvanpaşa Segmenti* (Master's Thesis). Afyon, Turkey: Afyon Kocatepe Üniversitesi.

Akyüz, H. S. (2002). Surface Rupture and Slip Distribution of the 12 November 1999 Duzce Earthquake (M 7.1), North Anatolian Fault, Bolu, Turkey. *Bulletin of the Seismological Society of America*, *92*(1), 61–66. 10.1785/0120000840

Akyüz, H. S., Kırkan, E., Basmenji, M., Aksoy, E., Akyüz, A. D., Uçarkuş, G., et al. (2019). Paleoseismological and morphotectonical characteristics of active faults in the vicinity of Muğla Area (SW Turkey). In N. Sundararajan, M. Eshagh, H. Saibi, M. Meghraoui, M. Al-Garni, & B. Giroux (Eds.), *On significant applications of geophysical methods* (pp. 253–256). Cham: Springer International Publishing. doi: 10.1007/978-3-030-01656-2_57

Alçiçek, M. C., Brogi, A., Capezzuoli, E., Liotta, D., & Meccheri, M. (2013). Superimposed basin formation during Neogene–Quaternary extensional Tectonics in SW-Anatolia (Turkey): Insights from the kinematics of the Dinar fault zone. *Tectonophysics*, *608*, 713–727. doi: 10.1016/j.tecto.2013.08.008

Altun, S., Sezer, A., & Burak Göktepe, A. (2012). A preliminary microzonation study on Northern Coasts of Izmir: Investigation of the local soil conditions. *Soil Dynamics and Earthquake Engineering*, *39*, 37–49. doi: 10.1016/j.soildyn.2012.02.006

Altunel, E. (1999). Geological and geomorphological observations in relation to the 20 September 1899 Menderes earthquake, Western Turkey. *Journal of the Geological Society*, *156*(2), 241–246. doi: 10.1144/gsjgs.156.2.0241

Altunkaynak, Ş., & Yilmaz, Y. (1999). The Kozak Pluton and its emplacement. *Geological Journal*, *34*(3), 257–274. doi: 10.1002/(SICI)1099-1034(199907/09)34:3<257::AID-GJ826>3.0.CO;2-Q

Ambraseys, N. N., & Tchalenko, J. S. (1972). Seismotectonic Aspects of the Gediz, Turkey, Earthquake of March 1970. *Geophysical Journal International*, *30*(3), 229–252. doi: 10.1111/j.1365-246X.1972.tb05811.x

Armijo, R., Meyer, B., Hubert, A., & Barka, A. (1999). Westward propagation of the North Anatolian fault into the northern Aegean: Timing and kinematics. *Geology*, *27*(3), 267. doi: 10.1130/0091-7613(1999)027<0267:WPOTNA>2.3.CO;2

Armijo, R., Meyer, B., Navarro, S., King, G., & Barka, A. (2002). Asymmetric slip partitioning in the Sea of Marmara pull-apart: A clue to propagation processes of the North Anatolian Fault? *Terra Nova*, *14*(2), 80–86. doi: 10.1046/j.1365-3121.2002.00397.x

Arpalıyiğit, İ. (2004). *Pliocene-quaternary geology of the Soma Graben, Western Turkey*. İzmir, Turkey: Dokuz Eylül University.

Avigad, D., Baer, G., & Heimann, A. (1998). Block rotations and continental extension in the central Aegean Sea: palaeomagnetic and structural evidence from Tinos and Mykonos (Cyclades, Greece). *Earth and Planetary Science Letters*, *157*(1–2), 23–40. doi: 10.1016/S0012-821X(98)00024-7

Aydin, S., & Tuncer, K. (2021). Bozdağ'da (Denizli) dolinlerin morfometrik özellikleri. *Türk Coğrafya Dergisi*, *78*, 33–48. 10.17211/tcd.1013232

Bağcı, U., & Parlak, O. (2009). Petrology of the Tekirova (Antalya) ophiolite (Southern Turkey): Evidence for diverse magma generations and their tectonic implications during Neotethyan-subduction. *International Journal of Earth Sciences*, *98*(2), 387–405. doi: 10.1007/s00531-007-0242-7

Bakliwal, P. C., & Ramasamy, S. M. (1987). Lineament fabric of Rajasthan and Gujarat. *Records of the Geological Survey of India*, *113*(7), 54–64.

Bally, A. W. (1981). Thoughts on the tectonics of folded belts. *Geological Society, London, Special Publications*, *9*(1), 13–32. doi: 10.1144/GSL.SP.1981.009.01.03

Barbot, S., & Weiss, J. R. (2021). Connecting subduction, extension and shear localization across the Aegean Sea and Anatolia. *Geophysical Journal International*, *226*(1), 422–445. doi: 10.1093/gji/ggab078

Basmenji, M., Akyüz, H. S., Kirkan, E., Aksoy, M. E., Uçarkuş, G., & Yakupoğlu, N. (2021). Earthquake history of the Yatağan Fault (Muğla, SW Turkey): Implications for regional seismic hazard assessment and paleoseismology in extensional provinces. *Turkish Journal of Earth Sciences*, *30*(2), 161–181. doi: 10.3906/yer-2006-23

Beaudoin, A., Augier, R., Laurent, V., Jolivet, L., Lahfid, A., Bosse, V., et al. (2015). The Ikaria high-temperature Metamorphic Core Complex (Cyclades, Greece): Geometry, kinematics and thermal structure. *Journal of Geodynamics*, *92*, 18–41. doi: 10.1016/j.jog.2015.09.004

Beccaletto, L., & Jenny, C. (2004). Geology and correlation of the Ezine zone: A Rhodope fragment in NW Turkey? *Turkish Journal of Earth Sciences*, *13*(2), 145–176.

Bello, O. A., Özgür, N., & Çalışkan, T. A. (2017). Hydrogeological, Hydrogeochemical and Isotope Geochemical Features of Geothermal Waters in Simav and Environs, Western Anatolia, Turkey. *Procedia Earth and Planetary Science*, *17*, 29–32. doi: 10.1016/j.proeps.2016.12.014

Benlest, A., Brun, J. P., Gorini, C., Crombez, V., Deschamps, R., Hamon, Y., & Smit, J. (2016). Interaction between trench retreat and Anatolian escape as recorded by neogene basins in the northern Aegean Sea. *Marine and Petroleum Geology*, *77*, 30–42. doi: 10.1016/j.marpetgeo.2016.05.011

Ben-Zion, Y., Peng, Z., Okaya, D., Seeber, L., Armbruster, J. G., Ozer, N., et al. (2003). A shallow fault-zone structure illuminated by trapped waves in the Karadere-Duzce branch of the North Anatolian Fault, western Turkey. *Geophysical Journal International*, *152*(3), 699–717. doi: 10.1046/j.1365-246X.2003.01870.x

Black, K. N., Catlos, E. J., Oyman, T., & Demirbilek, M. (2013). Timing Aegean extension: Evidence from in situ U–Pb geochronology and cathodoluminescence imaging of granitoids from NW Turkey. *Lithos*, *180–181*, 92–108. doi: 10.1016/j.lithos.2013.09.001

Bonev, N., Beccaletto, L., Robyr, M., & Monié, P. (2009). Metamorphic and age constraints on the Alakeçi shear zone: Implications for the extensional exhumation history of the northern Kazdağ Massif, NW Turkey. *Lithos*, *113*(1–2), 331–345. doi: 10.1016/j.lithos.2009.02.010

Bosworth, W. (1985). Geometry of propagating continental rifts. *Nature*, *316*(6029), 625–627. doi: 10.1038/316625a0

Bozkurt, E. (2003). Origin of NE-trending basins in western Turkey. *Geodinamica Acta*, *16*(2–6), 61–81. doi: 10.1016/S0985-3111(03)00002-0

Bozkurt, E., & Park, L. R. G. (1994). Southern Menderes Massif: An incipient metamorphic core complex in western Anatolia, Turkey. *Journal of the Geological Society*, *151*(2), 213–216. doi: 10.1144/gsjgs.151.2.0213

Bozkurt, E., & Satır, M. (2000). The southern Menderes Massif (western Turkey): Geochronology and exhumation history. *Geological Journal*, *35*(3–4), 285–296. doi: 10.1002/gj.849

Bozkurt, E., Satır, M., & Buğdaycıoğlu, Ç. (2011). Surprisingly young Rb/Sr ages from the Simav extensional detachment fault zone, northern Menderes Massif, Turkey. *Journal of Geodynamics*, *52*(5), 406–431. doi: 10.1016/j.jog.2011.06.002

Brogi, A., Capezzuoli, E., Alçiçek, M. C., & Gandin, A. (2014). Evolution of a fault-controlled fissure-ridge type travertine deposit in the western Anatolia extensional province: the Çukurbağ fissure-ridge (Pamukkale, Turkey). *Journal of the Geological Society*, *171*(3), 425–441. doi: 10.1144/jgs2013-034

Brooks, M., & Ferentinos, G. (1984). Tectonics and sedimentation in the Gulf of Corinth and the Zakynthos and Kefallinia channels, Western Greece. *Tectonophysics*, *101*(1–2), 25–54. doi: 10.1016/0040-1951(84)90040-4

Brun, J.-P., & Sokoutis, D. (2010). 45 m.y. of Aegean crust and mantle flow driven by trench retreat. *Geology*, *38*(9), 815–818. doi: 10.1130/G30950.1

Brun, J.-P., & Faccenna, C. (2008). Exhumation of high-pressure rocks driven by slab rollback. *Earth and Planetary Science Letters*, *272*(1–2), 1–7. doi: 10.1016/j.epsl.2008.02.038

Bulut, F., Özener, H., Doğru, A., Aktuğ, B., & Yaltırak, C. (2018). Structural setting along the Western North Anatolian Fault and its influence on the 2014 North Aegean Earthquake (Mw 6.9). *Tectonophysics*, *745*, 382–394. doi: 10.1016/j.tecto.2018.07.006

Burke, K., & Sengör, C. (1986). Tectonic escape in the evolution of the continental crust. In M. Barazangi & L. Brown (Eds.), *Geodynamics series* (Vol. *14*, pp. 41–53). Washington, DC: American Geophysical Union. doi: 10.1029/GD014p0041

Caglayan, A., Saber, R., & Isik, V. (2021). *Geomorphic evidence of active faulting in the Simav Halfgraben, Western Turkey*. Paper presented at the 11th Congress of the Balkan Geophysical Society doi: 10.3997/2214-4609.202149BGS78

Çakir, Z. (1999). Along-strike discontinuity of active normal faults and its influence on Quaternary travertine deposition; Examples from Western Turkey. *Turkish Journal of Earth Sciences*, *8*(1), 67–80.

Candan, O., Çetinkaplan, M., Oberhänsli, R., Rimmelé, G., & Akal, C. (2005). Alpine high-P/low-T metamorphism of the Afyon Zone and implications for the metamorphic evolution of Western Anatolia, Turkey. *Lithos*, *84*(1–2), 102–124. doi: 10.1016/j.lithos.2005.02.005

Capitanio, F. A., Zlotnik, S., & Faccenna, C. (2010). Controls on subduction reorganization in the Hellenic margin, eastern Mediterranean. *Geophysical Research Letters*, *37*(14), 2010GL044054. doi: 10.1029/2010GL044054

Catlos, E. J., Baker, C., Sorensen, S. S., Çemen, I., & Hançer, M. (2010). Geochemistry, geochronology, and cathodoluminescence imagery of the Salihli and Turgutlu granites (central Menderes Massif, western Turkey): Implications for Aegean tectonics. *Tectonophysics*, *488*(1–4), 110–130. doi: 10.1016/j.tecto.2009.06.001

Catlos, E. J., & Çemen, I. (2005). Monazite ages and the evolution of the Menderes Massif, western Turkey. *International Journal of Earth Sciences*, *94*(2), 204–217. doi: 10.1007/s00531-005-0470-7

Catlos, E. J., & Çemen, İ. (2023). A review of the dynamics of subduction zone initiation in the Aegean region. In E. J. Catlos & İ. Çemen (Eds.), *Compressional tectonics: plate convergence to mountain building* Geophysical Monograph Series (Vol. *1*, pp. 87–117). Wiley. doi: 10.1002/9781119773856.ch4

Cavazza, W., Okay, A. I., & Zattin, M. (2009). Rapid early-middle Miocene exhumation of the Kazdağ Massif (western Anatolia). *International Journal of Earth Sciences*, *98*(8), 1935–1947. doi: 10.1007/s00531-008-0353-9

Çelik, Ö. F., & Chiaradia, M. (2008). Geochemical and petrological aspects of dike intrusions in the Lycian ophiolites (SW Turkey): A case study for the dike emplacement along the Tauride Belt Ophiolites. *International Journal of Earth Sciences*, *97*(6), 1151–1164. doi: 10.1007/s00531-007-0204-0

Çelik, Ö. F., Delaloye, M., & Feraud, G. (2006). Precise $^{40}Ar/^{39}Ar$ ages from the metamorphic sole rocks of the Tauride Belt Ophiolites, southern Turkey: Implications for the rapid cooling history. *Geological Magazine*, *143*(2), 213–227. doi: 10.1017/S0016756805001524

Çemen, I., Catlos, E. J., Göğüs, O., & Özerdem, C. (2006). Postcollisional extensional tectonics and exhumation of the Menderes Massif in the Western Anatolia extended terrane, Turkey. In Y. Dilek & S. Pavlides (Eds.), *Postcollisional tectonics and magmatism in the Mediterranean Region and Asia* (pp. 353–379). Geological Society of America. doi: 10.1130/2006.2409

Çemen, İ., Göncüoglu, M. C., & Dirik, K. (1999). Structural evolution of the Tuzgölü Basin in Central Anatolia, Turkey. *The Journal of Geology*, *107*(6), 693–706. doi: 10.1086/314379

Cengiz, M., Karabulut, S., Özçep, F., Çabuk, B. S., & Heller, F. (2021). Paleomagnetic results from Western Anatolia: Evidence of microblock rotations after emplacement of the Lower Miocene Yuntdağ Volcanic Rocks. *Annals of Geophysics*, *64*(6), GM661. doi: 10.4401/ag-8699

Channell, J. E. T., & Kozur, H. W. (1997). How many oceans? Meliata, Vardar and Pindos oceans in Mesozoic Alpine paleogeography. *Geology*, *25*(2), 183. doi: 10.1130/0091-7613(1997)025<0183:HMOMVA>2.3.CO;2

Chousianitis, K., Ganas, A., & Evangelidis, C. P. (2015). Strain and rotation rate patterns of mainland Greece from continuous GPS data and comparison between seismic and geodetic moment release. *Journal of Geophysical Research: Solid Earth*, *120*(5), 3909–3931. doi: 10.1002/2014JB011762

Chousianitis, K., Ganas, A., & Gianniou, M. (2013). Kinematic interpretation of present-day crustal deformation in central Greece from continuous GPS measurements. *Journal of Geodynamics*, *71*, 1–13. doi: 10.1016/j.jog.2013.06.004

Çifçi, G., Pamukçu, O., Çoruh, C., Çopur, S., & Sözbilir, H. (2011). Shallow and deep structure of a supradetachment basin based on geological, conventional deep seismic reflection sections and gravity data in the Buyuk Menderes Graben, Western Anatolia. *Surveys in Geophysics*, *32*(3), 271–290. doi: 10.1007/s10712-010-9109-8

Çiftçi, N. B., & Bozkurt, E. (2010). Structural evolution of the Gediz Graben, SW Turkey: temporal and spatial variation of the graben basin. *Basin Research*, *22*(6), 846–873. doi: 10.1111/j.1365-2117.2009.00438.x

Çiftçi, N. B., Temel, R. Ö., & İztan, Y. H. (2010). Hydrocarbon occurrences in the western Anatolian (Aegean) Grabens, Turkey: Is there a working petroleum system? *AAPG Bulletin*, *94*(12), 1827–1857. doi: 10.1306/06301009172

Collins, A. S., & Robertson, A. H. F. (1998). Processes of Late Cretaceous to Late Miocene episodic thrust-sheet translation in the Lycian Taurides, SW Turkey. *Journal of the Geological Society*, *155*(5), 759–772. doi: 10.1144/gsjgs.155.5.0759

Coskun, B. (1997). Oil and gas fields—transfer zone relationships, Thrace basin, NW Turkey. *Marine and Petroleum Geology*, *14*(4), 401–416. doi: 10.1016/S0264-8172(96)00062-1

Crameri, F., Magni, V., Domeier, M., Shephard, G. E., Chotalia, K., Cooper, G., et al. (2020). A transdisciplinary and community-driven database to unravel subduction zone initiation. *Nature Communications*, *11*(1), 3750. doi: 10.1038/s41467-020-17522-9

Çubuk-Sabuncu, Y., Taymaz, T., & Fichtner, A. (2017). 3-D crustal velocity structure of western Turkey: Constraints from full-waveform tomography. *Physics of the Earth and Planetary Interiors*, *270*, 90–112. doi: 10.1016/j.pepi.2017.06.014

Dahlstrom, C. D. (1970). Structural geology in the eastern margin of the Canadian Rocky Mountains. *Bulletin of Canadian Petroleum Geology*, *18*(3), 332–406. 10.35767/gscpgbull.18.3.332

d'Atri, A., Zuffa, G. G., Cavazza, W., Okay, A. I., & Di Vincenzo, G. (2012). Detrital supply from subduction/accretion complexes to the Eocene–Oligocene post-collisional southern Thrace Basin (NW Turkey and NE Greece). *Sedimentary Geology*, *243–244*, 117–129. doi: 10.1016/j.sedgeo.2011.10.008

Davis, P. B., & Whitney, D. L. (2006). Petrogenesis of lawsonite and epidote eclogite and blueschist, Sivrihisar Massif, Turkey. *Journal of Metamorphic Geology*, *24*(9), 823–849. doi: 10.1111/j.1525-1314.2006.00671.x

Demirci, A., Özden, S., Bekler, T., Kalafat, D., & Pınar, A. (2015). An active extensional deformation example: 19 May 2011 Simav earthquake (Mw = 5.8), Western Anatolia, Turkey. *Journal of Geophysics and Engineering*, *12*(4), 552–565. doi: 10.1088/1742-2132/12/4/552

Denèle, Y., Lecomte, E., Jolivet, L., Lacombe, O., Labrousse, L., Huet, B., & Le Pourhiet, L. (2011). Granite intrusion in a metamorphic core complex: The example of the Mykonos laccolith (Cyclades, Greece). *Tectonophysics*, *501*(1–4), 52–70. doi: 10.1016/j.tecto.2011.01.013

Di Rosa, M., Farina, F., Marroni, M., Pandolfi, L., Göncüoğlu, M. C., Ellero, A., & Ottria, G. (2019). U-Pb zircon geochronology of intrusive rocks from an exotic block in the Late Cretaceous – Paleocene Taraklı Flysch (northern Turkey): Constraints on the tectonics of the Intrapontide suture zone. *Journal of Asian Earth Sciences*, *171*, 277–288. doi: 10.1016/j.jseaes.2018.11.017

Dikbaş, A., Akyüz, H. S., Basmenji, M., & Kırkan, E. (2022). Earthquake history of the Gökova fault zone by paleoseismologic trenching, SW Turkey. *Natural Hazards*, *112*(3), 2695–2716. doi: 10.1007/s11069-022-05284-0

Doğan, A. & Emre, O. (2006). *Ege Graben Sisteminin Kuzey Sınırı: Sındırgı-Sincanlı Fay Zonu (Northern boundary of Aegean Graben System: Sındırgı-Sincanlı fault zone)*. Paper presented at the 59th Türkiye Jeoloji Kurultayı, Ankara. https://www.jmo.org.tr/resimler/ekler/afc8fe9c61d6760_ek.pdf

Doğan, A., Kaygusuz, Ç., Tiryakioğlu, İ., Yigit, C. O., Sözbilir, H., Özkaymak, Ç., & Turgut, B. (2022). Geodetic evidence for aseismic fault movement on the eastern segment of the Gediz Graben system (western Anatolia extensional province, Turkey) and its significance for settlements. *Acta Geodaetica et Geophysica*, *57*(3), 461–476. doi: 10.1007/s40328-022-00389-1

Doğan, U., Koçyiğit, A., & Gökkaya, E. (2017). Development of the Kembos and Eynif structural poljes: Morphotectonic evolution of the Upper Manavgat River basin, central Taurides, Turkey. *Geomorphology*, *278*, 105–120. doi: 10.1016/j.geomorph.2016.10.030

Dooley, T., & McClay, K. (1997). Analog Modeling of Pull-Apart Basins. *AAPG Bulletin*, *81*(1997), 1804–1826. doi: 10.1306/3B05C636-172A-11D7-8645000102C1865D

Drahor, M. G., & Berge, M. A. (2017). Integrated geophysical investigations in a fault zone located on southwestern part of İzmir city, Western Anatolia, Turkey. *Journal of Applied Geophysics*, *136*, 114–133. doi: 10.1016/j.jappgeo.2016.10.021

Duran, İ., Sözbilir, H., Eski, S., Softa, M., Uytun, H., Yüksel, M., & Topaksu, M. (2021). Paleoseismic history of the Manisa fault zone, Western Anatolia. *Turkish Journal of Earth Sciences*, *30*(SI-1), 806–832. doi: 10.3906/yer-2109-3

Ece, Ö. I., Ekinci, B., Schroeder, P. A., Crowe, D., & Esenli, F. (2013). Origin of the Düvertepe kaolin–alunite deposits in Simav Graben, Turkey: Timing and styles of hydrothermal mineralization. *Journal of Volcanology and Geothermal Research*, *255*, 57–78. doi: 10.1016/j.jvolgeores.2013.01.012

Ece, Ö. I., & Schroeder, P. A. (2007). Clay mineralogy and chemistry of halloysite and alunite deposits in the Turplu area, Balikesir, Turkey. *Clays and Clay Minerals*, *55*(1), 18–35. doi: 10.1346/CCMN.2007.0550102

Emre, O. (1996). Geology and the Tectonics of the Gediz Graben. *Turkish Journal of Earth Sciences*, *5*(3), 171–185.

Emre, O. & Barka, A. (2000). *Active faults between Gediz Graben and Aegean Sea (İzmir region)*. Paper presented at the International Symposia on Seismicity of Western Anatolia.

Emre, Ö, Duman, T. Y., Özalp, S., Elmacı, H., Olgun, Ş., & Şaroğlu, F. (2013). *Active fault map of Turkey with explanatory text*. Turkey: General Directorate of Mineral Research and Exploration.

Emre, Ö., Duman, T. Y., Özalp, S., Şaroğlu, F., Olgun, Ş., Elmacı, H., & Çan, T. (2018). Active fault database of Turkey. *Bulletin of Earthquake Engineering*, *16*(8), 3229–3275. doi: 10.1007/s10518-016-0041-2

Engin, T., & Hirst, D. M. (1970). Serpentinisation of harzburgites from the alpine peridotite belt of Southwest Turkey. *Chemical Geology*, *6*, 281–295. doi: 10.1016/0009-2541(70)90029-X

Erkül, F. (2010). Tectonic significance of synextensional ductile shear zones within the Early Miocene Alaçamdağ granites, northwestern Turkey. *Geological Magazine*, *147*(4), 611–637. doi: 10.1017/S0016756809990719

Erkül, F., Helvaci, C., & Sözbilir, H. (2005). Stratigraphy and geochronology of the Early Miocene volcanic units in the Bigadiç Borate Basin, western Turkey. *Turkish Journal of Earth Sciences*, *14*(3), 227–253.

Erkül, F., Helvacı, C., & Sözbilir, H. (2006). Olivine basalt and trachyandesite peperites formed at the subsurface/surface interface of a semi-arid lake: an example from the Early Miocene Bigadiç basin, western Turkey. *Journal of Volcanology and Geothermal Research*, *149*(3–4), 240–262. doi: 10.1016/j.jvolgeores.2005.07.016

Erkül, F., Tatar Erkül, S., Manap, H. S., & Çolak, C. (2017). An extensional and transtensional origin of elongated magmatic domes and localised transfer faults in the northern Menderes metamorphic core complex, western Turkey. *Geodinamica Acta*, *29*(1), 139–159. doi: 10.1080/09853111.2017.1343523

Ersoy, E. Y., Çemen, İ., Helvacı, C., & Billor, Z. (2014). Tectono-stratigraphy of the Neogene basins in Western Turkey: Implications for tectonic evolution of the Aegean Extended Region. *Tectonophysics*, *635*, 33–58. doi: 10.1016/j.tecto.2014.09.002

Ersoy, E. Y., & Palmer, M. R. (2013). Eocene-Quaternary magmatic activity in the Aegean: Implications for mantle metasomatism and magma genesis in an evolving orogeny. *Lithos*, *180–181*, 5–24. doi: 10.1016/j.lithos.2013.06.007

Ersoy, E. Y., Palmer, M. R., Genç, Ş. C., Prelević, D., Akal, C., & Uysal, İ. (2017). Chemo-probe into the mantle origin of the NW Anatolia Eocene to Miocene volcanic rocks: Implications for the role of, crustal accretion, subduction, slab roll-back and slab break-off processes in genesis of post-collisional magmatism. *Lithos*, *288–289*, 55–71. doi: 10.1016/j.lithos.2017.07.006

Ersoy, Y., Helvacı, C., Sözbilir, H., Erkül, F., & Bozkurt, E. (2008). A geochemical approach to Neogene–Quaternary volcanic activity of western Anatolia: An example of episodic bimodal volcanism within the Selendi Basin, Turkey. *Chemical Geology*, *255*(1–2), 265–282. doi: 10.1016/j.chemgeo.2008.06.044

Ersoy, Y. E., Helvacı, C., & Palmer, M. R. (2011). Stratigraphic, structural and geochemical features of the NE–SW trending Neogene volcano-sedimentary basins in western Anatolia: Implications for associations of supra-detachment and transtensional strike-slip basin formation in extensional tectonic setting. *Journal of Asian Earth Sciences*, *41*(2), 159–183. doi: 10.1016/j.jseaes.2010.12.012

Ersoy, Y. E., Helvacı, C., & Sözbilir, H. (2010). Tectono-stratigraphic evolution of the NE–SW-trending superimposed Selendi basin: Implications for late Cenozoic crustal extension in Western Anatolia, Turkey. *Tectonophysics*, *488*(1–4), 210–232. doi: 10.1016/j.tecto.2010.01.007

Eski, S., Sözbilir, H., Uzel, B., Özkaymak, Ç., & Sümer, Ö. (2020). Gölmarmara Fayı'nın Morfotektonik Evriminin CBS Tabanlı Yöntemlerle Arastırılması, Gediz Grabeni, Batı

Anadolu. *Türkiye Jeoloji Bülteni / Geological Bulletin of Turkey*, 1–28. doi: 10.25288/tjb.679584

Etzel, T. M., Catlos, E. J., Ataktürk, K., Lovera, O. M., Kelly, E. D., Çemen, I., & Diniz, E. (2019). Implications for thrust-related shortening punctuated by extension from P-T Paths and geochronology of garnet-bearing schists, Southern (Çine) Menderes Massif, SW Turkey. *Tectonics*, *38*(6), 1974–1998. doi: 10.1029/2018TC005335

Etzel, T. M., Catlos, E. J., Cemen, I., Ozerdem, C., Oyman, T., & Miggins, D. (2020). Documenting exhumation in the central and northern Menderes Massif (western Turkey): new insights from garnet-based P-T estimates and K-feldspar 40Ar/39Ar geochronology. *Lithosphere*, *2020*(1), 8818289. doi: 10.2113/2020/8818289

Eyidoğan, H. (2020). Manisa Akhisar Civarındaki Deprem Aktivitesi ile Gutenberg-Richter Parametrelerinin Zamansal ve Mekânsal Özelliklerinin İncelenmesi. *Türk Deprem Araştırma Dergisi*, *2*(2), 138–159. doi: 10.46464/tdad.776186

Eyubagi, E.,. E., Solak, H. İ., Kavak, U. S., Tiryakioğlu, İ., Sözbilir, H., Aktuğ, B., & Özkaymak, Ç. (2021). Present day strike-slip deformation within the southern part of the İzmir-Balıkesir Transfer Zone based on GNSS data and implications for seismic hazard assessment in western Anatolia. *Turkish Journal of Earth Sciences*, *30*(2), 143–160. doi: 10.3906/yer-2005-26

Faccenna, C., Bellier, O., Martinod, J., Piromallo, C., & Regard, V. (2006). Slab detachment beneath eastern Anatolia: A possible cause for the formation of the North Anatolian fault. *Earth and Planetary Science Letters*, *242*(1–2), 85–97. doi: 10.1016/j.epsl.2005.11.046

Faruk Çelik, Ö., & Delaloye, M. F. (2003). Origin of metamorphic soles and their post-kinematic mafic dyke swarms in the Antalya and Lycian ophiolites, SW Turkey. *Geological Journal*, *38*(3–4), 235–256. doi: 10.1002/gj.954

Faulds, J. E., & Varga, R. J. (1998). The role of accommodation zones and transfer zones in the regional segmentation of extended terranes. In J. E. Faulds & J. H. Stewart (Eds.), *Accommodation zones and transfer zones; the regional segmentation of the Basin and Range Province*. Geological Society of America. doi: 10.1130/0-8137-2323-X.1

Fichtner, A., Saygin, E., Taymaz, T., Cupillard, P., Capdeville, Y., & Trampert, J. (2013). The deep structure of the North Anatolian fault zone. *Earth and Planetary Science Letters*, *373*, 109–117. doi: 10.1016/j.epsl.2013.04.027

Flerit, F., Armijo, R., King, G. C. P., Meyer, B., & Barka, A. (2003). Slip partitioning in the Sea of Marmara pull-apart determined from GPS velocity vectors. *Geophysical Journal International*, *154*(1), 1–7. doi: 10.1046/j.1365-246X.2003.01899.x

Fountoulakis, I., Evangelidis, C. P., & Ktenidou, O.-J. (2023). Coseismic and Postseismic Imaging of a Composite Fault System: The Samos 2020 Mw 7.0 Sequence. *Bulletin of the Seismological Society of America*, *113*(3), 997–1012. doi: 10.1785/0120220207

Fraser, J., Vanneste, K., & Hubert-Ferrari, A. (2010). Recent behavior of the North Anatolian Fault: Insights from an integrated paleoseismological data set. *Journal of Geophysical Research: Solid Earth*, *115*(B9), 2009JB006982. doi: 10.1029/2009JB006982

Frassi, C., Marroni, M., Pandolfi, L., Göncüoğlu, M. C., Ellero, A., Ottria, G., et al. (2018). Burial and exhumation history of the Daday Unit (Central Pontides, Turkey): Implications for the closure of the Intra-Pontide oceanic basin. *Geological Magazine*, *155*(2), 356–376. doi: 10.1017/S0016756817000176

Gamal, N., Yousef, M., Moustafa, A. R., & Bosworth, W. (2021). Spatiotemporal evolution of transfer structures and linked fault systems in an extensional setting: Southwest Gebel Akheider, Cairo-Suez District, Egypt. *Marine and Petroleum Geology*, *133*, 105260. doi: 10.1016/j.marpetgeo.2021.105260

Gawthorpe, R. L., & Hurst, J. M. (1993). Transfer zones in extensional basins: Their structural style and influence on drainage development and stratigraphy. *Journal of the Geological Society*, *150*(6), 1137–1152. doi: 10.1144/gsjgs.150.6.1137

Geçkin, B., Sözbilir, H., Özkaymak, Ç., & Softa, M. (2021). Active Tectonics of Gülbahçe fault zone (GBFZ) by Using Geomorphic Indices, İzmir Province, Western Anatolia, Turkey. *Afyon Kocatepe University Journal of Sciences and Engineering*, *21*(5), 1195–1209. doi: 10.35414/akufemubid.946298

Géli, L., Henry, P., Zitter, T., Dupré, S., Tryon, M., Çağatay, M. N., et al. (2008). Gas emissions and active tectonics within the submerged section of the North Anatolian Fault zone in the Sea of Marmara. *Earth and Planetary Science Letters*, *274*(1–2), 34–39. doi: 10.1016/j.epsl.2008.06.047

Gessner, K., Gallardo, L. A., Markwitz, V., Ring, U., & Thomson, S. N. (2013). What caused the denudation of the Menderes Massif: Review of crustal evolution, lithosphere structure, and dynamic topography in southwest Turkey. *Gondwana Research*, *24*(1), 243–274. doi: 10.1016/j.gr.2013.01.005

Gessner, K., Markwitz, V., & Güngör, T. (2018). Crustal fluid flow in hot continental extension: tectonic framework of geothermal areas and mineral deposits in western Anatolia. *Geological Society, London, Special Publications*, *453*(1), 289–311. doi: 10.1144/SP453.7

Gessner, K., Piazolo, S., Güngör, T., Ring, U., Kröner, A., & Passchier, C. (2001). Tectonic significance of deformation patterns in granitoid rocks of the Menderes nappes, Anatolide belt, southwest Turkey. *International Journal of Earth Sciences*, *89*(4), 766–780. doi: 10.1007/s005310000106

Gibbs, A. D. (1984). Structural evolution of extensional basin margins. *Journal of the Geological Society*, *141*(4), 609–620. doi: 10.1144/gsjgs.141.4.0609

Gokten, E., Havzoğlu, T., & Şan, Ö. (2001). Tertiary evolution of the central Menderes Massif based on structural investigations of metamorphics and sedimentary cover rocks between Salihli and Kiraz (western Turkey). *International Journal of Earth Sciences*, *89*(4), 745–756. doi: 10.1007/s005310000099

Göncüoglu, M. C. (2010). *Introduction to the geology of Turkey: Geodynamic evolution of the pre-alpine and alpine terranes*. Ankara: General Directorate, Mineral Research and Exploration.

Göncüoglu, M. C., Marroni, M., Pandolfi, L., Ellero, A., Ottria, G., Catanzariti, R., et al. (2014). The Arkot Dağ Mélange in Araç area, central Turkey: Evidence of its origin within the geodynamic evolution of the Intra-Pontide suture zone. *Journal of Asian Earth Sciences*, *85*, 117–139. doi: 10.1016/j.jseaes.2014.01.013

Göncüoğlu, M. C., Marroni, M., Sayit, K., Tekin, U. K., Ottria, G., Pandolifi, L., & Ellero, A. (2012). The Ayli Dag ophiolite sequence (central-northern Turkey); A fragment of Middle Jurassic oceanic lithosphere within the Intra-Pontide suture zone. *Ofioliti*, *37*, 77–92.

Görgün, E. (2014). Source characteristics and Coulomb stress change of the 19 May 2011 Mw 6.0 Simav–Kütahya earthquake, Turkey. *Journal of Asian Earth Sciences*, *87*, 79–88. doi: 10.1016/j.jseaes.2014.02.016

Günay, A., Çağlayan, A., Saber, R., & Işik, V. (2021). Fracture analysis of Çaysimav segment (West Anatolia) and its surroundings using remote sensing methods. In F. Chitea (Ed.), *Geoscience for society, education & environment* (pp. 156–164). Bucureşti, România: Editura Academiei Române. doi: 10.5281/ZENODO.4322617

Gürboğa, Ş. (2014). Structural analyses of Şaphane relay ramps and fault linkage evolution in active extensional regime, western Turkey. *Turkish Journal of Earth Sciences*, *23*, 615–626. doi: 10.3906/yer-1405-16

Gürboğa, Ş., Koçyiğit, A., & Ruffet, G. (2013). Episodic two-stage extensional evolutionary model for southwestern Anatolian Graben–horst system: New field data from the Erdoğmuş-Yenigediz Graben (Kütahya). *Journal of Geodynamics*, *65*, 176–198. doi: 10.1016/j.jog.2012.06.007

Gürer, Ö. F. (2022). A new look at the origin of N-S trending young basins of western Anatolia. *Bulletin of the Mineral Research and Exploration*, *22*, 1–34. doi: 10.19111/bulletinofmre.1066949

Haklıdır, F. S. T., & Şengün, R. (2020). Hydrogeochemical similarities and differences between high temperature geothermal systems with similar geologic settings in the Büyük Menderes and Gediz Grabens of Turkey. *Geothermics*, *83*, 101717. doi: 10.1016/j.geothermics.2019.101717

Hançer, M. (2013). Study of the structural evolution of the Babadağ-Honaz and Pamukkale fault zones and the related earthquake risk potential of the Buldan region in SW Anatolia, east of the Mediterranean. *Journal of Earth Science*, *24*(3), 397–409. doi: 10.1007/s12583-013-0333-2

Hancock, P. L., & Altunel, E. (1997). Faulted archaeological relics at Hierapolis (Pamukkale), Turkey. *Journal of Geodynamics*, *24*(1–4), 21–36. doi: 10.1016/S0264-3707(97)00003-3

Hancock, P. L., Chalmers, R. M. L., Altunel, E., & Çakir, Z. (1999). Travitonics: Using travertines in active fault studies. *Journal of Structural Geology*, *21*(8–9), 903–916. doi: 10.1016/S0191-8141(99)00061-9

Harrison, R. W., Tsiolakis, E., Stone, B. D., Lord, A., McGeehin, J. P., Mahan, S. A., & Chirico, P. (2013). Late Pleistocene and Holocene uplift history of Cyprus: implications for active tectonics along the southern margin of the Anatolian microplate. *Geological Society, London, Special Publications*, *372*(1), 561–584. doi: 10.1144/SP372.3

Helvaci, C. (2015). Geological features of Neogene basins hosting borate deposits: An overview of deposits and future forecast, Turkey. *Bulletin of the Mineral Research and Exploration*, *0*(151). doi: 10.19111/bmre.05207

Helvaci, C., & Alaca, O. (1991). Geology and mineralogy of the Bigadiç Borate Deposıts and vicinity. *Bulletin of the Mineral Research and Exploration*, *113*, 31–63.

Hetzel, R., Passchier, C. W., Ring, U., & Dora, Ö. O. (1995). Bivergent extension in orogenic belts: The Menderes massif (southwestern Turkey). *Geology*, *23*(5), 455. doi: 10.1130/0091-7613(1995)023<0455:BEIOBT>2.3.CO;2

Hetzel, R., Ring, U., Akal, C., & Troesch, M. (1995b). Miocene NNE-directed extensional unroofing in the Menderes Massif, southwestern Turkey. *Journal of the Geological Society*, *152*(4), 639–654. doi: 10.1144/gsjgs.152.4.0639

Hitchcock, C., Altunel, E., Barka, A., Bachhuber, J., Lettis, W., Helms, J., & Lindvall, S. (2003). Timing of Late Holocene Earthquakes on the Eastern Düzce Fault and Implications for Slip Transfer between the Southern and Northern Strands of the North Anatolian Fault System, Bolu, Turkey. *Turkish Journal of Earth Sciences*, *12*(1), 119–136.

Hou, Z., Şen, O., Gou, Y., Eker, A. M., Li, M., Yal, G. P., et al. (2015). Preliminary geological, geochemical and numerical study on the first EGS project in Turkey. *Environmental Earth Sciences*, *73*(11), 6747–6767. doi: 10.1007/s12665-015-4407-6

Hubert-Ferrari, A., Armijo, R., King, G., Meyer, B., & Barka, A. (2002). Morphology, displacement, and slip rates along the North Anatolian Fault, Turkey. *Journal of Geophysical Research: Solid Earth*, *107*(B10). doi: 10.1029/2001JB000393

İncı, U. (1998). Lignite and carbonate deposition in Middle Lignite succession of the Soma Formation, Soma coalfield, western Turkey. *International Journal of Coal Geology*, *37*(3–4), 287–313. doi: 10.1016/S0166-5162(98)00010-X

İnci, U. (2002). Depositional evolution of Miocene coal successions in the Soma coalfield, western Turkey. *International Journal of Coal Geology*, *51*(1), 1–29. doi: 10.1016/S0166-5162(01)00066-0

Inel, M., Ozmen, H. B., & Akyol, E. (2013). Observations on the building damages after 19 May 2011 Simav (Turkey) earthquake. *Bulletin of Earthquake Engineering*, *11*(1), 255–283. doi: 10.1007/s10518-012-9414-3

Işik, V., & Tekeli, O. (2001). Late orogenic crustal extension in the northern Menderes massif (western Turkey): evidence for metamorphic core complex formation. *International Journal of Earth Sciences*, *89*(4), 757–765. doi: 10.1007/s005310000105

Isik, V., Tekeli, O., & Seyitoglu, G. (2004). The $^{40}Ar/^{39}Ar$ age of extensional ductile deformation and granitoid intrusion in the northern Menderes core complex: Implications for the initiation of extensional tectonics in western Turkey. *Journal of Asian Earth Sciences*, *23*(4), 555–566. doi: 10.1016/j.jseaes.2003.09.001

Janssen, C., Bohnhoff, M., Vapnik, Y., Görgün, E., Bulut, F., Plessen, B., et al. (2009). Tectonic evolution of the Ganos segment of the North Anatolian Fault (NW Turkey). *Journal of Structural Geology*, *31*(1), 11–28. doi: 10.1016/j.jsg.2008.09.010

Jolivet, L., Baudin, T., Calassou, S., Chevrot, S., Ford, M., Issautier, B., et al. (2021). Geodynamic evolution of a wide plate boundary in the Western Mediterranean, near-field versus far-field interactions. In O. Lacombe, et al. (Eds.), *Orogen lifecycle: learnings and perspectives from Pyrenees, Western Mediterranean and analogues* (Vol. *192*, –48). doi: 10.1051/bsgf/2021043

Jolivet, L., & Faccenna, C. (2000). Mediterranean extension and the Africa-Eurasia collision. *Tectonics*, *19*(6), 1095–1106. doi: 10.1029/2000TC900018

Jolivet, L., Faccenna, C., Huet, B., Labrousse, L., Le Pourhiet, L., Lacombe, O., et al. (2013). Aegean tectonics: Strain localisation, slab tearing and trench retreat. *Tectonophysics, 597–598*, 1–33. doi: 10.1016/j.tecto.2012.06.011

Jolivet, L., Famin, V., Mehl, C., Parra, T., Aubourg, C., Hébert, R., & Philippot, P. (2004). Strain localization during crustal-scale boudinage to form extensional metamorphic domes in the Aegean Sea. In D. L. Whitney, et al. (Eds.), *Gneiss domes in Orogeny* (Vol. *380*, pp. 185–210). Boulder, CO: Geological Society of America. doi: 10.1130/0-8137-2380-9.185

Jolivet, L., Menant, A., Sternai, P., Rabillard, A., Arbaret, L., Augier, R., et al. (2015). The geological signature of a slab tear below the Aegean. *Tectonophysics, 659*, 166–182. doi: 10.1016/j.tecto.2015.08.004

Kadir, S., & Erkoyun, H. (2013). Genesis of the hydrothermal Karaçayır kaolinite deposit in Miocene volcanics and Palaeozoic metamorphic rocks of the Uşak-Güre Basin, western Turkey. *Turkish Journal of Earth Sciences, 22*(3), 444–468. doi: 10.3906/yer-1112-2

Kaiser, E. P. (1950). Structural significance of lineaments. *Geological Society of America Bulletin, 61*(12), 1475–1476.

Karabacak, V., Sançar, T., Yildirim, G., & Uysal, I. T. (2022). When did the North Anatolian fault reach southern Marmara, Turkey? *Geology, 50*(4), 432–436. doi: 10.1130/G49726.1

Karaoğlu, Ö. (2014). Tectonic controls on the Yamanlar volcano and Yuntdağı volcanic region, western Turkey: implications for an incremental deformation. *Journal of Volcanology and Geothermal Research, 274*, 16–33. doi: 10.1016/j.jvolgeores.2014.01.013

Karaoğlu, Ö., & Helvacı, C. (2012). Growth, destruction and volcanic facies architecture of three volcanic centres in the Miocene Uşak–Güre basin, western Turkey: Subaqueous–subaerial volcanism in a lacustrine setting. *Journal of Volcanology and Geothermal Research, 245–246*, 1–20. doi: 10.1016/j.jvolgeores.2012.06.028

Karaoğlu, Ö., & Helvacı, C. (2014). Isotopic evidence for a transition from subduction to slab-tear related volcanism in western Anatolia, Turkey. *Lithos, 192–195*, 226–239. doi: 10.1016/j.lithos.2014.02.006

Karaoğlu, Ö., Helvacı, C., & Ersoy, Y. (2010). Petrogenesis and $^{40}Ar/^{39}Ar$ geochronology of the volcanic rocks of the Uşak-Güre basin, western Türkiye. *Lithos, 119*(3–4), 193–210. doi: 10.1016/j.lithos.2010.07.001

Karasözen, E., Nissen, E., Bergman, E. A., Johnson, K. L., & Walters, R. J. (2016). Normal faulting in the Simav Graben of western Turkey reassessed with calibrated earthquake relocations. *Journal of Geophysical Research: Solid Earth, 121*(6), 4553–4574. doi: 10.1002/2016JB012828

Karayiğit, A. İ., Littke, R., Querol, X., Jones, T., Oskay, R. G., & Christanis, K. (2017). The Miocene coal seams in the Soma Basin (W. Turkey): Insights from coal petrography, mineralogy and geochemistry. *International Journal of Coal Geology, 173*, 110–128. doi: 10.1016/j.coal.2017.03.004

Kaya, O., Ünay, E., Göktaş, F., & Saraç, G. (2007). Early Miocene stratigraphy of central west Anatolia, Turkey: implications for the tectonic evolution of the eastern Aegean area. *Geological Journal, 42*(1), 85–109. doi: 10.1002/gj.1071

Keay, S., Lister, G., & Buick, I. (2001). The timing of partial melting, Barrovian metamorphism and granite intrusion in the Naxos metamorphic core complex, Cyclades, Aegean Sea, Greece. *Tectonophysics, 342*(3–4), 275–312. doi: 10.1016/S0040-1951(01)00168-8

Kent, E., Whittaker, A. C., Boulton, S. J., & Alçiçek, M. C. (2021). Quantifying the competing influences of lithology and throw rate on bedrock river incision. *GSA Bulletin, 133*(7–8), 1649–1664. doi: 10.1130/B35783.1

Kırkan, E., Akyüz, H. S., Basmenji, M., Dikbaş, A., Zabcı, C., Yazıcı, M., & Erturaç, M. K. (2023). Earthquake history of the Milas Fault: an active dextral fault in an extensional province (SW Anatolia, Türkiye). *Natural Hazards, 116*(2), 1639–1662. doi: 10.1007/s11069-022-05733-w

Kirkpatrick, J. D., Shipton, Z. K., Evans, J. P., Micklethwaite, S., Lim, S. J., & McKillop, P. (2008). Strike-slip fault terminations at seismogenic depths: The structure and kinematics of the Glacier Lakes fault, Sierra Nevada United States. *Journal of Geophysical Research: Solid Earth, 113*(B4), 2007JB005311. doi: 10.1029/2007JB005311

Koç-Taşgın, C., & Diniz-Akarca, C. (2018). Soft-sediment deformation structures related to tectonomagmatic activity: A case study from the borate-bearing lacustrine deposits of early Miocene Bigadiç Basin, NW Turkey. *Sedimentary Geology, 373*, 32–47. doi: 10.1016/j.sedgeo.2018.05.012

Koçyiğit, A., & Gürboğa, Ş. (2021). Active tectonics of Gülpınar-Tuzla area (Biga Peninsula, NW Turkey): The source of 6 February-24 March 2017 earthquake cluster. *Bulletin of the Mineral Research and Exploration*, 1–29. doi: 10.19111/bulletinofmre.984922

Koçyiğit, A., Yusufodlu, H., & Bozkurt, E. (1999). Evidence from the Gediz Graben for episodic two-stage extension in western Turkey. *Journal of the Geological Society, 156*(3), 605–616. doi: 10.1144/gsjgs.156.3.0605

Konstantinovskaya, E. A., Harris, L. B., Poulin, J., & Ivanov, G. M. (2007). Transfer zones and fault reactivation in inverted rift basins: Insights from physical modelling. *Tectonophysics, 441*(1–4), 1–26. doi: 10.1016/j.tecto.2007.06.002

Kumsar, H., Aydan, Ö., Şimşek, C., & D'Andria, F. (2016). Historical earthquakes that damaged Hierapolis and Laodikeia antique cities and their implications for earthquake potential of Denizli basin in western Turkey. *Bulletin of Engineering Geology and the Environment, 75*(2), 519–536. doi: 10.1007/s10064-015-0791-0

Kuşçu, İ. (2009). Cross-basin Faulting and Extinction of Pull-apart Basins in the Sea of Marmara, NW Turkey. *Turkish Journal of Earth Sciences, 18*(3), 331–349. doi: 10.3906/yer-0706-4

Lamont, T. N., Searle, M. P., Waters, D. J., Roberts, N. M. W., Palin, R. M., Smye, A., et al. (2020). Compressional origin of the Naxos metamorphic core complex, Greece: Structure, petrography, and thermobarometry. *GSA Bulletin, 132*(1–2), 149–197. doi: 10.1130/B31978.1

Larsen, P.-H. (1988). Relay structures in a Lower Permian basement-involved extension system, East Greenland. *Journal of Structural Geology, 10*(1), 3–8. doi: 10.1016/0191-8141(88)90122-8

Le Pichon, X., Şengör, A. M. C., & İmren, C. (2019). A new approach to the opening of the eastern Mediterranean

Sea and the origin of the Hellenic subduction zone. Part 2: The Hellenic subduction zone. *Canadian Journal of Earth Sciences*, *56*(11), 1144–1162. doi: 10.1139/cjes-2018-0315

Le Pourhiet, L., Huet, B., May, D. A., Labrousse, L., & Jolivet, L. (2012). Kinematic interpretation of the 3D shapes of metamorphic core complexes. *Geochemistry, Geophysics, Geosystems*, *13*(9), 2012GC004271. doi: 10.1029/2012GC004271

Leeder, M. R., & Gawthorpe, R. L. (1987). Sedimentary models for extensional tilt-block/half-Graben basins. *Geological Society, London, Special Publications*, *28*(1), 139–152. doi: 10.1144/GSL.SP.1987.028.01.11

Lips, A. L. W., Cassard, D., Sözbilir, H., Yilmaz, H., & Wijbrans, J. R. (2001). Multistage exhumation of the Menderes Massif, western Anatolia (Turkey). *International Journal of Earth Sciences*, *89*(4), 781–792. doi: 10.1007/s005310000101

Lister, G. S., Banga, G., & Feenstra, A. (1984). Metamorphic core complexes of Cordilleran type in the Cyclades, Aegean Sea, Greece. *Geology*, *12*(4), 221. doi: 10.1130/0091-7613(1984)12<221:MCCOCT>2.0.CO;2

Lister, G. S., Etheridge, M. A., & Symonds, P. A. (1986). Detachment faulting and the evolution of passive continental margins. *Geology*, *14*(3), 246. doi: 10.1130/0091-7613(1986)14<246:DFATEO>2.0.CO;2

Malandri, C., Soukis, K., Maffione, M., Özkaptan, M., Vassilakis, E., Lozios, S., & Van Hinsbergen, D. J. J. (2017). Vertical-axis rotations accommodated along the Mid-Cycladic lineament on Paros Island in the extensional heart of the Aegean orocline (Greece). *Lithosphere*, *9*(1), 78–99. doi: 10.1130/L575.1

Marroni, M., Frassi, C., Göncüoğlu, M. C., Di Vincenzo, G., Pandolfi, L., Rebay, G., et al. (2014). Late Jurassic amphibolite-facies metamorphism in the Intra-Pontide suture zone (Turkey): an eastward extension of the Vardar Ocean from the Balkans into Anatolia? *Journal of the Geological Society*, *171*(5), 605–608. doi: 10.1144/jgs2013-104

McClusky, S., Balassanian, S., Barka, A., Demir, C., Ergintav, S., Georgiev, I., et al. (2000). Global Positioning System constraints on plate kinematics and dynamics in the eastern Mediterranean and Caucasus. *Journal of Geophysical Research: Solid Earth*, *105*(B3), 5695–5719. doi: 10.1029/1999JB900351

Meulenkamp, J. E., Wortel, M. J. R., Van Wamel, W. A., Spakman, W., & Hoogerduyn Strating, E. (1988). On the Hellenic subduction zone and the geodynamic evolution of Crete since the late Middle Miocene. *Tectonophysics*, *146*(1–4), 203–215. doi: 10.1016/0040-1951(88)90091-1

Moix, P., Beccaletto, L., Kozur, H. W., Hochard, C., Rosselet, F., & Stampfli, G. M. (2008). A new classification of the Turkish terranes and sutures and its implication for the paleotectonic history of the region. *Tectonophysics*, *451*(1–4), 7–39. doi: 10.1016/j.tecto.2007.11.044

Morley, C. K. (1995). Developments in the structural geology of rifts over the last decade and their impact on hydrocarbon exploration. *Geological Society, London, Special Publications*, *80*, 1–32.

Morley, C. K., Nelson, R. A., Patton, T. L., & Munn, S. G. (1990). Transfer Zones in the East African Rift System and their relevance to hydrocarbon exploration in rifts. *AAPG Bulletin*, *74*. doi: 10.1306/0C9B2475-1710-11D7-8645000102C1865D

Morris, A., & Anderson, M. (1996). First palaeomagnetic results from the Cycladic Massif, Greece, and their implications for Miocene extension directions and tectonic models in the Aegean. *Earth and Planetary Science Letters*, *142*(3–4), 397–408. doi: 10.1016/0012-821X(96)00114-8

Moustafa, A. R. (2002). Controls on the geometry of transfer zones in the Suez rift and northwest Red Sea: Implications for the structural geometry of rift systems. *AAPG Bulletin*, *86*. doi: 10.1306/61EEDC06-173E-11D7-8645000102C1865D

Mozafari, N., Özkaymak, Ç., Tikhomirov, D., Ivy-Ochs, S., Alfimov, V., Sözbilir, H., et al. (2021). Seismic Activity of the Manisa fault zone in Western Turkey Constrained by Cosmogenic 36Cl Dating. *Geosciences*, *11*(11), 451. doi: 10.3390/geosciences11110451

Muthumaniraja, C. K., Anbazhagan, S., Jothibasu, A., & Chinnamuthu, M. (2019). Remote Sensing and Fuzzy Logic Approach for Artificial Recharge Studies in Hard Rock Terrain of South India. In *GIS and geostatistical techniques for groundwater science* (pp. 91–112). Elsevier. doi: 10.1016/B978-0-12-815413-7.00008-0

Mutlu, A. K. (2020). Seismicity, focal mechanism, and stress tensor analysis of the Simav region, western Turkey. *Open Geosciences*, *12*(1), 479–490. doi: 10.1515/geo-2020-0010

Naylor, M. A., Mandl, G., & Supesteijn, C. H. K. (1986). Fault geometries in basement-induced wrench faulting under different initial stress states. *Journal of Structural Geology*, *8*(7), 737–752. doi: 10.1016/0191-8141(86)90022-2

Nyst, M., & Thatcher, W. (2004). New constraints on the active tectonic deformation of the Aegean. *Journal of Geophysical Research: Solid Earth*, *109*(B11), 2003JB002830. doi: 10.1029/2003JB002830

Ocakoğlu, N., Demirbağ, E., & Kuşçu, İ. (2005). Neotectonic structures in İzmir Gulf and surrounding regions (western Turkey): Evidences of strike-slip faulting with compression in the Aegean extensional regime. *Marine Geology*, *219*(2–3), 155–171. doi: 10.1016/j.margeo.2005.06.004

Okay, A., Satır, M., Tüysüz, O., Akyüz, S., & Chen, F. (2001). The tectonics of the Strandja Massif: late-Variscan and mid-Mesozoic deformation and metamorphism in the northern Aegean. *International Journal of Earth Sciences*, *90*(2), 217–233. doi: 10.1007/s005310000104

Okay, A. I. (1984). Distribution and characteristics of the north-west Turkish blueschists. *Geological Society, London, Special Publications*, *17*(1), 455–466. doi: 10.1144/GSL.SP.1984.017.01.33

Okay, A. I. (1986). High-pressure/low-temperature metamorphic rocks of Turkey. In B. W. Evans & E. H. Brown (Eds.), *Geological Society of America Memoirs* (Vol. *164*, pp. 333–347). Geological Society of America. doi: 10.1130/MEM164-p333

Okay, A. I. (2002). Jadeite–chloritoid–glaucophane–lawsonite blueschists in north-west Turkey: Unusually high *P/T* ratios in continental crust. *Journal of Metamorphic Geology*, *20*(8), 757–768. doi: 10.1046/j.1525-1314.2002.00402.x

Okay, A. I. (2008). Geology of Turkey: A synopsis. *Der Anschnitt*, *21*, 19–42.

Okay, A. İ., Özcan, E., Cavazza, W., Okay, N., & Less, G. (2010). Basement types, Lower Eocene series, Upper Eocene olistostromes and the initiation of the southern Thrace Basin, NW Turkey. *Turkish Journal of Earth Sciences*, *19*(1), 1–25. doi: 10.3906/yer-0902-10

Okay, A. I., Satir, M., Maluski, H., Siyako, M., Monie, P., Metzger, R., & Akyüz, S. (1996). Paleo- and Neo-Tethyan events in northwestern Turkey: Geologic and geochronologic constraints. In A. Yin & T. M. Harrison (Eds.), *The tectonic evolution of Asia* (pp. 420–441). Cambridge: Cambridge University Press.

Okay, A. I., & Siyako, M. (1993). The new position of the İzmir-Ankara Neo-Tethyan suture between İzmir and Balıkesir. In *Proceedings of the Ozan Sungurlu Symposium* (pp. 333–355). Ankara, Turkey. Retrieved from https://web.itu.edu.tr/~okay/AralOkayPapersThematically.htm

Okay, A. I. (2001). Stratigraphic and metamorphic inversions in the central Menderes Massif: A new structural model. *International Journal of Earth Sciences*, *89*(4), 709–727. doi: 10.1007/s005310000098

Okay, A. I., & Satir, M. (2000). Coeval plutonism and metamorphism in a latest Oligocene metamorphic core complex in northwest Turkey. *Geological Magazine*, *137*(5), 495–516. doi: 10.1017/S0016756800004532

Okay, A. I., Sunal, G., Sherlock, S., Kylander-Clark, A. R. C., & Özcan, E. (2020). İzmir-Ankara suture as a Triassic to cretaceous Plate Boundary—Data from Central Anatolia. *Tectonics*, *39*(5), e2019TC005849. doi: 10.1029/2019TC005849

Okay, A. I., & Tüysüz, O. (1999). Tethyan sutures of northern Turkey. *Geological Society, London, Special Publications*, *156*(1), 475–515. doi: 10.1144/GSL.SP.1999.156.01.22

Oner, Z., & Dilek, Y. (2011). Supradetachment basin evolution during continental extension: The Aegean province of western Anatolia, Turkey. *Geological Society of America Bulletin*, *123*(11–12), 2115–2141. doi: 10.1130/B30468.1

Oner, Z., & Dilek, Y. (2013). Fault kinematics in supradetachment basin formation, Menderes core complex of western Turkey. *Tectonophysics*, *608*, 1394–1412. doi: 10.1016/j.tecto.2013.06.003

Oner, Z., Dilek, Y., & Kadioglu, Y. K. (2010). Geology and geochemistry of the synextensional Salihli granitoid in the Menderes core complex, western Anatolia, Turkey. *International Geology Review*, *52*(2–3), 336–368. doi: 10.1080/00206810902815871

Oudeika, M. S., İlkimen, E. M., Taşdelen, S., & Aydin, A. (2020). Distinguishing groundwater flow paths in fractured rock aquifers formed under tectonic stress using geophysical techniques: Cankurtaran Basin, Denizli, Turkey. *International Journal of Environmental Research*, *14*(5), 567–581. doi: 10.1007/s41742-020-00279-w

Oygür, V., & Erler, A. (2000). Metallogeny of the Simav graben (Inner-Western Anatolia, Turkey). *Geological Bulletin of Turkey*, *43*(1), 7–19.

Özalp, S., Emre, Ö., Şaroğlu, F., Özaksoy, V., Elmacı, H., & Duman, T. Y. (2018). Active fault segmentation of the Çivril Graben System and surface rupture of the 1 October 1995 Dinar earthquake (Mw 6.2), Southwestern Anatolia, Turkey. *Journal of Asian Earth Sciences*, *166*, 136–151. doi: 10.1016/j.jseaes.2018.07.037

Özbakır, A. D., Şengör, A. M. C., Wortel, M. J. R., & Govers, R. (2013). The Pliny–Strabo trench region: A large shear zone resulting from slab tearing. *Earth and Planetary Science Letters*, *375*, 188–195. doi: 10.1016/j.epsl.2013.05.025

Özdamar, Ş., Billor, M. Z., Sunal, G., Esenli, F., & Roden, M. F. (2013). First U–Pb SHRIMP zircon and $^{40}Ar/^{39}Ar$ ages of metarhyolites from the Afyon–Bolkardağ Zone, SW Turkey: Implications for the rifting and closure of the Neo-Tethys. *Gondwana Research*, *24*(1), 377–391. doi: 10.1016/j.gr.2012.10.006

Ozener, H., Dogru, A., & Acar, M. (2013). Determination of the displacements along the Tuzla fault (Aegean region-Turkey): Preliminary results from GPS and precise leveling techniques. *Journal of Geodynamics*, *67*, 13–20. doi: 10.1016/j.jog.2012.06.001

Özkaymak, Ç., & Sözbilir, H. (2008). Stratigraphic and Structural Evidence for Fault Reactivation: The Active Manisa fault zone, Western Anatolia. *Turkish Journal of Earth Sciences*, *17*(3), 615–635.

Özkaymak, Ç., & Sözbilir, H. (2012). Tectonic geomorphology of the Spildağı High Ranges, western Anatolia. *Geomorphology*, *173–174*, 128–140. doi: 10.1016/j.geomorph.2012.06.003

Özkaymak, Ç., Sözbilir, H., & Uzel, B. (2013). Neogene–Quaternary evolution of the Manisa Basin: Evidence for variation in the stress pattern of the İzmir-Balıkesir Transfer Zone, western Anatolia. *Journal of Geodynamics*, *65*, 117–135. doi: 10.1016/j.jog.2012.06.004

Özsayın, E., Dirik, K., Ocakoğlu, F., Cartigny, S. A., & Sağlam Selçuk, A. S. (2023). Tectonic geomorphology of Bozdoğan and Karacasu Grabens, Western Anatolia. *Geologica Acta*, *21*, 1–14. doi: 10.1344/GeologicaActa2023.21.1

Öztürk, S. (2015). A study on the correlations between seismotectonic b-value and Dc-value, and seismic quiescence Z-value in the Western Anatolian region of Turkey. *Austrian Journal of Earth Sciences*, *108*(2). doi: 10.17738/ajes.2015.0019

Özürlan, G., Candansayar, M. E., & Şahin, M. H. (2006). Deep resistivity structure of the Dikili-Bergama region, west Anatolia, revealed by two- dimensional inversion of vertical electrical sounding data. *Geophysical Prospecting*, *54*(2), 187–197. doi: 10.1111/j.1365-2478.2006.00525.x

Pamukçu, O., Gönenç, T., Uyanik, O., Sözbilir, H., & Çakmak, O. (2014). A microgravity model for the city of İzmir (Western Anatolia) and its tectonic implementations. *Acta Geophysica*, *62*(4), 849–871. doi: 10.2478/s11600-014-0203-z

Papadopoulos, G. A. (1997). On the interpretation of large-scale seismic tomography images in the Aegean sea area. *Annals of Geophysics*, *40*(1), 3. doi: 10.4401/ag-3933

Papanikolaou, D. I. (2021). *The geology of Greece*. Cham: Springer International Publishing. doi: 10.1007/978-3-030-60731-9

Papanikolaou, D. J., & Royden, L. H. (2007). Disruption of the Hellenic arc: Late Miocene extensional detachment faults and steep Pliocene-Quaternary normal faults—Or what happened at Corinth? *Tectonics*, *26*(5), 2006TC002007. doi: 10.1029/2006TC002007

Peacock, D. C. P., & Sanderson, D. J. (1991). Displacements, segment linkage and relay ramps in normal fault zones. *Journal of Structural Geology*, *13*(6), 721–733. doi: 10.1016/0191-8141(91)90033-F

Peacock, D. C. P., & Sanderson, D. J. (1994). Geometry and development of relay ramps in normal fault systems. *AAPG Bulletin, 78*. doi: 10.1306/BDFF9046-1718-11D7-8645000102C1865D

Pe-Piper, G., Piper, D. J. W., & Matarangas, D. (2002). Regional implications of geochemistry and style of emplacement of Miocene I-type diorite and granite, Delos, Cyclades, Greece. *Lithos, 60*(1–2), 47–66. doi: 10.1016/S0024-4937(01)00068-8

Perinçek, D. (1991). Possible strand of the North Anatolian fault in the Thrace Basin, Turkey—an interpretation. *AAPG Bulletin, 75*, 241–257. doi: 10.1306/0C9B2795-1710-11D7-8645000102C1865D

Perinçek, D., & Çemen, I. (1990). The structural relationship between the East Anatolian and Dead Sea fault zones in southeastern Turkey. *Tectonophysics, 172*(3–4), 331–340. doi: 10.1016/0040-1951(90)90039-B

Philippon, M., Brun, J.-P., Gueydan, F., & Sokoutis, D. (2014). The interaction between Aegean back-arc extension and Anatolia escape since Middle Miocene. *Tectonophysics, 631*, 176–188. doi: 10.1016/j.tecto.2014.04.039

Piper, J. D. A., Gürsoy, H., Tatar, O., Beck, M. E., Rao, A., Koçbulut, F., & Mesci, B. L. (2010). Distributed neotectonic deformation in the Anatolides of Turkey: A palaeomagnetic analysis. *Tectonophysics, 488*(1–4), 31–50. doi: 10.1016/j.tecto.2009.05.026

Plunder, A., Agard, P., Chopin, C., & Okay, A. I. (2013). Geodynamics of the Tavşanlı zone, western Turkey: Insights into subduction/obduction processes. *Tectonophysics, 608*, 884–903. doi: 10.1016/j.tecto.2013.07.028

Polat, O., Gök, E., & Yilmaz, D. (2008). Earthquake hazard of the Aegean extension region (West Turkey). *Turkish Journal of Earth Sciences, 17*(3), 593–614.

Pourteau, A., Candan, O., & Oberhänsli, R. (2010). High-pressure metasediments in central Turkey: Constraints on the Neotethyan closure history. *Tectonics, 29*(5) n/a-n/a. doi: 10.1029/2009TC002650

Pourteau, A., Oberhänsli, R., Candan, O., Barrier, E., & Vrielynck, B. (2016). Neotethyan closure history of western Anatolia: A geodynamic discussion. *International Journal of Earth Sciences, 105*(1), 203–224. doi: 10.1007/s00531-015-1226-7

Pourteau, A., Scherer, E. E., Schorn, S., Bast, R., Schmidt, A., & Ebert, L. (2019). Thermal evolution of an ancient subduction interface revealed by Lu–Hf garnet geochronology, Halilbağı Complex (Anatolia). *Geoscience Frontiers, 10*(1), 127–148. doi: 10.1016/j.gsf.2018.03.004

Pourteau, A., Sudo, M., Candan, O., Lanari, P., Vidal, O., & Oberhänsli, R. (2013). Neotethys closure history of Anatolia: insights from $^{40}Ar/^{39}Ar$ geochronology and P–T estimation in high-pressure metasedimentary rocks. *Journal of Metamorphic Geology, 31*(6), 585–606. doi: 10.1111/jmg.12034

Poyraz, F., Hastaoğlu, K. O., Koçbulut, F., Tiryakioğlu, I., Tatar, O., Demirel, M., et al. (2019). Determination of the block movements in the eastern section of the Gediz Graben (Turkey) from GNSS measurements. *Journal of Geodynamics, 123*, 38–48. doi: 10.1016/j.jog.2018.11.001

Purvis, M., & Robertson, A. (2005). Miocene sedimentary evolution of the NE–SW-trending Selendi and Gördes Basins, W Turkey: Implications for extensional processes. *Sedimentary Geology, 174*(1–2), 31–62. doi: 10.1016/j.sedgeo.2004.11.002

Rabayrol, F., & Hart, C. J. R. (2021). Petrogenetic and tectonic controls on magma fertility and the formation of post-subduction porphyry and epithermal mineralization along the late Cenozoic Anatolian metallogenic trend, Turkey. *Mineralium Deposita, 56*(2), 279–306. doi: 10.1007/s00126-020-00967-9

Rabayrol, F., Hart, C. J. R., & Creaser, R. A. (2019). Tectonic triggers for postsubduction magmatic-hydrothermal gold metallogeny in the late Cenozoic Anatolian metallogenic trend, Turkey. *Economic Geology, 114*(7), 1339–1363. doi: 10.5382/econgeo.4682

Rabillard, A., Arbaret, L., Jolivet, L., Le Breton, N., Gumiaux, C., Augier, R., & Grasemann, B. (2015). Interactions between plutonism and detachments during metamorphic core complex formation, Serifos Island (Cyclades, Greece): PLUTONISM- STRAIN LOCALIZATION. *Tectonics, 34*(6), 1080–1106. doi: 10.1002/2014TC003650

Rangin, C., Le Pichon, X., Demirbag, E., & Imren, C. (2004). Strain localization in the Sea of Marmara: Propagation of the North Anatolian Fault in a now inactive pull-apart. *Tectonics, 23*(2), 2002TC001437. doi: 10.1029/2002TC001437

Reilinger, R., McClusky, S., Vernant, P., Lawrence, S., Ergintav, S., Cakmak, R., et al. (2006). GPS constraints on continental deformation in the Africa-Arabia-Eurasia continental collision zone and implications for the dynamics of plate interactions. *Journal of Geophysical Research: Solid Earth, 111*(B5), 2005JB004051. doi: 10.1029/2005JB004051

Ring, U. (2001). Stacking of nappes with different pressure-temperature paths: An example from the Menderes nappes of western Turkey. *American Journal of Science, 301*(10), 912–944. doi: 10.2475/ajs.301.10.912

Ring, U., & Collins, A. S. (2005). U–Pb SIMS dating of synkinematic granites: timing of core-complex formation in the northern Anatolide belt of western Turkey. *Journal of the Geological Society, 162*(2), 289–298. doi: 10.1144/0016-764904-016

Ring, U., Gessner, K., Güngör, T., & Passchier, C. W. (1999). The Menderes Massif of western Turkey and the Cycladic Massif in the Aegean—do they really correlate? *Journal of the Geological Society, 156*(1), 3–6. doi: 10.1144/gsjgs.156.1.0003

Ring, U., Johnson, C., Hetzel, R., & Gessner, K. (2003). Tectonic denudation of a Late Cretaceous–Tertiary collisional belt: regionally symmetric cooling patterns and their relation to extensional faults in the Anatolide belt of western Turkey. *Geological Magazine, 140*(4), 421–441. doi: 10.1017/S0016756803007878

Robertson, A. H. F., & Dixon, J. E. (1984). Introduction: aspects of the geological evolution of the Eastern Mediterranean. *Geological Society, London, Special Publications, 17*(1), 1–74. doi: 10.1144/GSL.SP.1984.017.01.02

Robertson, A. H. F., & Ustaömer, T. (2004). Tectonic evolution of the Intra-Pontide suture zone in the Armutlu Peninsula, NW Turkey. *Tectonophysics, 381*(1–4), 175–209. doi: 10.1016/j.tecto.2002.06.002

Roche, V., Conand, C., Jolivet, L., & Augier, R. (2018). Tectonic evolution of Leros (Dodecanese, Greece) and correlations between the Aegean Domain and the Menderes Massif. *Journal of the Geological Society, 175*(5), 836–849. doi: 10.1144/jgs2018-028

Roche, V., Jolivet, L., Papanikolaou, D., Bozkurt, E., Menant, A., & Rimmelé, G. (2019). Slab fragmentation beneath the Aegean/Anatolia transition zone: Insights from the tectonic and metamorphic evolution of the Eastern Aegean region. *Tectonophysics*, *754*, 101–129. doi: 10.1016/j.tecto.2019.01.016

Rosendahl, B. R. (1987). Architecture of continental rifts with special reference to East Africa. *Annual Review of Earth and Planetary Sciences*, *15*(1), 445–503.

Rosendahl, B. R., Reynolds, D. J., Lorber, P. M., Burgess, C. F., McGill, J., Scott, D., et al. (1986). Structural expressions of rifting: Lessons from Lake Tanganyika, Africa. *Geological Society, London, Special Publications*, *25*(1), 29–43. doi: 10.1144/GSL.SP.1986.025.01.04

Rost, S., Houseman, G. A., Frederiksen, A. W., Cornwell, D. G., Kahraman, M., Altuncu Poyraz, S., et al. (2021). Structure of the northwestern North Anatolian fault zone imaged via teleseismic scattering tomography. *Geophysical Journal International*, *227*(2), 922–940. doi: 10.1093/gji/ggab265

Saccani, E., Photiades, A., Santato, A., & Zeda, O. (2008). New evidence for supra-subduction zone ophiolites in the Vardar zone of Northern Greece: Implications for the tectono-magmatic evolution of the Vardar oceanic basin. *Ofioliti*, *33*(1), 65–85.

Sadiki, N., Godfray, G., & Msabi, M. (2021). Characterization of the accommodation zones along restraining and releasing bends from analogue modelling simulating the seagap fault, off-shore Tanzania. *Petroleum Research*, *6*(4), 431–442. doi: 10.1016/j.ptlrs.2021.05.007

Sangu, E., Gürer, Ö. F., & Gürer, A. (2020). Fault kinematic and Plio-Quaternary paleostress evolution of the Bakırçay Basin, Western Turkey. *International Geology Review*, *62*(10), 1245–1261. doi: 10.1080/00206814.2019.1642148

Şaroğlu, F., Emre, Ö., & Kuşçu, İ. (1992). *Active Fault Map of Turkey*. Ankara, Turkey: General Directorate of Mineral Research and Exploration.

Satir, M., & Friedrichsen, H. (1986). The origin and evolution of the Menderes Massif, W-Turkey: A rubidium/strontium and oxygen isotope study. *Geologische Rundschau*, *75*(3), 703–714. doi: 10.1007/BF01820642

Sayit, K., Marroni, M., Göncüoglu, M. C., Pandolfi, L., Ellero, A., Ottria, G., & Frassi, C. (2016). Geological setting and geochemical signatures of the mafic rocks from the Intra-Pontide suture zone: Implications for the geodynamic reconstruction of the Mesozoic Neotethys. *International Journal of Earth Sciences*, *105*(1), 39–64. doi: 10.1007/s00531-015-1202-2

Seghedi, I., Helvacı, C., & Pécskay, Z. (2015). Composite volcanoes in the south-eastern part of İzmir–Balıkesir Transfer Zone, Western Anatolia, Turkey. *Journal of Volcanology and Geothermal Research*, *291*, 72–85. doi: 10.1016/j.jvolgeores.2014.12.019

Sengör, A. M. C., Satir, M., & Akkök, R. (1984). Timing of tectonic events in the Menderes Massif, western Turkey: Implications for tectonic evolution and evidence for pan-African basement in Turkey. *Tectonics*, *3*(7), 693–707. doi: 10.1029/TC003i007p00693

Şengör, A. M. C., Tüysüz, O., İmren, C., Sakınç, M., Eyidoğan, H., Görür, N., et al. (2005). The North Anatolian fault: A new look. *Annual Review of Earth and Planetary Sciences*, *33*(1), 37–112. doi: 10.1146/annurev.earth.32.101802.120415

Şengör, A. M. C., & Zabcı, C. (2019). The North Anatolian Fault and the North Anatolian Shear Zone. In C. Kuzucuoğlu, A. Çiner, & N. Kazancı (Eds.), *Landscapes and landforms of turkey* (pp. 481–494). Cham: Springer International Publishing. doi: 10.1007/978-3-030-03515-0_27

Şengör, A. M. C., & Yilmaz, Y. (1981). Tethyan evolution of Turkey: A plate tectonic approach. *Tectonophysics*, *75*(3–4), 181–241. doi: 10.1016/0040-1951(81)90275-4

Seyitoğlu, G. (1997). The Simav Graben: An example of Young E_W Trending Structures in the Late Cenozoic Extensional system of Western Turkey. *Turkish Journal of Earth Sciences*, *6*(3), 135–141.

Seyitoğlu, G., Esat, K., Kaypak, B., & Çivgin, B. (2022). Seismotectonics of the southern branch of North Anatolian fault zone along Bolu, Bursa, and İzmir cities and Değirmenlik (Milos) island in the Aegean Sea. *Yerbilimleri/Hacettepe Üniversitesi Yerbilimleri Uygulama ve Araştırma Merkezi Dergisi*, *43*(2), 138–159. doi: 10.17824/yerbilimleri.948130

Seyitoğlu, G., & Işik, V. (2015). Late Cenozoic extensional tectonics in Western Anatolia: Exhumation of the Menderes core complex and formation of related basins. *Bulletin of the Mineral Research and Exploration*, *151*(151), 47–106. doi: 10.19111/bmre.49951

Seyitoğlu, G., Işık, V., & Çemen, İ. (2004). Complete Tertiary exhumation history of the Menderes massif, western Turkey: An alternative working hypothesis. *Terra Nova*, *16*(6), 358–364. doi: 10.1111/j.1365-3121.2004.00574.x

Seyítoğlu, G., & Scott, B. C. (1994). Late Cenozoic basin development in west Turkey: Gördes basin tectonics and sedimentation. *Geological Magazine*, *131*(5), 631–637. doi: 10.1017/S0016756800012425

Sharp, I. R., & Robertson, A. H. F. (2006). Tectonic-sedimentary evolution of the western margin of the Mesozoic Vardar Ocean: Evidence from the Pelagonian and Almopias zones, northern Greece. *Geological Society, London, Special Publications*, *260*(1), 373–412. doi: 10.1144/GSL.SP.2006.260.01.16

Shin, T. A., Catlos, E. J., Jacob, L., & Black, K. (2013). Relationships between very high pressure subduction complex assemblages and intrusive granitoids in the Tavşanlı Zone, Sivrihisar Massif, central Anatolia. *Tectonophysics*, *595–596*, 183–197. doi: 10.1016/j.tecto.2012.07.012

Simão, N. M., Nalbant, S. S., Sunbul, F., & Komec Mutlu, A. (2016). Central and eastern Anatolian crustal deformation rate and velocity fields derived from GPS and earthquake data. *Earth and Planetary Science Letters*, *433*, 89–98. doi: 10.1016/j.epsl.2015.10.041

Sözbilir, H., İnci, U., Erkül, F., & Sümer, Ö. (2003). *An active intermittent transform zone accommodating N-S extension in Western Anatolia and its relation to the North Anatolian Fault system*. Paper presented at the International Workshop on the North Anatolian, East Anatolian and Dead Sea Fault Systems: Recent Progress in Tectonics and Paleoseismology, and Field Training Course in Paleoseismology, Ankara, Turkey.

Sözbilir, H., Özkaymak, Ç., Uzel, B., Sümer, Ö., Eski, S., & Tepe, Ç. (2016). Palaeoseismology of the Havran-Balıkesir fault zone: Evidence for past earthquakes in the strike-slip-dominated contractional deformation along the southern branches of the North Anatolian fault in northwest Turkey.

Geodinamica Acta, *28*(4), 254–272. doi: 10.1080/09853111.2016.1171111

Sözbilir, H., Sarı, B., Uzel, B., Sümer, Ö., & Akkiraz, S. (2011). Tectonic implications of transtensional supradetachment basin development in an extension-parallel transfer zone: the Kocaçay Basin, western Anatolia, Turkey. *Basin Research*, *23*(4), 423–448. doi: 10.1111/j.1365-2117.2010.00496.x

Spakman, W., Wortel, M. J. R., & Vlaar, N. J. (1988). The Hellenic Subduction Zone: A tomographic image and its geodynamic implications. *Geophysical Research Letters*, *15*(1), 60–63. doi: 10.1029/GL015i001p00060

Speciale, P. A., Catlos, E. J., Yıldız, G. O., Shin, T. A., & Black, K. N. (2012). Zircon ages from the Beypazarı granitoid pluton (north central Turkey): Tectonic implications. *Geodinamica Acta*, *25*(3–4), 162–182. doi: 10.1080/09853111.2013.858955

Stern, R. (2004). Subduction initiation: spontaneous and induced. *Earth and Planetary Science Letters*, *226*(3–4), 275–292. doi: 10.1016/S0012-821X(04)00498-4

Stern, R. J., & Gerya, T. (2018). Subduction initiation in nature and models: A review. *Tectonophysics*, *746*, 173–198. doi: 10.1016/j.tecto.2017.10.014

Stewart, I. S., & Hancock, P. L. (1991). Scales of structural heterogeneity within neotectonic normal fault zones in the Aegean region. *Journal of Structural Geology*, *13*(2), 191–204. doi: 10.1016/0191-8141(91)90066-R

Sümer, Ö., Sözbilir, H., & Uzel, B. (2020). Büyük Menderes Granbeni'nin Rolling Hinge (Yuvarlanan Reze) Modelinde Supra-Detachment (Sıyrılma Üstü) Havzadan Rift Havzasına Evrimi. *Türkiye Jeoloji Bülteni/Geological Bulletin of Turkey*, *63*, 1–36. doi: 10.25288/tjb.562552

Sümer, Ö., Uzel, B., Özkaymak, Ç., & Sözbilir, H. (2018). Kinematics of the Havran-Balıkesir fault zone and its implication on geodynamic evolution of the Southern Marmara Region, NW Anatolia. *Geodinamica Acta*, *30*(1), 306–323. doi: 10.1080/09853111.2018.1540145

Tan, O. (2013). The dense micro-earthquake activity at the boundary between the Anatolian and South Aegean microplates. *Journal of Geodynamics*, *65*, 199–217. doi: 10.1016/j.jog.2012.05.005

Tekin, U. K., Göncüoğlu, M. C., & Turhan, N. (2002). First evidence of Late Carnian radiolarians from the Izmir–Ankara suture complex, central Sakarya, Turkey: implications for the opening age of the Izmir–Ankara branch of Neo-Tethys. *Geobios*, *35*(1), 127–135. doi: 10.1016/S0016-6995(02)00015-3

Temiz, U., & Eikenberg, J. (2011). U/Th dating of the travertine deposited at transfer zone between two normal faults and their neotectonic significance: Cambazli fissure ridge travertines (the Gediz Graben-Turkey). *Geodinamica Acta*, *24*(2), 95–105. doi: 10.3166/ga.24.95-105

Tepe, Ç., & Sözbilir, H. (2017). Tectonic geomorphology of the Kemalpaşa Basin and surrounding horsts, southwestern part of the Gediz Graben, Western Anatolia. *Geodinamica Acta*, *29*(1), 70–90. doi: 10.1080/09853111.2017.1317191

Thomson, S. N., & Ring, U. (2006). Thermochronologic evaluation of postcollision extension in the Anatolide orogen, western Turkey. *Tectonics*, *25*(3), 2005TC001833. doi: 10.1029/2005TC001833

Thuizat, R., Whitechurch, H., Montigny, R., & Juteau, T. (1981). K-Ar dating of some infra-ophiolitic metamorphic soles from the Eastern Mediterranean: New evidence for oceanic thrustings before obduction. *Earth and Planetary Science Letters*, *52*(2), 302–310. doi: 10.1016/0012-821X(81)90185-0

Twiss, R. J., & Moores, E. M. (1992). *Structural geology*. New York: W.H. Freeman.

Ustaomer, T., Ustaömer, P. A., Robertson, A. H. F., & Gerdes, A. (2020). U-Pb-Hf isotopic data from detrital zircons in late Carboniferous and Mid-Late Triassic sandstones, and also Carboniferous granites from the Tauride and Anatolide continental units in S Turkey: implications for Tethyan palaeogeography. *International Geology Review*, *62*(9), 1159–1186. doi: 10.1080/00206814.2019.1636415

Uysal, I. T., Ünal-İmer, E., Shulmeister, J., Zhao, J.-X., Karabacak, V., Feng, Y.-X., & Bolhar, R. (2019). Linking CO_2 degassing in active fault zones to long-term changes in water balance and surface water circulation, an example from SW Turkey. *Quaternary Science Reviews*, *214*, 164–177. doi: 10.1016/j.quascirev.2019.04.029

Uzel, B. (2016). Field evidence for normal fault linkage and relay ramp evolution: the Kırkağaç fault zone, western Anatolia (Turkey). *Geodinamica Acta*, *28*(4), 311–327. doi: 10.1080/09853111.2016.1184778

Uzel, B., Kuiper, K., Sözbilir, H., Kaymakci, N., Langereis, C. G., & Boehm, K. (2020). Miocene geochronology and stratigraphy of western Anatolia: Insights from new Ar/Ar dataset. *Lithos*, *352–353*, 105305. doi: 10.1016/j.lithos.2019.105305

Uzel, B., Langereis, C. G., Kaymakci, N., Sözbilir, H., Özkaymak, Ç., & Özkaptan, M. (2015). Paleomagnetic evidence for an inverse rotation history of Western Anatolia during the exhumation of Menderes core complex. *Earth and Planetary Science Letters*, *414*, 108–125. doi: 10.1016/j.epsl.2015.01.008

Uzel, B., & Sözbilir, H. (2008). A first record of a strike-slip basin in Western Anatolia and its tectonic implication: The Cumaovası basin. *Turkish Journal of Earth Sciences*, *17*(3), 559–591.

Uzel, B., Sözbilir, H., Özkaymak, Ç., Kaymakcı, N., & Langereis, C. G. (2013). Structural evidence for strike-slip deformation in the İzmir–Balıkesir transfer zone and consequences for late Cenozoic evolution of western Anatolia (Turkey). *Journal of Geodynamics*, *65*, 94–116. doi: 10.1016/j.jog.2012.06.009

Van Hinsbergen, D. J. J., & Meulenkamp, J. E. (2006). Neogene supradetachment basin development on Crete (Greece) during exhumation of the South Aegean core complex. *Basin Research*, *18*(1), 103–124. doi: 10.1111/j.1365-2117.2005.00282.x

Van Noten, K., Topal, S., Baykara, M. O., Özkul, M., Claes, H., Aratman, C., & Swennen, R. (2019). Pleistocene-Holocene tectonic reconstruction of the Ballık travertine (Denizli Graben, SW Turkey): (De)formation of large travertine geobodies at intersecting grabens. *Journal of Structural Geology*, *118*, 114–134. doi: 10.1016/j.jsg.2018.10.009

Vandenberg, L. C., & Lister, G. S. (1996). Structural analysis of basement tectonites from the Aegean metamorphic core complex of Ios, Cyclades, Greece. *Journal of Structural*

Geology, *18*(12), 1437–1454. doi: 10.1016/S0191-8141(96)00068-5

Wakabayashi, J., & Dilek, Y. (2003). What constitutes 'emplacement' of an ophiolite?: Mechanisms and relationship to subduction initiation and formation of metamorphic soles. *Geological Society, London, Special Publications*, *218*(1), 427–447. doi: 10.1144/GSL.SP.2003.218.01.22

Walcott, C. R., & White, S. H. (1998). Constraints on the kinematics of post-orogenic extension imposed by stretching lineations in the Aegean region. *Tectonophysics*, *298*(1–3), 155–175. doi: 10.1016/S0040-1951(98)00182-6

Wells, D. L., & Coppersmith, K. J. (1994). New empirical relationships among magnitude, rupture length, rupture width, rupture area, and surface displacement. *Bulletin of the Seismological Society of America*, *84*(4), 974–1002. doi: 10.1785/BSSA0840040974

Westaway, R. (1994). Present-day kinematics of the Middle East and eastern Mediterranean. *Journal of Geophysical Research: Solid Earth*, *99*(B6), 12071–12090. doi: 10.1029/94JB00335

Westerweel, J., Uzel, B., Langereis, C. G., Kaymakci, N., & Sözbilir, H. (2020). Paleomagnetism of the Miocene Soma basin and its structural implications on the central sector of a crustal-scale transfer zone in western Anatolia (Turkey). *Journal of Asian Earth Sciences*, *193*, 104305. doi: 10.1016/j.jseaes.2020.104305

Whipp, P. S., Jackson, C. A.-L., Schlische, R. W., Withjack, M. O., & Gawthorpe, R. L. (2017). Spatial distribution and evolution of fault-segment boundary types in rift systems: Observations from experimental clay models. *Geological Society, London, Special Publications*, *439*(1), 79–107. doi: 10.1144/SP439.7

Wilcox, R. E., Harding, T. P., & Seely, D. R. (1973). Basic wrench tectonics. *AAPG Bulletin*, *74*(1), 74–96.

Wong, H. K., Lüdmann, T., Ulug, A., & Görür, N. (1995). The Sea of Marmara: a plate boundary sea in an escape tectonic regime. *Tectonophysics*, *244*(4), 231–250. doi: 10.1016/0040-1951(94)00245-5

Woodside, J. M., Mascle, J., Zitter, T. A. C., Limonov, A. F., Ergün, M., & Volkonskaia, A. (2002). The Florence Rise, the western bend of the Cyprus Arc. *Marine Geology*, *185*(3–4), 177–194. doi: 10.1016/S0025-3227(02)00194-9

Yang, X., Becker, F., Knitter, D., & Schütt, B. (2021). An overview of the geomorphological characteristics of the Pergamon micro-region (Bakırçay and Madra River Catchments, Aegean Region, West Turkey). *Landscape*, *10*(7), 667. doi: 10.3390/land10070667

Yiğitbaş, E., Kerrich, R., Yılmaz, Y., Elmas, A., & Xie, Q. (2004). Characteristics and geochemistry of Precambrian ophiolites and related volcanics from the Istanbul-Zonguldak Unit, Northwestern Anatolia, Turkey: Following the missing chain of the Precambrian South European suture zone to the east. *Precambrian Research*, *132*(1–2), 179–206. doi: 10.1016/j.precamres.2004.03.003

Yilmaz, H., Oyman, T., Arehart, G. B., Colakoglu, A. R., & Billor, Z. (2007). Low-sulfidation type Au–Ag mineralization at Bergama, Izmir, Turkey. *Ore Geology Reviews*, *32*(1–2), 81–124. doi: 10.1016/j.oregeorev.2006.10.007

Yılmaz, P. O., & Maxwell, J. C. (1984). An example of an obduction melange: The Alakır Çay unit, Antalya Complex, southwest Turkey. In L. A. Raymond (Ed.), *Geological Society of America special papers* (Vol. *198*, pp. 139–152). Geological Society of America. doi: 10.1130/SPE198-p139

Yılmaz, Y., Genç, Ş. C., Karacık, Z., & Altunkaynak, Ş. (2001). Two contrasting magmatic associations of NW Anatolia and their tectonic significance. *Journal of Geodynamics*, *31*(3), 243–271. doi: 10.1016/S0264-3707(01)00002-3

Yolsal-Çevikbilen, S., Taymaz, T., & Helvacı, C. (2014). Earthquake mechanisms in the Gulfs of Gökova, Sığacık, Kuşadası, and the Simav Region (western Turkey): Neotectonics, seismotectonics and geodynamic implications. *Tectonophysics*, *635*, 100–124. doi: 10.1016/j.tecto.2014.05.001

Yoshioka, T. (1996). Evolution of fault geometry and development of strike-slip basins: Comparative studies on the transform zones in Turkey and Japan. *Island Arc*, *5*(4), 407–419. doi: 10.1111/j.1440-1738.1996.tb00162.x

Yousefi-Bavil, K., Koçkar, M. K., & Akgün, H. (2022). Development of a three-dimensional basin model to evaluate the site effects in the tectonically active near-fault region of Gölyaka basin, Düzce, Turkey. *Natural Hazards*, *114*(1), 941–969. doi: 10.1007/s11069-022-05418-4

INDEX

Note: Page numbers in *italics* refer to *figures*; those in **bold** refer to **tables**.

Extensional Tectonics: Rifting and Continental Extension, Geophysical Monograph 278, First Edition.
Edited by İbrahim Çemen and Elizabeth J. Catlos.

DOI: 10.1002/9781119773757.index

Printed in the USA/Agawam, MA
September 16, 2026

910661.200